Medicinal Plant Biotechnology

Medicinal Plant Biotechnology

Ciddi Veeresham

M.Pharm., Ph.D., PDF (USA)

UNESCO BAC Fellow

Associate Professor in Pharmacy

Kakatiya University, Warangal (A.P.), India

Foreword by Prof. C.K. Kokate

CBSPD

CBS Publishers & Distributors Pvt Ltd

New Delhi • Bengaluru • Chennai • Kochi • Kolkata • Lucknow • Mumbai

Hyderabad • Jharkhand • Nagpur • Patna • Pune • Uttarakhand

Medicinal Plant Biotechnology

Copyright © Author and Publisher

ISBN-13: 978-81-239-1097-0
ISBN-10: 81-239-1097-5

First Edition: 2004
Reprint: 2006, 2008, 2011, 2023

Published by Satish Kumar Jain and produced by Varun Jain for

CBS Publishers & Distributors Pvt Ltd

4819/XI Prahlad Street, 24 Ansari Road, Daryaganj, New Delhi 110 002, India
Ph: 011-23289259, 23266861, 23266867   Website: www.cbspd.com
Fax: 011-23243014   e-mail: delhi@cbspd.com;
                                    cbspubs@airtelmail.in.

Corporate Office: 204 FIE, Industrial Area, Patparganj, Delhi 110 092, India
Ph: 011-4934 4934   Fax: 011-4934 4935   e-mail: publishing@cbspd.com;
                                                      publicity@cbspd.com

Branches

- **Bengaluru:** Seema House 2975, 17th Cross, KR Road, Banasankari 2nd Stage, Bengaluru 560 070, Karnataka, India
  Ph: +91-80-26771678/79   Fax: +91-80-26771680   e-mail: bangalore@cbspd.com
- **Chennai:** 7, Subbaraya Street, Shenoy Nagar, Chennai 600 030, Tamil Nadu, India
  Ph: +91-44-26680620, 26681266   Fax: +91-44-42032115   e-mail: chennai@cbspd.com
- **Kochi:** 42/1325, 1326, Power House Road, Opp KSEB, Power House, Ernakulum Kochi 682 018, Kerala, India
  Ph: +91-484-4059061-65,67   Fax: +91-484-4059065   e-mail: kochi@cbspd.com
- **Kolkata:** 147, Hind Ceramics Compound, 1st Floor, Nilgunj Road, Belghoria, Kolkata-700056, West Bengal, India
  Ph: +033-25633055, 033-25633056   e-mail: kolkata@cbspd.com
- **Lucknow:** Basement, Khushnuma Complex, 7 Meerabai Marg (Behind Jawahar Bhawan),Lucknow-226001, UP, India
  Ph: +0522-4000032   e-mail: tiwari.lucknow@cbspd.com
- **Mumbai:** PWD Shed, Gala no 25/26, Ramchandra Bhatt Marg, Next to JJ Hospital Gate no. 2, Opp. Union Bank of India, Noorbaug, Mumbai-400009, Maharashtra, India
  Ph: 022-66661880/89   e-mail: mumbai@cbspd.com

Representatives

• Hyderabad	0-9885175004	• Jharkhand	0-9811541605	• Nagpur	0-9421945513
• Patna	0-9334159340	• Pune	0-9623451994	• Uttarakhand	0-9716462459

Printed at   India Binding House, Noida, UP, India

Foreword

Foreword

Plant tissue culture is an important biotechnological tool for *in vitro* production of plant species and useful secondary metabolites. The feasibility of this technique for commercial production of pharmaceutically significant secondary products has been extensively studied in last two decades. The author, Dr. Veeresham, has exhaustively reviewed the literature in the field of plant tissue culture and made sincere efforts in projecting the milestones in the development of this branch of biosciences. The methodology of plant tissue culture, biogenesis of secondary metabolites in cell cultures, immobilized cell systems, cryopreservation, hairy root cultures, transgenic plants and related topics have been explained well. The chapters on large scale cultivation of plant cells using bioreactors, two-phase cultures and synthetic seeds are added features of the book useful for the students of biotechnology.

I congratulate Dr. Veeresham for his sincere efforts in bringing out this book. His wide experience in the field of plant tissue culture and his exposure to global trends in the field of biotechnology, including his interaction with international scientists have certainly added flavour to the contents of the book. It is highly informative and useful compilation of information for the students of pharmacy, botany, agriculture and related subjects.

(Prof. Chandrakant Kokate)

Preface

Preface

Plants are the sources for phytopharmaceuticals, flavours, fragrances, colours, pesticides and food. The availability of these plants became scarce because of ruthless exploitation and depletion of forest. Most of the pharmaceuticals are chemically complex in nature. There is no economically viable method of synthesis. The alternatives for the production of phytopharmaceuticals and allied products are met through the plant tissue culture. This biotechnology requires the knowledge of chemical engineering, life sciences and pharmaceutical sciences. This book has incorporated the requirements of all these subjects. Medicinal plant biotechnology has been successfully employed for the production of phytopharmaceuticals (Taxol and Shikonin), perfumes (Sandalwood oil), colours (Saffron), flavours (Vanilla) and biopesticides (Azadirachtin).

The book starts with an introduction of history followed by two chapters covering practical aspects namely, laboratory requirements and aseptic transfers. The chapter 4 deals with the nutritional requirements of plant cells. Chapter 5 was on methods of initiation and maintenance of callus and cell suspension cultures. This is followed by a chapter on production of phytopharmaceuticals and other value-added products from plant cell culture. The major disadvantage of plant cell culture is low productivity. This problem can be overcome by various strategies that are given in chapters 7, 8,11, 12 and 21. Plants are big chemical factories which can perform a various chemical reactions. This has been illustrated with many examples in chapter 10, so that plant cells can be used for the production of valuable products. Chapters 13, 14, 15, 16, 17, 18 and 19 deal with the fundamentals and application of plants in agriculture and biological sciences with their application towards the production of edible vaccines, antigens, antibodies and other pharmaceuticals. Chapter 20 has been described with how to produce the valuable products through plant cells using bioreactors. Some of the examples such as taxol, shikonin, rosmarinic acid, sanguinarine and indole alkaloids have been discussed in detail. Finally, the methods for the storage of high yielding cell lines have been given.

Each chapter is profusely illustrated with figures, tables and chemical structures and attempts to convey the practical as well as theoretical aspects of medicinal plant biotechnology has been done. It should, therefore, find application in the lecture hall and in the laboratory.

I have made a humble attempt to present the subject in lucid manner and have taken all possible care to avoid mistakes. Still, any mistake found in this book by the readers may please be brought to my notice for improvement.

I would like to thank Prof. M.L. Shuler (Cornell University, USA); Dr. V. Bringi (Phyton Inc., USA); Dr. V. Srinivasan (Phyton Inc., USA); and Dr. Harry (Phong Institute of Science and Technology. South Korea) for their help in learning the subject of medicinal plant biotechnology. I am also thankful to all my research scholars for their help in preparing this book.

I am especially indebted to my wife, Parameshwari, without whose cooperation and patience, I could not have seen the book to completion. I am also thankful to my children, Deepthi and Dileep, for their sacrifice of time for this cause.

I am thankful to my colleagues and friends for their help and encouragement. I am specially grateful to Dr. Chandrakant Kokate for his kind foreword to the book.

Ciddi Veeresham

Contents

Contents

Chapter 1

History of Plant Cell Cultures

The technique of plant tissue and cell culture has evolved over several decades. This technique combined with recent advances in developmental, cellular, molecular genetics, metabolic engineering, genetic transformation and using conventional plant breeding have turned plant biotechnology into an exciting research field with a significant impact on pharmaceutical industries, agriculture, horticulture and forestry.

Gottlieb Haberlandt accomplished the first successful plant tissue culture at the turn of the 20th century when he reported the culture of leaf mesophyll tissue and hair cells (Krikorian and Berquam 1969). Schwan (1839) expressed the view that each living cell of multicellular organism should be capable of independent development if provided with proper external conditions. A totipotent cell is one that is capable of developing by regeneration into a whole organism. Unfortunately, the cells that Haberlandt cultured did not come to successful. The possible reason for this failure may be lack of knowledge about plant growth regulators needed for cell division and embryo induction. Haberlandt became discouraged and worked on other physiological investigations. Hannig (1904) cultured nearly matured embryo excised from seeds of several species of crucifers. More ever, Haberlandt's lack of success was not invain because one of his students, Kotte (1922) from Germany reported the growth of isolated root tips on a medium consisting of inorganic salts. At the same time Robbins (1922) reported the similar success with root and stem tips and White (1934) reported that, not only could be cultured tomato root tips to grow, but they could be repeatedly subcultured to fresh medium of inorganic salts supplemented with yeast extract, which is a good source of vitamin B. White (1939) reported the growth promoting effects of thiamine isolated from tomato root tips. Went and Thimann (1937) discovered the indole acetic acid (IAA), a natural auxin. Duhamet (1939) reported the stimulation of growth of excised roots by IAA.

The avenue was now open for rapid progress in the successful culture of plant tissue cultures during 1930 with refined media, La Rue (1936) achieved better success at culturing embryos compared to Haning (1904). Guatheret (1934) described the successful culture of cambium of several species of trees to produce callus on medium containing IAA and vitamin B. Further research with other tissues led Nobecourt (1937) and Gautheret (1939) to obtain callus from carrot and White (1939) to obtain tobacco tumor (crown gall) tissue. Johannes Van Overbeek and his co-workers (1942) reported that they were able to obtain seedlings from heart shaped embryos by enriching culture media with coconut milk in addition to the usual salts, vitamins and other nutrients.

At the same time, Panchanan Maheshwari and co-workers from India were very active in angiosperm embryology research. Braun (1943) reported tumor induction related crown gall disease. Folke Skoog (1944) described the organ formation in cultured tissue and organs of tobacco. Camus· (1943) from Europe was the first to report grafting experiments in plant tissue cultures. Haberlandt and his associates (1946) made improvements on the medium for the growth of tobacco and sunflower tissue and Morel (1948) was well into applying tissue technique to study of parasites associated with plant tissues. Street (1950) and his associates began a series of extensive studies on the nutrition of excised tomato root tips.

Skoog and Co-workers (1951) worked on investigations on the nutritional requirements of tobacco cells and they discovered a compound (kinetin) which promotes cell division (Skoog and Carlos Miller, 1957). Stowe and Yamaki (1957) discovered gibberellins another plant growth regulator. Crocker et al (1935) were the first to propose that ethylene was involved in fruit ripening.

At the end of 1950 plant tissue culture methods had progressed to the point of making a major impact on research in plant biology. The progress in the plant tissue culture research was hastened by innovations in laboratory equipment and supplies and improved method of worldwide transportation and communication between the scientists around the world. In particular, the development of HEPA filters, which screens fungal spores and bacteria from air, made laminar flow transfer hoods possible. Laminar flow hoods became commonly available by the end of 1960's. At the same time the methods of obtaining plant callus from several sources were well developed by the middle of 1950's. The scientists involved in callus and cell cultures during 1950's are Bergman, Nitsch, Henderson, Murashige and Skoog.

Murashige and Skoog (1962) developed a culture medium for the rapid growth of tobacco callus. Morel (1965) reported the micropropagation of orchids. Guha and Maheshwari (1966) reported the first successful culturing of haploid cells of Datura. Cocking (1960) isolated the plant protoplast and began to be cultured in the media. This led to somatic hybridization. In early 1970's discovery of endonuclease enzymes, which led to rapid development of gene transformation. The prospects of success with the genetics of plants have created considerable public interest. Melchers, Sacristan and Holder (1978) produced somatic hybrid plants from the fusion of potato and tomato protoplasts. Several research groups have produced transformed tobacco plants following single cell transformation (i.e. gene insertion, Chilton 1983). The use of *Agrobacterium* mediated plant transformation, and the use of the so-called gene gun to shoot DNA into plant cells, has led to the development of novel plants. The calgene of USA is the first to commercialize the technique of production of transformed plants.

The important milestones in the development of plant tissue culture technology and the production of secondary metabolites is given in Tables 1.1 and 1.2 respectively.

Table 1.1. Brief history of plant cell culture technology

1902	Haberlandt	Culturing of single cells isolated from plant tissues in simple nutrient media
1904	Haning	Embryo culture
1934	White	Establishment of actively growing tomato roots
1937	White	Discovery of the importance of the vitamin B for the growth
1937	Went and Thimann	Role of auxin in plant tissue culture
1939	White, Guatheret	Establishment of Callus culture
1954	Muir, Haberlandt	Establishment of cell suspension culture

(Contd.)

1960	Bergmann	Single cell cloning
1960	Cocking	Isolation of protoplast
1962	Murashigae &Skoog	Medium for the growth of the Callus
1964	Morel	Microporpagation
1977	Street	Cultivation of single cell
1978	Melchers, Sacristan, Holder	Production of Somatic embryo
1983	Chilton	Transformed tobacco plants
1994	Calgene, USA	Production of Novel transformed Plants

Table 1.2. The milestones for the production of secondary metabolites from plant cell cultures

Year	Milestone
1934	Plant tissue culture obtained (Gautheret, 1934)
1939	Discovery of Auxin (Gautheret, 1939)
1942	Secondary metabolite (Diosgenin) produced in callus culture (Gautheret, 1942).
1954	Cell suspension culture (Muir, 1954)
1955	Production of secondary metabolites in cell cultures (Mothes and Kala, 1955; Routien and Nickell, 1956)
1959	Large scale cultivation of plant cells (Tulecke and Nickell, 1959)
1967	Yields of Phytopharmceuticals equal to intact plant (Kaul and Staba, 1967)
1977	Cultivation of tobacco cells in 20,000 litre bioreactor (Noguchi *et al.,* 1977)
1979	Immobilization of plant cells by sodium alginate (Brodelius et al., 1979)
1981	Use of hollow fiber reactor for secondary metabolite Production (Shuler, 1981)
1983	First Plant tissue culture process commercialized (Curtin, 1983)
1995	Taxol produced from plant cell culture on commercial scale (Phyton Newsletter)

REFERENCES

Bautheret, R.J. (1939). Sur la Possibilite de Realise la Culture Indefinite des Tissue de Tubercules de Tissues Vegetaux. *Cr. Seanc. Soc. Biol.* 130: 1270-1271.

Bergamann, L., (1960). Growth and division of single cells of higher plants in-vitro, *J.Gen.Physiol.* 43: 841-851.

Braun, A.C. (1943). Studies on tumor inception in the crown gall disease. *Am. J. Bot.* 30: 674-677.

Brodelius, P., Deus, B., Mosbach, K., and Zenk, M.H (1979). Immobilized Plants Cells for the Production and Transformation of Natural Products. *FEBS Letters* 103: 93-97.

Camus, G. (1943). Sur le greffage des bourgeons d'endive sur des fragments de tissues cultives in vitro. *C.R. Acad. Sci.* 137: 184-185.

Chilton , M.D., Tepfer, D.A, Petit, A., David, C., Casse-Delbart, F. and Tempe, J. (1982). *Agrobacterium rhizogenes* inserts T-DNA into the genomes of the bost plant root cells. *Nature,* 295 : 432-434.

Cocking, E.C. (1960). A method for the isolation of plant protoplasts and vacuoles. *Nature.* 187: 962-963.

Crocker, W., Hitchcock, A.E., and Zimerman, P.W., (1935). Similarities in the effects of ethylene and the plant auxins. *Contrib. Boyce Thompson Inst.* 7: 231-248.

Curtin, M.E. (1983). Harvesting profitable products from Plant Tissue Culture. *Biotechnology,* October 1983, 649-657.

Duhamet, L. (1939). Action de l'heteroauxine sur la croissance de racines isolees de *Lupinus albus. C.R. Acad. Sci. Paris.* 208: 1838-1840.

Gautheret, R.J. (1934). Culture de tissue cambial. *C.R. Acad. Sci.Paris.* 198: 2195-2196.

Gautheret, R.J. (1939). Sur la possibilite de realiser la culture indefinie des tissus de tubercules de carotte. *C.R. Acad. Sci. Paris.* 208: 118-120.

Gautheret, R.J. (1942). Hetero-Auxines et Cultures de Tissues vegetaux. *Bull. Soc. Ch. Biol.* 41(1-3): 13-47.

Guha, S. and Maheshwari, S.C. (1966). Cell division and differentiation in the pollen grains of *Datura* in vitro. *Nature.* 212: 97-98.

Haning, E. (1904). Zur physiologie pflanzlicher Embryonen. I. Uber die Kultur von cruciferen-embryonen Aussernalb des Embryosacks. *Bot. Zig.* 62: 45-80.

Haberlandt, G. (1902). Kultinversuche mit isolierten pflanzellen. *Sber.Akad.wiss.wien.* 111: 69-92.

Haberlandt, A.C., Riker, A.J., and Duggar, B.M., (1946). The influence of the composition of the medium on growth in vitro of excised tobacco and sunflower tissue culture. *Am. J. Bot.* 33: 591-597.

John Van Overbeck., Conklin, M.E., and Blakeslee, A.F., (1942) Cultivation in-vitro of small Datura embryo. *Am. J.Bot.* 29, 472-477.

Kaul, B., and Staba, E.J. (1967). *Amni Visnaga* (L) Tissue Cultures: Multiliter Suspension Growth and Examination for Furanocoumerins. *Planta Medica,* 15: 145-156.

Kotte, W. (1922). Kulturversuche mit isolierten Wurzelspitzen. *Beitr. Allg. Bot.* 2: 413-434.

Krikorian, A.D. and Berquam, D.L., (1969). Plant cell and tissue culture – the role of Haberlandt. *Bot. Rev.* 35: 58-88.

La Rue, C.D. (1936). The growth of plant embryos in culture. *Bull. Torrey Bot. Club.* 63: 365-382.

Maheshwari, P. (1950). *An Introduction to the Embryology of Angiosperms.* P. 453. Mcgraw-Hill Book Co., Inc., New York.

Melchers, G., Sacristan, M.D., and Holder, A., (1978) Somatic hybrid plants of potato and tomato regenerated from fused protoplasts, Carlsberg Res.Commun. 43: 203-218.

More, G. (1965). Clonal propagation of orchids by meristem culture. *Cymbidium Soc. News* 20: 3-11.

Morel, G. (1948). Recherches sur la culture associee de parasites obligatoires et de tissus vegetaux. *Ann. Epiphyt. N.S.* 14: 123-234.

Mothes, K., and Kala, H., (1955). Die Wurzel als Bildungsstaete fuer Cumarine. *Naturwiss,* 42: 159.

Muir, W.H., Hilderbrandt, A.C., and Riker, A.J., (1954) Plant tissue cultures produced from isolated plant cells. Science, N.Y. 119: 877-887.

Murashige, T. and Skoog, F., (1962). A revised medium for rapid growth and bioassays with tobacco tissue cultures. *Physiol. Plant.* 15: 473-497.

Nobecourt, P. (1937). Cultures en serie de tissus vegetaux sur milieu artificiel. *C.r. Seanc. Soc. Biol.* 205: 521-523.

Noguchi, M., Matsumato, T., Hirata, Y., Yamamato, K., Katsuyama, A., Kato, A., Azechi, S., and Kato, K., (1977). Improvement of Growth rates of plant cell cultures. In : *Plant Tissue Culture and its Biotechnological Application.* (W. Barz, E. Reinhard, M.H. Zenk, eds.), Springer-Verlag Berlin, pp.85-94.

Robbins, W.J. (1922). Cultivation of excised root tips and stem tips under sterile conditions. *Bot. Gaz.* 73: 376-390.

Routien, J.B., and Nickell, L.G., (1956). Cultivation of Plant Tissue. U.S. Patent No. 2: 747, 334.

Shuler, M.L., (1981) Production of secondary metabolites from plant tissue culture-problems and and prospects. In: Biochemical engineeringII. Constantinides, A., Vieth, W.R., and Venkatsubramanian, K., (eds). *Ann. N.Y. acad.Sci.* 369: 65-79.

Skoog, F. (1944).Growth and organ formation in tobacco tissue cultures. *Am.J.Bot.* 31: 19-24.

Skoog, F. and Tsui, C., (1951). Growth substances and the formation of buds in plant tissues. In : *Plant Growth Substances.* F. Skoog., Ed., pp. 263-285. University of Wisconsin Press, Madison.

Skoog, F. and C.O. Miller, (1957). Chemical regulation of growth and organ formation in plant tissues culture in vitro. *Symp. Soc. Exp. Biol.* 11: 118-131.

Stowe, B.B. and T. Yamaki. (1957). The history and physiological action of the gibberellins. *Annu. Rev. Plant Physiol.* 8: 181-216.

Street, H.E., and Lowe, J.S. (1950) The carbohydrate nutrition of tomato roots II. The mechanism of sucrose absorption by excised roots. Ann. Bot. 14: 307-325.

Tulecke, W., L.G. Nickell, (1959). Production of large amounts of plant tissue by submerged culture. *Science,* 130: 863-864.

Went, F.W. and Thimann. K.V. (1937). *Phytohormones.* p. 294. The Macmillan Co., New York.

White, P.R. (1934). Potentially unlimited growth of excised tomato root tips in liquid medium. *Plant Physiol.* 9: 585-600.

White, P.R. (1937). Vitamin B_1 in the nutrition of excised tomato roots. *Plant Physiol.* 12: 803-811.

White, P.R. (1939). Potentially unlimited growth of excised plant callus in an artificial medium. *Am. J. Bot.* 26: 59-64.

Chapter 2

Laboratory Facilities

A laboratory devoted to in vitro procedures with plant tissue cell organ cultures must have adequate space for the performance of several functions. According to White (1963) it must provide facilities for :

1. Media preparation, sterilization, cleaning and storage of supplies.

2. Aseptic manipulation of explant.

3. Growth of the cultures under controlled environmental conditions.

4. Examination and analysis of the cultures.

5. Assembling and filing of records.

An ideal design of Medicinal Plant Biotechnology Laboratory (MPB) is depicted in Fig. 2.1.

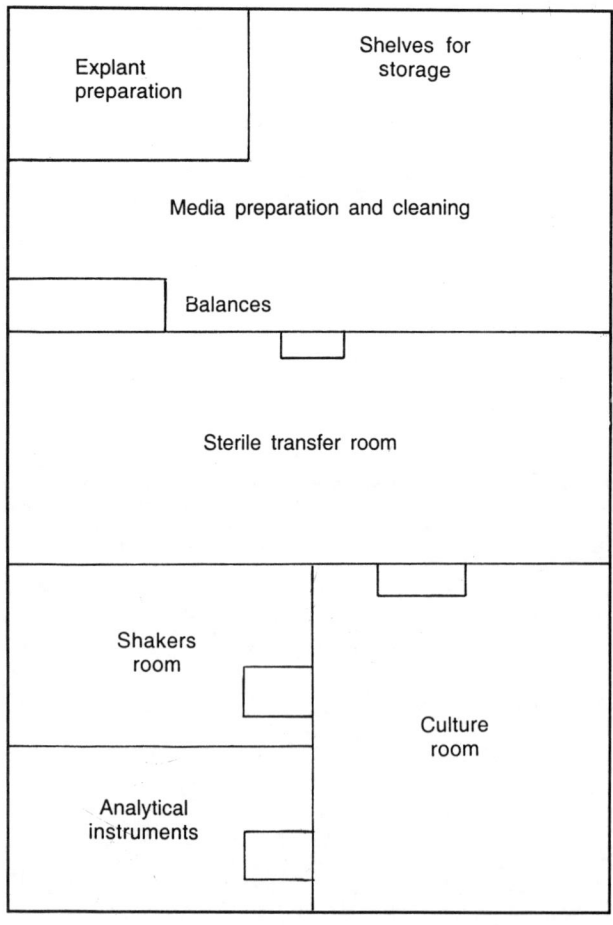

Fig. 2.1. A ideal design of Medicinal Plant Biotechnology Laboratory.

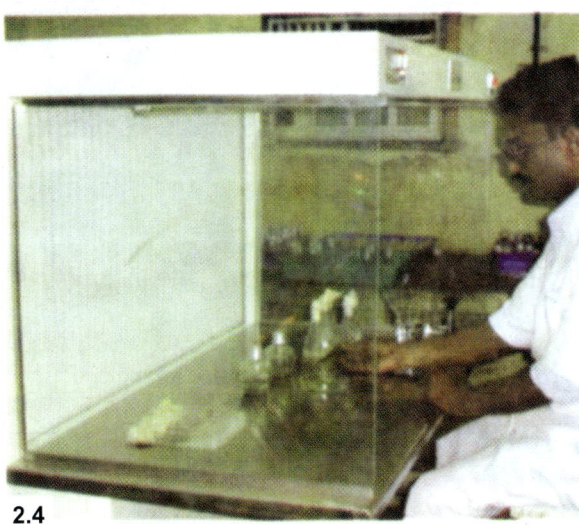

Fig. 2.2. Medicinal Plant Biotechnology Laboratory.

Fig. 2.3. Vacuum pump.

Fig. 2.4. Laminar flow hood.

Fig. 2.5. Centrifuge, table top.

Fig. 2.6. HPLC system.

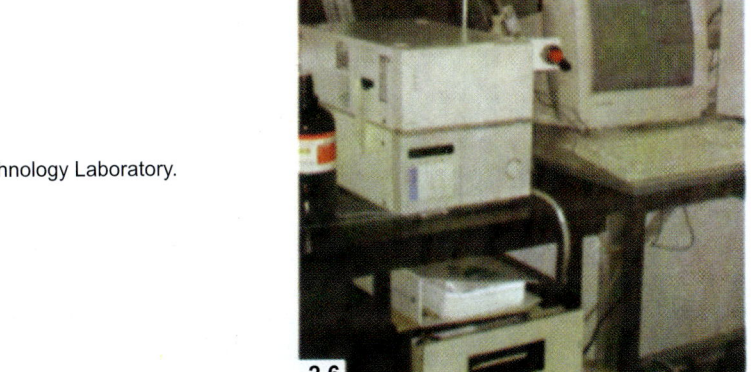

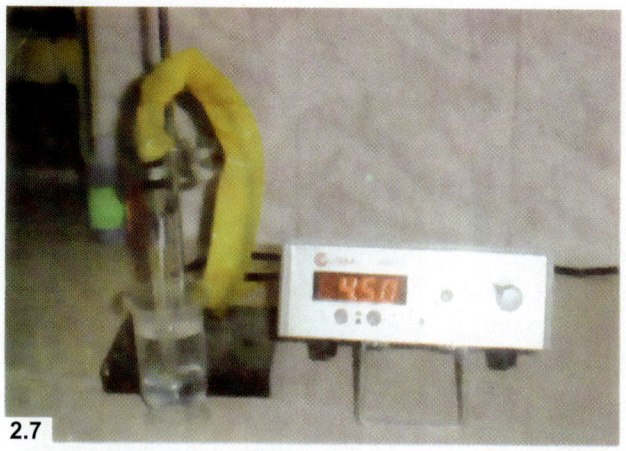

2.7

2.8

2.9

2.11

2.10

Fig. 2.7. pH meter.

Fig. 2.8. Hot-air oven.

Fig. 2.9. Ultrasonic bath.

Fig. 2.10. Horizontal autoclave.

Fig. 2.11. Vertical autoclave.

2.12

2.13

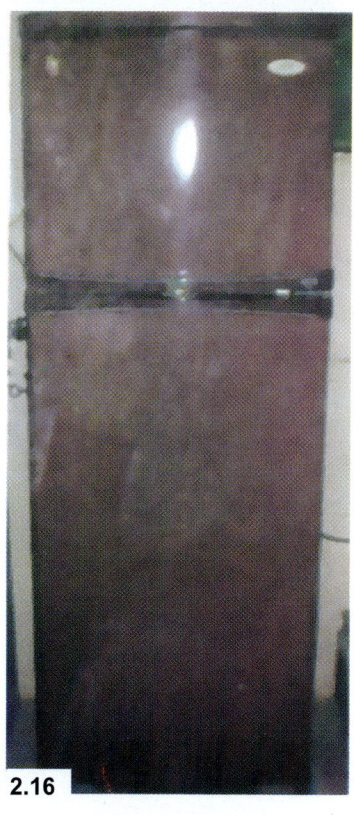

2.14

2.15

Fig. 2.12. Magnetic stirrer/hot plate.

Fig. 2.13. Filter unit.

Fig. 2.14. B.O.D. incubator.

Fig. 2.15. Refrigerator.

Fig. 2.16. Water filtration system.

2.16

2.17

2.18

2.19

2.20

Fig. 2.17. Rotary vacuum evaporator.

Fig. 2.18. Incubator shaker.

Fig. 2.19. Balance.

Fig. 2.20. Lyophilizer.

2.1. BASIC REQUIREMENTS FOR MEDICINAL PLANT BICTECHNOLOGY LABORATORY
(Fig. 2.2)

2.1.1. Equipment and Apparatus

1. Vacuum pump (Fig. 2.3)
2. Laminar flow hood (Fig. 2.4)
3. Centrifuge, table top (Fig. 2.5)
4. HPLC system (Fig. 2.6)
5. pH meter (Fig. 2.7)
6. Hot air oven (Fig. 2.8)
7. Ultrasonic bath (Fig. 2.9)
8. Autoclave or pressure cooker (Figs. 2.10 and 2.11)
9. Magnetic stirrer/hot plate (Fig. 2.12)
10. Filter units for filter sterilization (Fig. 2.13)
11. BOD incubator (Fig. 2.14)
12. Water filtration system (Fig. 2.15)
13. Refrigerator (Fig. 2.16)
14. Rotary Vacuum evaporator (Fig. 2.17)
15. Shaker, gyratory with platform for suspension cultures (Fig. 2.18)
16. Balance (Fig. 2.19)
17. Lyophilizer (Fig. 2.20)

2.1.2. Sterile Transfer Room

Laminar horizontal flow sterile transfer cabinets, which are available in 8, 6, 4 and 3 feet from many commercial sources. They should have horizontal airflow from the back to front and should be equipped with gas connection if gas burner has to be used.

2.1.3. Culturing facilities

Incubators : An ideal incubator should be having temperature control, light and humidity control. Most of the callus culture requires a temperature of $25 \pm 2°C$ (Fig. 2.14).

Shakers : Gyratory with plat form with clips for different size flasks for cell suspension cultures (Fig.2.18).

Media Preparation Room : This part is a central section where most of the activities are performed, except for sterile transfers and incubation of cultures. It should have direct access to sterile transfer room and the culturing section.

Cleaning Room : A separate room for cleaning and dish washing is very advantageous.

Instrument Room : In a larger operation a separate room should be set aside for housing and operation of analytical balances, Research microscopes, delicate instruments to avoid vibration caused by shaker etc.

Chapter 3

Aseptic Transfer

The application of plant, cell, tissue and organ culture requires sterile technique. The maintenance of aseptic or sterile conditions is essential for successful tissue culture procedures. A few simple precautions to avoid contamination will save valuable time in not repeating experiments. The best example of aseptic technique can be found in the operating room of a modern hospital.

The most important factor in the selection of suitable working area is the possible flow or unfiltered air over the disinfected working areas; air currents must be avoided, because of airborne spores of contaminating microorganisms. An interior room, similar in layout to a photographic dark room, is an excellent choice for aseptic procedures. Because opening the door creates a draft, post No admittance sign on the door during aseptic transfer.

Most plant scientists using plant tissue cultures procedures conduct sterile operations within some type of transfer chamber, bacteriological glove box or laminar flow cabinet. With a laminar flow cabinet (Fig. 2.4) air is forced through a dust filter and then passed through a high-efficiency - particular air (HEPA) filter. Depending on the type of cabinet, the air is directed either downward or outward over the working area (Torres 1989). The flow of sterile air is designed to present any spore - laden unfiltered air from entering the cabinet. Since ethanol is highly inflammable, flaming instruments (forceps, scalpel blade etc.) in a laminar flow cabinet should be done with caution. The air flow from the cabinet would direct a flash fire toward the worker (Wethereu, 1982). Before the start of any sterile procedure, the working areas should be thoroughly scrubbed with a tissue soaked with either ethanol or isopropanol (70% v/v). Dixon (1985) has indicated the use of a 20% aqueous solution of phenol for surface disinfection; volatile phenolics in the laboratory may pose a threat to the growth of tissue cultures.

Aseptic cabinets and transfer rooms are often equipped with germicidal lamps emitting ultraviolet (UV) light. This type of radiation is useful in eliminating airborne contaminants and for surface disinfection. The emission at 254 nm is slowly germicidal, but UV does not penetrate surfaces; dust and shadowed areas protect contaminants from its effects. Although the effectiveness of UV lamps in creating a sterile environment is questionable (Klein & Klein, 1970, Collins & Lyne, 1984), lamp performance can be tested with the Uvicide Germicidal Lamp Monitor (Vangard International, Inc., UK).

An extremely important point about aseptic procedure, and one of the leading causes of contamination, are unclean hands. Simply rinsing the hands with water is insufficient; it is necessary to scrub them *Vigorously* with soap and hot water for several minutes. Attention must be given to the fingernails and to any part of the forearm that extends into the working area. It is not advisable to use strong disinfectants that could produce a skin rash. The hands may be dipped in a dilute solution of ethanol or isopropanol, although this can cause skin dryness and must not be used indiscriminately around an open flame. Biondi and Thrope (1981) reported that the seed coat could be removed manually from surface-sterilized seeds without contamination by dipping the fingers repeatedly in ethanol (40-70% v/v) prior to the operation. As an alternative to bare hands, some workers prefer to use disposable sterile surgeon's gloves.

Several techniques are used for the sterilization of glassware, surgical instruments, liquids, and plant material. The term "Sterilization" is the total inactivation of all forms of microbial life in terms of the ability of the organisms to reproduce (Willett, 1988). "Disinfection, " means the reduction of bacterial numbers to some arbitrary "acceptable" level (Hamilton, 1973). Ethanol is a bacteriostatic agent and disinfects a working area, but the treated area is not sterile. The methods of sterilization can be classified as follows: dry heat, moist heat, microfiltration, and chemical.

3.1. DRY HEAT

This method is used only for glassware, metal instruments, and other materials that are not charred by oven temperatures. Strips of tape should not be on any glassware, and the prolonged oven heat may dull the cutting edges of surgical blades. Cotton, paper, and plastic should never be placed in the oven for sterilization. Although laboratory-drying ovens may be used, the oven of a gas or electric kitchen stove will serve the same purpose. It is highly important not to be used a laboratory oven that has previously been used for paraffin embedding.

Objects to be sterilized are wrapped in heavy-duty aluminum foil before being placed in the oven. Take care not to pack the foil wrapped packages in the oven too tightly, but to leave some space between them (Hamilton, 1973). After sterilization the wrapped packages are taken to the laminar flow cabinet. In calculating the time required for dry-heat sterilization, three time periods must be considered:

1. Approximately 1 hr (heating-up period) is allowed for the entire load to reach the sterilization temperature of 180°C (356°F).
2. A minimum of 2hrs at this temperature is required to kill all organisms including the spore formers (Willett, 1988).
3. Finally, a cooling-down period is advisable in order to prevent the glassware from cracking due to a rapid drop in temperature.

3.2. MOIST HEAT

This procedure employs an autoclave operated with steam under pressure (Fig 2.10 and 2.11). If the laboratory is not equipped with an autoclave, a pressure cooker can be used. For the sterilization of paper products, glassware, instruments, and liquids the standard procedure is for the autoclave to operate at a steam pressure of 15 lb./in.^{2} (103.4kPa) and a chamber temperature of 121°C (250°F). For anything other than liquids, the time required for sterility is 15 min after the chamber has reached the sterilization temperature of 121°C. The time required for the sterilization of liquids varies considerably depending on the volume. The greater the volume, the longer it will take for the content to reach sterilization temperature (see Table 3.1).

Table 3.1. Minimum autoclaving time for plants tissue culture media (Burger 1988)

Volume of medium Per vessel (ml)	Minimum autoclaving time (min)
25	20
50	25
100	28
250	31
500	35
1000	40
2000	48
4000	63

Minimum autoclaving time includes the time required for the liquid volume to reach the sterilizing temperature (121°C) and 15 min at this temperature. Times may vary due to differences in autoclaves.

Some heat-resistant species of Bacillus that can withstand recommended sterilization times as well as instrument flaming (Leifert & Waites, 1990). If a pressure cooker is used, do not close the escape valve until a steady stream of pure steam is evident. At the end of the sterilization period, the pressure must be permitted to return to the atmospheric level slowly because rapid decompression will cause the liquids to boil out of the vessels. Prolonged autoclaving must be avoided because it results in the degradation of certain components of the medium. For example, one group reported that 5% of the sucrose in liquid medium was hydrolyzed during autoclaving (Hagen et al., 1991). If the autoclave does not have a drying cycle, paper products should be placed in a drying oven (<60°C) briefly in order to evaporate the condensed moisture. Steam in the autoclave chamber must penetrate the materials; a temperature of 121°C will not by itself achieve sterilization. With the exception of the flasks, instruments and other materials should be wrapped in unwaxed kraft paper. Although aluminum foil is commonly used as a wrapping, it is impermeable to the steam vapors and therefore is not recommended (Hamilton, 1973). Demineralized water should be used in boilers of autoclaves that generate their own supply of steam, as well as in pressure cookers. Steam generated by external power plants often contains contaminates that may be absorbed by the materials in the autoclave (Kordan, 1965; Bonga, 1982). Caution should be used in subjecting plastic labware to steam heat (Biondi & Thorpe, 1981).

Another form of moist-heat sterilization is a boiling-water bath. The bath is filled with distilled water and heated to 100°C. Instruments, placed directly into the water, should remain in the boiling water for a minimum of 20 min. Although microorganisms in the vegetative state are destroyed by this treatment, spores will be unaffected (Hamilton, 1973; Collins & Lyne, 1984). Also, a boiling-water bath is ineffective as a device for sterilization above 6, 000 ft. elevation.

3.3. FILTER STERILIZATION

Filtration is the process of removing contaminants in the range of 0.025-10 μm from fluids by passage through a microporous medium, such as membrane filter. This is necessary for media components that are degraded during autoclaving and must be sterilized at room temperature. A relatively small volume can be sterilized by passage through a filtration unit attached to a hypodermic syringe. For example, the Swinney (stainless steel) and the Swinnex (polypropylene) are reusable units equipped with membrane filters, whereas the Millex units and disposable (Millipore Corporation). Millex units are available in 4-, 13-, 25-, and 50-μm diameters. The appropriate volume of the sterile liquid is added directly to the autoclaved medium with a graduated syringe. If an agar medium is used, this is done while the agar is still hot (approximately 45°C) and in the liquid (sol) state. For larger volumes special units are equipped for either vacuum or pressure operation.

Most of the membrane filters are screens consisting of a uniform continuous mesh of polymeric material with pore size precisely determined by the manufacturing process. It is important to select a filter with a low level of water extractable since the release of these substances can be inhibitory to cell cultures (Cahn, 1967). An excellent choice for plant tissue culture work is the Millipore Durapore (Hydrophilic) membrane filter, which contains less than 0.5% by weight of water extractables. For the sterilization of hydrophobic solvents, such as dimethyl sulfoxide solutions, a Fluoropore (Millipore) filter constructed of polytetrafluoroethylene is recommended. Disposable filtration units composed of polystyrene with 0.1 or 0.2 μm cellulose acetate membranes are practical for infrequent filtration's. These units have been presterilized by radiation and may be purchased with a variety of filter pore sizes and filling and receiving containers. Although some workers use a 0.45 μm pore size, the pore diameter should measure at least 0.22-μm for the complete removal of all bacteria, yeast's, and molds (Torres, 1989).

3.4. CHEMICALS

The working area is generally disinfected with either ethanol or isopropanol (70% v/v). Although acidified alcohol (70% v/v, pH 2.0) may be a more effective disinfectant, it has a corrosive effect on metal instruments. A higher concentration of ethanol (80% v/v), a more inflammable mixture, is used for flaming instruments. An ethanol dip can be assembled by inserting a large test tube (20 mm O.D. × 150 mm) filled with ethanol (80% v/v) into an empty metal can. After immersion in the alcohol, the instrument is then passed through the flame of a methanol lamp. Avoid prolonged heating of the instrument after the methanol has evaporated. When not in use, the test tube should be capped to retard evaporation of the ethanol.

The surface sterilization of plant material may be accomplished with an aqueous solution of either sodium hypochlorite (NaOCI) or calcium hypochlorite (Ca[OCI]₂). This concentration is adequate for the surface sterilization of pith parenchyma explants. There is no general agreement in regard to NaOCI concentrations, and higher levels. The sterilization may be enhanced by agitating the solutions and by the addition of a wetting agent such as Tween 20 or Tween 80. Because of the corrosive effect on metal instruments within the chamber, the bleach solution and the rinse water should be discarded immediately after use (See Table 3.2).

Table 3.2. Surface sterilization of plant parts (Constabel, 1984)

Disinfectant	Concentration (%)	Duration of treatment (min)
Calcium hypochlorite	9-10	10-40
Sodium hypochlorite	0.9-2	10-40
Mercuric chloride	0.1	10
Ethanol	70	2-3
Hydrogen peroxide	3-12	5-15
Silver Nitrate	1	5-30
Benzalkonium chloride	0.1-1	2-10
Bromine water	1-2	2-10

Several agents were tested by Sweet and Bolton (1979) for the surface sterilization of seeds. Calcium hypochlorite was one of the most effective and least injurious agents. Sodium ions (i.e., in sodium hypochlorite) can induce abnormal development in some seedlings. In addition to Ca (OCI)$_2$, the mixture contained a phosphate buffer giving a final pH of 6.0, and a 1.0% solution of either Triton or Tween 80 as a wetting agent. The seeds were immersed in the hypochlorite mixture for 10 min and then rinsed in three changes of sterile distilled water. Other workers have used 9-10% calcium hypochlorite solutions for periods ranging from 5 to 30 min (*Sigma Plant Cell Culture 1995-1997*).

Some workers prefer to treat seed surfaces with a combination of hypochlorite and ethanol. One worker suggests treating barley seeds in the following manner. Seeds were immersed in a 20% bleach solution containing a detergent and agitated 15-20 min with a magnetic stirrer. They were transferred aseptically to a Buchner funnel and rinsed with sterile Double distilled water (DDH$_2$O). Then the seeds were covered with 70% ethanol for 1 min, and finally rinsed with sterile DDH$_2$O (Stiff, 1991).

Some plant tissues pose a problem if they contain microorganisms within the tissue sample. Surface treatments will obviously be completely ineffective. If a fleshy organ examined for explant preparation shows any localized discoloration, it should be rejected. The goal in surface sterilization is to remove all of the microorganisms with a minimum of damage to the plant cells to be cultured. In some cases, the achievement of this goal is empirical and the worker must be flexible in the approach to the problem (de Fossard, 1976). In the case of seeds, the use of higher concentrations of chemicals or longer periods of treatment does not appear to improve the decontamination without reducing the percentage of germination (Sweet & Bolton, 1979). The different surface-sterilizing agents with contact time and concentration is shown in Table 3.3.

Table 3.3. Surface sterilization procedures for different plant organs (Yeoman and Macleod, 1977)

Tissue	Pre-sterilization	Procedure of sterilization	Post-sterilization
Seeds	Submerge in absolute ethanol for 10s and rinse in sterile distilled water	Submerge seeds with intact testae for 20-30 min in 10% w/v calcium hypochlorite or for 5 min in a 1% (w/v) solution of bromine water	Wash three times in sterile water. Wash five times with sterile distilled water and germinate on damp. Sterile filter paper
Fruits	Rinse briefly with absolute ethanol	Submerge for 10 min in 2% (w/v) sodium hypochlorite	Wash repeatedly with sterile water, remove seeds of interior tissue
Pieces of stem	Scrub clean under running tap water and rinse with pure ethanol	Immerse for 15-30 min in 2% (w/v) sodium hypochlorite, remove ends	Wash three times in sterile water
Storage organs	Scrub clean under running tap water	Submerge for 20-30 min in 2% (w/v) hypochlorite	Wash three times in sterile water dry with sterile tissue paper
Leaves	Rub surface briefly with absolute ethanol	Immerse for about 1 min 0.1% (w/v) mercuric chloride	Wash repeatedly with sterile water, dry with sterile tissue

REFERENCES

Biondi, S. and Thorpe, T.A. (1981). Requirements for a tissue culture facility. In: *Plant tissue culture:* Methods and applications in agriculture, Thorpe, T.A. (eds) pp. 1-20. Academic Press, New York.

Bonga, J.M. (1982). Tissue culture techniques. In: *Tissue culture in forestry,* Bonga, J.M and Durzan, D.J. (eds.) pp. 4-35. The Hague: Martinus Nijhoff Junk.

Burger, DW (1988). Guidelines for autoclaving liquid media used in plant tissue culture. *Hortic Sci.* 23(6): 1066.

Cahn, R.D. (1967). Detergents in membrane filters. *Science,* 155: 195-196.

Collins, C.H. and Lyne, P.M. (1984). Microbiological methods. 5th Ed. Butterworths, London.

Constable, F. (1984). Callus culture induction and maintenance. In: Cell Culture and somatic cell Genetics of plants, Vol. I.K. Vasil (ed.) Academic Press, New York, pp 27-35.

De Fossard, R.A. (1976). Tissue culture for plant propagators. Armidale (Australia): University of New England.

Dixon, R.A. (1985). Isolation and maintenance of Callus and Cells suspension cultures. In: Plant Cell Cultures: A practical approach, Dixon, R.A.(eds), Oxford; IRL press. pp 1-20

Hagen, S.R, Muneta, P., Augustin, J. and Letourneau, D. (1991). Stability and utilization of picloram, vitamins, and sucrose in a tissue culture medium. Plant Cell, tissue & Organ Cult. 25 : 45-48.

Hamilton, R.D. (1973). Sterilization. In: Handbook of Phycological methods, . Stein, J.R., (eds.) Cambridge University Press, Cambridge, pp 181-93,

Klein, R.M. and Klein, D.T. (1970). Research methods in plant science. Garden City, Natural History Press. New York..

Kordan, H.A. (1965). Fluorescent contaminants from plastic and rubber laboratory equipment. Science 149 : 1382-3.

Leifert, C. and Waites, W.M. (1990). Contaminants in plant tissue cultures. Newsl. IAPTC Newsletter, 60 : 2-13.

Sigma Plant Cell Culture 1995-97. St. Louis: Sigma Chemical Company.

Stiff, C.M. (1991). Barley tissue culture methodology media, culture technique, and approaches to biolistic transformation.

Sweet, H.C. and Bolton, W.E. (1979). The surface decontamination of seeds to produce axenic seedlings. Am. J. Bot. 66 : 692-8.

Torres, K.C. (1989). Tissue culture technique for Horticultural crops. New York: Van Nostrand, Reinhold.

Wetherell, D.F. (1982). Introduction to in vitro propagation. Wayen, N.J.: Avery Publishing Group.

White, P.R. (1963). The cultivation of animal and plant cells, 2nd Ed. New York, Ronald press.

Willett, H.P. (1988). Sterilization and disinfection. In Zinsser microbiology, Joklik, W.K., Willett, H.P. Amos, D.B and Wilert, C.M. (eds.) Norwalk, Conn.: Appleton & Lange, 19, pp. 161-71

Yeoman, M.M., and Macleod A.J. (1977) Tissue (Callus) cultures techniques In: Street, H.E., (eds.) Plant Tissue and Cell Culture. Botanical Monographs. Vol II. Blackwell, Oxford, p. 31.

Chapter 4

Nutritional Requirements of Plant Cell Cultures

INTRODUCTION

There are three essential sources of nutrition for plants growing in nature. The mineral nutrients are obtained, along with water, from the soil through the root system. Atmospheric carbondioxide is used in the process of photosynthesis to provide carbon as a source of basic energy. Lastly, the plant, particularly its meristematic regions and young organs such as leaves, using fixed carbon and minerals, synthesizes all of the vitamins and various plant growth substances that are critical and essential for normal growth and development of the plant.

The requirements of plant tissues grown *in vitro* are similar in general to those of intact plants growing in nature. The nature of the explant and the composition of the nutrient medium (White 1951) generally determine the successful establishment and growth of plant cells *in vitro*. In the first culturing attempts, either media that were known from nutrition experiments with intact plants (Knoop solution), or media consisting of juices and extracts of biological origin (Haberlandt 1902, Gautheret 1937, White 1934) were used.

Today, the only that is still commonly used is coconut milk, however only for monocotyledonous cultures. Mainly media of purely chemical composition are used (Rechigl 1977).

Media containing nutrients of plant origin that are chemically not precisely characterized are called complex or highly enriched media, while those containing exclusively chemically defined compounds are called synthetic or regular media.

To maintain the vital functions of a culture, the basic medium consisting of inorganic salts, organic compounds (amino acids, vitamins), growth regulators (phytohormones) and carbon sources recognized as essential (Gamborg et al 1968, Schenk and Hildebrandt 1972, Murashige 1973).

4.1. INORGANIC COMPONENTS

Basic media are solutions of inorganic salts in different concentrations, called macro and micro nutrients. Cultured plant tissues require a continuous supply of certain inorganic chemicals. The macro nutrients are the compounds containing N, S, P, K, Mg, K as chlorine and sodium, added in concentrations of more than 30 ppm. In contrast, the elements added in the concentrations of less than 30 ppm e.g. Boron, Iron, Manganese, Iodine and Molybdenum and rarely Copper and Zinc are called Micronutrients. The Micronutrients are necessary as cofactor or enzyme synthesis e.g. nickel is essential for urease synthesis. The role of various components are given in Table 4.1.

Table 4.1. Examples of the function of some nutritional elements (Adopted from Endress, 1994)

Element	Function
Fe	Co-factor of cytochromes, peroxidases
Cu	Co-factor of phenol oxidases
	Ascorbic acid oxidase
Zn	Co-factor of peptidase
Mo	Co-factor of nitrate-reductase, xanthine oxidase
Co	Co-factor of Vitamin B_{12}
	3-deoxy-D-arabinoso-heptulose-7-phosphate (DAHP) synthase of the cytoplasm
Mg	Co-factor of phosphatases, kinases,
	Activator of ribulose-1, 5-PP-carboxylase oxygenase
	Translation factor
Ni	Induction of urease synthesis
Bo	Chemical composition, permeability and integrity of the plasmalemma, Influences P_1 and glucose uptake,
	Enhances IAA oxidase activity
Ca	Stabilization (α-amylase) and activity (NAD-kinase), protein-kinase, (α-amylase) of metallo enzymes
	Supports phospholipids on and deposition in membranes
	Combines pectinic acids forming pectate

4.2. MACRO ELEMENTS

4.2.1. Sulfur

Sulfur is primarily supplied as sulphate (SO^{-2}_4). Usually, it is utilized for protein synthesis via sulphate respiration as soluble cysteine (99.9%) and a smaller proportion as soluble methionine (Giovanelli et al. 1980). The sulphate supply is directly incorporated only beneath a minimal concentration threshold. In a medium lacking Sulfur or in the presence of growth-limiting factors, e.g. the nonprotein amino acid djenkolic acid [HOOC-CH(NH_2)-CH_2-S-CH_2-S-CH_2-CH(NH_2)-COOH] (Bell 1981), this becomes a apparent due to a five-to ten-fold increase in the level of ATP-dependent sulphurylase (Klapheck et al. 1982). The sulphur requirements of a culture vary depending on the object (0.5 to 10 mM). Inorganic sulphur may be replaced by organically bound sulphur (DL-cysteine, DL-methionine, DL-homocysteine and glutathione).

4.2.2. Phosphorus

Phosphorus is commonly added as PO^{-3}_4 at concentrations of 1.1-1.25 mM (Murashige and Skoog 1962). Due to rapid uptake (Kato et al. 1977) and interactions with other components (Fe, K, Sucrose), deficiencies may rapidly arise in a medium. In addition, its uptake is influenced by the supply of other elements. For example, boron deficiency induces in *Daucus carota* cultures a reduction in the phosphorus uptake capacity (Goldbach 1985).

4.2.3. Nitrogen

Most standard media (Table 4.2) offer nitrogen as NH^+_4 and NO^-_3. Individual cultures (*Cannabis sativa, Ipomoea, Daucus carota*) prefer NH^+_4 under certain conditions. Utilization of NO^-_3 requires

Table 4.2. Composition of some media used in plant cell and tissue culture

Name Components	Concentration (mg/L)								
	Gamborg's B5	Heller's Salts	Linsmaier–Bednar & Skoog	Murashige & Skoog minimal organic	Nitsch's H	Schenk Hildebrandt	Takebe	White's Salts	WPM (Woody plant medium)
AlCl$_3$	–	0.03	–	–	–	–	–	–	–
CaCl$_2$.2H$_2$O	150	750	440	440	160	200	220	–	96
Ca(NO$_4$)$_2$.4H$_2$O	–	–	–	–	–	–	–	3000	556
CoCl$_2$.6H$_2$O	0.025	0.03	0.025	0.025	–	0.1	0.025	–	–
CuSO$_4$.5H$_2$O	0.025	0.03	0.025	0.025	0.025	0.2	0.025	–	0.25
FeCl$_3$.6H$_2$O	–	1.0	–	–	–	–	–	–	–
Fe$_2$(SO$_4$)$_3$	–	–	–	–	–	–	–	2.5	27.8
FeSO$_4$.7H$_2$O	–	27.8	–	–	–	–	–	–	99.00
H$_3$BO$_3$	3	30	–	–	–	–	–	–	6.2
KCl	–	7500	–	–	–	–	–	–	–
KH$_2$PO$_4$	–	–	170	170	68	–	680	68	170
KI	0.75	0.01	0.83	0.83	–	1	0.83	0.75	–
KNO$_3$	2500	–	1900	1900	950	2500	950	80	–
MgSO$_4$.7H$_2$O	246	250	370	370	185	400	1223	720	370
MnSO$_4$.H$_2$O	10	–	16.89	–	–	10	16.9	–	–
MnSO$_4$.4H$_2$O	–	0.1	–	22.3	25	25	–	7	22.3
NaH$_2$PO$_4$ anhydrous	–	–	–	–	–	–	300	–	16.5
NaH$_2$PO$_4$.H$_2$O	150	–	–	–	–	–	–	–	–
NaH$_2$PO$_4$.2H$_2$O	–	141	–	–	–	–	–	–	–
NaNO$_3$	–	600	–	–	–	–	–	–	–

(Contd.)

Name / Components	Gamborg's B5	Heller's Salts	Linsmaier–Bednar & Skoog	Murashige & Skoog minimal organic	Nitsch's H	Schenk Hildebrandt	Takebe	White's Salts	WPM (Woody plant medium)
Na_2EDTA	–	–	–	37.3	–	20	37.3	–	–
$Na_2EDTA.2H_2O$	–	–	37.25	–	–	–	–	–	37.3
$Na_2MoO_4. 2H_2O$	0.25	0.25	0.25	–	0.25	0.1	0.25	–	0.25
Na_2SO_4	–	–	–	–	–	–	–	200	–
NH_4NO_3	–	1650	1650	720	–	–	8.25	–	400
$(NH_4)_2SO_2$	134	–	–	–	–	–	–	–	–
$NiCl_2 6H_2O$	–	0.03	–	–	–	–	–	–	–
$ZnSO_4. 7H_2O$	2	1	10.58	8.6	10	1.0	8.6	3.0	8.6
1–Inositol	100	–	100	100	–	1000	100	–	100
Nicotinic acid	1	–	–	–	–	5	–	–	0.5
Pantothenic acid	0.4	–	–	–	–	–	–	–	–
Pyridoxine HCl	1	–	–	–	–	5	–	–	–
Riboflavin	0.0015	–	–	–	–	–	–	–	–
Thiamine HCl	10	–	0.4	0.1	–	5	10	–	16
Sucrose	20	20	30	20	–	25	30	20	20
pH	5.5	5.5	5.8	5.5	5.8	5.5	5.8	5.5	5.5

WPM needs K_2SO_4 6.2 mg/l in addition to the above composition.

functioning nitrate reductase, the presence of which has by now been described in numerous callus and suspension cultures (Bray 1983).

In a few cases, other sources of nitrogen may replace NO_3 or NH_4 as the only nitrogen sources or they may augment the existing supply. The supply already present from the medium solidifier is not considered here. (Table 4.3). In some cultures (*Nicotiana tabacum, Daucus carota*), use of amino acids (threonine, glycine, valine) for this purpose leads to diminished ammonia assimilation.

Table 4.3. Amount (%, mg/l) of amino acids and ions in different forms of agar (Sigma 1990)

Amino acids	Agar-Agar purified[a] (μM/10mg)	Ions	Agar[b] (%)	Type A[b] (%)	Type M[b] (%)	Washed[b] (%)	Bacto-agar[c] (%) (mg/l)		Noble agar[c] (%) (mg/l)		Purified agar[c] (%) (mg/l)	
ASP	0.37	Ba	–	–	–	–	0.01	–	0.01	–	0.01	–
Thr	0.162	Ca	0.05	0.01	0.29	0.15	0.13	–	0.23	–	0.27	–
Ser	0.237	Cl	–	–	–	–	0.43	–	0.18	–	0.13	–
Pro	0.197	Cu	–	–	–	–	–	5.00	–	7.50	–	20.00
Glu	0.372	Fe	–	–	–	–	–	11.00	–	11.00	–	11.00
Gly	0.380	K	–	0.1	–	0.07	–	–	–	–	–	–
Ala	0.375	Mg	–	0.01	0.14	0.09	–	285.00	–	260.00	–	695.00
Val	0.225	N	–	–	–	–	0.17	–	0.10	–	0.14	–
Ile	0.122	Na	1.3	1.8	1.4	0.43	–	–	–	–	–	–
Leu	0.227	P	–	0.17	0.01	–	–	–	–	–	–	–
Tyr	0.182	S	0.6	1.2	0.7	0.51	2.54	–	1.90	–	1.32	–
Phe	0.110	Si	–	0.01	0.02	–	0.19	–	0.26	–	0.09	–
Lys	0.115											
His	0.047											
Arg	0.090											
Cys	–											
Met	–											

– Below detection limit; [a] Specification Merck; [b] Specification Sigma; [c] Specification Difco.

4.2.4. Mg, K, Ca

The cations Mg, Ca and K play an essential role in cell metabolism. For example, Mg^{2+} is one of the essential factors in translation. It acts as a cofactor (e.g. glutamine synthase) and activator of various enzymes. Therefore, not least in photoautotrophic cultures, it is of central significance. K^+, and especially Ca^{2+}, inhibit enzymes such as the glycolysis enzyme pyruvate kinase, while others require Ca^{2+} to maintain their activity (NAD-kinase, protein-kinase, α-amylase) or stability (α-amy-lase). The Ca^{2+}- triggered binding of pectic acid, a polymerization product of galacturonic acid, to calcium pectate is an elementary step in cell wall formation. Ca^{2+} is also required for deposition of phospholipids and proteins on or within plasma membranes. Its importance is further demonstrated by the efforts of cells to maintain their intracellular concentration 10^{-6} to 10^{-8} M even against a concentration gradient using specific Ca^{2+} pumps and Ca^{2+} binding proteins (calmodulin) located in the cytoplasm and/or individuals organelles. The concentration increase to a value of 10^{-5} M induced by the sesquiterpenoid phytohormone abscisic acid (ABA), an apo-carotenoid and by light is only temporary and is the basis of its signaling effect in its function as a second messenger. In *Nicotiana tabacum* culture deficiency in nitrate reductase, an increased level of Ca^{2+} induces increased ammonium utilization. Chlorine plays a role by binding to positively charged histidine residues of proteins like the enzymes of the photosystem II and ATPases of the tonoplast and by influencing osmoregulation.

4.3. MICRONUTRIENTS

The Fe, Mn, Zn, Cu, Mo, B, Co, and Ni act as cofactors (Table 4.1) and as inducers of enzyme synthesis, as for example nickel in urease synthesis in tobacco, rice and soybean cell suspension cultures. Boron is essential for membrane function, permeability and integrity. Therefore, membrane-fixed processes like ATPases, membrane potential and iron-flow and phytohormone metabolism are influenced. Lack of iron results in increased contents of DNA and free amino acids, as well as a reduced RNA content (Table 4.4; Koblitz 1977). In order to maintain a minimum supply of Fe it is therefore usually added in complexes with EDTA or sequestrin. This also facilitates uptake over a broad pH range, which varies depending on the content of phosphate, NO_3^- and NH_4^+ in the medium. Although iodine in the form of KI is a constituent of several media, the necessity of this element remains questionable. A trace of cobalt is found in several media and yet this element is not known to have any function in higher plants.

Table 4.4. Effect of Lack of Iron in Cell Cultures of rice (Koblitz, 1977)

Parameter	Effect
Respiration	–50%
Growth	–50%
Protein /g Fresh weight	+20%
Glucose, amino acids	4 fold increase
RNA content	Reduction
DNA	Increase

4.4. ORGANIC COMPONENTS

4.4.1. Amino Acids

Amino acids are added for substitution or augmentation of the nitrogen supply. It is to be noted that threonine, glycine and valine reduce ammonium utilization by inactivating glutamate synthase located in chlorophasts and cytoplasm. Arginine is usually able to compensate this inactivation.

4.4.2. Vitamins

Plant cells are usually autotrophic with respect to vitamins. However, in most cases, the amount of vitamins synthesized even in photosynthetically active cells and tissues is insufficient to guarantee a sufficient supply. Thiamine (vitamin B_1) may be the only essential vitamin for nearly all plant tissue cultures, whereas nicotinic acid (niacin) and pyridoxine (Vitamin B_6) may stimulate growth (Gamborg et al 1976, Ohira *et al.,* 1976). Thiamine is added as thiamine hydrochloride in amounts ranging from 0.1-10 mg/l.

Some other vitamins that have been used in tissue culture media include p-aminobenzoic acid (PABA), ascorbic acid (vitamin C), tocopherol (vitamin E), biotin (vitamin H) choline chloride, cyanocobalamin (vitamin B_{12}), folic acid, riboflavin (vitamin B_2) and calcium pantothenate (Hung and Murashige 1977, Gamborg and Shyluk 1981).

4.5. CARBON SOURCE

All plant tissue culture media requires the presence of a carbon source and is added in the form of carbohydrates. Sucrose is usually added in concentration of 20-30g/l. Often, myoinositol is also used. Occasionally less common carbon sources such as lactose, galactose, glycerine and unrefined natural carbon sources are also employed (Table 4.5). The various carbon sources used in tissue culture medium are showed Table 4.6. However, some cultures can satisfy their energy requirements by assimilating carbon dioxide (Bergmann 1967) and they can survive photoautotrophically. In some cultures, the medium solidifier may also be used as carbon source.

Table 4.5. The plant cell cultures utilizing galactose or lactose

Lactose	Galactose
Coffea arabica	Daucus carota
Datura innoxia	
Daucus carota	Pentunia hybrida
Medicago sativa	
Petunia hybrida	
Vinca minor	

Table 4.6. The various carbon sources used in plant tissue culture media (Adopted from Endress, 1994).

Commonly used	Scarcely used
Glucose	Lactose
Sucrose	Galactose
Glycerol	Non-refined carbohydrates
Pentoses	Molasses
Uronic acid	Whey
	Potato starch
	Grain starch

4.6. PLANT GROWTH REGULATORS

The success of plant tissue, cell and organ culture will depends on the amount of plant hormones and plant growth regulators or growth substances added into nutrient medium. Plant growth regulators are organic compounds, which affect the morphological structure and/or physiological process of plants in low concentrations. Phytohormones or plant hormones are naturally occurring growth regulators, which in low concentration control physiological process in plants. More commonly, the term plant growth regulators is used, because it includes both the native (endogenous) and synthetic (exogenous) substances; which modify the plant growth.

Auxins, ethylene, abscisic acid, cytokinins and gibberellins are commonly recognized as the five main classes of naturally occurring plant hormones. Auxins, cytokinins, and auxin-cytokinin interactions are usually considered to be most important for regulating growth and organized development in plant tissue and organ cultures, as these two classes of hormones are generally required (Evan et al 1981, Vasil and Thrope 1994). However, absciscic acid, ethylene, gibberellins and other hormone-like compounds have regulatory roles, which must not be ignored in culture system. For instance, although one may not need to add abscisic acid, ethylene and gibberellins to cultured cells to ensure organogenesis or cell proliferation, this does not mean that these hormones are of no importance.

Synthetic compounds that act like natural plant hormones are called "Plant growth regulators" (Davies, 1995). Many such plant growth regulators have been discovered with biological activities, which equals or exceeds that of the equivalent endogenous hormones. In addition to these useful compounds, there are now quite a number of chemicals which interfere (generally inhibit) with the synthesis, transport, or action of endogenous hormones. These inhibitors are extremely helpful in the study of the role of plant hormones in *in vitro* cultures.

In addition to the classical plant hormones, new natural growth substances with regulatory roles in tissue cultures have been discovered in the last few years (Gross and Partheir, 1994). Examples of these are polyamines, jasmonates, brassinosteroids, oligosaccharins, sterols, phosphoinositosides, salicylic acid, and systemins.

The effects of natural and synthetic plant growth regulators are rarely specific in their ultimate influence on growth and development, and the responses of cells, tissues, and organs *in vitro* can vary with cultural conditions, the type of explant, and the genotype. Usually a combination of two or more growth regulators of different classes is required; either applied simultaneously or sequentially (Evans et al. 1981). One must be aware that besides exerting a direct effect on cellular mechanisms, many exogenously applied synthetic and natural regulators may modify the synthesis, destruction, activation, sequestration, transport, or sensitivity to endogenous growth substances of the same or other types (Beale and Sponsel, 1993; Davies, 1995).

4.6.1. Auxins

Auxins show a strong influence over processes such as cell growth expansion, cell wall acidification, initiation of cell division, and organization of meristems giving rise to either callus or defined organs (generally roots). In organized tissue, auxins cause, root formation affecting abscission, delaying leaf senescence, and fruit ripening (Addicott 1982, Sabater 1985, Tamas 1995, Aloni 1995, Liu and Reid 1992).

4.6.1.1. Naturally occurring auxins

The most commonly detected natural auxin is indole-3-acetic acid (IAA), but depending on the species, age of the plant, season, and the conditions under which it has been growing, other natural auxins have been identified such as 4-chloroindole-3-acetic acid, indole-3-acrylic acid, indole-3-butyric acid (IBA). Although it has been know for some time that IAA is commonly formed from L-tryptophan via indole-3-pyruvic acid or tryptamine, other pathways exist (Bandurski *et al.,* 1995 see Fig. 4.1).

Within plant tissues, IAA and other naturally occurring auxins combine with small molecules (alcohols, amino acids, sugars) to produce ester, amide, or glycoside conjugates (Bandurski et al., 1995). This appears to be a mechanism for storing auxins in cells and stabilizing the level of free auxin by metabolizing the excess. Auxin in conjugated molecules is protected from oxidative breakdown and may be later enzymatically released when required. Indole acetyl aspartic acid and its conjugates with sugars, such as glucobrassicin (in Crucifereae), are commonly found (Fig. 4.2.). There have been several reports on the successful use of IAA conjugates as growth regulators in plant cultures (Bandurski et al., 1995).

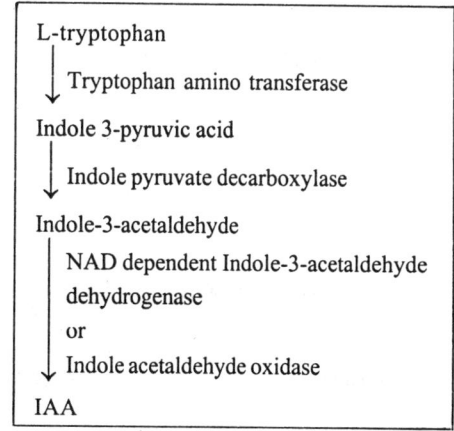

Fig. 4.1. Biosynthesis of Indole acetic acid.

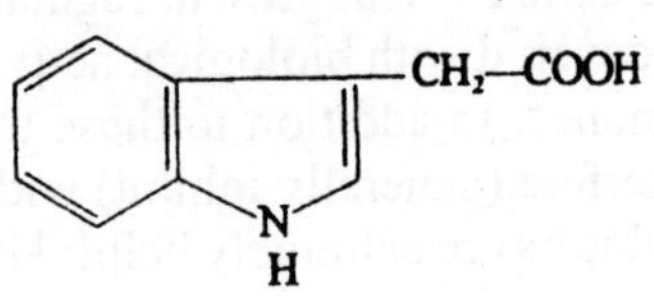

Indolyl-3-acetic acid (IAA)

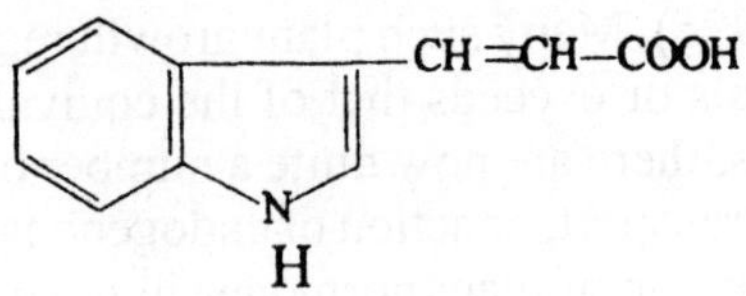

Indolyl-3-acrylic acid (IAcrA)

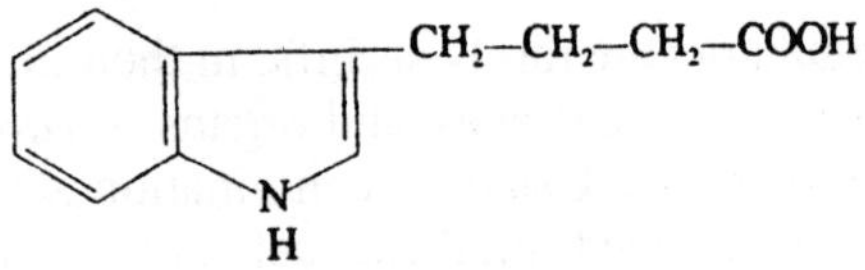

Indolyl-3-butyric acid (IBA)

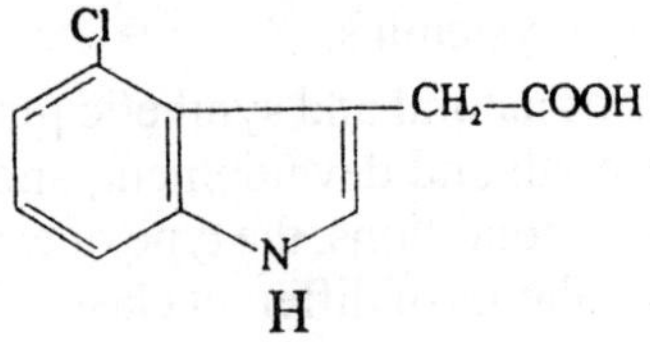

4-Cl-indolyl-3-acetic acid (4-Cl-IAA)

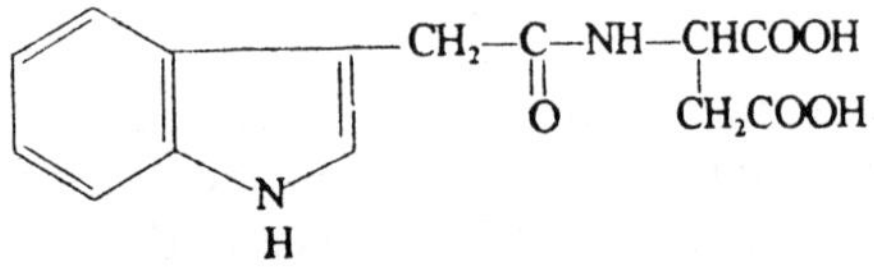

Indolyl-3-acetylaspartate

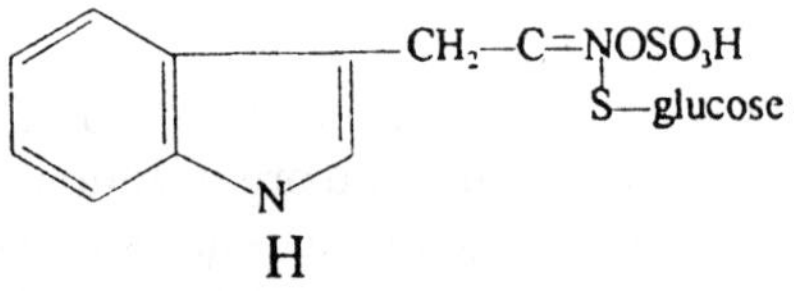

Glucobrassicin

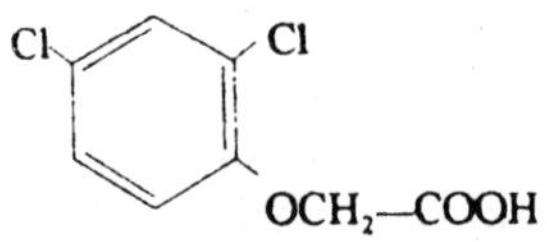

2,4-Dichlorophenoxyacetic acid (2,4-D)

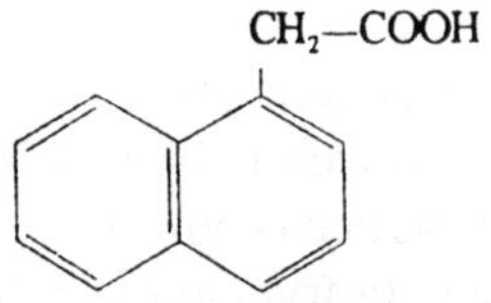

1-Naphthaleneacetic acid (NAA)

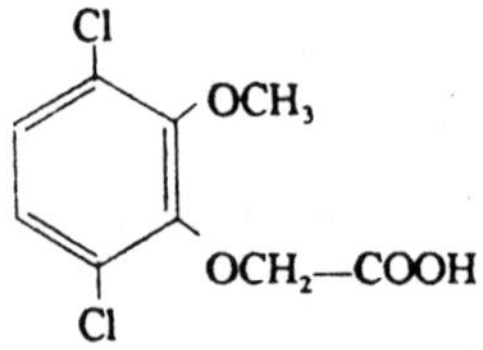

Dicamba

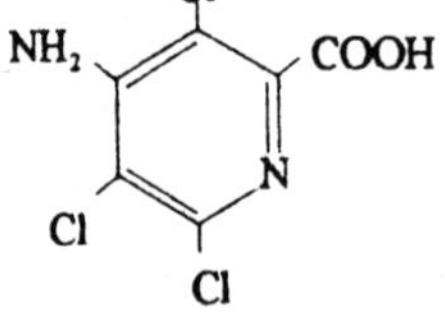

Picloram

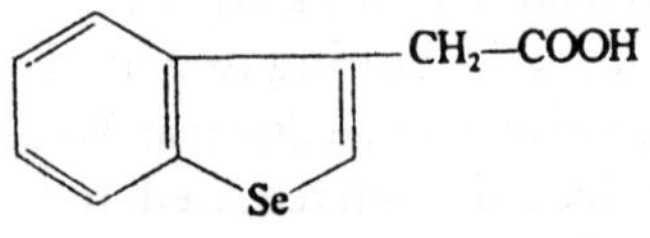

Benzo(b)selenienyl-3 acetic acid (BSAA)

Fig. 4.2. Some synthetic and natural auxins (Adopted from Gaspar et al., 1996).

4.6.1.2. Auxin-like growth regulators

Synthetically-prepared IAA and IBA are commonly used in plant culture media. They tend to be denatured in media and rapidly metabolized within plant tissues. These attributes can be useful when developmental phases in progress require less auxin (in rooting for instance, Gaspar et al., 1995), or in automatically changing the auxin : cytokinin ratio (organogenesis from some callus). Several indole derivatives, both naturally occurring and synthetic, are active in culture, e.g. inolde-3-acetaldehyde, indole-3-acetamide, indole-3-propionic acid, indole-3-pyruvic acid, and tryptophan, as well as phenylacetic acid have all been shown to support callus growth and shoot formation in tobacco callus cultures (Maeda and Thorpe, 1979).

Commonly used synthetic auxins in tissue culture are 2, 4-dichlorophenoxyacetic acid (2, 4-D; often used for callus induction and suspension cultures), and 1-naphthaleneacetic acid (NAA; when organogenesis is required; Fig. 4.2). Among others, dicamba (3, 6-dichloro-o-anisic acid) and picloram (4-amino-3, 5, 6-trichloropyridine-2-carboxylic acid) are often effective in inducing the formation of embryogenic tissue or in maintaining suspension cultures (Gray and Conger, 1985; Hagen et al., 1991). BSAA [benzo(b)selenienyl-3 acetic acid] is another synthetic auxin with powerful auxin-like activities (Lamproye et al., 1990; Gaspar, 1996).

4.6.1.3. Regulation of levels and activity of auxins

Auxin activity is dependent on the free availability of boron. In boron-deficient plants, both the translocation of IAA and nuclear RNA synthesis in response to auxin treatment can be inhibited. Boron may affect rooting by controlling IAA-oxidase activity (Jarvis et al., 1983). Other hormones such as cytokinins (Aloni. 1995) and ethylene can also modify the responsiveness of tissues to auxins.

For the regulation of at least some processes involving cell growth and cell division, it is thought that auxins need to become bound to auxin-binding proteins. Specific high-affinity binding proteins have been found (Venis and Napier, 1991). However, only if these proteins can be shown to be involved in the physiological action of auxin, can they be truly called receptors?

The levels of endogenous free auxins depend on the rate of their anabolism, catabolism, transport and conjugation (Bandurski et al., 1995). Increased endogenous concentrations of IAA, via synthesis, generally depend on the quantity of precursors. Auxin biosynthesis can also be influenced by treatments, which alter internal ethylene levels. However, modification of the oxidation or conjugation rate appears to be an important way that regulates internal IAA levels naturally. Phenolic inhibitors of IAA oxidase such as phloroglucinol, catechol, chlorogenic acid, and rutin have been shown to enhance auxin activities, being sometimes described as "auxin-synergists" (Bearder, 1980). Activators of IAA-oxidase can behave as auxin antagonists (Gaspar et al., 1996).

4.6.2. Cytokinins

Cytokinins are useful in culture for stimulation of cell division and release of lateral bud dormancy. They can also induce adventitious bud formation (Fabijan *et al.,* 1981; Krikorian 1995). Cell division is regulated by the joint actions of auxins and cytokinins, each of which influences different phases of the cell cycle. Auxins affect DNA replication, where as cytokinins seems to exert some control over the events leading to mitosis (Vesely et al 1994). Thus, auxin and cytokinin levels in cultures need to be carefully balanced and controlled. In intact plants, cytokinins promote lateral bud growth and leaf expansion, promote chlorophyll synthesis and enhance chloroplast development (Kuhnle et al. 1977; Nooden and Leopold 1988.)

4.6.2.1. Naturally occurring cytokinins

Cytokinins have been identified, either as free compounds, glucosides, or ribosides. The most commonly used cytokinins in plant tissue culture are zeatin, 2-iP, dihydro-zeatin, and zeatin riboside (Fig. 4.3). These natural cytokinins contain an isoprenoid side chain attached to the N^6-position of the adenine. Cytokinins with an aromatic ring substituting at N^6 (6-benzyladenine and its glycosides) have been identified recently.

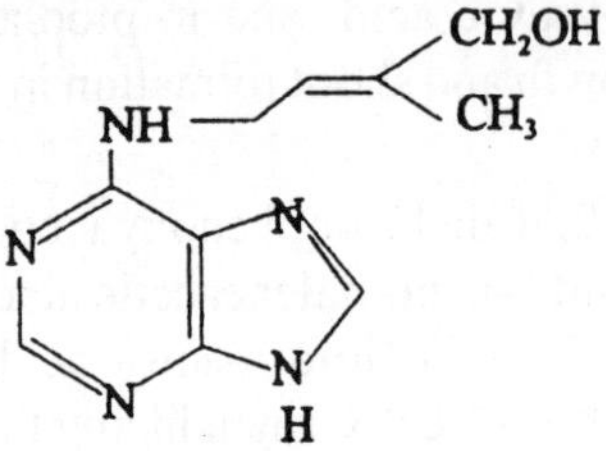

trans-Zeatin

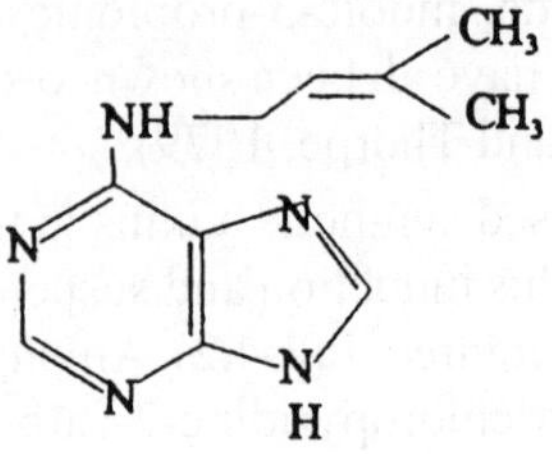

2-iP, N^6-(2-isopentyl)adenine

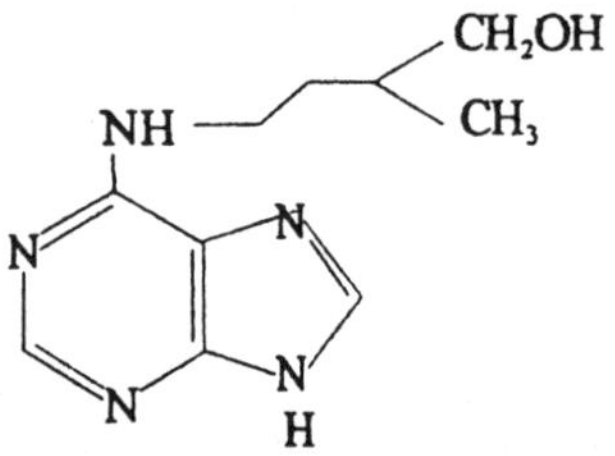

dihydro-Zeatin

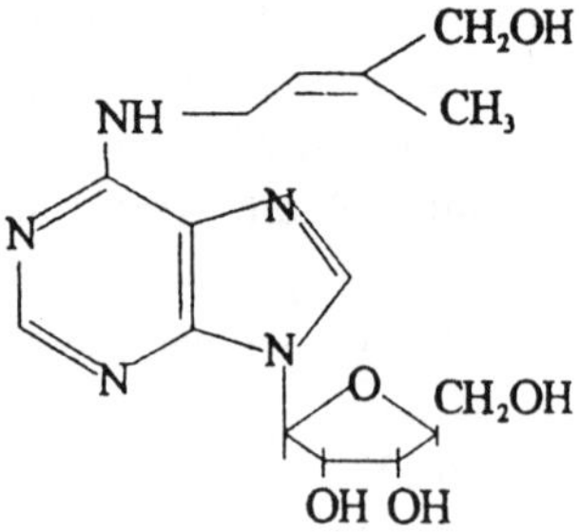

Zeatin riboside

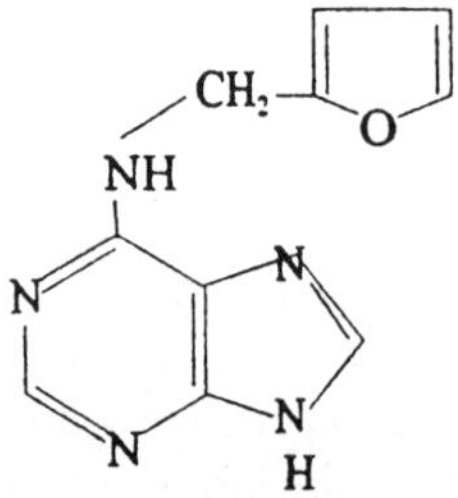

Kinetin 6-furfurylaminopurine

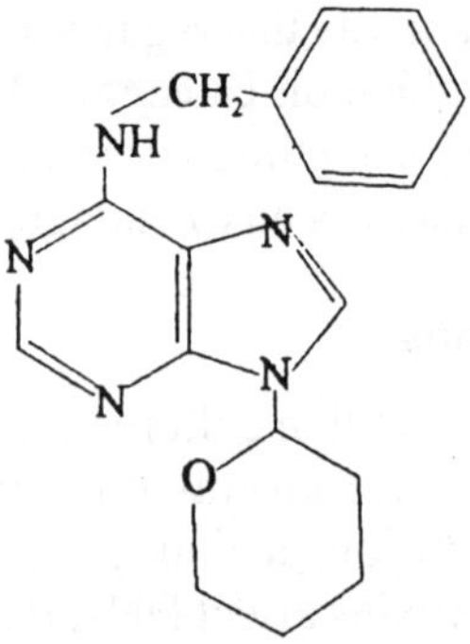

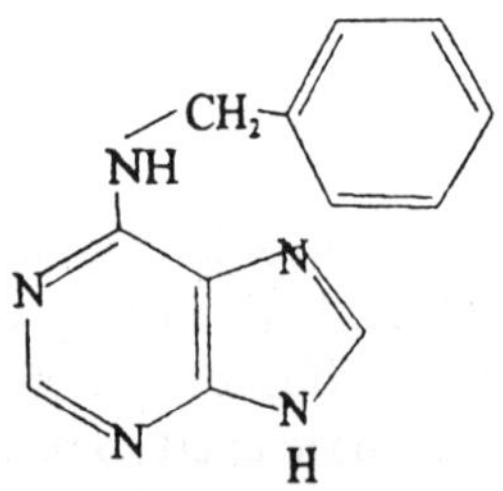

BA 6-benzylaminopurine
or benzyladenine

PBA 6-(benzylamino])-9-
(2-tetrahydropyranyl)-9H-purine

Fig. 4.3. Some synthetic and natural cytokinins (Adopted from Gaspar et al., 1996).

4.6.2.2. Cytokinin-like growth regulators

The most commonly used cytokinins are the substituted purines: kinetin and BA (Fig. 4.3). Adenine (vitamin B$_4$), adenosine, and adenylic acid may have cytokinin activity, although less than that of the cytokinins.

Many substituted ureas (Fig. 4.4) have cytokinin activity (Krikorian, 1995). In some plants, thidiazuron is more effective than adenine-based compounds for inducing axillary or adventitious shoots. Such shoots, however, tend not to elongate sufficiently and are susceptible to rapid hyperhydricity on repeated subcultures in the presence of this compound (Murthy et al., 1998).

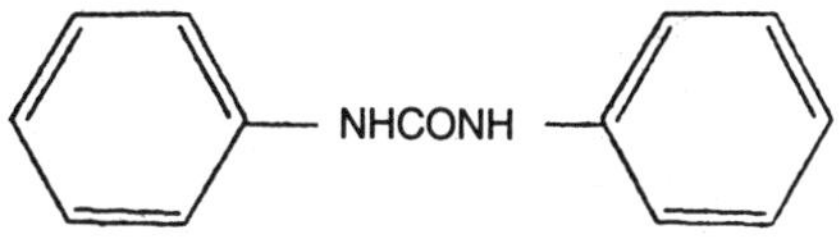

4.6.2.3. Regulation of activity and levels of cytokinin

The pool of active cytokinins present at any one time is a product of a number of dynamic processes: biosynthesis, formation or mobilization of storage forms (sequestered or O-glucoside forms), and deactivation (N-glucosylation, alanine conjugation and side chain removal by cytokinin oxidase (Mc Gaw and Burch, 1995).

Many aspects of cell differentiation, and organogenesis in tissue and organ cultures have been found to be controlled by an interaction between cytokinins and auxins. The requisite concentration of each phytohormone varies greatly according to the kind of plant being cultured, the cultural conditions, and the form of the phytohormone used. Although both auxin and cytokinin are usually required for growth and morphogenesis, auxin can inhibit cytokinin accumulation, whereas cytokinins can inhibit at least some of the actions of auxin.

1,3-diphenylurea

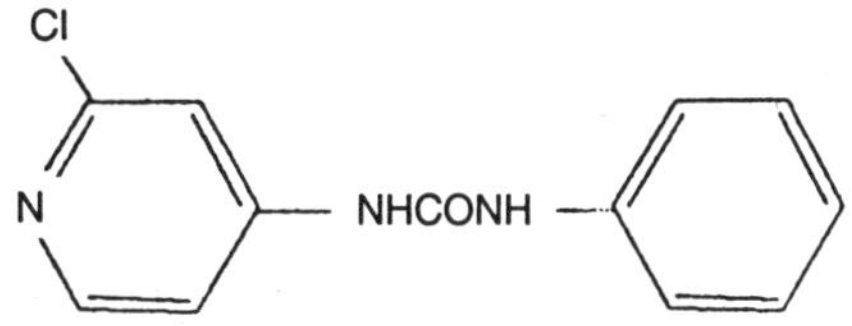

2 Cl-4PU or PPU
N-(2-chloro-4-pyridyl)-N'-phenylurea

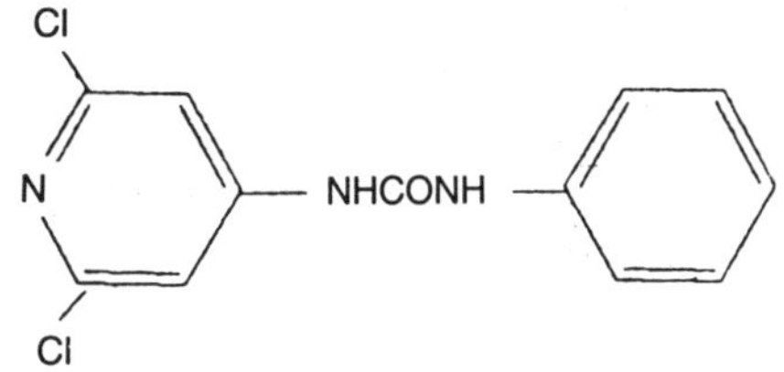

2,6 Cl-4PU
N-(2,6-dichloro-4-pyridyl)-N'-phenylurea

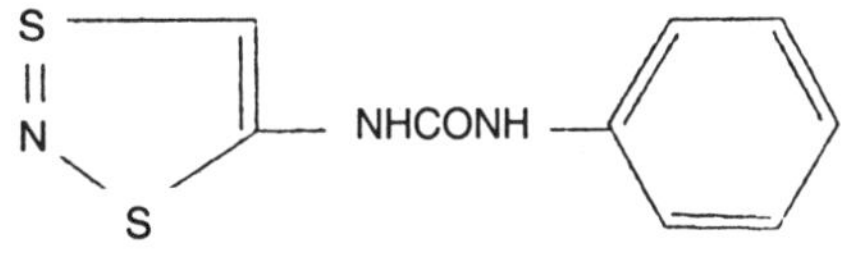

Thidiazuron (TDZ)
N-phenyl-N'-1,2,3-thiadiazol-5-ylurea

Fig. 4.4. Phenylureas with cytokinin activity.

Some of the effects of cytokinins on plant metabolism, physiology, and development can be antagonized by abscisic acid. In many tissues, including cultured tissues, cytokinins often promote ethylene biosynthesis (Gamborg and LaRue, 1971; Wright, 1979; Abeles et al., 1992).

The ability to stimulate cytokinin oxidase appears to be a property of a limited range of phenolic structures such as 2, 6-dimethoxyphenol and acetosyringone (3, 5-dimethoxy-4-hydroxy-acetophenone). The activity of acetosyringone is of particular interest because this compound is reported to serve as a signal compound in the interaction of the plant pathogen *Agrobacterium tumefaciens* with its higher plant hosts.

4.6.3. Gibberellins

Gibberellins (GAs) will promote flowering, cone initiation in some conifers, seed germination, and stem elongation (by increasing cell division and elongation). The ability to promote bolting and stem elongation can be exploited *in vitro* for the elongation of the shoots of woody species before rooting. Some GA effects are caused by increase or decrease in the biosynthesis and activity of specific enzymes (e.g., increase in levels of aleurone hydrolytic enzymes).

When GAs are added to plant tissue culture media, they often diminish or prevent the formation of roots, shoots, or somatic embryos, although the opposite has also been seen. Some endogenous GAs may be necessary for normal callus growth (Lance et al., 1976), and inhibition of GA biosynthesis can influence development of cells in liquid cultures (Ziv and Ariel, 1991). Some differences in the inhibition or promotion of adventitious root and shoot formation by GAs might be due to the fact that GAs inhibit, meristemoid initiation (Thorpe and Murashige, 1970) but are required for assisting the further growth and development of preformed organs. GAs may also alter the availability of endogenous auxin. Growth of shoots in meristem and shoot cultures may be enhanced by addition of GA (Fry and Street, 1980).

4.6.3.1. Naturally occurring gibberellins and regulation of levels of activity

About 90 naturally occurring GAs are known (Fig. 4.5). Cultured callus cells synthesize their own GA and can metabolize added GA (Lance et al., 1976). Because anti-GAs may normally influence growth of cultured cells.

GAs have numerous interactions with other hormones. GA-induced α-amylase activity is antagonized by ABA. Ethylene blocks (or in rice, promotes) the ability of stems to respond to GAs. GA antagonizes the senescence promoting effects of ABA and ethylene in leaves and petals. Light conditions (wavelength and photoperiod) affect GA metabolism; however, the precise mechanism still requires much study (Reid et al., 1991).

4.6.4. Abscisic Acid (ABA)

Abscisic acid (ABA) and other structurally related natural compounds with similar activity (Fig.4.6.) are most likely produced by the cleavage of xanthophyll. ABA is often regarded as being an inhibitor, as it maintains bud and seed dormancy, inhibits auxin-promoted cell wall acidification loosening, and slows cell elongation. ABA plays a key role in closing of stomatal apertures (reducing transpiration), control of water and ion uptake by roots (in part by increasing hydraulic conductivity) and, with other phytohormones, promoting leaf abscission and senescence. ABA is important in seed maturation, as it induces the synthesis of storage proteins in developing seeds. It acts antagonistically to GAs in many systems. ABA, along with ethylene and jasmonic acid, aids in defense against insect wounding. With ethylene, ABA is intimately involved with plant responses to a wide range of environmental stresses. (Gaspar et al., 1996).

In tissue cultures, exogenously applied ABA can affect (generally positively at low concentrations, while high concentrations inhibit) callus growth and organogenesis (buds, roots, embryos). Some ABA is essential for the maturation and normal growth of somatic embryos and only in its presence do they closely resemble zygotic embryos in their morphological and biochemical development (e.g., Roberts et al., 1990; Rock and Quatrano, 1995). Manipulation of endogenous and/ or exogenous ABA levels increases the frequency of embryos reaching maturity (Label and Lelu, 1994). ABA increases freezing tolerance of axenically grown plants and cell cultures.

4.6.4.1. Occurrence, activity, and antagonists

In nature, the most common form of ABA is (S)-(+)-abscisic acid (Fig. 4.6, formula 2). This is often called the cis isomer or simply ABA. The so-called trans isomer has the carboxyl group on the end

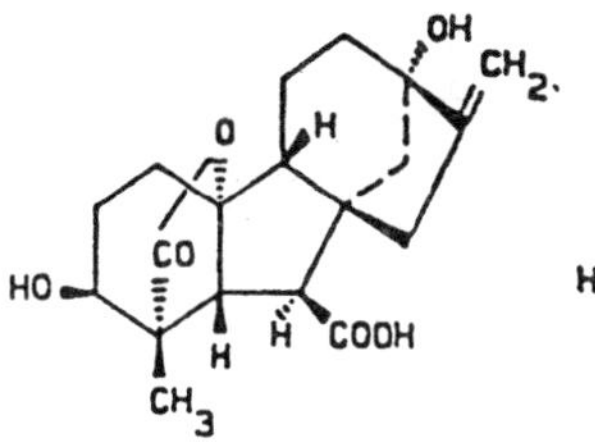

GA$_1$

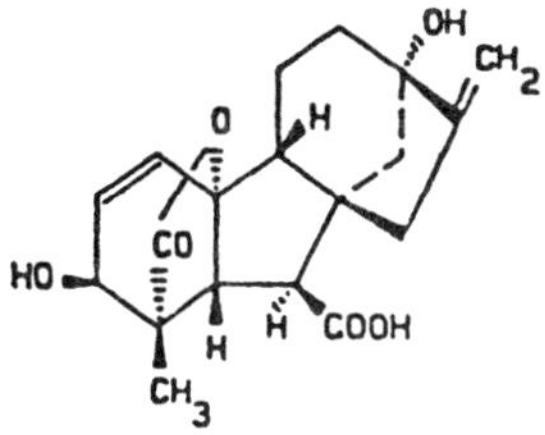

GA$_3$

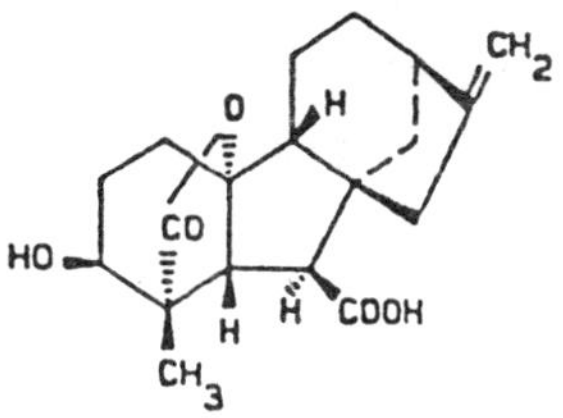

GA$_4$

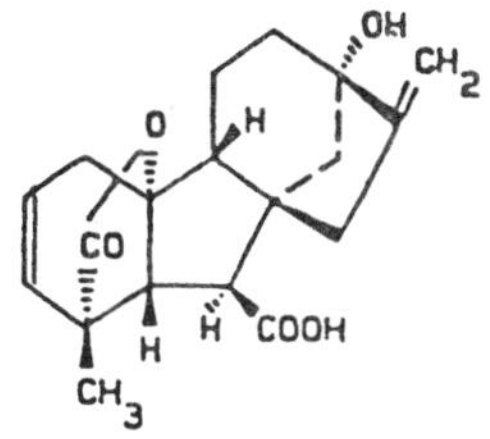

GA$_5$

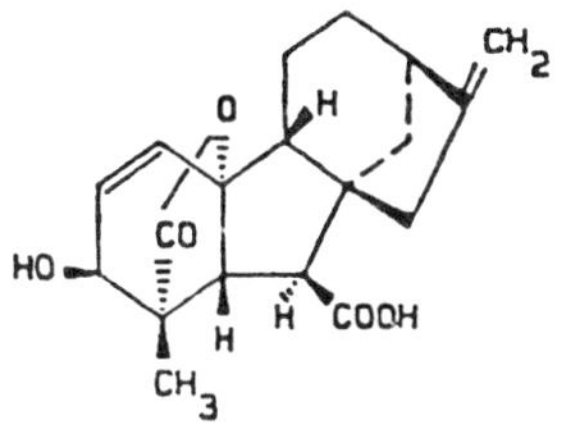

GA$_7$

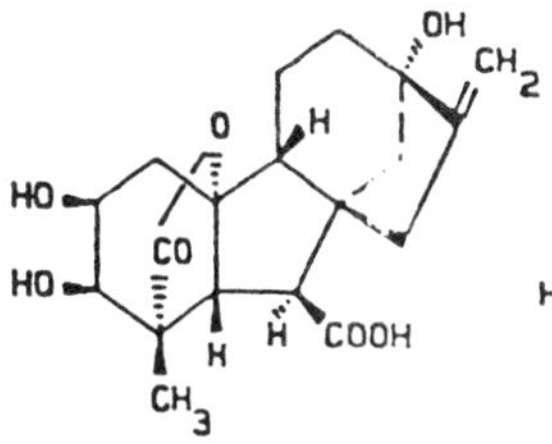

GA$_8$

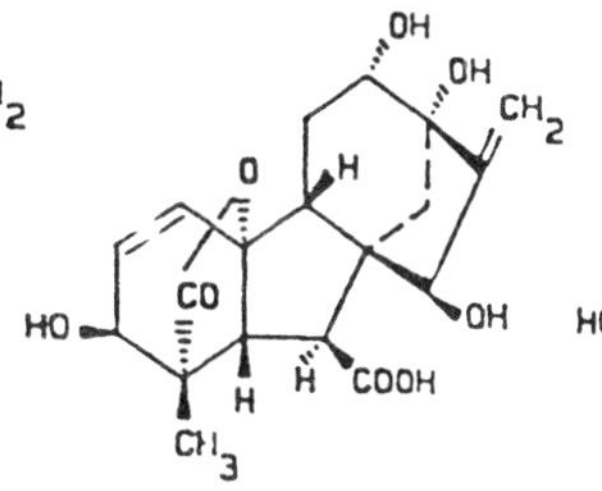

GA$_{32}$

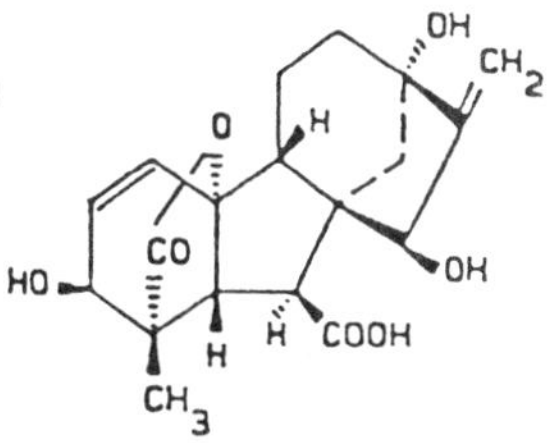

15β-OHGA$_3$

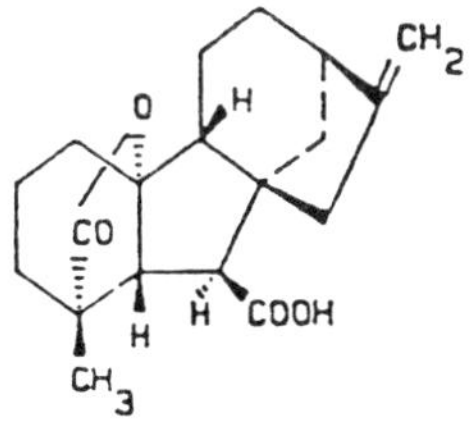

GA$_9$

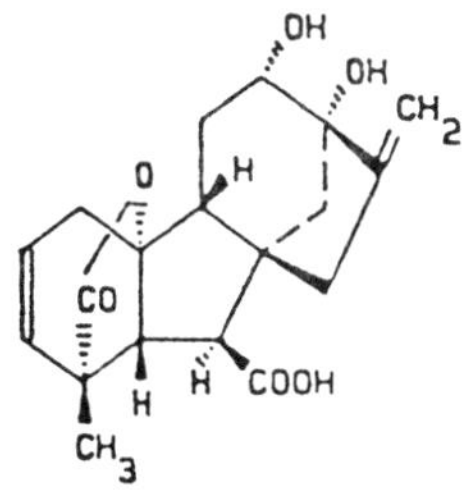

12α-OH GA$_5$

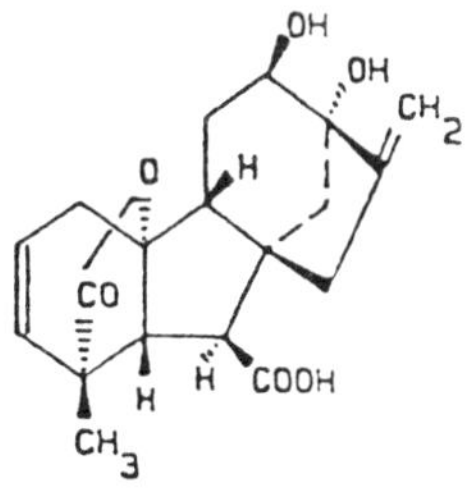

12β-OH GA$_5$

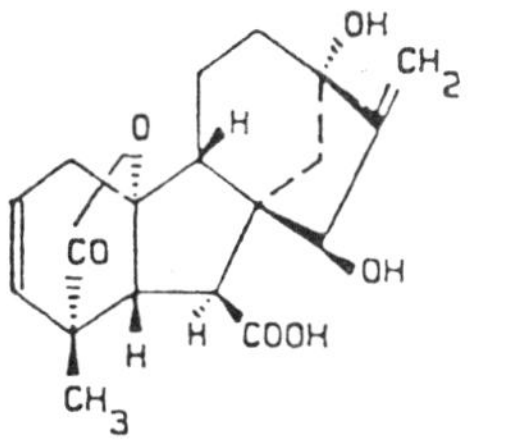

15β-OH GA$_5$

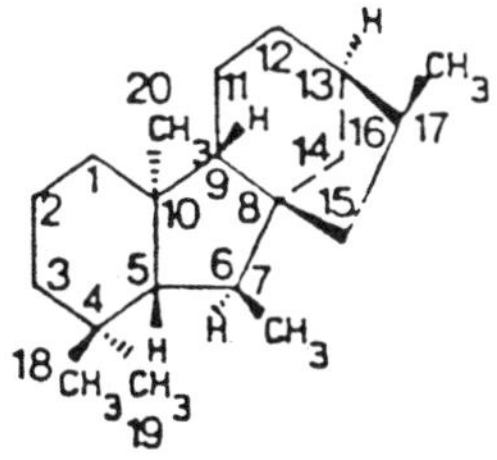

ent-gibberellane
skeleton

Fig. 4.5. Gibberellin Structures (Adopted from Gaspar et al., 1996).

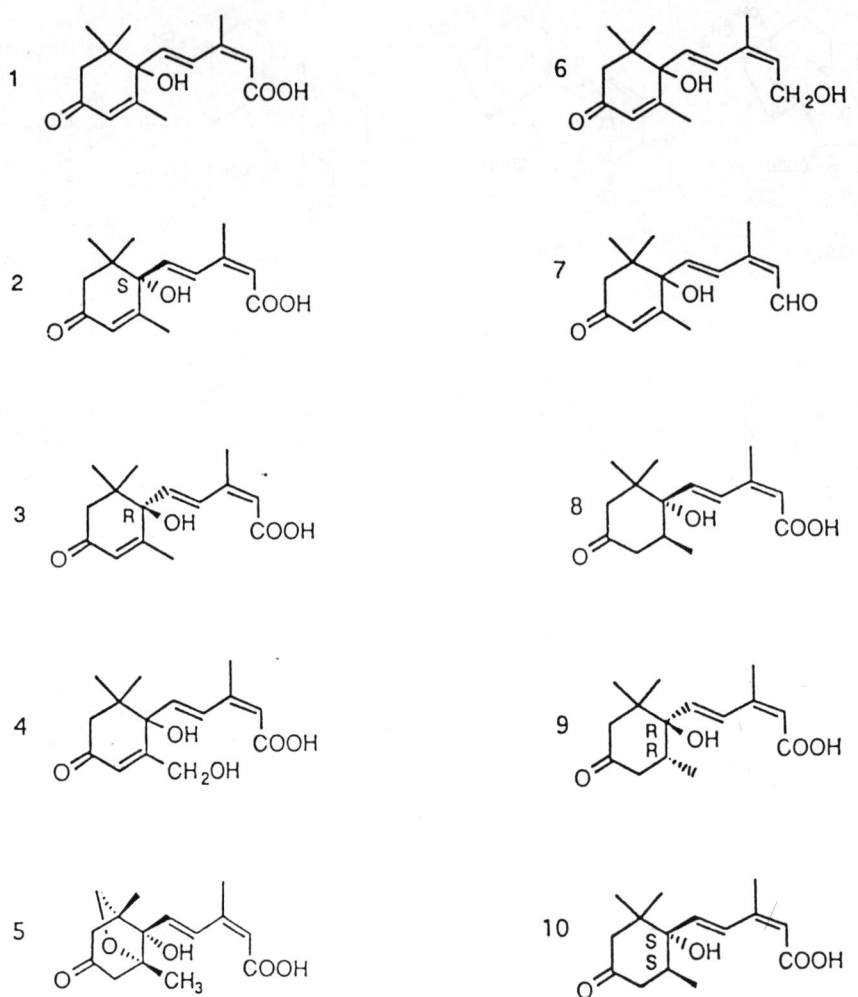

Fig. 4.6. Structural formulae of ABA and analogs (Adopted from Gaspar et al., 1996).

of the side chain oriented in the other direction. The cheaper ABA often sold commercially is a 1:1 mixture of cis- and the trans-ABA optical isomers. For rapid responses, such as stomatal closure, only cis-ABA is active, but for slower responses, both isomers are effective. Exposure of UV light converts some of the cis-ABA to the trans form.

Fluridone and norflurazon, inhibit ABA synthesis, but unfortunately their action is not specific to ABA, as they act as inhibitors of carotenoid biosynthesis (Henson, 1984; Gamble and Mullet, 1986). Wilen et al., (1983) have found that stereoisomeric analogs of ABA are effective specific antagonists of ABA action.

4.6.5. Ethylene

In conjunction with other phytohormones, this gas promotes fruit ripening, senescence, and leaf abscission. There is often a self-adjusting balance between natural auxin and ethylene levels. Some of the responses of plant to auxins may be caused by increased ethylene synthesis in response to auxin treatment. At higher concentrations the gas appears to alter microtubule and microfibril orientation which results in decreased cell elongation but increased cell expansion (Apelbaum and Burg, 1971; Steen and Chadwick, 1981). Depending upon the time after subculture, ethylene can stimulate or inhibit growth and organogenesis in *in vitro* cultures (Huxter et al., 1981). Ethylene can specifically affect growth of callus and suspension cultures, stem and root elongation, axillary and adventitious bud formation, rooting, and embryogenesis. The role of ethylene can be difficult to understand because its effects vary with developmental stage and because low concentrations can promote (or sometimes inhibit) a process, whereas higher levels have the opposite effect.

Ethylene is synthesized from methionine (Fig. 4.7) which in turn is converted to S-adenosyl-methionine (AdoMet) and then ACC. Adding ACC to plant cultures usually increases ethylene production because ACC often is the limiting factor. Adding methionine may have the same effect, and as methionine can be produced in plants from aspartic acid, asparagine, and cysteine, their presence may have a similar effect as well. Because the genes for ACC synthase and ACC oxidase have been cloned, it is possible to modulate ethylene synthesis through the use of mRNA antisense and similar techniques (Mckeon et al., 1995).

Useful inhibitors of specific steps in ethylene biosynthesis include: (a) aminoethoxyvinylglycine (AVG) which inhibits ACC-synthase; (b) α-aminobutyric acid (AIBA) and cobalt ions, which inhibit

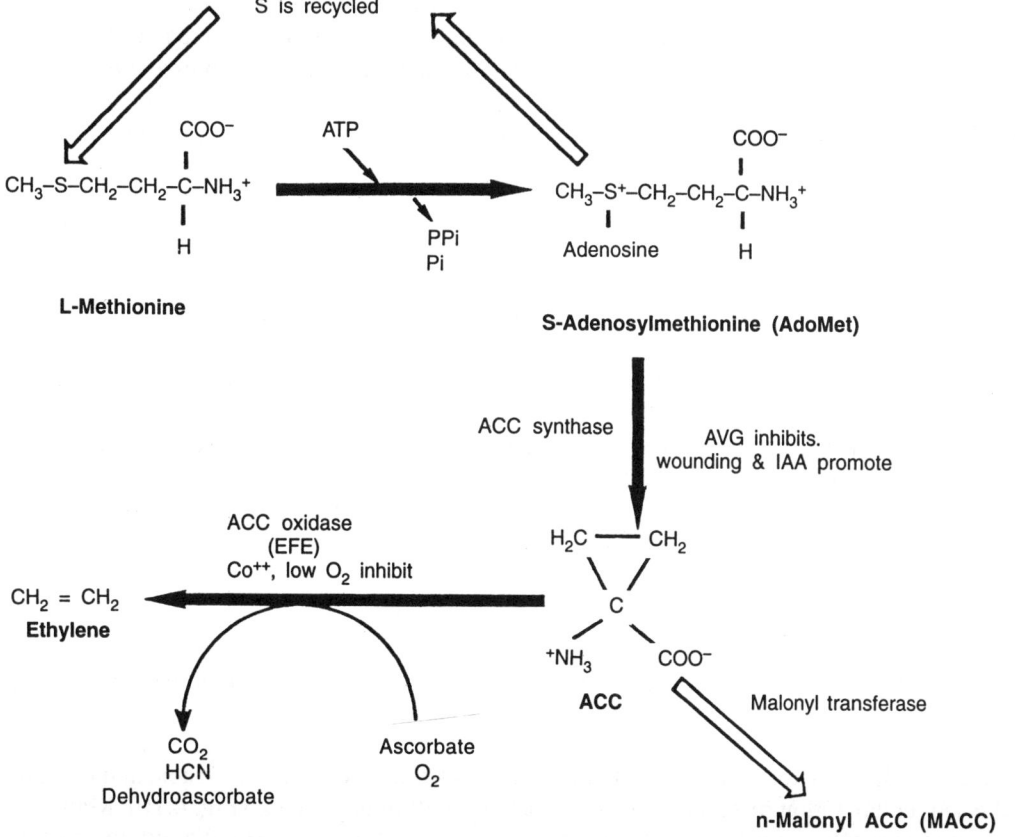

Fig. 4.7. Biosynthetic pathway of ethylene (Adopted from Gaspar et al., 1996). ACC = 1-aminocylo-propane-1-carboxylic acid; AVG = aminoethoxyvinylglycine; EFE = ethylene forming enzyme.

ACC oxidase (EFE) and prevent the conversion of ACC to ethylene; (c) N-propyl gallate, ioxynil, bramoxynil, 3,4,5-trichlorophenol, polyamine and salicylic acids, which can inhibit the conversion of ACC to ethylene.

4.6.6. Other plant growth regulators

4.6.6.1. Polyamines (PA)

Less is known about polyamines (PAs) than other phytohormones, and not all researchers are prepared to classify them as hormones. They require higher concentrations to produce an effect than do the more traditional plant hormones. However, PAs seem to be involved in a wide range of growth and developmental phenomena, as tissues that are deficient in PAs are abnormal. In particular, aliphatic PAs like putrescine, spermidine, and spermine have been shown to enhance morphogenesis in many cultured tissues (Bagni et al., 1993; Turbico et al., 1993).

At a physiological pH, PAs act as polycations and complexing agents. They bind strongly to phospholipid groups and other anionic sites on membranes (thus affecting membrane fluidity), cell wall polysaccharides and to RNA and DNA. Their association with PAs stabilizes nucleic acids. Free PAs can compensate for ionic deficiencies or the damaging effects of some stresses on membranes. Part of their function may be to act as buffers to minimize fluctuations in cellular pH, to modulate enzyme activities, and enhance DNA replication and transcription. PAs appear to be involved in cell division, cell elongation, and rooting and in certain circumstances can be used as a substitute for auxin treatment, which has led some to consider them as secondary messengers. (Evans and Malmberg 1989). The biosynthesis of PAs is influenced by the medium used for *in vitro* culture, biosynthesis is greater on ammonium based medium than when nitrogen is supplied as nitrate ions. The biosynthesis of polyamine is shown in Fig. 4.8.

4.6.6.2. Salicylates

Salicylic acid (SA) and its derivatives (glycosides of SA, esters of SA, glycosides of SA esters, and other salicyl alcohol derivatives) (Fig. 4.9) belong to the large group of plant phenolics and are probably ubiquitous in higher plants. SA probably is biosynthesized from cinnamic acid.

The most typical biological responses of SA are: (a) the induction of flowering (b) the induction of tuberization *in vitro* (c) induction or regulation of thermogenesis (heat production during flowering of thermogenic plants like Arum lilies), (d) systemic signal for the induction of disease resistance, in particular against necrotrophic pathogens [the action of SA is likely mediated by elevated amounts of H_2O_2 concentration *in vitro* (Chen et al., 1993)], and induce of certain pathogenesis-related (PR) proteins, and (e) inhibition of germination.

4.6.6.3. Jasmonates

Following early detection and chemical identification from *Jasminum, Rosarinus* and other plants containing fragrant essential oils, jasmonic acid (JA) and in particular its volatile methylester (MejA) have been detected in many species. The two physiologically active compounds, (–)-jasmonic acid and (+)-7-iso-jasmonic acid, are stereomeric forms normally found in a molar ratio of 9:1. Both can be metabolically transformed. The biosynthetic pathway starts with linolenic acid and is synthesized by the sequential action of several enzymes starting with lipoxygenase (Fig. 4.10). Many naturally occurring jasmonates have been identified (Sembdner and Pantheir, 1993). Wounding and pathogens can promote JA synthesis as follows: sytemin and oligouronides act as signals which eventually activate lipases, releasing linolenic acid and leading to JA formation. The jasmonates then activate

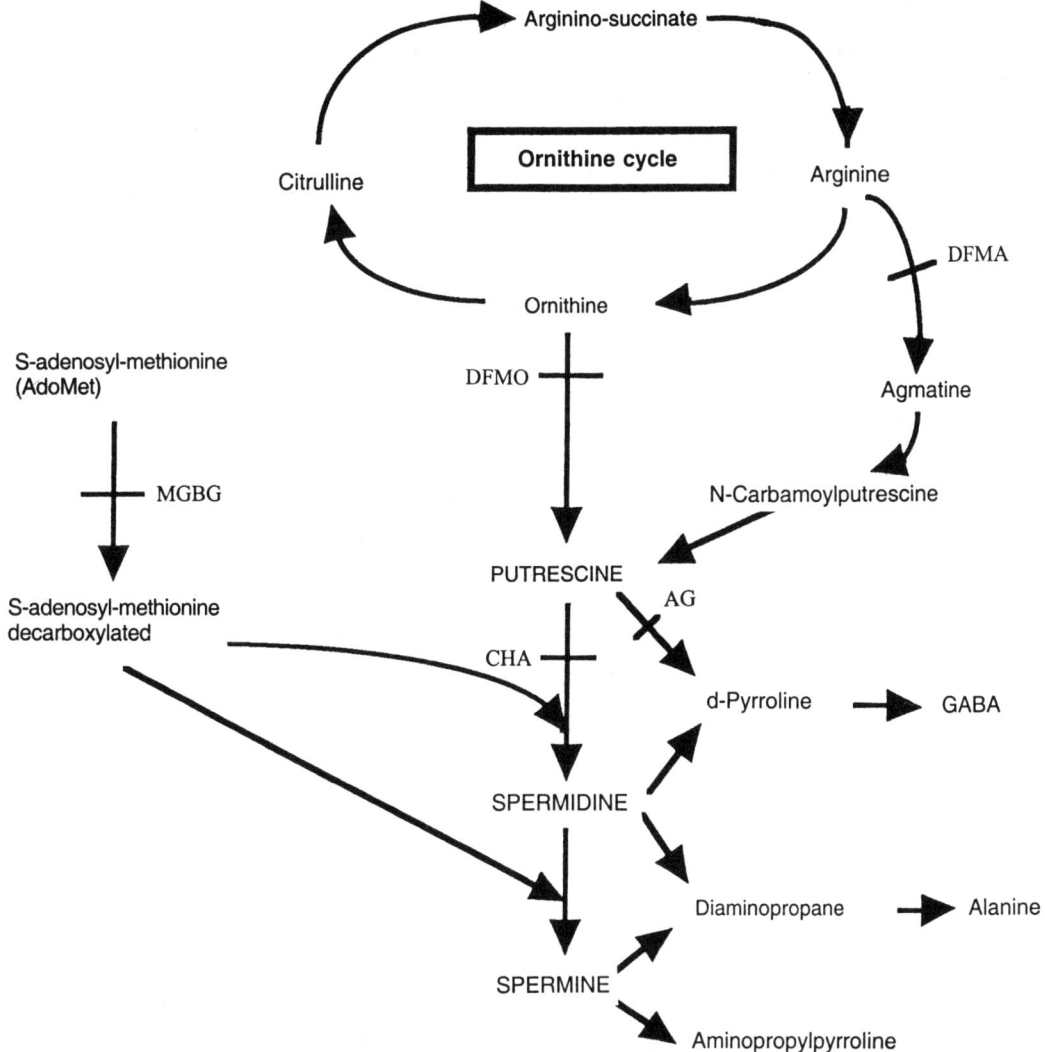

Fig. 4.8. Schematic outline of the pathways of polyamine synthesis and catabolism (Hausman et al., 1994). AG, aminoguanidine; CHA, cyclohexylamine; DFMA, α-difluoromethylarginine; DFMO, difluoromethylornithine; GABA, gamma aminobutyric acid; MGBG, methylglyoxal-bis-guanylhydrazone.

genes for the formation of stress proteins (different from the PR proteins) such as proteinase inhibitors. In some cases JA also promotes ethylene synthesis (Emery and Reid, 1996).

Jasmonates are thus involved in the cellular transduction processes between external stress (herbivore, pathogen, desiccation, mechanical, or osmotic stresses) and macromolecular stress responses involving the expression of "defense genes" and production of JA induced proteins (JIPS) (Reinbothe et al., 1994). Jasmonates display a multiplicity of effects in plants (Gross and Partheir, 1994) such as the promotion of leaf senscence, abscission, fruit ripening, tendril coiling, and tuber formation (tuberonic acid or 12-hydroxylated jasmonic acid is very effective here). Jasmonates also inhibit seed germination and stem elongation. As inducers of vegetative storage protein gene expression, jasmonates also stimulate bulb formation *in vitro*.

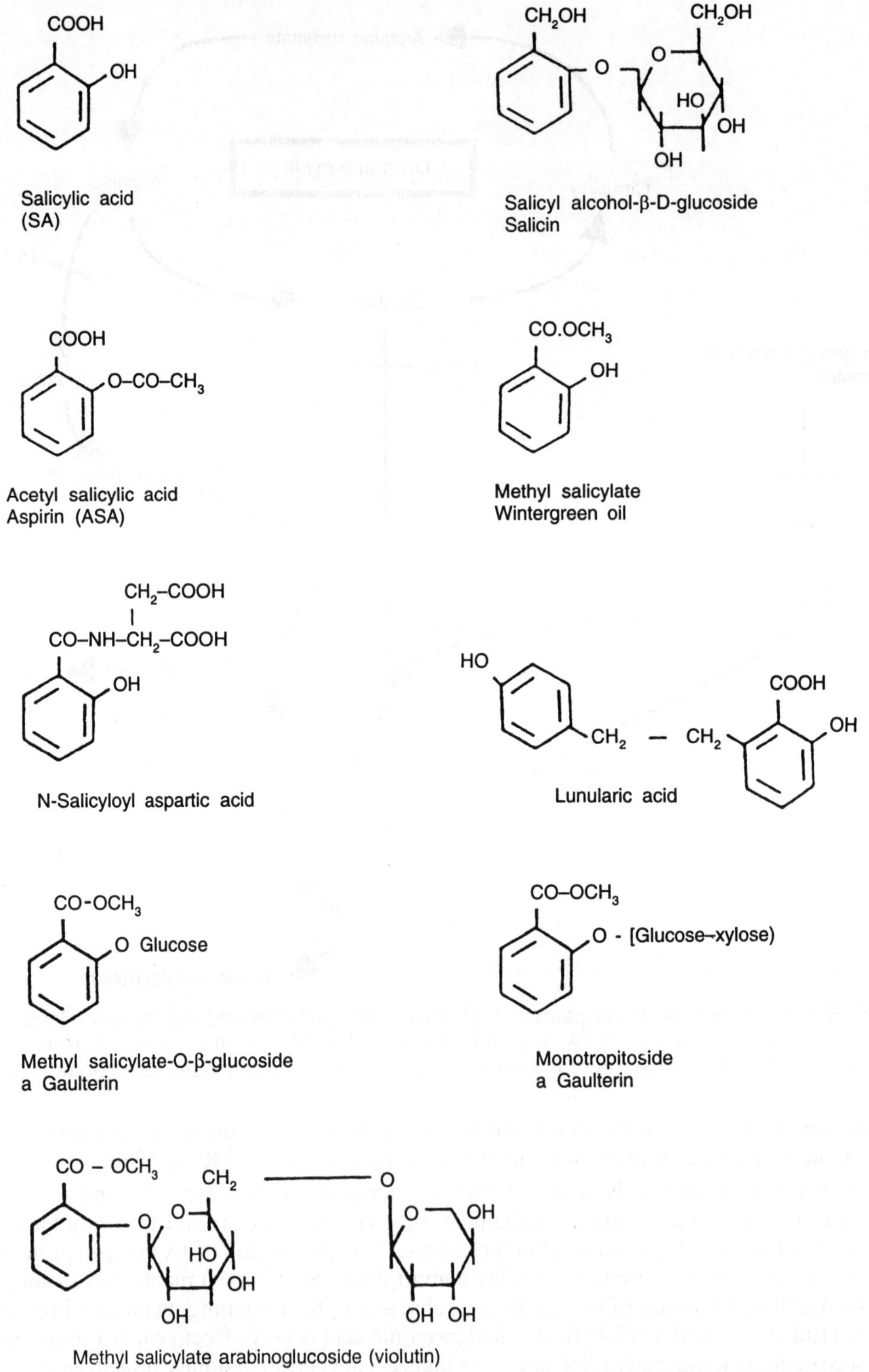
COOH
OH
Salicylic acid
(SA)
CH2OH
CH2OH
O
O
HO
OH
OH
Salicyl alcohol-β-D-glucoside
Salicin
COOH
O–CO–CH3
Acetyl salicylic acid
Aspirin (ASA)
CO.OCH3
OH
Methyl salicylate
Wintergreen oil
CH2–COOH
CO–NH–CH2–COOH
OH
N-Salicyloyl aspartic acid
HO
CH2 – CH2
COOH
OH
Lunularic acid
CO-OCH3
O Glucose
Methyl salicylate-O-β-glucoside
a Gaulterin
CO–OCH3
O - [Glucose–xylose)
Monotropitoside
a Gaulterin
CO – OCH3
CH2
O
HO
OH
OH
OH OH
Methyl salicylate arabinoglucoside (violutin)

Fig. 4.9. Some naturally occurring derivatives of salicylic acid (Adopted from Gaspar et al., 1996).

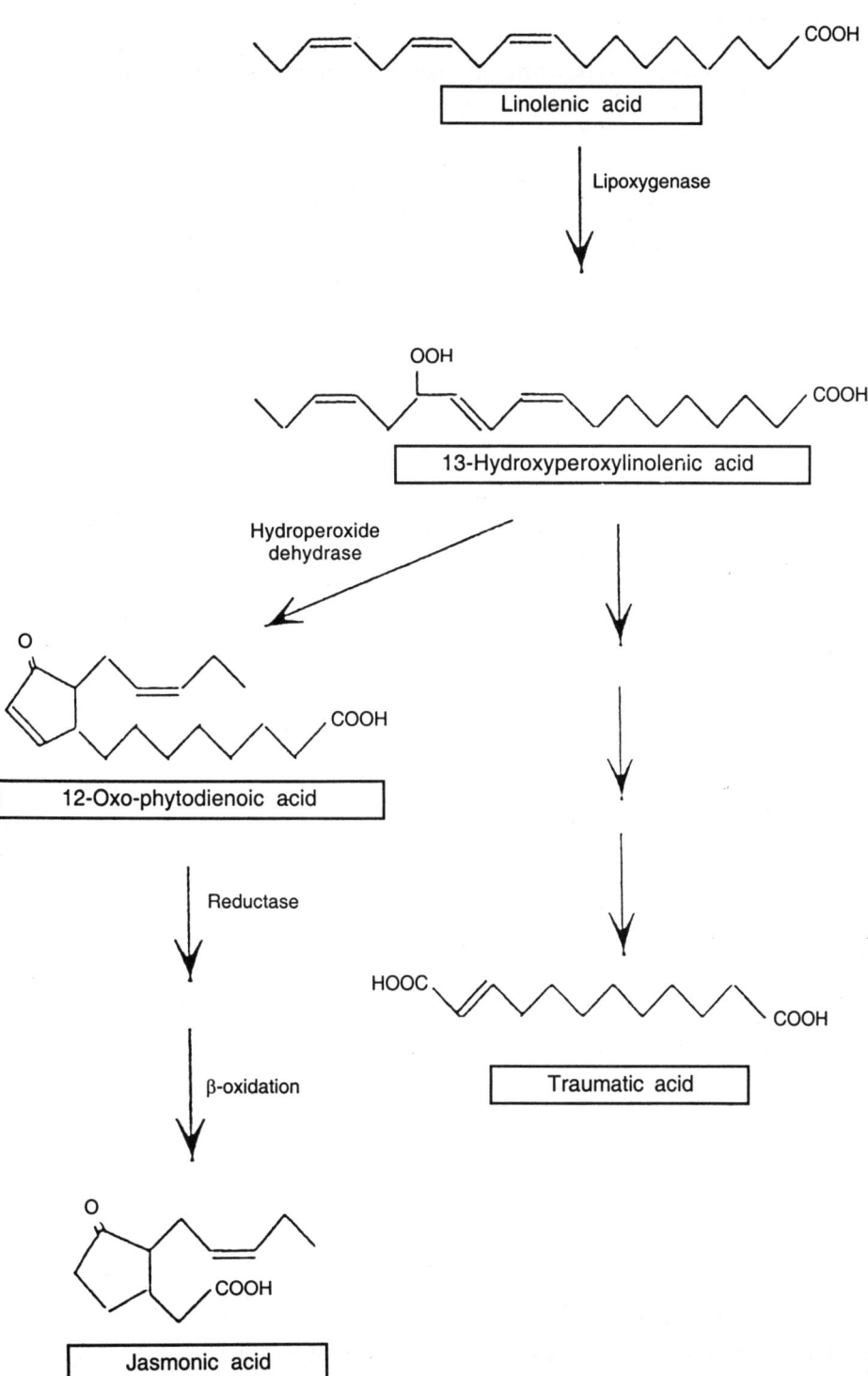

Fig. 4.10. Formation of jasmonic and traumatic acids from linolenic acid (Beale and Ward, 1998).

4.7. CHOICE OF MEDIUM

The various cell tissue and organ types differ in medium requirements. Therefore, a new combination of medium components must be tested for each new system. There are several media developed by Gamborg et al 1968, Murashige and Skoog 1962, White 1943 and Gautheret 1942. (Table 4.2).

The testing process usually begins with growth regulators experiencing the greatest variations, i.e. auxins and cytokinins. Beginning with five different NAA and five different BAP concentrations between 0 and 10 μmol, the optimum combination of hormones from the 25 combinations are chosen. Subsequently, the effect of other hormones at these concentrations is tested, while always varying the concentration of one of the two phytohormones. The salt concentrations are optimized by the systematic reduction of the concentrations in the chosen basic medium to one-half or one-fourth, while maintaining the optimum hormone combination. In media according to Murashige and Skoog, the ammonium concentration usually must be reduced substantially or substituted by nitrate or amino acids to prevent ammonium toxicity. Finally, in the media thus modified the nature and concentration of the optimum carbon source are determined.

If this screening method does not lead to success, the suitable medium combination must be determined by a broadband experiment. The various ingredients (sugar, amino acids, inositol, etc.) are each divided into low, medium and high concentration. Using the 81 combinations thus formed, the optimum combination can usually be determined even in intractable cases. The usefulness of special additives, e.g. organic acids, should be tested independently.

Modification of the Medium

The static cultures on solid media are preferred over liquid cultures because of improved oxygen supply and the developing chemical gradients. The various gelling agents used for this purpose are shown in Table 4.3.

The most commonly used substance for this purpose is Agar-Agar, which obtained from red algae (Gelidium, Gracilaria). Robert Kock (1882) first time described agar as gelling agent. The composition of agar is shown in Table 4.3. It contains amino acid and carbohydrates as major ingredients. Agar - Agar is used in the concentration 0.6 - 20% depending upon the quality of Agar-Agar.

REFERENCES

Abeles, F.B.; Morgan, O.W.; Salvert. M.E.(1992) Ethylene in plant biology, 2nd Ed. San Diego Academic press.

Addicott., F.T.(1982) Abscission. Berkeley: University of California Press; Along, R. The induction of vascular tissues by auxin and cytokinin. In: Davies, P.J., ed. Plant hormones. Dordrecht: Kluwer Academic Publishers: 1995:531-546.

Aloni, R.(1995).The induction of vascular tissue by auxin and cytokinin. In: Davies, P.J.(ed) Plant hormones. Dordrecht: Kluwer academic publishers. P531-546.

Apelbaum. A.; Burg. S.P. (1971) Altered cell microfibrillar orientation in ethylene treated *Pisum sativum stems. Plant Physiol.* 48:648-652.

Bagin, N.; Altamura, M.N.; Biondi, S., et al.(1993) Polyamines and morphogenesis in normal and transgenic plant cultures. In: Roubelaks-Angelakis, K.A.; Tran Thanh Van, K., Ed. Morphogenesis in plants: molecular approaches. New York: Plenum Press; 89-111.

Bandurski, R.S.; Cohen, J.D.; Slovin, J., et al. (1995) Auxin biosynthesis and metabolism. In: Davies, P.J. Ed. Plant hormones. Dordrecht: Kluwer Acaddemic Publishers: 39-65.

Beale, M.H.; Sponsel, V.M.(1993) Future directions in plant hormone research. *J. Plant Growth Regul.,* 12: 227-235

Beale, M.H. and Ward, J.L. (1998). Jasmonates : Key players in the plant defence. Natural Product Reports, 15, 533-548.

Bearder, J.R. (1980) substances and their background, structures and occurrence. In: MacMillan, J., ed. Ency. Plant Physiol. N.S., Vol. 9, Berlin: Springer-Verlag; 9-112.

Bell, E.A (1981) The on-protein amino acids occurring in plants. In: Reinhold, L. Harborne, JB. Swain T (eds) Progress in Phytochemistry 7. Pergamon Press Oxford. P. 171.

Bergmann L (1967) Wachstum gruner Suspensionskulturen von *Nicotiana tabacum* var. Samsun mit CO_2 als Kohlenstoffquelle. Planta 74:243.

Bray, C.M. (1983) Nitrogen metabolism in plants. Longman. London.

Chen. Z.; Silva, H.; Klessig. D.F.(1993) Active oxygen species in the induction of plant systemic acquired resistance by salicylic acid. *Science,* 262:1883.

Davies, P.J.(1995)(ed.) Plant hormones. Dordrecht: Kluwer Academic Publishers; 13-38.

Emery, R.J.N.; Reid, D..M.(1996) Methyl jasmonate effect on ethylene synthesis and organ-specific senescence in *Helianthus annuus* seedlings. *Plant Growth Regul.* 18:213-222.

Endress, R. (1994) In: Plant Cell Biotechnology. Springer-Verlag, Berlin, Heidelberg; 29-40.

Evans, P.T. Malmberg, R.L.(1989) Do polyamines have roles in plant development? *Ann. Rev. Plant Physiol. Plant. Mol. Biol.* 40:235-269.

Evans, D.A; Sharp, W.R.; Flick, C.E.(1981) Growth and behavior of cell cultures; embryogenesis and organogenesis. In: Thorpe, T.A. (ed.) Plant cell culture; methods and applications in agriculture. New York: Academic Press; 45-113.

Fabijan, D.M.; Plumb-Dhindsa, P.; Reid, D.M.(1981) Effects of two growth retardants on tissue permeability in *Pisum sativan* and *Beta vulgaris. Plant* 152:481-486.

Fry, S.C.; Street, H.E.(1980) Gibberellin-sensitive cultures. *Plant Physiol.* 65:472-477.

Gamble , P.E., Mullet, J.(1986) Inhibition of carotenoid accumulation and abscisic acid biosynthesis in fluridone-treated dark-grown barley *Eur. J. Biochem.* 160:117-121.

Gamborg, O.L.; LaRue T.A.G.(1971) Ethylene production by plant cell cultures. The effects of auxins, abscisic acid, and kinetin on theylene production in suspension cultures of rose and *Ruta* cells. *Plant Physiol.* 48:399-401.

Gamborg OL., Miller R.A. Ojima K (1968) Nutrient requirements of suspension cultures of soybean root cells. *Exp. Cell Res.* 50. 151.

Gamborg, O.L. Murashige, T. Thorpe, T.A and Vasil, I.K. (1976). Plant tissue culture media, *In vitro* 12, 473-478.

Gamborg, O.L and Shyluk, T.P. (1981) Nutrition, media and characteristics of plant cell and tissue cultures. In: Plant Tissue Culture Methods and Applications in Agriculture (ed.) Thrope, T.A., pp 21-44, New York, Academic Press.

Gaspar, T.; Kevers, C.; Hausman, J., et al. (1994) Peroxidase activity and endogenous free auxin during adventitious root formation. In: Lumsden, P.J.; Nicholas, J.R.; Davies, W.J. ed. Physiology, growth and development of plants in culture. Dordrecht: Kluwer Academic Publishers 289-298.

Gaspar, T. (1995) Selenieted forms of indolylacetic acid: new powerful synthetic auxins. *Across Organics Aeta* 1:65-66.

Gaspar, T.; Kevers, C.; Penel, C.; Greppin, H.; Reid, D.M. and Thrope, T.A. (1996). Plant hormones and plant growth regulators in plant tissue culture. *In vitro Cell Biol. plant,* 32 : 272-289.

Gautheret, R.J. (1942) Hetero-auxiners et cultures de tissues vegetaux. Bull Scotum Biol 41 (13).

Guatheret, R.J. (1937). Nouvelles recherches sur la culture du tissue cambial. *C.R. Hebd. Seances Acad. Sci.* 205, 572-574.

Giovanelli, J. Mudd, S.H. Datke, A.H (1980) Sulfur amino acids in plants. In:Stumpf PK. Conn 1:1 (eds in chief) The biochemistry of plants vol. 5. Miflin BF (ed) Amino Acids and derivatives Academic Press, New York. p. 453

Goldbach H (1985) Influence of boron nutrition on net uptake and efflux of ^{32}P and ^{14}C-glucose in *Heianthus annus* roots and cell cultures of *Daucus carota. J. Plant Physiol* 118.431.

Gray, D.J.; Conger, B.V.(1985) Influence of dicamba and casein hydrolysate on somatic embryo number and culture quality in cell suspension of *Dactylis glomerata* (Gramineae). Plant Cell Tissue Organ Cult. 4:123-133.

Gross, D.; Partheir, B.(1994) Novel natural substances acting in plant growth regulation. *J. Plant Growth Regul.* 13:93-114.

Haberlandt, G (1902) Kulturversuche mit isolierten Pflanzen zellen. Mat. Nat. Kais. Akad. Wiss 111, 69-92.

Hagen, S.R.; Muneta, P.; Augustin, J., et al. (1991) Stability and atilization of picloram, vitamins and sucrose in a tissue culture media. *Plant Cell Tissue Organ Cult.* 25:45-48.

Henson, I.E.(1984) Inhibition of abscisic acid accumulation in shoots of pearl millet (*Pennisetum americanum* L.) following induction of chlorosis by norflurazon. *Z. Pflanzenphysiol.* 114:35-43.

Huang L.C. & Murashige, T. (1977) Plant tissue culture media. major constituents, their preparation and some applications. Tissue culture Association manual 3, 539-548.

Huxter, T.J; Reid, D.M.; Thorpe, T.A.(1981) Shoot initiation in light-and dark grown tobacco callus: the role of ethylene. *Physiol. Plant.* 53:319-326.

Jarvis, B.C.; Ali, A.H.N.; Shaheed. A.I.(1983) Auxin and boron in relation to the rooting response and aging of mung bean cuttings. *New Phytol.* 95:509-518.

Kato, A. Fukasawa, A. Shimizu, Y. Soh, Y. Nagai, S (1977) Requirements of PO_4, NO_3, SO_4, potassium and calcium for the growth of tobacco cells in suspension culture. *J. Ferment. Technology* 55:207.

Krikorian, A.D.(1995) Hormones in tissue culture and micropropagation. In: Davies, P.J., ed. Plant hormones. Dordrecht: Kluwer Academic Publishers, 774-796.

Kuhnle, J.A.; Fuller, G.; Corse, J., et al. (1977) Antisenesent activity of natural cytokinins. *Plant Physiol.* 41: 14-21.

Klapheck, S.; Grobe W.; Bergmann L (1982) Effect of sulfur deficiency on protein synthesis and amino acid accumulation in cell suspension cultures of *Nicotiana tabacum.* Z Planzephysiol 108:235.

Koblitz H (1977) Zell-und Gewebzuchtung bei Planzen. Fischer, Stuttgart.

Label, P.; Lelu, M.A.(1994) Influence of exogenous abscisic acid on germination and plantlet conversion frequencies of hybrid larch somatic embryos (*Larix* x *leptoeuropaea).* Plant Growth Regul. 15.175-182.

Lamproye, A.; Hofinger, M.; Berthon, J.Y., *et al.*(1990*)*[Benzo(b)selenieny 1-3] acetic acid: a potent synthetic auxin in somatic embryogenesis. C.R. Acad. Sci. Paris, Ser. 111. 311:127-132.

Lance, B.; Durely, R.C.; Reid, D.M., Thorpe, T.A. (1976)Endogenous gibberellins and growth of tobacco callus cultures. *Physiol. Plant* 36:287-292.

Liu, J.H.; Reid, D.M.(1992) Auxin and ethylene-stimulated adventitious rooting in relation to tissue sensitivity to auxin and ethylene production in sunflower hypocotyls. *J.Exp. Bot.* 43:1191-1198.

Maeda, E.; Thorpe, T.A.(1979) Effects of various auxins on growth and shoot formation on tobacco callus. *Phytomorphology* 29:146-155.

McGaw, B.A. Burch, L.R. (1995) Cytokinin biosynthesis and metabolism. In: Davies, P.J., ed. Plant hormones. Dordrecht: Kluwer Academic Publishers, 98-117.

McKeon, T.A.; Fernandez-Maculet, J.C.; Yang, S.F.(1995) Biosynthesis and metabolism of ethylene, In: Davies, P.J., (ed.) Plant hormones. Dordrecht: Kluwer Academic Publishers, 118-139.

Murashige, T. Skoog, F. (1962) A revised medium for rapid growth and bioassays with tobacco tissue cultures: *Physiol Plant* 15. 473.

Murashige, T., (1973) Sample prepartion of media, C. plant cultures, In; Kruss, Jr. P.F and Patterson, Jr M.K. (Eds) Tissue culture methods and Applications. Academic Press. NewYork, PP 698-703.

Murthy, B.N.S.; Murch, S.J. and Saxena, P.K. (1998). Thidiazuron : A potent regulator of in-vitro plant morphogenesis. *In-vitro Dev. Biol. Plant*, 34 : 267-275.

Nooden, L.D.; Leopold, A.C. (1988) Senescence and aging in plants. San Diego: Academic Press.

Ohira, K. Ikeda, M and Ojima K (1976) Thiamine requirements of various plant cells, in suspension culture. *Plant and Cell Physiol.* 17, 583-588.

Rechcigl, M. Jr. (1977). Plant tissue culture media. In: Rechcigl, M. (ed.) CRC Handbook series in nutrition and food, vol. IV. Culture media for cells, Organelles and embryos, CRC Press, C'levand, USA, p. 605.

Reid, D.M.; Beall, F.D., Pharis, R.P.(1991) Environmental cues in plant development, In: Bidwell, R.G.S., Ed. Plant physiology, a treatise, Vol. X, Growth and development San Diego; Academic Press, 65-181.

Reinbothe, S.; Mollenhauer, B.; Reinbothe, C. Jips, and Rips (1994)The regulation of plant gene expression by jasmonates in response to environmental cues and pathogens. *Plant Cell* 6:1197-1209.

Roberts, D.R.; Flinn, B.S.; Webb, D.T., et al.(1990) Abscisic acid and indole-3-butyric acid regualtion of maturation and accumulation of storage proteins in somatic embryos of interior spruce. *Physiol. Plant.* 78:355-360.

Rock, C.D.; Quatrano, R.S. (1995)Hormones during seed development. In: Davies, P.J. ed. Plant hormones. Dordrecht: Kluwer Academic Publishers, 671-697.

Sabater, B. Hormonal regulation of senescence.(1985) In: Burohit, S.S., ed. Hormonal regulation of plant growth and development, Vol. 1. India: Agro. Botanical Publ.169-217.

Steen, D.A.; Chadwick, A.V.(1981) Ethylene effect in pica stem tissue. Evidence of microtubule mediation. *Plant Physiol.* 67: 460-466.

Schenk R.U, Hildebrandt AC (1972) Medium and techniques for induction and growth of monocotyledonous and dicotyledonous plant cell cultures. *Can. J. Bot* 50:199.

Sigma Chemical Gmbh (ed) (1990) Catalog: plant cell culture. Chemic Gmbh, Deisenhofen, p 28.

Tamas, I.A.(1995) Hormonal regulation of apical dominance. In: Davies, P.J., ed. Plant hormones. Dordrecht: Kluwer Academic Publishers, 572-597.

Thorpe, T.A.; Murashige, T.(1970) Some histochemical changes underlying shoot initiation in tobacco callus cultures. *Can. J. Bot.* 48:277-285.

Turbicio, A.F.; Campers, J.L.; Figueras, X., et al.(1993) Polyamines and morphogenesis in monocots. In: Roubelakis -Angelakis, K.A.; Tran Thanh Van, K., ed. Morphogenesis in plants: molecular approaches. New York: Plenum Press; 113-135.

Vasil, I.K.; Thorpe, T.A., (1994) Plant Cell and tissue culture. Dordrecht: Kluwer Academic Publishers.

Venis, M.A.; Napier, R.M.(1991) Auxin receptors: recent developments. *J. Plant Growth Regul.* 10:329-340.

Vesely, J.; Havlicek, L.; Strnad, M., et al.(1994) Inhibition of cyclin-dependent kinases by purine analogues. *Eur. J. Biochem.* 224:771-786.

Wilen, R.W.; Hays, D.B.; Mandel, R.M., et al.(1993) Competitive inhibition of abscisic acid-regulated gene expression by stereoisomeric acetylenic analogs of abscisic acid. *Plant Physiol.* 101:469-476.

Wright, S.T.C.(1979) The effect of 6-benzyladenine and leaf-aging treatments on the levels of stress-induced ethylene emanating from wilted wheat leaves. Planta 144:179-188.

White. P.R (1934) Potentially unlimited growth of excised tomato root tips in a liquid medium. *Plant physiol.* 9, 585-600.

White PR (1943) A Handbook of Plant Tissue Culture, Jaques Cattell Press. Lancaster, pp 80-100.

White P.R. (1951). Nutritional requirements of isolated plant tissues and organs. *Annu. Rev. Plant Physilogy* 2, 231-244.

Ziv, M.; Ariel, T.(1991) Bud proliferation and plant regeneration in liquid-cultured Philodendron treated with ancymidol and paclobutrazol. *J. Plant Growth Regul.* 10:53-57.

Chapter 5

Initiation and Maintenance of Callus and Suspension Cultures

INTRODUCTION

When an organ of a plant is damaged, a wound repair response is induced to bring about the repair of the damaged portion. This response consists intially of the induction of division in the undamaged cells adjacent to the lesion, thus sealing off the wound. This is then followed by the hardening of this layer through the deposition of lignin, suberin ,wax etc in order to regain the integrity of the protective outer barrier of the plant. If, however, wounding is followed by the aseptic culture of the damaged region on a chemically defined medium, the initial cell division response can be stimulated and induced to continue indefinitely through the exogenous influence of the chemical constitution of the culture medium. Wounding may not be essential to the *in vitro* callussing response although it is generally stimulatory to the rate of callus formation. The result is a continually-dividing mass of generally poorly differentiated and disroganised plant cell aggregates termed a callus.

Callus is generally grown in Petri dishes, glass tubes or extra-wide necked Erlenmeyer flasks on medium solidified with agar or one of its replacements (agarose, gelrite, etc). In morphological terms it can vary extensively, ranging from being very hard compact, where the cells have extensive and strong cell to cell contact, to being 'friable' where the callus consists of small, disintegrating aggregates of poorly-associated cells and has a rather crumbly or creamy appearance. Friable callus is generally most sought-after as it is usally the fastest growing and most uniform type and is best suited for the intiation of cell suspension cultures. Callus morphology is often explant-dependent but can usually be altered by the modification of the growth substance supplementaion of the culture medium.

Due to their size and nature, callus cultures have an inherent degree of heterogeneity. As there is a unidirectional supply of nutrients (from the medium below) and gases and light, chemical and physical gradients will be present within the callus mass. While, in some instances, this heterogeneity is a disadvantage (e.g. in the production of uniform biomass) it may also be an important influential factor in the developmental response of the callus in, for example, plant regeneration.

5.1. PLANT MATERIAL

The most commonly used starting materials are seeds (especially for monocots), young, asepctically-germinated seedlings and greenhouse-grown, young, healthy plants. The first two have the advantages that seed can generally withstand more severe sterilization conditions than plant tissues and thus it is easier to obtain sterile explants for culture. In addition, as all the plant material is at a very young stage, there is a high potential for cell division within the explants. Low germination frequencies, can however, be problematic. Greenhouse-grown plants provide a larger source of explant material but sterilization can prove difficult both due to sensitivity to the chemicals used and to the external morphology of the plant. If such greenhouse-grown plant material is to be used for callus initiation, it must be entirely free from infection (including viruses), devoid of any signs of insect attack and be in a well maintained state. Unlike for *in vitro* grown plants, for which constant conditions are maintained, the physiological state of greenhouse-grown plants can, without care, vary considerably and this can have a profound influence upon the response of cultured explants.

Essentially all organs can be used as explant sources. However, the degree of success with different tissues can vary and calluses with differing morphologies are frequently obtained. For dicotyledonous plants most commonly young leaves, petioles, stems and hypocotyls (from seedlings) are used. For monocots meristematic regions are usually chosen (leaf bases, young inflorescences etc). Prior to culture all material must be surface sterilized, for which there is a range of chemicals to choose from (e.g.sodium hypochlorite, mercuric chloride etc.), following sterilization the material must be washed thoroughly to remove all traces of the sterilant, cut into suitably-sized pieces (explants) and plated onto a solid medium. The choice of medium is determined by the species to be used and the aim of the experiment (i.e. biomass production, plant regeneration etc). Today with the extensive work which has already taken place in this field, a suitable medium to begin with can usally be found in the literature (see chapter 4). After plating out explants, it is important to seal petri plates with parafilm in order to prevent desiccation. Plates or conical flask should then be incubated at approximately 25°C either in dark or if beneficial under low level illumination. Callus initiation may be observed as an initial swelling at the areas of tissue damage. The complete callus formation takes palce within 1-5 weeks of incubation.

If sufficient callus for subculture has not formed within three to eight weeks, it will be necesssary to re-evaluate the culture medium and conditions. Subculture by removing newly formed callus with a sterile scalpel and transferring to fresh medium. Be careful not to squash the somewhat fragile callus too much when re-plating, and do not cut the callus into too many, very small pieces. The optimum inoculum size varies depending on the species, but it is best to err on the large side while the culture is in the early stages of establishment. Once well established, transfer approximately 2-10 mm³ of callus. The callus cultures will require regular subculture at approximately one month intervals. For newly initiated callus, it may be necessary to transfer the entire callus to the fresh medium for the first two or three subcultures. Fast growing callus cultures may require subculture at two to three week intervals.

In general suspension cultures form readily after transfer of callus to shake flasks of the culture medium minus agar. A large inoculum may be necessary to intiate the suspension, and agitation rates on orbital shakers should be in the range of 60-150 rpm. with an orbital motion stroke of 2-4 cm. If the callus culture is non-friable, very little will break off the callus clumps and the necessary inoculum size of freely suspended cells and small cell clusters may not be attained. In such cases, modification may have to be made to the callus, the approaches here are somewhat empirical.

When the newly established suspension culture has reached a suitable cell density for subculture, remove as much of the remaining unbroken callus material and large clumps as possible. This can be done by either transferring the single cells and small clumps with a sterile syringe (the orifice of which will exclude large clumps), by filtration, by allowing the large material to settle and pipetting off from the top of the culture, or simply by direct pouring. Remember, however, that there is a minimum inoculum size below which cell suspensions do not readily resume active growth following transfer, and that in some cultures most of the growth occurs on the surface of small clumps. The minimum density of cells required for cell suspension cultures depends on the rate of growth and the composition of the medium. Generally, ten percent of initial cell density to the total volume of the culture will be sufficient. Addition of medium in which the cell line had previously been growing (conditioned medium) can sometimes stimulate growth of cell suspensions at low inoculum density. The growth rate of the culture will often increase if the cells are transferred before the culture reaches stationary phase (Fig. 5.1). For some purposes, e.g. isolation of protoplasts which retain viability after electroporation, it is very importoant to have a vigorously growing culture.

5.2. MEASUREMENT OF GROWTH OF CELL CULTURES

Growth of callus and cell suspension cultures can be monitered by increase in fresh or dry weight or increase in cell number. For cell suspension cultures, increase in packed cell volume (pcv) is also a good indicator of growth.

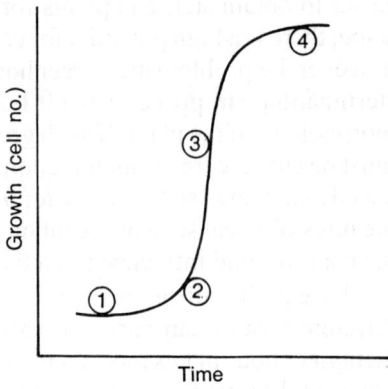

Fig. 5.1. Model growth curve for a plant cell suspension grown in a closed system. The four different growth phases are labelled. (1) Lag phase; (2) Exponential phase; (3) Linear phase; (4) Stationary phase.

5.2.1. Fresh and dry weight measurements

For callus culture, transfer the entire callus (scrape off the medium) to a pre-weighed weighing boat (or container) and determine the fresh weight. For cell suspension cultures, collect the cells on a pre-weighed nylon membrane and determine the fresh weight (Determine the weight of water retained by the membrane separately and subtract this amount from the measured fresh weight). Alternatively, transfer the entire contents of the cell culture flask to a pre-weighed centrifuge tube. Spin the tube for 5-10 minutes at 200 g. Examine the supernatant for cells. If cells are present in the supernatant, repeat centrifugation until the supernatant is free from cells. Then carefully pipette out the entire medium without disturbing the pellet. Weigh the centrifuge tube with cells to determine the fresh weight. After measuring the fresh weight, dry the samples in an oven at 60 °C until no change in dry weight is observed.

Increase in fresh weight can also be measured without sacrificing the samples at the beginning of an experiment. Transfer callus or cells to pre-weighed petri dishes or culture flask containing the medium. Weigh the petri dish or the flask again and determine the weight of callus or cells added. At the end of an experiment, remove the entire callus or cells from a suspension culture and determine the final weight.

$$\text{Growth index (GI)} = \frac{\text{Final weight of the callus or cells}}{\text{Initial weight of the callus or cells}}$$

5.2.2. Increase in cell number

A haemocytometer can be used for determining the cell numbers in a fine suspension culture. Callus cultures and suspension cultures consisting of cell aggregates have to be macerated prior to counting the cell numbers. An ideal macerating fluid for callus or cell aggregates from suspension cultures consists of equal volumes of 10% chromic acid and 10% nitric acid. Treat cell clumps for 5-30 minutes in this mixture at 60°C. After cooling, shake the container vigorously to loosen the cell clumps. Alternatively, add 1% (w/v) of macerozyme (Sigma Chemical Co. USA) to the culture medium and incubate flasks overnight on a shaker in the growth room under normal culture conditions.

For suspension culture filter the culture through miracloth (Baxter, USA) and use the same method as 5.2.1. and 5.2.2.

5.2.3. Packed Cell Volume (PCV)

This method will be useful for suspension cultures only. It is relatively simple and can be determined at any stage of growth. Small volume of (10 ml) suspension culture is aspectically sampled and placed in a graduated conical centrifuge tube. The total volume of a cell pellet is determined after centrifuging at a constant speed (e.g. 500 g) for a specific time (5 min.). The PCV is typically expressed as a percentage of the total volume in tube.

5.2.4. Molecular Protein and DNA

The amount of a cell component which is a constant proportion of the total cell dry weight, can be used to estimate biomass concentration during balanced growth. The components such as protein and DNA are usually determined by using well established techniques e.g. Bradford's reagent for protein.

5.2.5. Mitotic index

Mitosis is characterised by a number of stages called as prophase, metaphase, anaphase and telophase. Mitotic index is an estimate of the number of cells of the population in these stages (mitosis). It is simple but time consuming.

$$\text{Mitotic Index} = \frac{\text{Number of nuclei in mitosis}}{\text{Total number of nuclei scored}} \times 100$$

5.2.6. Medium component calibration

The depletion of medium components e.g glucose, nitrate, ammonium, phosphate can be used to estimate cell growth.

5.2.7. Conductivity of medium

The integrity of the plant cell membrane is reflected in the concentration of solutes which leak into the medium. Thus, the electrical conductivity of the medium can be used to analyze the cell growth. There is an inverse relationship between medium conductivity and growth.

5.2.8. Cellular Protein

The total protein present in cell or callus cultures will also reflect the growth of the culture. The total protein is directly proportional of the growth of the cells or amount of the cells. It can be estimated by Bradford's reagent.

5.3. GROWTH CURVE

The model growth curve for a plant cell suspention growth in a closed system is shown in Fig. 5.1. It is a sigmoid curve and this is useful for the determination of the time for subculture, effector addition for enhanced production of secondary metabolites, to know growth and production kinetics. The growth curve has lag phase, exponential phase, linear phase and stationary phase.

5.4. CELL VIABILITY

The staining techniques are useful for the determination of viability of cells from plant tissue cultures.

5.4.1. Fluorescein diacetate vital stain

The Fluorescein diacetate (FDA) vital stain is dependent upon the ability of esterases in viable cells to cleave the stain, which then fluoresces yellow/green under a UV microscope (Widholm, 1972).

Add a few drops of FDA (0.1% w/v in acetone) stain to cells on a microscope slide. Allow the stain to penetrate for a few minutes, place a cover-slip over the cells, and examine under a UV fluorescence microscope using a blue/violet filter. Count the number of fluorescing cells in a dark field of view and count the total number of cells in the same field under bright illumination.

$$\% \text{ Viability} = \frac{\text{Number of fluorescent cells}}{\text{Total number of cells}} \times 100$$

5.4.2. The triphenyl tetrazolium chloride (TTC) assay

This is a useful alternative for larger specimens (Steponkus and Lamphear, 1967). Dehydrogenase activity of viable cells reduces TTC to a red formazan product which is measured spectrometrically.

5.4.3. Evan's blue

The presence of a semi-permeable cytoplasmic membrane allows viable plant cells to be identified by their ability to either accumulate or prevent the uptake of certain stains. The non-viable cells stain with Evan's blue. The viability is the percentage of viable cells within the population of cells observed.

5.5. TYPES OF CULTURE

A plant consists of different organs and each organ is composed of different tissues, which in turn are made up of different individual cells. Plant tissue culture refers to the *in vitro* cultivation of plants, plant parts (organs, tissues, single cells, embryo and protoplasts) and seeds on nutrient media under aseptic conditions. Unlike animal cells, plant cells, even highly mature and differentiated, retain ability to change to a meristematic state and differentiate into a whole plant if it has retained an intact membrane system and a viable nucleus (totipotency).

In tissue culture, more often we use an explant (an excised piece of differentiated tissue or organ) to initate their growth in culture. The non-dividing, differentiated, quiescent cells of the explant when grown on a nutrient medium first undergo changes to achieve the meristematic state.The phenomenon of mature cells reverting to a meristematic state and forming undifferentiated callus tissue is termed as dedifferentation. Since the multicellular explant comprises cells of diverse types, the callus derived will be heterogenous. The ability of the component cells of the callus to differentiate into a whole plant or a plant organ is termed as redifferentiation. These two phenomena of dedifferentation and redifferentiation are inherent in the capacity of a plant cell, and their by giving rise to a whole plant is described as cellular totipotency. Generally, a callus phase is involved before the cells can undergo redifferentiation leading to the regeneration of a whole plant. The dedifferentiated cells can rarely give rise to whole plants directly without an intermediate callus phase.

5.5.1. Cytodifferentiation

The cells in a callus are parenchymatous in nature, the differentiation of these cells into a variety of cells is required during redifferentiation of cells into whole plants. This redifferentiation of cells is known as cytodifferentiation. *In- vitro* and *in-vivo*, the main emphasis in plant cytodifferentiation has been laid on vascular tissue differentiation (xylem, phloem), particularly the xylem elements. In an intact plant, tissue differentiation goes on in a fixed manner that is characteristic of the species and the organ, while callus cultures which lack vascular elements offer a valuable system for the study of the effect of various chemicals and physical factors on vascular tissue differentiation. The auxins and sucrose have a effect on vascular tissue differentiation. While cytokinins and gibberellins promote differentiation into xylem tissue (xylogenesis). There is an inverse relationship between the auxin concentration and degree of differentiation.

5.5.2. Organogenic differentiation

In nature, totipotentiality of somatic cells has been observed in several taxa where stem, leaf and root pieces are able to differentiate into shoots and roots. *In- vitro* studies have indicated that totipotentiality is not restricted to few species, most plant species if provided with appropriate conditions would show differentiation. For the regeneration of whole plant from a callus tissue or cell , cytodifferentiation is not enough and there should be differentiation leading to shoot bud or embryo formation. This may occur either through organogenesis or somatic embryogenesis (see chapter-13).

Plant tissue culture, which covers all type of aspetic plant culture should be used in a restricted sense and it is possible to distinguish it into various types of cultures.

Seed Culture : Culture of seeds *in vitro* to generate seedlings/plants.

Organ Culture : Culture of isolated plant organs. Different types can be distinguished e.g. meristem, shoot tip, root culture, anther culture, pollen culture.

Embryo Culture : Culture or excised mature or immature embryo from seeds.

Callus Culture : Culture of a differentiated tissue from explant allowed to differentiate *in vitro* and a so called callus tissue is produced.

Protoplast Culture : Culture of plant protoplasts i.e., cells devoid of their cell walls.

Cell Culture : Culture of isolated cells or very small aggregates remaining dispersed in liquid medium.

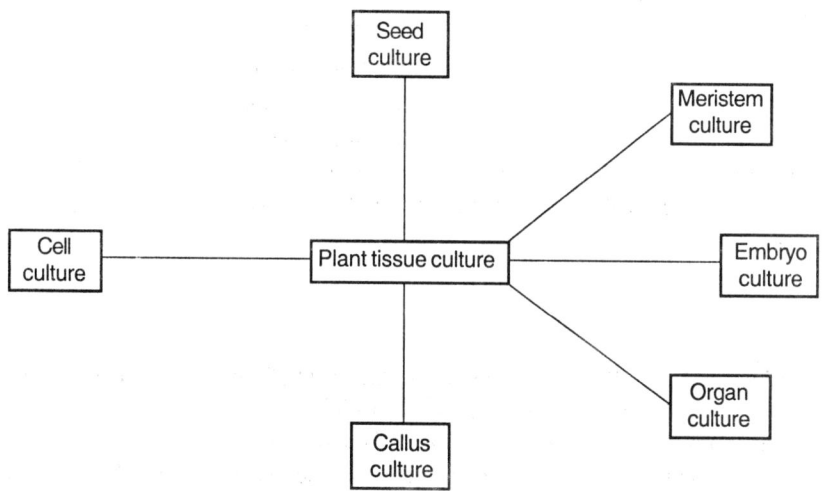

5.6. APPLICATION OF PLANT TISSUE CULTURES

The various applications of plant tissue cultures are

1. *Agriculture* : Production of high-yielding, drought, herbicide, salt resistant and insect resistant crops.
2. *Horticulture and Forestry* : Micropropagation of medicinal and aromatic plants (see chapter 16). Production of synthetic seeds (chapter 17).

3. Production of phytopharmaceuticals, food flavors, colors, etc from plant cell cultures (see Chapter 6).

   The yields of secondary metabolites are very poor in some species. The various strategies for enhancing the yields are shown in Fig. 5.2.

   (a) Biotransformation (see Chapter 10).

   (b) Immobilization (see Chapter 11).

   (c) Genetic transformation - transgenic plants (see Chapters 18 &19).

4. The valuable products from transgenic plants e.g., Antigen, antibodies, edible vaccines and monoclonal antibodies (see Chapter 18).

Strain improvement	Selection
	Screening
Medium variation	Nutrients
	Phytohormones
	Precursors
	Antimetabolites
Culture conditions	Inoculum size
	pH
	Temperature
	Light
	Agitation
Specialized techniques	Elicitors
	Immobilization
	Permeabilization
	Two-phase systems
	Two-stage systems

Fig. 5.2. Various strategies for the improvement of productivity of cultured plant cell (Dornenburg and Knorr 1995).

REFERENCES

Dornenburg, H and Knorr, D., (1995) stragies for the improvement of secondary metabolites production in plant cell culture. *Enzyme and microbial technology*, **17**: 674-684.

Steponkus, P.L., and Lamphear, F.O., (1967) *Plant physiology*, **42**: 1423.

Widholm, J.M. (1972) The Use of fluorescein diacetate and phenosafranine for the determining viability of cultured plant cell. *Stain technology* **47**: 189-194.

Chapter 6

Secondary Metabolic Products (Phytopharmaceuticals) from Plant Cell Cultures

6.1. INTRODUCTION

Secondary metabolism is the synthesis and metabolism of endogenous compounds by specialized proteins (Luckner, 1971, Luckner and Nover, 1977, Luckner *et al.*, 1977). The results of these processes are secondary metabolic products. They are an expression of cell specialization which is either triggered by the process of cell differentiation (Luckner and Nover 1977, Yeoman *et al.*, 1982) or represents an aspect of the process of plant development. Cell differentiation is itself a basic component of metabolic regulation by higher organisms. It includes all processes which differentiate cell with the same genetic composition. However, not all genetic information is actually used in the course of development of a cell or plant. Thus, the biosynthesis of secondary compounds is usually limited to (1) particular developmental stages and, or (2) specialized cells.

6.2. FUNCTION OF SECONDARY COMPOUNDS

The function of secondary compounds is as follows (Luckner 1986, Tables 6.1 to 6.3):
1. Detoxification of substances accumulated in primary metabolism.
2. Act as physiological effectors.
3. Provision of chemical signals to coordinate the metabolism of multicellular organisms.
4. Coordination of activities of different individuals of the same species, and
5. Development of ecological relationships.

6.3. ACCUMMULATION

6.3.1. Turnover

Depending on the object, secondary products are either secreted into the surrounding medium (Table 6.4) or stored, intracellularly. There they experience turnover processes with characteristic half–lives (Table 6.5). Their degradation was first proven in cell suspension cultures (Barz, 1975). Degradation and synthesis often occur simultaneously.

The extent of their accumulation is mainly determined by three cell capacities, synthetic capacity, storage capacity and the capacity to metabolize the compounds in transport and detoxification processes. Individual plant organs differ in their significance in this process. Often, synthesis and

Table 6.1. Examples of the protective function of secondary products (Adopted from Endress, 1994)

Compound	Stomach insecticide
Active intake L-DOPA (5%)	For the polyphagous larvae of *Proderma eridania*
Mimosine (0.1%) 3-Cyanoalanine (0.1%) Azetidinecarboxylic acid (0.1%) 5-Hydroxytryptophan (1%) N-Methyltyrosine (1%) Albizin (5%) S-Carboxycysteine (5%) L-DOPA (5%) Canavanine (5%)	Supplements a diet from seeds of *Vigna unguiculata* (cowpea). These compounds are lethal to the larvae of the snout beetle *Calloso-bruchus maculatus*
Flavonoid aglyca with two adjacent OH in the B-ring	For the larvae of the polyphagus butterfly *Heliothis zea*
Sambunigrin : a cyanogenic glycoside from *Sambucus niger* stored by plant lice	For plant lice feeders
Contact agent : Benzopyran From secretion Benzofuran Containers of the Asteraceae	Penetrate actively the cuticle; they are activated by UV (350 nm); affect membranes and DNA; interfere with the immune system

Table 6.2. Examples of the destructive effect of some secondary products for man and domesticated animals (Adopted from Endress, 1994)

Compound	Occurrence	Effect
Chlorogenic acid	Sunflower seeds	Inhibits digestive enzymes
β-N-methyl-amino-1-alanine (BMAA)	*Cycas circinalis*	Base for food-induced • Parkinson's disease • Alzheimer disease • Amyotrophic lateral sclerosis
β-Oxalyl-amino-L-alanine (BOAA)	*Cicer arietinum*	Lathyrism
Gossypol	*Gossypium arboreum*	Anaemia, Sperm decay
Thioglyceride	Seeds of rape	Struma formation
3-Cyanoalanine	*Vicia sativa*	Neurotoxic
Hypoglycin A and B [3-(methylene-cyclo-propyl)-alanine]	Seeds and fruits of *Bligleia sapida*	Hyperglycemia
Selenocystathionine	*Phaseolus lunatus*	Universal toxicity
Methyl-selenocysteine	*Astragalus bisulcatus, Neptunia amplexi-caulis, Stanleya pinnata, Haplopappus fremontii, Morinda reticulata, Lecythis ollaria*	Universal toxicity
Mimosine	*Mimosa* sp., *Leucaena* sp.	Liver damage, Loss of hair
Isoflavone	*Medicago sativa*	Disorder of the reproductive cycle

Table 6.3. Comparison of function and characteristics of primary and secondary metabolism in higher plants (Adopted from Luckner, 1986)

Primary metabolism		Secondary metabolism
	Plays a role in	
Growth and development		The interaction of an organism and its environment
	Characteristics	
Universal		Singular
Uniform		Manifold
Conservative		Adaptable
Indispensable	For growth and development	Dispensable
Indispensable	For existence and survival in the ecological system	Indispensable

Table 6.4. Selected examples of constitutively formed secondary products excreted into the extra-cellular compartment (Barz et al., 1990)

Atropine	Datura innoxia
Berberine	Coptis japonica
Berberine	Thalictrum minus
Caffeine	Coffea arabica
Caffeine	Coffea robusta
Theobromine	Coffea robusta
Quinine	Cinchona ledgeriana
Lupanine	Lupinus polyphyllus
Nicotine	Nicotiana tabacum
Protopine	Macleaya microcarpa
Sanguinarine	Papaver somniferum
Scopolamine	Datura innoxia
Shikonin	Lithospermum erythrorhizon
Phenolics	Lithospermum erythrorhizon
Capsaicin	Capsicum frutescens
Monoterpenes	Thuja occidentalis

storage occur in different sites (Table 6.6). In many cases their concentration varies according to geographical location (Table 6.7), climate, soil and fertilization, and is compartmentalized in both time and space (Nover *et al.* 1980).

6.3.2. Spatial compartmentalization

6.3.2.1. Synthesis

Localization. The synthesis of secondary metabolites is often bound to particular organs of reaction vessels, specific cell organelles, vesicles, or membrane bound enzyme or isoenzyme (Table 6.8) and / or it is characterized by temporal separation between reactions and by separate pools for primary and secondary metabolism (Table 6.9).

Table 6.5. Half-life of some plant secondary products (Adopted from Luckner, 1986)

Group of substances	Individual compounds	Half-life	Plant/organ
Isoprenoids	Menthol	Several h	*Mentha piperita*
	Mono- and diterpenes	170 days	*Pinus sylvestris*,
		46 days	Cortex, needles
	Marrubin	24 days	*Marrubium vulgare*
	α-Tomatine	6 days	*Solanum esculentum* fruits
Cyanogenic glycosides	Dhurrin	10 h	*Sorghum*, seedlings
Alkaloids	Gramine	80 h	*Hordeum vulgare*
	Nicotinic acid glucoside	24 h	*Glycine max*, cell culture
	Nicotine	22 h	*Nicotiana tabacum*
	Morphine	7.5 h	*Papaver somniferum*
	Ricinine	4h	*Ricinus communis*
Amino acid derivatives	Hordenine	42 h	*Hordeum vulgare*
Cinnamic acid derivatives	Chlorogenic acid	20 h	*Xanthium pensylvanicum*, leaves
	Coniferin	60-120 h	*Picea abies*, seedlings
Flavonoids	Delphinidin glycosides	25-31 h	*Petunia hybrida*, corolla
	Kaempferol and quercetin glycosides	7-12 days	*Leaves*
	Biochanin A	25-320 h	*Several organs, cell cultures*
	Formononetin	72 h	*Roots*
	Kaempferol, quercetin and isorhamnetin		*Cucurbita maxima*, seedlings
	glucosides	30-36 h	
	biosides	48 h	

Table 6.6. Some storage places of secondary metabolites (Adopted from Endress, 1994)

Compound	Species	Storage place
Quinolizidine alkaloids	Fabaceae	Throughout
Cardenolides	*Digitalis lanata*	Mesophyll cells of leaves
Quinine, Quinidine	*Cinchona succirubra*	Cortex
Shikonin derivatives	*Lithospermum erythrorhizon*	Cork crust
Gossypol	*Gossypium hirsutum*	Secretary spaces of seeds
	Gossypium arboreum	
Monoterpenes	*Pelargonium fragrans*	Specific glandular hairs
Benzofuran	*Asteraceae*	Secretory canals
Benzopyrane		Secretory containers

Table 6.7. Alkaloid content (%), depending on the latitude (Adopted from Endress, 1994)

Plant species	Origin	Content
Atropa belladonna	Crimea	1.3
	Leningrad	0.5
Aconitum napellus	Nova scotia	Atoxic
	Mediterranean region	One of the most poisonous plant
Anisodus luridus	Northern regions of Europe	Poor
Datura stramonium, Hyoscyamus niger	Southern regions of Europe	Rich
Scopolia carniolica	Caucasus	1
	Sweden	0.3

Table 6.8. Examples of the three groups of cinnamate CoA-ligases, characterised by their cofactor (group 1), their tissue specific localization (group 2) of their high substrate and/or product specificity (group 3) (Adopted from Endress, 1994)

a) Examples for group 1 : Cinnamate : CoA-ligase supply of leaves of *Petunia*

Enzyme	Substrate	Specific inhibitor
Isoenzyme 1	p-Coumaric acid Coffeic acid	Quinic acid Esters
Isoenzyme 2	p-Coumaric acid Sinapic acid	None
Isoenzyme 3	p-Coumaric acid Ferulic acid	Flavonoid Aglycons

b) Examples for group 2 and 3 : Cinnamate : CoA-ligase supply and concentration of phenolic compounds of different plants

b1) Example for group 2: stem of *Populus* Eur-americana

Enzyme origin	Substrate	End-product
Sclerenchyma cells	Sinapic acid p-Coumaric acid	Syringin Lignin
Xylem cells	Ferulic acid p-Coumaric acid	Guajacol Lignin
Parenchyma cells	Caffeic acid p-Coumaric acid	Soluble phenolic compounds

b2) Examples for group 3 : cell suspension cultures of *Petroselinum hortense* and *Glycine max*

P. hortense	p-Coumarin a cid	Flavonoids
Glycine max	Ferulic acid Sinapic acid p-Coumaric acid	Lignin-like compounds
Glycine max	p-Coumaric acid Caffeic acid	Flavonoids

Table 6.9. Examples of compartmentalization in the course of secondary product formation (Adopted from Endress, 1994)

1. Chronological compartmentalization

a) Annual rhythm

Atropa belladonna	April: 0.7% dry wt. June : 0.1% dry wt.	Alkaloids
Festuca arundinacea	Only from July 15th to September 15th	Perlodine
Genisia aethnensis	Spring Autumn stem flower	N-methyl-cystisine Anagyrine Cytisine
Impatiens balsamina	Flower bud differentiation	Anthocyanins
Lilium henrvi	Formation of anthers	Carotenoids flavonoids
Nigella damascna	Spermatogenesis	Damascenine

(Contd.)

Ranufuculus thora	Only during flowering	Contact induces irritation and pain
Salix purpurea		Phenolic glucosides

b) Diurnal rhythm

Atropa belladonna	Early in the morning (a.m.) Late afternoon	Atropine
Baptisia	Early noon ⎤	
Lupinus	Night ⎬	Leaf alkaloids
Sarothannus	Early afternoon ⎦	
Conium maculatum	Near 6 p.m. : increase 400%	γ-coniceine
Papaver somniferum	In the morning 12 p.m. 2 p.m.	Morphine Codeine Thebaine
Partulaca grandiflora	Dark	Catecholamines
Thymus vulgaris	2 p.m. : maximum	Thyme oil (thymol)

2. Local compartmentalization

a) Cellular

Catharanthus roseus	Synthetic capacity depends on the number of storage cells	Alkaloids
Datura innoxia	Synthesis in the roots and stored in the cytosol of leaf cells	L-hyoscyamine nicotine
Digitalis lanata	Mesophyll cells of leaves	Digitoxin
Macleaya cordata	Synthesis at pH 5, Storage at pH 3	Alkaloids
Pelargonium fragrans	Glandular cells	Monoterpenes
Rheum palmatum	Vascular rays of the roots	Anthracene derivatives
Ruta graveolens	Specific secretion spaces	Volatile oils

b) Sub-cellular

i) Cell structures

Coptis japonicum	Rough endoplasmic reticulum (ER)	Anthraquinones
Juniperus communis	Vacuoles	Lannins
Lithospermum erythror-hizon	Naphthoquinone vesicles rough endo-plasmic reticulum	Naphthoquinones
Morinda lucida	Plastid structure antutrophic nutrition, heterotrophic nutrition	Phylloquinones anthraquinones
Papaver somniferum	Alkaloid-vesicles	Alkaloids
Pinus elliotti	FR and Golgi vesicles	Tannins
Thalictrum minus	Rough ER anthraquinone vesicles	Anthraquinone

ii) Membrane formation multi-functional proteins / multi-enzyme complexes

Haplopappus gracilis	Endoplasmic reticulum	Naringenine eriodictyol
Sorghum bicolor	Endoplasmic reticulum Microsomes	Coumaric acid p-hydroxy-mandelic nitrile

iii) Isoenzymes

Glycine max	Methyltransferase	Cinnamic acid flavonoids
Haplopappus gracilis	Methyltransferase	Cinnamic acid flavonoids

iv) Different precursor pools

Malic acid	Mitochondria	Vacuoles

Isoenzymes. Isoenzymes involved in the transfer of precursors from primary metabolism are often characterized by their high specificity to particular substrates or inhibitors, inclusion of specific co–factors (group 1), tissue specificity (group 2), or specificity for a substrate and final product (group 3). Occassional deviations from this high substrate specificity occur primarily among peroxidases, some dehydrogenases and a few methyltransferases (Yeoman *et al.* 1990).

Local Differences. Cinnamic acid CoA–ligase, active in flavonoid metabolism, provides examples of all three groups (Table 6.8). The second group also includes isoenzymes of phenylalanine–ammonialyase from suspension cultures of *Quercus pendunculata.* Enzymes found in particular compartments (mitochondria, microbodies, microsomes) differ in their Michaelis–Menten constants (K_M), an expression of substrate affinity. Synthesis of aromatic amino acids occurring in the chloroplasts and cytosol of *Nicotiana tabacum* differs in the specificity of the 3–deoxy–D–arabinoso–heptulose–7–phosphate–(DAHP)–synthases and chorismate mutases involved (Jensen, 1986).

Synthetic Pathways. Tissue cultures of *Glycine max* are additionally characterized by the presence of a methyltransferase belonging to the third group. One of these methyltransferases methylates only substituted cinnamic acids (Caffeic acid, 5–hydroxyferulic acid) and is thus active during lignin biosynthesis. The other methylates only flavonoid derivatives (luteolin, quercetin) and this alters the flavonoid pattern. In addition, they are characterized by their dependence on Mg^{+2} ions.

The role of three strictosidine–synthase (SS) isoenzymes isolated from *Catharanthus roseus* is as yet unclear. SS from *Rauwolfia serpentina* exists only in a homogeneous form.

Initiation Enzymes. Particularly the initiation enzymes involved at the beginning of a specialized branch of a synthetic pathway (Fig. 6.1) are regulated by specific mechanisms (Hahlbrock *et al.* 1971, Noe and Berlin 1985, Rolfs *et al.* 1987). They open the way to specific groups, which very little in their structure. Methodological difficulties prevented the successful isolation of enzymes participating in secondary product formation until the 1960 (Heide and Tabata, 1987).

Autonomus Sites of Synthesis. Some synthesis are bound to particular cell organelles. Chloroplast represents such an autonomous site of synthesis. In addition to the biosynthesis of aromatic amino acids the synthesis of terpenoids from mevalonic acid and part of the synthesis of conine have been shown to take place here. The role of chloroplasts is confirmed by the localization of enzymes involved in the synthesis of lupanine and sparteine in chloroplast from *Lupinus polyphyllus* cell suspension culture (Wink and Hartmann, 1982). In *Capsicum* chromoplasts the stroma is the site of synthesis for the carotenoid phytoene.

Membrane Integrity. Binding to a membrane as demonstrated for most of the enzymes of carotenoid synthesis channelizes reaction processes and limits interference from the intracellular medium. The degree of biological activity of membrane dependent enzymes, enzyme complexes or multi-functional enzymes or the degree of cooperation among the individual components of a reaction pathway, depends at least partially on membrane integrity (Endress, 1994).

Microsomes. The importance of the membrane systems involved can be demonstrated using the example of the hydroxylation potential of oxidases dependent on NADPH and O_2 in *Haplopappus gracilis.* These mixed function oxidases transform naringenin into eriodictyol or di–hydroxy–kaempherol only in microsomes whose integrity has been destroyed. They can be transformed by treatment with dimethylsulphoxide in this "softened" – state which is especially suited for the deposition of foreign proteins. The membrane dependence of this reaction is also confirmed by the failure of cell–free extracts to convert taxifoline into cyanidin.

Stabilization of Synthesis. p–hydroxy–mandelo–nitrile synthesis in *Sorghum bicolor* process, microsomal membrane particle fraction transform tyrosine into the nitrile. Neither the exogenous addition of N–hydroxy–tyrosine nor p–hydroxy–phenylacetate can significantly affect its transformation into this nitrile or into p–hydroxy–phenyl–acetaldoxime, once transformation has begun (Table 6.10, Eilert *et al.* 1987).

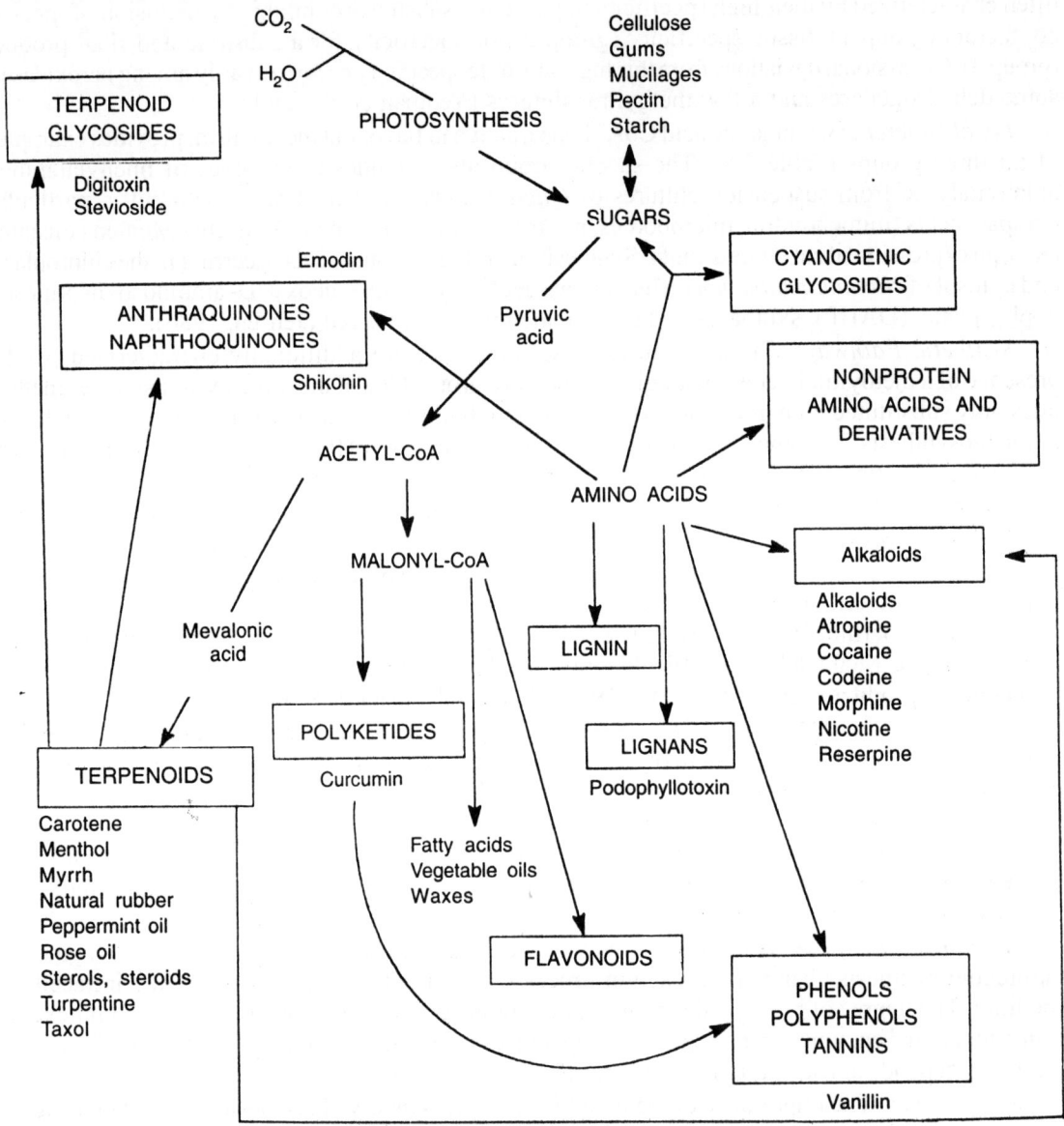

Fig. 6.1. Biosynthetic origin of some commercially important plant-derived compounds. Major groups of secondary metabolites are indicated by boxes (Balandrin et al., 1985).

Changes in Location : Intermediary products must often change location during such synthesis. For example, the intermediate products of berberine synthesis in *Coptis japonica* and *Berberis wilsonide* travel black and forth between the cytoplasm and specific vesicles (Amann *et al.,* 1986).

6.3.3. Storage

The sites of synthesis and storage are often separated spatially. However, the products of synthesis are generally not directly secreted into the surrounding medium. They must therefore be stored

Table 6.10. Compartments of secondary metabolite production in cells of suspension cultures of different species (Adopted from Endress, 1994)

Compartment	Formed product	Species
Endoplasmic reticulum	L–phenylalanine to o–, p–coumaric acid	*Haplopappus gracilis*
	Shikonin	*Lithospermum erythrorhizon*
	Berberine	*Coptis japonicum*
Cytoplasmic vesicles	Shikonin	*Lithospermum erythrorhizon*
	Berberine	*Coptis japonicum*
	Berberine	*Thalictrum* sp.
Vacuoles	Berberine	*Coptis japonicum*
Microsomes	Tyrosine to p–hydroxy–mandelo–nitrile	*Sorghum bicolor*

either within the synthesizing cells in separate storage cells, or in other secretory vessels. Often, single specialized cells, cell compartments or tissues, and sometimes even specialized parts of organs serve as storage sites.

6.3.3.1. Vacuole

A common site for short and or long–term storage is the aqueous interior of a vacuole (Table 6.11, Ryan and Walker–Simmons (1983). Usually, those components are stored which either have only one function within the vacuole or which are at least temporarily metabolically useless or harmful (Matile, 1978, 1984). By means of proton, pumps fixed in the tonoplast. Using ATP as energy source, an acidic pH (pH 5.3–5.6) relative to the cytoplasm (pH 7.1–7.6) is maintained. In *Catharanthus roseus* and *Macleaya cordata* this allows the differentiation between synthetic (pH 5) and storage (pH 3) cells. Neither cell type can be morphologically distinguished from other cells in the tissue (Luckner *et al.,* 1977).

6.3.3.2. Discrimination

In general, only two categories of secondary metabolites are stored in vacuoles, nitrogenous, usually non–glucosidated alkaloids and glucosides of various classes other than alkaloids. In roses, their toxic monoterpenes are glucosidated for this purpose. Accumulation of alkaloids in plant vacuoles has been demonstrated repeatedly (Table 6.11). *Catharanthus roseus* does not store the final metabolic product serpentine, but rather its precursor ajmalicine (Table 6.12). Some vacuoles even discriminate stereoisomers.

6.3.3.3. Product Trap

In this case, the vacuole functions as a trap for ions, distinct configurations or enantiomers. It acts as an ion trap for lipophilic compounds which diffuse into the acidic interior of the vacuole and are then protonated. Due to the induced change in electric charge, they lose their fat solubility and can no longer penetrate the tonoplast. Entering products can also be bound to intracellular materials such as phenols e.g *Cinchona ledgeriana*.

The vacuole acts as a configuration trap in compounds that undergo isomerization in the acidic medium of the vacuole (o–coumaric acid), or whose configuration is changed depending on the concentration of protons (apigenin cyanidin–ester).

6.3.3.4. Catabolism

In its interactions with the environment, a plant must always be capable of using its secondary products. Therefore, the required set of enzymes is often stored in the direct or close vicinity of the

Table 6.11. Secondary metabolites accumulated in vacuoles and some trapping methods (Renaudin and Guern, 1990)

Molecular species	Plant species	Mechanism of trapping
Serpentine	*Catharanthus roseus*	Precipitation with phenolics
Vindoline	*Catharanthus roseus*	
Chelidonin	*Chelidonium majus*	Binding to phenolics
Sanguinarine	*Macleaya microcarpa*	Crystallization
Berberine	*Thalictrum minus*	Crystallization
S-Reticuline	*Fumaria capreolata*	Enanticmer trapping
S-Scoulerine	*Fumaria capreolata*	Enantiomer trapping
Nicotine	*Nicotiana tabacum*	
Lupanine	*Lupinus polyphyllus*	
Betanin (betanidin-5-glucoside)	*Beta vulgaris*	
Purpurea glycoside A	*Digitalis lanata*	
Deacetyl lanatoside C	*D. lanata*	
Lanatoside A	*D. lanata*	Ionization (= ion trapping)
Lanatoside C	*D. lanata*	
Dhurrin	*Sorghum bicolor*	
Capsaicin	*Capsicum annuum*	
Glucosinolates	*Armoracia lapathifolia*	
Glucobrassicin	*Armoracia rusticana*	
Anthocyanin	*Hippeastrum, Tulipa*	
Cyanidin-ester[a]	*Daucus carota*	Conformational shift due to acidic pH
Apigenin-ester[b]	*Peteroselinum hortense*	Conformational shift due to acidic pH
cis glucoside of o-coumaric acid[c]	*Melilotus alba*	
Hydroxy-cinnamic acid ester	*Vitis* sp.	
Hydroxy-cinnamic acid-malic acid ester	*Raphanus sativus*	*cis-trans* isomerization
Isoflavone malonyl glycosides	*Cicer arietinum*	
Avenacoside	*Arena sativa*	

[a] Absorbed and accumulated only as cyanidin-(3-sinapyl-xylosyl-glucosyl-galactoside)-ester.
[b] Absorbed and accumulated only as apigenin-7-O-(â-D-6-O-malonyl-glycoside)-ester
[c] Absorbed only as trans-glucoside.

storage site. Protective mechanisms of varying elaboration have evolved in individual species in order to prevent toxic components from damaging the plant itself. Knowledge of these relationships can be important in choosing extraction methods for biotechnological purposes (Matile, 1984).

6.4. TRANSPORT MECHANISMS

6.4.1. Cardenolides

Both passive (diffusion) and active processes are involved in transport from the site of synthesis to storage. Uptake of cardenolides by vacuoles of *Digitalis lanata* mesophyll cells also serves to show the importance of glucosidation. Only primary glucosides remain inside the vacuole. Therefore, products such as β-methyl digitoxin that cannot be glycosilated cannot be stored. Upon completion of all possible chemical transformations, they are resecreted into the surrounding medium (Kreis, 1987).

Table 6.12. Mechanisms of the transport of secondary metabolites across the tonoplast in cells of various cell suspension cultures (Renaudin and Guern, 1990)

Compound	Mechanism	Occurrence	Related molecules not transported
Uncharged alkaloids	Diffusion	*Macleaya cordata*	Charged alkaloids
Ajmalicine	Carrier	*Catharanthus roseus*	Codeine, morphine, nicotine
Ajmalicine	Carrier	*Rauwolfia serpentina*	Codeine, morphine, nicotine
Acylated anthocyanin	Carrier	*Daucus carota*	Deacylated anthocyanin
Cardenolides			
Secondary glucosides	Diffusion	*Digitalis lanata*	--
Primary glycosides	Carrier	*Digitalis lanata*	Secondary glycosides
Catharanthine	Carrier	*Catharanthus roseus*	Codeine, morphine, nicotine
*trans–o–*coumaric acid	Diffusion	*Melilotus alba*	*cis–o–*coumaric acid
Dopamine	Carrier	*Papaver somniferum*	Coniceine, reserpine, sanguinarine
Esculine	Carrier	*Hordeum vulgare*	--
Lupanine	Carrier	*Lupinus polyhyllus*	Atropine
Nicotine	Carrier	*Nicotiana sylvestris*	Ajmalicine, vindoline, catharanthine
(S)–Reticuline	Carrier	*Fumaria capreolata*	(R)–Reticuline, ajmalicine, catharan-thine, nicotine

6.4.2. Chemical Form of Transport

The pyridine–alkaloid ricinine, which is formed with varying efficiency in all parts of the annual *Ricinus communis* which reaches a height of up to 10 m in the tropics, is demethylated for this purpose. It is transported via the xylem from yellow to green leaves or to seeds in this water–soluble form. There it is again stored as the methylated product.

6.4.3. Temporal Compartmentalization

6.4.3.1. Rhythmic Phenomena

Changes in Activity

Diurnal cycles (Wink and Hartmann, 1982) and annual rhythms characterize the accumulation of many compounds in differentiated plants. They are expressions of changes in enzymatic activities in the course of the day or year (Tissut and Egger, 1972).

These variations are probably caused by variations in temperature or light conditions in the course of the day or year. Influence by diurnally varying HR/DR conditions resupposes involvement of absorbent pigment systems, such as the phytochrome system. Dependence on annually varying day length causes changes in luminous flux. Light–sensitive synthetic processes that occur only in the dark are known (Endress *et al.,* 1984).

6.5. THE ROLE OF INDIVIDUAL DEVELOPMENTAL STAGES

6.5.1. Degree of Differentiation

The degree of differentiation of individual cells or plants is especially important (Niemann, 1976). Often, morphological and biochemical differentiation processes are linked (Wiermann, 1973; Yamamoto *et al.,* 1986). For example, young *Ammi visnaga* plants primarily accumulate rare furanochromes produced via polyketide metabolism, while blooming and fruiting plants accumulate pyranocoumarine esters from phenylpropane metabolism.

6.5.2. Pollen Maturation

In several objects, specific secondary products could be associated with particular stages of pollen maturation. The activity of corresponding enzymes increases prior to or simultaneously with their synthesis. The exine of the sporoderm of *Lilium henryi* pollen walls is characterized by a high sporopollenine content. This polymerization product is primarily synthesized following completion of tetrad formation by oxidative conversion of accumulated carotenoids. This synthesis is catalyzed by the peroxidase accumulated in parallel with the carotenoids (Shaw, 1971).

6.5.3. Germination

Ricinus communis is a leguminous plant with alkaloid–containing seeds. Formation of N–demethyl–ricinine during the first 2 days of germination may be regarded as one of the first metabolic processes triggered by growth and differentiation. This biologically active form of ricinine is obviously required only in the first few days of germination. By incorporation of radioactive nicotinic acid, it was demonstrated that this demethylated product of the mildly toxic pyridine–alkaloid ricinine is synthesized *de novo* from additionally provided ricinine. Thereafter, demethyl–ricinine disappears completely, which the ricinine content continues to increase. Seed germination of *Impatients balsamina* is characterized by a typical flavonoid pattern (Weissenbock and Reznik, 1970).

6.5.4. Tissue specificity

Phaseolus lunata, which occurs naturally in Costa Rica, accumulates the aceto–cyanhydrine–glucoside linamarine in its cotyledons. During germination, it is completely removed and gradually transported to other plant parts, by passing catabolic enzymes. Removal from leaves and transport to seeds have also been shown to occur with caffeine in *Coffea arabica*.

An example of metabolism linked with such transport is provided by differing nicotine contents in shoot and root of *Nicotiana tabacum*, *N. alata* and *N. sylvestris*. The difference is due to pronounced demethylation occurring only in the shoot, especially in the stage of axial elongation immediately following flower formation. Apparently, a developmental physiological process dependent on day length determines whether this potential is realized. Thus the capacity for demethylation diminishes if the plant remains a rosette under short day conditions (Endress, 1994).

6.5.5. Plastids

In many cases the presence of fully differentiated plastids is a necessary precondition for the production of secondary compounds (*e.g. Datura sps. Morinda citrifolia, Chenopodium rubrum, Lupinus sp.* and *Medicago sativa)* In *Hordeum vulgare* leaves, light regulated synthesis of 6–methyl salicylic acid from acetate in the presence of co–enzyme A and CO_2 is limited to the developmental period of active membranes and grana. i.e. of etioplast maturation (Lucker and Nover, 1977).

6.6. SIGNIFICANCE OF THE DEGREE OF CELL DIFFERENTIATION

In general the concentration of interesting compounds is relatively low and highly variable in fully differentiated plants. Therefore effective technical utilization and economic use are only possible using enormous amounts of biomass and costly processing methods. If this is accompanied by the extreme rarity of the wild plant, the basic substances can become very expensive. Therefore, the price of diosgenin, the basic material for the production of steroid hormones and cortisone derivatives, increased fivefold over the last several years.

The advantages of using dedifferentiated tissue (callus or cell cultures) are:

1. Cells in a tissue complex are constantly exposed to complex signals from surrounding tissue. The resulting endogenous chemical gradients can be experimentally induced in cultures of differentiated cells to fulfill set requirements.

2. *Decoupling* : The possibility of decoupling production and storage capacity led to a considerable increase in production e.g., *Macleaya cordata*, cell cultures, large amount of auxins addition led to decrease in number of storage cells with no adverse effect on alkaloid production. There is a difference in accumulative process in fully differentiated tissues compared to dedifferentiated tissue.

The synthetic capacity of dedifferentiated tissue often differs substantially from that of fully differentiated tissue, both quantitatively and qualitatively (Suzuki *et al.* 1987). In all *Lupinus, Cytisus* and other species studied the quinolizidine content of the calli is two to three orders of magnitude less than in fully differentiated plants. However, the alkaloid pattern was always characterized by lupanine as the major compound. By systematic selection, lines with a synthetic capacity exceeding that of corresponding organs were obtained e.g. *Dioscorea deltoidea* (diosgenin), *Ammi visnaga* (visnagine) and *Peganum harmala* (harmine).

Callus Heterogenicity

6.6.1. Distance from Meristematic Activities

Productive callus generally does not exclusively consist of parenchymatous cells of uniform morphology and biochemical activity. To a greater or lesser extent, the individual cells are differentiated. Production of secondary compounds is usually most efficient in tissue that is farthest from meristematic activity, and in which biochemical cell maturation is advanced. Nevertheless, the pattern of compounds is far more complex in calli with microscopic or even macroscopic differentiations that in completely dedifferentiated tissue. Therefore, callus cultures are usually more productive than cell suspension cultures of the same organ, cultivated on the same medium composition.

6.6.2. Shoot and Root Induction

Linked Expression

In *Hyoscyamus niger*, the expression of hyoscyamine 6β-hydroxylase (H6H) RNA occurs only in developmentally young roots. Valepotriate synthesis in *Cetrantus* and *Valeriana* is linked to the accumulation of cytochrome P_{450} and does not proceed independently from root differentiation (Table 6.13). It remains questionable whether this usually membrane–bound oxygenase is a biochemical marker of root differentiation. Certainly, the accumulation process of this enzyme is an indication of linked expression of morphological and biochemical processes.

Root Induction

The incorporation of phenylalanine into tropic acid during the synthesis of tropane alkaloids in *Duboisia myoporoides. Scopolia parviflora, S. japonica* and *Atropa belladonna* is specially linked to root induction. Formation of small roots induces the synthesis of forskolin, a labdenone–diterpenoid, in *Coleus forskohlii* cell aggregates, and an increase in the content of steroid alkaloids to 5.2% dry weight in *Solanum khasianum*. It is also essential for some components (disulphides) of the onion scent in *Allium cepa* callus cultures.

Shoot Induction

The major components of onion volatile oils were only formed following shoot induction. A noteworthy example of this dependence is the induction of cardenolide synthesis in *Digitalis lanata* callus and suspension cultures. In unspecialized cells of these cultures, traces of these economically important cardiac glycosides can be detected only by means of highly sensitive radioimmunoassays, if at all (Kartnig and Kobosil, 1977). The onset of synthesis is linked to shoot formation (Hirotani and Furuya,

Table 6.13. Examples of secondary product formation, depending on shoot and root induction (Adopted from Endress, 1994)

Shoot induction	Root induction
Monoterpenes	Monoterpenes
Lavendula angustifolia	*Eucalyptus citriodora*
Rosmarinus officinalis	*Coriandrum sativum*
Cyclic Monoterpene	
(menthol, menthone)	Labdenone diterpene (forskolin)
Mentha piperita	*Coleus forskohlii*
Tri-terpenes (cardenolides)	Sesquiterpenes
Digitalis lanata	*Pimpinella anisum*
Volatile oils	Onion fragrance
Pelargonium fragrans	*Allium cepa*
P. tomentosum	
Pyrethrines	*Valepotriates*
Tanacetum cineraiefolium	Centhranthus ruber
Pyrethrum cinerariefolium	*C. macrosiphon*
	Valeriana officinalis
Indole alkaloids (vindoline, vinblastine)	Tropane alkaloids
Catharanthus roseus	*Hyoscyamus niger*
	Scopolia parviflora
	Atropa belladonna
	Steroid alkaloids (solasodine, solasonine, solanidine)
	Solanum khasianum
Bupleurin	
Bupleurum falcatum	

1977). In contrast, other compounds such as steroids, steroid saponins and progesterone also accumulate in disorganized long–term cultures.

Feedback Effects

Conversely, in *Nicotiana tabacum* callus cultures, root induction is only induced by the addition of nicotine at a concentration of 50 mg/l to the medium.

6.6.3. Plastid Differentiation

Plastid Metamorphosis

The synthetic herbicide chlor–phenyl–thio–ether–tri–ethylamine (CPTA) induces *de novo* synthesis of the isoprenoid *trans*–lycopine in callus and suspension cultures of Paul's scarlet roses and tomatoes (Radin, 1986). This occurs in addition to a basic concentration resulting from constitutive gene expression, parallel to the transfer of proplastids into chromoplasts or parallel to the disappearance of chloroplasts and the simultaneous accumulation of chromoplasts (LL cultures). A non–specific effect of CPTA seems unlikely due to the lack of effect on culture growth at non–toxic concentrations (< 100 mg/l) and because of its high specificity. It specifically induces *trans*–lycopine, while its analogue, the di–chloro compound DCPTA induces. Formation of rudimentary membranes characteristic of developing chloroplasts has been shown to be essential for cardenolide synthesis in *Digitalis lanata* cell cultures.

Chlorophyll Synthesis

Functioning chloroplasts are essential for the synthesis of many different secondary compounds in a great variety of objects. In *Morinda citrifolia* cell cultures, the presence of plastidial lamellar system determines the utilization of the key compound 1,4-di-hydroxy-2-naphtholic acid for the conversion into Anthraquinone.

Morphogenesis

Accumulation of cardenolides in unspecialized isodiametric cells of *Digitalis lanata* suspension cultures occurs only following initiation of the programs for the formation of compact green tissue, for short formation or for the development of adventitious embryos (Garve *et al.,* 1980). In the absence of these specific induction conditions, they can only be detected by means of highly sensitive immunoassays, if at all (Kartnig and Kobosil, 1977). Only in those plastids formed after induction by blue light does steroid formation occur.

Table 6.14. Some selected compounds formed in plant tissue culture with a yield equal to or higher than that of the parent plant (Czygan, 1984).

Compound	Plant species	Culture condition*	A	B	C	F
Ajmalicine	*Catharanthus roseus*	S	1.0	0.26	0.3	3.0
Anthraquinone	*Cassia tora*[b]	C	6.0	--	0.6	10.0
	Galium aparine	S	3.8-5.7	0.43	0.28	13.5-20.0
	Morinda citrifolia	S	18.0	2.5	2.2	8
	Rubia fruticosa	S	20.0	--	2.5	8
Berberine	*Coptis japonicum*	S	13.0	1.4	7.5	1.7
Biscoclaurine	*Stephania cepharantha*	C	2.3		0.8	3
Caffeine	*Coffea arabica*	C	1.6	--	1.6	1
Diosgenin	*Dioscorea deltoidea*	S	7.8	--	2	3.9
Ginsengosides	*Panax ginseng*	C	27.0	--	4.5	6
Glutathione	*Nicotiana tabacum*	S	--	0.22	0.1	--
Nicotine	*N. tabacum*	C	2.5	--	0.7	3.6
Paniculide B	*Andrographis paniculata*	C	0.9	--	0	
Protopine	*Macleaya microcarpa*	C	0.4	--	0.32	1.25
Rosmarinic acid	*Coleus blumei*	S	15(23)	3.6	3.0	5 (7.5)
Serpentine	*Catharanthus roseus*	S	0.8	0.16	0.5	1.6
	C. roseus	C	0.5	--	0.5	1
Shikonin	*Lithospermum erythrorhizon*	C	14(23)	1.5	1.5	9(15)
Triptolide	*Trypterygium wilfordii*	S	0.05	0.004	0.001	50
Ubichinone-10	*N. tabacum*	S	0.19	0.015	0.003	63
Visnagin	*Ammi visnaga*	C	0.31	--	0.1	10

[a]C = Callus culture; S = suspension culture; A = % dry weight culture; B = g/l medium; C = % dry weight plant; F = culture/plant ratio; [b]From seeds

6.7. PHYTOPHARMACEUTICALS FROM SUSPENSION CULTURES

Plants are the source of large variety of phytopharmaceuticals which are metabolites of both primary and secondary metabolism. They are used as pharmaceuticals, agrochemicals, food flavors, colors and as cosmetics (Table 6.15). Many of these compounds have complex chemical structures which makes their chemical synthesis economically unattractive (Fig. 6.2) in addition to the depletion of forest causing decreased availability as wild source.

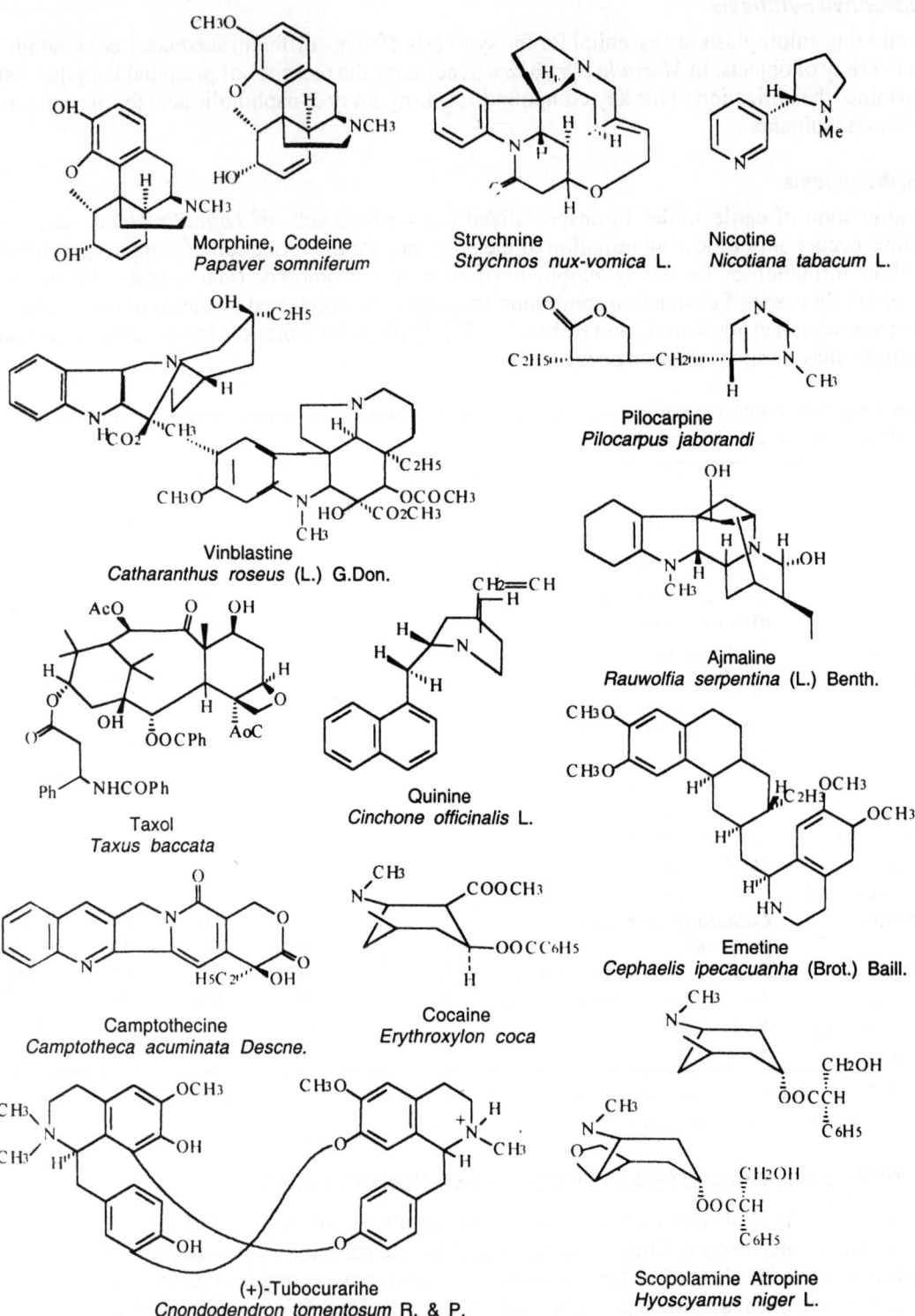

Fig. 6.2. Structures of secondary metabolites from plants.

Table 6.15. Substances reported from plant cell cultures

Alkaloids	Latex
Allergens	Lipids
Anthraquinones	
Antileukaemic agents	Naphthoquinones
Antitumor agents	Nucleic acids
Antiviral agents	Nucleotides
Aromas	Oils
Benzoquinones	Opiates, Organic acids
Carbohydrates (including polysaccharides)	
Cardiac glycosides	Peptides
Chalcones	Perfumes, Phenols
Dianthrones	Pigments, Plant growth regulators
Enzymes	Proteins
Enzyme inhibitors	
Flavanoids, flavones	Steroids and derivatives
Flavours (incluing sweeteners)	Sugars
Furanocoumarins	Tannins
Hormones	Terpenoids
Insecticides	Vitamins

6.7.1. Why Tissue Culture?

The pharmacologically active novel compounds are extracted from plants. In plant systems, they accumulate in leaves (nicotine in *Nicotiana*), roots (ajmalicine in *Catharanthus roseus*), bark (quinine in *Cinchona*) or in the whole plant (ephedrine in *Ephedra*). Sometimes these products are produced in specialized differentiated tissues such as resin in resin ducts and latex in laticifers. Except the herbaceous cultivated plants (e.g., *Papaver somniferum*), most of the secondary metabolites are accumulated after certain age or maturity of the plant. In the case of tree or shrub species e.g. Camptotheca, Cinchona, Rauwolfia, Ochrosia, etc. plants attain maturity in a few days before they accumulate the active principle in high amounts. It is difficult to increase the area under plantation for a particular species and growth of plants takes its own time. To meet the ever–increasing demand (e.g., vincristine) the natural resources are not sufficient. The world political scenario may also affect the supply of a particular raw material. To overcome all these hurdles, the industry requires alternative methods of assured supply of uniform material throughout the year. Harvesting of plants (except cultivated species) from natural forest resources is not only difficult, but also makes them endangered species; e.g., *Ephedra gerardiana* and several other Himalayan plants.

When plant material is not available throughout the year in a quantity sufficient for industrial production and chemical synthesis is not possible, particularly for large complex molecules, biotechnological methods offer an excellent alternative. But before implementing this approach, cost of the product and its demand should justify production by biotechnological means.

6.7.2. Secondary Product Formation in Suspension Cultures

Well–growing cell suspension cultures have been regarded for a long time as the only relevant system for biotechnological production processes. Most recently rapidly growing hairy root cultures with their root–specific production characteristics have also been grown in larger bioreactors (Curtis, 1993). There are also few groups who believe that immobilized plant cells are better suited for plant cell production processes than freely suspended cells (Tanaka, 1994). However, only cell suspension cultures of *Echinacea purpurea* have successfully been grown in bioreactors of up to 75 m³ (Westphal, 1990). Thus, without a doubt they remain the most attractive system from a technological point of view. The disadvantage of rapidly growing, morphologically undifferentiated callus cultures and suspension is that in such a cell state only a small part of the biosynthetic potential is expressed. Nevertheless, some products are found in suspension cultures at levels which exceed those of the differentiated plant.

6.7.2.1. Products Accumulating at High Levels in Supension Cultures

6.7.2.1.1.Cinnamic Acid Derivatives

Esters and amides of cinnamic acids, mainly of caffeic acid, form a major, widespread group of phenylpropanoid metabolites in plants. This group of compounds accumulates spontaneously, often at high levels, in cell suspension cultures. Two groups independently reported the accumulation of high levels of rosmarinic acid in *Coleus blumei* (Razzaque and Ellis, 1977) 2,4 –D and kinetin as phytohormones. These cultures were maintained in suspension for several years without loosing their capacity for synthesizing and accumulating rosmarinic acid. Razzaque and Ellis (1977) reported the accumulation of 8–11% rosmarinic acid (corresponding to 12-1.5 gL⁻¹) on the B5 growth medium. The synthesis of rosmarinic acid can be stimulated by increasing the sucrose levels to 5–8 %. Under these conditions, the content of rosmarinic acid increased up to 15 % (Zenk *et al.,* 1977a). The specific values exceed those of the various organs of the plants by a factor of 0.5. From their experiments in a volume of 25 mL Zenk's group calculated a yield 0f 3.6 gL⁻¹ rosmarinic acid within 13 days (0/3 gL⁻¹–d⁻¹). However, their first attempts at scaling up rosmarinic acid production in a 30 L airlift bioreactor showed that shake flask experiments are not necessarily transferable to larger–scale fermentations since productivity dropped to 15% of the shake flask experiments. As rosmarinic acid has recently been shown to have good antiphlogistic activity, the Nattermann Company, Cologne (Germany), investigated this culture further, they established new lines of *C. blumei* which spontaneously synthesized rosmarinic acid with yields of 200 mg L⁻¹ on the growth medium with an optimized concentration of sucrose as the main stimulator for maximum production (Ulbrich *et al.,* 1985). As expected from the experience with microbial fermentation's, these yields were further increased in 30 L airlift or stirred reactors up to 21 % or 5.6 g L⁻¹. The output of rosmarinic acid was thus raised to 0.93 gL⁻¹ d⁻¹ (Ulbrich *et al.,* 1985). This was at that time the highest amount of a defined secondary metabolite ever produced with plant cell culture.

1. Rosmarinic acid belongs to the group of compounds which is spontaneously accumulated at reasonably high levels in typical plant cell culture media. It is not only produced at high levels with *C.blumei* but also by several other plant species which are able to biosynthesize rosmarinic acid (Whitaker *et al.,* 1984). Cultures of *Anchusa officinalis* spontaneously accumulated 6% rosmarinic acid (De–Eknamkul and Ellis, 1984).

2. The production of rosmarinic acid can be greatly improved by very simple media variation (De–Eknamkul and Ellis 1985a, b).

3. As with most spontaneous high producer lines production is rather stable or can easily be re–established by controlled culture conditions.

It has recently been noted that the effect of sucrose depends on the carbohydrate level in the medium at the time when phosphate limitation occurs (Gertlowski and Petersen, 1993). This observation might explain why not all rosmarinic acid cell cultures react to the same extent to increased sucrose supply. At least two groups have used cultures of *A. officinalis* and *C. blumei* for identifying all enzymes involved in rosmarinic acid biosynthesis (Mizukami and Ellis, 1991; Petersen *et al.,* 1993) and some of these enzymes have been purified so that cloning of the corresponding genes could now proceed. Culture systems expressing a biosynthetic pathway quite well under all culture conditions may not be very suitable for the identification of regulatory factors controlling expression. Recently, it has been shown that rosmarinic acid and the enzymes involved in its biosynthesis can be induced in *Lithospermum erythrorhizon* and *Orthosiphon aristatus* by elicitors such as yeast and methyl jasmonate from nearly zero up to ca. 1.5% of dry mass (Mizukami *et al.*, 1993; Sumaryono *et al.*, 1991).

Ellis (1983) established various lines of *Syringa vulgaris* on B5 medium, all of which are produced spontaneously high levels of up to 1.4 g L^{-1} verbascoside. Rapidly growing suspension cultures of *S. vulgaris* were found to contain a higher specific content of verbascoside (15%) than callus cultures (5–8%) (Ellis 1983, 1985). Due to reports that verbascoside is a biologically active compound with antibacterial, antiviral, antihypertensive, and immunosuppressive properties, the interest in verbascoside producing tissue cultures has increased, especially because plant generally contain only low amounts of this compound (Inagaki *et al.*, 1991). The initial observation of Ellis that verbascoside belongs to the group of compounds whose production is favored under cell culture conditions has been confirmed. Cultures of *Hydrophila erecta* (Henry *et al.*, 1987) or *Leucosceptrum japonicum* (Inagaki *et al.*, 1991) yielded product levels in the range 2 g L^{-1} without optimization. The analysis of cultures of several other plant species, at the callus level indicate that many more highly effective systems for the production of verbascoside can be established (Dell *et al.*, 1989; Inagaki *et al.*, 1991).

The presence of various hydroxycinnamoyl putrescines in cultured cells of *Nicotiana tabacum* was first reported by Mizusaki *et al.* (1971). That these compounds have received some more attention during the last years is due to accidental finding. When selecting the *p*–fluorophenylalanine resistant cell line TX4from a widely distributed XD (TX1) Line (*N. tabacum* cv. Xanthi) Palmer and Widholm (1975) noted that the levels of phenolics were increased manifolds in TX4 cells. The phenolics were identified as hydroxycinnamoyl putrescines with caffeoyl putrescine as main component (Berlin *et al.*, 1982). TX1 cells accumulated between 0.6–1% hydroxycinnamoyl putrescines on the growth medium, while TX4 cells contained up to 10% of these compounds on a dry mass basis. This system has been used in two directions: optimization of product formation and comparison of the biochemical differences which lead to the different productivities. Growth limiting conditions (e.g., phosphate limitation) stimulated product formation (Schiel *et al.*, 1984a). It is a frequently observed phenomenon that growth and secondary metabolism are countercurrent processes in cultured cells. Thus, even the formation of products accumulating at high levels on the growth medium can often be strongly limiting conditions. The negative effect of accumulated phosphate on the synthesis of hydroxycinnamoyl putrescines suggested that the employment of fed batch fermentation would give highest yields. The productivity of the high producing variant TX4 was indeed increased to 1.5g L^{-1} by a 70L fed–batch fermentation with phosphate as a limiting nutrient (Schiel *et al.*, 1984b). Shake flask and batch fermentation of TX4 cells usually yielded 0.8-1.2 gl^{-1} (Berlin et al., 1982). The yield of TX1 cells was increased by the fed batch techniques from 160–200 mg L^{-1} to 300-400 mg L^{-1} This shows the importance of using the best possible line for product optimizations. TX4 cells were the first biochemically selected variant line which showed overproduction of secondary metabolites studies. It is also noteworthy that this highly productive variant line has maintained its production potential for more than 15 years (Meurer–Grimes *et al.*, 1989).

6.7.2.2. Naphthoquinones and Anthraquinones

Quinones comprise a large group of secondary metabolites that are widely distributed in the plant kingdom. Since they are colored and, therefore, visible compounds, they have been a favored target of tissue culture research, Naphthoquinones and anthraquinones sometimes accumulate in cultured cells at levels far exceeding the amounts found in the intact plant. Some of these structures represent the active components of drugs. They are also important as natural dyes.

The red shikonon pigments of the cork layer of the roots of *Lithospermum erythrorhizon* are derivatives of 1,4-naphthoquinones (Fig.6.3). These compounds have been used medicinally in Japan for the treatment of burns and skin disease and are now mainly used as a dye for lipsticks and for staining silk. The plants have to be grown for 3-4 years before a yield of 1–2% shikonin is achieved in the roots (Fujita, 1988). The total amount of *Lithospermum* roots used each year in Japan is 10000 kg. From this an annual demand of 150 kg shikonins can be calculated. As the plant cannot be grown in commercial quantities in Japan it has to be imported from Korea and China. It is often argued by Japanese scientists in industry (Komamine *et al.*, 1991) that due to the geographic situation the indigenous supply of plant material is not sufficient, and that plant tissue culture technology is, therefore, perhaps a more attractive alternative in Japan than elsewhere. This would explain the commercial production of shikonins from *L. erythrorhizon* cell suspension cultures of the Mitsui Company. Callus cultures of *L.erythrorhizon* were found to accumulate shikonin derivatives (Tabata *et al.*, 1974). By repeated analytical screening over a period of two years two highly productive strains containing 20–fold increased levels of 1 mg g^{-1} fresh mass were isolated (Mizukami et al, 1978). Fujita et al (1981a,b) investigated the effects of all media constitutents on growth and production. They developed a production medium yielding 1.4g/l shikonin derivatives within 23 d or 12 % on a dry mass basis. Combining the two media in a two stage process (1st stage 200 L growth medium, 2nd stage 750 L production medium) the yield was increased to 3.7 g g^{-1} dry mass inoculum within 23 d (Fujita *et al.*, 1982*)*. By screening protoplast derived clones they isolated lines with an improved growth and higher productivity. The best line had a specific content of 23.2% and yielded 6.45g Shikonin per g inoculum (Fujita and Hara, 1985; Fujita, 1988; Takahashi and Fujita, 1991). Since 1983 the Mitsui Company produces shikonins by this technology for cosmetics and dyes (Takahashi and Fujita, 1991). Comparison of the composition of shikonins extracted from cell cultures with that of various roots showed that the ratio of the individual shikonins was different (Fujita, 1988). The altered ratio of the various shikonin derivatives in the cell cultures is no problem if the mixture of compounds is to be used for cosmetic purposes or as a dye. However, if one wants to replace a known approved plant drug by a tissue culture extract extensive evaluation of the pharmacological properties and equivalence studies are required. Such studies have been initiated for *L. erythrorhizon* cultures by the Mitsui Company (Ozaki *et al.*, 1990, Suzuki *et al.*, 1991), Japan.

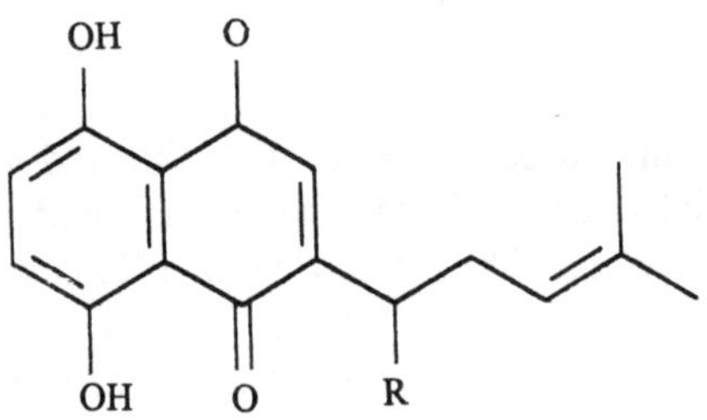

Shikonins R = H or aliphatic acid
Lithospermum erythrorhizon

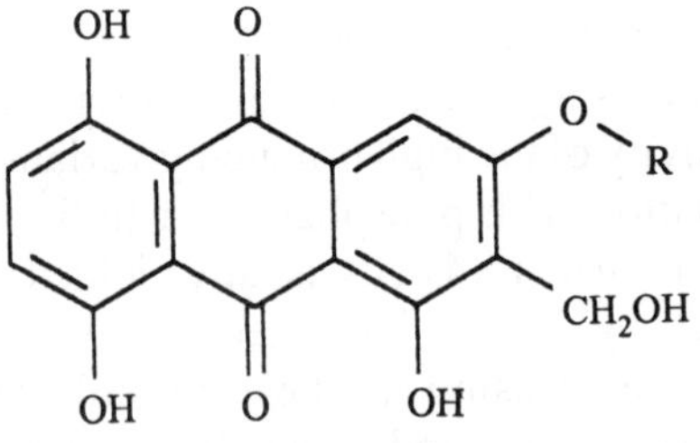

Anthraquinone (e.g. lucidin Primveroside R = Glucose)
Morinda citrofolia
Galium mollugo

Fig. 6.3. Napthaquinones and Anthraquinones from Plant Cell Cultures.

There are scattered reports in the literature of increased shikonin production in *L. erythrorhizon* cultures by *in situ* extraction, precursor feeding, elicitation, or media variation. Though the observed effects are well demonstrated their final levels remain far below those reported by Fujita *et al.* (1982). Such studies would deserve more attention if the production levels at the Mitsui Company could additionally be improved by the newly recommended techniques. This remark seems to be especially valid in view of the report of the Mitsui Company that two–phase cultures (see details in Chapter–21) did not improve productivity of their high yielding line (Deno *et al.*, 1987). In general, naphthoquinones and benzoquinones seem to belong to the groups of compounds which might readily be formed in cultured cells of various plant species (Fukui *et al.*, 1983; Inque *et al.*, 1984). It is evident that the highly expressed naphthoquinone biosynthetic pathway would be a good system for biochemical and mole-cular studies. The first results on the regulation of shikonin biosynthesis have been presented by Heide *et al.* (1989), showing that the ratio of *p*-hydroxybenzoic acid geranyltransferase and *p*-hydroxybenzoic acid glucosyltransferase activities is one of the regulatory controls of shikonin biosynthesis (see Fig. 20.24).

Anthraquinones in higher plants are formed either via the acetate mevalonate pathway or the *o*-succinylbenzoic acid pathway. Their production in cell cultures has been reviewed in great detail by Koblitz (1988). Highly productive suspension cultures of plant species (e.g., of *Cassia* sp. or *Rhamnus* sp.) producing anthraquinones via the acetate mevalonate pathway have never been reported (Van Den Berg *et al.*, 1988). However, it has been documented in numerous publications that the o-succinylbenzoic acid derived anthraquinones are well expressed in rapidly growing cell suspension cultures and accumulate sometimes at extraordinary levels. A good example is cell cultures of *Morinda citrifolia*. Zenk *et al.* (1975) tested a large variety of nutritional factors for their effect on anthraquinone production. They established a production medium yielding 2.5 gL^{-1} anthraquinones corresponding to more than 10% of dry mass which exceeds the concentration of the root by a factor of 10. Important for high production levels were the replacement of the phytohormone 2,4-D by NAA and an increase of the sucrose level to 7%. The anthraquinones of cell cultures of *M. citrifolia* consists of a mixture of at least 12 aglyca and glucosides (Leistner, 1975; Inque *et al.*, 1981). The main components are lucidin derivatives. Some of them have not yet been found in the intact plant (Inque *et al.*, 1981). The spontaneous high production of these anthraquinones was maintained in various bioreactors (Wagner and Vogelmann, 1977). The yields of anthraquinones in an airlift bioreactor were 30% higher than in experiments with shake flasks. An interesting example of manipulating the expression of pathways in cultured cells was reported for cell suspension cultures of *Morinda lucida*. In photoautotrophic cell cultures (chlorophyllous, no sugar in the medium) lipoquinones were the main components while anthraquinones were not found. When these cultures were transferred into the dark and sugar was added to the medium, anthraquinone biosynthesis was induced and lipoquinone formation was repressed (Igbavboa *et al.*, 1985).

The results achieved with *M. citrifolia* cultures can also be obtained with cultures of *Galium mollugo* (Bauch and Leistner, 1978). A B5-NAA medium with 7% sucrose gave the highest yields with 2 gL^{-1} within 14 days. Lucidin primveroside was the main component. While the pathway of anthraquinones was readily expressed in cultured cells, the biosynthesis of iridoids remained repressed under all culture conditions. Schulte *et al.* (1984) optimized cell suspension culture media of 19 different Rubiaceae species for optimal yields of *o*–succinylbenzoic acid derived anthraquinones. 17 of the cultures yielded anthraquinone levels higher than those found in the corresponding plants. It was shown that nutritional and hormonal requirements of the various anthraquinone producing cultures, even those of one family, may be quite different. Cultures of *Rubia cordifolia* maintained their high productivity when scaled upto 75L (Suzuki and Matsumoto, 1988) and cultures of *Rubia tinctorum* have been studied at San–Ei Chemical Industries, Japan (Odake *et al.*, 1991) for the development of a commercial production process of anthraquinone pigments (eg., alizarin, purpurin). It is clear that these culture systems are also suitable for biochemical and regulatory studies. The first enzymes involved in the biosynthesis of o–succinylbenzoic acid (Simantiras and Leistner, 1989) and its further metabolism (Sieweke and Leistner, 1992) have been studied in cell cultures of *Galium* sp. The cultures of *G. mollugo* have also been used to over produce shikimic acid. The addition of glyphosate (inhibitor) inhibits the formation of o–succinylbenzoic acid and anthraquinones in these cultures thus causing the accumulation of the biosynthetic precursor, shikimic acid (10% of dry mass, 1.2 gL^{-1}) (Steinrucken and Amrhein, 1980).

6.7.2.3. Protoberberines and Benzophenanthridine Alkaloids

Isoquinoline alkaloids are derived from tyrosine via (S)–norcoclaurine as a central intermediate (Rueffer and Zenk, 1987) and not, as initially thought, via (S)–norlaudanosoline synthase (Rueffer et al., 1981). Since the pathway is highly branched, a great variety of very different structures result from this central intermediate. Some of the branches are readily expressed in cultured cells, while others such as the morphinan alkaloids remain mostly repressed.

There are several reports showing that protoberberine alkaloids spontaneously accumulate at high levels in cultured cells. Various cultures of *Berberis* sp. accumulated between 0.2 and 1.7 g protoberberine alkaloids, mainly jatrorrhizine on a growth medium with 3.5% sucrose (Hinz and Zenk, 1981). Breuling et al. (1985) optimized the production of 3 g L^{-1} in a 20 L airlift bioreactor. Independently, two Japanese groups reported the accumulation of high levels of protoberberines in cultures of *Coptis japonica* (Fukui et al., 1982; Yamada and Sato, 1982). Fukui et al. (1982) established a line accumulating berberine, jatrorrhizine, palmatine, and coptisine at a ratio of 50 : 22 : 22 : 6. Interestingly, the content of alkaloids increased gradually during subculturing from 8–15% which was paralled by increased growth (Fukui et al., 1982). This shows that protoberberine formation is indeed a favored pathway in these rapidly growing cell cultures. Thus, alkaloid levels of up to 1.7 g L^{-1} were readily achieved in normal growth medium. The *Coptis* line described by Yamada and Sato (1982) reported contained mainly berberine and only trace amounts of other alkaloids. By screening small cell aggregates for high berberine (yellow) producing clones cell lines were established which produced more than 1 g berberine L^{-1} or 10% on a dry weight basis (Sato and Yamada, 1984).

Another important species spontaneously accumulating high levels of protoberberines is *Thalictrum*. While cell cultures of *Thalictrum minus* secrete most berberine into the medium (Nagakawa et al., 1984) those of *T. flavum, T. dipterocarpum* or *T. rugosum* accumulate the alkaloids within the cells (Suzuki et al., 1988, Piehl et al., 1988). Interestingly, when cells of *T. rugosum* were transferred to fresh medium lacking phosphate, protoberberines were released into the medium (Berlin et al., 1988a). There are other reports indicating that cultures of plant species with the capability to biosynthesize protoberberines will spontaneously yield cultures accumulating these colored alkaloids in high amounts (0.2-1 gL^{-1}). The levels of protoberberines are so high that the Mitsui Company has decided to develop a production process for berberine with *Coptis japonica* cell cultures (Matsubara and Fujita, 1991). As in the case of shikonins the medium was first optimized. For example, a 10–fold increase of Cu^{2+} increased the berberine content by 40% (Morimoto et al., 1988). The addition of very low levels of gibberellic acid increased the content by 30% (Hara et al., 1988). Since cell aggregate screening was regarded as very time–consuming and not ever efficient, protoplasts with high berberine content were isolated with a cell sorter (Hara et al., 1989). All selected lines contained higher levels than the original culture. The next step was to develop culture conditions for a high density culture (Matsubara et al., 1989) so that 70 g dry mass per L (corresponding ca. 700 g fresh weight) were obtained. This required increasing the nutrition and oxygen supply by fedbatch–perfusion cultivation (Matsubara and Fujita, 1991). Today, the Mitsui Company (Japan) produces berberines by a high–density continuous culture method over a period of several months with a yield of 0.65 g L^{-1}d^{-1}. This is 8 times more than the value obtained in batch cultures (Matsubara and Fujita, 1991).

In the case of protoberberines, an alternative production process seemed to be possible with the highly productive *T. minus* line. *T. minus* cells were probably tested for the Mitsui Company in an attempt to develop a new bioassay–based screening method for high berberine–producing cell colonies (Suzuki *et al.,* 1987). The advantage of *T. minus* cells can only be exploited if cells are immobilized for berberine production. Kobayashi *et al.* (1988) obtained a production rate of 50 mg L^{-1} d^{-1} in batch and semicontinuous bioreactors over a period of at least 60 days of cultivation. This was undoubtedly an exciting result. However, in view of the improvements obtained with the *C. japonica* culture, it is clear that *T. minus* cultures are not yet competitive for use in a commercial process. Berberine chloride is used in Japan as a medicine for intestinal disorders and treatments of abnormal zymosis and has previously been obtained by extraction from the roots of *C. japonica* and the cortex of *Phellodendron amurens* (Matsubara and Fujita, 1991). Berberine produced with tissue culture has not yet been approved as a drug in Japan. However, pharmacological studies towards this goal are underway (Suzuki *et al.,* 1993a, b).

Another group of isoquinoline alkaloids which accumulate spontaneously in cultured cells are benzophenanthridine (Fig. 6.4) and the related protopine alkaloids. Cell cultures of most Papaveraceae contain alkaloids of this group. Furuya *et al.* (1972) were the first to report the production of sanguinarines, protopines, and the aporphine magnoflorine in callus cultures of *Papaver somniferum.* When they analyzed the alkaloid pattern of callus cultures of 11 other species of Papaveraceae, almost identical alkaloid spectra were found (Ikuta *et al.,* 1974). Cell culture conditions seem to favor the synthesis of benzophenanthridine alkaloids, which are found in the corresponding intact plants often at rather low levels (Williams and Ellis, 1993). However, there is a significant difference between the production of benzophenanthridine and protoberberine alkaloids in cultured cells. While in *C. japonica,* e.g. protoberberine production increased with increasing growth rate (Fukui *et al.,* 1982), benzophenanthridine synthesis decreased with better growth (Berlin *et al.,* 1985). A suspension culture of *P. somniferum* with a growth cycle of 28d contained nearly 6% sanguinarines (360 mgL^{-1}) after the 5 passages. During further subcultivation the yield decreased to 200 mg L^{-1} after 20 subcultures and to 20 mg L^{-1} after 35 subcultures. Biomass production of this culture increased from 6g in 28d to 11g in 10d (Berlin *et al.,* 1985). The "biotechnological" interest in benzophenanthridine alkaloids increased only when it was noted that these alkaloids could be induced greatly by stress

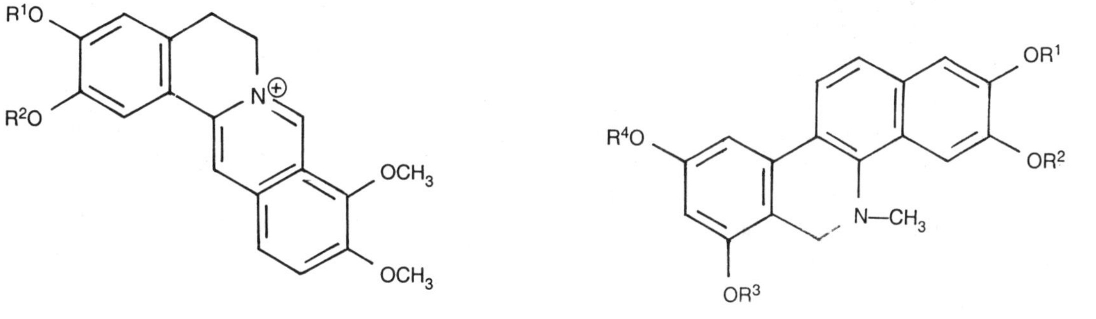

Protoberberine alkaloids

R^1 + R^2 = CH$_3$ Berberine
R^1 = CH$_3$; R^2 = H   Columbamine
R^1 = H; R^2 = CH$_3$ Jatrorhizine
e.g. *Coptis japonica*
    Berberis sp.
    Thalictrum rugosum

Benzophenanthridine alkaloids

R^1 + R^2 + R^3 + R^4 = CH$_3$ Dihydrosanguinarine
R^1 + R^2 = CH$_3$; R^3 + R^4 = CH$_3$ Dihydrochelerythrine
e.g. *Papaver somniferum*
Eschscholtzia californica

Fig. 6.4. Structure of Protoberberine and Benzophenanthridine alkaloids.

factors and elicitors. Thus, the yields of benzophenanthridine alkaloids of suspension cultures of *Eschscholtzia californica* were increased 10–fold to 150 mgL^{-1} by increasing the sucrose concentration in the medium to 8% (Berlin *et al.*, 1983). Eilert *et al.* (1985) reported that sanguinarine levels of *P. somniferum* were enhanced from 0.01% to 2.9% by treatment with fungal elicitors. The advantage of elicitor treatment in comparison with the altered production media is that the cells respond more rapidly, and the induced alkaloids in *Papaver* cell cultures are released into the medium (Eilert *et al.*, 1985, Cline and Coscia, 1988). In the case of *E. californica*, the alkaloids, induced by various fungal and yeast preparations evidently remained within the cells (Schumacher *et al.*, 1987).

Monoterpene Indole Alkaloids

About 1200 alkaloids derived from tryptophan have been isolated from higher plants, which corresponds to about one quarter to all alkaloids (Groger, 1980). Several of these alkaloids from *Rauwolfia, Catharanthus,* and *Cinchona* are used medicinally. Therefore, it was of great interest to see whether these alkaloids also accumulate in cultured cells. The dimeric monoterpene indole alkaloids, vinblastine and vincristine, used as antileukemia agents are only present in trace amounts in the whole plant.

More than 50 monoterpene indole alkaloids have been isolated from various Apocynaceae, e.g., *Catharanthus* and *Rauwolfia* sp. There has been one report that a cell suspension culture of *Rauwolfia serpentina* accumulated 1.6g raucaffricine per liter of medium (Schubel *et al.*, 1989). Cell cultures of *R. serpentina* were scaled upto 75 m^3 by the Diverse Company (Germany) and no alkaloids were found in the rapidly growing cell cultures (Westphal, 1990).

Callus cultures of *C. roseus* were found to contain low levels of monoterpene indole alkaloids (Carew, 1975). However, suspension cultures accumulated no or only trace amounts of these compounds when grown on a growth medium with 2,4-D as auxin. Zenk *et al.* (1977b) developed a production medium for *C. roseus* cell cultures. When the cells were transferred from the growth to a production medium, alkaloid formation was resumed after 3-5 d. Of 458 independently established cell lines, 312 (approx. 75%) produced alkaloids when transferred to Zenk's production medium (Kurz *et al.,* 1980). This production medium, however, had no special composition; the sole transfer of the cells into a 2,4-D free medium allows the accumulation of reasonable levels of indole alkaloids (Zenk *et al.*, 1977b, Knobloch and Berlin, 1980; Pareilleux and Vinas, 1984). The increase of sucrose to 5-8% had an additional beneficial effect. The phytohormone composition (indole acetic acid and benzylaminopurine) of Zenk's medium may have an additional stimulatory effect. The stimulatory effect of the production medium probably depends on the physiological state of the cells at the time of the transfer from the growth to the production medium. The level of phosphate accumulated in the cells seems to play an important role in the inducibility of alkaloid formation in a phosphate–free production medium (Knobloch and Berlin, 1983, Schiel *et al.,* 1987).

Cell suspension cultures of *C. roseus* grew rapidly, were easy to scale up in fermentors and yields of 30–70 mgL^{-1} serpentine were usually obtained within 20-30 days (Zenk *et al.*, 1977, Wagner and Vogelmann, 1977, Smart *et al.*, 1982). There have been several efforts to isolate higher producing cell lines using screening methods. The group at the NRC in Saskatoon (Canada) established more than 2000 individual lines from three *C. roseus* cultivars and screened them for alkaloid patterns and levels (Kurz *et al.*, 1985). They found a substantial variation among the lines, and one variety seemed to be more suitable for catharanthine formation. However, this tremendous effort did not evidently result in the isolation of an exceptionally high producing line. Zenk's group has screened established lines for high yielding clones using specific radioimmunoassays and the fluorescence microscope (Zenk *et al.*, 1977b, Deus and Zenk, 1982). Indeed, they found clones having accumulated high levels of serpentine or ajmalicine of which theoretical levels of up to 400 mg L^{-1} were calculated.

The isolated clones were, however, unstable, and alkaloid levels rapidly dropped to the values of the unscreened culture (Deus–Neumann and Zenk, 1984). The behaviour of the clones indicates that they were not true variants as claimed but instead were cell colonies in a transient physiological state.

Repeated analytical screening improved the specific catharanthine content from 0.1% to 0.7%, and these lines have maintained their superior productivity for at least 10 subcultures. Additionally, a biochemical selection for 5–methyl–tryptophan–tolerant cell lines was applied (Fujita *et al.,* 1990). There have been tremendous efforts to stimulate the production of ajmalicine and serpentine in cultures of *C. roseus*. Many approaches have been used such as the addition of various abiotic and biotic elicitors, the use of high and low density culture, precursor feeding and elicitation, alteration of media composition, immobilization, etc. (Van Den Heijden *et al.,* 1989; Moreno *et al., 1995).*

6.7.2.4. Anthocyanins and Betalains

Anthocyanins have been found in cultured cells of many plants (*Aralia, Catharanthus, Daucus, Euphorbia, Haplopappus, Mathiola, Perilla, Petunia, Vitis,* and many others). For some cultures the individual anthocyanin components have been identified, for others only the total amount of the pigments has been given (Ellis, 1988, Seitz and Hinderer, 1988, Yamamoto, 1991). In few culture systems anthocyanins are formed in dark grown cells, but in most cases light is required for optimal production (Sakamoto *et al.,* 1994). Due to their low toxicity, anthocyanins are widely used as food additives and also as a dye for silk (Yamamoto, 1991). While anthocyanin extracts are rather inexpensive the prices for individual anthocyanins seem to be very high which would justify the use of tissue culture based processes.

Since anthocyanin-producing cells are easily detected in callus cultures they have become an early target for visual screeing (Kinnersley and Dougall, 1980). Yamamoto *et al.,* (1982) established high yielding cell lines of *Euphorbia millii* by consecutive repeated screening. Only after 24 clonal selections were stable lines found; this required permanent screening over a period of many months. Nozue *et al.* (1987) were also successful in selecting high anthocyanin yielding cell lines from sweet potato by consecutive repeated screening. This result casts some doubt to claims of the instability of high producing variant lines if permanent screening was not applied over a very long period (Berlin, 1990). As for many other secondary metabolities, the culture medium can also be optimized for anthocyanin production. The optimization of *E. millii* cultures by the Nippon Paint Corporation (Japan) improved the productivity 4.5-fold to 32 mg $L^{-1}d^{-1}$ (Yamamoto *et al.,* 1989). The main component of the *E. millii* cell culture is cyanidin-3-arabinoside (Yamamoto, 1991). A great disadvantage of the *Euphorbia* system in commercial production is the requirement of light for pigment production. The same holds true for a highly productive culture of *Perrilla frutescens* (Zhong and Yoshida, 1995) which produces upto 5.8 gL^{-1} anthocyanins under optimized culture conditions and light. Therefore, the best line for the commercial production of anthocyanins may be cultures of *Aralia cordata,* which produce high levels in the dark. As with the cultures of *Euphorbia* and an *Ipomea,* continuous cell aggregate cloning led to a highly productive and fast growing line (Sakamoto *et al.,* 1994). 90% of all cells of the cloned line produced anthocyanins, a result which is only possible if product formation and growth are closely interrelated. Optimization of the basic inorganic salt concentration, the carbohydrate concentration, the ratio NO_3^-/NH_4^+, and the total nitrogen levels improved the anthocyanin content to 10.3% of dry mass. This culture was scaled up to 300 L in a 500 L jar fermentor (Kobayashi *et al.,* 1993). After 16 d of cultivation 69 kg fresh mass with 545 g anthocyanin (corresponding to 17.2% of dry mass) were obtained. The further improvement of productivity during scale–up was evidently due to the controlled supply of CO_2 to the cultures. Efforts to identify the anthocyanins present in this high yielding culture system are underway. As is typical in Japan the project is being done with close cooperation between researchers from university and industry (Kywa Hakko Kogyo Co. and Tonen Co., Japan).

Tissue culture systems may only become competitive for the production of individual anthocyanins and not for mixtures of anthocyanins. Therefore, it might be useful to look not only a total yields during selection and media optimization but also at whether such methods affect the percentage of any one anthocyanin in a complex mixture (Do and Cormier, 1991a & b).

Other pigments often found in cell cultures of centrospermae are the betalains which include red betacyanins and yellow betaxanthins (Bohm and Rink, 1988). Cultures of *Chenopodium rubrum* (Berlin *et al.,* 1986*), Phytolacca americana* (Sakuta *et al.,* 1987), *Amaranthus tricolor* (Bianco–Colomas and Hugues, 1990), *Beta vulgaris* (Girod and Zyrd, 1991), and *Portulaca* (Kishima et al., 1991) produce rather high levels of betacyanins under light conditions.

6.7.2.5. Steroidal Compounds

There are numerous reports on the occurrence of sterols, steroids, sapogenins, saponins, and steroidal alkaloids (Ellis, 1988) in cell cultures of plants. The root of *Panax ginseng* has widely been used as a tonic and medicine in oriental countries (China, Taiwan, Thailand and Korea) since ancient times. It is regarded as an adjuvant to prevent health disorders and is considered to be a "miraculous" drug for preserving health and promoting longevity (Misawa, 1994). The active components with proven pharmacological effects are saponins, e.g., ginsenoside–Rb and –Rg. Due to the worldwide increased demand for ginseng extracts and the fact that a cultivation period of 5–6 years is necessary before harvest, production of ginseng extracts or saponins using tissue cultures was regarded as an alternative source of supply. Furuya's group analyzed a number of callus cultures of *Panax ginseng* with different phytohormone requirements (Furuya, 1988), optimized their growth conditions, and compared their saponin patterns with those of *P. ginseng* roots. The patterns were quite comparable and levels of total saponins reached upto 0.7% of dry mass or 50 mgL^{-1} within 4 weeks in a 30 L jar fermenter (Ushiyama, 1991).

Since 1988 the use of the tissue culture material has been approved for the Nitto Denko Co. (Japan) by the Japan Ministry of Welfare and Health. The extracts from the tissue culture material have been added to tonic drinks, wines, soups, herbal liquours and other food preparations since 1989 (Ushiyama, 1991).

6.7.2.6. Immunologically Active Polysaccharides

There are a few reports that plant cell cultures sometimes release considerable amounts of polysaccharides were isolated and characterized (Prosksch and Wagner, 1987). The large amounts of these polysaccharides needed for further *in vivo* experiments were difficult to obtain from plant material. Therefore, cell cultures of *E. chinacea purpurea* were established. The culturees also produced two major active polysaccharides and released them into the culture medium; however, they were not identical with the polysaccharides of the intact plant (Wagner *et al.,* 1989).

6.8. PRODUCTS OF COMMERCIAL INTEREST DETECTED FROM SUSPENSION CULTURES

6.8.1. Morphinan Alkaloids

A number of reports have claimed that morphinan alkaloids are present in callus and cell suspension cultures of *Papaver somniferum* and *P. bracteatum* (Kamo and Mahlberg, 1988, Kamimura, 1991). However, there are also several reports where no morphinan alkaloids were detected (Berlin, 1986, Kamo and Mahlberg, 1988). The elicitation of poppy cell cultures was not initially planned for the

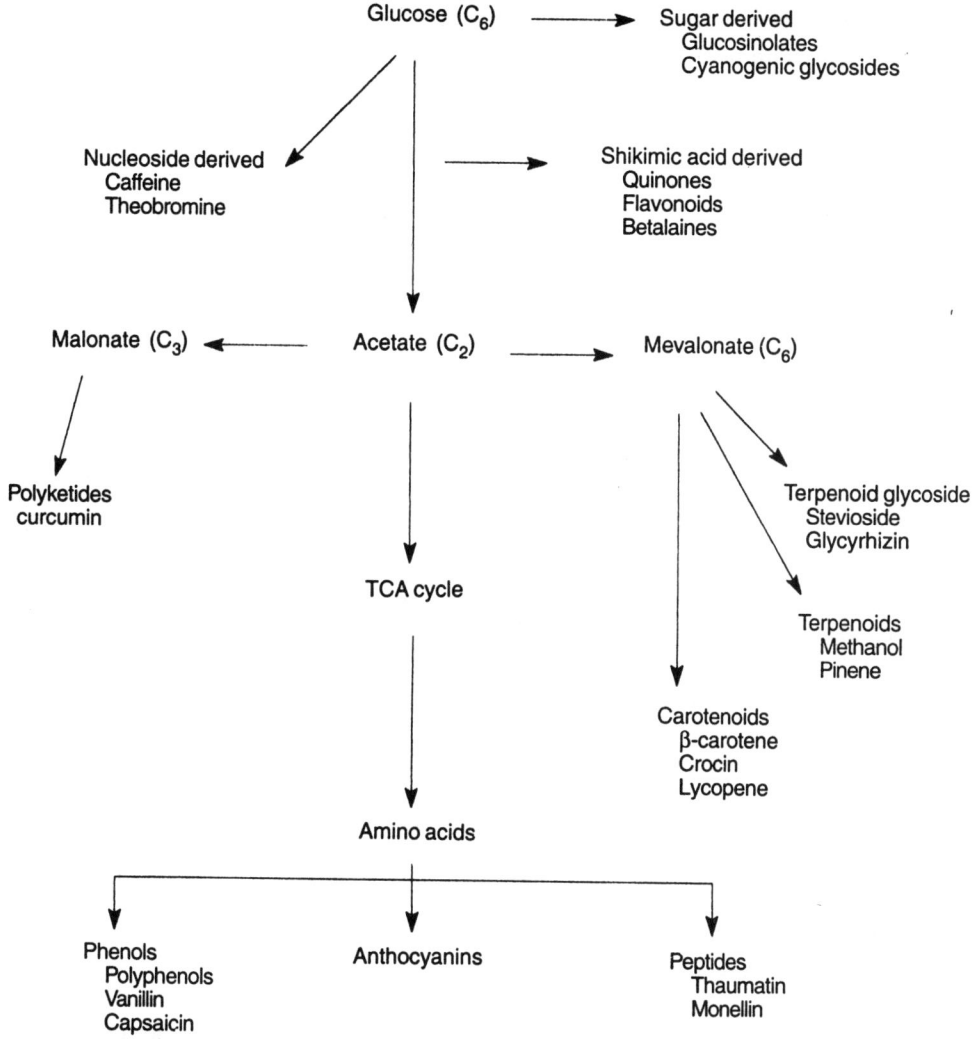

Fig. 6.5. Schematic of biosynthetic pathways for secondary metabolites with reference to the color and flavors (Payne *et al.*, 1991).

production of sanguinarines but instead for the induction of morphinan alkaloids. Eilert *et al.* (1985) clearly stated that morphinan alkaloids could not be induced by any of the elicitors tested. Claims in the literature of substantial morphinan alkaloid levels of upto 0.15% dry mass are of no biotechnological relevance since such data were not confirmed in further investigations. Evidently, some morphological differentiation is required for expression of the morphinan alkaloid pathway. Thus, freshly initiated cultures having retained the capacity to redifferentiate when stressed by elicitors or altered phytohormone composition, are able to synthesize and accumulate low levels of these alkaloids (Kamo and Mahlberg, 1988). The most recent report on morphinan alkaloids in cell cultures claims a production of 2.5 mg morphine and 3.0 mg codeine per g dry mass (Siah and Doran, 1991).

6.8.2. Tropane Alkaloids

Hyoscyamine and scopolamine are the most important tropane alkaloids. Hence, several groups have worked on the production of these compounds in cultured cells. However, the alkaloid levels were found to be extremely low in callus and suspension cultures of *Atropa, Datura, Duboisia, Hyoscyamus* and *Scopolia*. By using a squash technique in screening for alkaloid producing clones, Yamada and Hashimoto (1982) selected a cell line of *Hyoscyamus niger* containing just 0.01–0.02% hyoscyamine per g dry weight in suspension cultures. Scopolamine levels were 10–fold lower. Also media variation experiments were not successfully in increasing tropane alkaloid levels. Evidently, some root formation is required to express the tropane alkaloid pathway to a reasonable extent. There has only been one report (Ballica and Ryu, 1994) describing a selected cell suspension culture derived from stem cells of *D. stramonium* which seems to produce upto 80 mg L^{-1} total tropane alkaloids with in 25 days.

6.8.3. Quinoline Alkaloids

Cinchona have been grown in plantations since more than 130 years for the production of bark containing the antimalarial, antifever compound quinine. The related compound quinidine is used as a drug against cardiac arrhythmia. Though the compounds have repeatedly been found in cell suspension cultures of *Cinchona succirubra, C. pubescens* and *C. ledgeriana* at levels of upto 0.9% (Koblitz *et al.,* 1983), suspension cultures of *Cinchona* could not be further developed for biotechnological purposes. Attempts to increase the usually rather low levels by screening and media variation were not very successful (Harkes *et al.,* 1985). In addition, growth of cultures of this species is generally slow (Wjnsma and Verpoorte, 1988). Highest alkaloid production was obtained with cultures showing some degree of differentiation e.g., roots, shoots (Wijnsma and Verpoorte, 1988) and compact globular structures (Hoekstra *et al.,* 1990). Due to the difficulties of establishing reasonably productive culture systems, the progress in the elucidation of the biosynthetic pathway leading to quinoline alkaloids has remained slow (Wijnsma and Verpoorte, 1988).

6.8.4. Antitumor Compounds

The search for antitumor compounds in tissue culture has been a favoured goal of many researchers. The great interest in cell cultures of *C. roseus* resulted from the fact that this plant species synthesizes in very low amounts (0.0005% of dry mass) the expensive dimeric monoterpene indole alkaloids vinblastine and vincristine which are used in the treatment of certain forms of cancer (Misawa and Endo, 1988). Hirata *et al.* (1989) determined a vinblastine content of 1.4 mg g^{-1} of dry mass in leaves of *C. roseus*, a level which would reduce the need for a tissue culture derived process. It is now generally accepted that cell suspension cultures of *C. roseus* are not able to synthesize the dimeric alkaloids; vindoline, one of the monomeric precursors, is not formed in cultured cells but in leaves (Decarolis and Deluca, 1993). Research has gone into two directions – development of shoot culture systems and semi-syntheses from the two monomers. However, as with many shoot culture systems, productivity is presently 100-fold lower than in the leaves of intact plant (Hirata *et al.,* 1989). More efficient is the approach of coupling vindoline and catharanthine by chemical or enzymatic means to form 3', 4'-anhydrovinblastine and vinblastine (Goodbody *et al.,* 1988). The coupling of catharanthine and vinblastine with peroxidases of ferric ions and subsequent reduction yielded 50–70% anhydrovinblastine and 12-28% vinblastine (Dicosmo, 1990). The results obtained by Goodbody and Coworkers at Allelix Corporation (Canada) looked promising from a commerical point of view. Nevertheless, the project was discontinued. Evidently, there was no need for developing a tissue-culture-based commercial process. High levels of catharanthine and vindoline for chemical coupling are readily isolated as by-products from *C. roseus* plants. Since vindoline is not found in cell suspension cultures, tissue cultures are not needed for efficient synthesis of vinblastine.

Presently, there is a tremendous interest in developing a tissue culture process for the production of taxol (taxol is a registered trademark of Bristol Mtyers-Squibb for paclitaxel). Detected during the screening program of the National Cancer Institute (NCI), USA, it was considered the most interesting compound among 110000 tested structures (Nicolaou *et al.,* 1994). Taxol has been proven to be a very effective drug against several forms of cancer. It was approved by the US Food and Drug Administration (FDA) for the treatment of ovarian cancer in 1992. The chemistry and the mode of action of taxol has been reviewed by Nicolaou *et al.,*(1994) and Kingston (1994). A significant problem is a shortage in taxol supply (Cragg *et al.,* 1993). To isolate 1 kg taxol 10,000 kg of dried bark from 3000 *Taxus brevifolia* trees have to be extracted. As nearly 2g taxol are needed for the treatment of one patient it is evident that research has been initiated to improve taxol production in trees, to look for alternative sources of taxol, and to develop a total or semi–synthetic chemical synthesis (Nicolaou *et al.,* 1994; Kingston, 1994). Presently, taxol is produced commercially by semi–synthesis from 10-deacetyl baccatin III from needles of young *Taxus* tree grown in plantations. One alternative is to produce taxol by using tissue cultures and indeed, as pointed out by Edgington (1991) "if ever there was a plain target for plant tissue culture, this is it". In 1991, two American companies, Phyton Inc. (Ithaca, NY, USA) and ESCA genetics (California, USA), announced that the development of their tissue culture-based processes was to be completed within 2-5 years. It was impossible to estimate how realistic these claims were, because both companies did not release scientifically sound information about their processes. Thus, conclusions could only be drawn from the results published by researchers from universities and from the improvements of systems with similar productivity obtained when industry was involved. In the meantime, ESCA genetics dropped out of the race, while Phyton Incorporated USA licensed their tissue culture process to Bristol–Myers Squibb. It has to be seen whether the huge fermentation facilities of Phyton in Germany will be used for commercial taxol production. In contrast to the American companies, the Mitsui Company has published some of their research results on taxol production in *Taxus* cultures (Yukimune *et al.,* 1996). According to the published data, the Japanese company again is ahead of all others. Now, the Taxol is produced by Phyton using bioreactor in Germany. There is also another Korean Company is in race for production as Taxol using tissue culture (Samyoung Corporation).

Early publication on taxol production in tissue cultures of various *Taxus* species looked not very promising and did not support the taxol levels (153 mg L^{-1} after 6 weeks) claimed in the patent of Bringi and Kadkade (1993). Wickremesinhe and Arteca (1993) induced callus cultures of various *Taxus* sp. and screened them for taxol production. The levels ranged from 0.001–0.0131% of dry mass, and older browning callus produced more taxol than young pale callus. When they established suspension cultures from taxol producing callus, the specific taxol content could not be determined because the taxol peak was overlaid by other compounds (Wickremesinhe and Arteca, 1994a). From 28g dry mass 120 µg pure taxol (4.3×10^{-5} %) was extracted. The group of DiCosmo at Toronto University, Department of Botany (Canada), published several articles regarding the optimization of taxol production in *Taxus* callus and suspension cultures. They reported initial yields of 0.02% for a culture of *Taxus cuspidata* calli (Fett–Neto *et al.,* 1992). Subsequently, it was shown that the medium composition affects taxol production and growth. As only percent of controls were given, it is not clear whether the experiments resulted in really higher productivity (Fett-Neto *et al.,* 1993). When the cells were cultivated in shake flasks productivity went down by a factor of 10 in rapidly growing cultures (Fett–Neto *et al.,* 1993), and in a more productive suspension culture levels of 0.15 mg L^{-1} were detected (Fett–Neto *et al.,* 1994a). Taxol production could only be improved a little by feeding of potential precursors and thus, levels of 0.01% dry mass were found (Fett-Neto *et al.,*

1994b). Substantially higher levels were reported by two groups from Cornell University, Ithaca, USA during culture media optimization experiments they reached levels of up to 15 mgL^{-1} taxol using lines of *T. baccata, T. canadensis* and *T. cuspidata* (Hirasuna *et al.,* 1996). We reported the production and elicitation of *Taxus wallichiana* cell cultures for the production of taxanes and its analogs using six-well plate technique (Prasad Babu *et al.,* 2001). However, substantial variations of yields in individual experiments were observed making interpretations quite difficult. To overcome the oscillations of taxol production might be a severe problem in developing a large-scale process. In a bioreactor experiments with a working volume of 600 mL levels of upto 22mg L^{-1} or 1.1 mgL^{-1}d^{-1} taxol were found of which 90% were extracellular (Pestchanker *et al.,* 1996). A breakthrough may be was the finding that methyl jasmonate is an inducer of taxol biosynthesis in cultures of various *Taxus* species (Mirajalili and Linden, 1996). Addition of 10 µM methyl jasmonate increased taxol productivity 19-fold, from 0.2 mg L^{-1} to 3.4 mg L^{-1}. Yukimune *et al* (1996) added higher concentrations of methyl jasmonate (100 µM) used a cell line of *T. media* with a higher capacity for taxol biosynthesis, and observed taxol accumulation after induction for a longer period. Although taxol levels increased only 5–fold compared with controls, a productivity of 110 mg L^{-1} within 14 d was obtained. The specific content of taxol obtained was 0.6% and thus exceeded by far the specific contents found in any tissue of the intact plant.

Recently, transformed phytohormone independent callus cultures were found to contain taxol levels upto 16 µg g^{-1} dry mas (Han *et al.,* 1994). Embryo culture has also been tested for taxol production (Flores *et al.,* 1993). However, taxol and taxane yields were not very different from those reported initially for callus and suspension cultures. A recent publication suggests root of hydroponically grown *Taxus* plant as a source for taxol and related taxanes (Wickremesinhe and Afteca, 1994b), which would indicate that transformed root cultures should also be a suitable source of these compounds. Indeed, a recent patent application of Celex–Laboratoreis (1994) describes the production of taxol using hairy root cultures. Studying the published literature on taxol production in cell culture systems, one might conclude that most cell lines of all *Taxus* species contain low levels of taxol, that some more productive lines can be obtained by screening, that methyl jasmonate is presently the only unambigously proven enhancer of taxol synthesis, and that production levels of higher producing lines often oscillate to an unacceptable extent. Nevertheless, if the published production rates can reproducibly be obtained in large volumes of several thousand liters and if they are optimized further by biochemical engineers, culture systems may compete with semi–synthesis. More knowledge about the biochemistry, regulation and limiting steps of the taxol pathway may eventually help to engineer lines with improved productivity (Srinivasan *et al.,* 1996). They provide some evidence that taxol biosynthesis is located in plastids. They also conclude that the conversion of phenylalanine to phenylisoserine is a rate–limiting step of taxol biosynthesis. It might be important to stimulate the flux of the primary precursors into the taxane skeleton. Previously, it was assumed that mevalonate is the precursor of the isoprenoids. However, the findings of Ensenreich *et al.,* (1996) suggest that the isoprenoid moieties of the taxane molecule are derived from a yet unknown precursor which is likely to be derived from a novel pathway of isoprenoid biosynthesis (using triose phosphate–type compounds and activated acetaldehyde), recently detected by Rohmer *et al.,* (1993). Croteau's group (University of Washington, USA) demonstrated that taxa 4(5),11(12)–diene is the first intermediate in taxane biosynthesis, and recently they have purified the corresponding enzyme taxadiene synthase cyclizing the universal diterpene precursor geranylgeranyl pyrophosphate to taxa-4(5), 11(12)-diene in a single step (Hezari *et al.,* 1995). Due to the importance of taxol biosynthesis and its regulation will soon be made and that genetically engineered lines might be created in the near future.

Table 6.16. Some high yielding plant cell culture lines for secondary metabolites with yields equal to or higher than that of parent plant (George and Ravishankar, 1996)

Product	Species	Yield (% on dry wt. basis)
Anthraquinones	Morinda citrifolia	18.0
Berberine	Coptis japonica	13.2
Caffeine	Coffea arabica	7.6
Diosgenin	Dioscorea deltoidea	7.8
Ginsenosides	Panax ginseng	27.0
Morphinan alkaloids	Papaver somniferum	5.6
Nicotine	Nicotiana tabacum	3.4
Rosmarinic acid	Coleus blumei	27.0
Shikonin	Lithospermum erythrorhizon	14.0
Valeoptriates	Valeriana officinalis	10.3
Shikimic acid	Galium mollugo	10.0
Serpentine	Catharanthus roseus	2.2
Ajmalicine	C. roseus	1.8

Table 6.17. Natural Products yield from cell cultures and whole plants (Dougall 1980)

Natural Product	Species	Yield Cell culture	Whole plant
Anthraquinones	Morinda citrifolia	900 nmol g^{-1} dry wt.	Root 100 nmol g^{-1} dry wt.
Anthraquinones	Cassia tora	0.334% fresh wt.	0.209% seed, dry wt.
Ajmalicine and serpentine	Catharanthus roseus	0.3% dry wt.	0.26% dry wt.
Diosgenin	Dioscorea deltoidea	26 mg g^{-1} dry wt.	20 mg g^{-1} dry wt tuber
Ginseng saponins	Panax ginseng	0.38% fresh wt.	0.3, 3.3% fresh wt.
Nicotine	Nicotiana tabacum	3.4% dry wt.	2-5% dry wt.
Thebaine	Papaver bracteatum	130 mg g^{-1} dry wt	1400 mg g^{-1} dry wt less and 3000 mg g^{-1} dry wt. root
Uniquinone	Nicotiana tabacum	0.5 mg g^{-1} dry wt.	16 mg g^{-1} dry wt. leaf

Table 6.18. Antitumor compounds from cell cultures vs intact plant (Adopted from Misawa and Endo, 1988)

Plant	Antitumor compounds	Plant (dry wt. %)	Reference	Cultured cells (dry wt.%)	Reference
Baccharis megapotamica	Baccharin	2.0×10^{-2}	Gittermann (1980)	--	--
Brucea anidysenterica	Bruceantin	1.0×10^{-2}	Kupchan et al. (1973).	5.8×10^{-5}	Misawa et al. (1983)
Camptotheca acuminata	Camptothecine	5.0×10^{-3}	Misawa et al. (1974)	2.5×10^{-4}	Misawa et al. (1974)
Ochrosia moore	Ellipticine	3.2×10^{-5}	Kouadio et al. (1984)	2.7×10^{-4}	Kouadio et al. (1985)
Cephalotaxus harringtonia	Homoharringtonine	1.8×10^{-5}	Misawa et al. (1985)	5.5×10^{-8}	Misawa et al. (1985)

(Contd.)

Plant	Antitumor compounds	Plant (dry wt. %)	Reference	Cultured cells (dry wt.%)	Reference
Putterlickia verrucosa	Maytansine	2.0×10^{-5}	Douros et al. (1979)	5.0×10^{-7}	Masawa et al. (1985)
Podophyllum peltatum	Podophyllotoxin	6.4×10^{-1}	Kadkade (1981)	7.1×10^{-1}	Kadkade (1982)
Taxus brevifolia	Taxol	5.0×10^{-1}	Boettner et al. (1977)	--	--
Trypterygium wilfordii	Tripdiolide	1.0×10^{-3}	Kupchan et al. (1972a)	1.0×10^{-2}	Misawa et al. (1985)
Catharanthus roseus	Vinblastine, vincristine	5.0×10^{-3}	Furuya (1984)	5.0×10^{-4}	Govindachari & Viswanathan (1972)

Table 6.19. Establishment of Callus and Cell Cultures of Taxus sp

Taxus sp.	Explants	Basal medium	Phytohormone	Additives	Cultures	References
T. wallichiana	Young stems	MS	2,4-D/Kn/ NAA	Ascorbic acid	Callus	Banerjee et al. (1996)
T.x. media	Stem	B5	2,4-D	Casein hydrolysate	Callus	Wickramesinhe & Arteca (1991)
T.x. media var Hatfelda	Stem Needles	WR	Picloram + NAA	Methyl jasmonate	Callus	Furnanowa et al. (1997)
T. brevifolia	Bark, stem, needles	B5	2,4-D	Casamino acid	Callus	Gibson et al. (1993)
T. brevifolia	Young stem	SH	NAA./BA	--	Callus and Cel! cultures	Kim et al. (1995)
T. brevifolia	Cambium	B5 ABA, GA$_3$	Picloram, Kn, arginine, glycine proline	Aspartic acid	Cell	Ketchum et al. (1995)
T. cuspidata	Young stems	B5	2,4-D	PVP	Callus	Fett-Neto et al. (1992)
T. yunnanensis	Young stems	6,7V	IAA/2,4-D	--	Callus and Cell cultures	Chen et al. (1996)
T. floridana	Needles	MS	NAA/BA	Glutamine	Embryonic Cell cultures	Salandy et al. (1993)
T. chinensis	--	B5	NAA/BA	Glutamine	Cell cultures	Srinivasan et al. (1996)
T. canadensis	Stems, Needles	B5	2,4-D/BA/ Picloram	--	Callus	Ketchum and Gibson (1995)
T. baccata	Twigs and Needles	B5	NAA, Picloram	--	Callus	Furmanowa et al. (1995)
T. baccata	Hypocotyl, Leaf	WPM	NAA 2,4-D/Kn	--	Callus	Zhiri et al. (1995)
T. baccata	Stem	WPM	NAA/BA	--	Callus	Han et al. (1994)
T. baccata L.	Stem	B5	2,4-D/Kn	--	Callus	Monacelli et al. (1995)
T. baccata L. (Himalayan yew)	Leaf	B5	2,4-D/Kn	Casein hydroxylate	Callus	Jha and Jha (1995)
T. wallichiana	Leaf	MS	NAA/Kn	--	Cell	Prasad babu et al. (2001)

Table 6.20. Secondary Products detected in plant tissue culture

Compound	Plant source	Type of culture	Reference
Cinnamic acid	*Nicotiana tabacum*	C	Brown and Tenniswood (1974)
Caffeic acid	*N. tabacum*	C	Brown and Tenniswood (1974)
Ferulic acid	*N. tabacum*	C	Brown and Tenniswood (1974)
Ferulic acid	*Linum usitatissimum*	C	Liau and Ibrahim (1973)
p–Coumaric acid	*L. usitatissimum*	C	Liau and Ibrahim (1973)
Caffeoyl putrescine	*N. tabacum*	C	Mizusaki *et al.* (1971)
Feruloyl putrescine	*N. tabacum*	C	Mizusaki *et al.* (1971)
p–Coumaroyl putrescine	*N. tabacum*	C	Mizusaki *et al.* (1971)
Chlorogenic acid	*Haplopappus gracilis*	C	Stickland and Sunderland (1972a)
Chlorogenic acid	*Solanum tuberosum*	S	Gamborg (1967)
p–Hydroxybenzoic acid	*L. usitatissimum*	C	Liau and Ibrahim (1973)
Vanillic acid	*L. usitatissimum*	C	Liau and Ibrahim (1973)
Scopoletin	*N. tabacum*	C	Sargent and Skoog (1960, 1961)
Scopoletin	*N. tabacum*	C and CGC	Brown and Tenniswood (1974)
Scopoletin	*Ruta graveolens*	S	Reinhard *et al.* (1968); Steck *et al.* (1971)
Esculetin	*N. tabacum*	C and CGC	Brown and Tenniswood (1974)
Umbelliferone	*N. tabacum*	C	Brown and Tenniswood (1974)
Umbelliferone	*R. graveolens*	S	Reinhard *et al.* (1971), Steck *et al.* (1971)
Bergapten	*N. tabacum*	C	Brown and Tenniswood (1974)
Bergapten	*R. graveolens*	S	Reinhard *et al.* (1968)
Psoralen	*R. graveolens*	S	Reinhard *et al.* (1968), Steck *et al.* (1971)
Xanthotoxin	*R. graveolens*	S	Reinhard *et al.* (1968), Steck *et al.* (1971)
Herniarin	*R. graveolens*	S	Reinhard *et al.* (1968)
Rutaretin	*R. graveolens*	S	Reinhard *et al.* (1968)
3-(1,1-dimethylallyl)-scopoletin	*R. graveolens*	S	Brocke *et al.* (1971)
Isopimpinellin	*R. graveolens*	S	Steck *et al.* (1971)
Rutacultin	*R. graveolens*	S	Steck *et al.* (1971)
Rutamarin	*R. graveolens*	S	Steck *et al.* (1971)
Apigenin	*Petroselinum hortense*	S	Kreuzaler and Hahlbrock (1973); Hahlbrock and Wellmann (1970)
Apigenin	*Glycine max*	S	Hahlbrock (1971)
Luteolin	*P. hortense*	S	Kreuzaler and Hahlbrock (1973)
Chrysoeriol	*P. hortense*	S	Kreuzaler and Hahlbrock (1973)
Sinensetin	*Citrus aurantium*	C	Brunet and Ibrahim (1973)
Nobiletin	*C. aurantium*	C	Brunet and Ibrahim (1973)
Isorhamnetin	*P. hortense*	S	Kreuzaler and Hahlbrock (1973)
Quercetin	*P. hortense*	S	Kreuzaler and Hahlbrock (1973)

(Contd.)

Compound	Plant source	Type of culture	Reference
Quercetin	*Crotalaria juncea*	C	Jain and Khanna (1974)
2',4,4'-trihydroxychalcone	*Phaseolus aureus*	C	Berlin and Barz (1971)
Daidzein	*P. aureus*	C and S	Berlin and Barz (1971)
Daidzein	*G. max*	C	Miller (1969)
Coumestrol	*P. aureus*	C and S	Berlin and Barz (1971)
Soyagol	*P. aureus*	C and S	Berlin and Barz (1971)
Pisatin	*Pisum sativum*	C	Bailey (1970)
Cyanidin	*H. gracilis*	C	Ardenine (1965); Stickland and Sunderland (1972a)
Cyanidin	*L. usitatissimum*	C	Ibrahim *et al.* (1971)
Cyanidin	*Dimorphotheca auriculata*	C	Harborne *et al.* (1970)
Delphinidin	*D. auriculata*	C	Harborne *et al.* (1970)
Catechin	*Camellia sinensis*	C	Forrest (1969)
Catechin	Paul's Scarlet Rose	S	Davies (1972a)
Epicatechin	*C. sinensis*	C	Forrest (1969)
Epicatechin	Paul's Scarlet Rose	S	Davies (1972a)
D–glucogallin	Paul's Scarlet Rose	S	Davies (1972a)
Leucooanthocyanins	C. sinensis	C	Forrest (1969)
Digitolutein	*Digitalis lanata*	C	Furuya and Kojima (1971)
4–Hydroxydigitolutein	*D. lanata*	C	Furuya and Kojima (1971)
3–Methylpurpurin	*D. lanata*	C	Furuya *et al.* (1972)
3–Methylquinizarin	*D. lanata*	C	Furuya *et al.* (1972)
3–Methylalizarin	*D. lanata*	C	Furuya *et al.* (1972)
Pachybasin	*D. lanata*	C	Furuya *et al.* (1972)
Morindone	*Morinda citrifolia*	S	Leistner (1973)
Alizarin	*M. citrifolia*	S	Leistner (1973)
Damnacanthal	*M. citrifolia*	S	Leistner (1973)
Chrysophanol	*Cassia angustifolia*	C	Friedrich and Baier (1973)
Physcion	*Cassia tora*	C	Tabata *et al.* (1975)
Rheum emodin			
Aloe emodin	*C. angustifolia*	C	Friedrich and Baier (1973)
Rhein	*C. angustifolia*	C	Friedrich and Baier (1973)
Plumbagin	*Plumbago zeylanica*	C	Heble *et al.* (1974)
Lindenenol	*Lindera strychnifolia*	C	Tomita *et al.* (1969)
Lindenenol acetate	*L. strychnifolia*	C	Tomita *et al.* (1969)
Linderane	*L. strychnifolia*	C	Tomita *et al.* (1969)
Linderalacton	*L. strychnifolia*	C	Tomita *et al.* (1969)
Lindesterene	*L. strychnifolia*	C	Tomita *et al.* (1969)
Caryophyllene	*L. strychnifolia*	C	Tomita *et al.* (1969)
Paniculides A, B and C	*Andrographics paniculata*	C and S	Allison *et al.* (1968); Butcher and Connolly (1971)
Trans, trans and *cis trans* farnesols	*A. paniculata*	S	Overton and Roberts (1974a, b)

(Contd.)

Compound	Plant source	Type of culture	Reference
γ–bisabolene	A. paniculata	S	Overton and Roberts (1974a, b)
β–sitosterol	N. tabacum	C	Benveniste et al. (1966)
Stigmasterol	Dioscorea tokoro	C	Tomita et al. (1970)
Campesterol	Withania somnifera	C	Yu et al. (1974)
	Paul's Scarlet Rose	C	Williams and Goodwin (1965)
	Tylophora indica	C	Benjamin and Mulchandani (1973)
	Helianthus annuus	C	Butcher et al. (1974)
Cholesterol	H. annuus	C and CGC	Butcher et al. (1974)
Isofucosterol	H. annuus	C and CGC	Butcher et al. (1974)
Cycloartenol	N. tabacum	C	Benveniste et al. (1966); Benveniste (1968)
2,4–methylene cycloartenol	N. tabacum	C	Benveniste et al. (1966); Benveniste (1968)
Citrostradienol	N. tabacum	C	Benveniste et al. (1966); Benveniste (1968)
Citrostradiol	N. tabacum	C	Benveniste et al. (1966); Benveniste (1968)
Cycloeucalenol	N. tabacum	C	Benveniste et al. (1966); Benveniste (1968)
Obtusifoliol	N. tabacum	C	Benveniste et al. (1966); Benveniste (1968)
2,4–ethylidene cholesterol	W. somnifera	C	Yu et al. (1974)
2,4–methylene cholesterol	W. somnifera	C	Yu et al. (1974)
β–amyrin	Paul's Scarlet Rose	C	Williams and Goodwin (1965)
β–amyrin	T. indica	C	Benjamin and Mulchandani (1973)
Arundoin	Oryza sativa	C	Yanagawa et al. (1972)
Tomatine	Lycopersicum esculentum	C	Roddick and Butcher (1972a)
Solasodine	Solanum xanthocarpum	C	Heble et al. (1971)
Voilaxanthin	Paul's Scarlet Rose	C	Williams and Goodwin (1965)
Zeaxanthin	R. groveolens	C	Scharlemann and Czygan (1971)
Neoxanthin			
Auroxanthin	Paul's Scarlet Rose	C	Williams and Goodwin (1965)
β–carotene	R. graveolens	C	Scharlemann and Czygan (1971)
Lutein	R. graveolens	C	Scharlemann and Czygan (1971)
Lutein–5,6 epoxide	R. graveolens	C	Scharlemann and Czygan (1971)
Antheraxanthin	R. graveolens	C	Scharlemann and Czygan (1971)
Sterculic acid	Malva pariflora	C	Yano et al. (1972a)
Dihydrosterculic acid			
Malvalic acid	Malva sylvestris	C	Yano et al. (1972a)
Dihydromalvalic acid			
Hydnocarpic acid	Hydnocarpus anthelmintheca	C	Spener et al. (1974)
Chaulmoogric acid	H. anthelmintheca	C	Spener et al. (1974)
Gorlic acid	H. anthelmintheca	C	Spener et al. (1974)

(Contd.)

Compound	Plant source	Type of culture	Reference
Erucic acid	*Crambe abyssinica*	C	Jones (1974)
2–undecanone	*R. graveolens*	C	Reinhard *et al.* (1971); Corduan and Reinhard (1972)
2–undecanyl acetate	*R. graveolens*	C	Reinhard *et al.* (1971); Corduan and Reinhard (1972)
2–nonanone	*R. graveolens*	C	Reinhard *et al.* (1971); Corduan and Reinhard (1972)
2–nonanyl acetate	*R. graveolens*	C	Reinhard *et al.* (1971); Corduan and Reinhard (1972)
2–nonanol	*R. graveolens*	C	Reinhard *et al.* (1971); Corduan and Reinhard (1972)
2–undecanol	*R. graveolens*	C	Reinhard *et al.* (1971); Corduan and Reinhard (1972)
Aliphatic alkanes C_{17}–C_{18}	*N. tabacum*	C	Weete *et al.* (1971)
Benzylglucosinolate	*Tropeaolum majus*	S	Kirkland *et al.* (1971)
2–phenylethyl glucosinolate	*T. majus*	S	Kirkland *et al.* (1971)
2-hydroxy-2-phenyl-ethylglucosinolate	*Reseda luteola*	S	Kirkland *et al.* (1971)
Glucobrassicin	*Brassica rapa*	C	El–Tigani (1972)
Neoglucobrassicin	*B. rapa*	C	El–Tigani (1972)

Key : C = callus; CGC = crown–gall callus; S = suspension culture

Table 6.21. Phytopharmaceuticals from plant tissue culture

(C = Callus; S = suspension)

Compound	Plant species	Culture type	Reference
Ajmalcine	*Catharanthus roseus*	S	Zenk *et al.* (1977), Knobloch and Berlin (1980)
Anisodamine (6-hydroxy-hyoscyamine) (bioconversion)	*Anisodus tanguticus*	S	Cheng *et al.* (1987)
Anthocyanins	*Euphorbia mill*	S	Yamamoto *et al.* (1982)
Anthraquinones	*Cassia angustifolia*	C	Friedrich & Baier (1973)
	Cassia obtusifolia	C	Takahashi *et al.* (1978)
	Cassia tora	C	Tabata *et al.* (1975)
	Galium mollugo	S	Bauch & Leistner (1978)
	Morinda citrifolia	S	Zenk *et al.* (1975)
	Rubia species	S	Shulte *et al.* (1984)
Berberine	*Coptis japonica*	C & S	Furuya *et al.* (1972), Sato and Yamada (1984)
Betacyanins	*Chenopodium rubrum*	S	Berlin *et al.* (1986)
Caffeine	*Coffea arabica*	C	Keller *et al.* 91972)
Carboline alkaloids	*Phaseolus vulgaris*	S	Veliky (1972)
	Peganum harmala	S	Sasse *et al.* (1987)

(Contd.)

Compound	Plant species	Culture type	Reference
Cardenolides	*Digitalis purpurea*	S & C	Buechner & Staba (1964), Kartnig *et al.* (1976)
	Digitalis lanata	S	Luckner and Diettrich (1988)
Cinnamoyl putrescines	*Nicotiana tabacum*	S	Schiel *et al.* (1984a & b)
Codeine	*Papaver somniferum*	S	Heinstein (1985)
Digoxin (Biotransformation)	*Digitalis lanata*	S	Reinhard *et al.* (1975); Ohlsson *et al.* (1983)
Diosgenin	*Dioscorea composita*	C	Mehta and Staba (1970)
	Dioscorea deltoidea	S	Kaul *et al.* (1969), Tal *et al.* (1983)
	Solanum xanthocarpum	S	Heble *et al.* (1968)
	Trigonella foenum–graecum	S	Khanna, P. and John (1973)
L. dopa	*Mucuna pruriens*	S	Brain (1974)
Ginsenosides	*Panax ginseng*	S	Furuya *et al.* (1973)
Glycyrrhizin	*Glycyrrhiza glabra*	S	Tamaki *et al.* (1973)
Harringtonine and homo-harringtonine	*Cephalotaxus* spp.	S	Luckner and Diettrich (1985, 1988), Heinstein (1985)
Hyoscyamine	*Hyoscyamus niger*	S	Hashimoto *et al.* (1982)
Indole alkaloids	*Ipomoea violacea*	C & S	Dobberstein & Staba (1969)
	Rivea corymbosa		
Isoprenoids (vol. Oil)	*Pelargonium fragrans*	C	Charlwood *et al.* (1988)
Morphine	*Papaver somniferum*	S	Heinstein (1985)
Naphthoquinone	*Lithospermum erythrorhizon*	C & S	Tabata *et al.* (1974)
Nicotine	*Nicotiana tabacum*	S	Ogino *et al.* (1978)
Papain	*Carica papaya*	C	Medora *et al. (1973)*
Phenolics	*Pinus resinosa*	C	Jorgensen & Balsillie (1969)
Protoberberines	*Thalictrum* sps.		
	Coptis japonica & Berberis sp.	S	Nagakawa *et al.* (1984); Hinz and Zenk (1981); Breuling *et al.* (1985)
Psoralen	*Ruta graveolens*	S	Austin & Brown (1973)
Quinoline alkaloids	*Cinchona ledgeriana*	C & S	Hoekstra *et al.* (1987) Staba & Chung (1981); Rhodes *et al.* (1986)
Reserpine	*Rauwolfia serpentina*	S	Yamamoto & Yamada (1986)
	Alstonia constricta	C	Carev (1965)
Rosamarinic acid	*Coleus blumei*	C & S	Zenk and Deus (1982); Ulbrich *et al.i (1985)*
Serpentine & other mono-meric alkaloids	*Catharanthus roseus*	C & S	Doeller *et al.* (1976)
Shikonin	*Lithospermum erythrorhizon*	C & S	Kutney *et al.* (1986); Maeda *et al.* (1983); Fujita *et al.* (1985)
Steroidal glycoalkaloids	*Solanum acculeatissimum*	C	Kadkade & Madrid (1977)
	Solanum xanthocarpum	C	Heble *et al.* (1971)
Trigonelline	*Trigonella foenum–graecum*	S	Radwan & Kokate (1980)
Tropane alkaloids	*Datura innoxia*	S	Tabata *et al.* (1971)

(Contd.)

Compound	Plant species	Culture type	Reference
	Datura innoxia (regeneration)	S	Hiraoka & Tabata (1974)
	Datura stramonium	C & S	Chan & Staba (1965); Stohs (1969)
	Hyoscyamus niger	S	Hashimoto *et al.* (1982)
	Scopolia parviflora	C	Hiraoka *et al.* (1973)
Ubiquinone-10	*Nicotiana tabacum*	S	Matsumoto *et al.* (1982)
Undecanone & other volatile components	*Ruta gravealens*	C	Corduan & Reinhard (1972)
Verbascoside	*Syringa vulgaris*	S	Ellis (1983)
Vinblastine	*Catharanthus roseus*	C	Miura *et al.* (1987)
Vindoline	*Catharanthus roseus*	S	Chan *et al.* (1969)
Visnagin	*Ammi visnoga*	S	Kaul & Staba (1967)
Xanthotoxin	*Ruta graveolens*	S	Reinhard *et al.* (1968)
Rosmarinic acid	*Lithospermum erythrorhizon*	S	Mizukarni *et al.* (1993)
	Coleus blumei	S	Petersen *et al.* (1995)
	Ocimum basilicum	S	Tada *et al.* (1996)
	Heliotropium perunianum	S	Motoyata *et al.* (1996)
10–methoxy camptothecine	*O. mungos*	C	Roja and Heble (1991); Tafuri *et al.* (1975)
Anthraquinone	*Ophiorrhiza pumila*	S	Kitajima *et al.* (1998)
Artemisinin	*Artemisia annua*	C & S	Liu *et al.* (1992)
Azadirachtin	*Azadirachta indica*	C	Veeresham *et al.* (1998)
Camptothecine (CPT) 10–OH CPT	*Camptothecia accuminata*	C	Wiedenfeld *et al.* (1997)
Capsaicin	*Capsicum annuum*	C & S	Veeresham *et al.* (1991)
Castanospermine	*Castanospermum austrate*		Roja and Heble (1995)
Cathin-6-one alkaloid	*Ailthanthus altissima*	S	Romero and Roberts (1996)
Cephaeline	*Cephaelis ipecacuanha*	C & S	Veeresham *et al.* (1994)
Cerebroside	*Lycium chinense*	S	Janga *et al.* (1998)
Colchicine	*Colchicum autumnale*	S	Hayashi *et al.* (1988)
CPT	*N. foetida*	S	Veeresham and Shuler (2002)
CPT	*Eravatamia heyeneana*	C	Sumalatha. (2001)
Crocetin	*Crocus sativus*	C	Dufresne *et al.* (1997)
Ellipticine	*Ochrosia elliptica*	S	Kouadio *et al.* (1984)
Ephedrine	*Ephedra* sp.	C	O'Dowd *et al.* (1993)
Forskolin	*Coleus forskohlii*	C	Tripathi *et al.* (1995)
Forskolin	*Coleus forskohlii*	S	Mersinger *et al.* (1988)
Galanthamine	*Narcissus confusus*	C & S	Selles *et al.* (1999)

(Contd.)

Compound	Plant species	Culture type	Reference
Harringtonine Isoharringtonine Homoharringtonine	*Cephalotaxus harringtonia*	C	Delfel and Rothfus (1977)
Hypericin	*Hypericum perforatum*	C	Rani *et al.* (2001)
Khellin, visnagin	*Ammi visnaga*	C	Fiky *et al.* (1989)
Plumbagin	*Droscophyllum lusitanium*	S	Nahalka *et al.* (1996)
Podophyllotoxin	*Linum album*	S	Smollny *et al.* (1998b);
	Linum flavum	C	Berlin *et al.* (1988)
Ribosome inactivating proteins	*Trichosanthes kirilowii*	S	Shih *et al.* (1998)
Rosmarinic acid	*Saliva officinalis*	S	Hippolyte *et al.* (1992)
Rosmarinic acid	*Anchusa officinalis*	S	De–Eknamkul and Ellis (1985a, b)
Scopolamine Hyoscyamine	*Datura innoxia*	S	Gontier *et al.* (1994)
Sennosides	*Rheum palmatum*	C	Ohshina *et al.* (1988)
Solasodine	*Solanum aviculare*	S	Kittipong Pastana *et al.* (1998)
Taxol	*T. cuspidata*	C	Fett–neto *et al.* (1992)
Triptotide Tripdiolide	*Tripterygium wilfordii*	S	Kutney *et al.* (1993)
Valetpotriaes	*Valeriana wallichi*	S	Becker and Chavadej (1985)
Vanillic acid	*Vanilla planifolia*	S	Funk and Brodelius (1990)
Withaferin A	*Withania somnifera*	C & S	Veeresham & Shuler (2004)

Another important, commercially used antitumor compound is etoposide, a semisynthetic podophyllotoxin. Production of podophyllotoxin by tissue cultures of *Podophyllum peltatum* was first reported by Kadkade (1982). However, the levels are very low. The interest in podophyllotoxins was revived when root cultures of *Linum flavum* were found to contain high levels (1% of dry mass) of 5–methoxypodophyllotoxin and its glucoside (Berlin *et al.*, 1988b). Van Uden *et al.* (1990), Wichers *et al.* (1991) reported that the suspension cultures of *L. flavum* also contained some podophyllotoxins. However, reasonable productivity required some morphological differentiation, e.g., root formation (Van Uden *et al.*, 1991). Other podophyllotoxin producing cell cultures are those of *L. album* (Smollny *et al.*, 1993) and *P. hexandrum* (Woerdenbag *et al.*, 1990). The levels of podophyllotoxins vary from 0.05 to 0.3% of dry mass depending on the culture conditions and the tendency to differentiate. Thus root cultures (producing at least 1% podophyllotoxins in dry weight) seem to be most suitable for future research efforts e.g., for biochemical studies (Oostham *et al.*, 1993).

Misawa and Endo (1988) described that the productivity of cell culture systems for other potential antitumor compounds produced by plants and plant cell cultures. These include camptothecin, homoharringtonine, and maytansine. We have reported the production of CPT and MCPT from callus and cell cultures of *Nothapodytes foetida* (Ciddi and Shuler, 2000; Sundaravelan *et al.*, 2002). It must be concluded that the levels found in the culture systems are far too low to be attractive from a biotechnological point of view. The same holds true for tripdiolide and triptolide produced at 0.01% of dry mass by *Tripterygium* species (Takavama. 1994)

6.8.5. Cardiac Glycosides

The formation of cardiac glycosides in undifferentiated cell cultures of various *Digitalis* species was found to be very low or even lacking in rapidly growing suspension cultures. In slow growing green callus cultures or in embryogenic culture some cardiac glycoside accumulation was observed (Luckner and Diettrich, 1988). However, low- or non-producing plant cell suspension cultures of *Digitalis lanata* are effective in the biotransformation of cardenolides and for this reason they are still rather attractive from a biotechnological point of view.

6.8.6. Vanillin and Vanilla Aroma

Vanilla is probably the most widely used flavor in food industries with a worldwide annual consumption of 1200–1500 tonnes in 1993 (Havkin-Frenkel, 1994). The prices for one kg natural vanillin from cured beans of *Vanilla planifolia* amount to $3000-4000 while the price for synthetic vanillin is less than $20 per kg. The flavor of natural vanilla extracts from beans of different origin varies. Though vanillin is the most important constituent of the aroma other compounds seem to affect substantially the taste of this flavor. Due to the high price of natural vanilla flavor plant tissue cultures, mainly of *V. planifolia*, have been regarded as an alternative source for the production of vanilla aroma and vanillin. The first publication of phenylpropanoid metabolism in *V. planifolia* showed that vanillin was not produced under normal culture conditions (Funk and Brodelius, 1990). Formation of vanillic acid, however, was observed when 3,4–(methylenedioxy)–cinnamic acid, an inhibitor of *p*–coumarate CoA ligase, was added to the suspension cultures. This led to a shift from lignin to benzoate biosynthesis. After feeding potential cinnamic acid precursors there were indications that the cell culture must contain enzymes involved in benzoate biosynthesis. Finally, it was found that kinetin is an efficient elicitor of vanillic acid formation (Funk and Brodelius, 1992). Vanillic acid levels of up to 0.1% of dry mass were detected. Thus, according to the literature, one would assume that cultures of *V. planifolia* are of no commercial interest. It was, therefore, somewhat surprising that ESCA genetics (Knuth and Sahai, 1991a,b) filed a patent in 1988 (issued in 1991) for the production of vanilla aroma (PhytoVanilla™) and biosynthesized vanillin (Phytovanillin™) by suspension cultures of *V. fragrans*. The aroma compounds are secreted into the medium and bind to adsorbents. The patent states that 16–18 mg vanillin per L of medium were produced with in ca. 45 days. A comparison of the vanilla flavor profiles of beans with that of tissue cultures as presented in the patents reveals that vanillin is the main component in both extracts, but that otherwise the composition of the two extracts is quite different. Thus one can conclude that the suspension cultures produce a new vanilla aroma for which a market may or may not exist. ESCA–genetics announced that they would be able to produce tissue culture flavors such as vanilla at an incredibly low price of $50-100 kg^{-1} (Moshy *et al.,* 1989). Stahlhut (1993) reported that the vanilla flavor from suspension cultures of *V. fragrans* "exited the lab and entered commercial development". Unfortunately, the company was not willing to support their announcement by providing any scientific data or comments of their "marketed" product for this review. Thus, some doubts remained as to whether "Phyto Vanilla™ is considerably more economic to produce than natural vanilla extract (Goldstein, ESCA genetics, USA). It has to be seen whether another company will continue the process of vanilla aroma production after the shutting down of ESCA genetics.

Based on the literature and the patent it is now clear that vanillin production can be induced and optimized in cultured cells of *Vanilla* sp. by media composition, selection, elicitation, and feeding of suitable precursors. The highest specific yields (0.16% vanillin on a dry mass basis) were reported for an embryo culture of *V. planifolia* grown in small bioreactors (Knorr *et al.,* 1993; Hjavkin Fraenkel, 1994). A comparison of an HPLC extract of the embryo culture and a Bourbons vanilla bean extract exhibited much more similarities of the components that shown in the patents of ESCA genetics. Havkin–Frenkel (1994) also compared the production costs. If the cultures produce 2% vanillin in large vessels, the estimated costs for the production of 1 kg vanillin would be between $ 500–1000 (investment for bioreactors not included). Although the embryo culture system contains the highest published content of vanillin reported upto now it is commercially not yet competitive. In this context it should be mentioned that alternative biotechnological approaches for the production of vanillin have been investigated (Cheetham, 1993). For example, root tissue of *V. planifolia* converts ferulic acid to vanillin at production rates of 400 mg kg^{-1} d^{-1} roots were obtained (Westcott *et al.,* 1994). This suggests that hairy root cultures should be tested for the production of vanillin by biotransformation. The production of vanillin and related compouns may also be improved in the near future on the basis of present studies on the enzymology of the formation of benzoic acids from cinnamic acids (Loscher and Heide, 1994).

6.9. COLORS AND FLAVORS FROM PLANT CELL AND TISSUE CULTURES

Many important food colors and flavors are of plant origin. In most cases these compounds can be obtained from field-grown plants (Table 6.22 and 6.23). However, plants are seasonal and the quality of a flavor or color varies due to uncontrolled fluctuations in growing conditions and geographical variations. The use of plant cell and tissue cultures could ensure a continuous supply of uniform quality. The product would have the important advantage of being naturally produced. The production of colors and flavors by plant cell and tissue cultures offers the possibility of quality control and availability independent of environmental changes.

Table 6.22. Example of colorants from botanical sources

Colorant	Source	Colour
Anthocyanins	Fruits, mainly grape; red cabbage	Red to violet blue
Betalains	Beetroot	Red to red purple
Bixin	Seeds of *Bixa orellana*	Yellow to orange
β–carotene	Carrot and palm oil	Yellow to orange
caspsanthin, capsorubin	Red pepper	Orange to red
crocetin , crocin	Saffron stigmas gardenia fruits	Yellow to orange-red
lycopene	Tomatoes	Orange and red
lutein	Alfalfa, Marigold petals	Yellow to orange
zeaxanthin	Corn, Marigold petals	Yellow to orange
Chlorophyll	Plants	Green
Santalan	Red sandlwood tree heartwood	Red
Turmeric	Root of *Curcuma longa*	Yellow to yellow green

Color and Pigment Development in Plant Cell Cultures

There has been much interest over the last decade in the use of biotechnology for the production of natural colors and pigments. The accumulation of colors and pigments in cell cultures is easily recognized. In addition to attracting attention to this expression of secondary metabolism, the pigmentation can facilitate the direct selection of high–producing cell lines. The four groups of compounds comprising the good colors include benzopyrans (anthocyanins, flavones and flavanones), betalaines (betacyanin and betaxanthines), carotenoids (carotenes and xanthophylls), chlorophylls and quinones.

Table 6.23. Flavor Compound Production by Plant Cell and Tissue Cultures (Dorenburg and Knorr, 1996)

Plant cell cultures	Products	References
Allium schoenoprasum	1–Propenyl disulfide	Misawa et al. (1973)
Anethum graveolens	Carvone	Dornenburg et al (1990)
Coriandrum ledgeriana	Quinone	Rhodes et al . (1992)
Coriandrum sativum	Terpenes	Sardesai and Tipnis (1969)
Eucalyptus citriodora	Monoterpenes	Guptha and Mascarenhas (1983)
Jasminum officinale	Monoterpenes	Banthrope et al . (1986)
Malus sylvestris	Ethanol, ethylacetate, ethylbutyrate	Ambib and Fallot (1981)
Martricaria chamomilla	Volatile oils	Prince (1991)
Lavandula angustifolia	Monoterpenes	Webb et al (1984)
Ocimum basillicum	Monoterpene phenylpropanoids	Purohit and Khanna (1983)
Passifolia eduils	Alcohols, aldehydes	Jensen et al. (1990)
Picrasma quassiodies	Quassin	Scragg and Arias Castro (1992)
Pimpinella anisum	Anethole	Reichling et al (1985)
Pridium guajava	Acetates, alcohols, esters	Prabha et al. (1990)
Quassia amara	Quassin	Scragg and Alan (1994)
Ribes nigrum	Black current flavor	Enevoldsen (1994)
Rosmarinus officinalis	Alcohols, esters	Webb et al. (1984)
Thuja occidentails	Thujon	Berlin et al. (1984)
Zingiber officinale	Gingerols, shoagols	Charlwood et al (1988)

6.9.1. Anthocyanins

Anthocyanins are common compounds and are widely found in various plant species. Because they are natural pigments and are safe, they have a high potential value as food additives. Obtaining stable pigments in large quantities throughout all seasons is difficult. Plant cell cultures have become an important biotechnological tool for producing such secondary metabolites. Seitz and Hinderer (1988) noted that 27 different species have been shown to produce anthocyanins in plant cell cultures. Seven different anthocyanidins were detected in these cultures.

Examples of culture conditions and process strategies leading to improved anthocyanin production in plant cell cultures are given in Table 6.24

6.9.2. Betalaines

Betalaines are a class of water-soluble pigments produced by plants and some higher fungi. They consists of two major subgroups: red and purple betacyanins, and betaxanthin pigments, which are yellow to orange. Betalaines are heat, light and oxygen sensitive and their use has been limited to foods with short shelf life or within low pH.

Bohm and Rink (1988) have given a detailed review on betalaine production by plant tissue cultures. Five of the ten betalain–producing plant families (*Amaranthus, Beta, Chenopodium, Phytolacca* and *Portulacca*) also produce similar compounds in plant cell culture. Betacyanins, the red pigments, as well as betaxanthins, the yellow ones, have been produced in culture. Both pigments were accumulated inside the cell vacuole.

The highest concentration of betacyanins was obtained in cell cultures of *Chenopodium rubrum* (Berlin *et al.,* 1986). This culture could accumulate upto 1% (dry weight) betacyanins under optimized

Table 6.24. Factors Affecting Anthocyanin Production in Various Plant Cell Cultures
(Dorenburg and Knorr, 1996)

Cell cultures	Factors	References
Aralia cordata	Cell aggregate cloning, nutrients	Sakamoto et al . (1993)
	Large scale	Kobayashi et al (1993)
	Selection	Hiraoka et al (1986)
Daucus carota	Selection (aggregate size)	Kinersley and Dougall (1980)
	Serial cloning plus selection	Dougall and Vogelien
	Limiting phosphate chemostat conditions	Dougall and Weyrach (1980)
	Inoculum density, zeatin, sucrose	Ozeki and Komamime (1985),
	Sinapic and (precursor)	Dougall (1989)
	Mutation, Large scale	Nagarajan et al. (1989)
	Light, 2,4 dichlorophenoxyacetic acid	Takeda (1990)
	Glucose, fructose	Zwayyed et al (1991)
Haplopappus gracilis	UV light	Welmann et al. (1976)
Frgaria ananassa	Cultivars	Hong et al. (1988)
	Cultured explants	Mori et al. (1993)
	Sugar, ammonium:nitrate ratio, nitrogen	Mori and Sakurai (1994)
	Auxin, cytokinin	Mori et al. (1994)
Euphorbia millii	Selection	Kino–Oka et al (1994)
	Culture conditions	Hamada et al. (1994)
Malus sylvestris "Star king"	Blue light	Anon. (1989)
Perilla frutescens	Light irradiation	Zhong et al. (1991)
	Jar fermentor, 2000 lx	Koda et al. (1992)
Solanum tuberosum	Gamma irradiation, selection	Zubko et al (1993)
Vitis vinifera	Vitis species⁻	Yamakawa et al. (1983),
	Osmotic potential	Do and Cormier (1990),
	Low nitrate, high sugar	Do and Cormier (1991a)
	High ammonium	Do and Cormier (1991b)
	Osmotic stress	

growth conditions with tyrosine as precursor. The presence of different betacyanins (amaranthin, celosian, betanin, and the corresponding isoforms) associated with two betaxanthins (vulgaxanthin I and vulgaxanthin II) was also demonstrated in *C. rubrum*.

In contrast to product released from *C. rubrum* cultures (Dornenburg and Knorr, 1993; 1995), *Beta vulgaris* cultures (Kilby and Hunter, 1990; 1991), or *B. vulgaris* "hairy roots" (Dilorio *et al.,* 1993) by various permeabilization procedures (e.g., electric field pulses, ultra high pressure, sonication, and heat treatment), it was found that in adventitious roots that were induced from red beet (*B. vulgaris* L. cv. Detroit dark red) by infecting the plant with *Agrobacterium rhizogenes*, significant amounts of pigment (mainly betanin and vulgaxanthin I) were released merely by the cessation of culture shaking. Repeated batch culturing of the hairy roots was performed with the cell growth phases for 9 or 10 d and with phases of pigment leakage upto 2d after termination of shaking. More than 20% of the total intracellular accumulated pigments were recovered from the culture broth after 35d of cultivation. The released pigments were confirmed to be substantially identical to those extracted from the hairy root and the original plant cells of red beets (Taya *et al.,* 1992). Misawa *et al.* (1973) have applied for a patent for the isolation of betanin from *Phytolacca americana* callus and suspension cultures. They obtained 32 mg of crude pigment from 1 g of the dried tissue. In *Amaranthus tricolor* cultures the presence of kinetin is essential for the induction of amaranthin and isoamaranthin synthesis in darkness. Additionally, a synergistic effect of light and kinetin suggests an interaction of these two factors on *in vitro* betacyanin production (Bianco–Colomas and Hugues, 1990).

6.9.3. Carotenoids

Carotenoids and xanthophylls belong to the class of polyenes and are probably the most widely distributed group of pigments in nature. Carotenoids are found in bacteria, fungi, yeasts and plants. The advantages of carotenoids as colorants are their good pH stability, their red to yellow color, and their insensitivity to reducing agents such as ascorbic acid.

(a) β–carotene

The production of β-carotene was possible by high-density cultures of *Daucus carota* using an air-lift column. In these carrot cells the productivity was higher in the logarithmic phase of cell growth than in the stationary phase. Semicontinuous cultures that maintained the logarithmic phase was performed for continuous production of β-carotene. The production rate of 6.26 µg/d (7.5 times higher than the productivity in batch culture) was possible for 45 d under replacement of 25% of the medium per day (Matsushita *et al.,* 1994).

b) Crocin

Commercial saffron contains crocin and crocetin (colors), picrocrocin (bitter principle), and safranal (flavor). Cell cultures of *Crocus sativus* are expected to lead to the development of a biotechnological process for *in vitro* saffron production. Cell cultures could be induced to form red globular callus and red filamentous structures that were able to produce crocin and crocetin as well as picocrocin and safranal. Quantificiation of the desired metabolites showed that the crocin content in the red filamentous structures of *C. sativus* cultures (0.13 mg/g dry tissue) was two times the concentration found in red globular callus. The safranal content of the red globular callus was almost equal to that of red dry saffron stigmata. The picrocrocin level in both tissue cultures was five–to six fold higher than that in stigmata (Vishvanth *et al.,* 1990). Sarma *et al.* (1991) described a sensor analysis of tissue culture derived saffronins in which the sensory profile was ten times lower than the natural one.

   Crocin is a main secondary metabolite in matured fruits of *Gardenia jasminoides.* Callus development from ovary, skin, and stems of gardenia and subsequently crocin production has been reported (Mamera *et al.,* 1988). Indoleacetic acid seemed to be the most effective auxin for formation of pigments in *G. jasminoides* callus. A young generation of fruit callus contained substantial amounts of crocin, but the content of the pigment decreased upon subculturing. "Fresh callus" retained a relatively constant ability to produce crocin, but the crocin content was below 10% of the plant tissue (Nawa and Ohtani, 1992).

c) Lycopene

The production of carotenoids was reported in cell cultures of tomato (Fosket and Radin, 1983). Dark-grown tomato suspension cultures contain low levels of carotenoids (mainly lycopene and β-carotene). The addition of a bioregulator resulted in a 60–fold increase in carotenoid production over a cultivation time of 14 d. The major component of the increase was lycopene. In another investigation, a herbicide was used to enhance the accumulation of lycopene in cell cultures of *Daucus carota* (Nishi *et al.,* 1974).

Table 6.25. Physiological functions of natural pigments

Pigment groups	Function in plant	Value to herbivores and carnivores
Carotenoids	Photosynthesis, Pollination	Vital precursors of vitamin A which cannot be synthesized
Porphyrins	Photosynthesis	None known; hemoglobin can be synthesized
Flavonoids	Growth control, Pollination	No physiological function known
Quinones	Respiratory enzymes	Vital; includes vitamins K_1, K_2 which cannot be synthesized
Indoles	Unknown	None known; melanins can be synthesized
Flavins	Quench molecular oxygen	Vital vitamin B_2 cannot be synthesize
Betalanis	Pollination ? Virus protection ?	No physiological function known.

6.9.4. Quinones

a) Anthraquinones

A large number of *Rubiaceae* species are able to accumulate a variety of anthraquinones *in vitro*, for example, *Morinda, Rubia, Cinchona* and *Galium*. Attempts to raise anthraquinone levels produced by plant cell cultures by variation of nutritional and hormonal conditions have been remarkably successful (Bauch and Leistner, 1978; Khouri *et al.,* 1986; Schulte *et al.,* 1984; Zenk *et al.,* 1975).

In the case of *Morinda citrifolia* it has been demonstrated that the yields of anthraquinones (upto 2.5 g/l) exceeded under optimum culture conditions those of differentiated root tissues by a factor of ten on a dry weight basis (Zenk *et al.,* 1975). Stored in the cell vacuole, anthraquinones are released only to a low extent under physiological conditions.

Cell suspension cultures of *Galium vernum* are also able to produce anthraquinones. Strobel *et al.* (1990) investigated the effects of medium components and supplements such as sucrose, phosphate, growth regulators and elicitors on growth and anthraquinone production. A maximum anthraquinone production (200 mg/l) was reached by a combined increase of sucrose and phosphate concentration. Lux–Pfister *et al.* (1995) observed a growth–associated production of anthraquinones in *G. vernum* suspension cultures.

b) Naphthoquinones

The commercially most important of the quinone pigments, some of which also possess pharmacological activity, is the naphthoquinone shikonin produced by *Lithospermum erythrorhizon*. The cell cultures can be induced to synthesize high levels of shikonin derivatives by optimization of the growth medium and environmental conditions. Callus growing on an ammonium–rich medium requires indoleacetic acid rather than 2,4-dichlorophenoxyacetic acid as exogenous auxin for shikonin production (Tabata and Fujita, 1985).

6.9.5. Other Pigments

a) Blue Pigments

Cultured cells of *Lavandula vera*, which had been screened as a potent biotin producer, were found to synthesize and excrete blue pigments, whose formation was induced by the precursor L-cysteine (Watanabe *et al.,* 1985). Addition of L-cysteine to the medium was essential for the induction of pigment synthesis by the cells, but this precursor severely inhibited cell growth. Various trials led to the successful repeated use of alginate-entrapped cells for the production of the pigments. The immobilized cells were alternately incubated for growth in the absence of L-cysteine and for pigment production in the presence of the inducer. The productivity fluctuated fairly and the calcium alginate-entrapped cells were alive and active even after 7 months of incubation. However, freely suspended cell cultures of *L. vera* lost viability only after one incubation cycle (Nakajima *et al.*, 1985).

Trichotomine, an attractive pigment for the food color industry, is present at very low concentration in the fruit of *Clerodendron trichotomum*. Plant cell cultures have been used as an effective tool for producing trichotomine. The pigments isolated from plant cells were examined for heat and light stability, which varied a little from those of the commercial pigment. The heat and light stabilities were inferior to gardenia blue pigment, but could be further improved by using stabilizers (Koda *et al.*, 1992b).

b) Red Pigments

Cultured cells of *Carthamus tinctorius* released into the culture medium a red pigment that is different from the red pigment, called carthamin, produced in the petals of the intact plant (Hanagata *et al.*, 1992; Saito *et al.*, 1988). Kobayashi *et al.* (1992) determined the structure of this red pigment and named it kinobeon A. An optimum production medium for red pigment formation was developed. In this medium no cell growth occurred and pigment was released from the cell cultures (Hanagata *et al.*, 1993). The total pigment production level of *C. tinctorius* cultures was 9.4 mg/l, which was 1.3 times that of a normal batch culture (Hanagata and Karube, 1994).

6.10. FLAVOR COMPOUND PRODUCTION BY PLANT CELL AND TISSUE CULTURES

Special flavors are complex and partially characterized mixtures of aroma compounds in definite proportions. Certain flavors consists of one or few related compounds, for example, benzaldehyde (cherry), methylethylcinnamates (strawberry), methylanthranilate (grape), menthol (mint), safranal (saffron), and vanillin (vanilla). While monoterpenes are often major components of volatile oils, hydrocarbons, esters, aldehydes and alcohols are also important in fruit and vegetable flavors.

In contrast to colors, a limited number of studies have been carried out on the production of flavor compounds via plant cell cultures. Plant cell tissue culture has been attempted to produce a wide variety of fruit flavors. Dziezak (1986) and Sahai (1994) presented tables listing examples (apple, apricot, banana, blueberry, cherry, cocoa, coconut, grape, lime mango, melon, orange, peach, pineapple, raspberry and strawberry). In addition to these flavors, work has been reported on vegetable flavors (chilli pepper, celery and tomato) and other flavors (vanilla, chocolate and licorice) as well as essential oil production (Mulder–Krieger *et al.*, 1988).

6.10.1. Chilli Flavor

The production of capsaicin, the pungent flavor compound of peppers, by tissue cultures have been investigated mainly in cultures of *Capsicum frutescens* Mill cv. *annuum*. Yeoman *et al.* (1980) demonstrated that the production of capsaicin by plant cell cultures of *Capsicum* is inversely related to the culture growth, and that secondary metabolite production can be increased by manipulating culture conditions to limit primary metabolism (Lindsey, 1985).

Capsaicin from cell cultures of *C. annuum* have been reported (Veeresham *et al.*, 1994). Cultures of *C. annuum* have also been immobilized in calcium alginate beads (Ravishankar *et al.*, 1988).

6.10.2. Celery Flavor

Suspension cultures of *Apium graveolens* have been induced to accumulate terpenoids (in particular, the phthalides 3-butylphthalide and sedanenolide) and limonene, the major components of its flavor.

Cultures grown in the absence of phytohormones tend to differentiate, becoming embryoidal, with a phthalide content similar to that of the plant. Phthalides and limonen were released from the cell by a suitable temperature regime, but the growth of producing cells is very low (Watts *et al.*, 1984).

6.10.3. Citrus Flavor

The accumulation of neohesperidin and naringin as major flavonoids was observed in callus cultures of *Citrus aurantium* (bitter orange). The levels of neohesperidin were higher than those of naringin in callus cultures, as they are in immature fruit. High concentrations of both compounds were found in young tissues such as immature fruits and in the outer zone of callus tissues (del Rio *et al.*, 1992). Creswell *et al.* (1990) demonstrated that sabinene and octanol are key elements produced in callus and suspension cultures of *C. limon*. It has been found that citrus fruit callus has an altered metabolism compared to that of intact fruit tissues. In different *Citrus* species naringin and limonin production was found in callus cultures and in regenerated shoots (Barthe *et al.*, 1987). Tisserat *et al.* (1989) showed that key lemon flavonoids characteristic of tree–grown lemons (hesperidin, eriocitrin, and diosmin) were produced by differentiated cultures. Juice vesicles of lemon fruit cultivated on a solid medium were suggested as a tool for improving juice yield and quality.

6.10.4. Cocoa, Coffee and Tea Flavors

a) Cocoa Flavor

Limited success has been obtained in producing flavor components in dedifferentiated cell cultures of *Theobroma cacao*. Townsley (1974) found that Cocoa aroma can be produced by senescent cells. The production of flavour processors was not observed in the cell cultures, but the components of cocoa aroma were produced when the cultures were maintained at roasting temperature. Jalal and Collin (1977) showed that the polyphenol compounds were different and somewhat fewer in number in callus cultures than in the explant tissues.

b) Coffee Flavor

Caffeine production of *Coffea arabica* was strongly stimulated by stress such as high light intensity or high sodium chloride concentration (Frischknecht and Baumann, 1983) and inhibited by preaddition of caffeine in high concentrations (Frischknecht *et al.*, 1977). Ethylene is considered to be a powerful natural regulating substance in plant hormones. The ethylene concentration in the culture medium is usually regulated by 2-chloro-ethylene-phosphonic acid and increases caffeine and theobromine production in *C. arabica*. Cell suspension cultures of *C. arabica* with large aggregates show a higher production than cultures with finely suspended cells. The production of the purine alkaloids was increased upto 13–fold by immobilization in alginate due to the organization of cells through physiochemical interactions between the alginate and the plant cell wall (Haldimann and Brodelius, 1987).

c) Tea Flavor

Theanine (γ-glutamethylamide), an amide that enhances the taste of infused green tea, will be required because of the increasing demand for natural food additives and because of its ability to inhibit stimulation by caffeine. The accumulation of theanine in callus cultures of tea (*Camellia sinensis*) was greatly enhanced by an increase of growth rate (Matsuura and Kakuda, 1990). Studies on the effects of the concentrations of major inorganic constituents on cell growth and theanine formation led to the development of a culture medium suitable for theanine production from suspension cultures of *C. sinensis* yielding approximately 1% theanine on fresh weight basis (Matsuura *et al.,* 1992). Tea theanine production in callus cultures was greatly increased by supplementation of the culture medium with a precursor, ethylamine (Matsuura and Kakuda, 1990; Matsuura *et al.,* 1992).

6.10.5. Garlic and Onion Flavors

a) Garlic Flavor

Garlic (*Allium sativum*) cell cultures ("unorganized white" as well as "organized green" callus) revealed a total of five amino acid precursors (methyl, propyl, allyl and cysteine sulfoxides) and two unidentified ninhydrin–positive compounds. All five compounds could be hydrolyzed to pyruvic acid by alliin lyase. The total flavor substrate index in globular white callus and semi–differentiated green callus was 4 and 13% respectively, of the original explant. In contrast to the substrate levels, the specific activity of alliin lyase enzyme was half the original activity of the garlic bulb explant (Madhavi *et al.,* 1991).

b) Onion Flavor

Plant cell cultures of onion (*Allium cepa*) have been developed by a number of research groups. Selby *et al.* (1979) showed that their cultures failed to produce onion flavors but could be stimulated by feeding with immediate flavor precursors. The presence of alliinase in the cultures, but absence of the precursors, appears a common feature of many onion cultures.

Prince (1991) has examined the enhancement of onion flavor compounds in root cultures. Through the addition of the amino acids cysteine, methionine and glutathione, final product levels could be greatly enhanced. This means that earlier steps in sulfate reduction are probably limiting in the absence of precursors. Different treatments elevated different flavor components to different concentration. Incorporation of thiol intermediates also led to the production of novel flavor composites with garlic and onion flavor components (Payne *et al.,* 1991).

6.10.6. Peppermint Flavor

The characterizing flavor components of peppermint oil include menthol and menthone. Minor components include isomenthol, isomenthone, menthylacetate, pulegone, piperitone and limonene. Undifferentiated callus cultures of *Mentha piperita* lacked the ability to produce essential oils, and structural differentiation was needed to induce product synthesis (Lin and Staba, 1961; Becker, 1970). The addition of colchicine to *M. piperita* cultures was effective in increasing yields of mint oil constituents. This was closely linked to an increase in the number of neoformed secretory glands in the callus material (Bricout *et al.,* 1978). These are the cells in which the cytotoxic substances were accumulated. Metabolism of secondary products seems to be correlated with organized cell structures.

6.10.7. Vanilla Flavor

Vanillin, 4-hydroxybenzaldehyde and 4–hydroxybenzyl methylether are the most abundant constituents of the vanilla flavor. These C_6-C_1 compounds are formed as products of the phenylpropanoid pathway and stored as glucosides within the vanilla pods.

Knuth and Sahai (1991a) showed that a hormone mix containing 2,4-D and benzyl adenine was necessary and useful in initiation of callus production and growth in *Vanilla fragrans* and other species. The influence of growth regulators on cell growth and formation of phenolic substances has also been studied by Funk and Brodelius (1990a). The phytohormones had a drastic influence on phenylpropanoid metabolism. 2,4-D suppressed secondary metabolism in this suspension culture. When naphthylacetic acid was substituted for 2,4-D, a significant increase in total extractable phenolics was observed, and combination with cytokinins resulted in production of relatively large amounts of extractable phenolics (mainly coumaric acid and sinapic acid). Under normal growth conditions the cultured cells did not produce any detectable benzoate derivatives (C_6-C_1 compounds). Kinetin was used to initiate vanillic acid synthesis in cell suspension cultures of *V. planifolia*.

Conditioning factors, that is, substances, released from living plant cells, seem to have a regulatory function. Paulmann *et al.* (1990) used "conditioned media" for culturing *V. planifolia* callus and detected a two fold increase in vanillin production of upto 15 µg vanillin per gram fresh weight. Feeding *V. planifolia* callus with 1 mM ferulic acid, a probable precursor of vanillin, resulted in an increase in vanillin concentration by a factor of 1.7 compared with an untreated callus, but precursor addition did not improve the ratio of the flavor components tested (Romagnoli and Knorr, 1988).

Knuth and Sahai (1991a) found that the nature and concentration of vanilla component precursors added to the culture medium was a factor influencing flavor production in *V. fragrans* cultures. Phenylalanine and ferulic acid resulted in little enhancement of vanillin production, whereas addition of vanillyl alcohol resulted in a significant increase in vanillin output.

Fine suspension cultures of *V. planifolia* have also been used to study the effects of precursor feeding as well as metabolic inhibitors on phenylpropanoid metabolism. Funk and Brodelius (1990b) postulated that cinnamic acid, and not ferulic acid, was a precursor of vanillic acid. In this case, the branching point of the pathway leading to vanillic acid must be located before ferulic acid in the general phenylpropanoid pathway.

Knorr *et al.* (1993) demonstrated a non-linear pathway in *V. planifolia* cluster cultures. A more complex pathways for the formation of vanillin and its precursors was postulated. The effect of different light sources on the production of vanillin and precursors was examined, and vanillin could be detected after 5 and 14d of incubation, with the greatest increase of accumulation in blue light (Havkin–Frenkel *et al.,* 1991).

In *V. fragrans* cultures, which were able to secrete the flavor components into the culture medium, vanillin production was improved by adsorption onto activated charcoal (Knuth and Sahai, 1991b). The vanillin yield obtained was 2.2% on dry weight basis, comparable to the 1 to 3% vanillin content in cured vanilla beans. Subsequent optimization improved the cell doubling time from 106 to 50h and the yield of vanillin from 100 to over 1000 mg/l (8% on dry weight basis; Sahai, 1994).

A process for producing natural vanillin from a ferulic acid precursor with aerial roots and charcoal as a product reservoir has been developed by Westcott *et al.* (1994). Organized root tissues from *V. planifolia* plants were shown to be active biocatalysts transforming ferulic acid to vanillin in the presence of charcoal. Productivity of 400 mg/kg tissue per day amounts to 40% of the concentration found in vanilla beans, producing a final product with a ratio of *p*–hydroxybenzaldehyde (the second most important component of vanilla flavor) to vanillin of 7.8 : 1, similar to that found in vanilla beans (12.8:1).

6.11. PRODUCTION OF SWEETENERS IN PLANT CELL CULTURES

There is a need to find non-nutritive and safe sweeteners for various food and pharmaceutical applications. The important non-nutritive sweeteners are stevioside from *Stevia rebaudiana*, thaumatin from *Thaumatococcus danielli,* monellin from *Dioscoreophyllum cumminsii* and miraculin from fruits of *Synsepalum dulcificum*. These sweeteners are potential products of plant cell cultures (Sahai and Knuth, 1985). Production of the glucoside stevioside from *Stevia rebaudiana* cell cultures has been reported, while products such as thaumatin, miraculin, and monellin, which are proteins are also targets for genetic engineering and could be made less expensively by using microorganisms.

6.11.1. Glycyrrhizin

The roots of *Glycyrrhiza glabra* contain the natural sweetener glycyrrhizin, which is widely used in industry and is regarded as being about 70 to 93 times sweeter than sucrose, depending on the concentration. Tamaki *et al.* (1973) have reported glycyrrhizin production by callus and suspension cultures of *Glycyrrhiza* species in substantial quantities at levels approaching 3 to 4% of dry weight. Hayashi *et al.* (1988) were unable to produce detectable amounts of glycyrrhizin in callus and cell suspensions of *G. glabra*. A suspension culture has been developed from seeds of *G. glabra* that were grown in an air–lift bioreactor with a volume of up to 80 litres. Growth of the cell culture was some what reduced in this reactor and "reasonable" levels of glycyrrhizin were obtained in the cultures (Scragg and Arias–Castro, 1992).

6.11.2. Hernandulcin

Hernandulcin, the sweetening agent from the flowers and leaves of the plant *Lippia dulcis*, is a sesquiterpenoid oil that exceeds the sweetening power of saccharose 1000-fold. The sweet sesquiterpene hernandulcin (0.25 mg/g dry weight) together with 20 other mono- and sesquiterpenes could be produced by green hairy root cultures of *L. dulcis*, where no terpenes were detected in nontransformed culture. The growth and production were influenced by the addition of auxins to the culture medium. Hernandulcin production could be enhanced fivefold in *L. dulcis* root cultures by the addition of chitosan as an elicitor to the culture medium (Sauerwein *et al.*, 1991).

6.11.3. Steviosides

The possibility of obtaining the sweet compounds (100 to 400 times sweeter than sucrose) from *Stevia rebaudiana* directly in culture was investigated in callus and cell suspensions. Komatsu *et al.* (1976) patented a process to produce stevioside from a callus cultured in a basal medium with indoleacetic acid (10^{-6} M). Kotani (1980) developed another process, a biotransformation of steviol into stevioside using calli cultured in a basal medium containing indoleacetic acid (1 mg/l), kinetin (0.1 mg/l) and steviol as a precursor at a concentration of 5 mg/l medium.

The occurrence of stevioside and rebaudioside A was observed in callus cultivated in a medium supplemented with naphthalene acetic acid (2 mg/l) and kinetin (2 mg/l). The stevioside content in the cells varied with the age of the calli. A maximum concentration of 36.4% of dry weight was attained after 38 d of incubation, which is four times higher than the stevioside content in *Stevia* leaves (Handro and Ferreira, 1989).

REFERENCES

Allison, A.J., Butcher, D.N., Connolly, J.D., Overton, K.H. (1968). Paniculides A, B and C Bisabolenoid lactones from tissue cultures of *Andrographis paniculata*. *Chem. Commun.,* **23** : 1493.

Amann, M., Wanner, G. and Zenk, M.H. (1986). Intracellular compartmentation of two enzymes of berberine biosynthesis in plant cell cultures. *Planta,* **167(3)** : 310.

Ambib, C. and Fallot, J. (1981). Role of gaseous environment on volatile compound production by fruit cell suspension cultures *in vitro*. In : *Flavour,* 81, pp. 529–538. Schreier, P., (eds). Walter de Gruyter, Berlin.

Ambib, C. and Fallot, J. 1981. Role of gaseous environment on volatile compound production by fruit cell suspension cultures *in vitro*. In : *Flavour* '81. Pp. 529–538. Schreier, P., Ed., Walter de Gruyter, Berlin.

Anon (1989). Mass production of red pigment. Takeda Chemical Industries, Japan. *Food Manuf. Int.,* **7/8** : 9.

Anon. 1989. Mass production of red pigment. Takeda Chemical industries, Japan. *Food Manuf. Int.,* **7/8** : 9.

Aredenne, R., Von (1965). Bestimmung der Natur der Anthocyane in Gewbekulturen von *Haplopappus gracilits*. *Z. Naturforsch.,* **20b** : 186–187.

Atul R. Mehta and Staba, E. J (1970). Presence of Diosgenin in Tissue–Cultures of *Discorea composita* Hemsl. and Related Species. *J. Pharm. Sci.*, **59(6):** 864-865.

Austin, D.J. and Brown, S.A. (1973). Furanocoumarin biosynthesis in *Ruta graveolens* cell cultures. *Phytochemistry,* 12 : 1657–1669.

Bailey, J.A. (1970). Pisatin production by tissue culture of *Pisum sativum* L. *J. Gen. Microbiol.,* 61 : 409–415.

Balandrin, M.F., Kloche, J.A., Wurtele, E.S. and Bollinger, W.H. (1985). Natural plant chemicals : sources of industrial and medicinal material. *Science,* **228** : 1154.

Ballica, R. and Ryu, D.D.Y. (1994). Tropane alkaloid production from *Datura stramonium* : An integrated approach to bioprocess optimization of plant cultivation, in : *Advances in Plant Biotechnology. Studies in Plant Sciences,* vol. 4 (Ryu, D.D.Y., Furasaki, S. eds.), pp. 221–254. Amsterdam : Elsevier.

Banerjee, S., Upadhyay,N., Kukreja, A.K. and Ahuja, P.S. (1996). Taxanes from *in–vitro* cultures of the Himalayan yew *Taxus wallichiana. Planta Med.,* **62:** 333–335.

Banthorpe, D., Branch, S., Njar, V., Osborne, M. and Watson, D. (1986). Ability of plant callus cultures to synthesize and accumulate lower terpenoids. *Phytochemistry,* 25 : 629–636.

Barthe, G.A., Jourdan, P.S., McIntosh, C.A. and Mansell, R.L. 1987. Naringin and limonin production in callus cultures and regenerated shoots from *Citrus* sp. *J. Plant Physiol.,* 127 : 55–65.

Barz, W. (1975). Abbau von Aromastoffen und heterozyklischen Pflanzenihaltsstoften durch Zellsuspensionskulture. *Planta Med. Suupl. Thieme Verlag, Stuttgart,* p. 128.

Barz, W., Beimer, A., Drager, B., Jaques, U., Otto, Ch., Super, E., Upmeier, B. (1990). Turnover and storage of secondary products in cell culture. In : Charlwood, B.A. and Rhodes, M.J.C. (eds.), Secondary Products from plant tissue culture. *Proc. Of the Phytochemical Socity of Europe* 30. Oxford Science Publication, Calderon Press, Oxford, p. 79.

Bauch, H.J. and Leistner, E. (1978). Aromatic metabolites in cell suspension cultures of *Galium mollugo. Plant Med.,* 33 : 105–127.

Becker, H. and Chavadej, S. (1985). Valeporiate production of Normal and Colchicine treated cell suspension cultures of *Valeriana wallichii. J. Nat. Products,* 48(1) : 17–21.

Becker, U.H. 1970. Untersuchungen zur Frage der Bildung fluchtiger Stoffwechselprodukte in Callus Kulturen. *Biochem. Physiol. Pflanzen.,* **161** : 425–441.

Benjamin, B.D., Mulchandani, N.B. (1973). Studies in biosynthesis of secondary constituents in tissue cultures of *Tylophora indica. Planta Med.,* **23** : 394–397.

Benveniste, P. (1968). La biosynthese des sterols dans les tissus de tabac cultives, *in vitro.* Mise en evidence du cycloeucalenol et del'obtusifoliol. *Phytochemistry,* **7** : 951–953.

Benveniste, P., Hirth, L., Ourisson, G. (1966). La biosynthese des phytosterols des tissus de tabac cultive *in vitro.* II. Particularities de la biosynthese des phytosterols des tissus de tabac cultives *in vitro. Phytochemistry,* **5** : 45–58.

Berlin, J. (1986). Secondary products from plant cell cultures, in : *Biotechnology,* vol. 4, Ist edn. (Rehm, H.J., Reed, G. eds), pp. 630–658.

Berlin, J. (1990). Screening and selection for variant cell lines with increased levels of secondary metabolites, in: Secondary Products from Plant Tissue Culture (*Charlwood, B.V., Rhodes, M.J.C.* eds.), pp. 119–137, Oxford: Clarendon.

Berlin, J. and Sasse, F. (1985). Selection and screening techniques for plant cell cultures. *Adv. Biochem. Eng.,* **31:** 99–132.

Berlin, J., Barz, W. (1971). Stoffwechsel von Isoflavonen und Cumostanen in Zell– un Callussus–pensionskulture von *Phaseolus aureus* Roxb. *Planta,* **98** : 300–314 (1971).

Berlin, J., Bedorf, N., Mollenschott, C. Wray, V., Sasse, F. and Hofle, G. (1988b). On the podophyllotoxins of root cultures of *Linum flavum. Planta Med.,* **54** : 204–206.

Berlin, J., Beier, H., Fecker, F., Forche, E., Noe, W., Sasse, F., Schiel, O. and Wray, V. (1985). Conventional and new approaches to increase alkaloid production of plant cell cultures. In : *Primary and Secondary Metabolism of Plant Cell Cultures* (Neumann, K.H., Barz, W., Reinhard, E., eds). pp. 272–280, Berlin : Springer–Berlin.

Berlin, J., Forche, E., Wray, V., Hammer, J. and Hosel, W. (1982). Biochemical characterization of two tobacco cell lines with high and low yields of cinamoyl putrescines. *J. Nat. Prod.,* **45** : 83–87.

Berlin, J., Forche, E., Wray, V., Hammer, J. and Hosel, W. (1983). Formation of benzophenanthridine alkaloids by suspension cultures of *Eschscholtzia californica Z. Naturforsch.,* **38c** : 346–352.

Berlin, J., Mollenschott, C., Wray, V. (1988a). Triggered efflux of protoberberine alkaloids from cell suspension cultures of *Thalictrum rugosum. Biotechnol Lett.,* **10** : 193–198.

Berlin, J., Sieg, S., Strack, D., Bokern, M., Harms, H. (1986). Production of betalains by suspension cultures of *Chenopodium rubrum. Plant Cell Tissue Organ Cult.,* **5** : 163–174.

Berlin, J., Witte, L., Schubert, W. and Wray, V. (1984). Determination and quantification of monoterpenoids secreted into the medium of cell cultures of *Thuja occidentalis. Phytochemistry,* **23** : 1277–1279.

Bianco–Colomas, J. and Hugues, M. (1990). Establishment and characterization of a betacynin producing cell line of *Amaranthus tricol :* Inductive effects of light and cytokinins. *J. Plant Physiol.,* **136** : 734–739.

Boetner, F.E., Mattes, T.D., Kopec, L., Sheng, M. and Halpern, B.D. (1977). Isolation of ten grams of pure taxol from AJ–8 concentrate preparation Rep. 2. In: NSc 125973.

Bohm, H. and Rink, E. (1988). Betalains, In : *Cell Culture and Somatic Cell Genetics of Plants,* vol. 5 : *Phytochemicals in Plant Cell Cultures* (Constabel, F., Vasil, I.K. eds.), pp. 449–463. San Diego : Academic Press.

Breuling, M., Alfermann, A.W., Reinhard, E. (1985). Cultivation of cell cultures of *Berberis wilsonae* in 20–L airlift bioreactors. *Plant Cell Rep.,* **4** : 220–223.

Brian, K.R. (1974). Accumulation of L–DOPA in cultures from *Mucuna pruriens. 3rd Intern. Congr. Plant Tissue and Cell Culture,* Abstract, 73, Leicester, Univ. Leicester.

Bricout, M.J., Garcia–Rodriguez, M.J. and Paupardin, C. (1978). Action de la colchicine sur la synthese de Phule essentielle par de tissus de *Mentha piperita* L. cultivees *in vitro C.R. Hebd. Seances Acad. Sci. Ser. D* **286** : 1585–1588.

Bringi, V., Kodkade, P. (1993). Enhanced production of taxol and taxanes by cell cultures of *Taxus* species. WO 93/17121.

Brocke, W., Von, Reinharde, E., Nicholson, G., Konig, W.A. (1971). Uber das Vorkommen von 3–(1',1'–dimethylallyl)–scopletin in Gewebekulturen von *Ruta graveolens. Z. Naturforsch.,* **266** : 1252–1255.

Brown, S.A., Tenniswood, M. (1974). Aberrant coumarin metabolism in crown gall tissues of tobacco. *Can. J. Botany,* **52** : 1091–1094 (1974).

Brunet, G., Ibrahim, R.K. (1973). Tissue culture of *Citrus* peel and its potential for flavonoid synthesis. *Z. Pflanzenphysiol.,* **69** : 152–162 (1973).

Butcher, D.N., Connolly, J.D. (1971). An investigation of factors which influence the production of abnormal terpenoids by callus cultures of *Andrographis paniculata. Nees. J. Exp. Botany,* **22** : 314–322 (1971).

Butcher, D.N., Phillips, R., Powell, R.G., Sogeke, A. (1974). Lipid components of membranes from normal and tumour tissues. 3rd Intern. Congr. Plant Tissue and Cell Culture. Abstr. 170, Leicester : Univ. Leicester.

Carew, D.P. (1965). Reserpine in a tissue culture of *Alstonia constricta* F. Muell. *Nature,* **207** : 89.

Carew, D.P. (1975). Tissue culture studies of *Catharanthus roseus.* In : *The Catharanthus Alkaloids* (Taylor, W.I., Fransworth, N.R. Eds.), pp. 193–208, New York : Marcel Dekker.

Chan, W.M. and Staba, E.J. (1965). Alkaloid production by Datura callus and suspension tissue cultures. *J. Nat. Products,* **28** : 55–62.

Charlwood, B.V. and Moustou, C. (1998). Essential oil accumulation in shoot proliferation cultures of *Pelargonium* sp. in : Manipulating secondary metabolism in culture (Robins, R.J. and Rhodes, M.J.C. ed.), Cambridge University Press, Cambridge, p. 187–194.

Charlwood, K.A., Brown, S. and Charlwood, B.V. (1988). The accumulation of flavour compounds by cultures of *Zingiber officinale.* In : *Manipulating Secondary Metabolism in Culture.* PP. 195–200. Robins, R.J. and Rhodes, M.J.C., (eds)., Cambridge University Press, Cambridge.

Cheetham, P.S.J. (1993). The use of biotransformations for the production of flavours and frangrances. *TIBTECH,* **11** : 478–488.

Chen, M., Stons, S.J., Staba, E.J. (1969). The biosynthesis of Visnagin from 2 [^{14}C] acetate by *Ammi visnaga* suspension cultures and the metabolism of [^{14}C]–visnagin and ^{14}C Khellin by *A. visnaga* and *A. majus*. *J. Nat. Products,* **32** : 339–346.

Cheng, K., Fang, W., Yang, Y., Xu, H., Meng, C., He, W. and Fang, Q. (1996) ^{14}C–oxygenated taxanes from *Taxus yunnanensis* cell cultures. *Phytochemistry,* **42**: 73–75.

Cho, G.H., Kim, D.J., Pederse, H. and Chin, C.K. (1988). Etephen enhancement of secondary metabolite synthesis in plant cell cultures. *Biotechnol. Prog.,* **4(3)** : 184.

Ciddi, V. and Shuler, M.L. (2000). Camptothecine from callus cultures of *Nothapodytes foetida. Biotechnol Lett,* **22**: 129–132.

Cline, S.D. and Coscia, C.J. (1988). Stimulation of sanguinarine production by combined fungal elcitation and hormonal deprivation in cell suspension cultures of *Papaver bracteatum. Plant Physio.,* **86** : 161–165.

Corduan, G., Reinhard, E. (1972). Synthesis of volatile oils in tissue cultures of *Ruta graveolens. Phytochemistry,* **11** : 917–922.

Cragg, G.M., Schepartz, S.A., Suffness, M. and Grever, M.R. (1993). The taxol supply crisis. New NCI policies for handling the large–scale production of novel anticancer and anti–HIV agents. *J. Nat. Prod.,* **56** : 1657–1668.

Cresswell, R. 1990. The production of flavor components from citrus tissue cultures. Abstract from the 7th International Congress of Plant Cell and Tissue Culture. International Association for Plant Cell Culture, Amsterdam.

Curtis, W.R. (1993). Cultivation of roots in bioreactors. *Curr. Opin. Biotechnol.,* **4** : 205–210.

Czygan, F.C. (1975). Moglichkeiten zur production von Arzneipflazen durch pflanziche. Gewebekulturen. *Planta. Med. Suppl.,* 169.

Czygan, F.C. (ed) (1984). Biogene Arzneistoffe Vieweg Braunschweig.

Davies, M.E. (1972a). Polyphenol synthesis in cell suspension cultures of Paul;s Scarlet Rose. *Planta,* **104** : 50–65.

DeCarolis, E. and De Luca, V. (1993). Purifiation characterization and kinetic analysis of a 2–oxo–glutarate dependent dioxygenase involved in vindoline biosynthetis from *Catharanthus roseus. J Biol Chem* **268**: 5504–5511.

De–Eknamkul, W. and Ellis, B.E. (1984). Rosmarinic acid production and growth characteristics of *Anchusa officinalis* cell suspension cultures. *Planta Med,* **51**: 346–350.

De–Eknamkul, W. and Ellis, B.E. (1985a). Effects of auxins and cytokinins on growth and rosmarinic acid formation in cell suspension cultures of *Anchusa officinalis. Plant Cell Rep.* **4** : 50–53.

De–Eknamkul, W. and Ellis, B.E. (1985b). Effects of macronutrients on growth and rosmanic acid formation in cell suspension cultures of *Anchusa officinalis. Plant Cell Rep.,* **4** : 46–49.

del Rio, J.A., Ortuna, A., Marin, F.R., Garcia Puig, D. and Sabater, F. 1992. Bioproduction of neosperidin and naringin in callus cultures of *Citrus aurantium. Plant Cell Rep.,* **11** : 592–596.

Delfel, N.E. and Rothfus, J.A. (1977). Antitumor alkaloids in callus cultures of *Cephalotaxus harningtonia. Phytochemistry,* **16** : 1595–1598.

Dell, B., Elsegood, C.L., Ghisalberti, E.L. (1989). Production of verbascoside in callus tissue of *Eremophila* spp., *Phytochemistry,* **28** : 1871–1872.

Deno, H., Suga, C., Morimoto, T., Fujita, Y. (1987). Production of shikonin derivatives by cell suspension cultures of *Lithospermum erythrorhizon.* VI. Production of shikonin derivatives by a two layer culture containing an organic solvent. *Plant Cell Rep.,* **6** : 197–199.

Deus, B. and Zenk, M.H. (1982). Exploitation of plant cells for the production of natural compounds. *Biotechnol. Bioeng.,* **24** : 1965–1974.

Deus–Neumann, B. and Zenk, M.H. (1984). Instability of alkaloid production in *Catharanthus roseus* cell suspension cultures. *Plant Med.,* **50** : 427–431.

Dicosmo, F. (1990). Strategies to improve yields of secondary metabolites to industrially interesting levels. In : *Progress in Plant Cellular and Molecular Biology* (Nijkamp, H.J.J., Van Der Plas, L.H.W., Van Aartrijk, J. eds), pp. 717–724, Dordrecht : Kluwer.

Dilorio, A.A., Weathers, P.J. and Cheetham, R.D. (1993). Non–lethal secondary product release from transformed root cultures of *Beta vulgaris. Appl. Microbiol. Biotechnol.,* **39** : 174–180.

Do, C.B. and Cormier, F. (1990). Accumulation of anthocyanins enhanced by a high osmotic potential in grape (*Vitis vinifera* L.) cell suspensions. *Plant Cell Rep.,* **9** : 143–146.

Do, C.B. and Cormier, F. (1991). Effects of high ammonium concentrations on growth and anthocyanin formation in grape (*Vitis vinifera* L.) cell suspension cultured in a production medium. *Plant Cell Tissue Organ Cult.,* **27**: 169–174.

Do, C.B. and Cormier, F. (1991a). Effects of low nitrate and high sugar concentrations on anthocyanin content and composition of grape (*Vitis vinifera* L.) cell suspension. *Plant Cell Rep.,* **9** : 500–504.

Do, C.B. and Cormier, F. (1991b). Effects of high ammonium concentrations on growth and anthocyanin formation in grape (*Vitis vinifera* L.) cell suspension cultured in a production medium. *Plant Cell Tissue Organ Cult.,* **27**: 169–173.

Do, C.B. and cormier, F. (1991c). Accumulation of peonidin–3–glucoside enhanced by osmotic stress in grape (*Vitis vinifera* L.) cell suspension. *Plant Cell Tissue Organ Cult.,* **24** : 49–54.

Dobberrstein, R.H., Staba, E. (1969). Ipomoea, Rivea and Argyeia Tissue cultures : Influence of various chemical factors on Indole alkaloid production and growth. *J. Nat. Product.,* **32** : 141–146.

Dornenburg, H. and Knorr, D. (1993). Cellular permeabilization of cultured plant tissues by electric field pulses and high pressure for the recovery of secondary metabolites. *Food Biotechnol.,* **7** : 35–48.

Dornenburg, H. and Knorr, D. (1995). Strategies for the improvement of secondary metabolite production in plant cell cultures. *Enzyme Microb. Technol.,* **17** : 674–684.

Dornenburg, H. and Knorr, D. (1996). Generation of colours and flavours in plant cell and tissue cultures. Critical Reviews in Plant Science, 15(2) : 141-168.

Dornenburg, H., Wermann, U. and Knorr, D. (1990). Improvement of secondary metabolite production by precursor and elicitor addition to cell cultures of *Anethum graveolens. Proc. Int Conf. Biotechnol. Food,* February 20–24, 1989, Poster Presentation, *Food Biotechnol,* **4** : 477.

Dougall, D.K. (1980). The use of tissue cultures in studies of metabolism. In: The Biochemistry of Plants. Vol 2. Davies, D.D. (ed) Academic Press, New York. Pp. 627–542.

Dougall, D.K. (1989). Sinapic acid stimulator of anthocyanin accumulation in carrot cell cultures. *Plant Sci.,* **60** : 259–262.

Dougall, D.K. and Vogelien, D.L. (1990). Anthocyanin yields of clonal wild carrot cell cultures. Effects of serial cloning of clonal wild carrot cell cultures. *Plant Cell Tissue Organ Cult.,* **23** : 79–91.

Douros, J., Suffness, M., Chiuten, D. and Adamson, R. (1979). Basis for cancer therapy I. In: Fox, B.W. (ed) Advances in medical oncology, research and education. Pergamon, Oxford New York pp. 59–73.

Dufresne, C., Cormier, F. and Dorion, S. (1997). *In–vitro* formation of Crocetin Glucosyl esters by *Crocus sativus. Planta Medica,* **63** : 150–153.

Dziezak, J.D. (1986). Biotechnology and flavor development : plant tissue culture. *Food Technol.,* **40** : 122–129.

Edgington, S.M. (1991). *Taxol*–out of the woods. *Bio/Technology,* **9** : 933–938.

Eilert, U., Kurz, W.G.W. and Costabel, F. (1987). Ultrastructure of *Catharanthus roseus* cells cultured *in vitro* and exposed to conditions for alkaloid accumulation. *Protoplasma,* **140** : 157.

Eilert, U., Kurz, W.G.W., Constable, F. (1985). Stimulation of sanguinarine accumulation in *Papaver somniferum* cells by fungal homogenates. *J. Plant Physiol.,* **119** : 65–76.

Eisenreich, W., Menhard, B., Hylands, P.J., Zenk, M.H. and Bacher, A. (1996). Studies of the biosynthesis of taxol : the taxane carbon skeleton is not of mevalonoid origin. *Proc. Natl. Acad. Sci. Usa,* **93** : 6431–6436.

Ellis, B.E. (1983). Production of hydroxyphenylethanol glycosides in suspension cultures of *Syringa vulgaris. Phytochemistry,* **22** : 1941–1943.

Ellis, B.E. (1985). Metabolism of caffeoyl derivatives in plant cell cultures in : Pharmacy and Secondary Metabolism of Plant Cell Cultures (Neumann, K.H., Barz, W., Reinhard, E., eds), pp. 164–173, Berlin : Springer–Verlag.

Ellis, B.E. (1988). Natural products from plant tissue culture. *Nat. Prod. Rep.,* **5** : 581–612.

El–Tigani, S. (1972). The role of plant hormones in the club root disease of *Brassica rapa* L. Ph.D. Thesis, Univ. Cambridge.

Endress, B.R., Jager, A. and Kreis, W. (1984). Catecholamine biosynthesis dependent on the dark in betacyanin forming *Partulaca* callus. *J. Plant Physiol.,* **115** : 291.

Endress, R. (1994). In: Plant Cell Biotechnology. Springer-Verlag, Berlin, Heidelberg, New York, pp. 121-140.

Enevoldsen, K. (1994). *Ribes nigrum* L. (Black–currant) : *In vitro* culture and the production of flavor compounds. In : *Biotechnology in Agriculture and Forestry,* vol. 26, *medicinal and Aromatic Plants VI,* pp. 327–338. Bajaj, Y.P.S., Ed., Springer–Verlag, Berlin.

Fett–Neto, A.G., Dicosmo, F., Reynolds, W.F. and Sakata, K. (1992). Cell culture of *Taxus* as a source of the antineoplastic drug taxol and related taxanes. *Bio/Technology,* **10:** 1572–1575.

Fett–Neto, A.G., Melanson, S.J., Sakata, K. and Dicosmo, F. (1993). Improved growth and taxol yield in developing calli of *Taxus cuspidata* by medium composition modification. *Bio/Technology,* **11** : 731–734.

Fett–Neto, A.G., Zhang, W.Y., Dicosmo, F. (1994a). Kinetics of *taxol* production, growth and nutrient uptake in cell suspensions of *Taxus cuspidata. Biotechnol. Bioeng.,* **44** : 205–210.

Fett–Neto, A.G., Melanson, S.J., Nicholson, S.A. Pennington,J.J and Dicosmo, F. (1994b). Improved taxol yield by aromatic and amino acid feeding to cell cultures *Taxus cuspidata. Biotechnol. Bioeng.,* **44** : 967–971.

Fiky, E.K.F., Remmel, R.P. and Staba, E.J. (1989). *Ammi Visnaga* : Somatic Embryo induction and Furanochrome production in embryo, seedlings and plants. *Planta Medica,* **55** : 446–451.

Flores, T., Wagner, L.J. and Flores, H.E. (1993). Embryo culture and taxane production in *Taxus* spp., *In Vitro Cell Dev. Biol.,* **29P:** 160–165.

Forrest, G.I. (1969). Studies on the polyphenol metabolism of tissue cultures derived from the tea plant (*Camellia sinensis* L.). *Biochem. J.,* **113** : 765–772.

Fosket, D. and Radin, D. (1983). Induction of carotenogenesis in cultured cells of *Lycopersicon esculentum. Plant Sci. Lett.,* **30** : 165–175.

Friedrich, H., Baier, S. (1973). Anthracene–derivative in Kalluskulturen aus *Cassia angustifolia. Phytochemistry,* **12** : 1456–1462.

Frischknecht, P.M. and Baumann, T.W. (1983). Stress induced formation of purine alkaloids in plant tissue culture of *Coffea arabica. Phytochemistry,* **24** : 2255–2257.

Fujita, Y. (1988). Shikonin : Production by (*Lithospermum erythrorhizon*) cell cultures, in *Biotechnology in Agriculture and Forestry,* vol. 4 (Bajaj, Y.P.S., Ed.), pp. 225–236, Berlin : Springer–Verlag.

Fujita, Y., Hara, Y. (1985). The effective production of shikonin by cultures with an increased cell population. *Agric. Biol. Chem.,* **49** : 2071–2075.

Fujita, Y., Hara, Y., Morimoto, T. and Misawa, M. (1990). Semisynthetic production of vinblastine involving cell cultures of *Catharanthus roseus,* In :*Progress in Plant Cellular and Molecular Biology.* Nijkamp, H.J.J., Van Der Plas, L.H.W., VanAartrijk, J. (eds.), pp. 763–768, Dordrecht : Kluwer.

Fujita, Y., Tabata, M., Nishi, A., Yamada, Y. (1982). New medium and production of secondary compounds with the two stage culture methods, In : *Plant Tissue Culture 1982* Fujiwara, A., (eds.), pp. 399–400, Tokyo : Maruzen Press.

Fujita, Y., Takahashi, S. and Yamada, Y. (1985). Selection of cell lines with high productivity of Shikonin derivatives by protoplast culture of *Lithospermum erythrorhizon* cells. *Agric. Biol. Chem.,* **49** : 1755–1759.

Fukui, H., Nagakawa, K., Tsuda, S. and Tabata, M. (1982). Production of isoquinoline alkaloids by cell suspension cultures of *Coptis japonica,* in : *Plant Cell Culture 1982* Fujiwara, A., (eds.), pp. 313–314, Tokyo : Maruzen Press.

Fukui, H., Tsukada, M., Mizukami, H., Tabata, M. (1983). Formation of stereoisomeric mixtures of naphthoquinone derivatives in *Echiumlycopsis* callus cultures. *Phytochemistry,* **22** : 453–456.

Funk, C. and Brodelius, P. (1990a). Influence of growth regulators and elicitor on phenylpropanoid metabolism in suspension cultures of *Vanilla planifolia*. *Phytochemistry,* **29** : 845–848.

Funk, C. and Brodelius, P. (1992). Phenylpropanoid metabolism in suspension cultures of *Vanilla planifolia*. IV. Induction of vanillic acid formation. *Plant Physiol.,* **99** : 256–262.

Funk, C. and Brodelius, P.E. (1990b). Phenyl propanoid metabolism in suspension cultures of *Vanilla planifolia Andr.* II. Effects of precursor feeding and metabolic inhibitors. *Plant Physiol.,* **94** : 95–101.

Furmanova, M., Glowniak, K., Barnek, K.S., Zgorka, G. and Jozefezyk, A. (1997). Efect of picloram and methyl jasmonate on growth and taxane accumulation in callus cultures of *Taxus media* var. Hatfieldii. *Plant Cell Tiss Org Cult.,* **49**: 75–79.

Furmanova, M., Glowniak, K., Zobel, A., Guzewska, J., Zgorka, G., Rapczewska, L., and Jozefezyk, A. (1995) Taxol in *Taxus baccata* L. var. elegantissima organs and in tissue culture. *Med Fac Landbouww Univ Gent.,* **60**: 2215–2118.

Furuya, T. (1984). Recent advances in plant tissue culture, *Oil Chem,* 33:666–671.

Furuya, T. (1988). Saponins (Ginseng saponins), In : *Cell Culture and Somatic Cell Genetics of Plants,* vol. 5. *Phytochemicals in Plant Cell Cultures.* Constabel, F., Vasil, I. K. (eds). PP. 213–236. San Diego: Academic Press.

Furuya, T. and Ishii, T. (1988). The manufacturing of panax plant tissue culture containing crude saponins and crude sapogenins which are identifical with those of natural panax roots. Jap. Patent No. 48–31917.

Furuya, T., Ikuta, A. and Syono, K. (1972). Alkaloids from callus tissue of *Papaver somniferum. Phytochemistry,* **11** : 3041–3044.

Furuya, T., Kojima, H. (1971). 4–Hydroxydigitolutein, a new anthraquinone from callus tissues of *Digitalis lanata. Phytochemistry,* **10** : 1607–16111.

Furuya, T., Syono, K., Ikuta, A. (1972). Isolation of berberine from callus tissue of *Coptis japonica. Phytochemistry,* **11** : 175.

Gamborg, O.L. (1967). Aromatic metabolism in plants V. The biosynthesis of chlorogenic acid and lignin in potato cell cultures. *Can. J. Biochem.,* **45**: 1451–1457.

Garve, R., Luckner, M., Vogel, E., Tewes, A. and Nover, L. (1980). Growth morphogenesis and cardenolide formation in long–time cultures of *Digitalis lanata. Planta Med.,* **40** : 92.

George,J., and Ravishankar, G.A., (1996). Harnesing high value metabolites from plant cells, *Ind.J.Pharm.Edn.* **30**:120–130.

Gertlowski, C. and Petersen, M. (1993). Influence of the carbon source on growth and rosmarinic acid production in suspension cultures of *Coleus blumei. Plant Cell Tissue Organ Cult.,* **34** : 183–190.

Gibson, D.M., Ketchum, R.E.B., Vance, N.C. and Christen, A.A. (1993) Initiation and growth of cell lines of *Taxus brevifolia* (Pacific Yew). *Plant Cell Rep.,* **12**: 479–482.

Girod, P.A. and Zyrd, J.P. (1991). Secondary metabolism in cultured red beet (*beta vulgaris* L.) cells. Differential regulation of betaxanthins and betacyanin biosynthesis. *Plant Cell Tissue Organ Cult.,* **25** : 1–12.

Gittermann,A (1980) Univ.Microfilms Int. 8013629. North carolina, USA. PP 1–115.

Gontier, E., Sangwan, B.S. and Barbotin, J.N. (1994). Effects of calcium, alginate and calcium alginate immobilization on growth and tropane alkaloid levels of a stable suspension cell line of *Datura innoxia* Mill. *Plant Cell Reports,* **13** : 533–536.

Goodbody, A.E., Endo, T., Vukovic, J., Kutney, J.P., Chol, L.S.L. and Misawa, M. (1988). Enzymic coupling of catharanthine and vindoline to form 3',4'–anhydrovinblastine by horseradish peroxidase. *Planta Med.,* **54** : 136–140.

Govindachari, T.R., and Vishwanthan, N (1972) 9–Methoxy camptothecine a new alkaloid from *Mappia foetida* Miers. *Ind.J.Chem.* 10:453–454.

Groger, D. (1980). Alkaloids derived from tryptophan and anthranilic acid, In : *Secondary Plant Products. Encyclopedia of Plant Physiology,* vol. 8 (Bell, E.A. Charlwood, B.V.eds.), pp. 128–159, Berlin : Springer–Verlag.

Gupta, P.K. and Mascarenhas, A.F. (1983). Essential oil production in relation to organogenesis in tissue cultures of *Eucalyptus citriodora* Hook. *Basic Life Sci., 22* : 299–308.

Hahlbrock, K., Kuhlen, E. and Lindl, T. (1971). Anderungen vol Enzymativitaten wahred des Wachstums von Zellsuspendionskulture von *Glycine max* :Phenylalain ammonium lyase und p–cumarat CoA Ligase. *Planta,* **99** : 311.

Hahlbrock, K., Wellmann, E. (1970). Light–induced flavone biosynthesis and activity of phenylalanine–ammonia–lyase and UDP–apiose synthetase in cell suspension cultures of *Petroselinum hortense. Planta,* 94 : 236–239.

Haldimann, D. and Brodelius, P. (1987). Redirecting cellular metabolism by immobilization of cultured plant cells : a model study with *Coffea arabica. Phytochemistry,* **26** : 1431–1434.

Hamada, R., Kinoshita, Y., Yamamoto, Y., Tanaka, M. and Yamada, Y. (1994). Anthocyanin production in cultured *Euphorbia millii* cells using film culture vessels. *Biosci. Biotechnol. Biochem.,* b58 : 1530–1533.

Han, K.H., Fleming, P., Walker, K., Loper, M., Chilton, W.S., Mocek, U., Gordon, M.P. and Floss, H.G. (1994). Genetic transformation of mature *Taxus* : an approach to genetically control the *in vitro* production of the anticancer drug taxol. *Plant Sci.,* **95** : 187–196.

Hanagata, N. and Karube, I. (1994). Red pigment production by *Carthamus tinctorius* cells in a two–stage culture system. *J. Biotechnol.,* 37 : 59–65.

Hanagata, N., Ito, A., Fukuju, Y. and Marata, K. (1992). Red pigment formation in cultured cells of *Carthamus tinctorius* L. *Biosci. Biotechnol. Biochem.,* **56** : 44–47.

Hanagata, N., Ito, A., Uehara, H., Asari, F., Takeuchi, T. and Karube, I. (1993). Behaviour of cell aggregate of *Carthamus tinctorius* L. cultured cells and correlation with red pigment formation. *J. Biotechnol.,* **30** : 259–269.

Handro, W. and Ferreira, C.M. (1989). *Stevia rebaudiana* (Bert.) Bertoni : production of natural sweeteners. In: *Biotechnology in Agriculture and Forestry,* vol. 7, *Medicinal and Aromatic Plants II,* pp. 468–487, Bajaj, Y.P.S., (eds.)., Springer–Verlag, Berlin.

Hara, Y., yamagata, H., Morimoto, T., Hiratsuka, J., Yoshioka, T., Fujita, Y. and Yamada, Y. (1989). Flow cytometric analysis of cellular berberine contents in high– and low– producing cell lines of *Coptis japonica. Plant Med.,* **55** : 151–154.

Hara, Y., Yoshioka, T., Morimoto, T., Fujita, Y. and Yamda, Y. (1988). Enhancement of berberine production in suspension cultures of *Coptis japonica* by gibberellic acid treatment. *J. Plant Physiol.,* **133** : 12–15.

Harborne, J.B., Arditti, J., Ball, E.A. (1964). The anthocyanins of callus culture from the stem of *Dimorphotheca auriculata* (Cape marigold, Compositae). *Am. J. Botany,* **57** : 763.

Harkes, P.A.A., Krijbolder, L.,Libbenga, K.R., Wjnsma, R., Nsengiyaremge, T. and Verpoorte, R. (1985). Influence of various media constituents on the growth of *Cinchona ledgeriana* tissue cultures and the production of alkaloids and anthraquinones therein. *Plant Cell Tissue Organ Cult.,* 4 : 199–214.

Havkin–Frenkel, D., Dorn, R. and Knorr, D. (1991). Influence of light on flavor component synthesis in *Vanilla planifolia* cluster cultures. Paper presented at World Congress n Cell and Tissue Culture, Anaheim, CA, June 16–20, 1991.

Hashi, H. (1991). Verbascoside production by plant cell cultures. *Plant Cell Rep.,* 9 : 484–487.

Hashimoto, T. and Yamada, Y. (1983). Scopolamine production in suspension cultures and redifferentiated roots of *Hyoscyamus niger. Plantamed.,* **47**: 195–199.

Hashimoto, T., Sato, F., Mino, M. and Yamada, Y. (1982). Production of tropane alkaloids from cultured solanaceae cells. In : Fujiwara, A. (ed.). *Plant Tissue Culture,* Maruzen, Tokyo, p. 305–306.

Havkin–Frenkel, D. (1994). Vanilla flavor production : Tissue Culture and the hole plant. *VIII IAPTC Congress,* Firenze (Abstract), S18–

Hayashi, H., Fukui, H. and Tabata, M. (1988). Examination of triterpenoids produced by callus and suspension cultures of *Glycyrrhiza glabra. Plant Cell Rep.,* 7 : 508–511.

Hayashi, T., Yoshida, K. and Sano, K. (1988). Formation of alkaloids in suspension–cultured *Colchicum autmnale. Phytochemistry,* **27(5)** : 1371–1374.

Heble, M.R., Narayana Swami, S., Chanda (1968). Diogenin and β–sitosterol : Isolation from *Solanum xanthocarpum* tissue cultures. *Phytochemistry,* 10 : 2393–2394.

Heble, M.R., Narayanaswamy, S., Chadha, M.S. (1971). Hormonal control of steroid synthesis in *Solanum xanthocarpum* tissue cultures. *Phytochemistry,* 13 : 2393–2394.

Heble, M.R., Narayanaswamy, S., Chadha, M.S. (1974). Tissue differentiation and plumbagin synthesis in variant cell strains of *Plumbago zeylanica. Plant Sci. Letters,* 2 : 405–409.

Heide, L. and Tabata, M. (1987). Enzyme activities in cell–free extracts of shikonin–producing *Lithospermum erythrorhizon* cell suspension cultures. *Phytochemistry,* 26(6) : 1645.

Heide, L.Nishioka,N., Fukai,H., and Tabata, M. (1989). Enzymatic regulation of shikonin biosynthesis in *Lithospermum erythrorhizon* cell cultures. *Phytochemistry,* 28 : 1873–1877.

Heinstein, P.F. (1985). Future approaches to the formation secondary natural products in plant cell suspension cultures. *J. Nat. Products,* **48(1)** : 1–9.

Henry, M., Roussel, J.L. and Andary, C. (1987). Verbascoside production in callus and suspension cultures of *Hygrophila erecia. Phytochemistry,* 26 : 1961–1963.

Hezari, M., Lewis, N.G. and Croteau, R. (1995). Purification and characterization of taxa–4 (5), 11(12)–diene synthase from Pacific yew (*Taxus brevifolia*) that catalyzes the firs committed step of taxol biosynthesis. *Arch. Biochem. Biophys.,* **322** : 437–444.

Hinz, H. and Zenk, M.H. (1981). Production of protoberberine alkaloids by cell suspension cultures of *Berberis* species. *Naturwissenschaften,* 68 : 620–621.

Hippolyte, I., Marin, B., Baccou, J.C., Jonard, R. (1992). Growth and rosmarinic acid production in cell suspension cultures of *Salvia officinalis. Plant Cell Rep.,* 11 : 109–112.

Hiraoka, N., Kodama, T. and Tomita, Y. (1986). Selection of *Bupleurum falcatum* callus line producing anthocyanins in darkness. *J. Nat. Prod.,* **49** : 470–474.

Hiraoka, N., Tabata, M. (1974). Formation of acetyltropine in *Datura* callus cultures. *Phytochemistry,* **13** : 1671.

Hiraoka, N., Tabata, M. and Konoshima, M. (1973). Formation of acetyltropine in Datura callus cultures. *Phytochemistry,* 12 : 795–799.

Hirasuna, T.J., Pestchanker, L.J., Srinivasan, V. and Shuler, M.L. (1996). Taxol production in suspension cultures of *Taxus baccata. Plant Cell Tiss. Org. Cult.,* 44 : 95–102.

Hirata, K., Kobayashi, M., Miyamoto, K., Hashi, T., Okazaki, M. and Miura, Y. (1989). Quantitative determination of vinblastine in tissue cultures of *Catharanthus roseus* by radioimmunoassay. *Planta Med.,* 55 : 262–264.

Hirotani, M. and Furuya, T. (1977). Restoration of cardenolide synthesis in redifferentiated shoots from callus of *Digitalis purpurea. Phytochemistry,* **16** : 610.

Hoekstra, S.S., Harkes, P.A.A., Verpoorte, R. and Libbenga, K.R. (1990). Effect of auxin oncytodifferentiation and production of quinoline alkaloid in compact globular structures of *Cinchona ledgeriana. Plant Cell rep.,* **8** : 571–574.

Hong, Y.C., Read, P.E., Harlander, S.K. and Labuza, T.P. (1988). Development of a tissue culture system from immature strawberry fruits. *J. Food. Sci.,* **54**: 388–392.

Ibrahim, R.K., thakur, M.L., Permanand, B. (1971). Formation of anthocyanins in callus tissue cultures. *Lloydia,* **34** : 175–182.

Igbavboa, U., Sieweke, H.J., Leistner, E., Rower, I., Husemann, W. and Barz, W. (1985). Alternative formation of anthraquinones and lipoquinones in heterotrophic and photoautotrophic cell suspension cultures of *Morinda lucida. Planta,* **166** : 537–544.

Ikuta, A., Syono, K. and Furuya, T. (1974). Alkaloids of callus tissues and redifferentiated plantlets in the Papaveraceae. *Phytochemistry,* **13** : 2175–2179.

Inagaki, N., Nishimura, H., Okada, M., Hashi, H. (1991). Verbascoside production by plant cell cultures. *Plant Cell Rep.,* **9** : 484–487.

Inque, K., Nayeshiro, H., Inouye, H. and Zenk, M.H. (1981). Anthraquinones in cell suspension cultures of *Morinda citrifolia. Phytochemistry,* **20** : 1693–1700.

Inque, K., Ueda, S., Nayeshiro, H., Moritome, N. and Inouye, H. (1984). Biosynthesis of naphthoquinones and anthraquinones in *Streptocarpus dunnii* cell cultures. *Phytochemistry,* 23 : 313–318.

Jain, S.C., Khanna, P. (1974). Quercetin from *Crotalaria juncea* Linn. Tissue cultures. *Indian J. Exp. Biol.,* 12: 466 (1974).

Jalal, M.A.F. and Collin, H.A. (1977). Polyphenols of mature plant, seedling and tissue cultures of *Theobroma cacao. Phytochemistry,* 16 : 1377–1380.

Janga, Y.P., Lee, Y.J., Kim, Y.C. and huh, H. (1998). Production of a hepatoprotective cerebroside from suspension cultures of *Lycium chinense. Plant Cell Reports,* 18 : 252–254.

Jensen, I., Werrmann, U. and Knorr, D. (1990). Use of *Passiflora* sp. plant tissue cultures in food biotechnology. *Proc. Int. Conf. Biotechnol. Food,* February 20–24, 1989, Poster Presentation. *Food Biotechnol.,* 4 : 482.

Jensen,R.A.(1986).The shikimate arogenate pathway link between carbohydrate metabolism and secondary metabolism.*Physiol. Plant,* 66 : 164.

Jha, S. and Jha, T.B. (1995). Improved taxol yield in cell suspension cultures of *Taxus wallichiana* (Himalayan yew). *Planta Med,* 64: 270–272.

Jones, L.H. (1974). Plant cell culture and biochemistry : studies for improved vegetable oil production. In : Industrial Aspects of Biochemistry, *Fed. European Biochem. Soc.,* Spencer, B. (eds) PP. 813–833.

Kadkade, P.G. (1981). Formation of podophyllotoxin by *Podophyllum peltatum.*tissue culture, *Naturwissenschaften,* 68: 481–482.

Kadkade, P.G. (1982). Growth and podophyllotoxin production callus tissues of *Podophyllum peltatum. Plant Sci. Lett.,* 25 : 107–115.

Kadkade, P.G., Kremer, B.P. (1977). Glycoalkaloids in tissue cultures of *Solanum acculeatissimum* (ii) Rotalgen– Chloroplaten also fun–Kationelle Endosymbioten in Einem marinen Opisthobranchier. *Naturwissenschaften, Z.,* 64 : 147.

Kamimura, S. (1991). Production of morphinan alkaloids, in : *Plant Cell culture in Japan. Progress in Production of Useful Plant Metabolites by Japanese Enterprises using Plant Cell Culture Technology* (Komamine, A., Misawa, M., DiCosmo, F. eds.), pp. 27–38, Tokyo : CMC Co.

Kamo, K.K., Mahlberg, P.G. (1988). Morphinan alkaloids : Biosynthesis in plant (*Papaver* spp.) tissue culture, in : *Biotechnology in Agriculture and Forestry,* vol. 4 (Bajaj, Y.P.S. ed.), pp. 251–263, Berlin : Springer– Verlag.

Kartnig, Th. and Kobosil, P. (1977). Beobachtungen uber das Vorkommen und die binding von Cardenoliden in Gewebekulturen aus *Digitalis purpurea* and *Digitalis lanata. Plant Med.,* 31 : 221.

Kaul, B. and Staba, E.J. (1967). *Ammi visnage* L. Lam. Tissue cultures : Multiliter suspension growth and examination for furanochromones. *Planta Med.,* 15 : 145–156.

Kaul, B., Stohs, S.J. and Staba, E.J. (1969). Dioscorea tissue culture. III. Influence of various factors on diosgenin production by *Dioscorea deltoidea* callus and suspension cultures. *J. Nat. Products.,* 32 : 347–359.

Keller, H., Wanner, B., Baumann, T.W. (1972). Kaffeinsynthese in fruchten and gewebekulturen von *Coffea arabica. Planta,* 108 : 339–350.

Ketchum, R.E.B., Gibson, D.M., Crouteau, R.B. and Shuler, M.L. (1999). The kinetics of taxoid accumulation in cell suspension cultures of *Taxus* following elicitation with methyl jasmonate. *Biotechnol Bioeng,* 62: 97– 105.

Ketchum, R.E.B. and Gibson, D.M. (1995). Novel method of isolating taxanes from cell suspension cultures of yew. *J Liq Chromatogr,* 18: 1093–1111.

Khanna, P and Jain, S.C. (1973). Diosgenin, Gitogenin and Tigogenin from *Trigonella Foenum–graecum* tissue cultures. *J. Nat. Products.,* 36 : 96–97.

Khouri, H.E., Ibrahim, R.K. and Rideau, M. (1986). Effects of nutritional and hormonal factors on growth and production of anthraquinone glucosides in cell suspension cultures of *Cinchona succirubra. Plant Cell Rep.,* 5 : 423–426.

Kilby, N.J. and Hunter, C.S. (1990). Repeated harvest of vacuole–located secondary product from *in vitro* grown plant cells using 1.02 MHz ultrasound. *Appl. Microbiol. Biotechnol.,* 33 : 448–451.

Kim, J.H., Yun, J.H., Hwang Y.S., Byun, S.Y. and Kam, D.I. (1995). Production of taxol and related taxanes in *T. brevifolia* cell cultures: effect of sugar. *Biotechnol Lett*, **17**: 101–106.

Kingston, D.G.I. (1994). Taxol : The chemistry and structure relationships of a novel anticancer agent. *TIBTECH* **12** : 222–227.

Kinnersley, A.M. and Dougall, D.K. (1980). Increase of anthocyanin yield from wild–carrot cell cultures by a selection system based on cell aggregate size. *Planta,* **148** : 200–204.

Kino–Oka, M., Mine, K., Taya, M., Tone, S. and Ichi, T. (1994). Production and release of anthraquinone pigments of madder (*Rubia tinctorum* L.) under improved culture conditions. *J. Ferment. Bioeng.,* **77** : 103–106.

Kirkland, D.F., Matsuo, M., Underhill, E.W. (1971). Glucosinolates and myrosinase in plant tissue culture. *Lloydia,* **34** : 195–198.

Kishima, Y., Nozali, K., Akashi, R. and Adachi, T. (1991). Light–inducible pigmentation in *Portulaca* callus; selection of a betalain producing cell line. *Plant Cell Rep.,* **10** : 304–307.

Kitajima, M. Fischer, U. Nakamura, M. Ohsawa, M. Veno, M. Takayama, H. Unger, M. Stockigt, J. and Aimi, N. (1998). Anthraquinones from *Ophirrhiza pumila* tissue and cell cultures, *Phytochemistry*, **48 (1):** 107–111

Kittipongpatana, N., Hock, R.S. and Porter, J.R. (1998). Production of solasodine by hairy root, callus and cell suspension cultures of *Solanum aviculare* Forst. *Plant Cell, Tissue and Organ Culture,* **52** : 133–143.

Knobloch, K.H. and Berlin, J. (1980). Influence of the medium composition on the formation of secondary compounds in cell suspension cultures of *Catharanthus roseus. Z. Naturforsch.,* **35c** : 551–556.

Knobloch, K.H. and Berlin, J. (1983). Influence of phosphate on the formation of the indole alkaloids and phenolic compounds in cell suspension cultures of *Catharanthus roseus.* I. Comparison of enzyme activities and product accumulation. *Plant Cell Tissue Organ Cult.,* **2** : 33–340.

Knorr, D., Caster, D., Domenburg, H., Dorn, R., Graf, S., Havkin–Frenkel, D., Podstolski, A. and Wermann, U. (1993). Biosynthesis and yield improvement of food ingredients from plant cell and tissue culture. *Food Technol.,* **47** : 57–63.

Knuth, M.E. and Sahai, O.P. (1991b). Flavor composition and method. U.S. Patent No. 5,068,184, November 26, 1991.

Knuth, M.E. and Sahai, O.P., (1991a). Flavor composition and Method : U.S. Patent No. 5,057,424, October 15, 1991.

Kobayashi, A., Kawazu, K., Hisasaka, K., Wakayama, Y., Nakgawa, N. and Terui, S. (1992). Abstract, Annual Meeting of the Japanese Society for Bioscience, Biotechnology and Agrochemistry, April 1–3, Tokyo. Japan.

Kobayashi, Y., Akita, M., Sakamoto, K., Liu, H., Shigeoka, T., Koyano, T., Kawamura, M. and Furuya, T. (1993). Large–scale production of anthocyanin by *Aralia cordata* cell suspension cultures. *Appl. Microbiol. Biotechnol.,* **40** : 215–218.

Kobayashi, Y., Fukui, H. and Tabata, M. (1988). Berberine production by batch and semi–continuous cultures of immobilized *Thalictrum* cells in an improved bioreactor. *Plant Cell rep.,* **7** : 249–252.

Koblitz, H. (1988). Anthraquinones, in : *Cell Culture and Somatic Cell Genetics of Plants,* vol. 5 : *Production of Phytochemicals in Plant Cell Cultures.* Constabel, F., Vasil, I.K., (eds.) pp. 113–142, San Diego : Academic Press.

Koblitz, H., Koblitz, D., Schmauder, H.P. and Groger, D. (1983). Studies on tissue culture of genus *Cinchona* L: Alkaloid production in cell suspension cultures. *Plant Cell Rep.,* **2** : 122–125.

Koda, T., Ichi, T. and Sekiya, J. (1992a). Properties of blue pigment produced by cultured plant cells of *Chlorodendron trichotomum* as food colors. *Nippon Shokuhin Kygyo Gakkai–Shi,* **39** : 850–855.

Koda, T., Ichi, T., Yoshimitu, M., Nihongi, Y. and Sekiva, J. (1992b). Production of perilla pigment in cell cultures of *Perilla frutescens. Nippon Shokulin Kygyo Gokkai–Shi,* **39** : 839–844.

Komamme, A., Misawa, M., DiCosmo, F. (1991). *Plant Cell Cultures in Japan. Progress in Production of Useful Plant Metabolites by Japanese Enterprises Using Plant Cell Culture Technology,* (eds.) CMC Co. Tokyo.

Komatsu, K., Nozaki,W., Takemura,M and nakaminami,M (1976). Japn.patent no. 51–19169.

Kotani, C (1980). Steveosides. Japn.Patent no. 55–19009.

Kouadio, K.,Chenieux, J.C. Rideau, M. and Viel, C., (1984). Antitumour alkaloid production in callus cultures of *Ochrosia elliptica*. *J. Nat. Produ.* **47**:872–874.

Kupchan,S.M., Court,W.A., Bailey, R.G. Jr. Gilmore, C.J., and Bryan, R.F. (1972 a). Triptolide and tripdiolide, novel antileukamic diterpenoid from *Tripterygium wilfordii, J.Am.Chem. Soc.* **94**: 7194–7195.

Kupchan,S.M., Britton, R.W., Ziegler,M.F. and Sigel, C.W., (1973). Bruceantin, a new potent antileukemic simaroubolide from *Brucea antidysentrica. J.Org.Chem.* **38**: 178–179.

Kouadio, K. Creche, J. Chenieux, J.C. Rideau, M. and Viel, C(1985). Alkaloid production by *Ochrosia elliptica* cell suspesnion cultures, *J Plant. Physiol.* **118**, 227–283

Kreuzaler, F., and Hahlbrock, K. (1973). Flavonoid glycosides from illuminated cell suspension cultures of *Petroselinum hortense*. *Phytochemistry,* **12** : 1149–1153.

Kreis, W. (1987). Untersuchungen zur Kompartimentierung der Cardenolid–Biotransformation in *Digitalis lanata* Zellkulture. Ph.D. thesis, Universitat Tubingen.

Kurz, W.G.W., Chatson, K.B., Constabel, F., Kutney, J.P., Chol, L.S.L., Kolodziejczyk, P., Sleigh, S.K., Stuart, K.L. and Worthy, B.R. (1980). Alkaloid production in *Catharanthus roseus* cell cultures : Initial studies on cell lines and their alkaloid content. *Phytochemistry,* **19** : 2583–2587.

Kurz, W.G.W., Chatson, R.B. and Costabel, F. (1985). Biosynthesis and accumulation of indole alkaloids in *Catharanthus roseus* cultivars, in : *Primary and Secondary Metabolism in Plant Cell Cultures*. Neumann, K.H., Barz, W., Reinhard, E. (eds.), pp. 143–153, Springer–Verlag. Berlin

Kutney, J.P., Aweryn, B., Chatson, K.B., Choi, L.S.L. and Kurz, W.G.W. (1985). Alkaloid production in *Catharanthus roseus* (L.) G. Don Cell Cultures. XIII. Effects of Bioregulators in indole alkaloid biosynthesis. *Plant Cell Reports,* **4** : 259–262.

Kutney, J.P., Samija, M.D., Hewitt, G.M., Bugnade, E.C. and Gu, H. (1993). Anti–inflammatory oleanane triterpenes from *Tripterygium wilfordii* cell cultures by fungal elicitation. *Plant Cell Reports,* **12** : 356–359.

Leistner, E. (1973). Biosynthesis of morindone and alizarin in intact plants and cell suspension cultures of *Morinda citrifolia*. *Phytochemistry,* **12** : 1669–1674.

Leistner, E. (1975). Isolierung, identifizierung und biosynthese von Anthrachinonen in Zellasuspensions kulturen von *Morinda citrifolia. Plant Med.* (Suppl.) 214–224.

Liau, S., Ibrahim, R.K. (1973). Biochemical differentiation in flax tissue culture. Phenolic compounds. *Can. J. Botany,* **51** : 820–824.

Lin, M.L. and Staba, E.J. (1961). Peppermint and spearmint tissue cultures. I. Callus formation and submerged culture. *Lloydia,* **24** : 139–145.

Lindsey, K. (1985). Manipulation, by nutrient limitation, of the biosynthetic capacity of immobilised cells of *Capsicum frutescens* Mill. Cv. *annuum. Planta,* **165** : 126–133.

Liu, C.C.S.K., Yang, S.L., Roberts, M.F., Elford, B.C. and Phillipson, J.D. (1992). Antimalarial activity of *Artemisia annua* flavanoids from whole plants and cell cultures. *Plant Cell Reports,* **11** : 637–640.

Loscher, R.,and Heide, L. (1994). Biosynthesis of *p*–hydroxybenzoate from *p*–coumarate and *p*–coumarate-coenzyme A in cell–free extracts of *Lithospermum erythrorhizon* cell cultures. *Plant Physiol.,* **106** : 271–279.

Luckner, M. (1986). Secondary metabolism in microorganisms, plants, and animals, Springer-Verlag, Berlin, Heidelberg, New York.

Luckner, M. and Diettrich, B. (1988). Cardenolides, In : *Cell Culture and Somatic Cell Genetics of Plants,* vol. 5: *Phytochemicals in Plant Cell Cultures* Costabel, F., Vasil, I.K., (eds), pp. 193–l212, San Diego: Academic Press.

Luckner, M. and Nover, L. (1977). Expression of secondary metabolism. An aspect of cell specialization microorganisms, higher plants and animals. In : Luckner, M. , Nover L., Bohm, H. (eds.) Secondary metabolism and cell differentiaton. *Mol. Biol. Biochem. Biophys.,* **23** : 1.

Luckner, M.(1971). Was ist Sekundarstoffwechsel ?. *Pharmazie,* **26** : 717.

Luckner, M., Nover, L. and Bohm, H. (1977). Secondary metabolism and cell differentiation. *Mol. Biol. Biochem. Biophys,* 23.

Lux–Pfister, C., Dornenburg, H. and Knorr, D. (1995). Bioreaktoren fur pflanziliche Zellkulturen – Kultivierung von *Galium vernum* Suspensionskulturen im Hubstrahlbioreaktor. *Bio Tec ,* 7 : 28–32.

Madhavi, D.L., Prabha, T.N., Singh, N.S. and Patwardhan, M.V. (1991). Biochemical studies with garlic (*Allium sativum*) cell cultures showing different flavour levels. *J. Sci. food Agric.,* **56** : 15–24.

Matile, P. (1978). Biochemistry and function of vacuoles. *Annu. Rev. Plant Physiol.,* **29** : 193.

Matile, P. (1984). Das toxische Kompartiment der Pflanzenzelle. *Naturwissenschaften,* **71** : 18.

Matsubara, K., Fujita, Y. (1991). Production of berberine in : *Plant Cell Culture in Japan. Progress in Production of useful Plant Metabolites by Japanese Enterprises using Plant Cell Culture Technology.* Komanne, A., Misawa, M., Dicosmo, F. (eds.), pp. 39–44, Tokyo: CMC Co.

Matsubara, K., Kitani, S., Yoshioka, Y., Morimoto, T., Fujita, Y. and Yamada, Y. (1989). High density culture of *Coptis japonica* cells increases berberine production. *J. Chem. Technol. Biotechnol.,* **46** : 61–69.

Matsumoto, T., Ikeda, T., Okimura, L., Obi, Y., Kisaki, T. and Noguchi, M. (1982). Production of ubiquinone–10 by highly producing strains selected by a cell cloning technique. In : Fujiwara, A. (eds.). *Plant Tissue culture,* 1982, Maruzen, Tokyo, pp. 275–276.

Matsushita, T., Koga, N., Ogawa, K., Fujino, K. and Funatsu, K. (1994). High–density culture of plant cells using an air–life column for production of valuable metabolites. In : *Studies in Plant Science,* 4, *Advances in Plant Biotechnology,* pp. 339–353. Ryu, D.D.Y. and Furusaki, S. (eds)., Elsevier, Amsterdam.

Matsuura, T. and Kakuda, T. (1990). Effects of precursor, temperature and illumination on theanine accumulation in tea callus. *Agric. Biol. Chem.,* **54**: 2283–2286.

Matsuura, T., Kakuda, T., Kinoshita, T., Takeuchi, N. and Sasaki, K. (1992). Theanine formation by tea suspension cultures. *Biosci. Biotechnol. Biochem.,* **56** : 1179–1181.

Medora, R.S., Campbell, J.M. and Mell, G.P. (1973). Proteolytic enzymes in Papaya tissue cultures. *J. Nat. Products,* **36** : 214–215.

Mersinger, R., Domauer, H., Reinhard, E. (1988). Formation of forskolin by suspension cultures of *Coleus forskohlii. Planta Medica,* **54** : 200–204.

Meurer–grimes, B., Berlin,J., and Strack, D., (1989) Hydroxy cinnamoyl–CoA: Putrescine hydroxy cinnamyl transferase in tobacco cell cultures with high and low levels of caffeoyl putrescine. *Plant Physiology,* **89**: 488–492.

Miller, C.O. (1969). Control of deoxyisoflavone synthesis in soybean tissue. *Planta,* **87** : 26.

Mirjalili, N. and Linden, J.C. (1996). Methyl jasmonate induced production of taxol in suspension cultures of *Taxu cuspidata* : ethylene interaction and industion models. *Biotechnol. Prog.,* **12** : 110–118.

Misawa, M., Sakato,K., Tanaka,H., and Hayashi,M., and Samejima,H., (1974) Production of physiologically active substances by plant cell suspension cultures.In:Tissue culture and Plant Science, Street, H.E. (eds) Academic Press. London, Newyork, PP 405–432.

Misawa, M.,Hayashi,M.,and Takayama,S.,(1983) Production of antineoplastic agents by plant tissue cultures. *Planta Medica,* **49**: 115–119.

Misawa,M.,Hayashi,M.,andTakayama,S.,(1985)Accumulation of antineoplastic agents by plant tissue cultures. In: Primary and Secondary metabolism of plant cell cultures, Neumann, K.H., Barz, W., and Reinhard, E., (eds) Springer, Berlin, Hiedelberg, PP235–246.

Misawa, M. (1994). Plant tissue culture : an alternative for production of useful metabolites. *FAO Agric. Serv. Bull.,* **108** Rome FAO.

Misawa, M. and Endo, T. (1988). Antitumor compounds, In : *Cell Culture and Somatic Cell Genetics,* vol. 5: *Phytochemicals in Plant Cell Cultures* Constabel, F., Vasil, I.K., (eds.), PP 553–568, San Diego : Academic Press.

Misawa, M., Hayashi, M., Nagano, Y. and Kawamoto, Z. (1973). *Jpn. Patent* (Kokai) 73–6153.

Mizukami, H. and Ellis, B.E. (1991). Rosmarinic acid formation and differential expression of tyrosine aminotransferase isoforms in *Anchusa officinalis* cell suspension cultures. *Plant Cell Rep.,* 10 : 321–324.

Mizukami, H., Konoshima, M. and Tabata, M. (1978). Variation in pigment production in *Lithospermum erythrorhizon* cell suspension cultures. *Plant Cell Rep.,* 12 : 706–709.

Mizukami, H., Tabira, Y. and Ellis, B.E. (1993). Methyl jasmonate induced rosmarinic acid biosynthesis in Lithospermum erythrorhizon cell suspension cultures. *Plant Cell Rep.,* 12 : 706–709.

Mizusaki, S., Tanabe, T., Noguchi, M., Tamaki, E. (1971). *p*–Coumaroyl–putrescine, caffeolyl–putrescine and feruloylputrescine from callus tissue culture of *Nicotiana tabacum. Phytochemistry,* 10 : 1347–1350.

Monacelli, B., Pasqua, G., Cuteri, A., Varusi, A., Botta, B. and Monache, G.D. (1995). Histological study of callus formation and optimization of cell growth in *Taxus baccata. Cyto bios,* 81: 159–170.

Moreno, P.R.H., Van Der Heijden, R., Verpoorte, R. (1995). Cell and tissue culture of *Catharanthus roseus* : a literature survey II. Updating from 1988–1993, *Plant Cell Tiss. Org. Cult.,* 42 : 1–25.

Mori, T. and Sakurai, M. (1994). Production of anthocyanin from strawberry cell suspension cultures; effects of sugar and nitrogen. *J. Food Sci.,* 59: 588–593.

Mori, T., Sakurai, M., Seki, M. and Furusaki, S. (1994). Use of auxin and cytokinin to regulate anthocyanin production and composition in suspension cultures of strawberry cells. *J. Sci. Food Agric.,* 65 : 271–276.

Mori, T., Sakurai, M., Shigeta, J., Yoshida, K. and Kondo, T. (1993). Formation of anthocyanins from cells cultured from different parts of strawberry. *J. Food Sci.,* 58 : 788–792.

Morimoto, M., Hara, Y., Kato, Y., Hiratsuka, J., Yoshioka, T., Fujita, Y. and Yamada, Y. (1988). Berberine productio by cultured *Coptis japonica* cells in a one stage culture using medium with a high copper concentration *Agric. Biol. Chem.,* 52 : 1835–1836.

Moshy, R.J., Nieder, M.H. and Sahai, O.P. (1989). Biotechnology in the flavor and food industry of the USA, in : *Biotechnology. Challenges for the Flavor and Food Industry* Lindsey, R.L., Willis, B.J., (eds.), PP 145–163. London : Elsevier.

Motoyama, E., Tada, H., Shimomura,K., Yoshihira, K. and Ishimaru, K. (1996). Caffeic acid esters in tissue cultures of *Heliotropium pervianum. Plant Tiss. Cult. Lett.,* 13 : 73–74.

Mulder–Krieger, Th., Verpoorte, R., Baerheim Svendsen, a. and scheffer, J.J.C. (1988). Production of essential oils and flavours in plant cell and tissue cultures. A review. *Plant Cell Tissue Organ Cult.,* 13 : 85–154.

Nagakawa, K., Konagai, A., Fukui, H., Tabata, M. (1984). Release and crystallization of berberine in liquid medium of *Thalictrum minus* cell suspension cultures. *Plant Cell Rep.,* 3 : 254–257.

Nagarjan, R.P., Kshavarz, E. and Gerson, D.F. (1989). Optimization of anthocyanin yield in a mutated carrot cell line (*Daucus carota*) and its implication in large scale production. *J. Ferment. Bioeng.,* 68 : 102–106.

Nahalka, J., Nahalkova, J., Gemeiner, P. and Blanarik, P. (1998). Elicitation of plumbagin by Chitin and its release into the medium in *Drosophyllum lusitanicum* link. Suspension cultures. *Biotechnology Letters,* 20(9) : 841–845.

Nakajima, H., Sonomoto, K., usui, N., Sato, F., Ichimura, K., Yamada, Y., Tanaka, A. and Fukui, S. (1985). Entrapment of *Lavandula vera* cells and production of pigments by entrapped cells. *J. Biotechnol.,* 2 : 107–117.

...wa, Y. and Ohtani, T. (1992). Induction of callus from flesh of *Gardenia jasminoides* fruit and formation o yellow pigment in the callus. *Biosci. Biotechnol. Bioeng.,* 56 : 1732–1736.

Nicolaou, K.C., Dai, W.M. and Guy, R.K. (1994). Chemie und Biologie von Taxol. *Angew. Chem.,* 106 : 38–69.

Niemann, G.J. (1976). Phenolics from *Larix* needles. XII. Seasonal variation of main flavonoids in leaves of *Larix leptolepis. Acta. Bot. Neerl.,* 25 : 349.

Nishi, A., Yoshida, A., Mori, M. and Sugano, N. (1974). Isolation of variant carrot cell lines with altered pigmentation. *Phytochemistry,* 13 : 1653–1656.

Noe, W. and Berlin, J. (1985). Induction of de–novo synthesis of tryptophan decarboxylase in cell suspensions of *Catharanthus roseus. Planta,* 166 : 500.

Nover, L., Lynen, F. and Mothes, K. (1980). Cell compartmentation and metabolic channeling. *Fischer. Jena.*

Nozue, M., Kawai, J. and Yoshitama, K. (1987). Selection of a high anthocyanin–producing cell line of sweet potato cell cultures and identification of the pigment. *J. Plant Physiol.,* **129** : 81–88.

O'Dowd, A.N., McCauley, P.G., Richard, H.S.D. and Wilson, G. (1993). Callus production, suspension culture and *in vitro* alkaloid yields of Ephedra. *Plant Cell Tissue and Organ Culture,* **34** : 149–155.

Odake, K., Ichi, T. and Kusuhara, K. (1991). Production of madder colorants, in : *Plant Cell Culture in Japan. Progress in the Production of Useful Plant Metabolites by Japanese Enterprises using Plant Tissue Culture Technology.* Komamaine, A., Misawa, M., Dicosmo, F., (eds.), PP 138–146, Tokyo : CMC Co.

Ogino, T., Hiraoka, N. and Tabata, M. (1978). Selection of high nicotine producing cell lines of tobacco callus by single cell cloning. *Phytochemistry,* **17** : 1907.

Ohlsson, A.B., Bjork, L., Gatenbeck, S. (1983). Effect of light on cardenolide production by *Digitalis lanata* tissue cultures. *Phytochemistry,* **22** : 2447–2450.

Ohshima, Y., Takahashi, K. and Shibata, S. (1988). Tissue culture of Rhubarb and isolation of sennosides from the callus planta medica, PP 20–24.

Oostham, A., Mot, J.N.M., Van Der Plas, L. H.W. (1993). Establishment of hairy root cultures of *Linum flavum* producing the lignan 5–methoxypodophyllotoxin. *Plant Cell Rep.,* **12** : 474–477.

Ozaki, Y., Suga, C., Yoshioka, T., Morimoto, T., Harada, M. (1990). Evaluation of equivalence on pharmacological properties between natural crude drugs and their cultured cells based on their components. Accelerative effect of *Lithospermi radix* and inhibitory effect of *Coptidis rhizoma* on proliferation and granulation tissue. *Yakugaku Zasshi,* **110** : 268–272.

Ozeki, Y. and Komamine, A. (1985). Effects of inoculum density, zeatin and sucrose on anthocyanin accumulation in a carrot suspension culture. *Plant Cell Tissue Organ Cult.,* **5** : 45–53.

Palmer, J.E.,and Widholm, J.M. (1975). Characterization of carrot and tobacco cell cultures resistant to *p*–fluorophenylalanine. *Plant Physiol.,* **56** : 233–238.

Pareilleux, A. and Vinas, R. (1984). A study on the alkaloid production of resting cell suspensions of *Catharanthus roseus* in a continuous flow reaction. *Appl. Microbiol. Biotechnol.,* **19** : 316–320.

Paulmann, H., Dornenburg, H. and Knorr, D. (1990). Effect of medium conditioning on vanillin production in *Vanilla planifolia* callus cultures. Berlin University of Technology.

Payne, G.F., Bringi, V., Prince, C. and Shuler, M.L. Eds. (1991). *Plant Cell and Tissue Culture in Liquid Systems.* Hanser Publishers, Munich.

Pestchanker, L.J., Roberts, S.C. and Shuler, M.L. (1991).Kinetics of taxol production and nutrient use in suspension cultures of *Taxu cuspidata* shake flasks and a Wilson–type reactor. *Enzyme Microb. Technol.,* **19** : 256–260 **19** : 256–260.

Petersen, M., Hausler, E., Karwatzki, B. and Meinhard, J. (1993). Proposed biosynthetic pathway for rosmarinic acid in cell suspension cultures of *Coleus blumei. Planta,* **189** : 10–14.

Piehl, G.W., Berlin, J., Mollenschott, C. and Lehmann, J. (1988). Growth and alkaloid production of a cell suspension culture of *Thalictrum rugosum*n shake flasks and membrane stirrer reactors with bubble free aeration. *Appl. Microbiol. Biotechnol.,* **29** : 456–461.

Prabha, T.N., Narayana, M.S. and Patwardhan, M.V. (1990). Flavour formation in callus cultures of guava (*pridium guajava*). *J. Sci. Food. Agric.,* **50** : 105–110.

Prasad Babu, Ch., Mamatha, R., Kokate, C.K. and Veeresham, C. (2001). Elicitation of *Taxus wallichiana* (Himalayan Yew) cell cultures for the production of Taxanes. *Indian Drugs,* **38(10)** : 502–505.

Prince, C. (1991). In : *Plant Cell and Tissue Culture in Liquid Systems,* Payne, G.F., Bringi, V. Prince, C. and Shuler, M.L., (eds) PP 256 Hanser Publishers, Munich.

Proksch, A. and Wagner, H. (1987). Structural analysis of a 4–*O*–methylglucuronarabinoxylan with immuno–stimulating activity from *Echinacea purpurea. Phytochemistry,* **26** : 1989–1993.

Purohit, P.v. and Khanna, P. (1983). Production of essential oil from callus cultures of *Ocimum basiclicum* L. *Basic Life Sci.,* **22** : 377–380.

Radin, D.N. (1986). A modell cell culture system to study isoprenoid regulation in plants. *Curr. Top. Plant Biochem. Physiol.,* **5** : 153.

perforatum. Ind. J. Pharm. Sci., 431–433.

Ravishankar, G.A., Sarma, K.S., Venkatamaran, L.V. and Kadyan, A.K. (1988). Effect of nutrial stress on capsaicin production in immobilised cell cultures of *Capsicum annuum. Curr. Sci.,* **57** : 381–383.

Razzaque, A. and Ellis, B.E. (1977). Rosmarinic acid production in *Coleus blumei. Planta,* **137** : 287–291.

Reichling, J., Bisson, W. and Becker, H. (1985). Vergleichende Untersuchungen zur Bildung und Akkumulation von etherischem Ol in der intakten Pflanze und in Zellkulturen von *Pimpinella anisum. Z. Naturforsch.,* **40**: 465–468.

Reinhard, E., Cordaun, G., Volk, O.H. (1968). Uber Gewebekulturen von *Ruta graveolens. Planta Med.,* **16** : 8–16.

Reinhard, E., Corduan, G., Brocke, W. Von (1971). Untersuchungen iiber das atherische OI und die Cumarine in Gewebekulturen von *Ruta graveolens.* Herba Hungarica, **10** : 9–26.

Reinhard, E., Corduan, G., Volk, O.H. (1968). Nachweis von Harmin in Gewebekulturen von *Peganum harmala. Phytochemistry,* **7** : 503–504.

Reinhard, E., Corduan, G., Volk, O.H. (1968a). Uber Gewebekulturen von *Ruta graveolens. Plant Med.,* **16** : 8–16.

Renaudin, J.P. and Guern, J. (1990). Transport and vacuolar storage of secondary metabolites in plant cell cultures. In: Charlwood, B.A., Rhodes, M.J.c. (eds.). Secondary products from plant tissue culture. *Proc. Phytochemical Society of Europe.* 30. Oxford Science Publications, Calderon Press, Oxford, PP 59.

Rhodes, M.J.C., Payne, J. and Robins, R.J. (1986). Cell suspension cultures of *Cinchona ledgeriana,* II. The effect of a range of auxins and cytokinins on the production of quinoline alkaloids. *Planta Mediea,* 226–229.

Roddick, J.G., Butcher, D.N. (1972a). Isolation of tomatine from cultured excised roots and callus tissues of tomato. *Phytochemistry,* **11** : 2019–2024.

Rohmer, M., Knani, M., Simonin, P., Sutter, B. and Sahm, H. (1993). Isoprenoid biosynthesis in bacteria : a novel pathway for the early steps leading to isopentenyl diphosphate. *Biochem. J.,* **295** : 517–524.

Roja, G. and Heble, M.R. (1994). The quinoline alkaloids camptothecin and 9–methoxy camptothecin from tissue cultures and mature trees of *Nothapodytes foetida. Phytochemistry,* **36(1)** : 65–66.

Roja, G. and Heble, M.R. (1995). Castanospermine, an HIV inhibitor from tissue cultures of Castanospermum australe. *Phytotheraphy Research,* **9** : 540–542.

Rolfs, C.H., Schon, H., Steffens, M. and Kindle, H. (1987). Cell suspension culture of *Arachis hypogea* L: model system of specific enzyme induction in secondary metabolism. *Planta,* **172** : 238.

Romagnoli, L.G. and Knorr, D. (1988). Effects of ferulic acid treatment on growth and flavor development of cultured *Vanilla planifolia* cells. *Food Biotechnol.,* **2** : 93–104.

Romero, R.M. and Roberts, M.F. (1996). Anthranilase synthase from *Ailanthus altissima* cell suspension cultures. *Phytochemistry,* **41(2)** : 395–402.

Rueffer, M. and Zenk, M.H. (1987). Distant precurssors of benzylisoquinoline alkaloids and their enzymatic formation. *Z. Naturforsch.,* **42c** : 319–332.

Rueffer, M., El–Shagi, H., Nagakura, N. and Zenk, M.H. (1981). *S*–Norlaudanosoline synthase : the first enzyme in the benzylisoquinoline biosynthetic pathway : *FEBS Lett.,* **129** : 5–9.

Ryan, C.A., and Walker–Simmons, M. (1983). *Plant Vacuoles Methods Enzymol.,* **96** : 580.

Sahai, O.P. (1994). Plant tissue cultures. In : *Bioprocess Production of Flavor, Fragrance and Color Ingredients,* pp. 239–275. Gabelman, A., (eds). John Wiley and Sons, New York.

Sahai, O.P. and Knuth, M. (1985). Commercializing plant tissue culture processes : economics, problems and prospects. *Biotechnol. Prog.,* **1** : 1–9.

Saito, K., Daimon, E., Kusaka, K., Wakayama, S. and Sekino, Y. (1988). Accumulation of a novel red pigment in cell suspension cultures of floral meristem tissues from *Carthamus tinctorius* L. *Z. Naturforsch.,* **43c** : 862–870.

Sakamoto, K., Iida, K., Sawamura, K., Hajiro, K., Asada, Y., Yoshikawa, T., Furuya, T. (1994). Anthocyanin production in cultured cells of *Aralia cordata. Plant Cell Tissue Organ Cult.,* **36** : 21–26.

Sakamoto, K., Iida, K., Sawamura, K., Hajiro, K., yashikawa, T. and Furuya, T. (1993). Effects of nutrients on

Sakamoto, K., Iida, K., Sawamura, K., Hajiro, K., Asada, Y., Yoshikawa, T., Furuya, T. (1994). Anthocyanin production in cultured cells of *Aralia cordata. Plant Cell Tissue Organ Cult.,* **36** : 21–26.

Sakamoto, K., Iida, K., Sawamura, K., Hajiro, K., yashikawa, T. and Furuya, T. (1993). Effects of nutrients on anthocyanin production in cultured cells of *Aralia cordata. Phytochemistry,* 33 : 357–360..

Sakuta, M., Takagi, T., Komamine, A. (1987). Effects of nitrogen source on betacyanin accumulation and growth in suspension cultures of *Phytolaca americana. Physiol. Plant.,* **71** : 459–463.

Salandy, A., Grafton, L., Uddin, M.R. and Shafri, M.I. (1993). Establishing an embryogenic cell suspension culture system in Florida yew (*Taxus floridana*). *In vitro Cell Dev Biol,* 29: 75A.

Sardesai, D.L. and Tipnis, H.P. (1969). Production of flavouring principles by tissue culture of *Coriandrum sativum . Curr. Sci.,* **38** : 545.

Sargent, J.A., Skoog, F. (1960). Effects of indoleacetic acid and kinetin on scopoletin–scopolin levels in relation to growth of tobacco tissues *in vitro. Plant Physiol.,* **35** : 934–941.

Sargent, J.A., Skoog, F. (1961). Scopoletin glycosides in tobacco tissue. *Physiol. Plantarum,* **14** : 504–519.

Sarma, K.S., Sharada, K., Maesato, K., Hara, T., and Sonoda, Y. (1991). Chemical and sensory analysis of saffron produced through tissue culture of *Crocus sativus. Plant Cell Tissue Organ Cult.,* **26** : 11–14.

Sasse, F., Witte, L., Berlin, J. (1987). Biotransformation of tryptamine to serotonin by cell suspension cultures of *Peganum harmala. Planta Med.,* **53** : 354–359.

Sato, F., Yamada, Y. (1984). High berberine producing cultures of *Coptis japonica* cells. *Phytochemistry,* **23** : 281–285.

Sauerwein, M., Yamazaki, T. and Shimomura, K. (1991). Hernandulcin in hairy root cultures of *Lippia dulcis. Plant Cell Rep.,* **9** : 579–581.

Scharlemann, W., Czygan, F.C. (1971). Zusammensetzung and Stoffwechsel der Tetraterpene und Chlorophylle in Kalluskulturen von *Ruta graveolens* L. Herba Hungarica, **10** : 43–48.

Schiel, O., Jarchow–Redecker, K., Piehl, G.W., Lehmann, J., Berlin, J. (1984a). Increased formation of cinnamoyl putrescines by fed–batch fermentation of cell suspension cultures of *Nicotiana tabacum. Plant Cell Rep.,* **3** : 18–20.

Schiel, O., Martin, B., Piehl, G.W., Nowak, J., Hammer, J., Sasse, F., Schaer, W., Lehmann, J., Berlin, J. (1984b). Some technological aspects on the production of cinnamoyl putrescines by cell suspension cultures of *Nicotiana tabacum,* in *3ʳᵈ Eur. Congr. Biotechnol.,* Vol. 1, pp. 167–172, Weinheim : VCH.

Schiel, O., Witte, L. and Berlin, J. (1987). Geraniol–10–hydroxylase activity and its relation to monoterpene indole alkaloid accumulation in cell suspension cultures of *Catharanthus roseus. Z. Naturforsch.,* **42c** : 1075–1081.

Schubel, H., Ruyter, C.M., Stockigt, J. (1989). Improved production of raucaffricine by cultivated *Rauwolfia* cells. *Phytochemistry,* **28** : 491–494.

Schulte, U., El–Shagi, H., Zenk, M.H. (1984). Optimization of 19 Rubiaceae species in cell culture for the production of anthraquinones. *Plant Cell Rep.,* **3** : 51–54.

Schumacher, H.M., Gundlach, H., Fielder, F. and Zenk, M.H. (1987). Elicitation of benzophenanthridine alkaloid biosynthesis in *Eschscholtzia californica* cell cultures. *Plant Cell Rep.,* **6** : 410–413.

Scragg, A.H. and Alan, E.J. (1994). *Quassia amara* (Surinam Quassia) : *In vitro* culture and the production of quassin. In : *Biotechnology in Agriculture and Forestry,* vo. 26, *Medicinal and Aromatic Plants,* VI, pp. 316–326. Bajaj, Y.P.S., (eds)., Springer–Verlag, Berlin.

Scragg, A.H. and Arias–Castro, C. (1992). Bioreactors for industrial production of flavours : use of plant cells. In : *Biotransformation of Flavours.* PP. 131–154. Paterson, R.L.S., Charlwood, B.V., MacLeod, G. and Williams, A.A., Eds. Royal Society of Chemistry, Cambridge.

Seitz, H.U. and Hinderer, W. (1988). Anthocyanins, in : *Cell Culture and Somatic Celle Genetics of Plants,* vol. 5 : *Phytochemicals in Plant Cell Cultures.* Constabel, F., Vasil, I.K. (eds.), pp. 49–76. San Diego : Academic Press.

Selby, C., Galpin, I.J. and Collin, H.A. (1979). Comparison of the onion plant (*Allium cepa*) and onion tissue culture I. Allinase activity and flavour precursor compounds. *New Phytol.,* **83** : 351.

Shaw, G. (1971). The chemistry of sporopollenin. In : Luckner M. Nover, L., Bohm, H. (eds.). Secondary metabolism and cell differentiation. *Mol. Biol. Biochem. Biophys.,* **23** : 72.

Shih, N.J.R. Mc. Donald, K.A. Dandekar, A.M. Girbes, T. Iglesias, R. and jackman, A.P. (1998) A novel type – 1ribosome in activating protein isolated from the supperrnatant of transformed suspension cultures of *Trichosanthes Kirilowi* Plant cell reports, **17** : 531–537

Siah, C.L. and Doran, P.M. (1991). Enhanced codeine and morphine production in suspended *Papaver somniferum* cultures after removal of exogenous hormones. *Plant Cell Rep.,* **10** : 349–353.

Sieweke, H.J. and Leistner, E. (1992). *O*–succinylbenzoate : CoA ligase from anthraquinone producing cell cultures of *Galium mollugo. Phytochemistry,* **31** : 2329–2335.

Simantiras, M. and Leistner, E. (1989). Formation of *o*–succinylbenzoic acid from isochorismic acid in protein from anthraquinone–producing plant cell suspension cultures. *Phytochemistry,* **28** : 1381–1382.

Smart, N.J., Morris, P. and Fowler, M.W. (1982). Alkaloid production by cells of *Catharanthus roseus* grown in airlift fermenter systems, in : *Plant Tissue Culture* 1982 Fujiwara A., (eds.), pp. 397–398, Tokyo : Maruzen Press.

Smollny, T., Wichers, H., De Ruk, T., VanZwam, A., Shasavari, A. and Alfermann, A.W. (1993). Formation of lignans in suspension cultures of *Linum album* 3rd Workshop on "*Primary and Secondary Metabolism of Plants and Plant Cell Cultures*". Abstract P13, Leiden University.

Smollny, T., Wichers, H., Kalenberg, S., Shahsavari, A., Petersen, M. and Alfermann, A.W. (1998). Accumulation of podophyllotoxin and related lignans in cell suspension cultures of *Linum album. Phytochemistry,* **48(6)** : 975–979.

Spener, F., Staba, E.J., Mangold, H.K. (1974). Lipids in plant tissue cultures. II. Unusual fatty acids in lipids of *Hydnocarpus anthelminthica* cultures. *Chem. Phys. Lipids,* **12** : 344–350.

Srinivasan, V., Ciddi, V., Bringi, V., Shuler, M. (1996). Metabolic inhibitors, elicitors and precursors as tools for probing yield limitation in taxane production by *Taxus chinensis* cell cultures. *Biotechnol. Prog.,* **12** : 457–466.

Stahlhut, R. (1993). ESCAgentics Corporation, *IAPTC Newslett.,* **73** : 12–15.

Steck, W., Bailey, B.K., Shyluk, J.P., Gamborg, O.L. (1971). Coumarins and alkaloids from cell cultures of *Ruta graveolens. Phytochemistry,* **10** : 191–198.

Steinrucken, H.C., Amrhein, N. (1980). The herbicide glyphosate is a potent inhibitor of 5–enolpyruvylshikimic acid 3–phosphate synthase, *Biochem. Biophys. Res. Commun.,* **94** : 1207–1212.

Stickland, R.G.,and Sunderland, N. (1972a). Production of anthocyanins, flavonols and chlorogenic acids by cultured callus tissues of *haplopappus gracilis. Ann. Botany,* **36** : 443–457.

Stohs, S.J. (1969). Production of Scopolamine and Hyoscyamine by *Datura stramonium* L. suspension cultures. *J. Pharm. Sci.,* **58** : 703.

Strobel, J., Bieke, M., Gebaurer, E., Wind, E. and Groger, D. (1990). The influence of organic and inorganic chemical factors on cell growth and anthraquinone formarion in suspension cultures of *Galium vernum. Biochem. Physiol., Pflanzen,* **186** : 117–124.

Sumalatha, D.V. (2001). Studies on tissue culture of *Eravatamia heyeneana.* M.Pharm. Thesis, Kakaktiya University, Warangal, A.P., India.

Sumaryono, W., Proksch, P., Hartmann, T., Nimtz, M. and Wray, N. (1991). Induction of rosmarinic acid accumulation in cell suspension cultures of *Orthosiphon aristants* after treatment with yeast extract. *Phytochemistry,* **30** : 3267–3271.

Sundaravelan, D.C., Brahmareddy, D. and Ciddi Veeresham (2002). Effects of precursors and elicitors on the production of camptothecine and its analogs in cell cultures of *Nothapodytes foetida.* Ethiopian J. Pharmaceutical, 20, 39-46.

Suzuki, H., Matsumoto, T. (1988). Anthraquinone : Production by plant cell culture, in : *Biotechnology in Agriculture and Forestry,* vol. 4 (Bajaj, Y.P.S. ed.), Pp. 237–250. Berlin : Springer–Verlag.

Suzuki, H., Ozaki, Y. and Satake, M. (1993b). Changes of components – dissolution from prescription containing *Coptis* rhizome or cultured cells of *Coptis japonica* into the detection. *Shoyakugaku Zasshi,* **47** : 396–401.

Suzuki, H., Ozaki, Y., Suga, C., Morimoto, T., Satake, M., Harada, M. (1993a). Dissolution tests of *Coptis* rhizome

Suzuki, H., Ozaki, Y. and Satake, M. (1993b). Changes of components – dissolution from prescription containing *Coptis* rhizome or cultured cells of *Coptis japonica* into the detection. *Shoyakugaku Zasshi*, **47** : 396–401.

Suzuki, H., Ozaki, Y., Suga, C., Morimoto, T., Satake, M., Harada, M. (1993a). Dissolution tests of *Coptis* rhizome and cultured cells of *Coptis japonica*. *Shoyakugaku Zasshi,* **47** : 311–315.

Suzuki, H., Suga, C., Morimoto, T., Harada, M. (1991). Quantitative analysis of plant hormones, auxins, in biotechnologically cultured products of medicinal plants. *Shoyakugaku Zasshi,* **45** : 137–141.

Suzuki, M., Makagawa, K., Fukui, H. and Tabata, M. (1987). Relationships of berbeine–producing capability between. *Thalictrum* plants and their tissue cultures. *Plant Cell Rep.,* **6** : 260.

Suzuki, M., Nakagawa, K., Fukui, H., Tabata, M. (1988). Alkaloid production in cell suspension cultures of *thalictrum flavum* and *T. dipterocarpum. Plant Cell Rep.,* **7** : 26–29.

Suzuki, T., Yoshioka, T., Hara, Y., Tabata, M. and Fujita, Y. (1987). A new bioassay system for screening high berberine–production in cell suspension cultures of *Thalictrum flavum* and *T. dipterocarpum. Plant Cell Rep.,* **7** : 26–29.

Tabata, M. and Fujita, Y. (1985). Producing of shikonin by plant cell cultures. In: *Biotechnology in Plant Science : Relevane to Agriculture in the Eighties.* PP 207–218. Zaitlin, M., Day, P. and Hollaender, A. (eds) Academic Press, New York.

Tabata, M., Hiraoka, N., Ikenoue, M., Sano, Y., Konoshima, M. (1975). The production of anthraquinones in callus cultures of *Cassia tora. Lloydia,* **38** : 131–134.

Tabata, M., Mizukami, H., Hiraoka, N., Konoshima, M. (1974). Pigment formation in callus cultures of *Lithospermum erythrorhizon. Phytochemistry,* **13** : 927–932.

Tabata, M., Yamamoto, H. and Hiraoka, N. (1971). Alkaloid production in the tissue cultures of some solanaceous plants. In : Les cultures de tissus de plantlets. *Colloq. Intern.* 193, pp. 390–420, Paris, L.N.R.S.

Tafur, S. Nelson, J,D. Delong, C.D. nad Svoboda, G.H. (1975). Antiviral compounds of *Ophiorrhiza mungos.* Isolation of campththecin and 10–methoxy camptothecin J. Nat. products **39 (4)** : 261–262

Takahashi, S., Fujita, Y. (1991). Production of shikonin, In : *Plant Cell Culture in Japan. Progress in the Production of Useful Plant Metabolites by Japanese Enterprises Using Plant Cell culture Technology* .Komanine, A., Misawa, M., DiCosmo, F., (eds.), PP 72–78, CMC Co. Tokyo :

Takahashi, S., Kitanaka, S., Takido, M., Ebizuka, Y., Sankawa, U., Hoson, M., Kobayashi, M. and Shibata, S. (1978). Formation of anthraquinones by the tissue cultures of *Cassia obtusifolia. Planta Med.,* **33** : 389–392.

Takayama, S. (1994). *Tripteryium wilfordii : In vitro* culture and the production of anticancer compounds tripdolide and tripdolide. In : *Biotechnology in Agriculture and Forestry,* vol. 28 Bajaj, Y.P.S. (eds.), pp. 457–468, Springer–Verlag. Berlin

Takayama, S., Takizawa, N., Kuroyanagi, M. (1994). The method and system for large–scale culture of plant roots using aeration–agitation bioreactor. Strategies for scale–up. VIII. *IAPTC Congr.,* Firenze (Abstract) S20–22.

Takeda, J. (1990). Light–induced synthesis of anthocyanin in carrot cells in suspension II. Effects of light and 2,4–D on induction and reduction of enzyme activities related to anthocyanin synthesis. *J. Exp. Bot.,* **41** : 749–754.

Tal, B., Rokem, J.S. and Goldberg, I. (1983). Factors affecting growth and product formation in plant cells growing in continuous culture. *Plant Cell Rep.,* **2** : 219–222.

Tamiki, E., Morishita, I., Nishida, K., Kato, K. and Matsumoto, T. (1973). Process for preparing liquorice extract like material for tobacco flavouring. U.S. Patent No. 3710512.

Tanaka, A. (1994). Immobilization of plant cells. In : *Advances in Plant Biotechnology. Studies in Plant Science,* vol. 4, (Ryu, D.D.Y., Furasaki, S. eds.), pp. 209–220, Amsterdam : Elsevier.

Taya, M., Mine, K., Kino–Oka, M. Tone, S. and Ichi, T. (1992). Production and release of pigments by culture of transformed hairy root of red beet. *J. Ferment. Bioeng.,* **73** : 31–36.

Tisserat, B., Vandercook, C.E. and Berhow, M. (1989). Citrus juice vesicle culture : A potential research tool for improving juice yield and quality. *Food. Technol.,* **43** : 95–100.

Tissut, M. and Egger, K.K. (1972). Les glycosides flavoniques fohares de quelques arbres au cours du cycle vegetatif. *Phytochemistry,* **11** : 631.

Tomita, Y., Uomori, A., Minato, H. (1969). Sesquiterpenes and phytosterols in tissue cultures of *Lindera strychnifolia. Phytochemistry,* **8** : 2249–2252.

Tomita, Y., Uomori, A., Minato, H. (1970). Steroidal sapogenins and sterols in tissue cultures of *Dioscorea tokoro. Phytochemistry,* **9** : 111–114.

Townsley, P.M. (1974). Chocolate aroma from plant cell cultures. *J. Inst. Can. Sci. Technol. Aliment.,* **7** : 76–78.

Tripathi, C.K.M. Basu, S.K, Jain, S. and Tandon, J.S. (1995). Production of coleonol (Forskolin) by root callus cells of plant *Coleus for skohlii,* **17 (4)** : 423–426.

Ulbrich, B., Wiesner, W., Arens, H. (1985). Large–scale production of rosmarinic acid from plant cell cultures of *Coleus blumei,* In : *Primary and Secondary Metabolism of Plant cell Cultures.* Neumann, K.H., Barz, W., Reinhard, E., (eds.) PP 293–303, Berlin : Springer–Verlag.

Ushiyama, K. (1991). Large–scale culture of ginseng, In : *Plant Cell Culture in Japan. Progress in the Production of useful Plant Metabolites by Japanese Enterprises using Plant Tissue Culture Technology.* Komamine, A., Misawa, M., DiCosmo, F. (eds), pp. 92–98, CMC Co. Tokyo

Van Den Berg, A.J.J., Radema, M.H., Labadie, R.P. (1988). Effects of light on anthraquinone production in *Rhamnus purshiana* suspension cultures. *Phytochemistry,* **27** : 415–417.

Van Den Heijden, R., Verpoorte, R., Ten Hoopen, H.J.G. (1989). Cell and tissue cultures of *Catharanthus roseus. Plant Cell Tissue Organ Cult.,* **18** : 231–280.

Van Der Kroi, A.R., Mur, L.A., Beld, M., Mot, J.N.M. and Stuitji, A.R. (1990). Flavonoid genes in *Petunia* : addition of a limited number of gene copies may lead to a suppression of gene expression. *Plant Cell.,* **2** : 218–221.

Van Uden, W., Pras, N., Homan, B., Malingre, T.M. (1991). Improvement of the production of 5–methoxypodophyllotoxin using a new selected root culture of *Linum flavum. Plant Cell Tissue Organ Cult.,* **27** : 115–121.

Van Uden, W., Pras, N., vossebeld, E.M., Mot, J.N.M., Malingre, T.M. (1990). Production of 5–methoxypodophyllotoxin in cell suspension cultures of *Linum flavum. Plant Cell Tissue Organ Cult.,* **20** : 81–87.

Veeresham, C. and Shuler, M.L. (2000). Camptothecine from callus cultures of *Nothapodytes foetida. Biotechnology Letters,* **22** : 129–132.

Veeresham, C., Kokate, C.K. and Venkateshwarlu, V. (1994). Influence of precursors on production of isoquinoline alkaloids in tissue cultures of *Cephaelis ipecacuanha. Phytochemistry,* **35(4)** : 947–949.

Veeresham, C., Kokate, C.K., Venkateshwarlu, V. and Apte, S.S. (1991). Enhanced capsaicin production in immobilized cell cultures of *Capsicum annuum. Indian Drugs,* **29(1)** : 12–14.

Veeresham, C., Kokate, C.K., Venkateshwarlu, V. and Apte, S.S. (1991). Enhanced capsaicin production in immobilized cell cultures of *Capsicum annuum. Indian Drugs,* **29(1)** : 1–4.

Veeresham, C., Raj Kumar, M., Sowjanya, D., Kokate, C.K. and Apte, S.S. (1998). Production of azadirachtin from callus cultures of *Azadirachta indica. Fitoterapia,* **LXIX (5)** : 423–425.

Veeresham, C. and Shuler, M.L. (2004). Production of Withaferin A from cell cultures of *Withania somnifera* in shake flasks vs six well plates. *Plata Medica* (communicated).

Veliky, I.A. (1972). Synthesis of carboline alkaloids by plant cell cultures. *Phytochemistry,* **11** : 1405–1406.

Visvanath, S., Ravishankar, G.A. and Venkatraman, L.V. (1990). Induction of crocin, crocetin, picrocrocin and safranal synthesis in callus cultures of saffron *Crocus sativus* L. *Biotechnol. Appl. Biochem.,* **12** : 336–340.

Wagner, F. and Vogelmann, H. (1977). Cultivation of plant cell cultures in bioreactors and formation of secondary metabolites. In : *Plant Tissue Culture and its Biotechnological Application,* Barz, W., Reinhard, E., Zenk, M.H. (eds.) PP 245–252, Springer–Verlag, Heidelberg

Wagner, H., Stuppner, H., Puhlmann, J., Brummer, B., Deepe, K. and Zenk, M.H. (1989). Gewinnung von immunologisch aktiven Polysacchariden aus *Echinacea*–Drogen und Gewebekulturen. *Z. Phytother.,* **10** : 35–38.

Watasnabe, K., Sato, F., Furuta, M. and Yamada, Y. (1985). Induction of pigment production by S–containing compounds in cultured *Lavandula vera* cells. *Agric. Biol. Chem.,* **49** : 533–534.

Watts, M.J., Galpin, I.J. and Collin, H.A. (1984). The effect of growth regulators, light and temperature on flavor production in celery cultures. *New Phytol.,* **98** : 583–591.

Webb, J.K., Banthorpe, D.V. and Watson, D.G. (1984). Monoterpene synthesis in shoots regenerated from callus cultures. *Phytochemistry,* **23** : 903–904.

Weete, J.D., Venketeswaran, S., Laseter, J.L. (1971). Two populations of aliphatic hydrocarbons of teratoma and habituated tissue cultures of tobacco. *Phytochemistry,* **10** : 939–945.

Weissenbock, G. and Reznik, H. (1970). Anderung des flavonoidmusters wahrend der samenkeimung von *Impatiens balsamina* L. *Z. Pflanzenphysiol.,* **63** : 114.

Wellmann, E., Harazdina, G. and Grisebach, H. (1976). Induction of anthocyanin formation and enzymes related to its biosynthesis by UV light in cell cultures of *Haplopappus gracilis. Phytochemistry,* **15** : 913–915.

Westcott, R.J., Cheetham, P.S.J. and Barraclough, A.J. (1994). Use of organized viable vanilla plant aerial roots for the production of natural vanillin. *Phytochemistry,* **35** : 135–138.

Westphal, K. (1990). Large–scale production of new biologically active compounds in plant cell cultures, In : *Progress in Plant Cellular and Molecular Biology.* Nijkamp, H.J.J., Van Der Plas, L.H.W., Van Aartruk, J. (eds.), PP 601–608, Kluwer. Dordrecht

Whitaker, R.J., Hashimoto, T., Evans, D.A. (1984). Production of secondary metabolite rosmarinic acid by plant cell suspension cultures. *Ann. N.Y. Acad. Sci.,* **435** : 364–366.

Wichers, H.J., Versluis–De Haan, G.G., Marsman, J.W., Harkes, M.P. (1991). Podophyllotoxins in plants and cell cultures of *Linum flavum. Phytochemistry,* **30** : 3601–3604.

Wickremesinhe, E.R.M.and Arteca, R.N. (1991). Habituated callus culture of *Taxus media* cv. Hicksii as a source of taxol. *Plant Physiol,* **96**: 96.

Wickremesinhe, E.R.M.and Arteca, R.N. (1993). *Taxus* callus cultures : Initiation, growth, optimization, characterization and taxolproduction. *Plant Cell Tissue Organ Cult.,* **35** : 181–193.

Wickremesinhe, E.R.M.and Arteca, R.N. (1994a). *Taxus* cell suspension cultures : optimization of growth and production of taxol. *J. Plant Physiol.,* **144** : 183–188.

Wickremesinhe, E.R.M.,and Arteca, R.N. (1994b). Roots of hydroponically plants as a source of taxol and related taxanes. *Plant Sci.,* **101** : 125–135.

Wiedenfeld, H., Furmanowa, M., Roeder, E., Guzewska, J. and Gustowski, W. (1997). Camptothecin and 10–Hydroxy comptothecin in callus and plantlets of *Camptotheca accuminata, Plant Cell Tissue and Organ Culture,* **49** : 213–218.

Wiermann, R. (1973). Uber die Beziehung zsischen Flavonol aufbauenden enzymen, cinem Flavonol unwandelnden enzym und der Akkumulation phenylpropanoider Verbindungen wahrend der Antherenentwicklung. *Planta (Berl),* **110**: 353.

Williams, B.L., and Goodwin, T.W. (1965). The terpenoids of tissue cultures of Paul's Scarlet Rose. *Phytochemistry,* **4** : 81–88.

Williams, R.D.,and Ellis, B.E. (1993). Alkaloids from *Agrobacterium rhizogenes*–transformed *papaver somniferum* cultures. *Phytochemistry,* **32** : 719–723.

Wink, M.,and Hartmann, Th. (1982). Diurnal fluctuation of quinolizidine alkaloid accumulation in legume plants and photomixotrophic cell suspension cultures. *Z. Naturforsch.,* **37c** : 369.

Wjnsma, R. and Verpoorte, R. (1988). Quinoline alkaloids of *Cinchona*, In : *In Cell Culture and Somatic Cell Genetics of Plants,* vol. 5 : *Phytochemicals in Plant Cell Cultures,* Constabel, F., Vasil, I.K. (eds.) PP 337–356. Academic Press. San Diego

Woerdenbag, H.J., Van Uden, W., Frijlink, H.W., Lerk, C.F., Pras, N., Malingre, T.M. (1990). Increased podophyllotoxin production in *Podophyllum hexandrum* cell suspension cultures after feeding coniferyl alcohol as a â–cyclodextrin complex. *Plant Cell Rep.,* 9 : 97–100.

Yamada, Y.,and Hashimoto, T. (1982). Production of tropane alkaloids in cultured cells of *Hyoscyamus niger. Plant Cell Rep.,* 1 : 101–103.

Yamada, Y.,and Sato, F (1982). Production of berberine in cultured cells of *Coptis japonica. Phytochemistry,* 20 : 545–547.

Yamakawa, T., Ishida, K., Kato, S., Komada, T. and Minoda, Y. (1983). Formation and identification of anthocyanins in cultured cells of *Vitis* sp. *Agric. Biol. Chem.,* 47 : 997–1001.

Yamamoto, H., Nakagawa, K., Fukui, H. and Tabata, M. (1986). Cytological changes associated with alkaloid production in cultured cells of *Coptis japonica* and *Thalictrum minus. Plant Cell Rep.,* 5 : 65.

Yamamoto, O. and Yamada, Y. (1986). Production of reserpine and its optimization in cultured *Rauwolfia serpentina* Benth. Cells. *Plant Cell rep.,* 5 : 50–53.

Yamamoto, Y. (1991). Anthocyanin production in plant cell cultures in : *Plant Cell Culture in Japan. Progress in the Production of Useful Plant Metabolites by Japanese Enterprises using Plant Cell Culture Technology.* Komamine, A., Misawa, M., Dicosmo, F. (eds.), PP 114–126CMC Co. Tokyo

Yamamoto, Y., Kinoshita, Y., Watanabe, S. and Yamada, Y. (1989). Anthocyanin production in suspension cultures of high producing cells of *Euphorbia millii. Agric. Biol. Chem.,* 53 : 417–423.

Yamamoto, Y., Mizuguchi, R. and Yamada, Y. (1982). Selection of a high and stable–pigment producing strain in cultured *Euphorbia millii* ells. *Theor. Appl. Genet.,* 61 : 113–116.

Yanagawa, H., Kato, T., Kitahara, Y., Kato, Y. (1972). Chemical components of callus tissues of rice. *Phytochemistry,* 11 : 1893–1899.

Yano, I., Nichols, B.W., Morris, L.J., James, A.T. (1972a). The distribution of cyclopropane and cyclopropene fatty acids in higher plants (Malvaceae). *Lipids,* 7 : 30–34.

Yeoman, M.M, Lindsey, K., Miedzybrodzka, M.B. and McLauchalan, W.R. (1982). Accumulation of secondary products as effect of differentiation in plant cell and tissue cultures. In : Yeomann, M.M., Truman, D.E.S. (eds). Differentiation *in vitro* vol. 4, British Society for cell Biology Symp. PP 65 Univ. Press, Cambridge.

Yeoman, M.M., Holden, M.A., Corchet, P., Holden, P.R., Goy, J.G. and Bobbs, M.C. (1990). Exploitation of disorganized plant cultures for the production of secondary metabolites. In : Charlwood, B.V., Rhodes, M.J.C. (eds.) Secondary products from plant tissue culture. *Proc. Of the Phytochemical Society of Europe,* vol. 30, PP 139 Oxford Science Publications, Calderon Press.

Yeoman, M.M., Miedzybrodzka, M.B., Lindsey, K. and McLauchlan, W.R. (1980). The synthetic potential of cultured plant cells. In : *Plant Cell Cultures : Results and Prespectives.,* pp. 327-343. Sala, F., Parisi, B., Cella, R. and Ciferri, O., Eds. Elsevier, Amsterdam.

Yu, P.L.C., El–Olemy, M.M., Stohs, S.J. (1974). A phytochemical investigation of *Withania somnifera* tissue cultures. *Lloydia,* 37 : 593–597.

Yukimune, Y., Tabata, H., Higashi, Y., Hara, Y. (1996). Methyl jasmonate–induced overproduction of paclitaxel and baccatin III in *Taxus* cell suspension cultures. *Nt. Biotechnol.,* 14 : 1129–1132.

Yun, D.J., Hashimoto, T. and Yamada, Y. (1992). Metabolic engineering of medicinal plants : Transgenic *Atropa belladonna* with an improved alkaloid composition. *Proc. Natl. Acad. Sci. USA,* 89 : 11799–11803.

Zenk, M.H., El–Shagi, H. and Schulte, U. (1975). Anthraquinone production by suspension cultures of *Morinda citrifolia. Plant Med. Suppl.,* 79–101.

Zenk, M.H., El–Shagi, H., Arens, H.,Stockigt, J., Weiler, E.W. and Deus, B. (1977b). Formation of indole alkaloids serpentine and ajmalicine in cell suspension cultures of *Catharanthus roseus,* In : *Plant Tissue Culture and its Biotechnological Application* , Barz, W., Reinhard, E., Zenk, M.H., (eds.) PP 27–43, Springer–Verlag, Heidelberg

Zenk, M.H., El–Shagi, H., Schulte, U. (1975). Anthraquinone production by cell suspension cultures of *Morinda citrifolia. Planta Med.* (Suppl.), 79–101.

Zenk, M.H., El–Shagi, H., Ulbrich, B. (1977a). Production of rosmarinic acid by cell suspension cultures of *Coleus blumei. Naturwissenschaften,* **64** : 585–586.

Zheng, G.Z., He, J.B. and Wang, S. (1983). Cryopreservation of calli and their suspension culture cells of *Anisodus acutangullus. Acta Bot. Sin.,* **25(6)** : 512–517.

Zhiri, A., Maciejewska, K., Jaziri, M., Homes, J. and Vanhaelen, M. (1995). Establishment of *Taxus baccata* callus cultures and evaluation of taxoid production. *Med Fac Lanbouww Univ Gent* **60**: 2111–2113.

Zhong, J., Seki, T., Kinoshita, S. and Yoshida, T. (1991). Effect of light on anthocyanin production by suspended culture of *Perilla frutescens. Biotechnol. Bioeng.,* **38** : 653–658.

Zhong, J.J.,and Yoshida, T. (1995). High density cultivation of *Perilla frutescens* cell suspensions for anthocyanin production : Effects of Sucrose concentration and inoculum size. *Biotechnol. Bioeng.,* **38** : 653–658.

Zubko, M.K., Schmeer, K., Glabgen, W.E., Bayer, E. and Seitz, H.U. (1993). Selection of anthocyanin–accumulating potato (*Solanum tuberosum* L.) cell lines by gamma–irradiated seeds. *Plant Cell Rep.,* **12** : 555–558.

Zwayyed, S.K., Frazier, G.C. and Dougall, D.K. (1991). Growth and anthocyanin accumulation in carrot cell suspension cultures growing on fructose, glucose on their mixtures. *Biotechnol. Prog.,* **7** : 288–290.

Chapter 7

Effect of Cultural Practices on Production of Secondary Metabolites

INTRODUCTION

The phenomenon of altered secondary metabolism in cell cultures has been attributed to various factors. The first of these is the lack of tissue differentiation in some callus and cell suspension cultures. For example root differentiating calluses of *Atropa belladonna* are capable of producing tropane alkaloids whilst non-differentiated cultures of the same plant material not (Thomas and Street 1970). Similarly, differentiated tissue cultures of *Papaver bracteatum* and *Nicotiana tabacum* tend to produce more thebaine and nicotine respectively than that of undifferentiated calluses (Pearson 1978). This may be due to lack of specialized cell structures in some cultures may be further reason for the absence of accumulated secondary metabolites (Krikorian and Steward 1969). The decreased levels of secondary metabolites yields have some times been reported for cultures of consequence of organogenesis e.g. undifferentiated cultures of *Agave wightii* and *Dioscorea deltoidea* yield 1-2-% (Dry weight basis) steroidal sapogenins, but when differentiate to produce bulbils or roots respectively, only trace amount of sapogenins are produced (Kaul and Staba 1968, Sharma and Khanna 1980). The second factor is structural rearrangements of the genomes of cultured cells caused by endoreduplication and/or nuclear fragmentation processes(D'Amato, 1977, Yeoman and Forche, 1980) might lead to significant alterations in the genotypes of a portion of a cell population, thereby causing altered secondary metabolism in these cells.

The third factor in decreased production of secondary metabolite from plant cell culture is epigenetic variability. This phenomena may result in the accumulation of auxin of these cells may drastically reduce the proportion of productive cells in cultures (Mein and Binn 1978, Dougall 1980). Improvements in the secondary metabolite productivity of plant cell cultures can be made by taking explants from parent plants, which accumulate high levels of a particular secondary metabolite. Culture derived from parent plants of *C. roseus* and *N. tabacum* which produce high levels of serpentine and nicotine respectively, produce higher amounts of these secondary metabolites than do comparable cultures derived from parent plants which produce low levels of these metabolites (Zenk *et al* 1977, Kinnersley and Dougall 1980).

7.1. EXPERIMENTAL SYSTEMS

Three major types of *in vitro* cultures have been regularly used in studies on secondary plant metabolites. These are organs, callus and liquid cell suspension cultures. The latter, through containing aggregates of anything up to 200 cells, normally consist of dispersed cell and in this state it has been possible for these to be maintained in large fermentor vessels (King, 1980). For metabolic studies, cell suspension cultures are generally used only after a stabilization period of culture (5-10 generations) in order to avoid the possibility of carry-over of secondary metabolites from parent tissue. Qualitative and quantitative analysis of secondary metabolites is normally carried out on culture media as well as on the cells since some metabolites (particularly alkaloids) are thought to be excreted from the cells into the media. Metabolite yields are usually compared on a product weight per unit weight of cells or volume of medium basis.

7.2. INFLUENCE OF PRODUCTION CONDITIONS

The factors such as external culture conditions (light, temperature and agitation) and internal culture conditions (medium components, pH, degree of mixing and aeration) will influence the production of secondary metabolites from plant cell cultures.

7.2.1. External Culture conditions

7.2.1.1. Light

The characteristics of radiation which influence plant development *in-vitro* are also those which affect plant tissues and cells *in vitro*. The behavior of cultures is influenced by photoperiodicity, light quality and light intensity (Seibertes and Kadkade 1980). For e.g. The activities of the Group 1 and Group 2 enzymes of the flavonoid pathway in cultured cells of *Petroselinum hortense* (parsley) shows an increase when cells are exposed to light for 2 and 4 h respectively. The activities of the Group 1 enzymes, phenylalanine ammonia lyase (PAL) included, can be increased independently of light by transferring dark-grown parsley cells to distilled water. Further, the extent of this PAL-activity change is dependent on the degree of cell dilution. A second increase in PAL activity is observed 5 h after dilution but this is light induced. In determining the active spectral region of PAL induction by light, Stickland and sunderland (1972) have shown that blue light increases both the activity of this enzyme and subsequent anthocyanin production in callus and cell cultures more than 20 flavone and flavone glycosides are produced after cool white light treatment (Grisebach and Hahl brock 1974). Interestingly, Alfermann and Reinhard (1971) reported that they were able to replace light requirement for anthocyanin biosynthesis with auxin treatment in carrot cell cultures. Upon illumination tea callus cultures showed several fold increase in catechin, epicatechin, and leuco anthocyanin synthesis.

Volatile-oil accumulation in cultures grown under continuous Cool White light (250 lx) resembled that of photosynthesizing parent plants (Corduan & Reinhard, 1972) However, the relative composition of the oils could be significantly altered by administering different spectral treatments to culture. Cells grown under continuous red or far-red light ($1.25 \, Wm^{-2}$) produced the same major oil components as those found in dark-grown cultures but others grown under blue light produced an oil composition comparable to that grown under coolwhite fluorescent light (250 lx) on long days (either 15-or 24-h photoperiods). On short days (6-h photoperiod) in the same light quality conditions, culture produced oil accumulations which were a mixture of those obtained separately under continuous light and continuous dark conditions (Nagel & Reinhard, 1975). 'Cool White' light also stimulates the biosynthesis of numerous other secondary metabolites including steroidal sapogenins (e.g. diosgenin in tuber-derived callus and cell suspensions of *Dioscorea*), steroidal alkaloids (e.g. solasodine in *Solanum* (Seibert & Kadkade, 1980) and some alkaloids (e.g. serpentine in *C. roseus* cell cultures (Roller,

1978). In these cases, the promotory effects of light may possibly be due to its known influence on the rate of uptake of sugars and nutrients into plant cells. Enhanced uptake of both (^{14}C) sucrose and nitrate is known to be stimulated by light in etiolated plant tissues, both of these responses being under the control of phytochrome (Goren and Galston, 1966, Jones and Sheard, 1975) through its mediation of increases in intracellular ATP levels that are necessary for the active uptake of nutrients by cells (White and Pike, 1974).

In contrast to the triggering effects of light on secondary metabolism in tissue cultures, several reports indicate that light can have inhibitory effect (Tabata et al., 1972). Higher levels of accumulation of secondary metabolites under dark rather than light conditions suggest that photo degradation of certain metabolites and/or enzymes may occur although there is no direct evidence to substantiate this (nicotine photodegradation, Doumery and Chouteau 1975).

7.2.1.2. Temperature

In general, a temperature range of $25 \pm 2^{\circ}$C has been found to be optimal for growth. Above 30°C, especially above 53°C and below 21°C growth usually diminishes rapidly. Nettleship and Slaytor (1974) showed that optimal growth of callus occurred at 30°C while maximum alkaloid production was attained of 25°C with levels of production decreasing rapidly at higher temperatures. For instance, the rate of both sucrose and amino-nitrogen utilization in *Ipomoea* cell suspension cultures is maximal between 30 and 32°C and both of these rates decline by about 25% when cells are moved from 30 to 25°C whereas growth rate declines little (Rose and Martin, 1975). The fatty acid content of cultured plant cells was increased in cultures grown at suboptimal growth temperatures such as 15°C (MacCarthy and Stumpf, 1980). The production of Taxol from cell cultures of *Taxus SP* the maximum production was at 22°C (Personal communication with Shuler, M.L., Cornell University, USA).

7.2.1.3. Culture vessel agitation

In the rotary shake-flask cultures, the rotation speed of a shaker (normally 90-120 rpm) can have an important effect on growth and metabolite accumulation. Rajasekhar et al (1971) in a study of the effect of the shaking rate on the growth of *Atropa* and *Acer* cultures, concluded that reduced growth at suboptimal shaking speeds was neither due to oxygen deficiency nor to accumulation of carbon dioxide but rather to either an unknown volatile toxic factor (not ethylene) or to restricted nutrient uptake resulting from a stationary liquid-phase boundary surrounding cells. Pearson (1978) in a study on tobacco cells indicated that raised culture agitation (150 rpm) induces high nicotine yields, where as normal agitation (110 rpm) causes slight inhibition of growth and an almost total suppression of nicotine production.

7.2.2. Internal cultural environment

7.2.2.1. Inoculum and preculture

Size of the Inoculum. The size of the inoculum not only exerts a critical influence on the survival of an induced culture, it is also important for the determination of the production capacity e.g. serpentine production in *C.roseus* cultures, the prevailing cell density is decisive, while only those *Tagetes patula* aggregates between 1 × 12 cm diameter synthesize and excrete thiophenes.

Effects on Subcultures. The size dependence of inocula may also be one of the reasons for the variations in secondary compound content frequently observed in subcultures of originally highly productive lines. If the size or number of cells of transferred cultures is not standardized, it varies due to subjective reasons, as manifested in variable secondary compound contents.

Influences from the Preculture. In addition, the mode of secondary compound accumulation is heavily influenced by preliminary treatment of the inoculum. Thus, although anthocyanin synthesis is always inducible in *Haplopappus gracilis* (syn. *Machaeranthera gracilis*) suspension cultures by exposure to blue light, the duration of the lag phase before accumulation begins and the elapsed time from the onset of irradiation to maximum production depend on the duration of preculturing in darkness (Fritsch *et al.,* 1971).

7.2.2.2. Medium components

Components of the basic medium exert a substantial influence on callus and suspension cultures. The medium most widely used is that MS (1962), which was devised originally for growth of tobacco callus cultures. Generally, medium conditions which most frequently support active secondary metabolism are those which limit rapid cell division and lead to a comparatively early cessation of exponential growth, since, the production of secondary metabolites is inversely proportional to the growth of plant cells in culture.

7.2.2.3. Phytohormones

Plant growth regulators are effective triggers of secondary metabolism *in vivo* (Bohm,1980). Similarly, *in vitro* both the quality and quantity of auxins initially present in media or administered during the course of culture development have a marked effect on primary (Everett, 1978,1981) and secondary metabolism (Gamborg *et al.,*1971). Plant cell cultures require the addition of growth regulators, i.e. auxins and cytokinins, to media for consistent growth by cell division. Growth by differentiation or morphogenesis, on the other hand, can usually be induced by lowering the auxin concentration or by supplying less active growth substances. Since the production of secondary metabolites in plant cell cultures is a function of both cell multiplication and division, growth regulators have a major role in determining the potential productivity of a given culture (Kurz and Constabel 1979, Staba 1980). The effect of auxin type on secondary metabolite synthesis has been investigated in cultures of numerous species. For e.g. the production of thebaine from cell cultures of *Papaver bracteatum,* IAA was found to be better then NAA or 2,4-D. Furuya *et al.* (1971) found that in callus cultures of *N. tabacum* cv. Bright yellow cultured for five years in the presence of 2,4-D no alkaloids were detected, while nicotine, anatabine and anabasine were readily found in callus growing in media supplemented with IAA. 2,4-D has generally been found to be less suitable for triggering secondary metabolism in tissue cultures than either IAA or NAA. There are only a few reports of metabolite levels being increased by raising auxin levels, e.g. maximum carotenoid and ubiquinone contents of carrot cell cultures were induced by the comparatively high 2,4-D level of 10mg $^{-1}$ (Ikeda *et al.,* 1976).

In most plant tissue culture studies auxin has been employed in combination with cytokinins. In callus cultures of *Datura tatula,* kinetin showed no noticeable effect on growth yet it was inhibitory to alkaloid production at high concentrations (Tabata *et al.,* 1971). In *Scopolia maxima* cultures, a comparatively high concentration of kinetin promotes alkaloid production. In *N. tabacum* callus and suspension cultures, kinetin levels in excess of 10^{-5} M totally suppress nicotine production (Shiio and Ohta, 1973, Pearson, 1978). It should be stressed that the combined effects of auxins and cytokinins on secondary metabolite production are difficult to assess particularly since there is a lack of data on relative endogenous levels of these growth regulators in the cultured cells.

Hara *et al.* (1988) reported that GA$_3$ (Gibberellic acid) result in a marked increase in berberine production in *Coptis japonicum* tissue culture, from 0.22 to 0.68 mg/l. The increased production of berberine is because of the stimulation of sugar uptake and inhibition of starch deposition. The resulting increased carbohydrate supply is therefore available for unhindered secondary metabolite production. Smith *et al.,* (1997) reported that the GA$_3$ (0.01 mg/l) increased the growth rate of hairy roots of *A. annua* by 25%, with a slight increase in Artemisin in compared to controls. However, Marshall and Staba (1976) were unable to detect any effects of a single concentration (0.3 mg/l) of GA on diosgenin production in *Discorea* cultures.

Abscisic acid (ABA) inhibits the growth of embryos. ABA inhibited the growth and production of phenolic compounds by tobacco callus cultures (Li *et al.,* 1970). Ethylene concentration in the medium is usually regulated by 2-chloroethylene phosphoric acid (ethephon). Ethylene, considerably increases secondary metabolite production in *Coffea arabica* (Caffeine, theobromine) and *Thalictrum rugosum* (berberine).

7.2.2.4. Macro and Micronutrients

Generally, increased levels of nitrate, potassium, ammonium and phosphate tend to support rapid cell growth while depletion or deficiency of some of these nutrients is associated with growth limitation and concomitant secondary metabolism. Significantly, lack of phosphate more than any other nutrient stimulates secondary metabolite biosynthesis for e.g. *Peganum* callus cultures (Nettleship and Slaytor 1974). The influence of intracellular phosphate concentrations on secondary compound production is known in microorganisms. Thus, the activity of certain enzymes is regulated by the cell energy level (AMP, ADP, and ATP). Similar influences have been observed in suspension cultures of higher plants (Knobloch and Berlin 1983). However, Carew and Krueger (1977) found that raised phosphate levels increased the yield of indole alkaloids in the medium of *C. roseus* and Zenk *et al.,* (1975) obtained a 50% increased in anthraquinones in *M. citrifolia* cultures when phosphate was increased to 50mm. Nitrogen in the plant tissue culture is two types. 1) organic nitrogen, usually; casein hydrolysate and peptone are used as organic nitrogen sources. The Nitrate (2-4.5 g/l) induced the production of *Morinda citrifolia* (anthraquinone) and *Lithospermum erythrorhizon* (Shikonin) tissue cultures.

Except for Aspartic acid, amino acids are generally not accepted as sources of organic nitrogen. The absolute amount of inorganic nitrogen, usually supplied as ammonium and/or nitrate, is usually of minor importance. Production capacity usually depends on the ratio of NH_4^+ to Nitrate.

7.2.2.5. Carbon sources

Carbon may be supplied either CO_2 in a photoautotrophic culture, or as carbohydrates in a heterotrophic cultures. *Vitis vinifera, Chenopodium, Morinda* and *Nicotiana* accumulated CO_2-typical secondary products (Acyclicnerol, Anthraquinone and Nicotine respectively). In general, raising the initial sucrose level leads to an increase in the secondary metabolite yields of cultures. The influence on secondary compounds depends on the source of carbon employed, its concentration, as well as on the biosynthetic process studied. The optimum concentration seems to vary according to the plant species. A high sucrose concentration has been found to be necessary for a high yield of shikonin derivatives in *Lithospermum erythrorhizon* (Mizukami *et al.,* 1978), and anthraquinone in *Morinda citrifolia* (Zenk *et al.,* 1975). In contrast, Ikeda *et al.* (1976) found that the ubiquinone content in tobacco cultures tended to decrease with an increase in the sugar concentration. Carbon sources other than sucrose and glucose have been tested for their suitability for supporting secondary metabolite accumulation in cultures. Zenk *et al.* (1975) tested 14 carbohydrates at 2% (w/v) levels and found that sucrose gave the highest yields of anthraquinones in *M. citrifolia* cultures. Kim *et al.,* (1995) reported that the high levels of a sugar increase the osmotic potential, though the role of osmotic pressure on the synthesis of secondary metabolite is not clear. For the production of *Catharanthus* alkaloids, 8% sucrose solution without any other nutrients was used as a production medium (Knobloch and Berlin 1983). Of all the carbon sources, the best carbon source for taxol production in cell cultures of *Taxus brevifolia* was found to be fructose 6% (Kim *et al.,* 1995). The various carbon sources on growth and Taxanes production from cell cultures of *T. brevifolia* is shown in Table 7.1 and 7.2. The possible mode of action of sucrose in cell cultures is :

Table 7.1. Effect of various carbohydrates on the growth of T. brevifolia cell cultures in growth medium. (Kim et al., 1995)

Sugar	DCW (g/l)
sucrose	15.9
lactose	14.8
galactose	16.7
glucose	15.3
fructose	15.7
mannitol	7.5
sorbitol	6.8

Table 7.2. Effect of various 8% sugar solution without any other nutrients on the production of taxol. (Kim et al., 1995)

8% sugar solution	Taxol (mg/L)
sucrose	0.012
lactose	0
maltose	0.01
fructose	0.172
galactose	0.123
glucose	0.061
sorbitol	0.135
mannitol	0.049
2% sucrose + 6% sorbitol	0.11
2% sucrose + 6% mannitol	0.142

1. Extension of the stationary growth phase (for example upto 30-45 days in cultures of Paul's scarlet roses)
2. Inhibition of particle-fixed RNAse in connection with increasing activities of soluble RNAse
3. Inhibition of endogenous auxin synthesis by the sucrose-cleaving product glucose, produced by cell exudates
4. Influence on differentiation an essential role of all carbohydrates characterized by increased activities of enzymes of the pentose-phosphate pathway.

The effects of other carbohydrates (galactose, glucose, raffnose) are less impressive.

7.2.2.6. Culture medium pH

Optimal growth in plant cell cultures usually occurs in media with initial pH values in the ranges 5-6. Media containing undefined organic components like casein hydrolysate and yeast extract are usually well buffered so that the pH changes relatively little during the course of culture development. However, in media without these substances, shifts in pH during culture can be dramatic. Veliky

(1977) found that *Ipomoea* cell culture, which transform tryptophan into a variety of indole metabolites such as tryptophol, yield double the amount of this metabolite when cells are grown at pH 6.3 in a pH-stat compared with yields obtained when cells are grown in uncontrolled culture. When the pH drops to 4.8, tryptophan accumulation is completely inhibited.

7.2.2.7. Aeration and culture mixing

In the most widely used small-scale liquid culture systems, i.e. 40 to 200 ml rotary shake flask cultures, upon which most current data on secondary metabolism *in-vitro* are based, culture agitation and aeration are interdependent cultural components. These can influence the yield of secondary metabolites *in-vitro*. In conical flask cultures, culture volume (with its obvious influence on oxygen absorption coefficients, 'OAC' associated with the area of culture medium having an air liquid interface) has a marked effect on nicotine biosynthesis in tobacco cultures. At low culture volumes (conditions which raise OAC values), nicotine production is enhanced. Wagner and Vogelmann (1977) have also reported the importance of aeration and culture agitation in studies on large-scale fermentors. (see chapter 20).

REFERENCES

Alfermann, W. and Reinhard, E. (1971). Isolierung anthocyanhaltzer and anthocyanfreir Gewebestamme von iDaucus carota. *Einfluss von Auxinen* auf die Anthocyanbildung. *Experimentia,* **27** : 353.

Alfermann, A.W. and Reinhard, E. (1978). Possibilities and problems in production of natural compounds by cell culture methods. In *Production of Natural Compounds by Cell Culture Methods,* ed. A.W. Alfermann & E. Reinhard, pp. 3-15, BPT Report, Gesellschaft fur Strahlen und Umweltforschung, Munich.

Bohm, H (1980) The formation of secondary metabolites in plant tissue and cell cultures. *International Review of cytology, Supplement* **11B,** 183-208.

Carew, D.P. and Krueger, R.J. (1977). *Catharanthus roseus* tissue culture : the effects of medium modifications on growth and alkaloid production. *Lloydia,* **40** : 326.

Corduan, G. and Reinhard, E. (1972). Synthesis of volatile oils in tissue cultures of *Ruta graveolens*. *Phytochemistry,* **11** : 917-22.

D'Amato, F. (1977). Cytogenetics of differentiation in tissue and cell cultures. In : *Plant Cell tissue and Organ Culture,* ed. J. Reinert and Y.P.S. Bajaj, pp. 343-57, Berlin, Heidelberg and New York : Springer-Verlag.

Dougall, D.K. (1980). Nutrition and metabolism. In *Plant Tissue Culture as a Source of Biochemicals,* ed. E.J. Staba, pp. 21-58, Boca Raton, Florida, CRC Press.

Doumery, B. and Chouteau, J. (1975). Photodegradation of chlorophyll pigments and nicotine. *Annales du Tabac.* Section 2, 183-200.

Everett, N.P., Wang, T.L., Gould, A.R. and Street, H.E. (1981). Studies on the control of the cell cycle in cultured plant cells. 2. Effects of 2,4-dichlorophenoxyacetic acid (2,4-D). *Protoplasma,* **106** : 15-22.

Everett, N.P., Wang, T.L. and Street, H.E. (1978). Hormone regulation of cell growth and development *in vitro*. In *Frontiers of Plant Tissue Culture,* ed. T.A. Thorpe, pp. 307-16, Calgary : University of Calgary.

Fritsch, H., Hahlbrock, K. and Grisebach, H. (1971). Biosyfnthese von Cyanidin in Zell suspensionkulturen von *Haplopappus gracilis. Zeitschrift fur Naturforschung,* **266** : 581.

Furuya, I., Kojima, H. and Syono, K. (1971). Regulation of nicotine biosynthesis by auxins in tobacco callus tissues. *Phytochemistry,* **10** : 1529-32.

Gamborg, G.L., Constabel, F., La Rue, I.A.G., Miller, R.A. and Steck, W. (1971). The influence of hormones on secondary metabolite formation in plant cell cultures. In :*les Cultures de Tissus de Plantes.* Pp. 335-44, Paris: Centre National de la Recherche Scientifique.

Goren, R. and Galston, A.W. (1966). Control by phytochrome of ^{14}C-sucrose incorporated into buds of etiolated pea seedlings. *Plant Physiology,* **41** : 1055-64.

Grisebach, H. and Hahlbrock, K. (1974). Enzymology and regulation of flavonoid and lignin biosynthesis in plants and plant cell suspension cultures. In: *metabolism and Regulation of Secondary Plant Products,* ed. V.C. Runeckles & E.E. Conn., pp. 21-52, New York : Academic Press.

Hara,Y.,Yoshioka, T., Morimoto,T., Fujita, Y., and Yamada,Y.,(1988) Enahancement of berberine production in suspension cultures of *Coptis japonica* by gibberellic acid treatment, *J.Plant.Physiol.* 133, 12-15.

Ikeda, T., Matsumoto, T. and Noguchi, M. (1976). Formation of ubiquinone by tobacco plant cells in suspension culture. *Phytochemistry,* **15** : 568-9.

Ikeda, T., Matsumoto, T. and Noguchi, M. (1977). Effects of inorganic nitrogen sources and physical factors on the formation of ubiquinone by tobacco plant cells in suspension culture. *Agricultural and Biological Chemistry,* **41** : 1197-201.

Jones, R.W. and Sheard, R.W. (1975). Phytochrome, nitrate movement and induction of nitrate reductase in etiolated pea terminal buds. *Plant Physiology,* **55** : 954-9.

Kaul, B. and Staba, E.J. (1968). *Dioscorea* tissue cultures. 1. Biosynthesis and isolation of diosgenin from *Dioscorea deltoidea* callus and suspension cells. *Lloydia,* **31** : 171-9.

Kim, J.H., Yun, J.H., Hwang, Y.S., Byun, S.Y. and Kim, D.I. (1995). Production of *Taxol* and related Taxanes in *Taxus brevifolia* cell cultures : Effect of sugar, *Biotechnology Letters,* **17(1)** : 101-106.

King, P.J., (1980) Cell proliferation and growth in suspension cultures. *International review of cytology supplement* **11A,** 25-54.

Kinnersley, A.M. and Dougall, D.K. (1980). Correlation between the nicotine content of tobacco plants and callus cultures. *Planta,* **149** : 205-6.

Knobloch, K.H., and Berlin, J., (1983) Influence of phosphate on the formation of the indole alkaloids and phenolic compounds in cell suspension cultures of *Catharanthus roseus.* 1. Comparison of enzyme activities and products accumulation. *Plant cell tissue organ culture.* **2,** 33-340.

Krikorian, A.D. and Steward, F.C. (1969). Biochemical differentiation the biosynthetic potentialities of growing and quiescent tissue. In *Plant Physiology,* vol. 1 B, ed. F.C. Steward, pp. 227-326, New York : Academic Press.

Kurz, W.E.W. and Constabel, F. (1979). Plant cell cultures, a potential source of pharmaceuticals. *Advances in Applied Microbiology,* **25** : 209-40.

Li, H.C., Rice, E.L., Rohrbaugh, L.M. and Wender, S.H. (1970). Effects of abscisic acid on phenolic content and lignin biosynthesis in tobacco tissue culture. *Physiologia Plantarum,* **23** : 928-36.

MacCarthy, J.J. and Stumpf, P.K. (1980). Effect of different temperatures on fatty-acid synthesis and polyunsaturation in cell suspension cultures. *Planta,* **147** : 389-95.

Marshall, J.G., and Staba,E.J., (1976). Hormonal effects on diosgenin biosynthesis and growth in Discorea deltoidea tissue cultures. *Phytochemistry,* **15,** 53-55.

Meins, F., Jr & Binns, A.N. (1978). Epigenetic clonal variation and the requirement of plant cells for cytokinins. In *The Clonal Basis for Development,* ed. S. Subtelny & I.M. Susses, pp. 185-201, New York : Academic Press.

Mizukami,H., Konoshima, M and Tabata, M (1978) Variation in pigment production in *Lithospermum erythrorhizon* callus cultures. *Phytochemistry,* **17,** 95-97.

Murashige, T. and Skoog, F. (1962). A revised medium for rapid growth and bioassays with tobacco tissue cultures. *Physiologia Plantarum.* **15** : 473-97.

Nagel, M. and Reinhard, E. (1975). Das atherische Ol der Calluskulturen von *Ruta graveolens* II. Physiologic zur Bildung des atherischen Oles. *Planta Medica,* **27** : 264.

Nettleship, L. and Slaytor, M. (1974). Adaptation of *Peganum harmata* callus to alkaloid production. *Journal of Experimental Botany,* **25** : 1114-23.

Pearson, D.W. (1978). Nicotine production by tobacco tissue cultures. Ph.D. thesis, Nottingham University.

Rajasekhar, E.W., Edwards, M., Wilson, S.B. and Street, H.E. (1971). Studies on the growth in culture of plant cells XI. The influence of shaking rate on the growth of suspension cultures. *Journal of Experimental Botany,* **22** : 107-17.

Roller, U. (1978). Selection of plants and plant tissue culture of *Catharanthus roseus* with high content of serpentine and ajmalicine. In *Production of Natural Compounds by Cell Culture* Methods, ed. A.W. Alfermann and E. Reinhard, pp. 95-108, Munich : Gesellschaft fur Strahlen und Umwelt-forschung MBH.

Rose, D. and Martin, S.M. (1975). Growth of suspension cultures of plant cells (*Pomoea* sp.) at various temperatures. *Canadian Journal of Botany,* **53** : 315-20.

Seibert, M. and Kadkade, P.G. (1986). Environmental factors : A light In *Plant Tissue Culture as a Source of Biochemicals*, ed. E.J. Staba, pp. 123-42, Boca Raton, Florida, CRC Press.

Sharma, O.P. and Khanna, P. (1980). Studies on steroidal sapogenins from tissue cultures of *Agave wightii*. *Lloydia,* **43** : 459-62.

Shiio, I. and Ohta, S. (1973). Nicotine production by tobacco callus tissues and effect of plant growth regulators. *Agricultural and Biological Chemistry,* **37** : 1857-64.

Smith, T.C., Weathers, P.J., Cheetham, R.D. (1997). Effects of gibberellic acid on hairy root cultures of *Artemisia annua* : growth and artemisinin production. *In vitro Cell Dev. Biol.,* **33** : 75-79.

Staba, E.J. (1980). *Plant Tissue Culture as a Source of Biochemicals.* Boca Raton, Florida : CRC Press.

Stickland, R.G., and Sunderland, N., (1972) Production of anthocyanins, flavanols and chlorogenic acids by cultured callus tissue of *Haplopappus gracilis. Annals of Botany,* **36,** 443-457.

Tabata, M., Yamamoto, H., Hiraoka, N. and Konoshima, M. (1972). Organization and alkaloid production in tissue cultures of *Scopolia parviflora. Phytochemistry,* **11** : 949-55.

Tabata, M., Yamamoto, H.,and Hiraoka, N.(1971)Alkaloid production in the tissue cultures of some solanaceous plants. In: *les cultures de tissus de plantes,* Paris: Centre national de la recherhe Scientifique, PP 390-402.

Thomas, E. and Street, H.E. (1970). Organogenesis in cell suspension cultures of *Atropa belladonna* L. and *Atropa belladonna* cultivars *lactea* Doll. *Annals of Botany,* **34** : 657-69.

Veliky, I.A. (1977). Effect of pH on tryptophol formation by cultured *Ipomoea* sp. plant cells. *Lloydia,* **40** : 482.

Wagner, F. and Vogelmann, H. (1977). Cultivation of plant tissue cultures in bioreactors and formation of secondary metabolites. In *Plant Tissue culture and its Biotechnological Applications,* ed. W. Barz, E. Reinhard and M.H. Zenk, pp. 245052, Berlin, Heidelberg and New York, Springer-Verlag.

White, J.M. and Pike, C.S. (1974). Rapid phytochrome-mediated changes in adenosine 5'-triphosphate content of etiolated beans bugs. *Plant Physiology,* **53** : 76.

Yeoman, M.M. and Forsche, E. (1980). Cell proliferation and growth in callus cultures. *International Review of Cytology,* Supplement, HA. 1-24.

Zenk, M.H., El-Shagi, H., Arens, H., Stockigt, J., Weiler, E.W. and Deus, B. (1977). Formation of the indole alkaloids, serpentine and ajmalicine in cell suspension cultures of *Catharanthus roseus.* In *Plant Tissue Culture and its Biotechnological Application,* ed. W. Barz, E. Reinhard and M.H. Zenk, pp. 27-43, Berlin, Heidelberg and New York, Springer-Verlag.

Zenk, M.H., El-Shagi, H. and Schulte, U. (1975). Anthraquinone production by cell suspension cultures of *Morinda citrifolia. Planta Medica Supplement,* pp. 79-101.

Chapter 8

Selection and Screening of Cultured Plant Cells

INTRODUCTION

A major limitation to the exploitation of plant cell cultures for the bioproduction of valuable secondary metabolites for use as therapeutic agents, flavors, food additives and pharmaceuticals etc. is the low and variable rate of production in many cultures. Plant cell cultures are genetically heterogeneous (Change in chromosomal and intra chromosomal rearrangement). In addition, epigenetic stability causes genetic instability leading to product accumulation only in some population of cells. Productivity in such mixed populations of cells is liable to be unstable since change any factor, environmental or nutritional which favors the growth of a particular sub population of cells will change the composition of the culture. The variation in culture offers the opportunity of isolating from the mixed population for variant cells with high productive capacity and developing from them cell lines of high productivity.

The overall production of a secondary metabolite in a cell culture depends on the rate of accumulation with in the productive cells and the proportion of such cells in the culture. The reasons for low production in cultures may be due to :

1. Competition between primary and secondary pathways for common intermediates.
2. Low levels of expression of key enzymes at rate limiting steps in a pathway or lack of gene expression.

Screening and selection are often used as interchangeable terms. Screening is a passive technique by which a great number of cells alone analyzed for a certain trait and those showing the desired features are further cultivated and screened. In the screening program unchanged wild type cells are not killed but only eliminated by preferential isolation of cell clumps with the desired trait. Selection is an active process, which deliberately favors only the survival of the desired variant while wild type cells are killed.

8.1. CULTURE SYSTEMS FOR SCREENING AND SELECTION

Screening and selection for wanted traits can be done with callus or cell cultures, with plated cell, or protoplasts. The selection of culture systems depends upon the purpose of program for e.g. protoplasts are used for whole plant regeneration.

8.1.1. Callus Cultures

The easiest system for screening and selection program is callus cultures. Several cell lines, which accumulate high levels of various colored compounds, have been selected over the years by visual selection. Callus which show the desired coloration is picked out and sub cultured until a pure cell line is established. Eichenberger (1951) reported the isolation of a *Daucus carota* callus line high in β-carotene. Differences in anthocyanin content in *Zea mays* endosperm callus led to the isolation of stable lines with different pigment levels. Stable cell lines, which accumulated high levels of red-betalain pigments, were selected from the white callus of *Beta vulgaris* by visual selection. The callus can be checked visually for altered pigmentation by the naked eye, for colored spots with UV lamps at a different intervals. Screening for pigments (Chlorophyll, anthocyanins, napthoquinone, betalanins, and carotenoids) can easily and successfully done with calli.

8.1.2. Suspension Cultures

The selection program with suspension cultures can be performed with a fine, rapidly growing suspension consisting of aggregates up to 50 cells or with a rather lump culture. Fine suspension culture is suitable in selection for resistance. The selective agent (a pesticide or metabolite) is added to the culture medium at a concentration, which kills all sensitive cells. The toxic compound equally affects all cells and thus the sensitive one should be eliminated. The disadvantage, however, is that surviving resistant cell population, will most likely be a mixture of cells being resistant for several biochemical reasons. The selection program for resistance in cell culture, one could normally detect growth of resistant population after 4 weeks, even some times only after 12 weeks.

8.1.3. Plating of Cells and Protoplasts

To avoid inherent problems in the use of callus or cell culture for selection and screening, the plating of cells or protoplasts on solid media or matrices may be an ideal alternative. In the plating of cell the first step is to prepare single cells or small sized aggregates for plating. The cells have to be plated at low densities allowing growth of individual, clearly separated colonies that can be isolated and sub cultured. Unlike microbial cells, cultured plant cells require a minimum inoculum density for growth.

8.1.3.1. Preparation of Single Cells and Protoplasts for Plating

The easiest way to obtain a suspension of single cells is to filter successively cells of a fine suspension culture through sieves of decreasing mesh-width from 500-75 μm. The obtained suspension will consist of single cells and small aggregates, which most likely originated from a single cell. If the cell cultures contain large aggregates, adding pectinase to the culture medium 1 or 2 days before sieving can disrupt it (Morris and Fowler, 1981). Morris *et al.* (1983) described a method for rapidly growing rather fine suspensions from immobilization of the cells in alginate. Single cells were then released from the alginate beads.

Single cells for plating can also be prepared from differentiated tissues of intact plants. One should start screening and selection program from cells freshly prepared from the plant rather from very old suspension culture. The most ideal tissue for the isolation of single cell is leaf and mesophyll cells. Mesophyll cells can be isolated mechanically or enzymaticaly (Colman and Mawson 1979, Schwenk 1981, Jensen et al 1971). In order to be sure that a selected cell line is indeed derived from a single cell selection, schemes should usually be started from protoplasts of cultured cells or mesophyll cells.

8.1.3.2. Plating Technique

Bergmann (1960) introduced the technique for plating of cells on solid media. In order to select a single cell, cell suspension must be diluted to densities at which plant cells do not grow without special precautions.

$$\text{Plating efficiency} = \frac{\text{Number of colonies/plate}}{\text{Number of viable cellular units per plate}} \times 100$$

A colony can be formed from every plated cellular unit. It is desirable to obtain highest possible plating efficiency of low densities.

8.1.4. Micro culture of Single Cells

This technique is not useful for the initial stages of selection program. The problem of minimum effective density will be overcome by culturing microdroplet culture.

8.1.5. Replica Technique

This technique was successfully used for bacteria and can be applied for plant cell culture also. Cells are placed on a nylon net with a mesh width of 500 μm on the master agar plate after the cells had grown for 10-14 days. The colonies grew through the net. After 20 days the net with the colonies was transferred to another petri dish and the net was removed from *Morinda citrifolia* cells. (Schulte and Zenk, 1977).

8.2. METHODS OF SCREENING AND SELECTION

8.2.1. Analytical Screening

The principle involved in the analytical screening technique is that the cells of wild type population are separately analyzed for the desired trait. This is mainly used for selection of high productive variant cell lines (Table 8.1). This technique is divided into:

(a) Direct Analytical Screening.
(b) Indirect Analytical Screening.

Table 8.1. The variant lines with enhanced yields established by analytical screening

Plant species	Culture system	Secondary Metabolite	Yield of wild type	Variant	References
Lavandula vera	Callus	Biotin	60 μg*	150 μg/g	Watanabe et al 1982
Lithospermum erythrorhizon	Callus	Shikonin	50 μg*	1mg	Mizukami et al (1978)
Euphorbia millii	Callus	Anthocyanin	$1,1 A_{530}$ per mg*	7,5	Yamamoto et al (1982)
Coptis japonica	Plated clones	Berberine	50 mg g^{-1}$	82	Sato and Yamada (1984)
Nicotiana tabacum	Plated clones	Nicotine	7 mg g^{-1}$	25	Ogino et al (1978)
Daucus carota	Plated clones	Anthocyanins	$1,2 A_{530}$ per mg$ 3,5	3,5	Kinnersley and Dougall (1980)

* Fresh weight basis; $ Dry weight basis

8.2.1.1. Direct Analytical Screening

Some secondary metabolites have an intense color allowing direct visual detection. Altered pigmented areas of callus can easily be detected. Alferman *et al.* (1975) isolated cell clones from callus of

Daucus carota capable and non-capable of producing anthocyanins. Yamamoto, et al. (1982) described a simple method for screening process high and stable anthocyanin producing cultures of *Euphorbia millii*. The initiated callus divided into many segments and placed the various segments separately on agar medium. One half of the segments were analyzed and the other half was sub cultured. The red pieces or the plates were selected, divided and analyzed before. The process is continued and after 23 passages the stable cell line was selected, there was a 7-fold increase compared to wild type cells. Shikonin derivatives, napthoquinone pigments of commercial interest for pharmaceutical and cosmetic industry are produced by callus cells of *Lithospermum erythrorhizon*. Tabata and his co-workers (1978) established high producing strains by repeatedly sub culturing only the red areas of callus. The high producing cell line yielded shikonin derivatives only on MS solid medium but not in LS liquid medium. The direct screening method not only used for colored compound but also for fluorescent compounds. Sasse et al (1982) screened callus of *Peganum harmala* under UV light for fluorescent spots. By sub-culturing only these areas, the culture produced 10 fold increase in harman alkaloids. Deus and Zenk (1982) used this method to isolate colonies from *Catharanthus roseus* with high fluorescence levels, which arose mainly from the accumulation of indole alkaloids.

8.2.1.2. *Indirect Analytical Screening*

In this method analysis of cell extracts is done to know the quality of clones. The clones under investigation have to be divided for subculture and chemical analysis. Ogino *et al.* (1978) used this technique for selection of high yield tobacco cell. He kept a small samples of tobacco callus on filter paper and squeezed them thoroughly between two glass plates. The cell sap was absorbed by the filter paper, which was then sprayed with Dragendroff's reagent. The alkaloid content of samples was estimated from the color reaction. By this they selected a cell line capable of producing 2.5% nicotine.

Screening of large number of clones by the individual evaluation of their cell extracts by measuring absorption, fluorescence and color reaction of chromatographic technique is very time consuming and tedious. So Zenk et al (1977) used Radio immunoassay (RIA) technique to screen for variant lines. They successfully employed this for the selection of high yielding cell lines from *C.roseus* using production media. RIA is very sensitive as it can detect in nanomolar or picomolar levels quantitatively.

8.2.2. Positive Selection System

Analytical screening for cell strains with altered biosynthetic potential is a passive process and will only give positive results when a trait is sufficiently expressed in the wild type population. The cells are made to adapt to the selective compound, then the selected characteristic is only maintained in the presence of the selective agent. By this you can select resistant cell lines for e.g. salt resistance, herbicide or phytotoxin.

Plant cell cultures synthesize biotin from the precursors of pimelic acid and alanine. High concentration pimelic acid is toxic to cells while they can tolerate high levels of biotin. Watanabe et al. (1982) used pimelic acid for developing high producing cell line of *Lavanduala vera* over producing strain of biotin they used pimelic acid as selective agent. The over producing strain rapidly detoxifies the pimelic acid rather than wild type cells (Table 8.2).

Many resistant cell lines have been selected by using anti metabolites, which are analogues of primary metabolites, can also yield high yielding cell lines. Resistance to 5 MT could be accounted for by an altered feed back sensitivity of anthranilate synthase (Widholm, 1972; Carlson and Widholm, 1978). This allows the resistant cells to over produce L tryptophan thus diluting the toxic effect of 5 MT.

Table 8.2. Examples of resistant cell lines selected from plant tissue cultures

Type of resistance	Plant	Culture system	Reference
i) Physical stress			
UV light	*Rosa damascena*	S	Murphy et al (1979)
Water stress	*Lycopersicon esculentum*	S	Bressan et al (1981)
ii) Inorganic ions			
NaCl	*Nicotiana tabacum*	S	Nabors et al (1975, 1980)
iii) Antibiotics			
Chloramphenicol	*Nicotiana sylvestris*	C	Dix (1981)
Lincomycin	*Nicotiana plumbaginifolia*	P	Cseplo and Maliga (1982)
Kanamycin	*Nicotiana sylvestris*	S	Dix et al (1977)
Streptomycin	*Nicotiana tabacum*	C	Maliga et al (1973, 1975)
iv) Nucleic Acid Base Analogues			
5-Fluorouracil	*Daucus carota*	S,P	Sung and Jacques (1980)
5-Bromodesoxyuridine	*Nicotiana tabacum*	C	Marton and Maliga (1975)
v) Amino acids			
L-Ornithine	*Nicotiana tabacum*	S	Berlin et al (1981)
L-Phenylalanine	*Nicotiana tabacum*	S	Berlin et al (1981)
L-Threonine	*Nicotiana tabacum*	S	Heimer and Filner (1970)
vi) Amino acid Analogues			
4-Methyltryptophan	*Catharanthus roseus*	S	Sasse et al (1983)
5-Methyltryptophan	*Datura innoxia*	S	Ranch et al (1983)
p-Fluoreophenylalanine	*Daucus carota*	S	Palmer and Widholm (1975)
m-Fluoreophenylalanine	*Daucus carota*	S	Berlin et al (1981)
Azatidine-2-carboxylic acid	*Daucus carota*	S	Cella et al (1982)
vii) Herbicides			
Glyphosate	*Corydalis sempervirens*	S	Amrhein et al (1983)
Picloram	*Nicotiana tabacum*	S	Chaleff and Parsons (1978)
2,4D	*Citrus sinensis*	C	Spiegel-Roy et al (1983)
viii) Miscellaneous compounds			
Pimelic acid	*Lavandula vera*	S	Watanabe et al (1982)

C= callus, S=suspension culture, P=Protoplasts

Para-fluoro phenylalanine (PFP) is an analogue of L-phenylalanine. PFP was found to block mitosis (Sisken 1973). This metabolite (PFP) is very widely used for bacterial selection cell rather than plant cells.

Glyphosate a broad-spectrum herbicide, is an inhibitor of 5 enol pyruvyl shikmic acid 3 phosphate (EPSP) synthase. Treatment of cultured cells with glyphosate resulted in increased levels of shikmic acid (Table 8.3). Therefore it is clear that auxotrophs can only be selected for primary biosynthetic pathway are absolutely necessary for survival.

8.2.3. Nutritional or Auxotrophic Mutants

Auxotrophic mutants are nutritional or deficiency mutants as they cannot grow on minimal medium due to a block in a biosynthetic pathway. Only if the lack of essential biosynthetic intermediates caused by the block is overcome by supplementing the medium with suitable compounds would auxotrophs grow on minimal medium. Carlson (1970) was the first to screen for auxotrophic mutants with cultured plant cells.

Table 8.3. Auxotrophic cell lines isolated by different methods

Selection method	Plant species	Nutritional requirement	Reference
Chlorate resistance	Rosa damascena	Nitrate reductase	Murphy and Imbrie (1981)
	N. tabacum	"	Murphy and Imbrie (1981)
Arsenate selection	D. innoxia	"	Horsch and King(1983 & 1984)
Negative selection	N. tabacum	Biotin	Carlson et al (1970)
Non-selective screening	H. muticus	Nitrate reductase	Gebhardte et al (1981)

8.2.3.1. Chlorate Resistance

This system, as it does not necessarily require haploid cells, can be used for the selection of nitrate reductase deficient mutants. By selecting for chlorate resistance cells many nitrate reductase minus lines have been established. The nitrate reductase deficient cell lines cannot reduce the nitrate analogue chlorate into toxic chlorite.

8.2.3.2. Non-selective Total Isolation System

Auxotrophic mutants must be isolated either by non-selective procedures or by methods which preferentially kill wild type cells during a reversible growth inhibition of auxtrophic cells. Beadle and Tatum (1945) first described non-selective, total isolation system for the isolation of auxotrophic cells of *Neurospora crassa*. Gebhardte et al (1981) prepared protoplasts from mesophyll cells of haploid

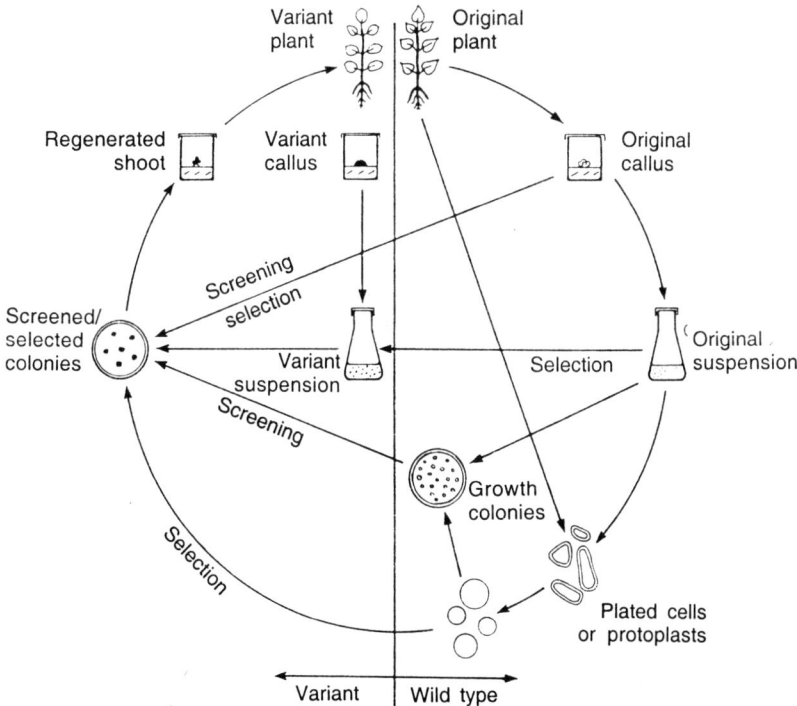

Fig. 8.1. The various culture systems, which can be used for screening and selction program. (Berlin and Sasse, 1993)

leaves of *Hyoscyamus muticus,* mutagenized the isolated protoplasts and plated the cells on minimal medium supplemented with amino acids, vitamins and nucleic acid bases.

8.2.3.3. Isolation by Enrichment /Selection Methods

Puck and Kao (1967) introduced the 5-bromodesoxyuridine light selection method for the isolation of auxotrophic mammalian cells. This technique can also be successfully used with cultured plant cells.

Polacco (1979) has demonstrated arsenate as a potential negative selection agent for auxotrophs in cultured plant cells. Sodium arsenate (1-2 mM) kill virtually all growing cells within 24h.

8.3. APPLICATIONS

1. The selection and screening program is useful for selecting high yielding strains, e.g. selection for enhanced activity of the enzyme in *C. roseus* led increased levels of serotonin
2. To understand the physiological basis for the application of selection pressure for yield improvement, e.g. Nicotiana.
3. By positive selection system use of pimelic acid, a toxic precursor of biotin synthesis in *Lavandula vera* to select higher biotin yielding cell lines.

REFERENCES

Alfermann, A.W. Merz, D., Reinhardt, E.: *Planta Med. Suppl.* 70 (1975).

Amrhein, N., et al *FEBS Letters.* 157, 191, (1983).

Barz, W., Reinhardt, E., Zenk, M.H. (eds.), p. 27, Springer-Verlag, Berlin, Heidelberg, New York, 1977.

Beadle, G.W. Tatum, E.L.: *Am. J. Bot.* 32, 678 (1945).

Bergmann, L.: *J. Gen. Physiol.* 43, 841 (1960).

Berlin, J., Kukoschke, K.G., Knobloch, K.H.: *Planta Med.* 42, 173 (1981).

Berlin, J. and Sasse, F. (1993). Selection and screening techniques for plant cell cultures. In : Plant Cell Line Selection, Procedures and Applications, Philip, J.D. (eds.) pp. 99-131, VCH, New York.

Bourgin, J.P., Chupeau, M.C., Missonier, C.: In: Variability in plants regenerated

Bourgin, J.P.: In: Genetic Engineering in Eukaryotes. (Lurquin, P.F., Kleinhofs, A.eds.), p. 195, Plenum Publ. Corp., New York 1983.

Bressan, R.A., Hasegawa, P.M., Handa, A.K.: *Plant Sci. Lett.* 21, 23 (1981)

Buchanan, R.J., Wray, J.L.: *ibid.* 188, 228 (1982).

Carlson, J.E., Widholm, J.M.: *Physiol. Plant.* 44, 251 (1978).

Carlson, P.S.: *Science* 168, 487 (1970).

Cella, R., Parisi, B., Nielsen, E.: *Plant Sci. Lett.* 24, 125 (1982).

Chaleff, R.S., Parsons, M. F.: *Proc. Natl. Acad. Sci.* USA 75, 5104 (1978).

Colman, B., Mawson, B.T.: *Can. J. Bot.* 57, 1505 (1979).

Cseplo, A., Maliga, P.: *Current Genet.* 6, 105 (1982).

Deus, B., Zenk, M.H.: *Biotech. Bioeng.* 24, 1965 (1982).

Dix, P.J. Joo, F., Maliga, P.: *Mol. Gen. Genet.* 157, 285 (1977).

Dix, P.J.: *Ann. Bot.* 48, 315 (1981).

Eichenberger, M.E., Sur une mutation survenue dans une culute de tissus de carotte C.R. Acad. Sci., 145, 239 (1951).

Evola, S.V.: *ibid.* 189, 447 (1983).

Gebhardte, C., Schnebli, V., King, P.J.: *Planta* 53, 81 (1981).

Hashimoto, T. et al.: as Ref. (61), p. 305 (1982).

Heimer, Y.M., Filner, P.: *Biochim. Biophys. Acta* 215, 152 (1970).

Horsch, R.B., King, J.: *ibid.* 159, 12 (1983) and 160, 168 (1984).

Jensen, R.G., Francki, R.I.B., Zaitlin, M.: *Plant Physiol.* 48, 9 (1971).

Kinnersley, A.M., Dougall, D.K.: *Planta* 149, 200 (1980).

Lawyer, A.L., Berlyn, M.B., Zelitch, I.: *Plant Physiol.* 66, 334 (1980).

Maliga, P., Marton, L., Sz-Breznovits, A.: *Plant Sci. Lett.* 1, 119 (1973).

Maliga, P., Sz-Breznovits, A., Marton, L.: *Nature* 255, 401 (1975).

Maliga, P., Sz-Breznovits, A., Marton, L.: *Nature New Biol.* 244, 299 (1973).

Marton, L., Maliga, P.: *Plant Sci. Lett.* 5, 77 (1975).

Meczel, L. et al.: *Theor Appl. Genet.* 59, 191 (1981).

Mizukami, H., Kanoshina, M., Tabata, M.: *Phytochemistry* 17, 95 (1978).

Morris, P., Fowler, M.W.: *Plant Cell Tissue Org. Cult.* 1, 15, (1981).

Morris, P., Smart, N.J. Fowler, M. W.: *ibid.* 2, 207 (1983).

Muller, A.J., Grafe, R.: *Mo,. Gen. Genet.* 161, 67 (1978).

Muller, A.J.: *Mo. Gen. Genet.* 192, 275 (1983).

Murphy, T., Imbrie, C.W.: *Plant Physiol.* 67, 910 (1981).

Murphy, T.M., Hamilton, C.M., Street, H.E.: *Plant Physiol.* 64. 936 (1979).

Nabors, M. W. et al.: *ibid.* r, 155 (1975).

Nabors, M.W. et al.: *Z. Pflanzenphysiol.* 97, 13 (1980).

Ogino, T., Hiraoka, N., Tabata, M.: *Phytochemistry* 17, 1907 (1978).

Palmer, J.E., Widholm, J.M.: *ibid.* 56, 233 (1975).

Polacco, J.C.: *Planta* 146, 155 (1979).

Puck, T., Kao, F.: *Proc. Natl. Acad. Sci. USA* 58, 1227 (1967).

Ranch, J.P., et al.: *Plant Physiol.* 71, 136 (1983).

Sasse, F., Buchholz, M., Berlin, J.: *Z. Naturforsch.* 38c, 916 (1983).

Sasse, F., Heckenberg. U., Berlin. J.: *Plant Physiol.* 69, 400 (1982).

Sato, F., Yamada, Y.: *Phytochemistry* 23, 281 (1984).

Schulte, U., Zenk, M.H. : *Physiol. Plant.* 39, 139 (1977)

Schwenk, F.W.: *Plant Sci. Lett.* 23, 147, (1981).

Sisken, J.E.: *Chromosoma* 44, 91 (1973).

Spiegel-Roy, P., Kochba, J., Shoshana, S.: *Z. Pflanzenphysiol.* 109, 41 (1983).

Strauss, J., Spontaneous changes in corn endosperm tissue culutre, Science, 128, 537 (1958).

Sung, Z.R., Jacques, S.: *Planta* 148, 389 (1980).

Tabata, M. *et al.* In : Frontiers of plant tissue culture (1978). *Proc. 4th Intl. Cong. Plant Tissue Cell Culture* (Thorpe, T.A. ed.), p. 381, Univ. Calgany, 1978.

Umberck, P.F., Gengenbach, B.G.: *Crop Science* 23, 584 (1983).

Vunsh, R., Aviv, D., Galun, E.: *Theor. Appl. Genet.* 64, 51 (1982).

Watanabe, K., Yano, S.I., Yamada, Y.: *Phytochemistry* 21, 513 (1982).

Widholm, J.M.: *Biochim. Biophys. Acta* 279, 48 (1972).

Yamamota, Y., Mizuguchi, R., Yamada, Y.: *Theor. Appl. Genet.* 61, 113 (1982).

Zenk, M.H. et al.: In: Plant Tissue Culture and its Biotechnological Application(Barz, W., Reinhardt, E and Zenk,M.H. eds) Springer- Verlag, Berlin, Heidelberg, New York . pp 27, (1977)

Chapter 9

Biogenesis of Phytopharmaceuticals

INTRODUCTION

All organisms need to transform and interconvert a vast number of organic compounds to enable them to live, grow and reproduce. They need to provide themselves with energy in the form of ATP, and a supply of building blocks to construct their own tissues. The energy required for this process is met through the catabolism. Catabolism is the intracellular process of degrading a compound into smaller and simpler products. Anabolism is involved in the synthesis of more complex compounds (secondary metabolites).

An integrated network of enzyme-mediated and carefully regulated reactions is used for this purpose, are known as intermediary metabolism, and the pathways involved are called as metabolic pathways. Some of the important molecules of life are proteins, carbohydrates, fats and nucleic acids. These molecules are synthesized in all the living organisms through the biosynthetic pathways. These processes demonstrate the fundamentals unity of all living matter and are collectively described as primary metabolism, with the compounds involved in the pathways being termed as primary metabolites. Thus degradation of carbohydrates and sugars generally proceeds via the glycolysis and krebs/citric acid/tricarboxylic acid cycle, which release energy from the organic compounds by oxidative reactions (Fig. 9.1 to 9.4).

In contrast to these primary metabolic pathways, which synthesize, degrade and generally interconvert compounds commonly encountered in all organisms, there also exists an area of metabolism concerned with compounds which have a much more limited distribution in nature. Such compounds are known as secondary metabolites. These are found in only specific organisms or groups of organisms, and are an expression of the individuality of species. They may represent chemical adoptions to environmental stresses, or they may serve a defensive, protective or offensive chemicals against microorganisms, insects and higher herbivorous predators. They are some times considered to be waste or secondary products of plant metabolism. The building blocks for secondary metabolites are derived from primary metabolism as shown in Fig. 9.5.

The photosynthetic process in plants is essential for all animal life on the earth, since it converts the solar energy into organic compounds which in turn are useful for the production of essential foods. The various biosynthetic reactions occurring in plant cells are enzyme-dependent, wherein

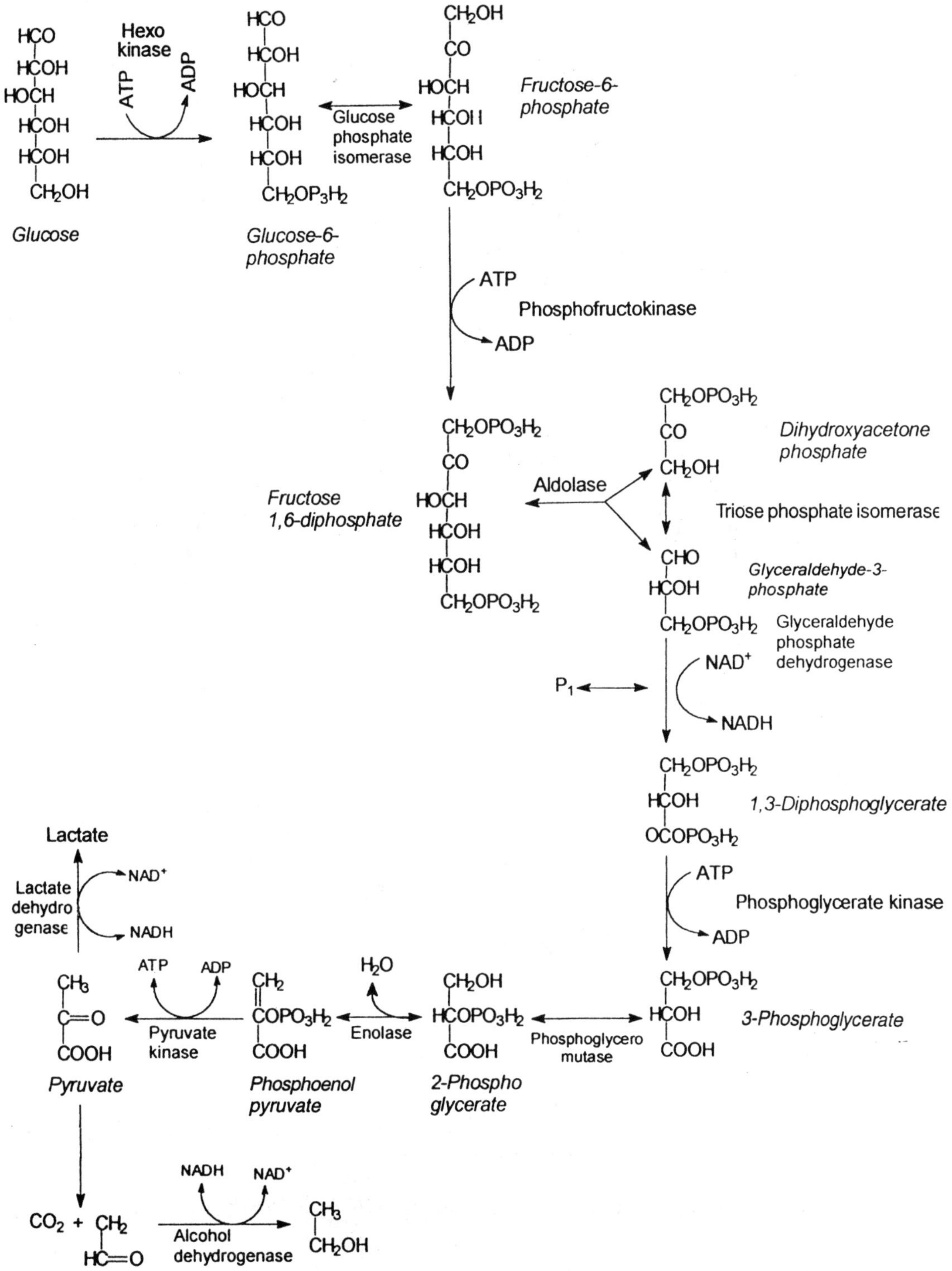

Fig. 9.1. Glycolysis.

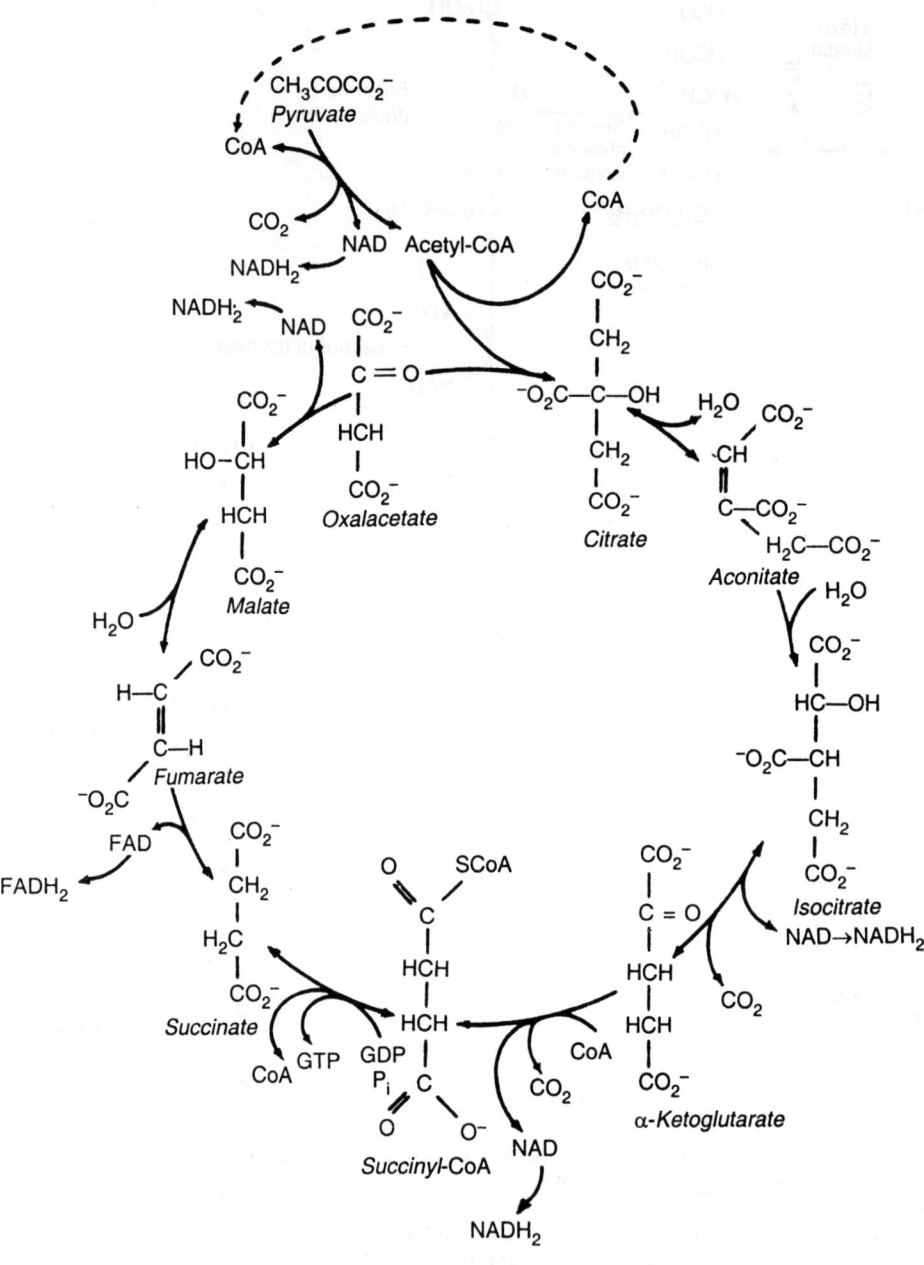

Overall reaction :   Pyruvate + 4NAD + FAD → 3CO$_2$ + 4NADH$_2$ + FADH$_2$
                        GDP + Phosphate → FTP
                        GTP + ADP → GDP + ATP $\Big]$
Oxidative phosphorylation :   4NADH$_2$ ≡ 12ATP $\Big\}$ 15 ATP
                                 FADH$_2$ ≡ 2ATP

Fig. 9.2. Glycolysis and TCA cycle.

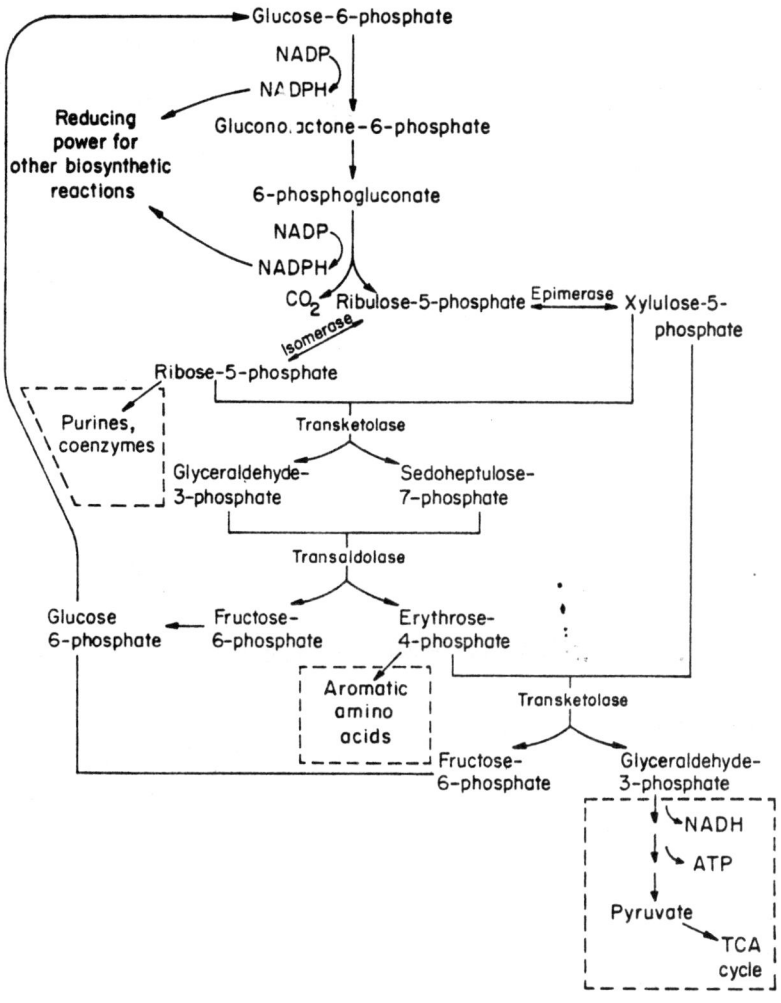

Fig. 9.3. Pentose-phosphate pathway (hexose-monophosphate).

enzymes act as catalysts of metabolism. It is through the control of enzymatic activity that plant metabolism is directed into specific biosynthetic pathways.

The course of biosynthetic pathways in plants can be investigated by means of following main techniques.

1. Tracer technique.
2. Use of isolated organs and tissues.
3. Grafting methods.
4. Use of mutant strains.

9.1. SHIKIMIC ACID PATHWAY

The shikimate pathway provide an alternative route to aromatic compounds, especially the aromatic amino acid such as L-phenyl alanine, L-tyrosine and L-tryptophan. This pathway is only employed in microorganisms and plants, but not by animals (Fig. 9.6-9.7).

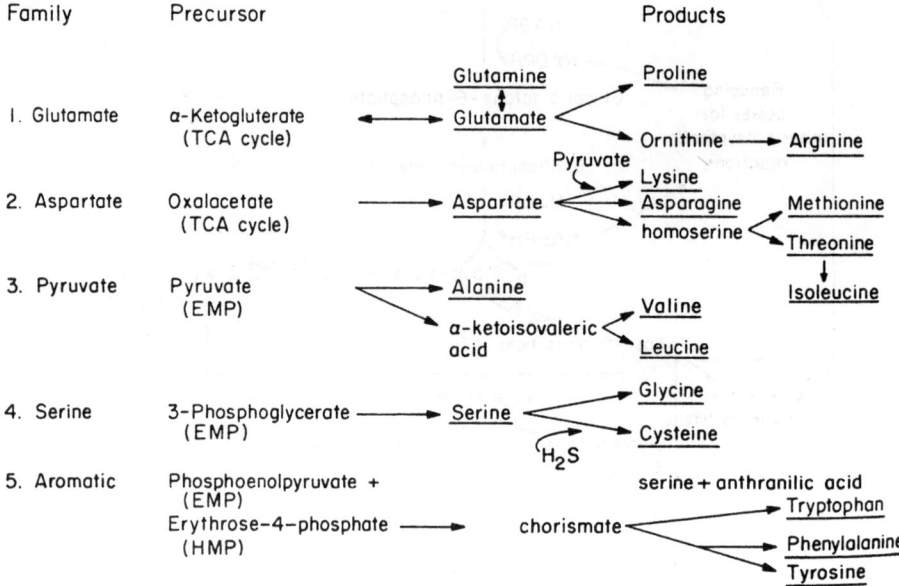

Fig. 9.4. Synthesis of Amino acids from intermediates of the glycolysis (EMP), TCA and HMP pathways.

9.2. THE ACETATE PATHWAY (POLYKETIDES)

Polyketides constitute a large class of natural products grouped together on the basis of biosynthetic grounds, they are derived from poly-β-keto chains, formed by the coupling of acetic acid (C_2 units) via condensation reactions, e.g. fatty acids, polyacetylenes, prostaglandins, macrolide antibiotics, tetracyclines and Anthraquinone.

9.2.1. Biogenesis of Anthraquinones

A number of natural anthraquinone derivatives are excellent examples of acetate – derived structures. The key intermediate in biosynthesis of anthraquinone, poly-β-keto methylene acid is identified through microbial cultures, produced a metabolite known as Islandicin structurally related to anthraquinones. The intermediate is probably produced from 8 acetate units which on intramolecular condensation form anthraquinones (Fig. 9.8).

Emodin, physcion, chrysophanol, aloe-emodin and rhein form the basis of range of purgative anthraquinone derivatives found in Senna, Cascara, Frangula, Rhubarb and Aloes.

Hypericin, a napthodianthrone, is an antidepressant, anti-HIV compound found in *H. perforatum* also produced through the same key intermediate (Fig. 9.9).

9.3. BIOSYNTHESIS OF FLAVONOIDS

Flavonoids are mostly yellow coloured substances and are the derivatives of phenol and present in the form of glycoside.

Cinnamic acids, as their coenzyme A esters, may also function as starter units for chain extension with malonyl-CoA units, thus combining elements of the shikimate and acetate pathways. Enzymes chalcone synthase and stilbene synthase couple a cinnamoyl Co-A unit with 3 Malonyl-CoA units, giving stilbenes, e.g. resveratrol or chalcone e.g. Naringenin-Chalcone, respectively (Fig.9.10). Chalcone act as precursor for a wide range of flavonoid derivatives found in the plants.

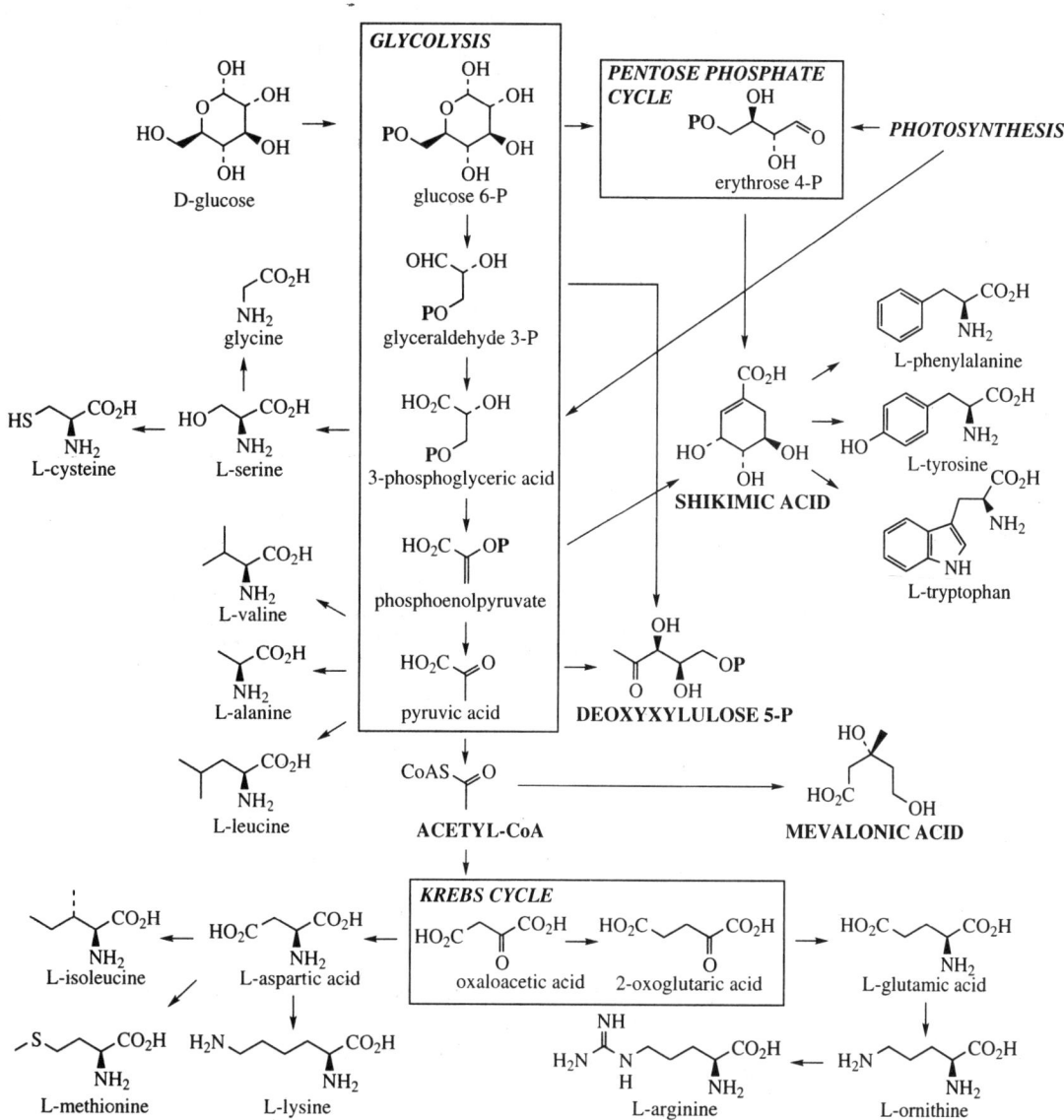

Fig. 9.5. Various metabolic pathways for primary metabolites.

Flavonones (Naringenin) gave rise to flavones (apigenin, luteolins), flavonols (Kaempferol, quertcetin), catechins (afzalechin) and anthocyanidins (pelargonidin). The biosynthetic pathway for these compounds is shown in Fig. 9.11-9.12. The flavonol glycoside rutin from (*Fagopyrum esculentum*, Polygonaceae) buck wheat and *Ruta graveolens* (Rutaceae) and flavonone glycoside, hesperidin from citrus peels, Neohesperidin from bitter orange (*Citrus aurantium*, Rutaceae) and naringin from grape fruit peel (*Citrus paradisi*) and the chemical structures are shown in Fig 9.13.

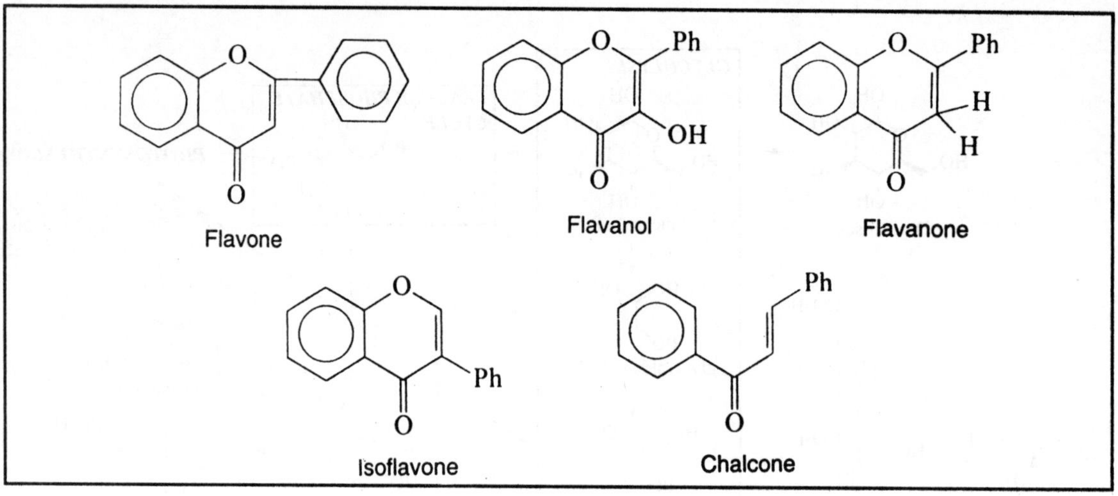

Flavone

Flavanol

Flavanone

Isoflavone

Chalcone

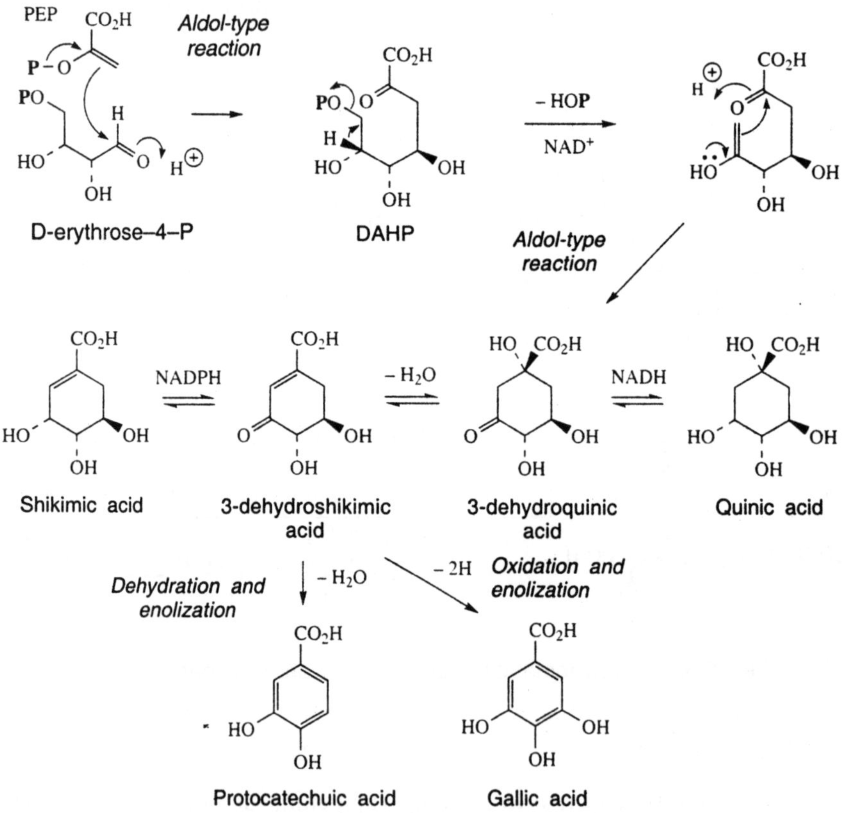

PEP

D-erythrose–4–P

DAHP

Shikimic acid

3-dehydroshikimic acid

3-dehydroquinic acid

Quinic acid

Protocatechuic acid

Gallic acid

Fig. 9.6. Shikimate pathway.

9.3.1. Flavanolignans

The flavanolignans are compounds formed as result of flavanoid and liganin structure. They arise by oxidative coupling processes between a flavonoid and a phenyl propanoid. Taxifolin a dihydroflavonol, generates a free radical which on combination with coniferyl alcohol (a phenyl propanoid) forms an

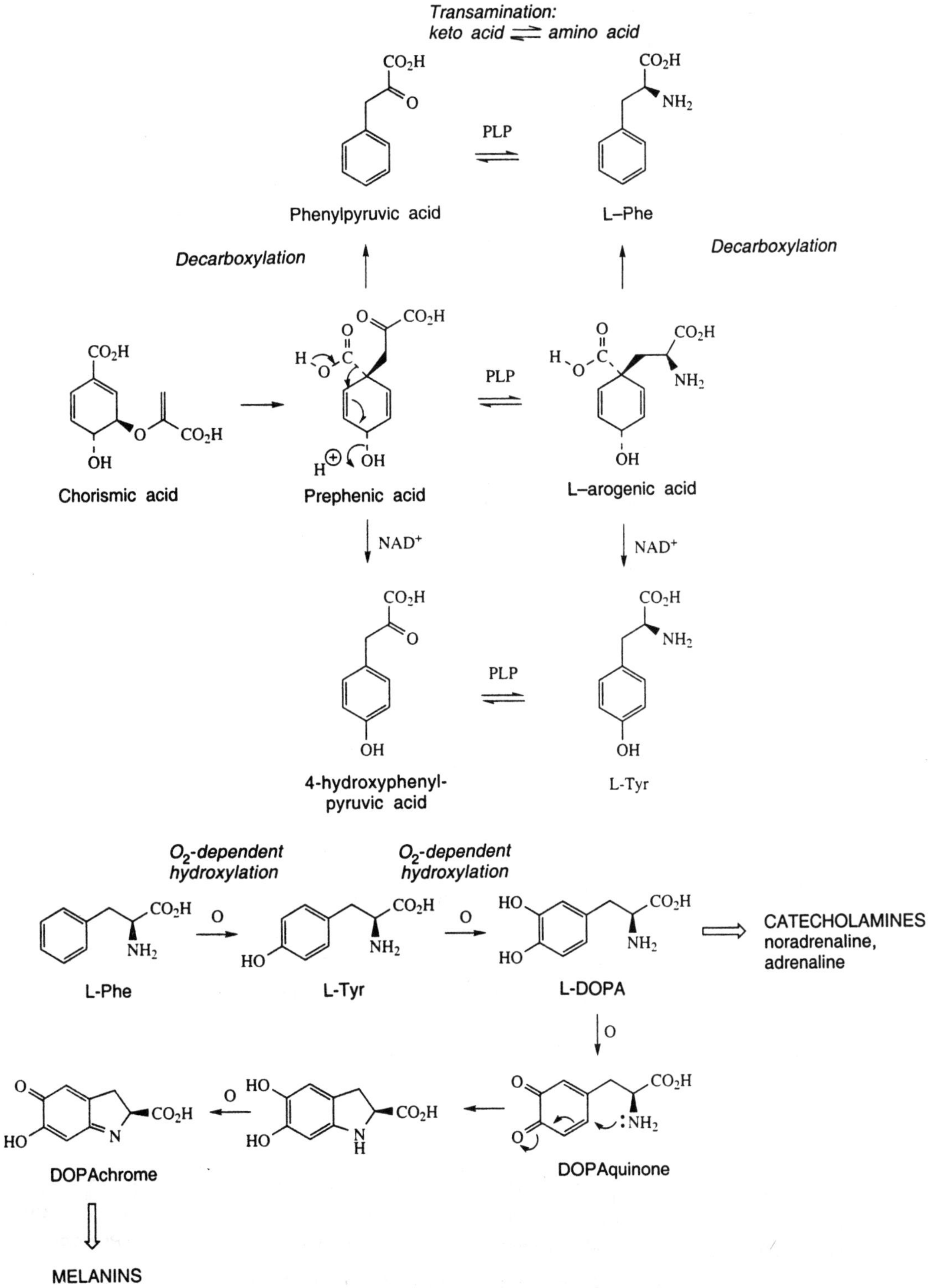

Fig. 9.7. Shikimic acid pathway for L-Tyrosine and Catecholamines.

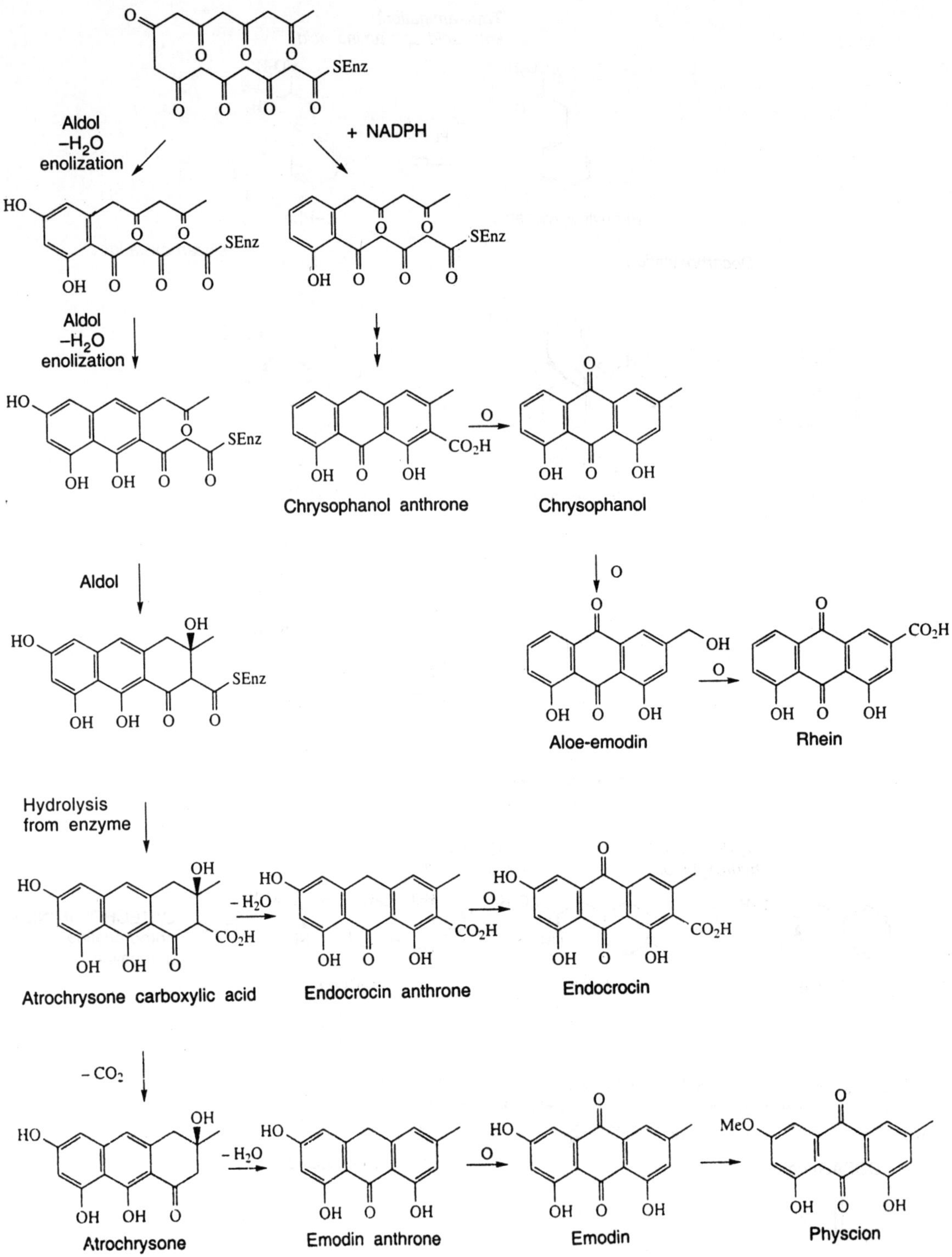

Fig. 9.8. Biosynthetic pathway for anthraquinones.

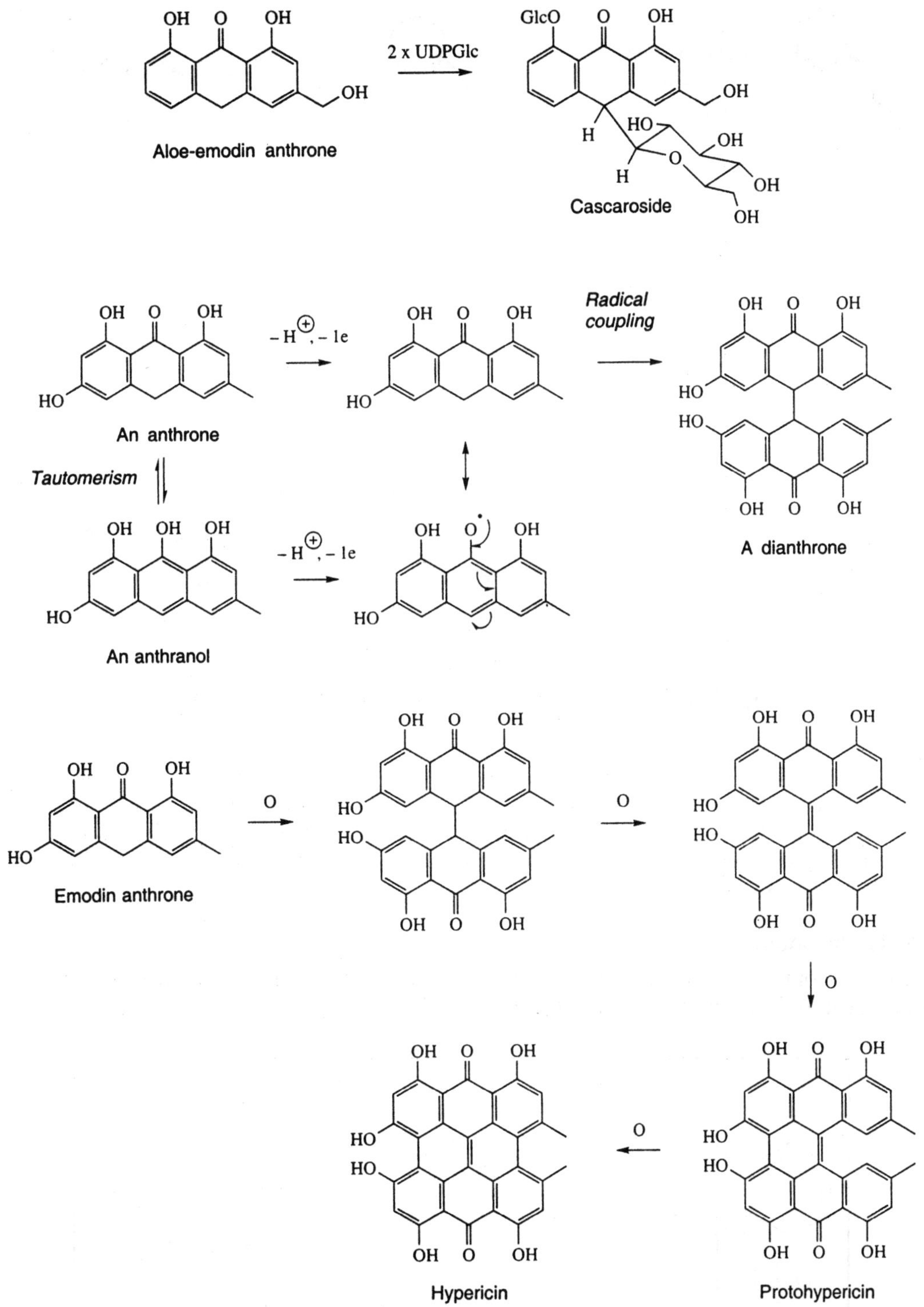

Fig. 9.9. Biosynthetic pathway of hypericin, cascaroside, dianthrane/anthranol.

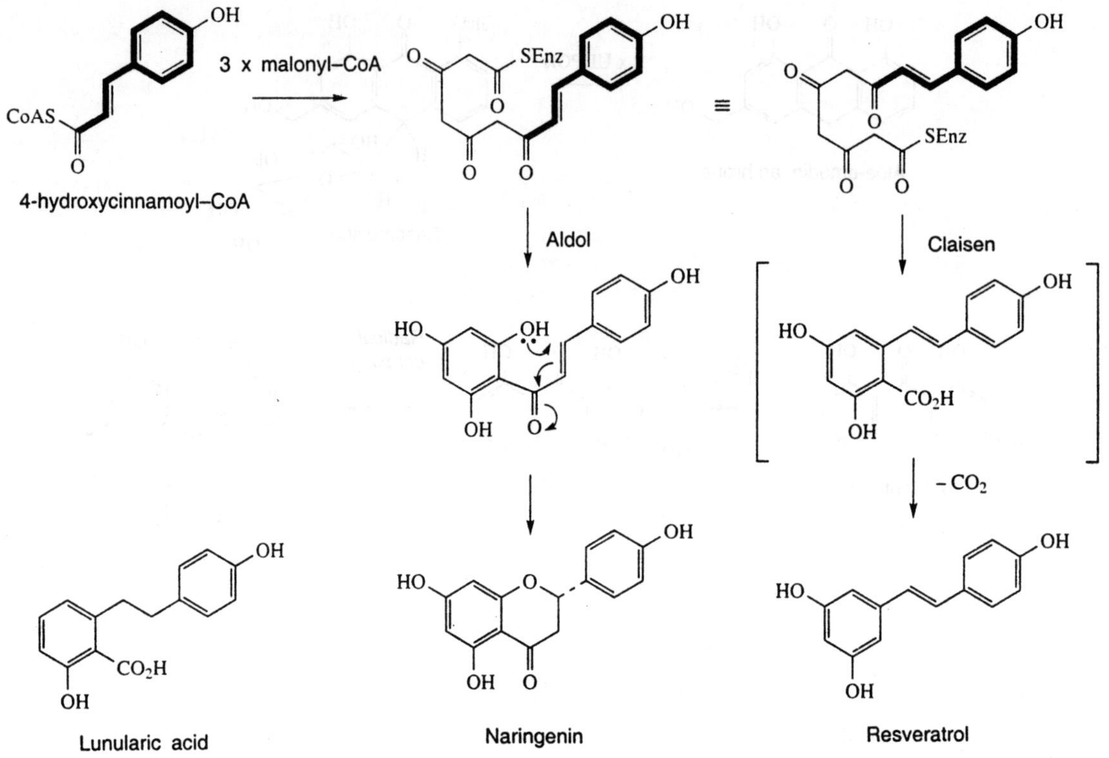

Fig. 9.10. Biosynthetic pathway of a flavanoid and resveratrol.

adduct which could generate a product known as Silybin (Fig. 9.14-9.15). *Silybum marianum* (Compositae) is a biennial thistle-like plant (milk thistle), seeds of this plant contains 1.5–3% of flavanolignans collectively termed silymarin. This mixture contains mainly silybin, together with silychristin, silydianin, and small amounts of isosilybin (Fig 9.14). Both silybin and iso silybin are a mixture of two trans diastereoisomers.

9.3.2. Isoflavonoids

The isoflavonoids are structural variants of flavonoids in which the shikimate derived aromatic ring has migrated to the adjacent carbon of the heterocycle.

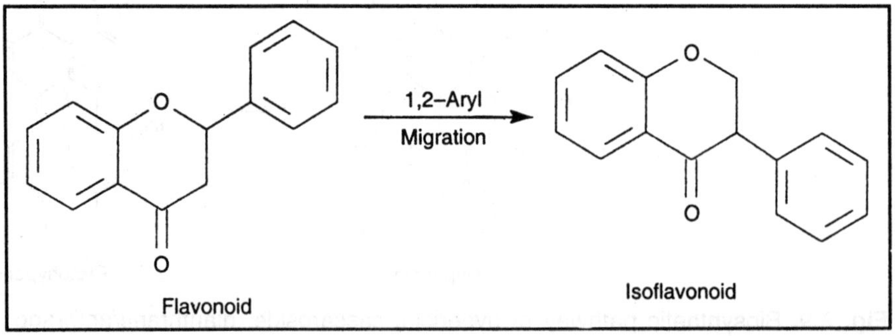

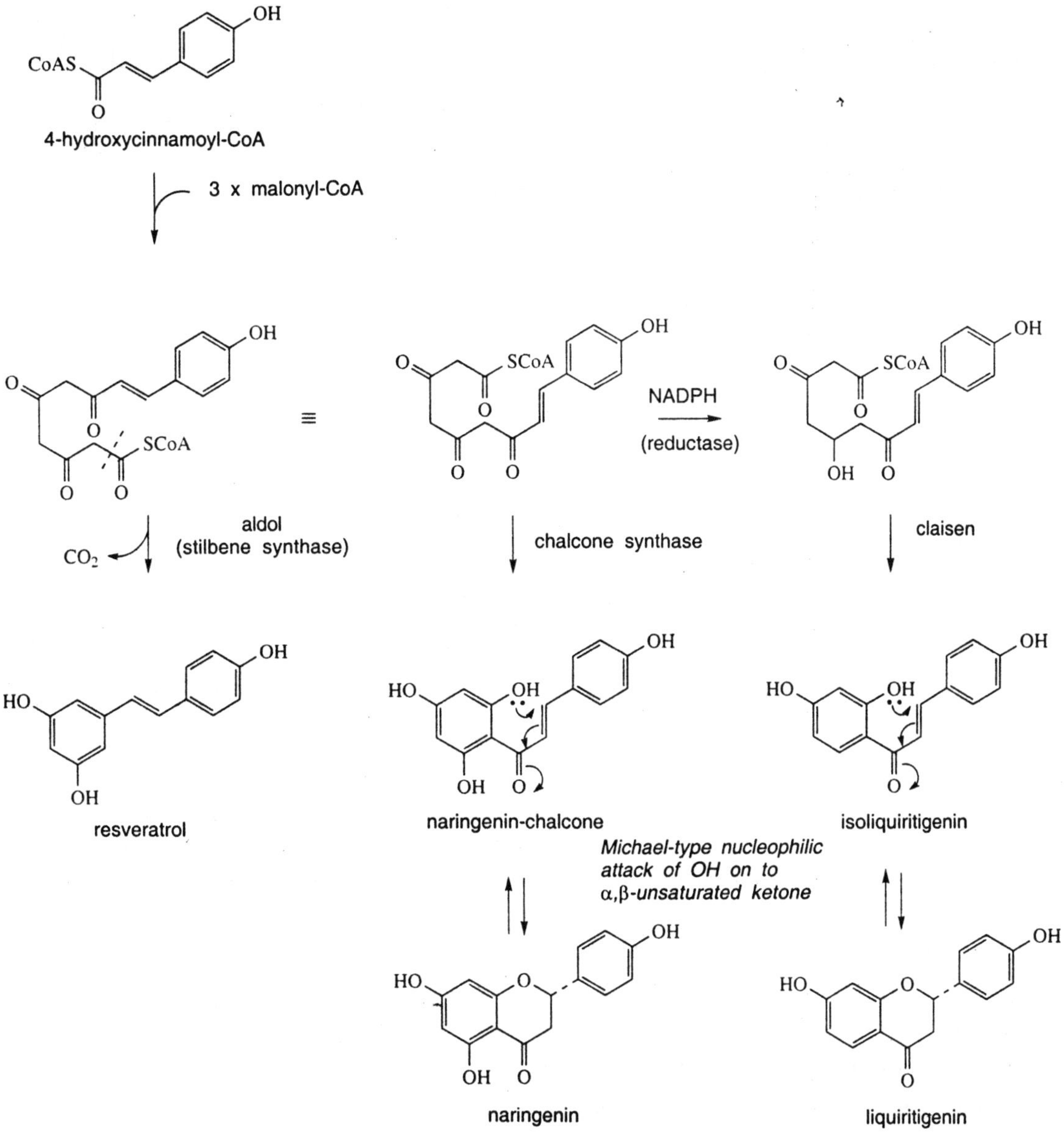

Fig. 9.11. Biosynthetic pathway of Naringenin and liquiritigenin.

The migration is brought about by a cytochrome P-450 dependent enzyme NADPH and O_2 cofactors, which transform the flavanones liquiritigenin or naringenin into the isoflavones daidzein or genistein (Fig. 9.16).

9.4. SAPONINS

Saponins are glycosides which even at low concentrations, produce a frothing in aqueous solution, because they have surfactant and soap like properties, e.g.

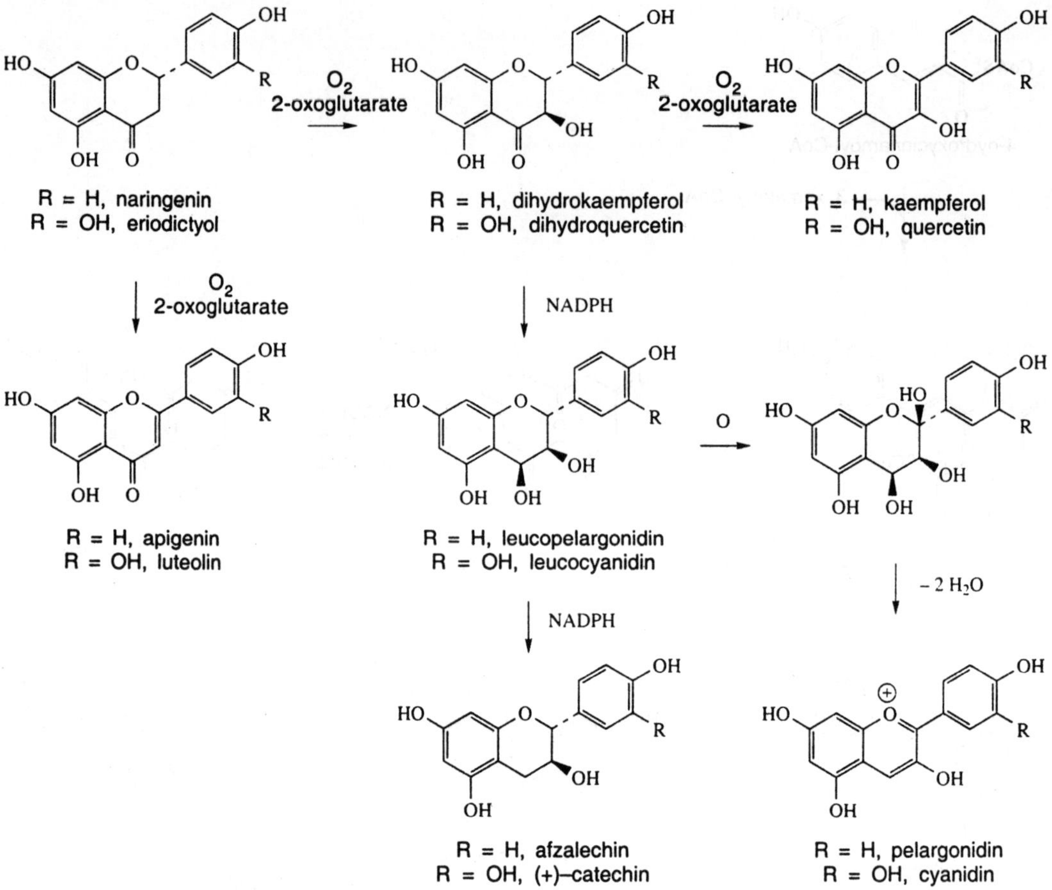

Fig. 9.12. Biosynthetic pathway of flavones, catechins and anthocyanidins.

Saponaria officinalis	Caryophyllaceae
Quillaja saponaria	Rosaceae
Smilax sp	Liliaceae
Polygala senega	Polygdeceae
Panax ginseng	Araliaceae
Glycyrrhiza glabra	Leguminosae
Dioscorea deltoidea	Dioscoreaceae

Saponins are divided into two categories based on the chemical constituents, these are :

1. Triterpenoid saponins
2. Steroidal saponins.

The name comes from the latin Sapo : Soap, and the plant materials containing saponins were originally used for cleaning clothes e.g. Soapwort, (*Saponaria officinalis,* Caryophyllaceae) and soap bark, *Quillaja saponaria* – Rosaceae). These materials also cause haemolysis, lysing red

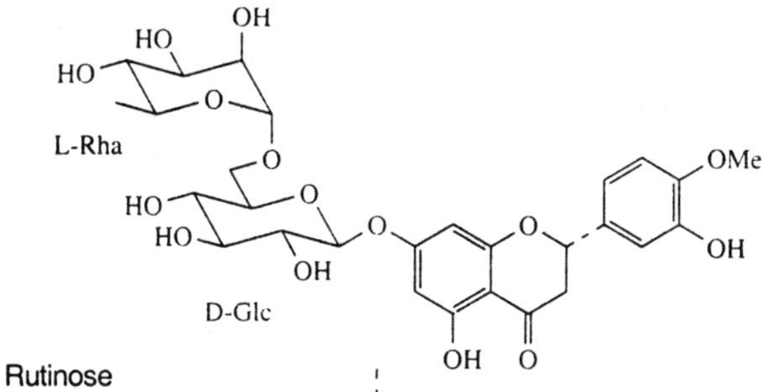

Rutinose
= rhamnosyl(α1→6)glucose | Hesperetin

Hesperidin

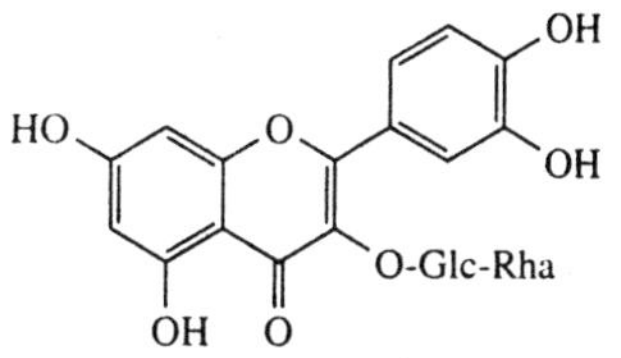

Quercetin | Rutinose

Rutin

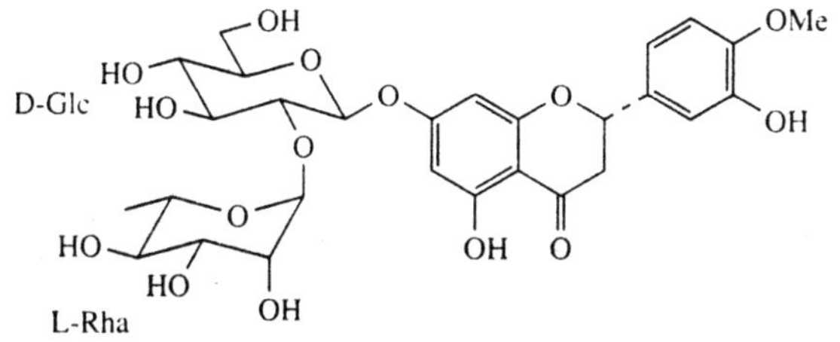

Neohesperidose
= rhamnosyl(α1→2)glucose | Hesperetin

Neohesperidin

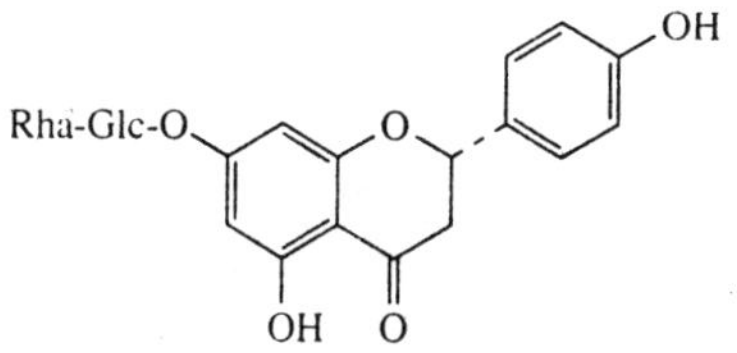

Neohesperidose | Naringenin

Naringin

Fig. 9.13. The structures of hesperidin/neohesperidin/rutin/naringin.

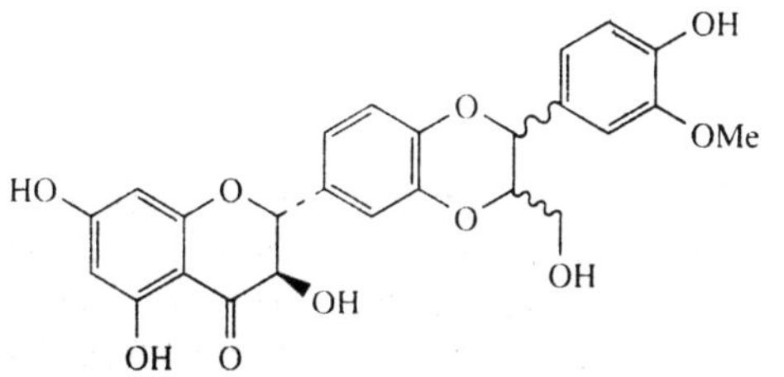

isosilybin

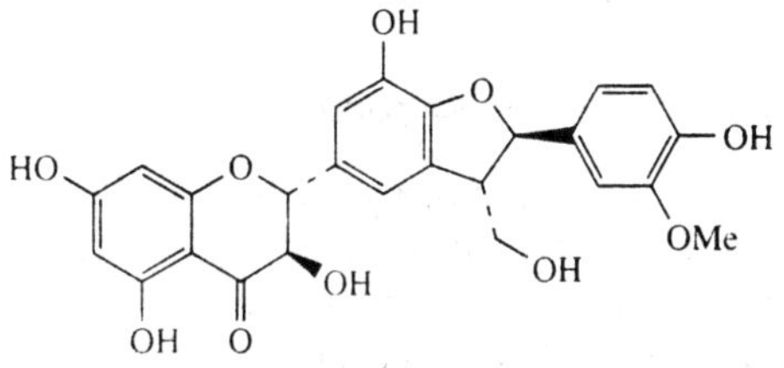

silychristin

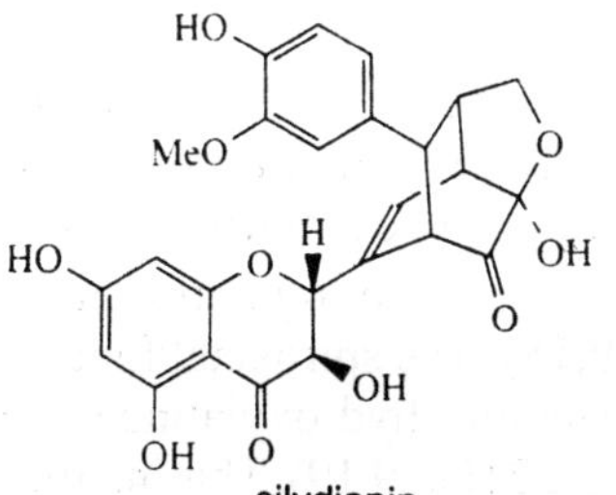

silydianin

Fig. 9.14. The structures of silydianin, isosilybin and silychristin.

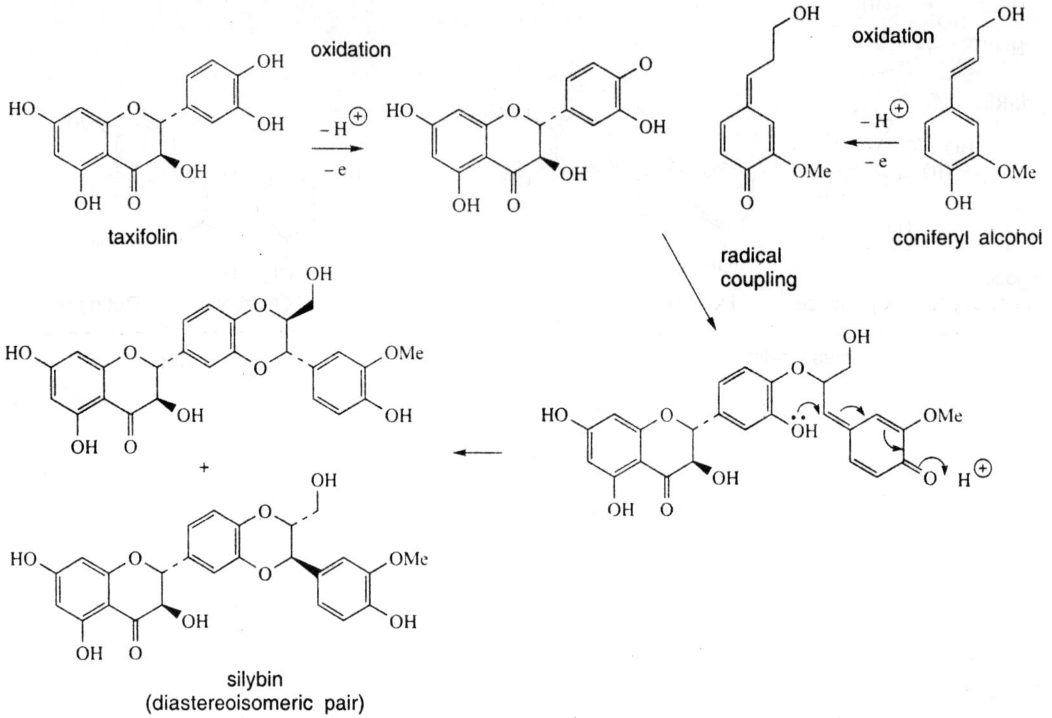

Fig. 9.15. Biosynthesis of Silybin.

blood cells by increasing the permeability of the plasma membrane, and thus they are highly toxic when injected into blood stream. They are also used as arrow poisons.

9.4.1. Triterpenoid Saponins

These are rare in monocotyledons, but abundant in many dicotyledons families. Quillaia bark contains a saponin mixture with quillaic acid as the principal aglycone and the roots of liquorice (*Glycyrrhiza glabra*) contain glycyrrhizin (Fig 9.17). The seeds of *Aeseulus hippocastrum* (Hippocastanaceae, Horse chest nut) contains a complex mixture of saponins termed aescin (anti-inflammatory) based on the polyhydroxylated aglycone protoasecigenin and barringtogenol (Fig 9.17).

9.4.2. Steroidal Saponins

They are C_{27} sterol and distributed in the families of Dioscoreaceae (*Dioscorea* sps.) Agavaceae (*e.g. Agave, Yucca*) and the liliaceae (*Smilax, Trillium*). The side chain of cholesterol has undergone modification to produce a spiroketal, e.g. Dioscin from Dioscorea.

All the steroidal saponins have the same configuration at the spiro centre C-22, but stereo isomers at C-25 exist e.g. Yamogenin (Fig.9.18). The spiroketal function is derived from the cholesterol side chain by a series of oxygenation reactions, hydroxylating C-16 and one of the terminal methyl, and then producing a ketone function at C-22 (Fig.9.19). This intermediate is transformed into the hemiketal and then the spiroketal. The chirality at C-22 is fixed by the stereospecificity in the formation of the ketal while the different possible.

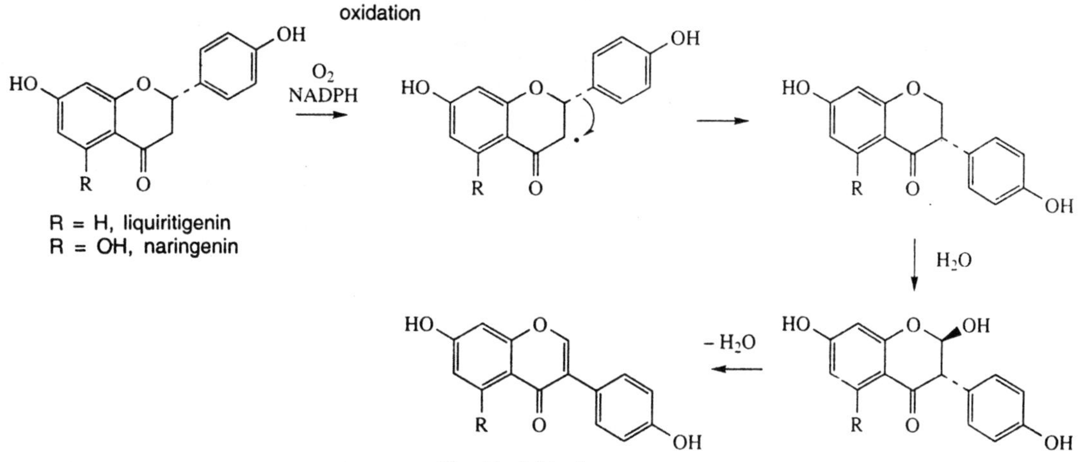

R = H, liquiritigenin
R = OH, naringenin

R = H, daidzein
R = OH, genistein

Fig. 9.16. Biosynthesis of isoflavonoids.

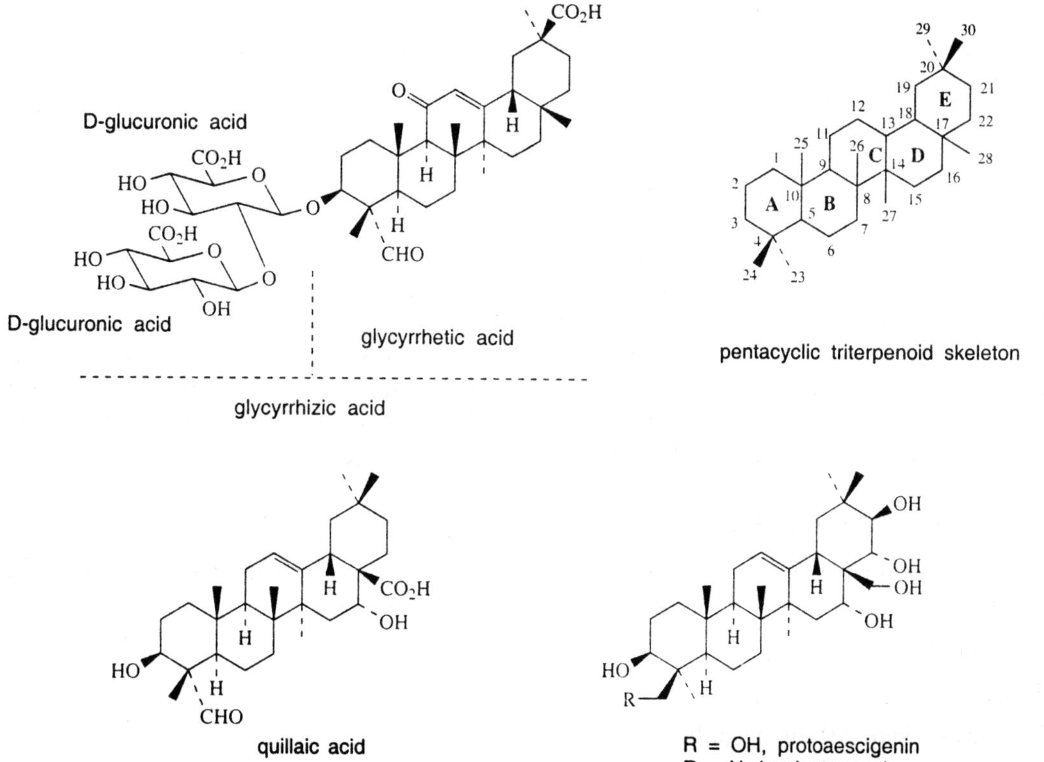

D-glucuronic acid

D-glucuronic acid

glycyrrhetic acid

glycyrrhizic acid

pentacyclic triterpenoid skeleton

quillaic acid

R = OH, protoaescigenin
R = H, barringtogenol

Fig. 9.17. The chemical structure of quillaic acid, glycyrrhizic acid protosecigenin and barringtogenol.

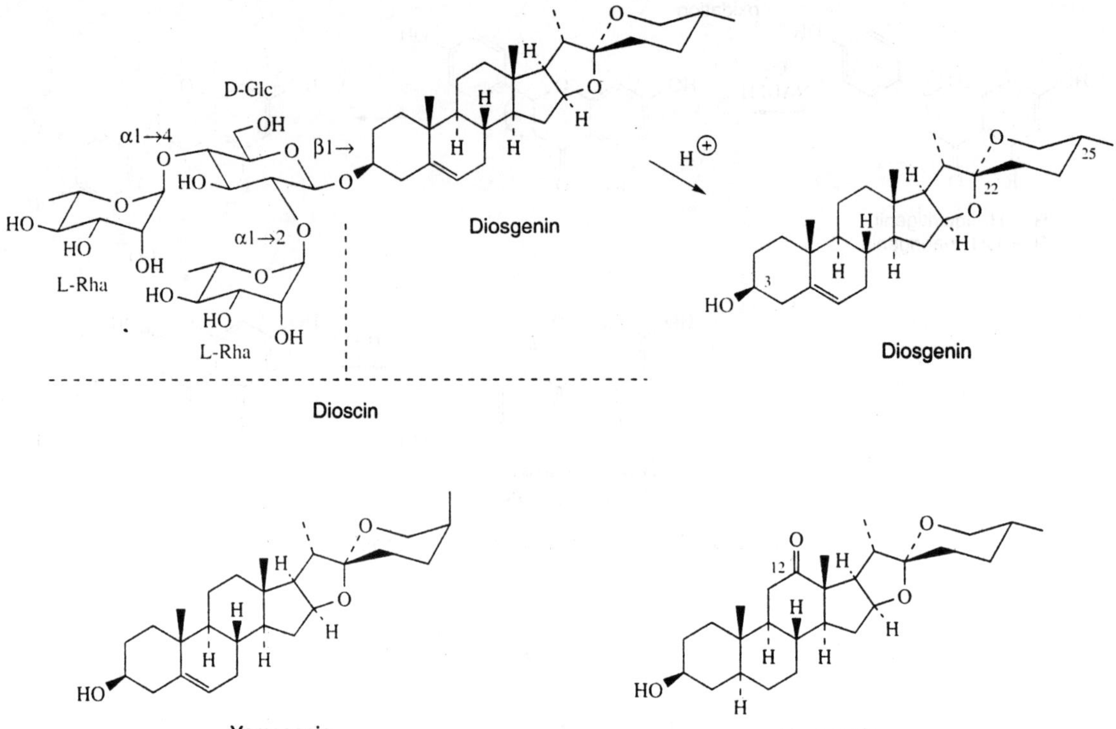

Fig. 9.18. The chemical structure of Dioscin, diosgenin, yamogenin and hecogenin.

Stereochemistries at C-25 are dictated by whether C-26 or C-27 is hydroxylated in the earlier step. Protodiscin is also isolated from the plants(Fig.9.20).

Diosgenin is the principal example of steroidal saponin and is obtained from Mexican yams (Dioscroea sp.) Fenugreek (*Trigonella foenum-graecum*), sisal (*Agave sisalana*) for yielding hecogenin. Solasonine (Fig 9.21) is another steroidal saponins found in many plants of the genus *Solanum* (Solanaceae), and tomatine from tomato *Lycopersicon esculenta*, (Solanaceae) all these are a rich source for the synthesis of steroidal drugs. Smilagenin and sarsasapogenin found in sarsaparilla are reduced forms of diosgenin and yamogenin respectively (Fig. 9.21).

9.5. CARDIAC GLYCOSIDES

Many of the plants known to contain cardiac glycosides, have long been used as arrow poisons (e.g. strophanthus). The following plants contain cardiac glycosides.

1. Digitalis sp. (Fox glove)
2. *Urginea maritima* (Squill)
3. *Urginea indica* (Indian squill)
4. *Strophanthus kombe*
5. *S. gratus* (Ouabain)
6. *Helleborus niger* (black hellebore)

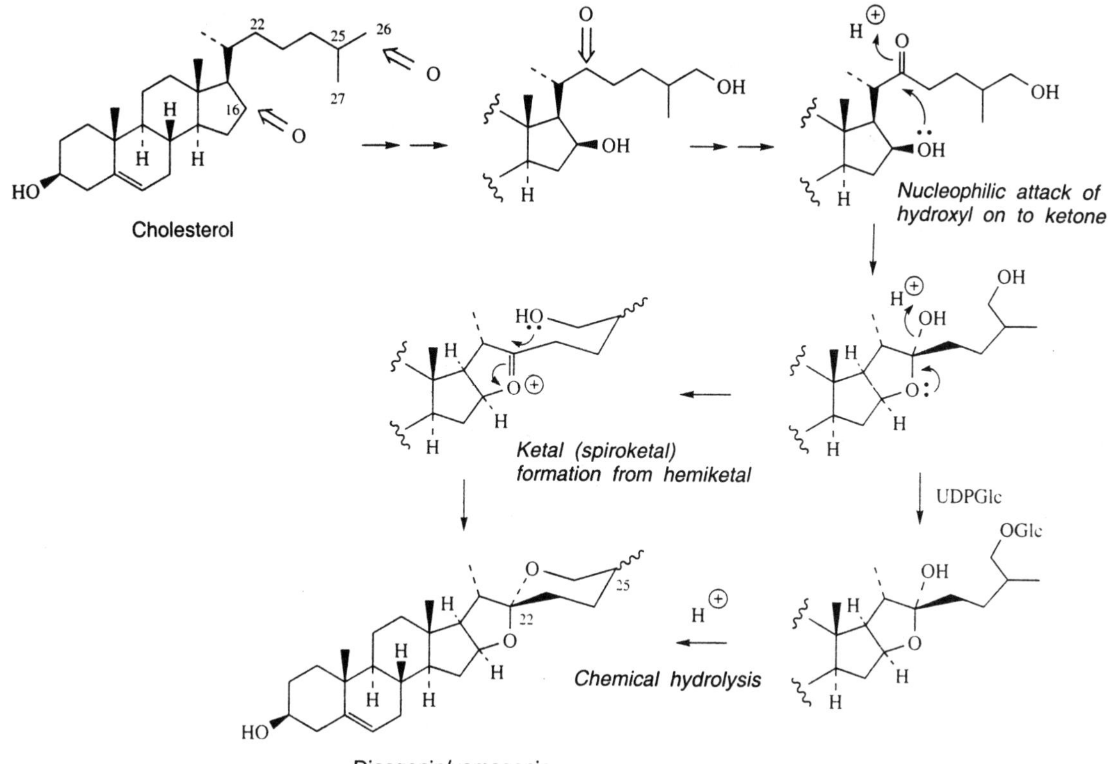

Fig. 9.19. Biosynthetic pathway of diosgenin/yamogenin.

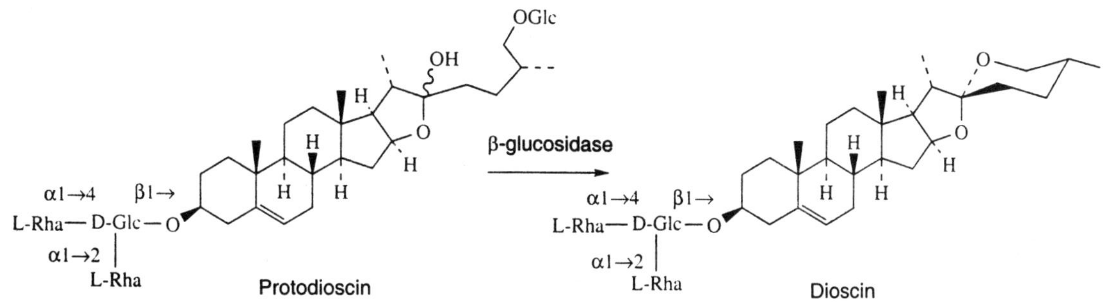

Fig. 9.20. Enzymatic conversion of protodioscin to dioscin.

7. *Thevetia nerifolia*
8. *Adonis vernalis*

The therapeutic action of cardioactive glycoside depends on the structure of the aglycone, and on the type and number of sugar units attached. Two types of aglycone are recognised, cardenolide e.g. Digitoxigenin from *D. purpurea* C-23 compounds, and bufadienolides, e.g. hellebrigenin from *Helleborus niger,* which are C-24 (Fig 9.22) compounds.

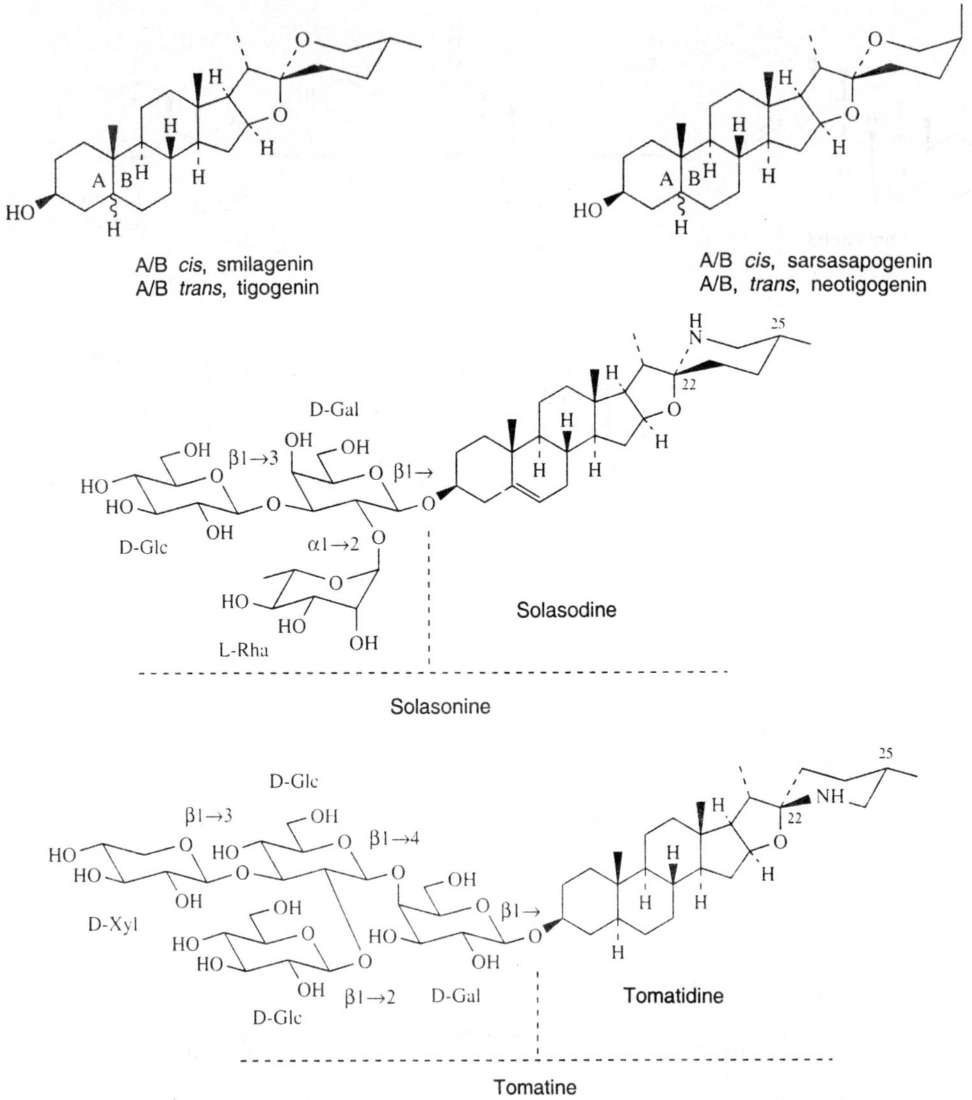

Fig. 9.21. The structure of tomatine, smilagenin, tigogenin, solasonine, sarsasapogenin and neotigo-genin.

The cholesterol side chain is shortened by stepwise hydroxylation at C-22 and then C-20, then cleavage as C-20/22 band, giving pregnenolone, which is then oxidized in ring A to give progesterone (Fig.9.23a). This upon reduction to give the cis β-fused A/B system as in pregnanolone for 14β hydroxylation, followed C-21, then hydroxy ketone. Then lactone ring is created by incorporating two carbons from acetate giving the digitoxigenin. Alternatively 3 carbons from oxaloacetate, will give rise the bufalin. The bufalin upon 5β-hydroxylation and oxidation at C-19 will form hellebringenin. On the other hand digitoxigenin upon hydroxylation (12β), 16β will form digitoxigenin and gitoxigenin respectively (Fig 9.23b). The biosynthetic pathway of side chain present in cardenolide and bufadienolide are shown in Fig 9.24.

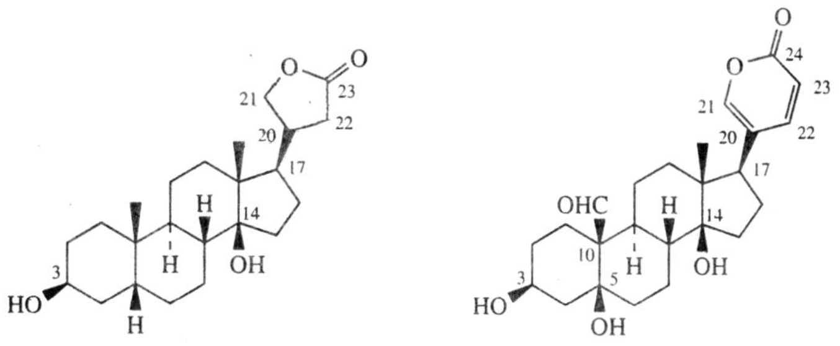

Cardenolide – digitoxigenin   **Bufadienolide** – hellebrigenin

Fig. 9.22. The chemical structure of cardenolide and bufadienolide.

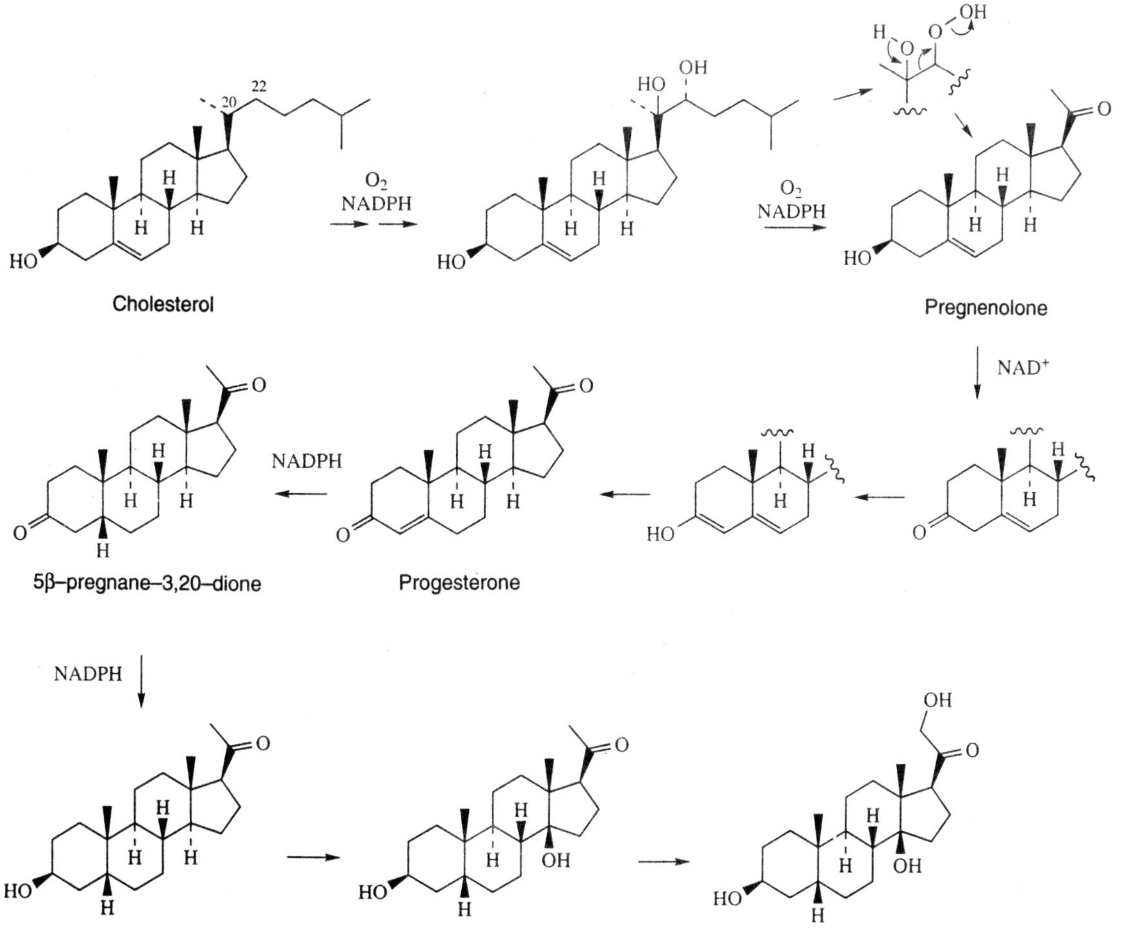

Fig. 9.23a. Biosynthesis of cardiacglycosides.

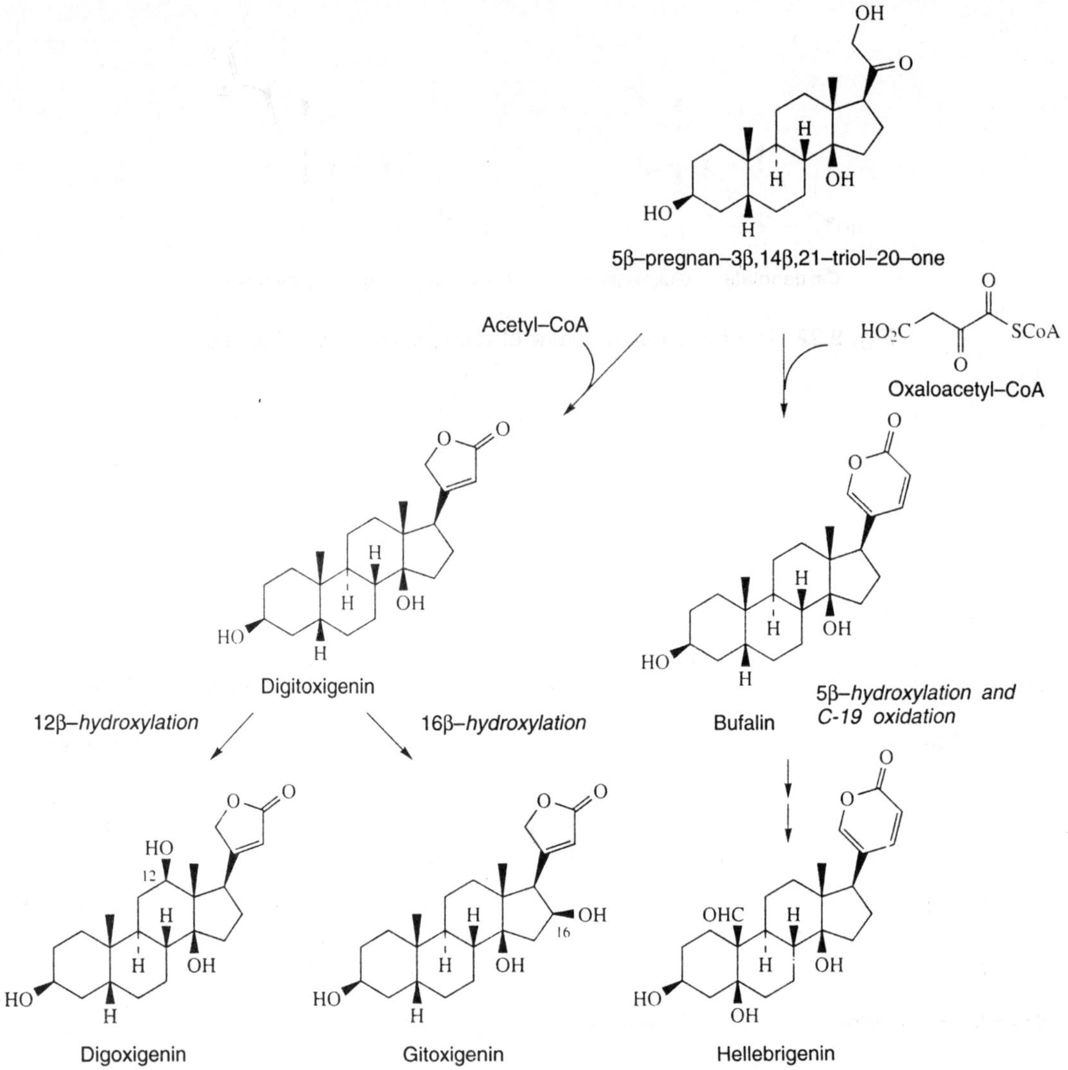

5β–pregnan–3β,14β,21–triol–20–one

Acetyl–CoA

Oxaloacetyl–CoA

Digitoxigenin

Bufalin

5β–hydroxylation and
C-19 oxidation

12β–hydroxylation

16β–hydroxylation

Digoxigenin

Gitoxigenin

Hellebrigenin

Fig. 9.23b. Biosynthesis of cardiac glycosides.

9.6. ACETATE-MEVALONATE PATHWAY (TERPENOIDS AND STEROIDS)

The terpenoids form a large and structurally diverse family of natural products derived from C_5 isoprene units joined in head to tail fashion(Fig 9.26) They are classified into 7 classes based on the number of isoprene units $(C_5)_n$.

1. Hemiterpenes (C_5)
2. Monoterpenes (C_{10})
3. Sesquiterpenes (C_{15})
4. Diterpenes (C_{20})
5. Triterpenes (C_{30})
6. Tetraterpnes (C_{40})

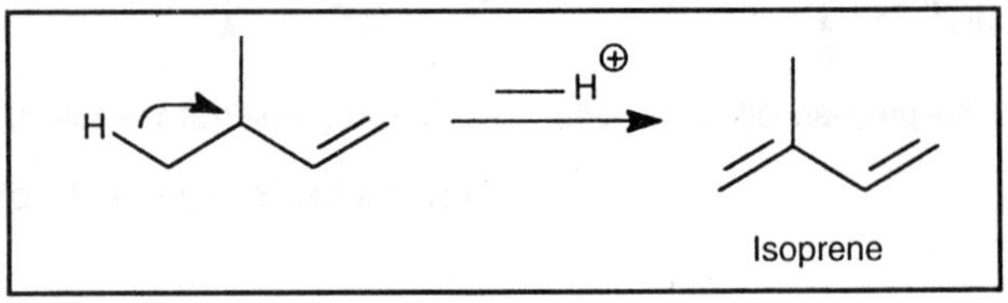

Isoprene

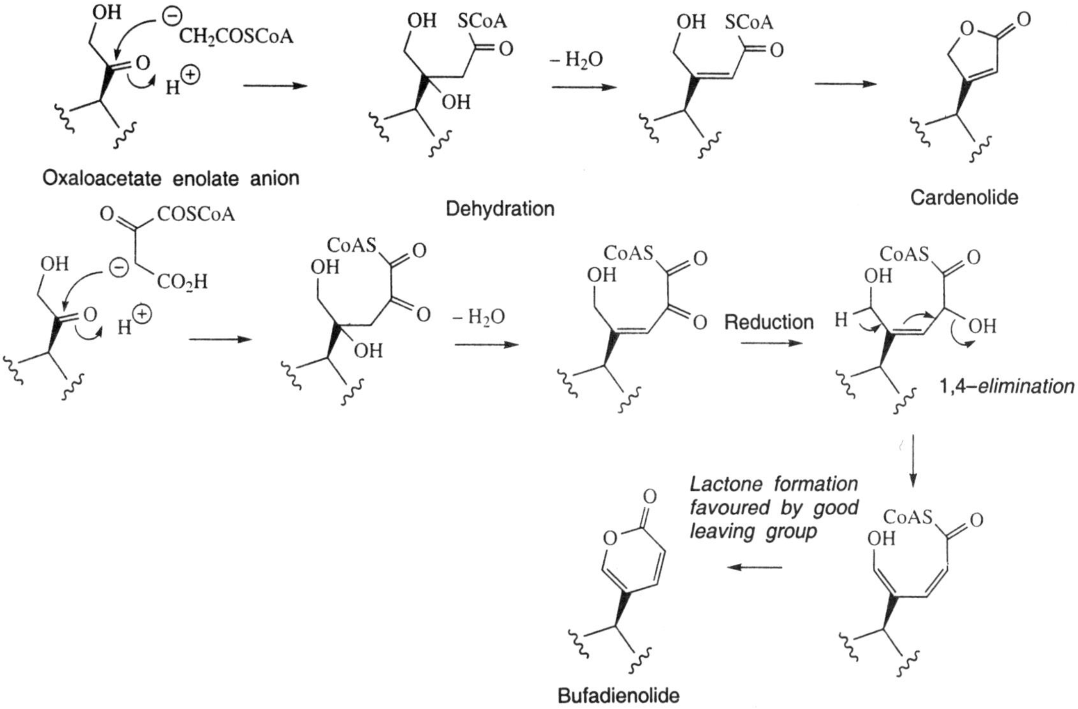

Fig. 9.24. Biosynthesis of side chain of cardiac glycosides.

Three molecules of acetyl coenzyme A are used to form mevalonic acid (MVA). MVA had been established as precursor of the sterol cholesterol. Two molecules of acetyl CoA undergo claisen type of condensation to give rise to aceto acetyl CoA to which a third molecule is incorporated via a stereospecific aldol addition giving the branched chain ester β-hydroxy-β-methyl glutaryl CoA (HMG-CoA). This upon two-step reduction will give rise MVA the six-carbon compound MVA is transformed into the five-carbon phosphorylated isoprene units in a series of reactions, beginning with phosphorylation of the primary alcohol group. Two different ATP-dependent enzymes are involved, resulting in MVA diphosphate and decarboxylation / dehydration then follow to give isopentenyl pyrophosphate (IPP) which upon isomerization will form dimethyl allyl pyrophosphate (DMAPP) (Fig.9.27). Combination of DMAPP and IPP via the enzyme prenyl transferase yields geranyl diphosphate (GPP) (Fig 9.28). The voltaile oils containing principally isoprenoid compounds is shown in Table 9.1.

Linalyl PP and neryl PP are isomers of GPP, and are likely to be formed from GPP by ionization to the allylic cation, which can thus allow a change in attachment of diphosphate group (to the tertiary carbon in linalyl PP) or a change in the stereochemistry of the double bond (to Z in neryl PP) (Fig 9.28). These 3 compounds, by modest changes can give rise to a range of monoterpenes found which are components of volatile oils used in flavoring and perfumery (Fig. 9. 29). The formation of monocyclic and bicyclic terpenes are shown in (Fig. 9.30 and 9.31). Most of the monterpenes are optically active; and there are many examples known where enantiomeric forms of the same compound can be isolated from different sources e.g. (+) camphor from *Salvia officinalis* (labiatae) and – camphor from *Tanacetum vulgare* (Compositae) or (+) – carvone from *Carum curvi* (umbelliferae)

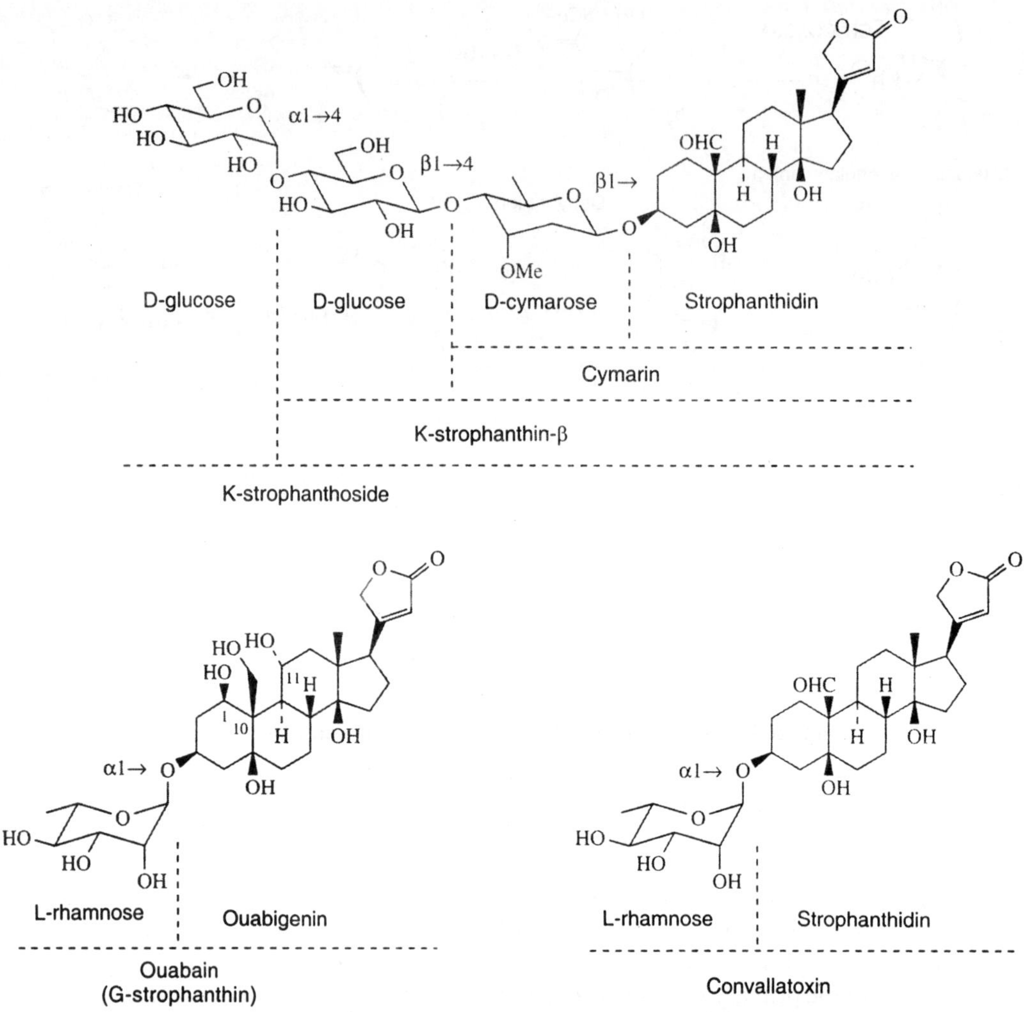

Fig. 9.25. Cardiac glycosides from Strophanthus and convallaria.

and – Carvone from *Mentha spicata* (labiatae). The biogenetic pathway for camphor, α-pinene, thymol is shown in Fig. 9.31. Similarly for thymol and related compounds is shown in Fig. 9.32.

Chrysanthemic acid and pyrethric acid found in ester form as pyrethrins (pyrethrins, cinerins and jasmolins) (Fig.9.33), are valuable insecticidal components in Pyrethrum flowers (*Chrysanthemum cinerariaefolium* (compositae). The formation of chrysanthemic acid is given in Fig. 9.34. The α-linolenic acid formed via 12-oxophytodienoic acid could be the precursor of jasmolone, which with β-oxidation and then followed by decarboxylation will give rise jasmonic acid (Fig 9.35).

9.7. BIOSYNTHESIS OF ALKALOIDS

The alkaloids are organic nitrogenous bases found mainly in plants, but also to a lesser extent in microorganisms and animals. They are often classified according to the nature of the nitrogen-containing structure, e.g. pyrrolidine, piperidine, quinoline, isoquinoline indole, etc. Though the structural

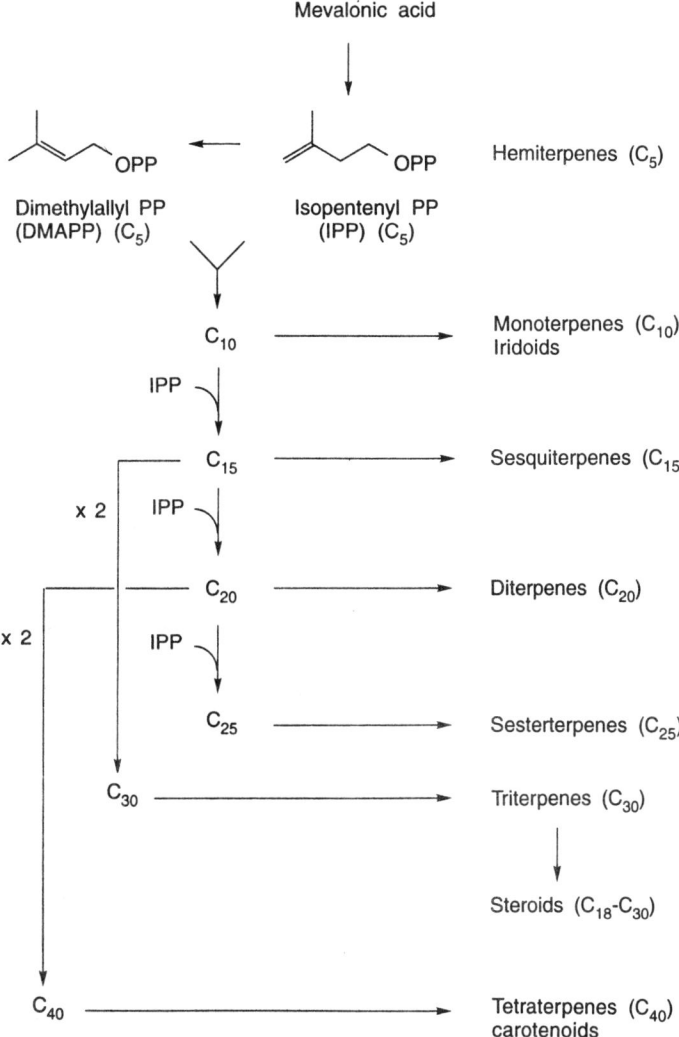

Fig. 9.26. The different classes of terpenoids.

complexity of some examples rapidly expands the number of sub-divisions. The nitrogen atoms in alkaloids originate from an amino acid, and in general, the carbon skeleton of the particular aminoacid precursor is also largely retained intact in the alkaloid structure, though the carboxylic acid carbon is often lost through decarboxylation. The principal precursor for alkaloids are ornithine, nicotinic acid, tyrosine, tryptophan, anthranilic acid and histidine. Building blocks from the acetate, shikimate and mevalonate pathways are also frequently incorporated into the alkaloid structures.

9.7.1. Alkaloids Derived From Ornithine (Pyrrolidine and Tropane Alkaloids)

These alkaloids are found in the solanaceae family members such as hyoscyamine or cocaine. The following plants contain the tropane alkaloids.

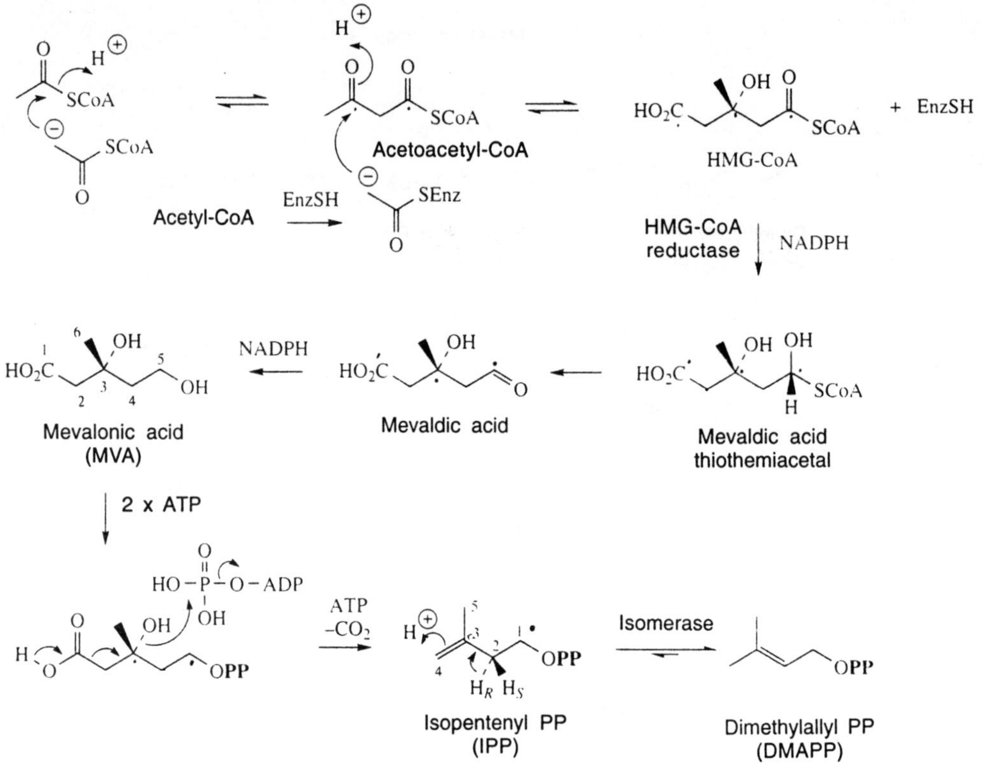

Fig. 9.27. Biosynthesis of dimethylallyl pyrophosphate.

1. *Atropa belladonna*
2. *Datura stramonium*
3. *Brugmansia* sp.
4. *Hyoscyamus niger*
5. *Duboisa* sp.
6. *Scopolia* sp.

The pyrrolidine ring system is formed initially as a Δ¹pyrrolinium - cation pyridoxal 5'-phosphate (PLP) dependent decarboxylation of ornithine gives putrescine, which is methylated to give N-methylputrescine, the oxidative deamination of N-methyl putrescine gives the aldehyde and form N-methyl-Δ¹pyrrolinium cation (Fig. 9.38). An alternative sequence to puterescine is from L-arginine, upon decarboxylation followed by hydrolysis to give putrescine. The extra carbon atoms required for hygrine formation are derived from acetate-MVA pathway. The biogenetic pathway for (−) Hyoscyamine and Cocaine are shown in Fig. 9.39. The N-methyl Δ¹ pyrrolinium cation in combination with acetyl CoA on S and R configuration leads the formation of cocaine and hyoscyamine. The intramolecular mannich reaction on the R enantiomer accompanied by decarboxylation will form tropinone, followed by stereospecific reduction of the carbonyl group will yield tropine. The tropine on combination with phenyl lactic acid (which orginates from L-phenylalanine) forms hyoscyamine. The hyoscyamine on 6β-hydroxylation followed by epoxidation will give rise to hyoscine (Scopolamine) (Fig. 9.40).

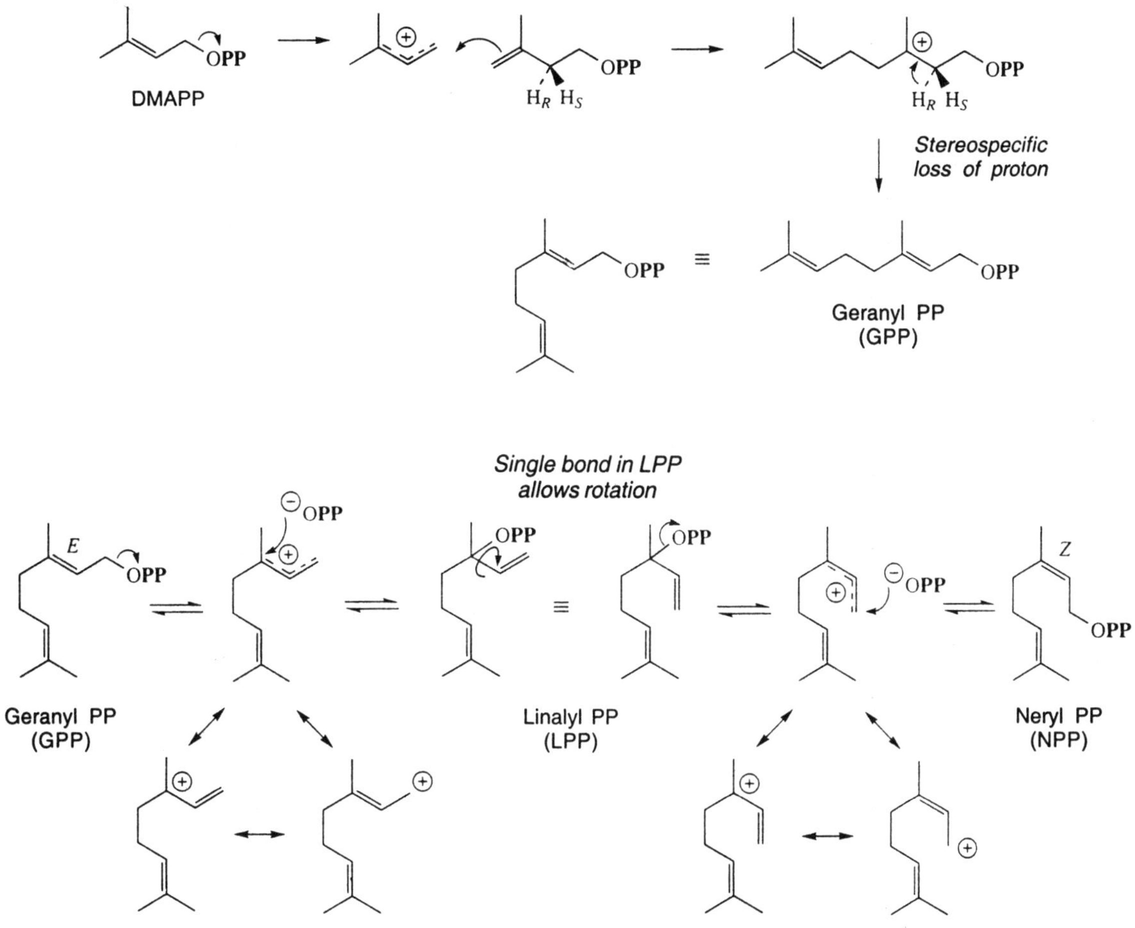

Fig. 9.28. Acetate-Mevalonate pathway (Biosynthesis of GPP/NPP).

9.7.2. Alkaloids Derived from Tryosine

9.7.2.1. *Phenylethylamine and Simple Tetrahydro Isoquinoline Alkaloids*

The PLP-dependent decarboxylation of L-tyrosine gives the simple phenyl ethylamine derivative pyramine which on methylation forms hordenine alkaloid obtained from barley (*Hordeum vulgare,* Graminae). The 3,4-di or trihydroxy derivatives of phenethylamines are derived from via dopamine, followed by decarboxylation yields L-DOPA (L-dihydroxy phenylanine). The aromatic hydroxylation, followed by O-methylation yields a alkaloid known as mescaline (*Lophora williamsii* Cactaceae) (Fig. 9.41).

The simple tetrahydroisoquinoline derived alkaloids such as anhalamine, anhalonine, anhalodine are also formed through the DOPA (Fig. 9.42). Lophocerine an alkaloid obtained from *Lophophora schotii* is also formed from DOPA and iPP (MVA) combination. The resulted compound undergoes methylation to yield alkaloid (Fig. 9.43). Salsolinol is an alkaloid found in plants e.g. Corydalis sp. (Papaveraceae), but can also be detected in the human urine. This alkaloid also formed from DOPA and acetaldehyde (Fig. 9.44).

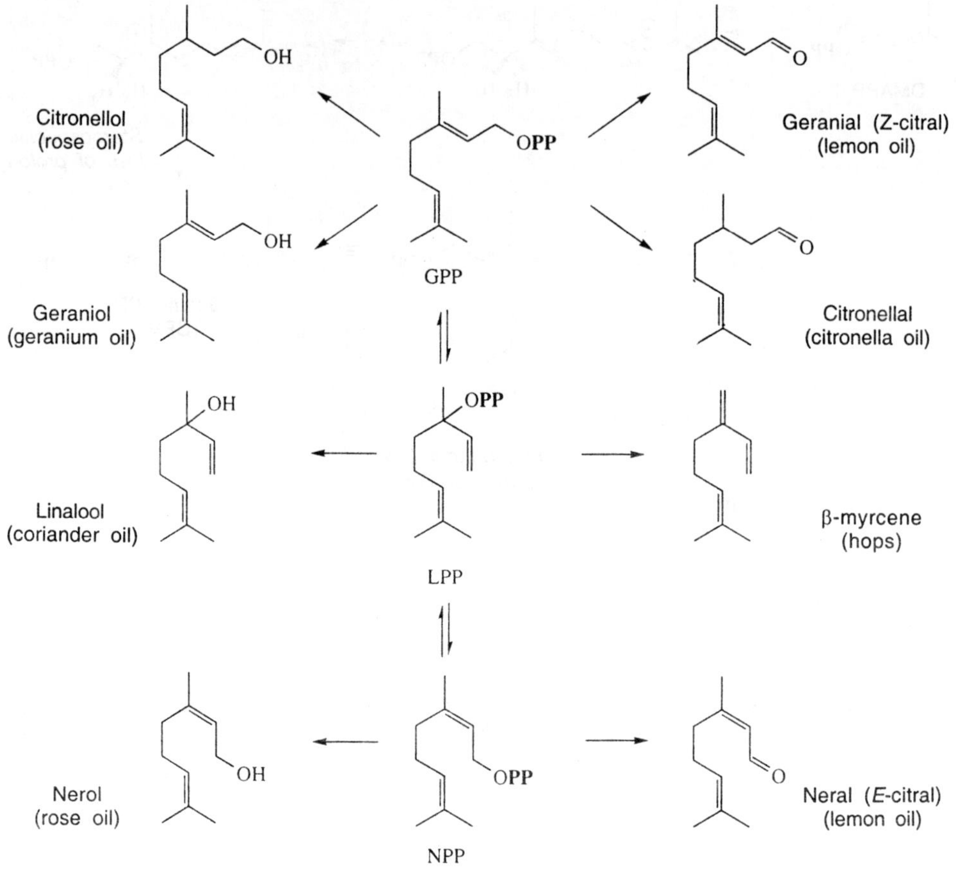

Fig. 9.29. The biosynthesis of monoterpenes.

Benzyl tetrahydroisoquinoline skeleton is formed by the incorporation of a phenylethyl unit into the phenyl ethylamine. This upon several modifications produces a wide range of alkaloids (Fig. 9.45). Two tryosine molecules are used in the biosynthetic pathway of norcoclaurine (S). One phenethylamine fragment of tetrahydroisoquinoline ring system is formed via DOPA, the remaining from tyrosine via 4-Hydroxyphenyl acetaldehyde, norcoclaurine (s) upon methylations will form (s) N—methyl coclaurine which further methylation forms(s) reticuline / (L) reticuline and morphinan (Fig. 9.45). Tetradrine a bis benzyl tetrahydroisoquinoline alkaloid isolated from *Stephania tetrandra* (Menispermaceae) is formed by the coupling of two molecules of (s)-N-methylcoclaurine (Fig. 9.46). Papaverine a benzyl isoquinoline alkaloid obtained from *Papaver somniferum* (Opium) is formed from N-nor-reticuline by O-methylations and oxidation on isoquinoline nucleus (Fig. 9.45).

Tubocurarine, is an alkaloid used as arrow poison obtained from *Chondrodendron tomentosum* (Meninspermaceae). The biosynthetic pathway is shown in Fig 9.47.

9.7.2.2. *Modified Benzyl Tetrahydrosiquinoline Alkaloids*

The opium alkaloids morphine, codeine and thebaine are derived by the coupling of tetrahydro-isoquinoline skeleton, followed by reduction. The (R)-reticuline is established as the precursor of the

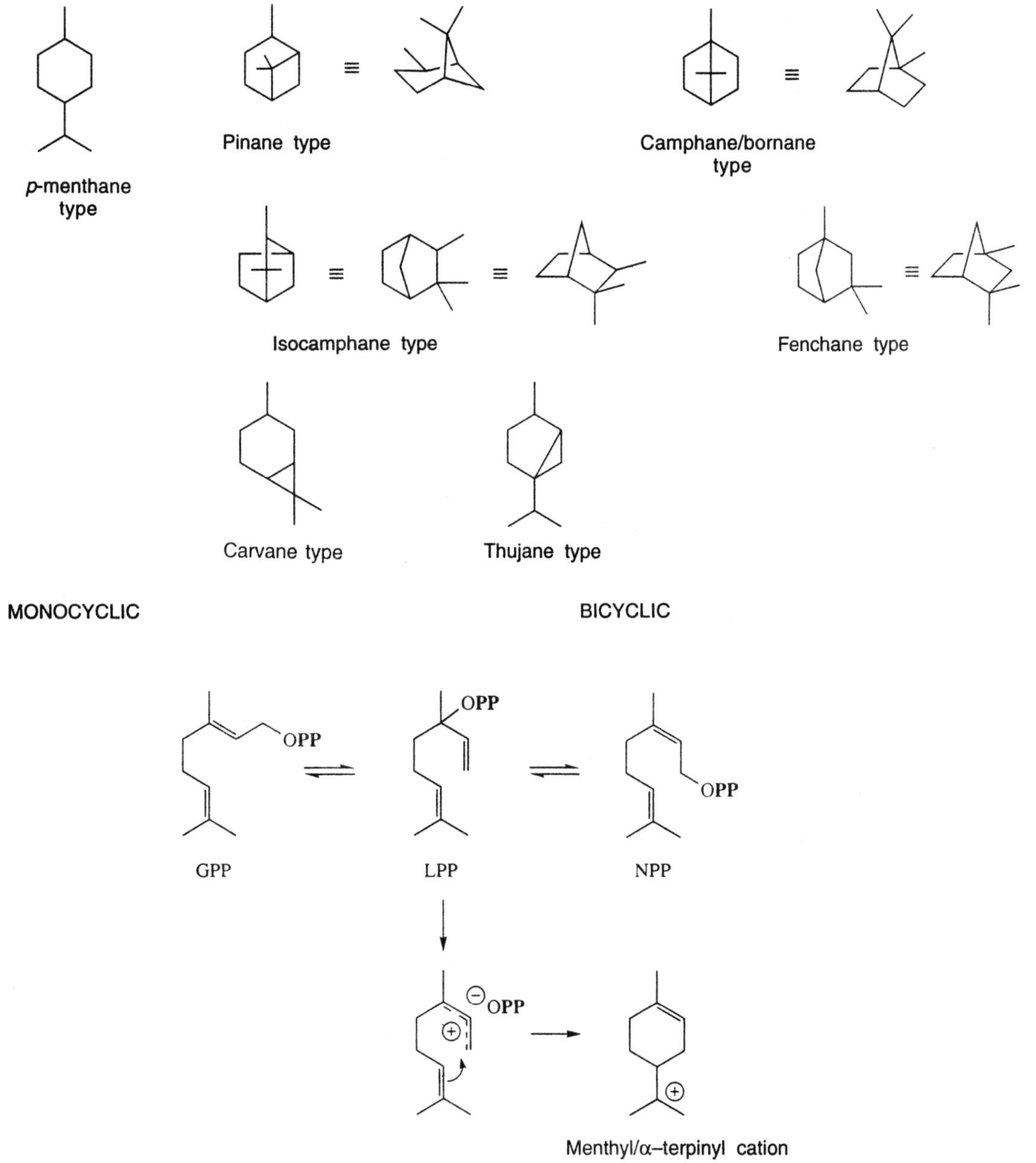

p-menthane
type

Pinane type

Camphane/bornane
type

Isocamphane type

Fenchane type

Carvane type

Thujane type

MONOCYCLIC

BICYCLIC

GPP

LPP

NPP

Menthyl/α–terpinyl cation

Fig. 9.30. Biosynthesis of monocyclic/bicyclic terpenes.

morphinan alkaloids (Fig. 9.48). Coupling ortho to the phenol group in the tetrahydroisoquinoline, and para to the phenol in the benzyl substituent then yields salutaridine (Minor alkaloid in opium). This upon reduction will forms salutaridinol which on combination with acetyl CoA forms thebaine. The oxidation, followed by hydroxylation and reduction yields codeine. Then demethylation forms morphine (Fig. 9.48). Isoboldine is a aporphine alkaloid and is a minor constituent in opium and biosynthesized from S-reticuline (Fig. 9.49).

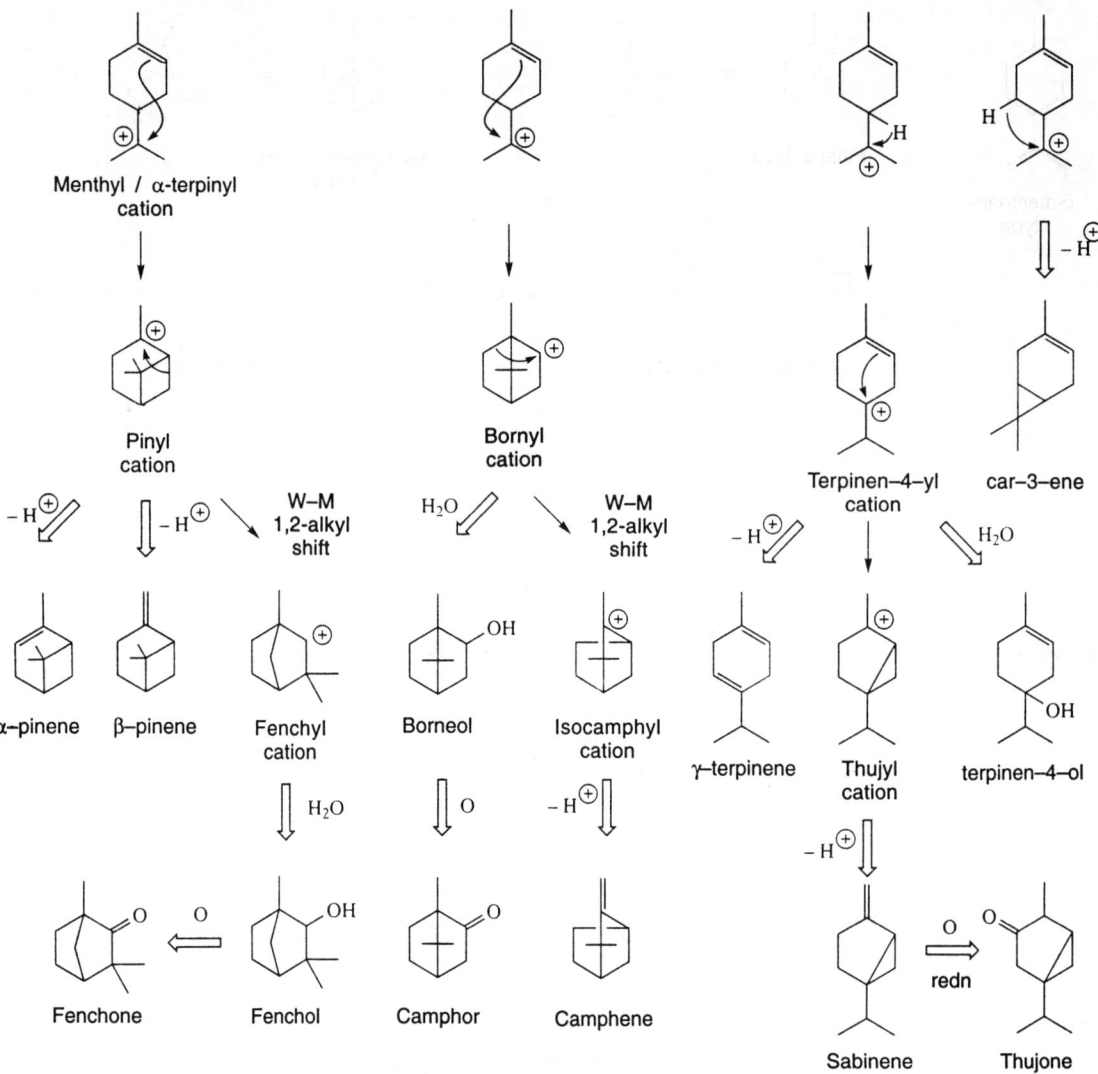

Fig. 9.31. Biosynthesis of camphor, fenchone and thujone.

Aristolochic acid is an active constituent present in many species of Aristolochia (*Aristolo-chiaceae*). The (R)-orientaline is a precursor for the biosynthesis of aristolochic acid via stephanine (*Stephania* sp. menispermaceae) (Fig. 9.50).

Berberine is the principal constituents of members of the Berberidaceae family (Berberis sp.), the Rananculaceae (e.g. Hydrastis sp.) and other families. The biosynthetic pathway is depicted in Fig 9.52. The intermediate scoulerine is the precursor for biosynthesis of hydrastine.

9.7.2.3. Phenethylisoquinoline Alkaloids

The liliaceae family members are found to contain several analogues of the benzyltetra hydro-isoquinoline alkaloids e.g., autumanaline. This alkaloid is formed by the condensation of dopamine

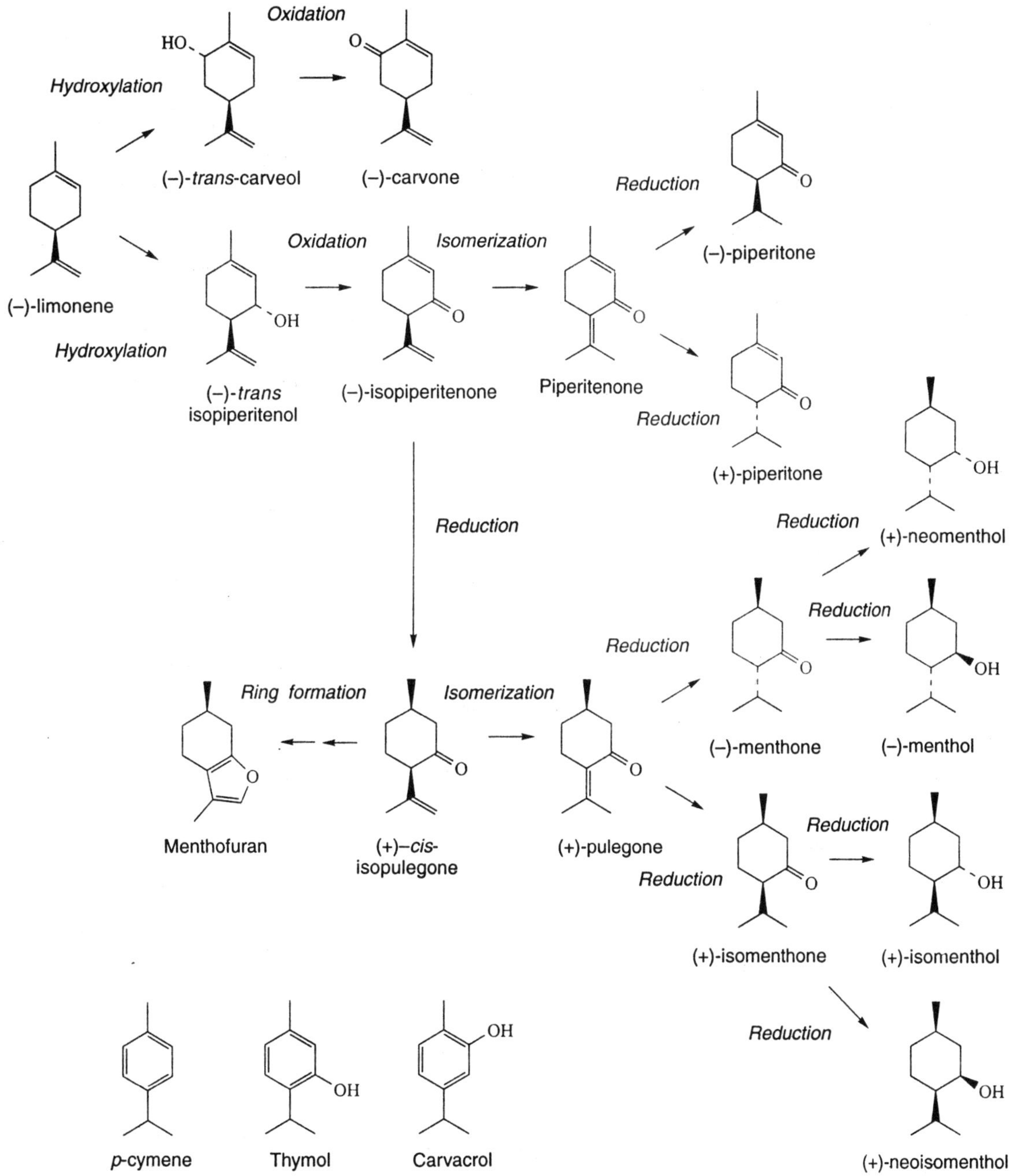

Fig. 9.32. Biosynthesis of thymol, carvacrol, p-cymene, menthol, neomenthol and neoisomenthol.

(from L-tyrosine) and 4-hydroxy cinnamaldehyde (from L-phenylalanine) the oxidative coupling of which yields floramultine, which upon methylation give rises Kreysigine (*Kreysigia multiflora* liliaceae) (Fig. 9.53). Colchicine found in species of Colchicum (e.g. Colchicum autumnale) are also biosynthesized from (s)-autmnaline (Fig 9.53).

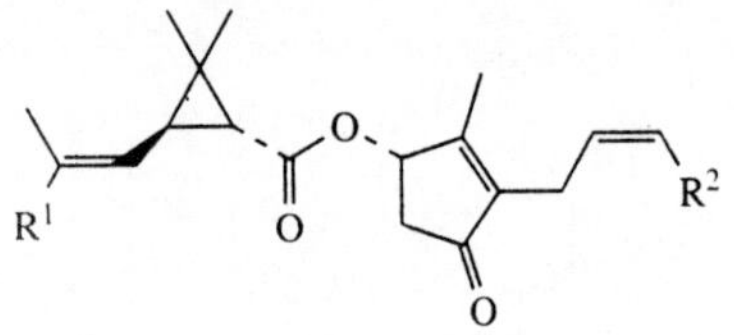

	R^1	R^2
Pyrethrin I	Me	CH=CH$_2$
Pyrethrin II	CO$_2$Me	CH=CH$_2$
Cinerin I	Me	Me
Cinerin II	CO$_2$Me	Me
Jasmolin I	Me	Et
Jasmolin II	CO$_2$Me	Et

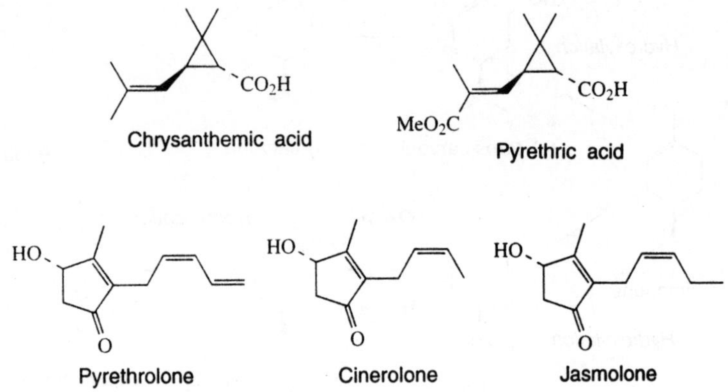

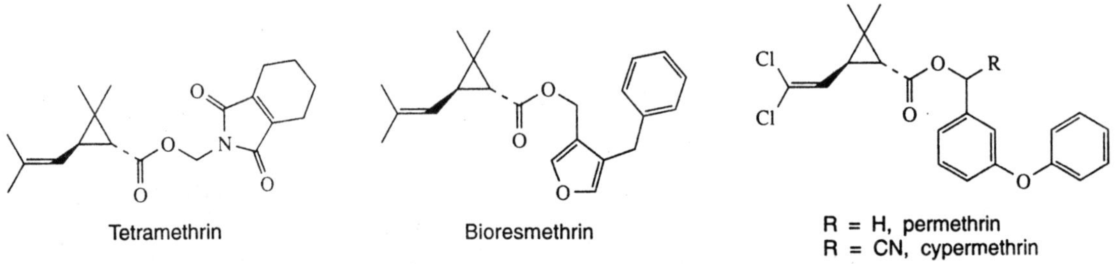

Chrysanthemic acid

Pyrethric acid

Pyrethrolone   Cinerolone   Jasmolone

Tetramethrin   Bioresmethrin   R = H, permethrin
R = CN, cypermethrin

Fig. 9.33. The active chemical constituents of Pyrethrum flowers.

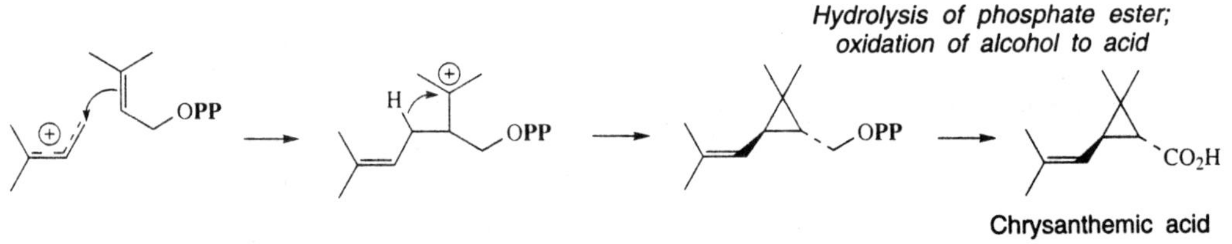

Hydrolysis of phosphate ester;
oxidation of alcohol to acid

Chrysanthemic acid

Fig. 9.34. Biosynthesis of chrysanthemic acid.

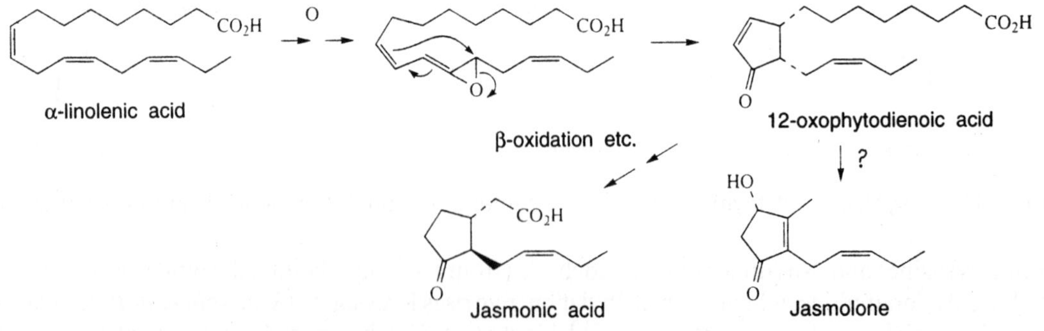

α-linolenic acid

β-oxidation etc.

12-oxophytodienoic acid

Jasmonic acid   Jasmolone

Fig. 9.35. Biosynthetic pathway of Jasmonic acid.

Biosynthesis of Triglycerides

$$\text{Acetyl CoA} \xrightarrow[\substack{\text{Biotin} \\ \text{ATP} / CO_2}]{Mn^{+2}} \text{Malonyl CoA} + \text{Acetyl CoA}$$

$$CH_3COCH\begin{matrix} \diagup COOH \\ \diagdown CO-SCoA \end{matrix}$$

$$+$$

$$CoA-SH$$

$$\Big| NADPH_2$$

$$CH_3CH_2CH_2-COSCoA$$
$$\text{Butyl CoA} +$$
$$CO_2 + H_2O$$

$$\text{Butyl CoA} + \text{Malonyl CoA}$$

$$CH_3CH_2CH_2COCH\begin{matrix} \diagup COOH \\ \diagdown CO-S-CoA \end{matrix} \quad + \quad CO_2 + H_2O$$

$$CH_3CH_2CH_2CH_2CO-S-CoA$$
$$\text{Carproyl CoA} + CO_2 + H_2O$$
$$\text{Caproyl CoA} + \text{Malonyl } (C_2)$$

Chain expansion for higher fatty acids

Fig. 9.36. Biosynthesis of fatty acids.

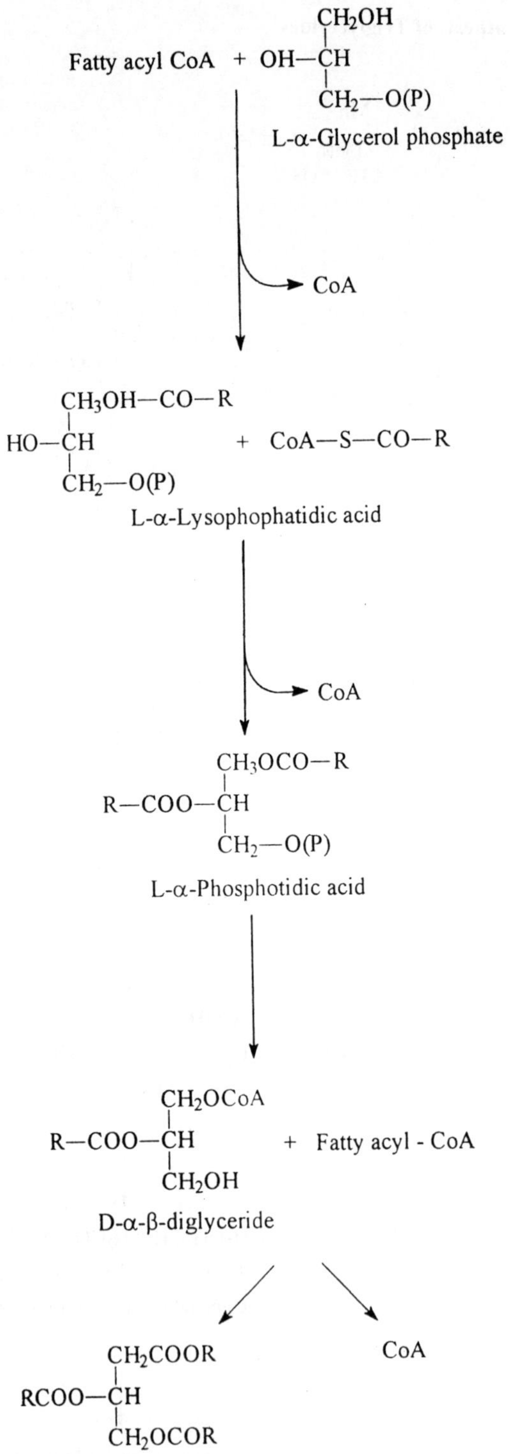

Fatty acyl CoA + OH—CH
$$\begin{array}{c} CH_2OH \\ | \\ OH{-}CH \\ | \\ CH_2{-}O(P) \end{array}$$

L-α-Glycerol phosphate

→ CoA

$$\begin{array}{c} CH_3OH{-}CO{-}R \\ | \\ HO{-}CH \\ | \\ CH_2{-}O(P) \end{array} \quad + \quad CoA{-}S{-}CO{-}R$$

L-α-Lysophophatidic acid

→ CoA

$$\begin{array}{c} CH_3OCO{-}R \\ | \\ R{-}COO{-}CH \\ | \\ CH_2{-}O(P) \end{array}$$

L-α-Phosphotidic acid

$$\begin{array}{c} CH_2OCoA \\ | \\ R{-}COO{-}CH \\ | \\ CH_2OH \end{array} \quad + \quad \text{Fatty acyl - CoA}$$

D-α-β-diglyceride

CoA

$$\begin{array}{c} CH_2COOR \\ | \\ RCOO{-}CH \\ | \\ CH_2OCOR \end{array}$$

Fig. 9.37. Biosynthesis of Triglycerides.

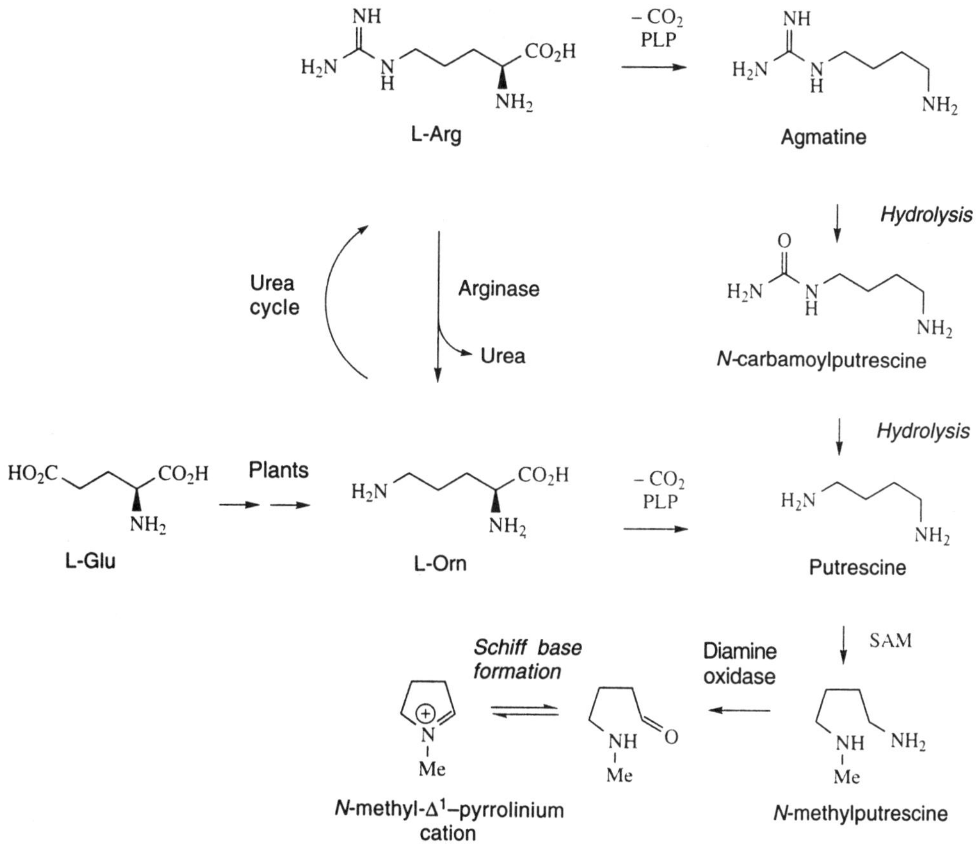

Fig. 9.38. Biosynthesis of methylputrescine.

9.7.2.4. Terpenoid tetrahydro isoquinoline alkaloids

The dried roots and rhizomes of ipecacuanha (*Cephaelis ipecacuanha*) found to contain terpenoid tetrahydro isoquinoline alkaloids e.g. Emetine, Cephaeline. The secologanin is an aldehyde and can condense with dopamine to give the N-deacetyl ipecoside (Tetrahydroisoquinoline). The biosynthetic pathway is shown in Fig. 9.54. Ipecoside itself found in the ipecac. The ipecoside an hydrolysis, followed by reduction and condensation with another mole of dopamine, followed by methylation forms cephaeline or emetine.

9.7.3. Alkaloids Derived from Tryptophan

9.7.3.1. Simple indole alkaloids

Psilocybin and simple indole alkaloid found mushroom (*Psilocybe* sp.) and biosyntheiszed from L-tryptophan. The biosynthetic pathway is depicted in Fig. 9.55. L-tryptophan upon decarboxylation forms tryptamine which undergoes N-methylation, hydroxylation and forms psilocin which upon phosphorylation yields Psilocybin. Gramine, simple indole alkaloid found in barley (*Hordeum vulgare*, Graminae) are also biosynthesized from L-tryptamine (Fig. 9.56).

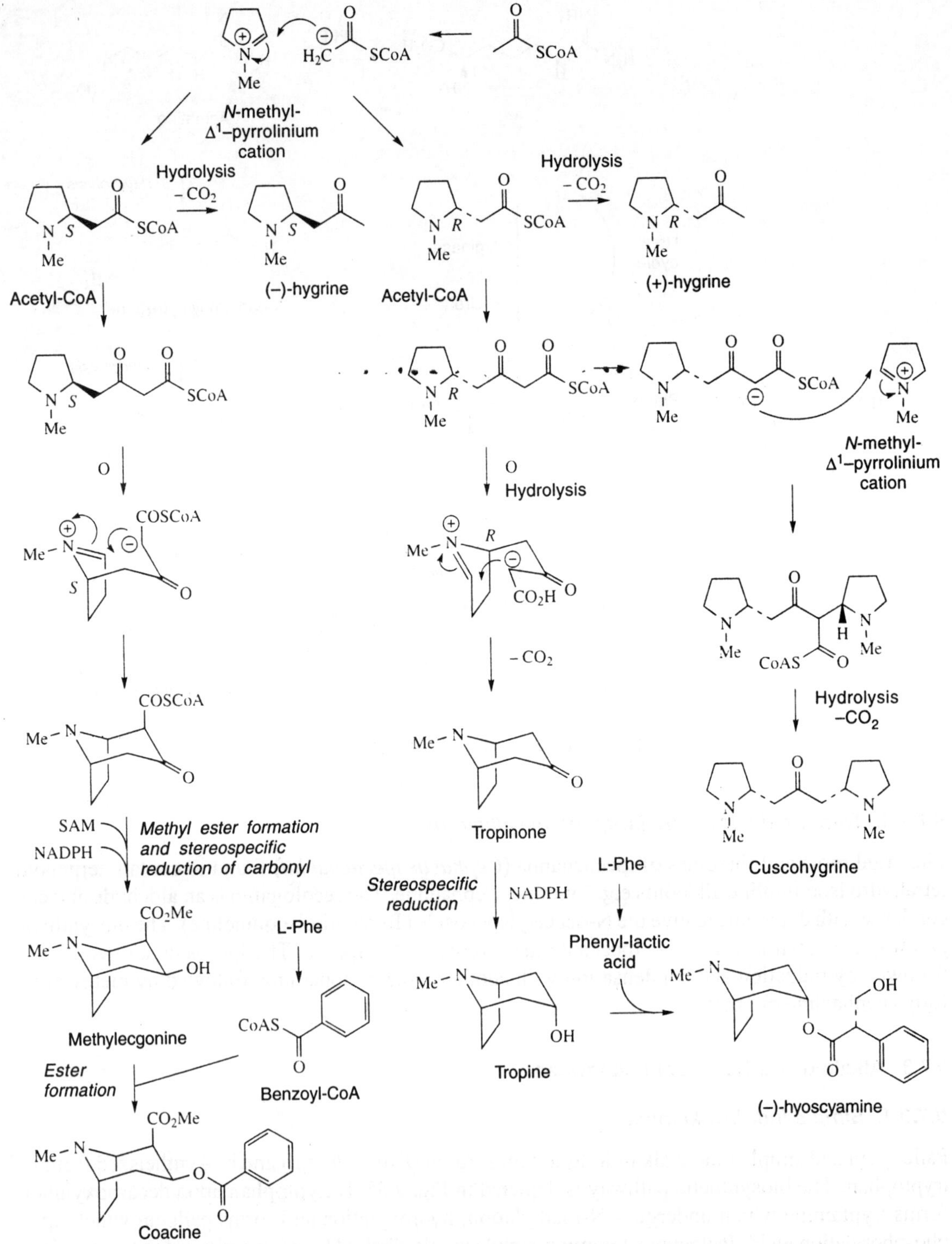

Fig. 9.39. Biosynthetic pathway of hyoscyamine and cocaine. (SAM = S-adenosylmethionine,

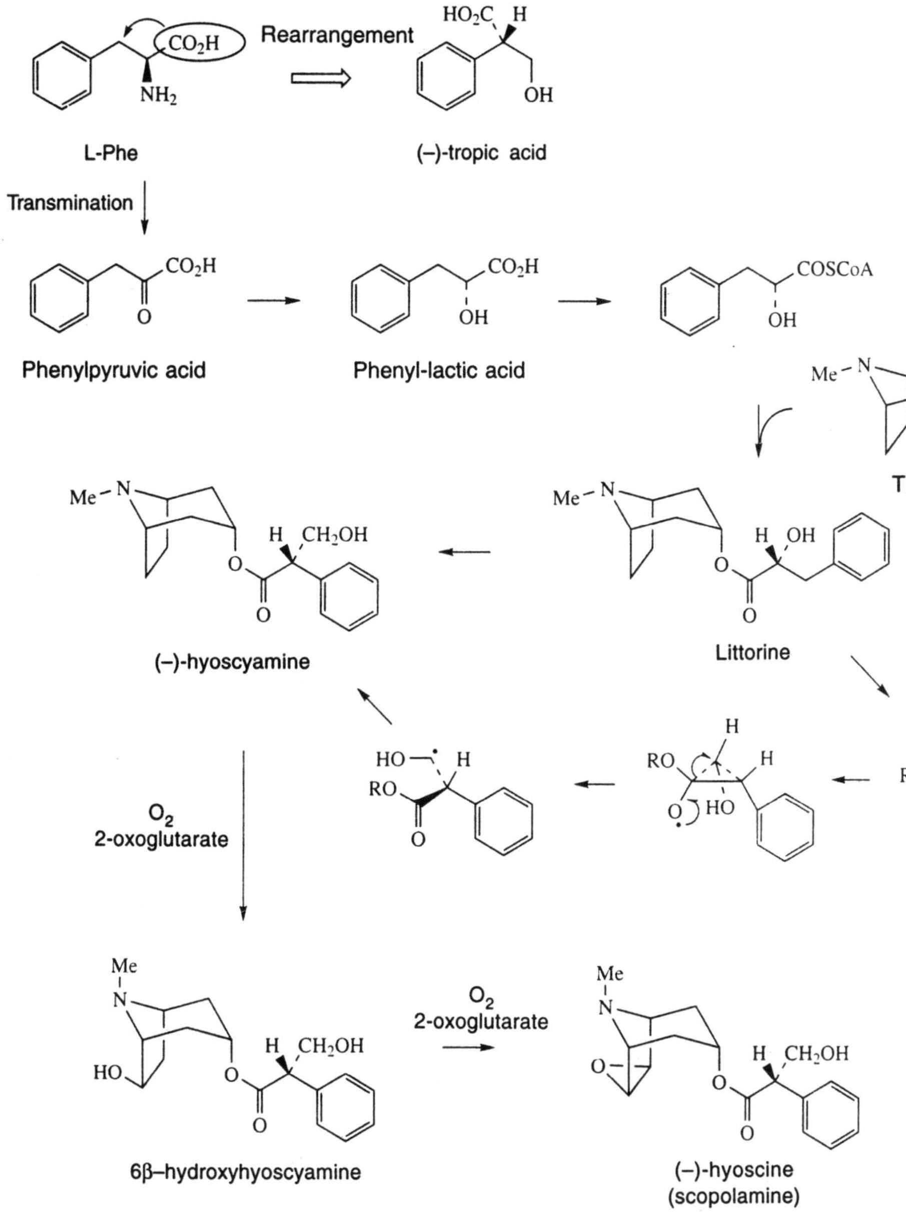

Fig. 9.40. Biosynthetic pathway for scopolamine.

9.7.3.2. Simple β-Carboline Alkaloids

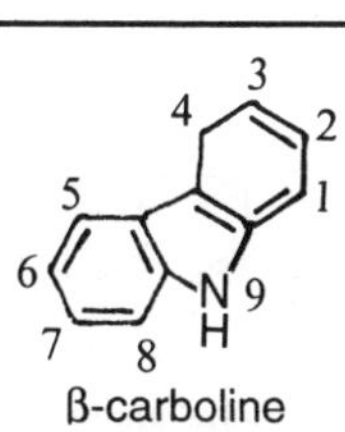

The β-Carboline alkaloids such as harmine (*Peganum harmala*, Zygophyllaceae), Eleagnine (*Eleagnus angustifolia*, Eleagnaceae) and tetrahydroharmine (*Banisteriopsis caapi*, Malphighiaceae) are biosynthesized from tryptamine and pyruvic acid (Fig. 9.57).

β-carboline

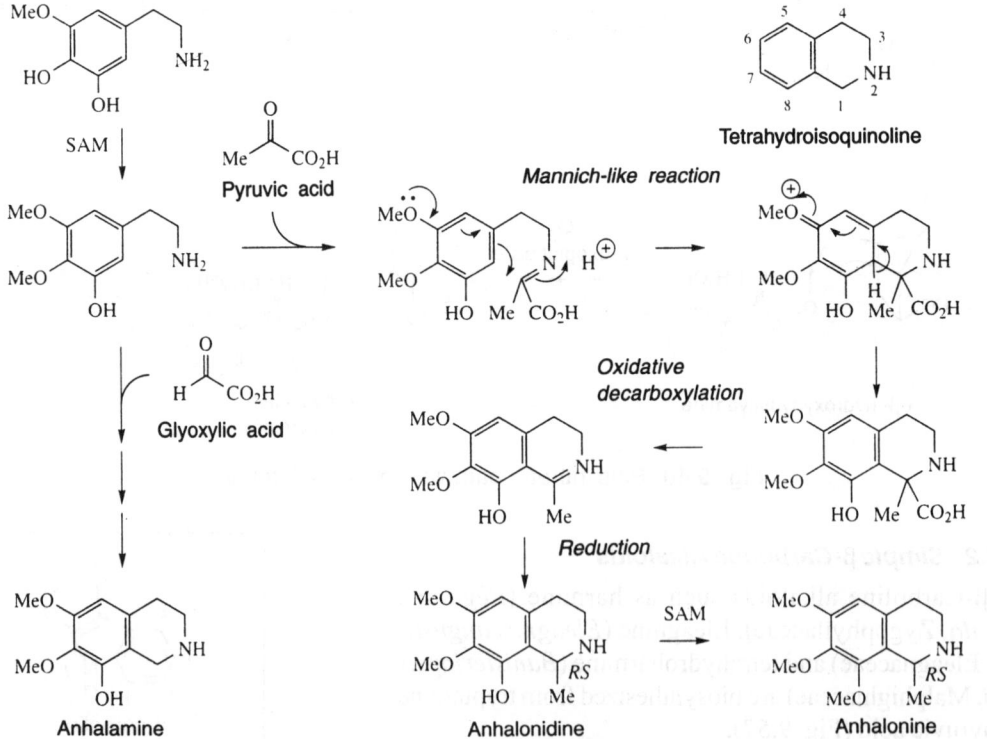

Fig. 9.41. Biosynthesis of hordenine, adrenaline and mescaline.

Fig. 9.42. Biosynthesis of tetrahydroisoquinoline alkaloids.

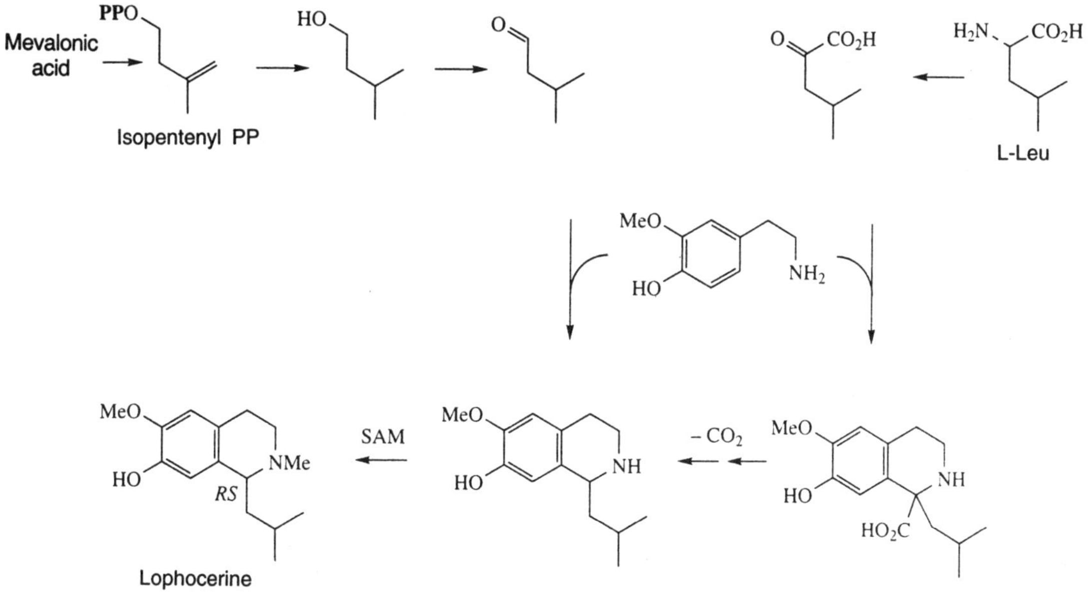

Fig. 9.43. Biosynthesis of lophocerine.

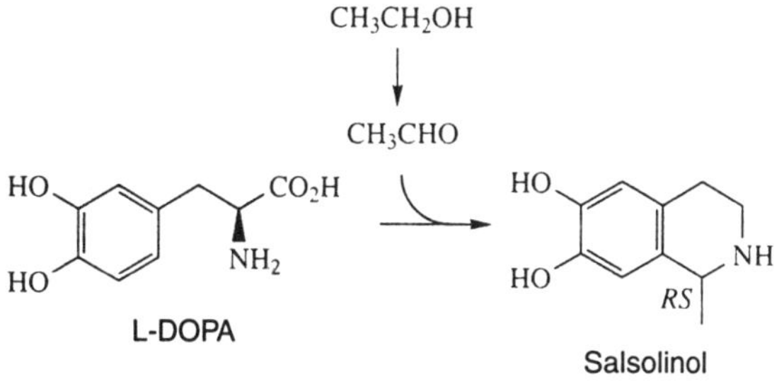

Fig. 9.44. Biosynthesis of salsolinol

9.7.3.3. Terpenoid Indole Alkaloids

Terpenoid Indole Alkaloids are mainly found in eight plant families of which Apocyanaceae, Loganiaceae, Rubiaceae provide the best sources. Tryptophan and its decarboxylation product tryptamine, serve as the precursor for the biosynthesis of indole alkaloids. The non-tryptophan portions of the alkaloids are however derived from monoterpenoid precursors which are designated as the coryanthe type (Ajmalicine & Akuammicine), the Aspidosperma type (Tabersonine), the Iboga type (Catharanthine).

The condensation of tryptamine with secologanin produces strictosidine, which upon hydrolysis of the glycoside function allows the opening of the hemiacetal, and exposure of an aldehyde group which can react with amine function. This intermediate upon allylic isomerization yields a product known as dehydrogeissoschizine and cyclization to cathenamine, which in turn reduction gives a compound known as ajmalicine (Fig. 9.58) present in *C. roseus*.

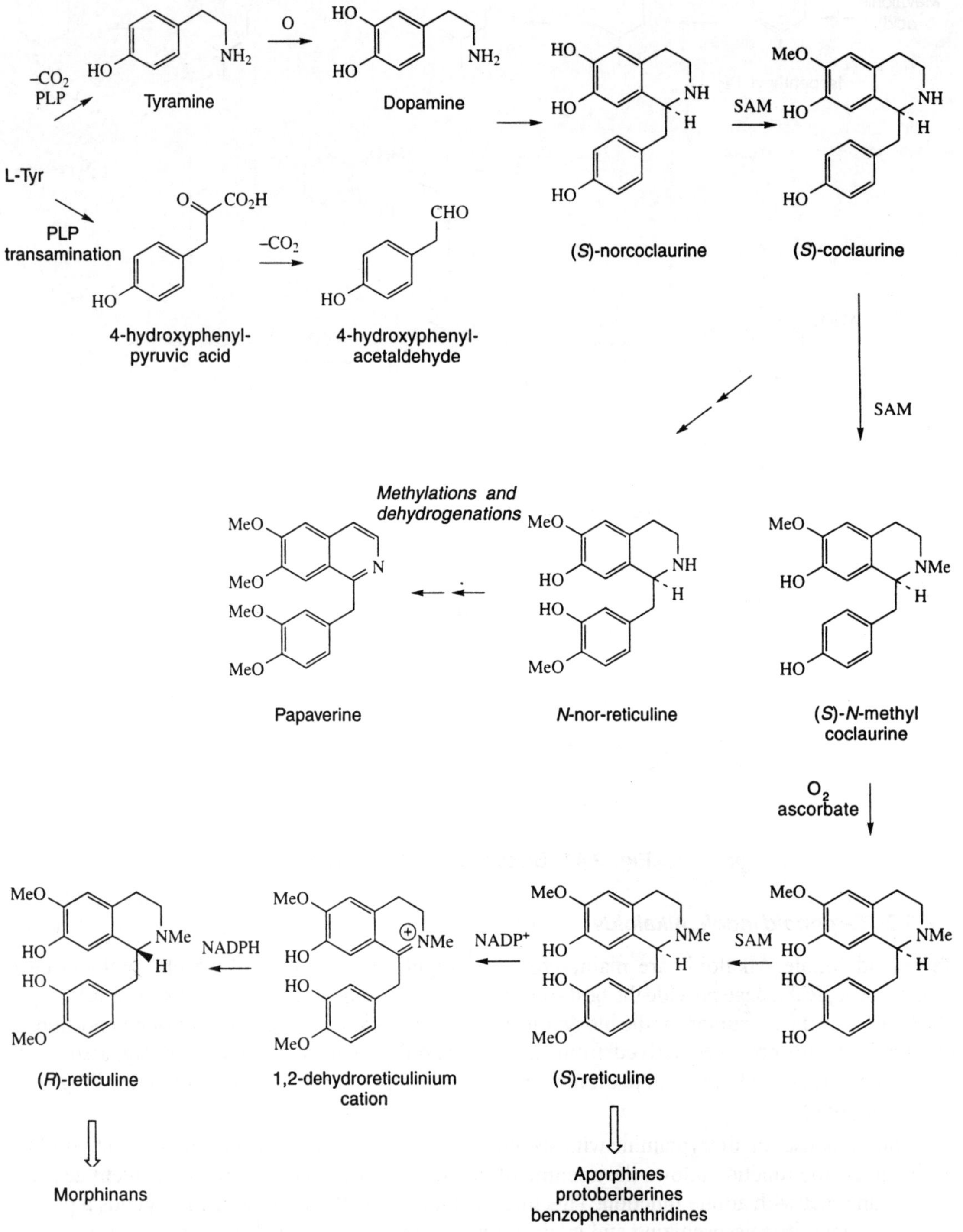

Fig. 9.45. Biosynthesis of benzyltetrahydroisoquinoline alkaloids.

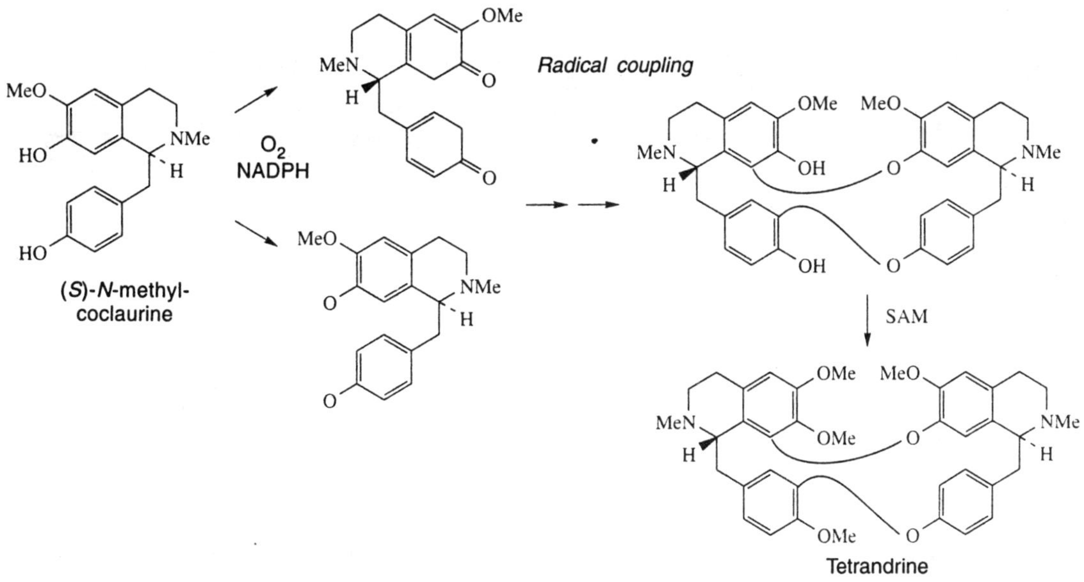

Fig 9.46. Biosynthesis of Tetrandrine.

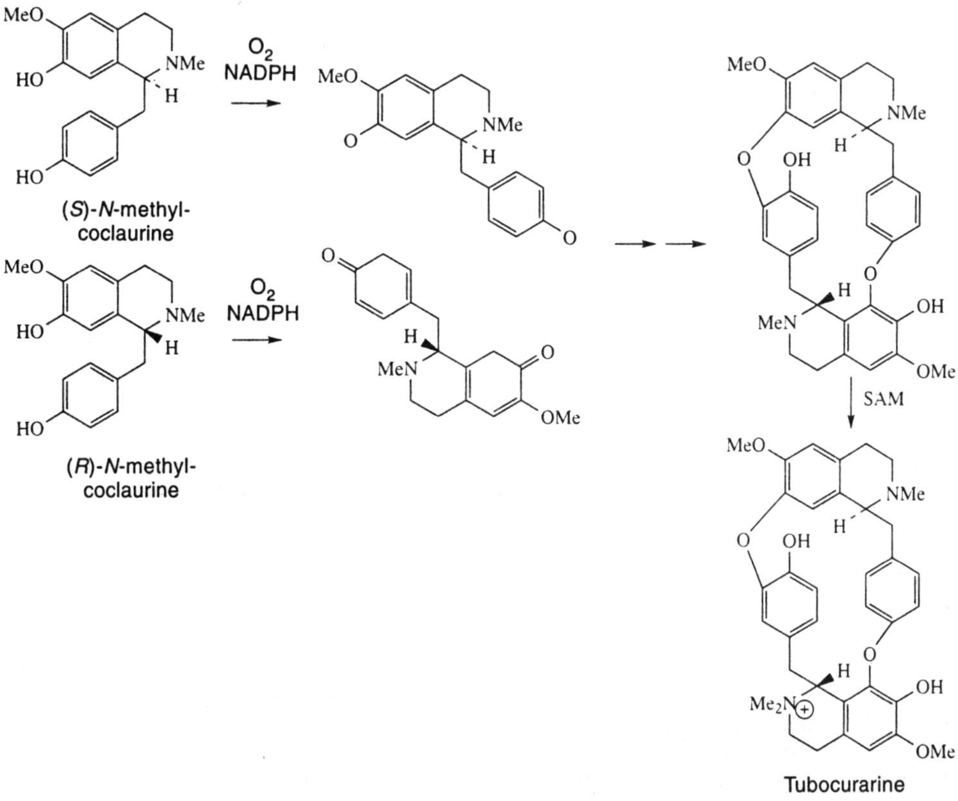

Fig. 9.47. Biosynthesis of tubocurarine.

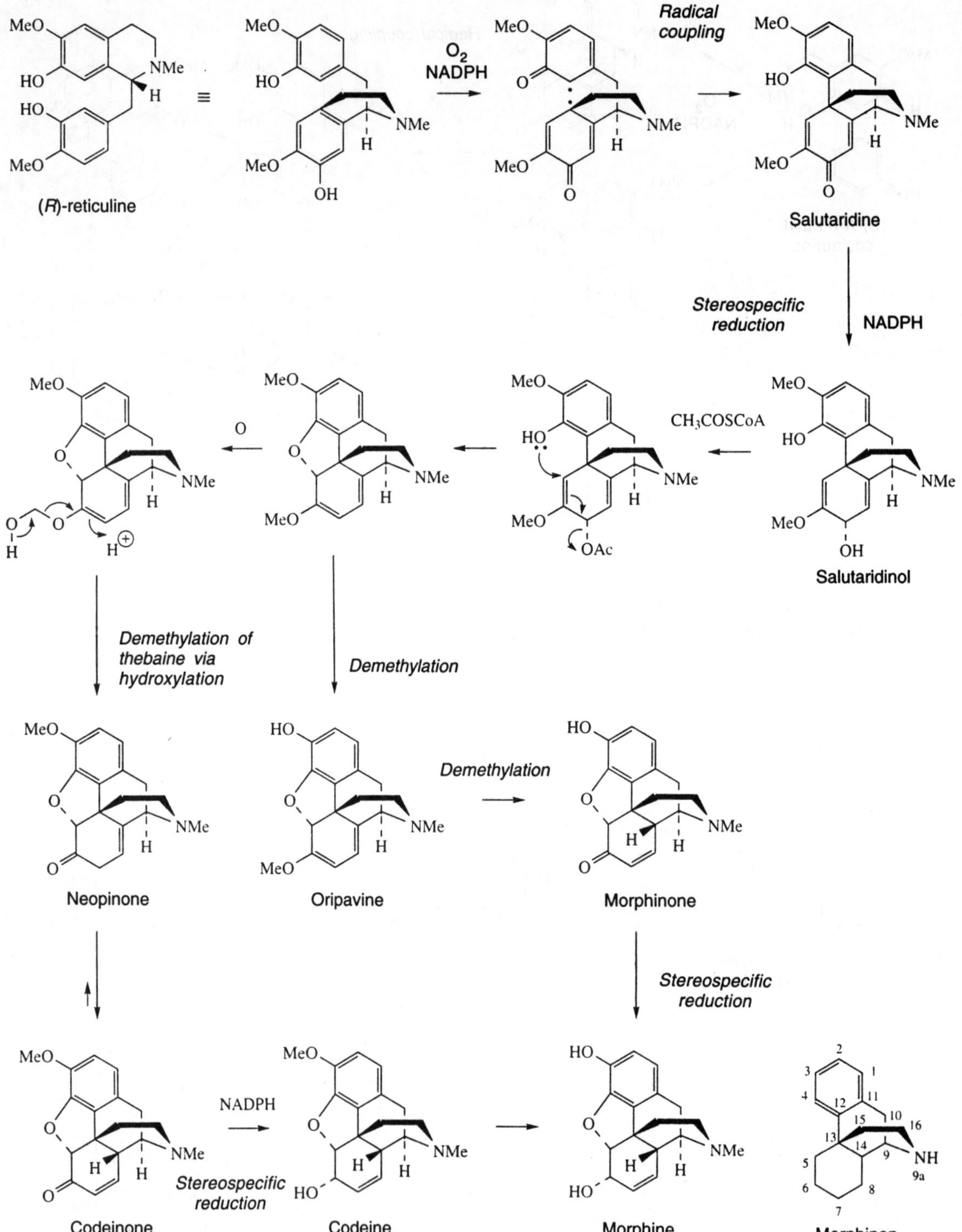

Fig. 9.48. Biosynthesis of morphine, codeine and other opium alkaloids.

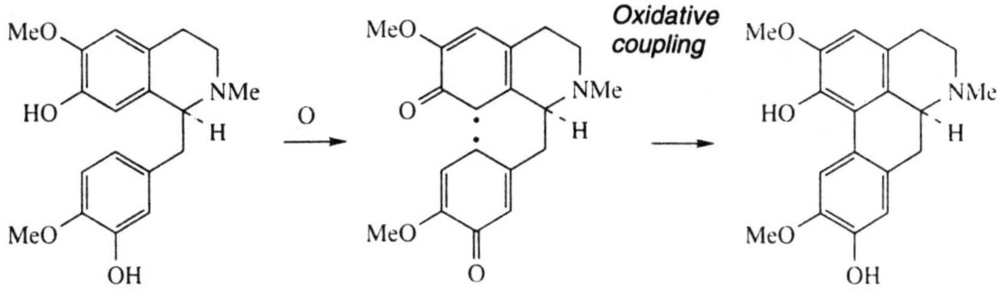

(*S*)-reticuline (*S*)-isoboldine

Fig. 9.49. Biosynthesis of (S)-isoboldine.

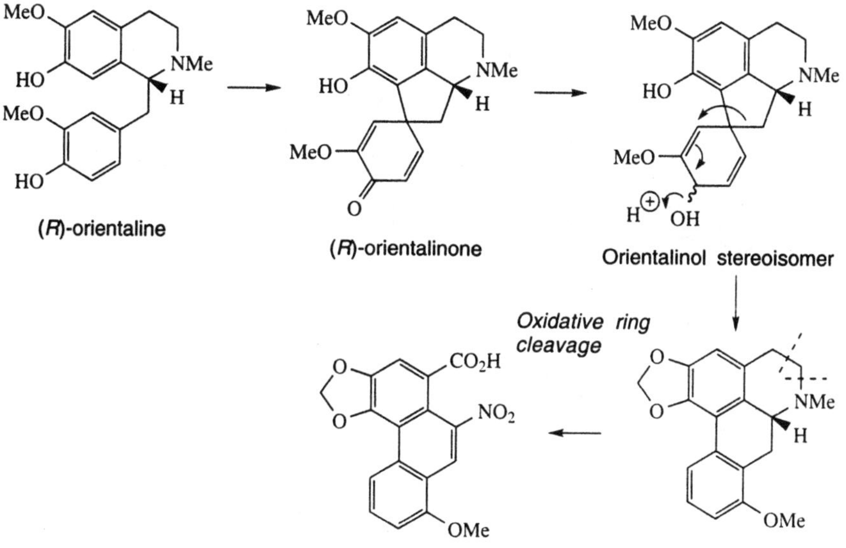

(*R*)-orientaline

(*R*)-orientalinone

Orientalinol stereoisomer

Aristolochic acid

Stephanine

Fig. 9.50. Biosynthesis of aristolochic acid.

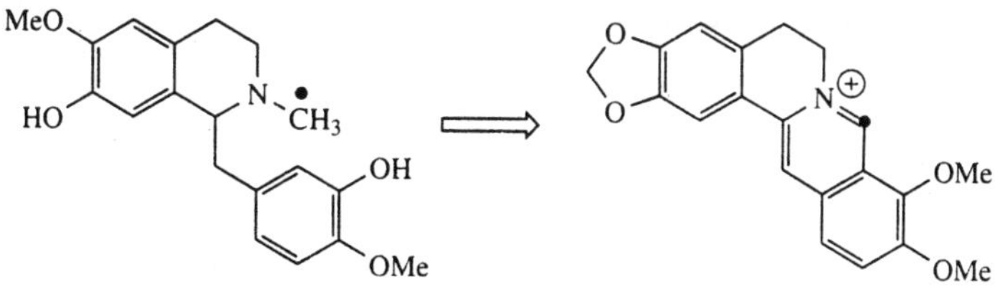

Reticuline Berberine

Fig. 9.51. Biosynthesis of berberine.

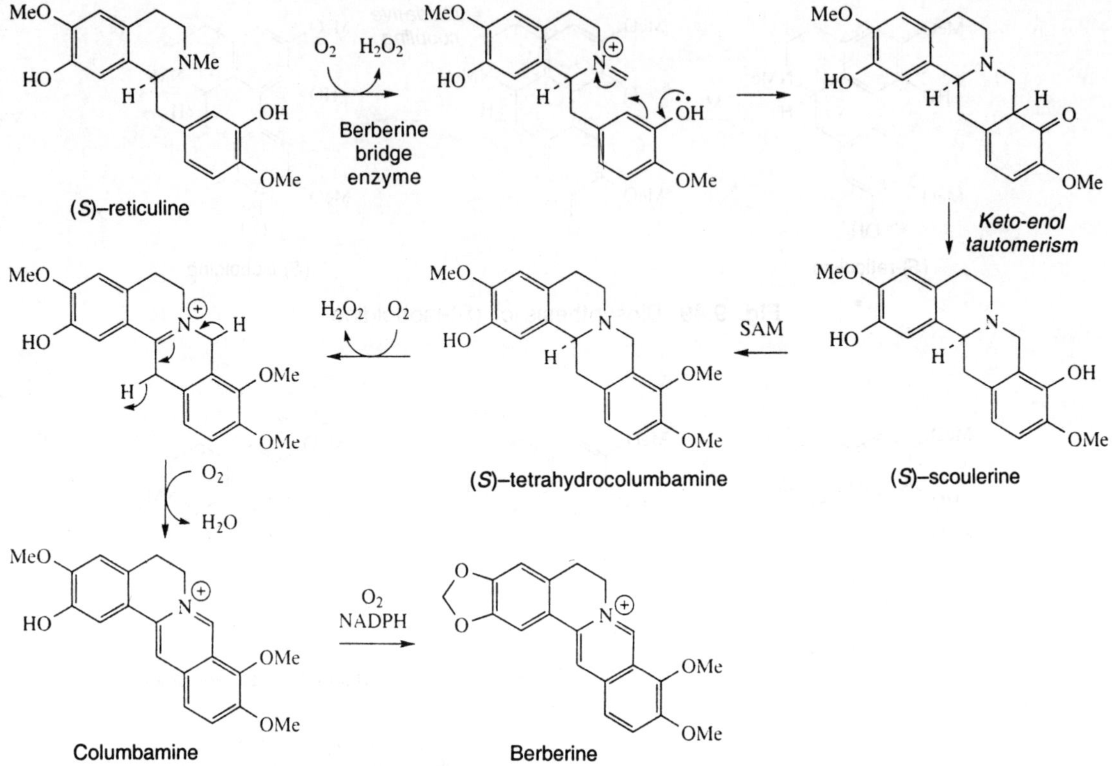

Fig. 9.52. Biosynthesis of berberine via scoulerine.

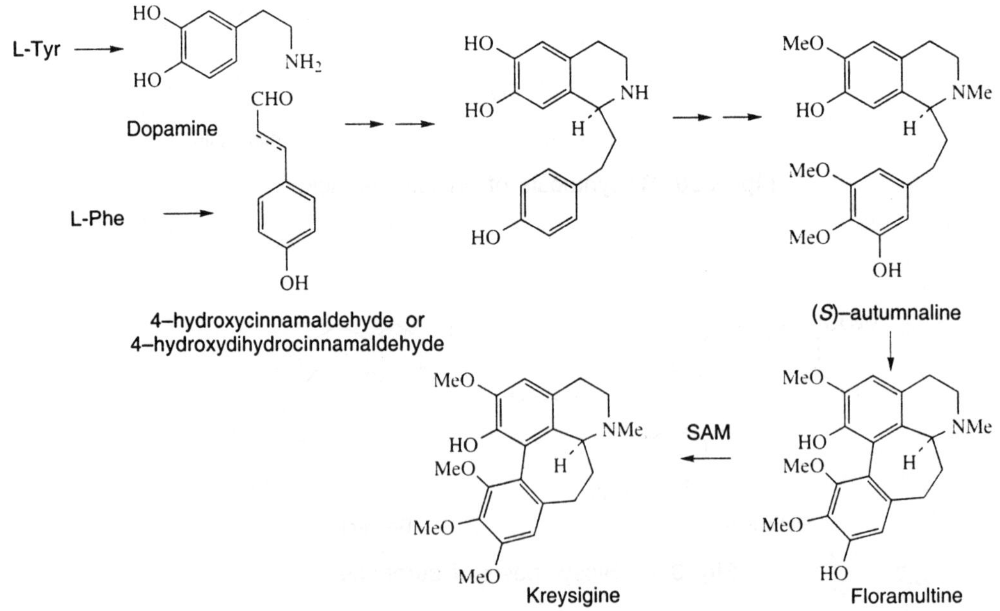

Fig. 9.53. Biosynthesis of phenethyl isoquinoline alkalods.

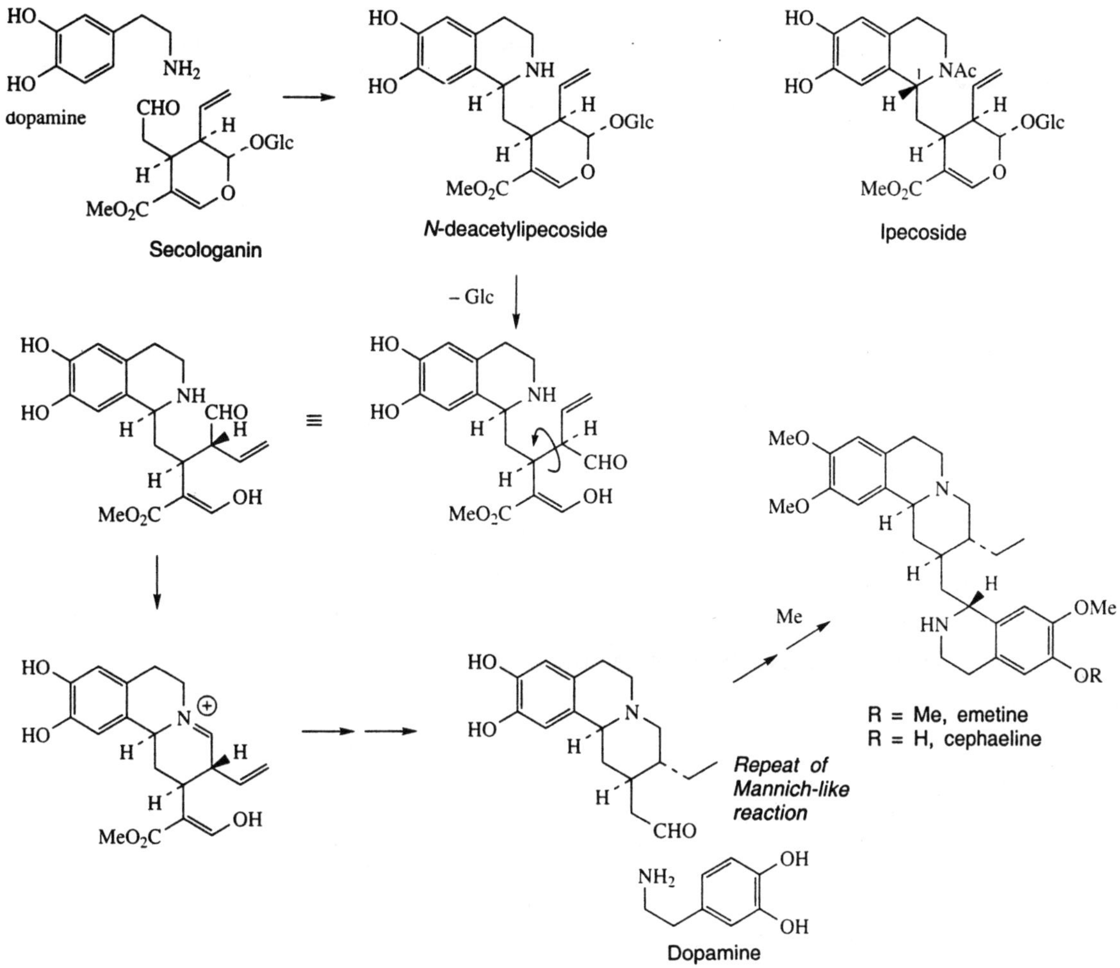

Fig. 9.54. Biosynthesis of ipecac alkaloids.

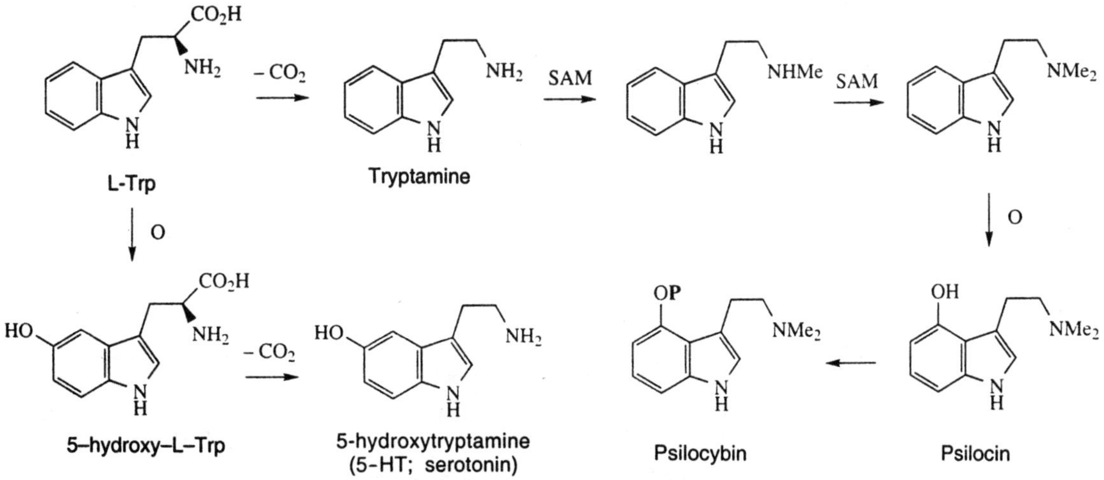

Fig. 9.55. Biosynthesis of psilocybin.

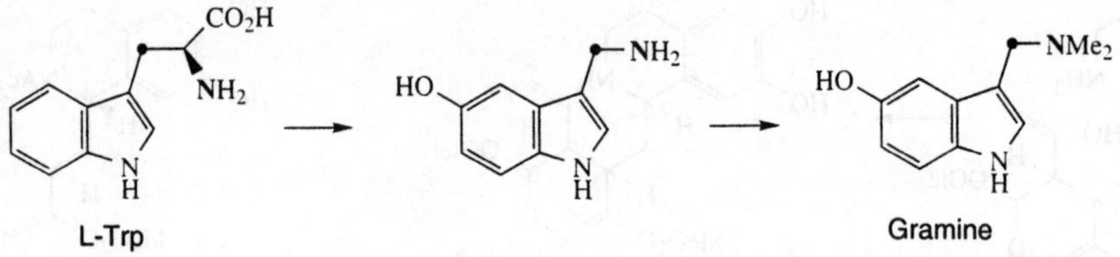

Fig. 9.56. Biosynthesis of gramine.

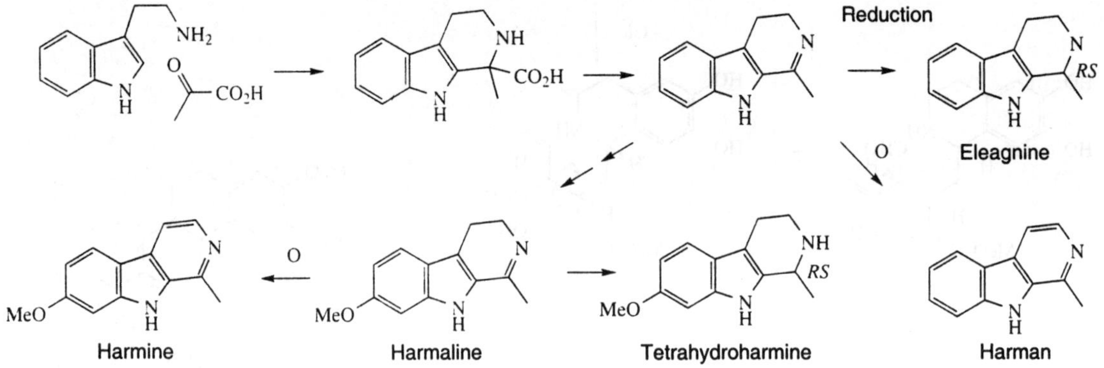

Fig. 9.57. Biosynthesis of harmine, harman and tetrahydroharmine.

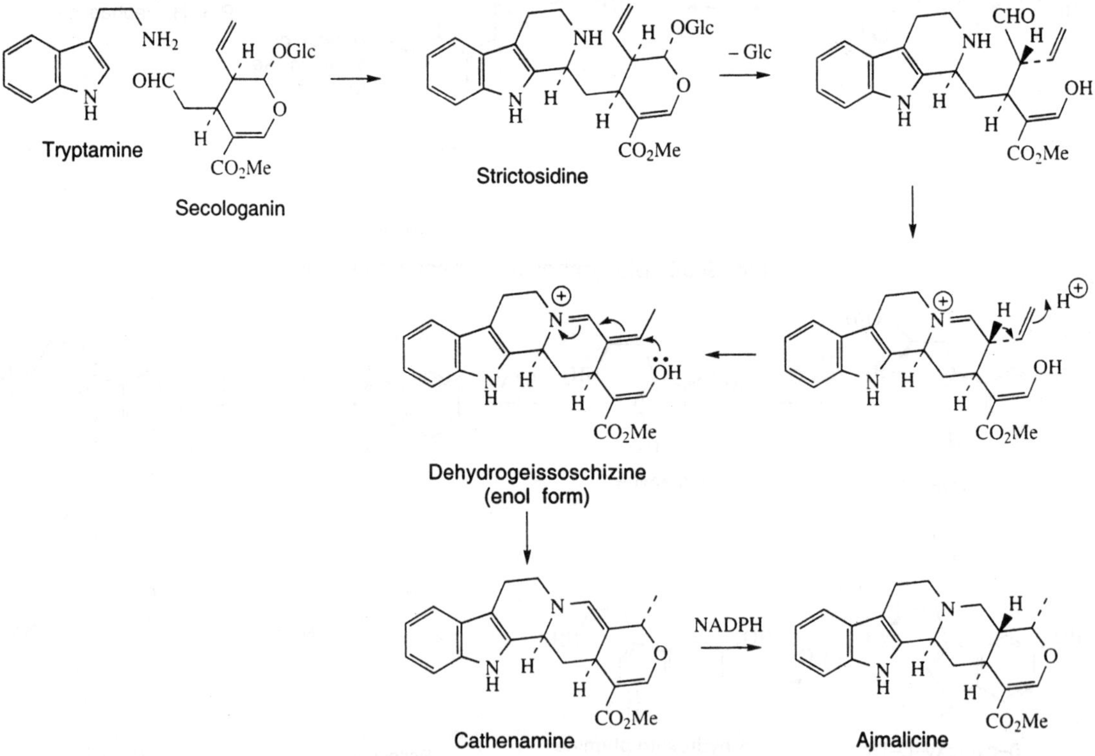

Fig. 9.58. Biosynthesis of ajmalicine.

Yohimbine is an indole alkaloid found in the bark of *Pausinystalia yohimbe* (Rubiaceae), Aspidosperma bark (*Aspidosperma* sp. *Apocyanaceae*) and Rauwolfia sp. (Apocyanaceae). Yohimbine arises from dehydrogeissoschizine (Fig. 9.59). Reserpine and deserpidine are trimethoxy benzoyl esters of Yohimbine like alkaloids, whilst rescinnamine is a trimethoxy cinnamoyl ester. All these alkaloids are present in *Rauwolfia* sp. (Fig. 9.60).

The structural changes involved in converting the corynanthe-type skeleton into those of the Aspidosperma and Iboga groups are quite complex, and are shown in Fig. 9.61. Preakuammicine arises from dehydrogeissoschizine, which inturn forms stemmadenine, catharthine, tebersonine and vindoline and the biosynthetic pathway is depicted in Fig. 9.61 The vinblastine and vincristine are anticancer indole alkaloids found in Madagascar periwinkle (*Catharanthus roseus*, Apocynaceae). The catharanthine is believed to involve firstly an oxidative reaction on catharanthine, catalyzed by a peroxidase, produces, vindoline which inturn reduction and hydroxylation forms vinblastine (Fig. 9.63).

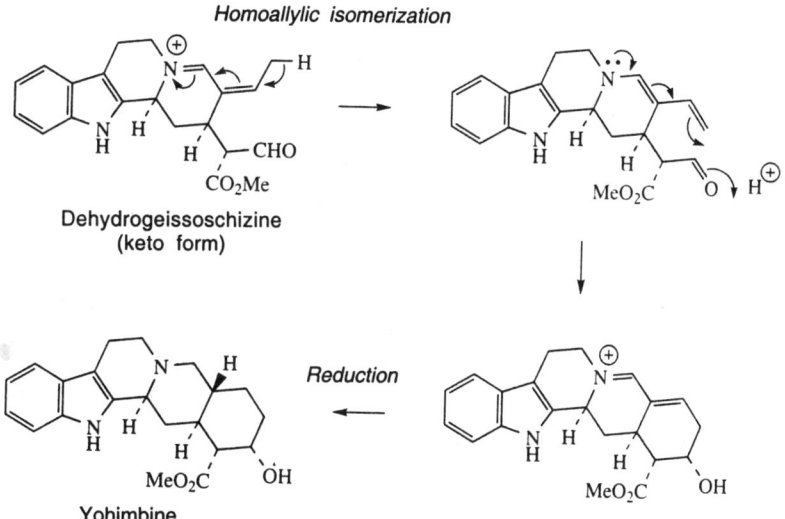

Fig. 9.59. Biosynthesis of yohimbine.

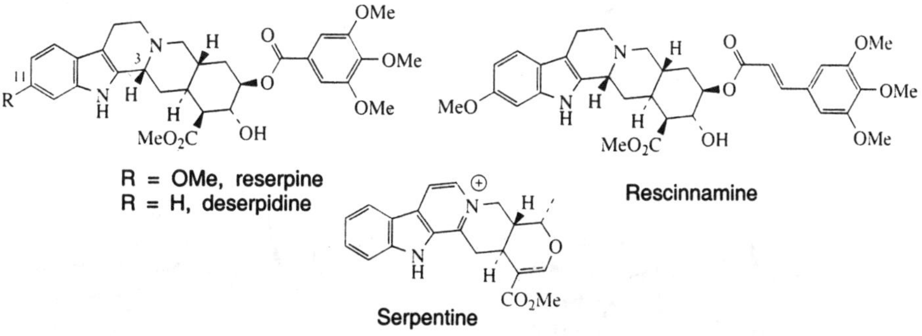

Fig. 9.60. The chemical structures of Rauwolfia alkaloids.

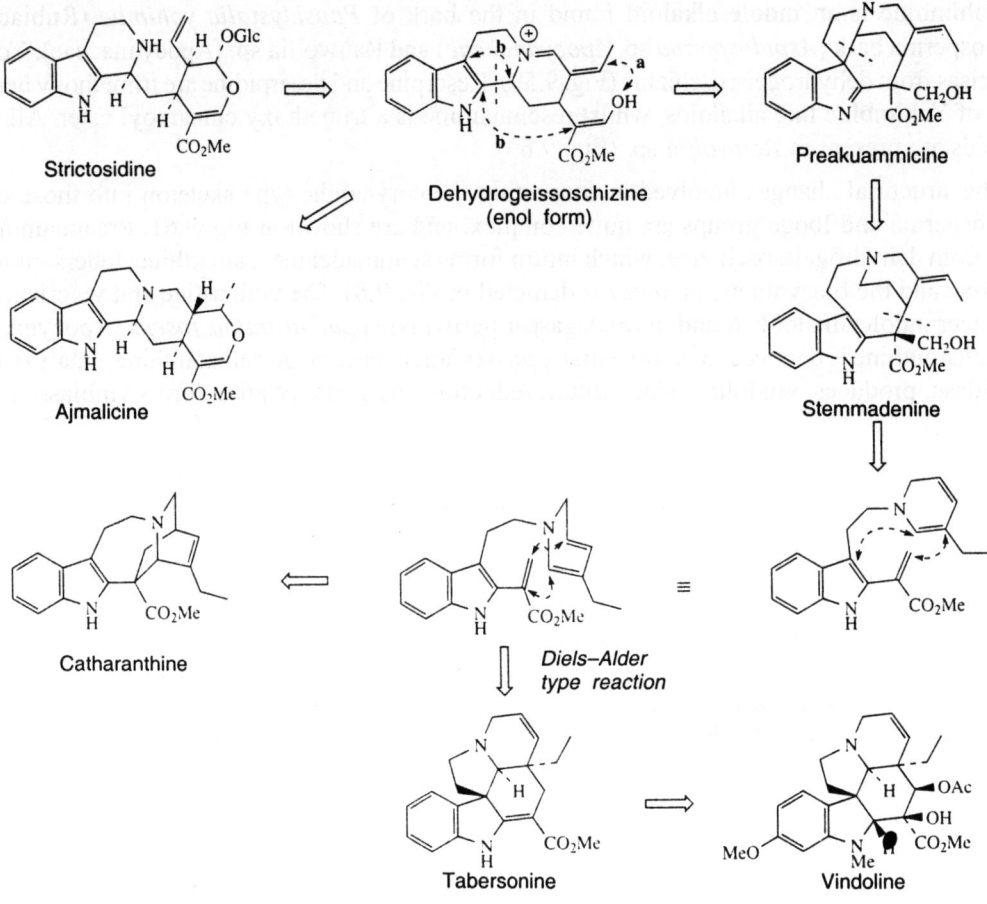

Fig. 9.61. Biosynthesis of vindoline, catharanthine, vindoline.

R = Me, vinblastine
R = CHO, vincristine

Vindesine

Vinorelbine

Anhydrovinblastine

Fig. 9.62. The structures of Catharanthus alkaloids.

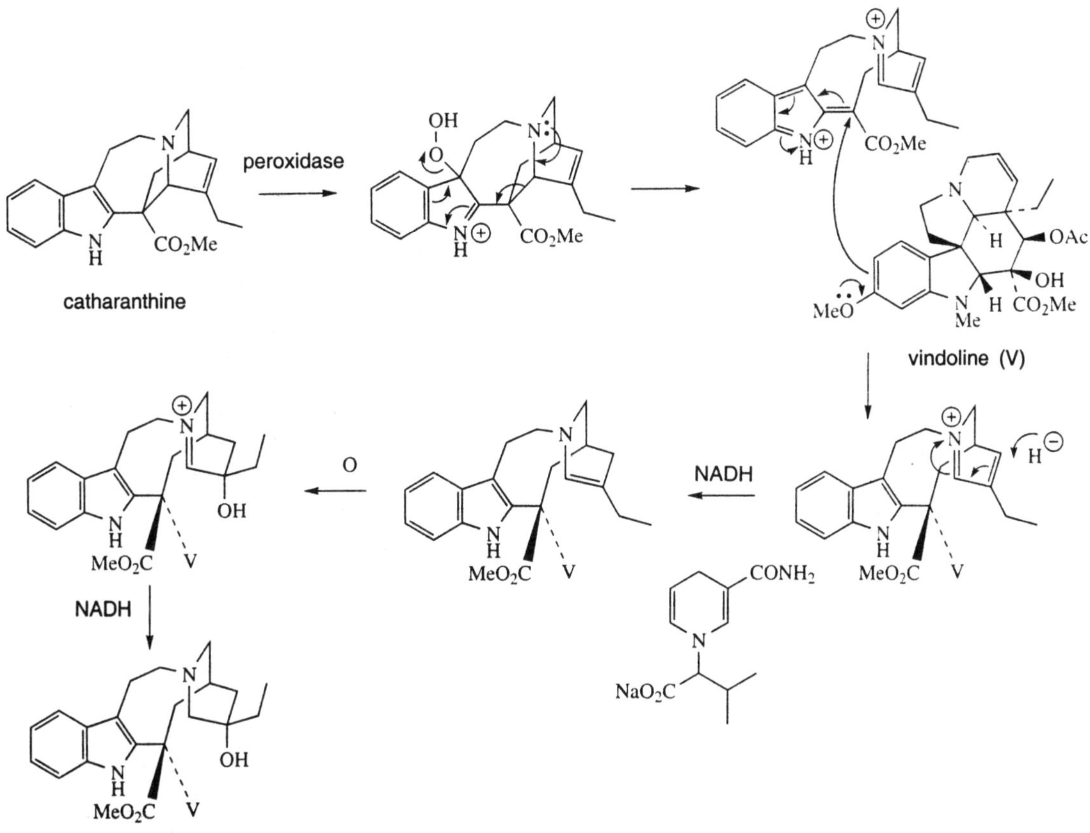

Fig. 9.63. Biosynthesis pathway of vinblastine.

The vincamine from *Vinca minor* and ajmaline from *Rauwolfia serpentina* are also biosynthesized from dehydrogeissoschizine (Fig 9.64). Ajmaline contains a C-9 corynanthe-type unit whereas vincamine C10 aspidosperma unit, and it orginates from tabersonine.

Nux-vomica consists of the dried ripe seeds of *Strychnos nux-vomica* (Loganiaceae), contains strychnine and its dimethoxy analogue brucine. The non-tryptamine portion of these compounds contains eleven carbons, and is derived from an iridoid-derived C9 unit, plus two carbons from acetate. The preakuammicine like structure upon hydrolysis/decarboxylation and addition of acetyl CoA form a hemiacetal in the weiland-gumlich aldehyde (Fig. 9.65).

9.7.3.4. Pyrroloindole alkaloids

Pyrroloindole type alkaloids are present in *Physostigma venenosum* (Leguminosae) Physostigmine (eserine), chimonanthine from *Chimonanthus fragrans* (Calycanthaceae) respectively. Physostigmine has anticholinesterase activity. A suggested pathway to physostigmine is by C-3 methylation of tryptamine, followed by ring formation involving attack of the primary amine function onto the minium ion. Further substitution produces the phytostigmine (eserine) (Fig. 9.66).

9.7.3.5. Ergot Alkaloids

Ergot is a fungal disease commonly found on many wild and cultivated grasses, and it caused by *Claviceps purpurea*. The disease is eventually characterized by the formation of hard, seed, like

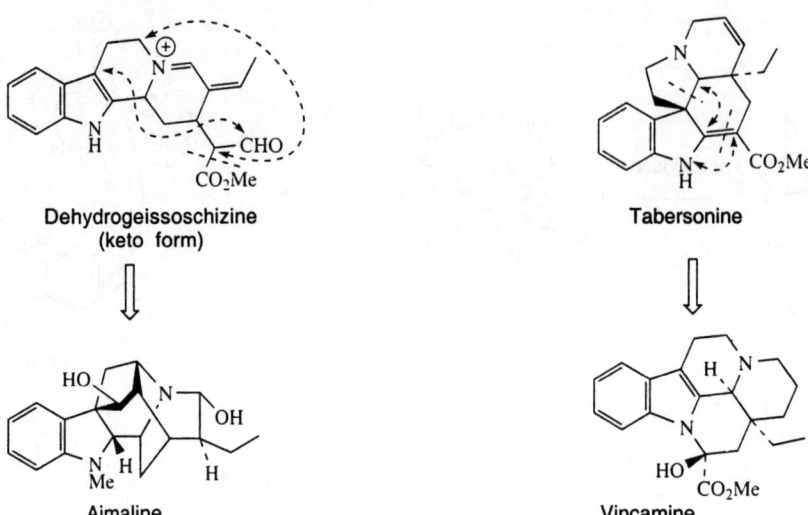

Dehydrogeissoschizine
(keto form)

Tabersonine

Ajmaline

Vincamine

Fig. 9.64. Biosynthesis of ajmaline and vincamine.

Preakuammicine

Acetyl–CoA

Wieland–Gumlich aldehyde

Strychnine

Wieland–Gumlich aldehyde
(hemiacetal form)

Brucine

Fig. 9.65. Biosynthesis of Nux-vomica alkaloids.

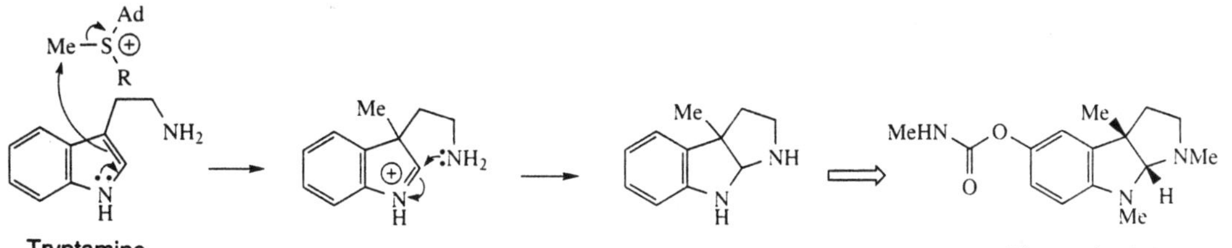

Fig. 9.66. Biosynthesis of physostigamine.

'ergots' instead of normal seeds, these structures, called sclerotia, forming the resting stage of the fungus. The group of alkaloids found in this fungus are called as Ergot alkaloids. These alkaloids are derived from (+) lysergic acid, which includes ergometrine, ergotamine. The structures of these compounds are shown in Fig. 9.68.

The building blocks for lysergic acid are tryptophan and an isoprene unit. Alkylation of tryptophan with DMAPP gives 4-dimethyl-allyl-L-tryptophan, which then undergoes N-methylation. Formation of the tetracyclic ring system of lysergic acid is known to form via chanoclavine-I and agroclavine, though the mechanistic details are not clear. The agroclavine hydroxylation to produce elymoclavine and further oxidation of the primary alcohol to give rises to paspalic acid. The paspalic acid upon isomerization yields the lysergic acid (Fig. 9.69).

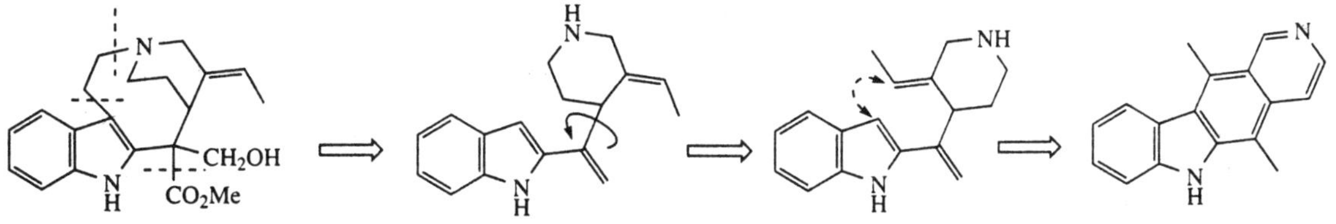

Fig. 9.67. Biosynthesis of ellipticine.

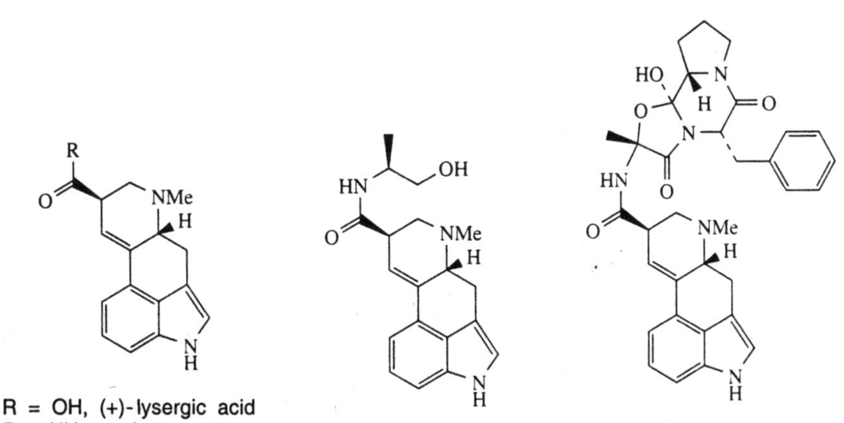

R = OH, (+)- lysergic acid
R = NH₂, ergine

Ergometrine

Ergotamine

Fig. 9.68. The chemical structures of ergot alkaloids.

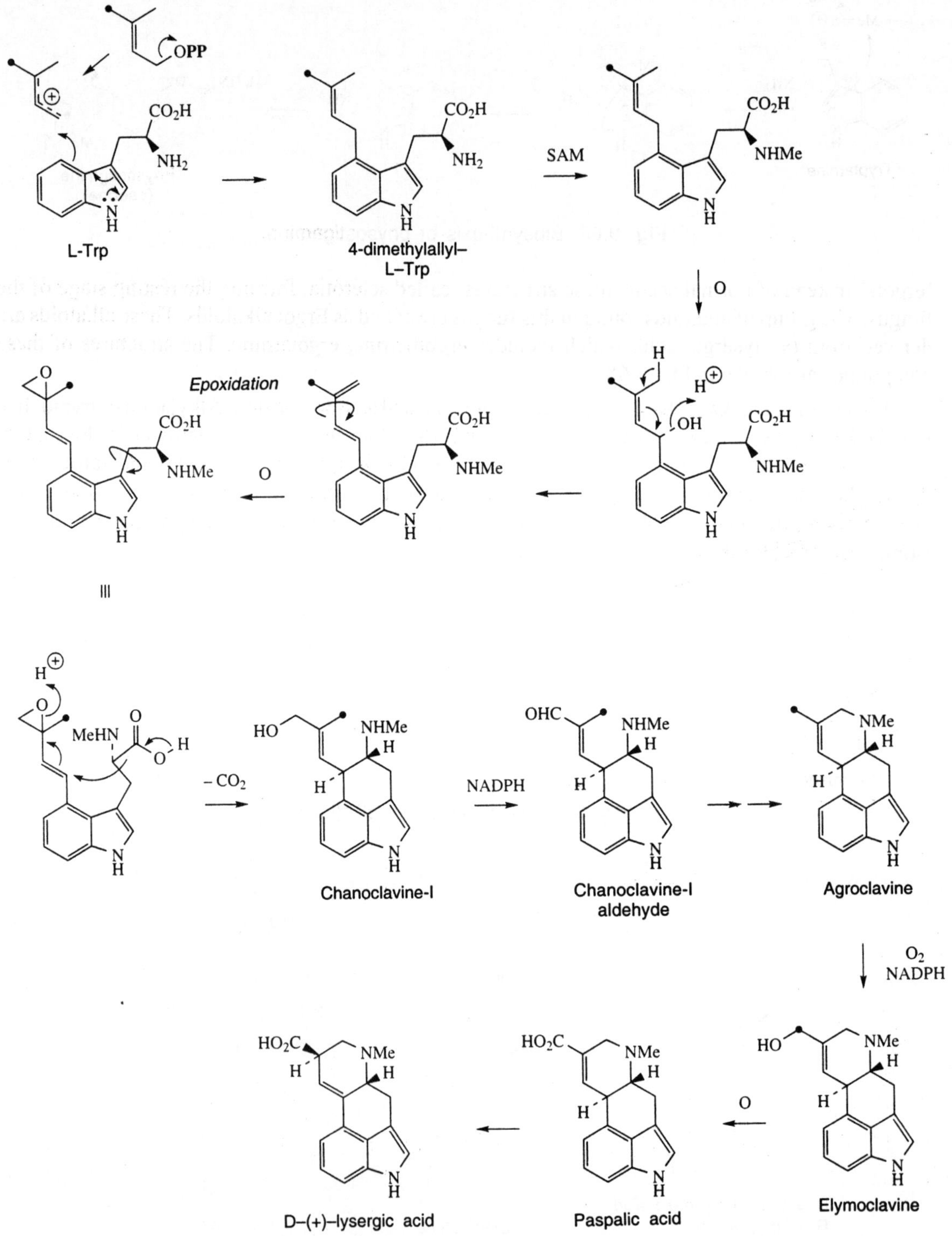

Fig. 9.69. Biosynthesis of lysergic acid.

Ergine is an ergot alkaloid found in Rivea and Ipomoea species contained lysergic acid amide, while ergometrine from *Claviceps purpurea* is the amide with 2-amino propanol. The peptide alkaloids of ergot are derived by adding amino acid residues to lysergl CoA, giving a linear tripeptide covalently bound to enzyme complex (Fig. 9.70).

9.7.3.5.1. Ellipticine

Ellipticine is found in *Ochrosia elliptica* (Apocyanaceae) and related species and has anticancer properties. The ellipticine is formed from a tryptamine-terpenoid precursor. Stemmadenine undergoes transformations that effectively remove the two-carbon bridge originally linking the indole and nitrogen tryptamine (Fig. 9.67).

9.7.3.6. Quinoline alkaloids

The modified terpenoid indole alkaloids are found to be in Cinchona sp. (Rubiaceae) e.g. quinine, quinidine, cinchonine and cinchonidine. Strictosidine upon hydrolysis followed decarboxylation produces a corynanthe-type indole alkaloid to cinchonidine (Fig. 9.71).

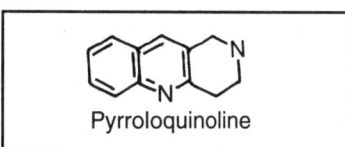
Pyrroloquinoline

Camptothecine is a pyrroloquinoline alkaloid found in *Camptotheca acuminata* (Nyssaceae). *Nothopodytes foetida* (Icacinaceae), *Merilliodendron megacarpum* (Icacinaceae), *Ophiorrhiza mungos* (Rubiaceae), and *Ervataemia heyneana* (Apocyenaceae). Strictosidine is a precursor for the biosynthesis of camptothecine (Fig. 9.72). Pumiloside is a potential intermediate which has been isolated from *C. acuminata* as well as in *Ophiorrhiza mungos* (Rubiaceae).

9.7.4. Alkaloids Derived from Anthranilic Acid

Anthranilic acid is a key intermediate in the biogenesis of L-tryptophan and which inturn is a precursor for indole alkaloids.

Anthranilic acid

9.7.4.1. Quinazoline Alkaloids

The quinazoline alkaloids are found in *Peganum harmala* Zygophyllaceae (Peganine co-occurs with the β-carboline alkaloid, harmine) and in *Adhatoda vasica* (Acanthaceae) (Vasicine, useful as a bronchodilator). The anthraniloyl CoA on condensation with pyrrolidine ring (orginated from L-Ornithine) produces vasicine (Peganine). Alternatively biosynthetic route is operative in *A. vasica*, that is from N-acetyl anthranilic acid and aspartic acid (Fig. 9.73).

9.7.4.2. Quinoline and Acridine Alkaloids

The quinoline ring system is formed by the combination of anthranilic acid and acetate / malonate, and an extension of this process also yields acridine ring system.

Quinidine   Acridine

The 4-Hydroxy-2-quinoline is formed by the condensation of anthraniloyl CoA with malonyl CoA, which inturn reacts with DMAPP, followed by methylation yields platydesmine. The platydesmine oxidative cleavage of side chain produces dictamnine which hydroxylation/methylation forms skimmianine. These furoquinoline alkaloids are found in the plants of *Dictamus albus* and *Skimmia japonica* (Rutaceae) (Fig. 9.74).

Fig. 9.70. Biosynthetic pathway of ergotamine.

The Acridine alkaloids are found in the plants of *Melicope fareana* (Melicopicine), *Acronychia baueri* (Acronycine), and *Ruta graveolens* (Rutacridone) and their biosynthetic pathway is shown in Fig. 9.75. N-methyl anthranilyl-CoA incorporate 3 acetate / malonate units, a polyketide is produced, the polyketide generates melicopicine, acronycine and rutacridone.

9.7.5. Alkaloids Derived From Histidine (Imidazole alkaloids)

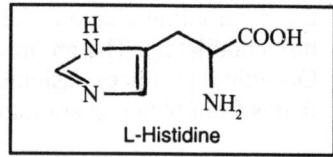

L-Histidine

The imidazole alkaloids found in *Pilocarpus microphyllus* and *P. jaborandi* (Rutaceae). The active constituents found are pilocarpine and pilosine. Pilocarpine is used for the treatment of glaucoma. Histidine is a proven precursor of dolichotheline in *Dolichothele sphaerica* (Cactaceae) (Fig. 9.76), the remaining carbon atoms originating from leucine via isovaleric acid [see biosynthesis of fatty acids (Fig. 9.36)]. The imidazole alkaloids found in leaves of *Pilocarpus* sp. are also probably derived from histidine, but experimental data are lacking. The author speculate that the histidine and alanine could serves as a precursor for pilocarpus alkaloid biogenetic pathway (Fig. 9.77).

The two possible pathways for biosynthesis of pilocarpine are shown in Fig. 9.78. Pathway A, suggested by Robinson, assumes the combination of 4 imidazolyl acetol phosphate (III) and either two molecules of acetate or one molecule of acetoacetate. Another precursor, according to pathway B, could be histidyl analogue of cinnamic acid (Urocanic acid) (IV) and α-ketobutyrate, a metabolite of threonine. However, there is no clear evidence for the biosynthesis of pilocarpine, so far.

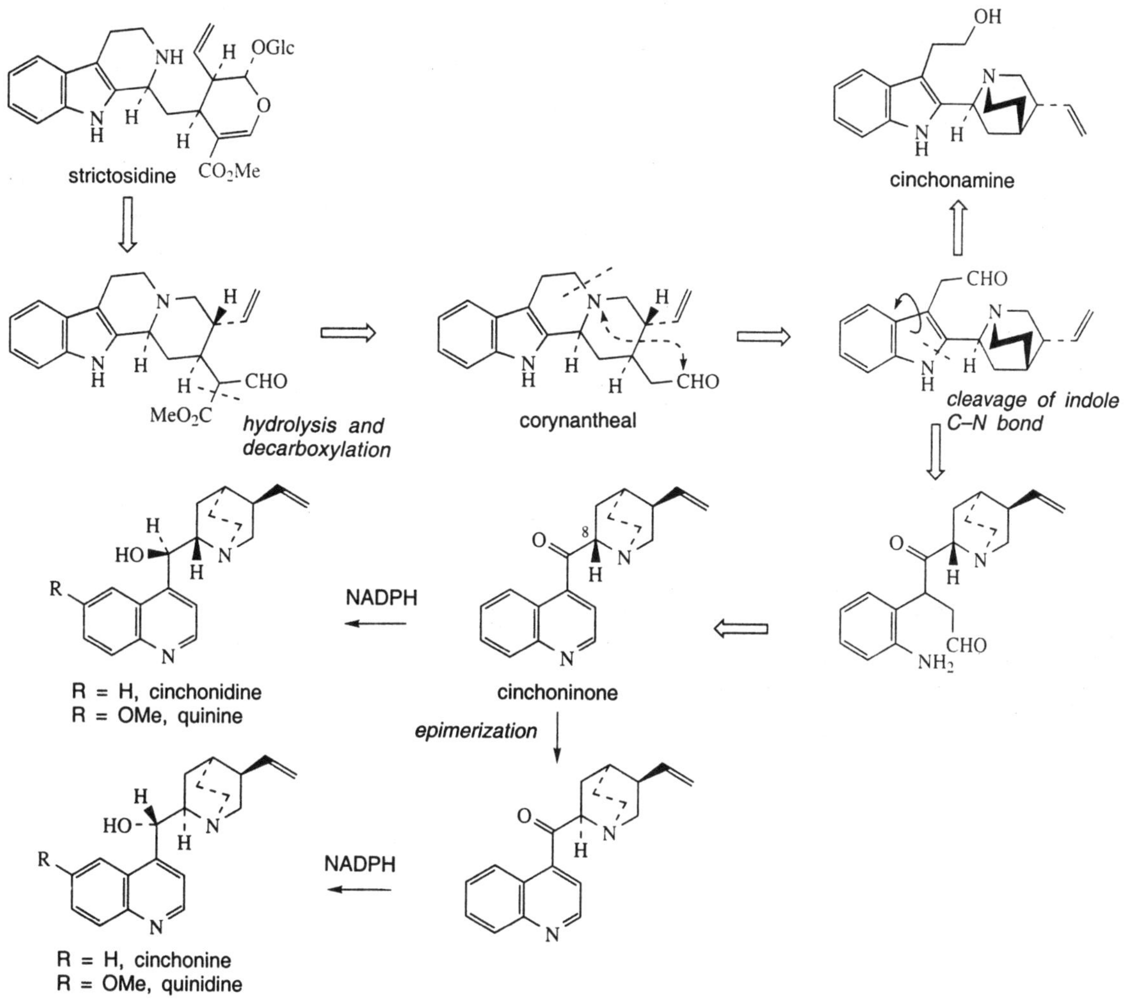

Fig. 9.71. Biosynthesis of Cinchona alkaloids.

9.8. PRECURSOR FEEDING

Any compound, whether endogenous or exogenous, that can be converted by an organism or living system into the investigated product or secondary metabolite or useful compounds known as precursor, for e.g L-Ornithine for tropane alkaloids. Intermediate compound is one which is both formed and further converted by the organism under identical conditions. Intermediate can be classified into a) Natural intermediate, a compound formed by the organism independently from the investigated biosynthetic pathway b) Obligatory intermediate, a member of a path that is the only one by which an organism can synthesize a given product from given source materials. The major disadvantage of plant cell cultures on large scale is low productivity. The precursor feeding is one of the strategies for improvement of secondary metabolites in plant cell culture (Veeresham and Kokate, 1997). The content of precursor and especially intermediate (Davis 1955) is usually lower in callus and cell cultures than in fully differentiated tissue.

Enrichment of the medium with precursors has given some degree of success in enhancement of metabolite production. Failure of the desired result may be due to lack of uptake, precipitation,

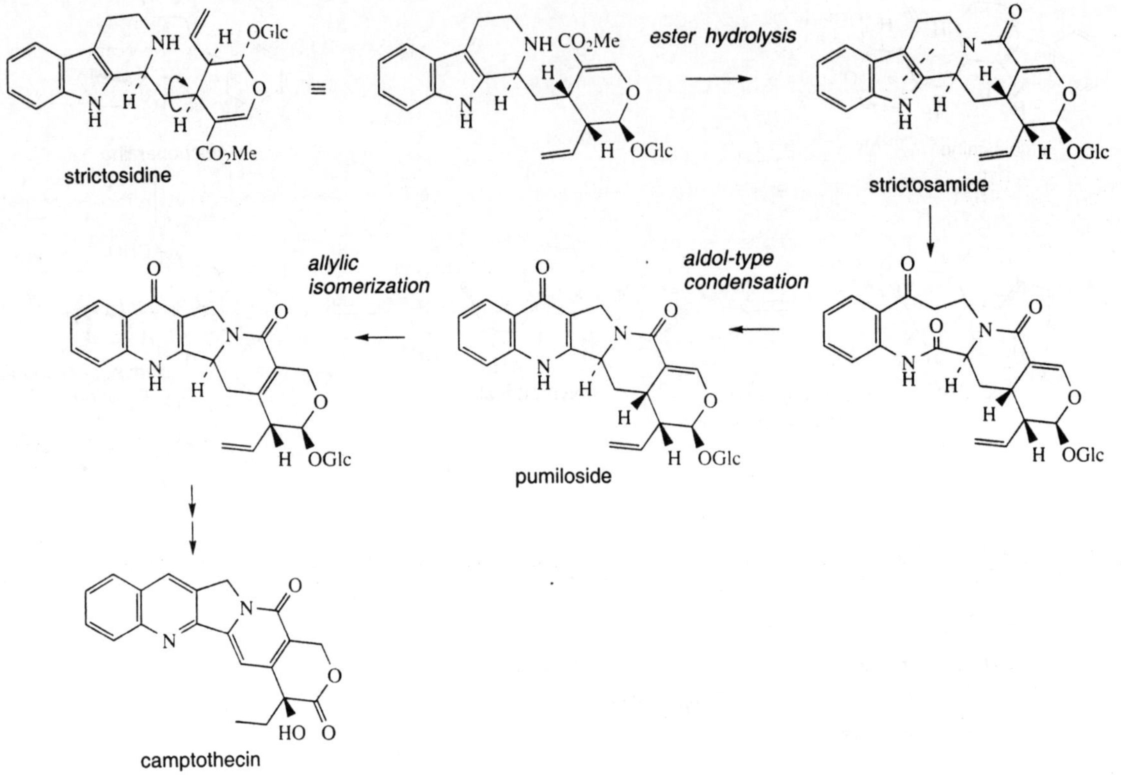

Fig. 9.72. Biosynthesis of pyrroloquinoline alkalods.

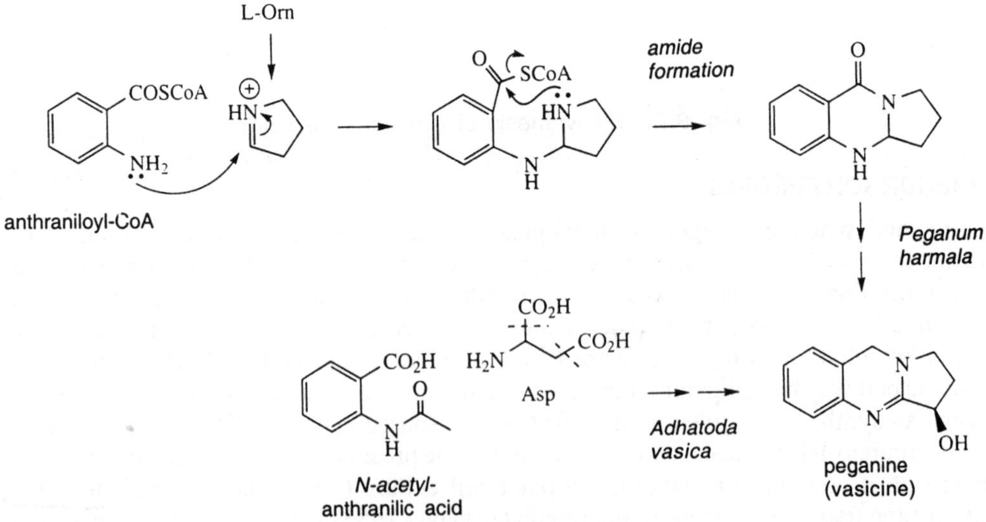

Fig. 9.73. Biosynthesis of vasicine.

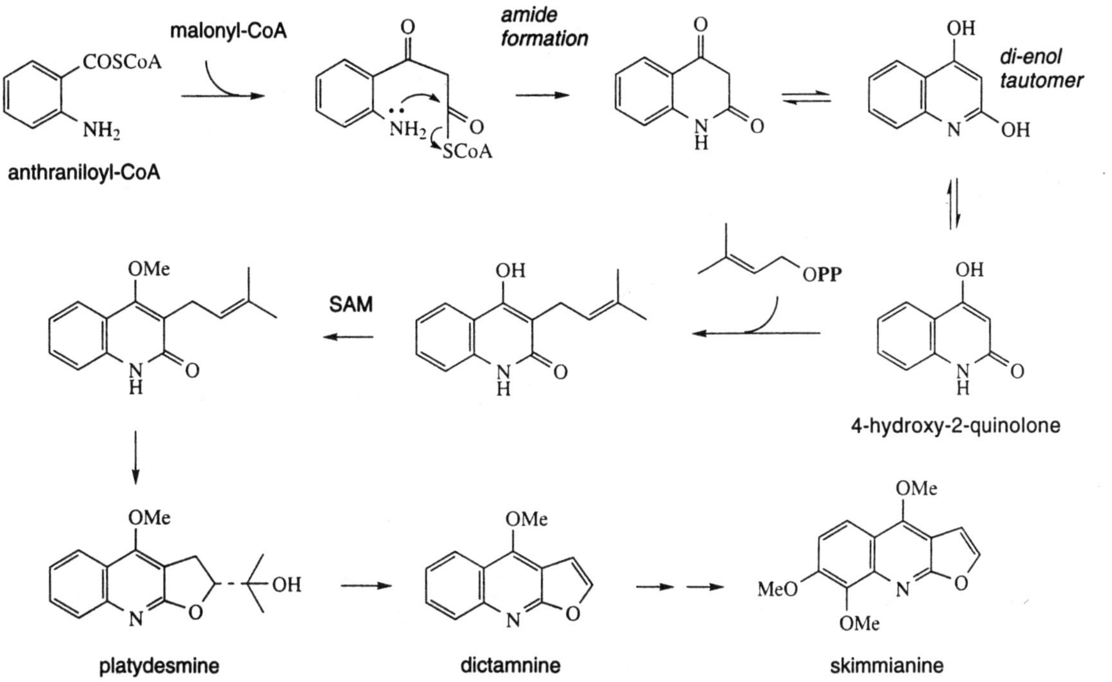

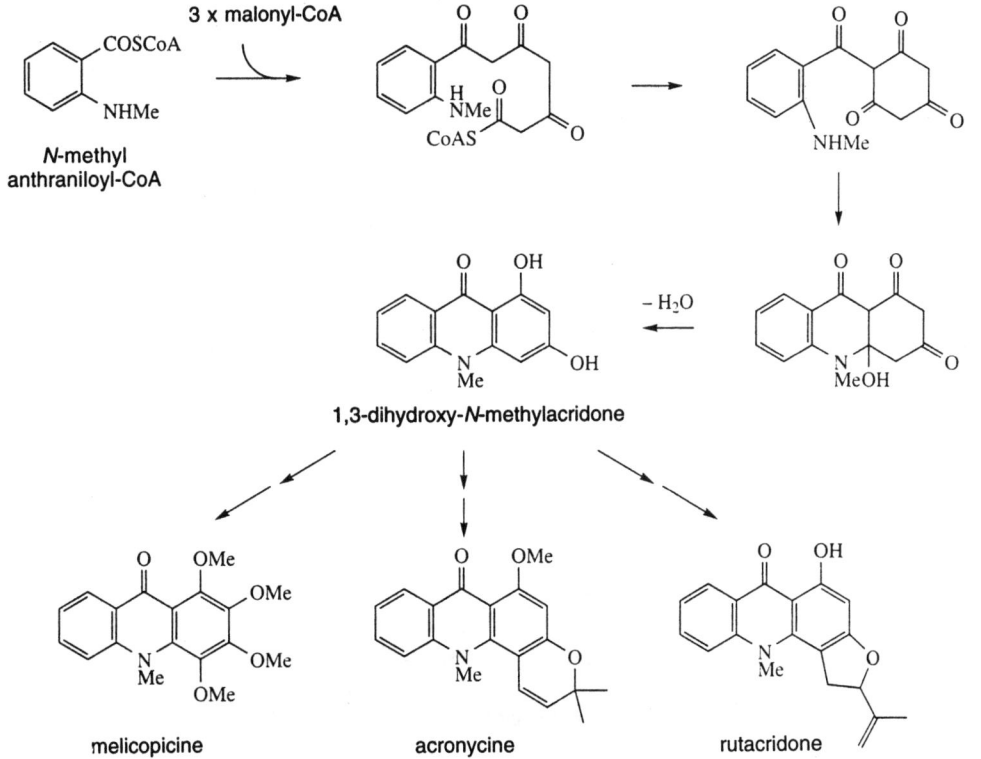

Fig. 9.74. Biosynthesis of skimmianine.

Fig. 9.75. Biosynthesis of acridone alkaloids.

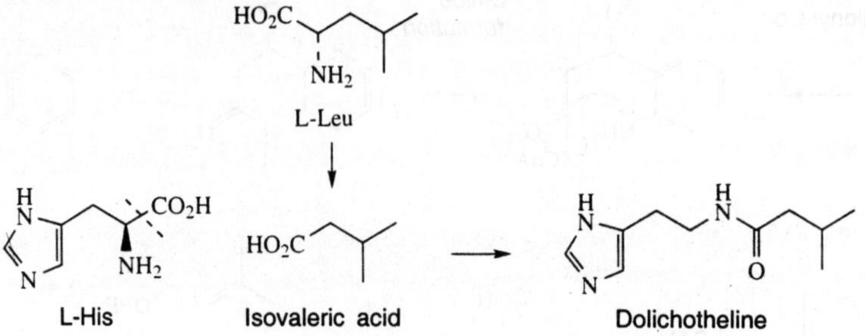

Fig. 9.76. Biosynthesis of doliochotheline.

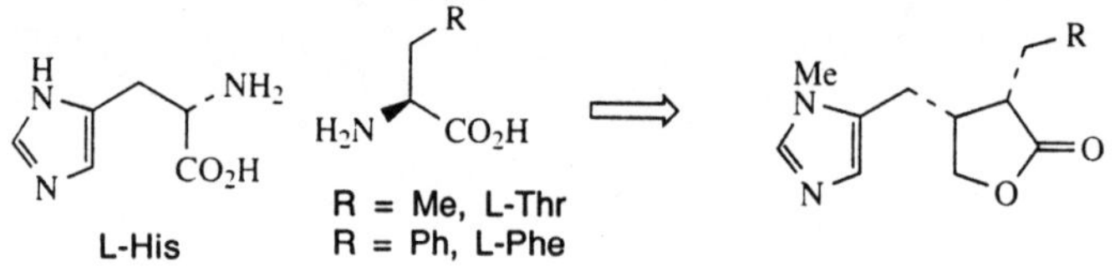

R = Me, L-Thr
R = Ph, L-Phe

Fig. 9.77. Biosynthesis of Pilocarpus alkaloids.

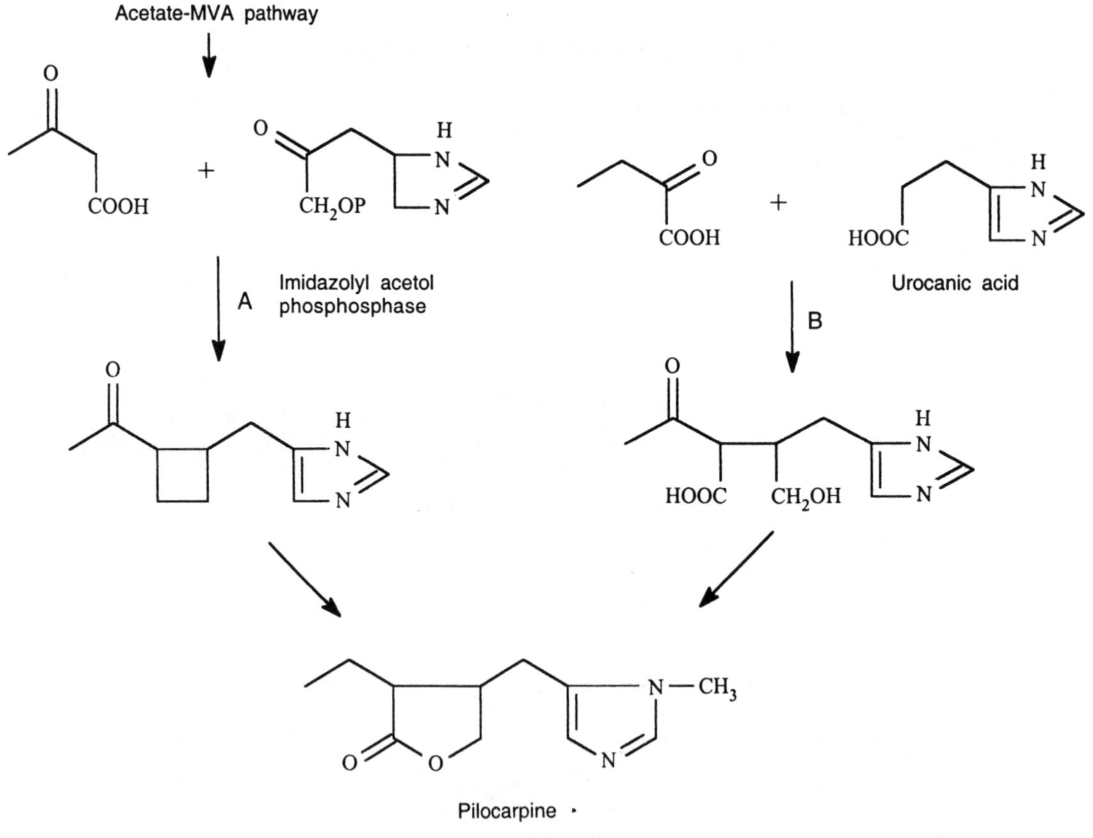

Pilocarpine ·

Fig. 9.78. Biosynthetic pathway of Pilocarpine.

conjugation, diversion into alternative pathways, or lack of one of the enzymes participating in bioproduction. The addition of presursor may influence spatial orientation of enzymes, compartmentation of enzymes, substrates, and reservoir sites for production and accumulation in the biosynthesis of secondary products by specialized tissues. The effects of precursors on secondary products in plant cell culture systems are generally diverse and often contradictory. There are two distinct methods of increasing the precursor supply within the cell. Firstly, by its addition to the medium, in which case the uptake mechanism may /may not be limiting. Secondly, by selecting for resistance to precursor analogues (or by treating with anti-metabolites), in, which case the intracellular level may be modified. The examples of the increasing effects of precursors and intermediates on secondary metabolite production in different suspension cultures are shown in (Table 9.2) Sasse *et al.* (1982) have reported that *Peganum harmala* cell cultures increased the bioproduction of serotonin, when supplied with tryptamine. Doeller (1978) demonstrated that the cell cultures of *C. roseus* increased alkaloid production with addition of L-tryptophan. Similar results were observed with cell cultures of *Cinchona ledgeriana* supplemented with precursor L-tryptophan (Harker *et al.,* 1986). Chan and Staba (1965) reported that L-phenylalanine stimulated alkaloid production in *D. stramonium* leaf suspension cultures. Sairam and Khanna (1971) also reported that the alkaloid content in *D. tatula* seed callus was greatly increased by phenylalanine.

Tabata *et al,* (1971,1972) showed that tropic acid was the most effective precursor for stimulating alkaloid production in callus cultures of *D. tatula, Scopolia japonica* and *Scopolia parviflora.* The biosynthetic pathway for capsaicin is shown in Fig. 9.79. Lindsey (1986) reported the incorporation of phenylalanine and cinnamic acid into cultured cells of *Capsicum frutescens.* Yeoman *et al* (1980) reported the effect of vanillylamine and isocapric acid on production of capsaicin in cell cultures of *C. frutescens.* Sudhakar and Ravishankar (1998) reported that increased yields of capsaicin and dihydrocapsaicin were obtained when immobilized placental tissue of *C.frutescens* was treated with phenylalanine, cinnamic acid, coumaric acid, caffeic acid, ferulic acid and vanillylamine in combination with valine. We reported the enchanced production of capsaicin from callus cultures of *C. annuum* upon supply of DL-DOPA and L-tyrosine (Veeresham *et al.,* 1994).

Allium cepa calli lack the major precursor of onion scent (S.-trans-prop-1-enyl-L-cysteine). Addition of precursors reveals the cells potential to produce S-trans-prop-1-enyl-L-cysteine sulphoxide (the tear factor) which degrades in water producing H_2SO_4 (Musker *et al* 1988). In *Chenopodium rubrum,* the addition of tyrosine triples betacyain accumulation. Zenk *et al* (1975) reported the two fold increased production of anthraquinones by the supply of precurser O-succinyl-benzoic acid to the cell culture of *Morinda citrifolia.*

The roots of *Cephaelis ipecacuahna* known as 'ipecac' are used as an expectorant, emetic and amoebicide. They contain isoquinoline alkaloids, as shown in Fig 9.54. The exogenous feeding of precursors e.g. L-phenylalanine and shikimic acid, L-tyrosine to callus, cell and immobilized cell cultures enhanced the production of isoquinoline alkaloids (Veeresham *et al.,* 1993, 1994).

Paclitaxel is a highly derivatized diterpenoid, isolated from *Taxus brevifolia* and other species of *Taxus.* It is useful for the treatment of ovarian, breast and lung cancer. The cell cultures of *Taxus* sp. reported to produce paclitaxel and its analogs (Hirasuna *et al.,* 1996, Fett-neto *et al* 1994a, 1994b). Fig. 9.80 shows a biogenetic scheme constructed from the results of Croteau *et al* (1995) and Fleming *et al* (1994). Fett-Neto *et al* (1994a, 1994b) showed the increased accumulation of paclitaxel in the presence of precursors such as phenylalanine, benzoic acid, serine, benzoyl glycine and glycine in cell and callus cultures of *Taxus cuspidata.* A study of *in-vitro* production of radio-labeled taxol by pacific yew, phenylalanine and leucine were demonstrated to be the best precursors, out of various compounds tested (Strobel *et al.,* 1992). Hirasuna *et al.* (1996) also claimed that feeding of acetate enhanced the formation of taxol like metabolites, and addition of fructose showed similar positive effect. The similar kind of effect was observed with the addition of phenylalanine,

Table 9.1. Volatile oils containing terpenoid compounds

Oil	Plant source	Major constituents with typical (%) composition
Bergamot	*Citrus aurantium* sp.	Limonene (42) Linalyl acetate (27) γ-terpinene (8) linalool (7)
Camphor oil	*Cinnamomum camphora* (Lauraceae)	Camphor (27-45) Cineole (4-21) Safrole (1-18)
Caraway	*Carum carvi* (Umbelliferae)	(+)-carvone (50-70) limonene (47)
Cardamom	*Elettaria cardomomum* (Zingiberaceae)	α-terpinyl acetate (25-35) cineole (25-45) linalool (5)
Camomile (Roman camomile)	*Chamaemelum nobile* (Anthemis nobilis) • (Compositae)	Aliphatic esters of angelic, tiglic, iso-valeric, and isobutyric acids (75-85) small amount of monoterpenes
Citrnoella	*Cymbopogon winterianus C. nardus* (Graminae)	(+)-citronellal (25-55) geraniol (20-40) (+)-citronellol (10-15) geranyl acetate (8)
Coriander	*Coriandrum sativum* (Umbelliferae)	(+)-linalool (60-75) γ-terpinene (5) α-pinene (5) camphor (5)
Dill	*Anethum graveolens* (Umbelliferae)	(+)-carvone (40-56)
Eucalyptus	*Eucalyptus globulus* E- smithii E. polybractea (Myrtaceae)	Cineole (= eucalyptol) (70-85) α-pinene (14)
Eucalyptus (lemon-scented)	*Eucalyptus citriodora* (Myrtaceae)	Cintronellal (65-85)
Ginger	*Zingiber officinale* (Zingiberaceae)	Zingiberene (34 β-sesquiphellandrene (12) β-phellandrene (8) β-bisabolene (6)
Juniper	*Juniperus communis* (Cupressaceae)	α-pinene (20) limonene (9) myrcene (9) borneol (8)
Lavender	*Lavandula angustifolia L. officinalis* (Labiatae)	Linalyl acetate (25-45) Linalool (25-38)
Lemon	*Citrus limon* (Rutaceae)	(+)-limonene (60-80) β-pinene (8-12) γ-pinene (8-12) citral (- geranial + neral) (2-3)

(Contd.)

Oil	Plant source	Major constituents with typical (%) composition
Lemon-grass	*Cymbopogon ciratus* (Graminae)	Citral (= geranial + nearl) (50-85)
Matricaria (German camomile)	*Matricaria chamomilla* (*Chamomilla recutica*) (Compositae)	(-)-α-bisabolol (10-25) bisabolol oxides A and B (10-25) chamazulene (1-15)
Orange (bitter)	*Citrus aurantium* ssp. *Amara* (Rutaceae)	(+)-limonene (92-94) myrcene (2)
Orange (sweet)	*Citrus sinensis* (Rutaceae)	(+)-limonene (90-95) myrcene (2)
Orange flower (Neroli)	*Citrus aurantium* ssp. *Amara* (Rutaceae)	Linalool (36) β-pinene (16) limonene (12) linalyl acetate (6)
Peppermint	*Mentha x piperita* (Labiatae)	Menthol (30-50) Menthone (15-32) Menthyl acetate (2-10) Menthofuran (1-9)
Pine	*Pinus palustris* or other *Pinus* species (Pinaceae)	α- and β-phellandrene (60)
Pumilio pine	*Pinus mugo* spp. *Pumilio* (Pinaceae)	α- and β-pinene (10-20) bornyl acetate (3-10)
Rose (attar of rose, otto or rose)	*Rosa damascena*, *R. gallica, R. alba,* and *R. centifolia* (Rosaceae)	Citronellol (36) Geraniol (17) 2-phenylethanol (3) C_{14}-C_{23} straight-chain Hydrocarbons (25)
Rosemary	*Romarinus officinalis* (Labiatae)	Cineole (15-45) α-pinene (10-25) camphor (10-25) β-pinene (8)
Sage	*Salvia officinalis* (Labiatae)	Thujone (40-60) Camphor (5-22) Cineole (5-14) β-caryophyllene (10) limonene (6)
Sandalwood	*Santalum album* (Santalaceae)	Sesquiterpenes; α-santalol (50) β-santalol (21)
Spearmint	*Mentha spicata* (Labiatae)	(-)-carvone (50-70) (-)-limonene (2-25)
Thymen	*Thymus vulgaris* (Labiateae)	Thymol (40) p-cymene (30) linalool (7) carvacrol (1)
Turpentine oil	*Pinus palustris* and other *Pinus* species (Pinaceae)	(+)-and (-)-α-pinene (35:65) (60-70) β-pinene (20-25)

Table 9.2. Examples of the increasing effects of precursors on Secondary metabolite production in different cell cultures

Culture	Precursor	Product	Reference
Azadirachta indica	Sodium acetate Squalene IPP GPP	Azadirachtin	Balaji (2001)
Capsicum frutescens	Phenylalanine Valine Vanillyl amine	Capsaicin	Lindsey (1986) Yeoman (1980)
	Isocapric acid Coumaric acid Cinnamic acid Caffeic acid Ferulic acid	Capsaicin	Sudhakar and Ravishankar (1998)
Capsicum annuum	L-tyrosine DL-DOPA	Capsaicin	Veeresham *et al.* (1984)
Chenopodium rubrum	Tyrosine	Betacyanin	
Cephaelis ipecacuanha	L-Tyrosine L-phenylalanine Shikimic acid	Cephaeline	Veeresham *et al.* 1994; 1993
Coleus blumei	Phenylalanine	Rosmaric acid	Sairam and Khanna, (1971)
Datura stramonium	L-phenylalanine	Tropane alkaloid	
D. talula	Tropic acid	Tropic acid	Tabata *et al.*1971,1972
Dioscorea deltoidea	Cholesterol	Diosogenin	
Lithospermum erythrorhizon	L-phenylalanine	Rosmarnic acid	
Morinda citrifolia	O-succinyl benzoic acid	Anthraquinones	Zenk *et al.* (1975)
Nicotiana tabacum	Ornithine	Nicotine	
Ruta graveolens	4-Hydroxy 2-quinoline	Dictamine	
Scopolia japonica	Tropic acid	Scopolamine	Tabata *et al.*1971,1972
Taxus cuspidata	Phenylalanine Benzoic acid Serine Benzoyl glycine Glycine	Taxol	Fett-neto *et al.* (1994a, b)
T. chinensis	Phenylalanine Benzaylglycine	Taxol	Srinivasan *et al.* (1996)
T. wallichiana	Sodium acetate Benzoyl glycine Leucine Phenylalanine	Taxol, baccatin DBA	Mamatha (2001)

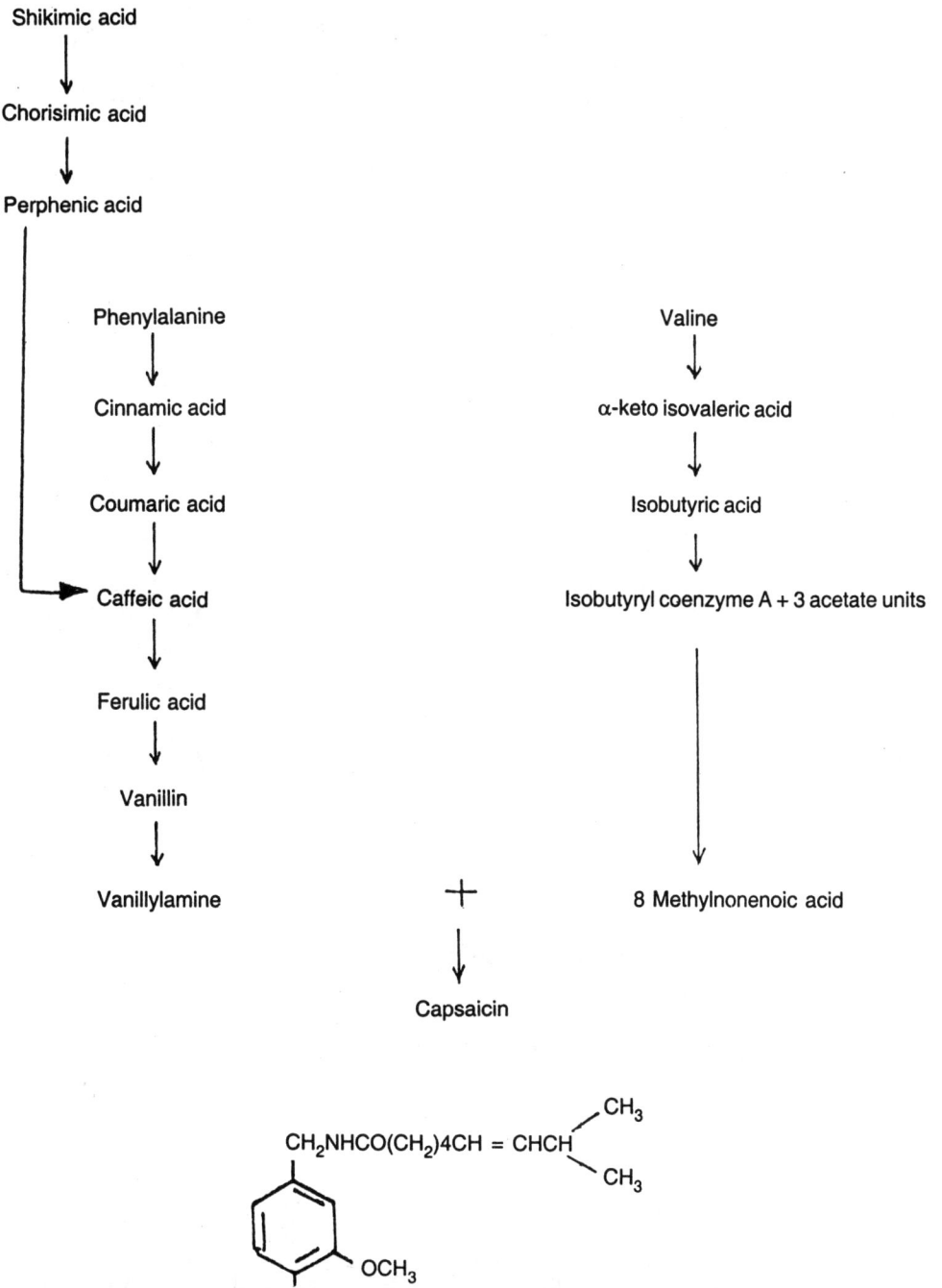

Fig. 9.79. Biosynthetic pathway of capsaicin.

Fic. 9.80. Biosynthesis of Taxol.

benzoylglycine in cell cultures of *T.chinensis* (Srinivasan *et al.*, 1996). The addition of precursors such as sodium benzoate, hippuric acid (benzoyl glycine), leucine, phenylalanine to the cell cultures of *T. wallachiana* significantly improved the production of paclitaxel, baccatin and 10-deacetyl baccatin (Mamatha, 2000). Similar kind effect was also observed with the cell cultures of *T. canadensis*.

Tropane alkaloids (hyoscyamine and scopolamine etc) are present in the plants belongs to solanaceae family e.g. *Datura, Atropa, Hyoscyamus niger* (Fig. 9.81). The addition of ornithine and phenylalanine and other tropane alkaloids from callus cultures of *D. innoxia* (Petri, 1992). Similar kind effect was also observed with cell cultures of *D. stramonium*.

Azadirachtin a biopesticide produced from seeds of *Azadirachta indica* (neem). A. Juss. Meliaceae. The callus and cell cultures are known to produce this secondary metabolite (Allan *et al.*, 1994; Veeresham *et al.* 1998). However, the yields are poor. The precursor feeding such as sodium acetate, squalene, isopentenyl pyrophosphate, geranyl pyrophosphate to the cell cultures of *A. indica* significantly enhanced the production of Azadirachtin (Balaji, 2001) (Fig. 9.82). *Catharanthus roseus* is a source for the indole alkaloids of pharmaceutically important compounds such as vincristine, vinblastine and ajmalicine (Fig. 9.83). The production of these alkaloids in tissue cultures has been detected. The effect of various precursor on the yield of alkaloid is shown in Table 9.3. An exogenous supply of a precursor to the culture medium may improve secondary metabolite accumulation if the endogenous level of these precursors is a limited factor of the flux. Feeding ornithine, arginine, putrescine, agmatine, tropine and tropane moiety had no influence on hyoscyamine accumulation (Robin *et al.*, 1991). The feeding of phenylalanine to the cell culture of *Taxus chinensis* had no influence on the production of taxanes (Srinivasan *et al.*, 1996).

Similarly there is no effect on the production of indole alkaloids upon feeding tryptophan. On other hand, Yatazowa *et al.* (1965) observed that the root callus tissue of *D. stramonium* developed alkaloid reaction when supplied simultaneously with ornithine and tropic acid but the reaction was negative when the callus was cultured with ornithine and phenylalanine.

Table 9.3. Effect of alkaloid precursors on cell yield and alkaloid production in suspension cultures of *Catharanthus roseus* (Zenk et al., 1977)

Precursor added	Precursor concentration (mg/l)	Cell yield (%)	Alkaloid yield (%)
Control	0	100	100
Indole componenet			
Shikimic acid	50	100	122
Quinic acid	250	100	65
Anthranilic acid	50	100	55
Indole	50	110	98
Tryptamine	100	75	67
L-Tryptophan	500	175	284
D-Tryptophan	500	65	55
Monoterpene component			
L-Methionine	100	100	50
Geraniol	50	100	55
Loganin	250	110	188
Loganic acid	250	110	121
Secologanin	50	100	106
Secologanic acid	500	100	75

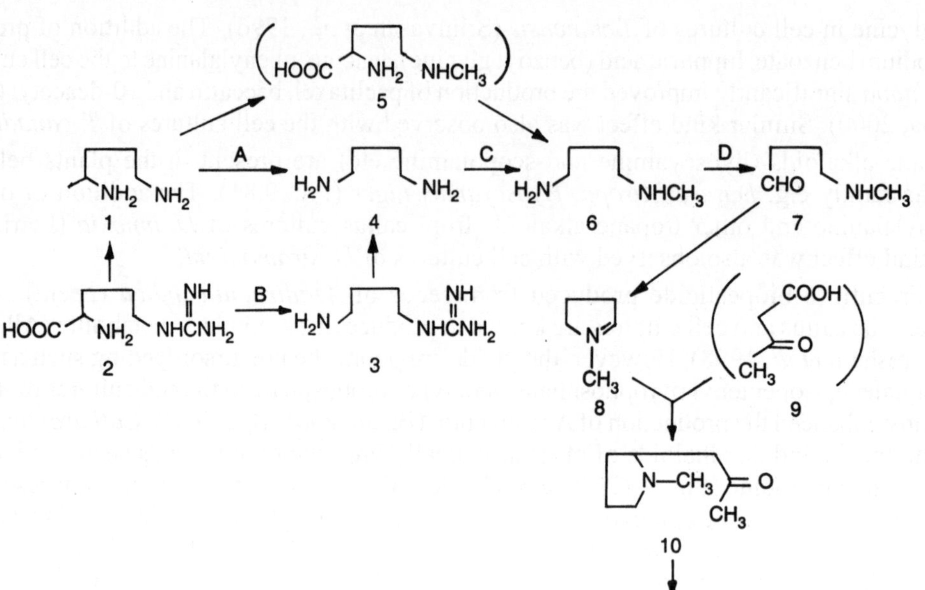

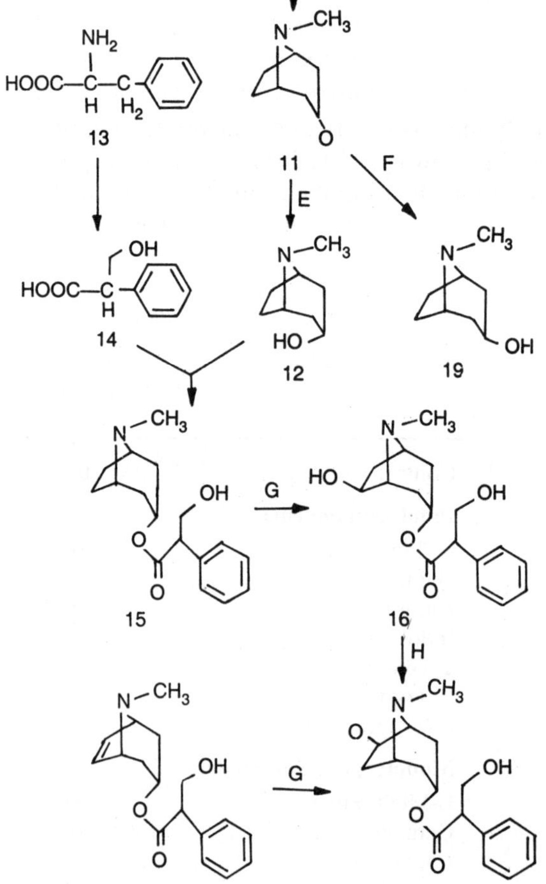

1. L-Ornithine
2. L-arginine
3. Agmatine
4. Putrescine
5. N-methyl-L-ornthine
6. N-methyl-putrescine
7. 4-methyl-aminobutanal
8. N-methyl-l-pyrrolinium cation
9. Acetoacetic acid
10. Hygrine
11. Tropinone
12. Tropine
13. Phenylalanine
14. (s)-tropic acid
15. (s)-hyoscyamine
16. 6 Beta-hydroxyl hyoscyamine
17. Scopolamine
18. 6,7-dehydro hyoscyamine
19. -tropine
A = Ornthine decarboxylase
B = Arginine decarboxylase
C = Putrescine N-methyltransferase
D = Diamine oxidase
E = Tropinone reductase i
F = Tropinone reductase ii
G = Hyoscyamine 6-B-hydroxylase
H = 6 B-(OH) hyoscyamine epoxidase

Fig. 9.81. The metabolic pathway from L-ornithine and L-arginine to scopolamine (adopted from Yamada et al, 1990).

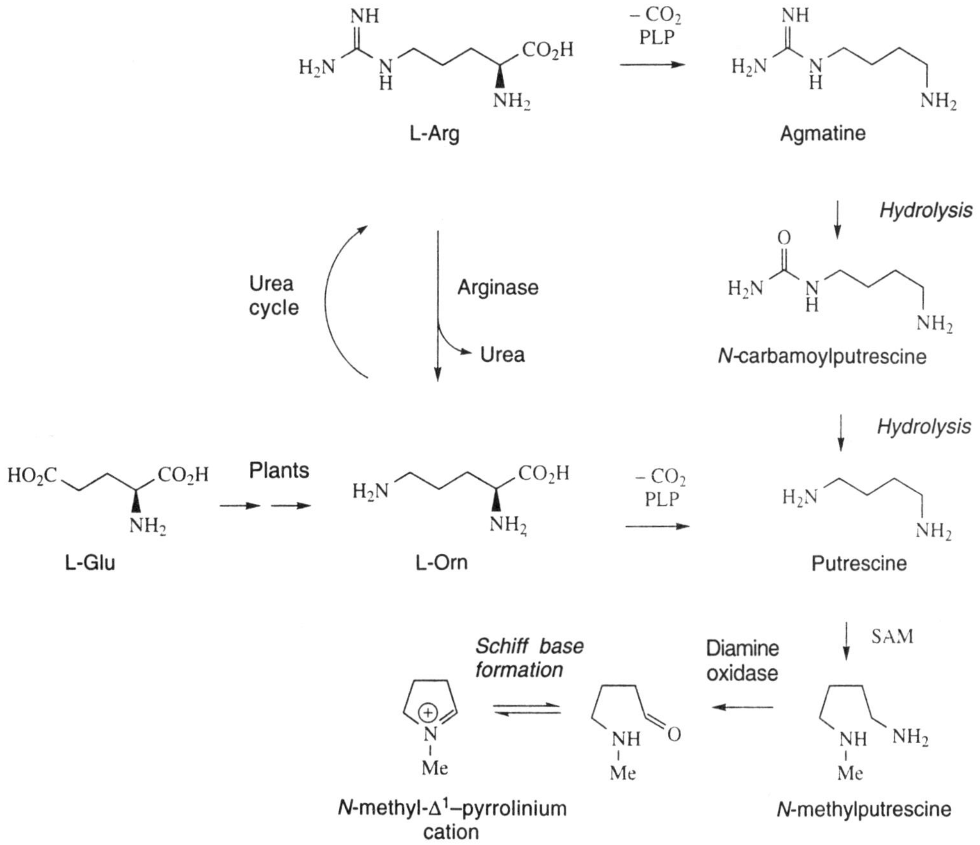

Fig. 9.38. Biosynthesis of methylputrescine.

9.7.2.4. *Terpenoid tetrahydro isoquinoline alkaloids*

The dried roots and rhizomes of ipecacuanha (*Cephaelis ipecacuanha*) found to contain terpenoid tetrahydro isoquinoline alkaloids e.g. Emetine, Cephaeline. The secologanin is an aldehyde and can condense with dopamine to give the N-deacetyl ipecoside (Tetrahydroisoquinoline). The biosynthetic pathway is shown in Fig. 9.54. Ipecoside itself found in the ipecac. The ipecoside an hydrolysis, followed by reduction and condensation with another mole of dopamine, followed by methylation forms cephaeline or emetine.

9.7.3. Alkaloids Derived from Tryptophan

9.7.3.1. *Simple indole alkaloids*

Psilocybin and simple indole alkaloid found mushroom (*Psilocybe* sp.) and biosyntheiszed from L-tryptophan. The biosynthetic pathway is depicted in Fig. 9.55. L-tryptophan upon decarboxylation forms tryptamine which undergoes N-methylation, hydroxylation and forms psilocin which upon phosphorylation yields Psilocybin. Gramine, simple indole alkaloid found in barley (*Hordeum vulgare*, Graminae) are also biosynthesized from L-tryptamine (Fig. 9.56).

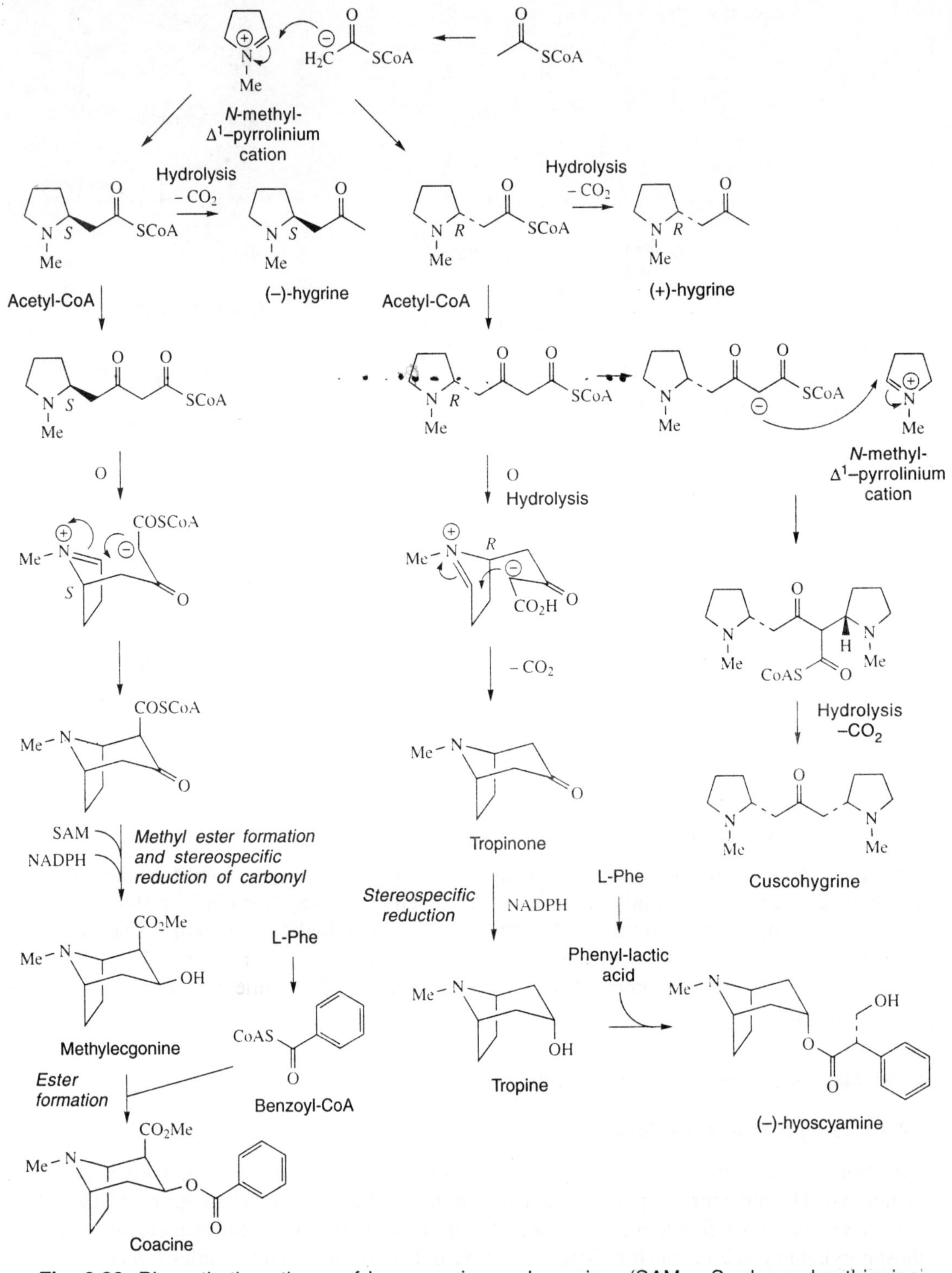

Fig. 9.39. Biosynthetic pathway of hyoscyamine and cocaine. (SAM = S-adenosylmethionine,

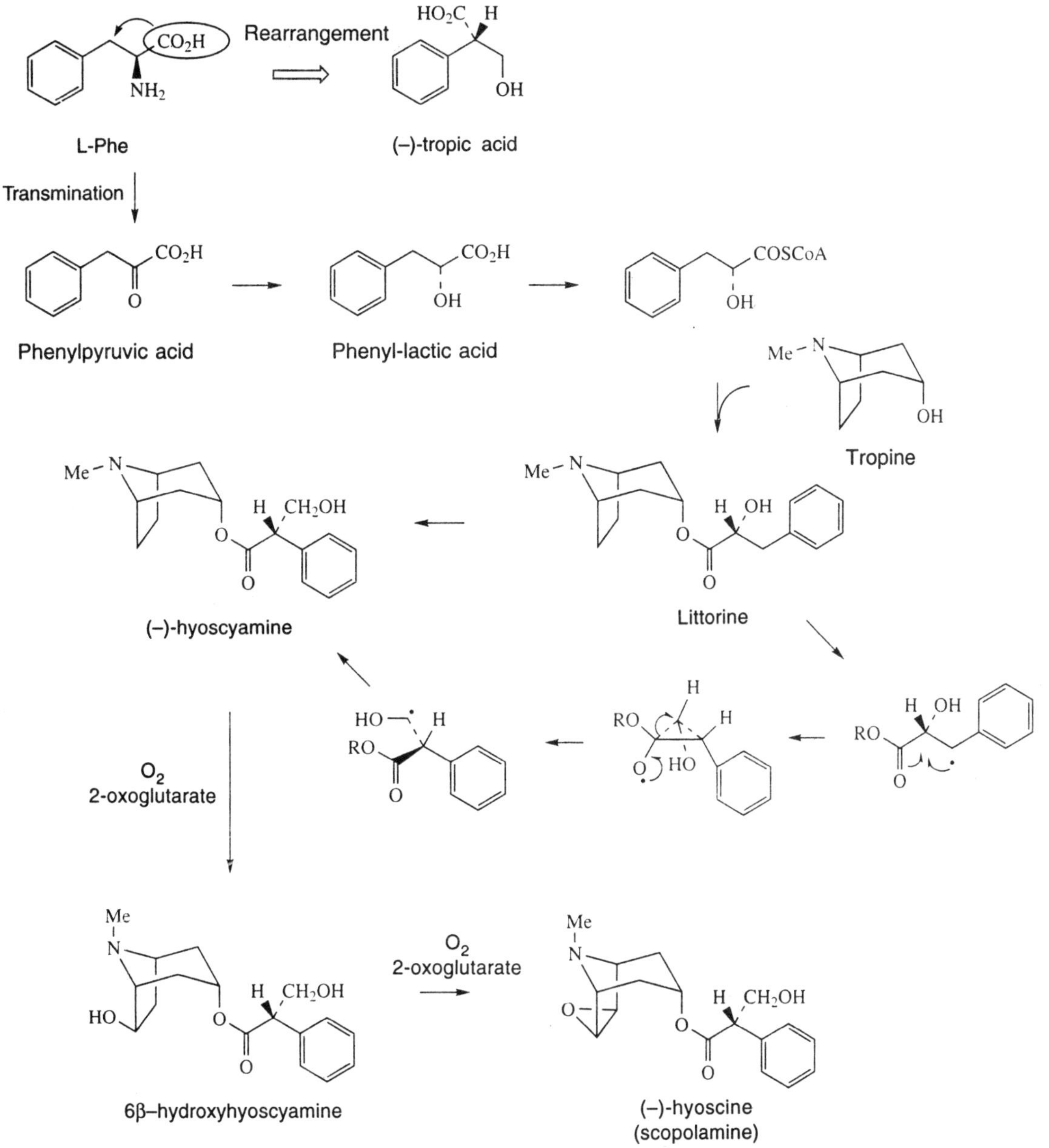

Fig. 9.40. Biosynthetic pathway for scopolamine.

9.7.3.2. Simple β-Carboline Alkaloids

The β-Carboline alkaloids such as harmine (*Peganum harmala*, Zygophyllaceae), Eleagnine (*Eleagnus angustifolia*, Eleagnaceae) and tetrahydroharmine (*Banisteriopsis caapi*, Malphighiaceae) are biosynthesized from tryptamine and pyruvic acid (Fig. 9.57).

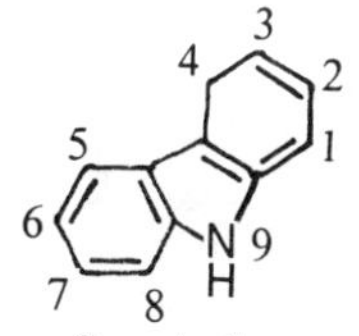

β-carboline

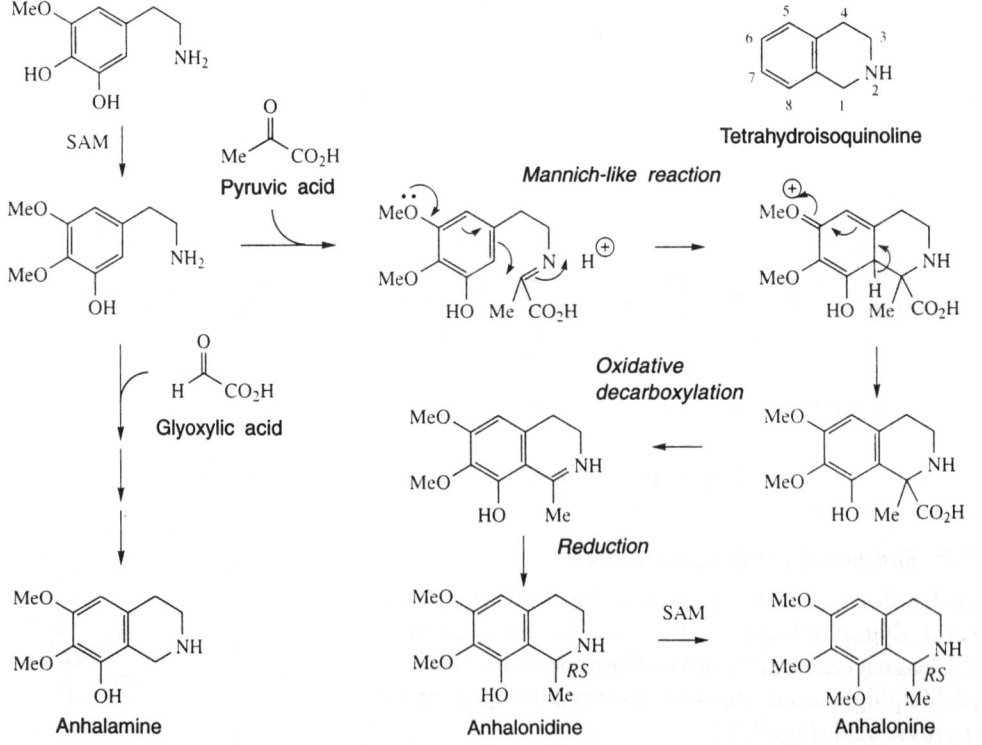

Fig. 9.41. Biosynthesis of hordenine, adrenaline and mescaline.

Fig. 9.42. Biosynthesis of tetrahydroisoquinoline alkaloids.

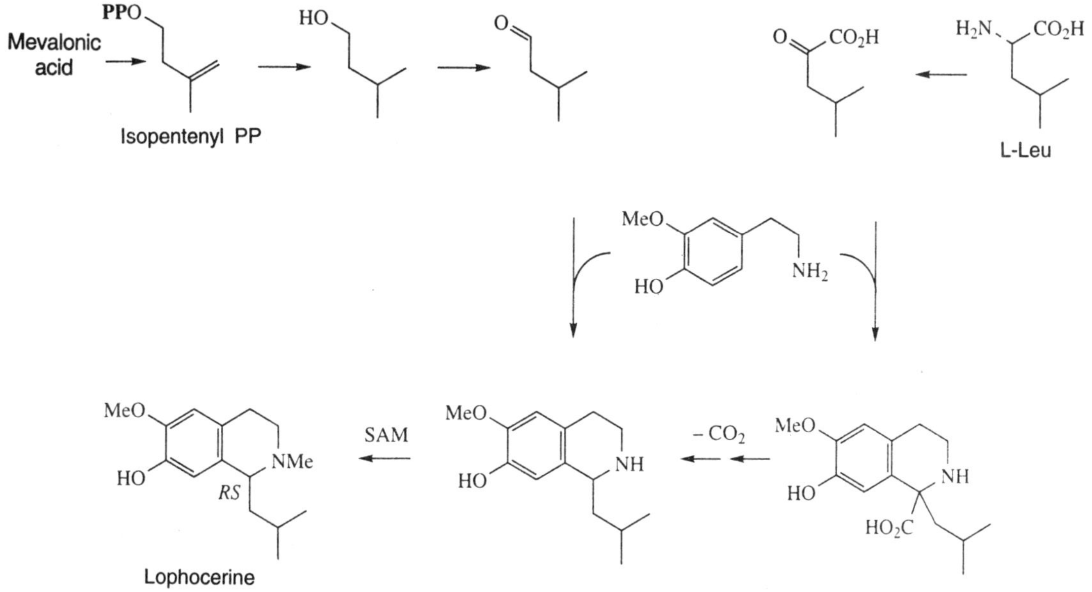

Fig. 9.43. Biosynthesis of lophocerine.

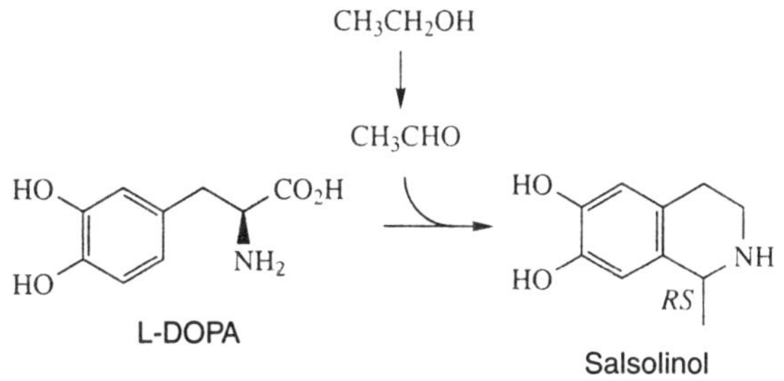

Fig. 9.44. Biosynthesis of salsolinol

9.7.3.3. Terpenoid Indole Alkaloids

Terpenoid Indole Alkaloids are mainly found in eight plant families of which Apocyanaceae, Loganiaceae, Rubiaceae provide the best sources. Tryptophan and its decarboxylation product tryptamine, serve as the precursor for the biosynthesis of indole alkaloids. The non-tryptophan portions of the alkaloids are however derived from monoterpenoid precursors which are designated as the coryanthe type (Ajmalicine & Akuammicine), the Aspidosperma type (Tabersonine), the Iboga type (Catharanthine).

The condensation of tryptamine with secologanin produces strictosidine, which upon hydrolysis of the glycoside function allows the opening of the hemiacetal, and exposure of an aldehyde group which can react with amine function. This intermediate upon allylic isomerization yields a product known as dehydrogeissoschizine and cyclization to cathenamine, which in turn reduction gives a compound known as ajmalicine (Fig. 9.58) present in *C. roseus*.

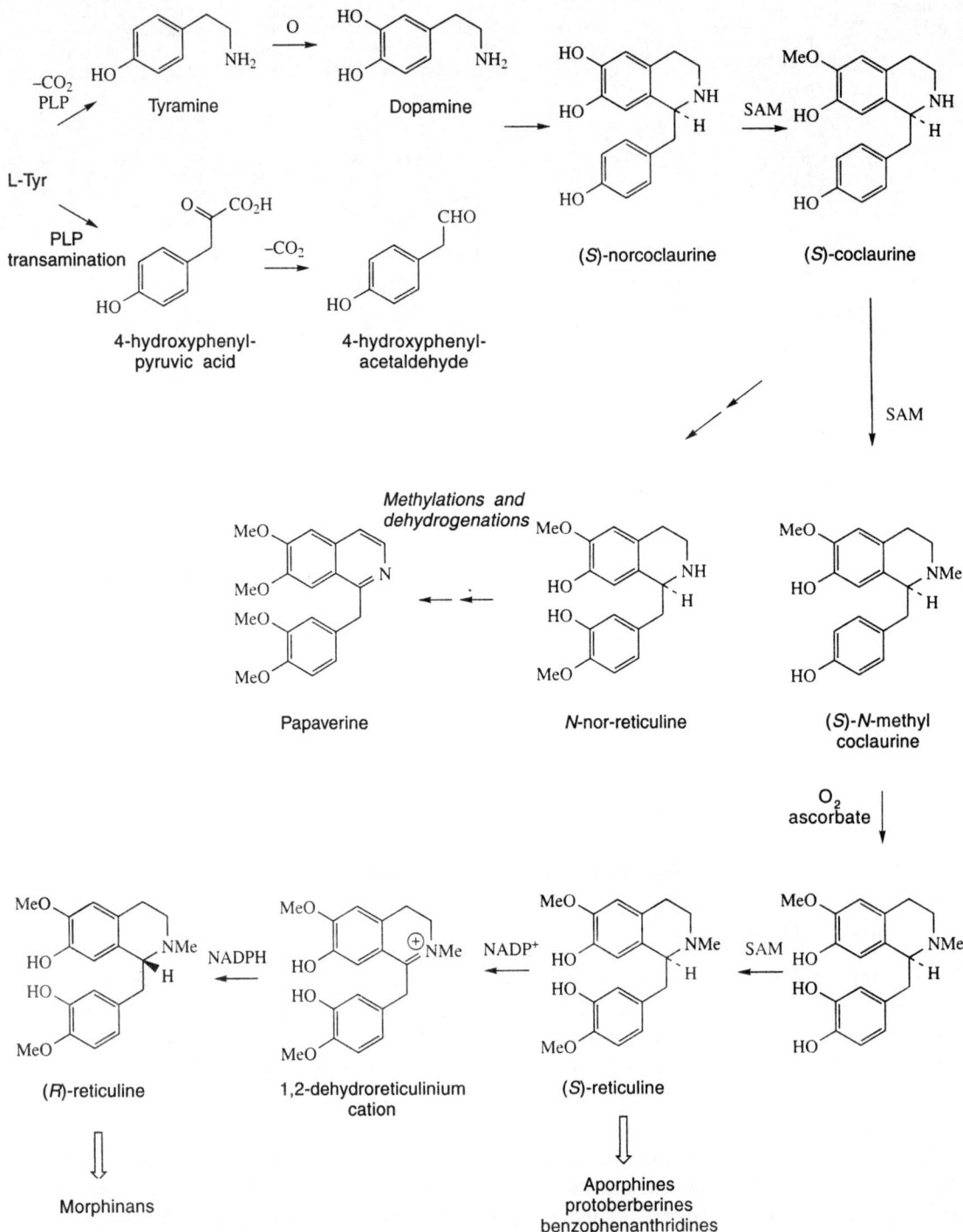

Fig. 9.45. Biosynthesis of benzyltetrahydroisoquinoline alkaloids.

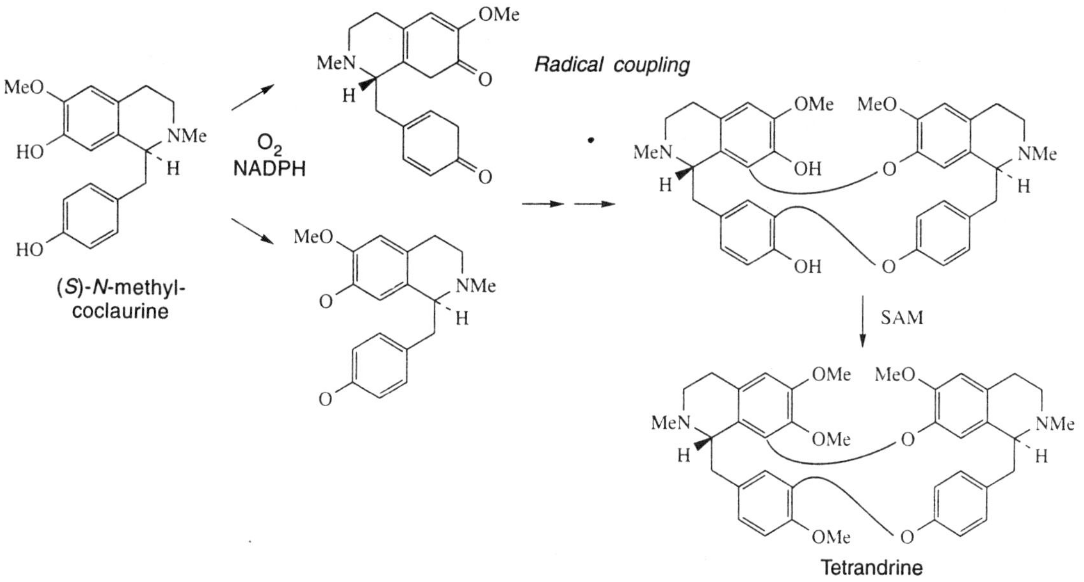

Fig 9.46. Biosynthesis of Tetrandrine.

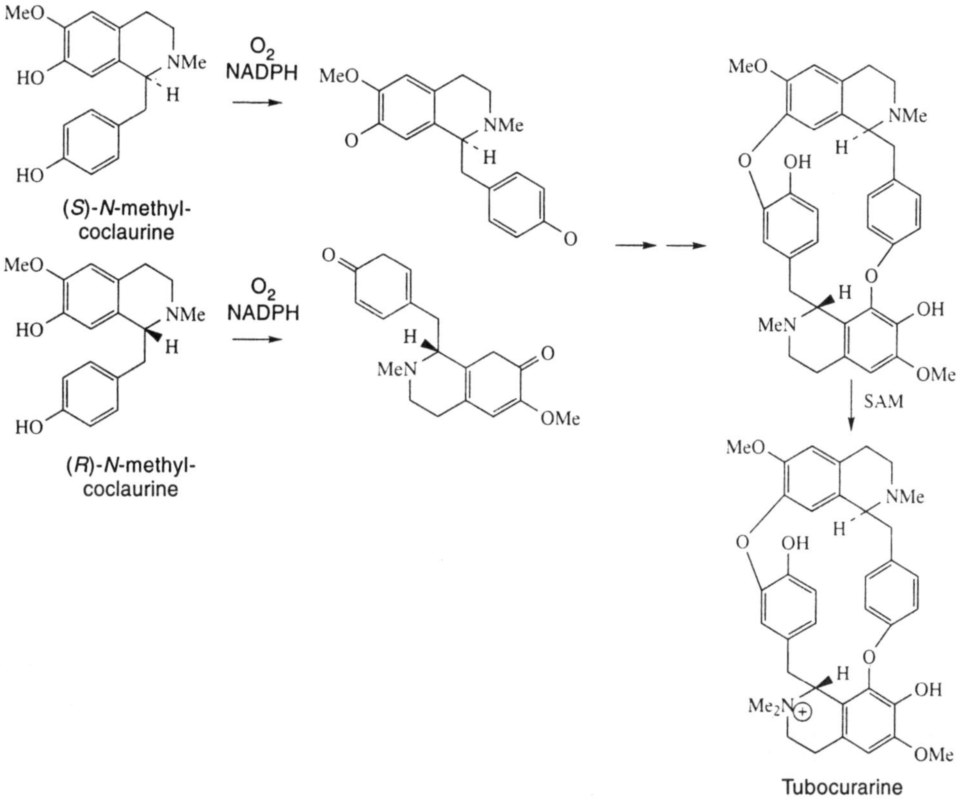

Fig. 9.47. Biosynthesis of tubocurarine.

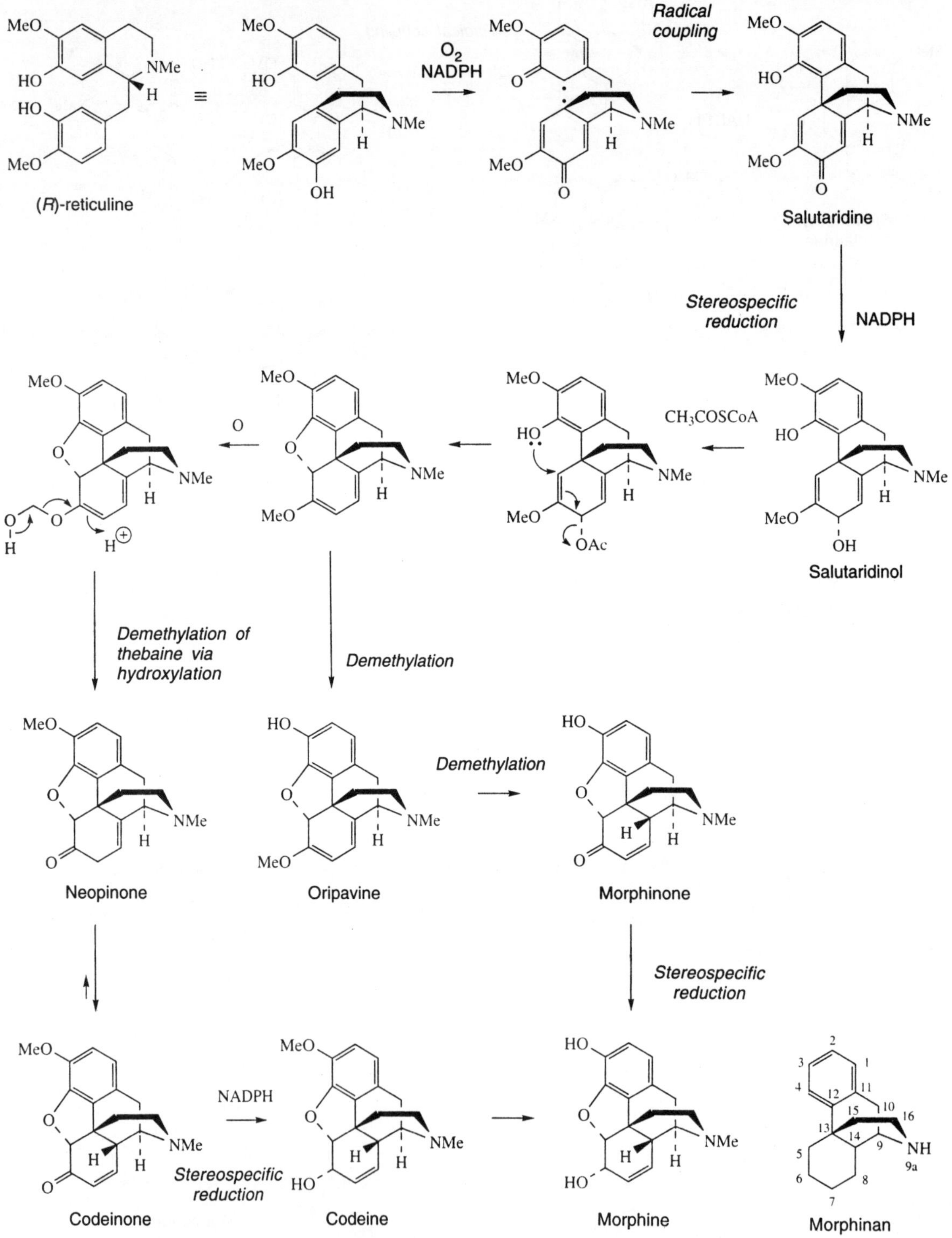

Fig. 9.48. Biosynthesis of morphine, codeine and other opium alkaloids.

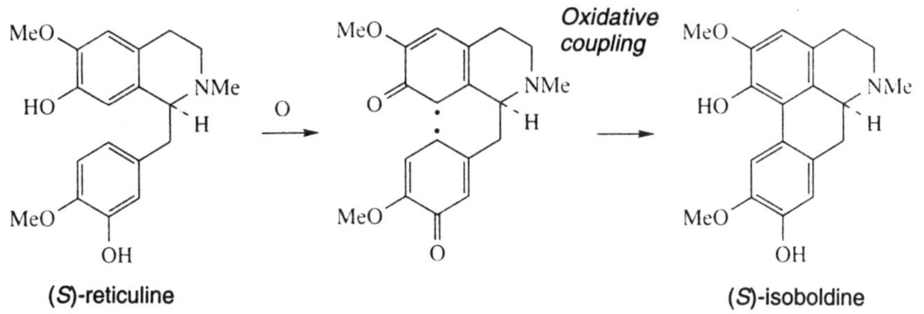

Fig. 9.49. Biosynthesis of (S)-isoboldine.

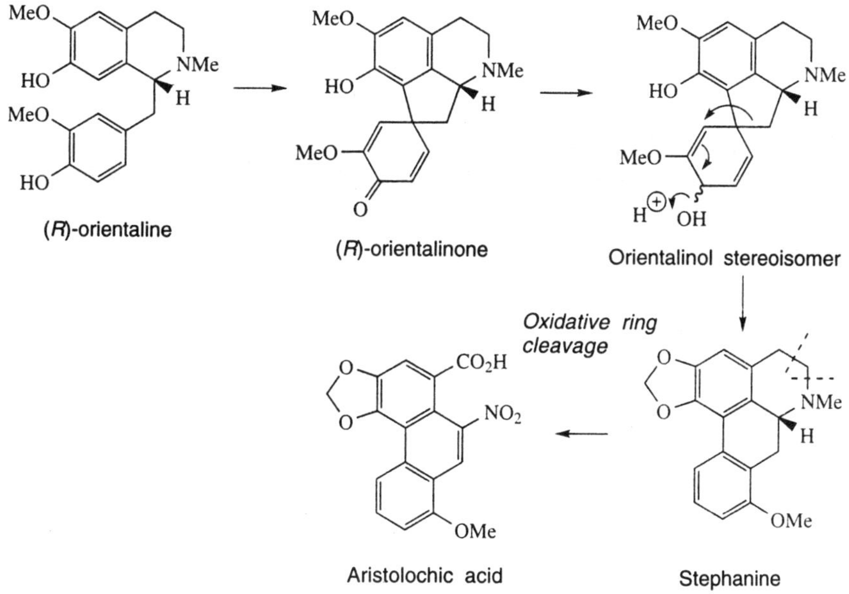

Fig. 9.50. Biosynthesis of aristolochic acid.

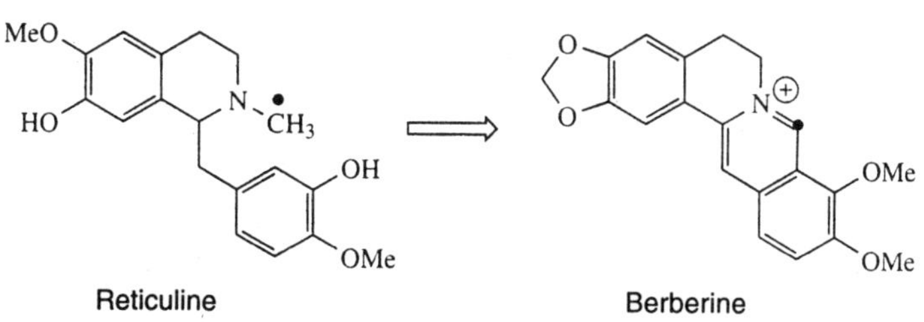

Fig. 9.51. Biosynthesis of berberine.

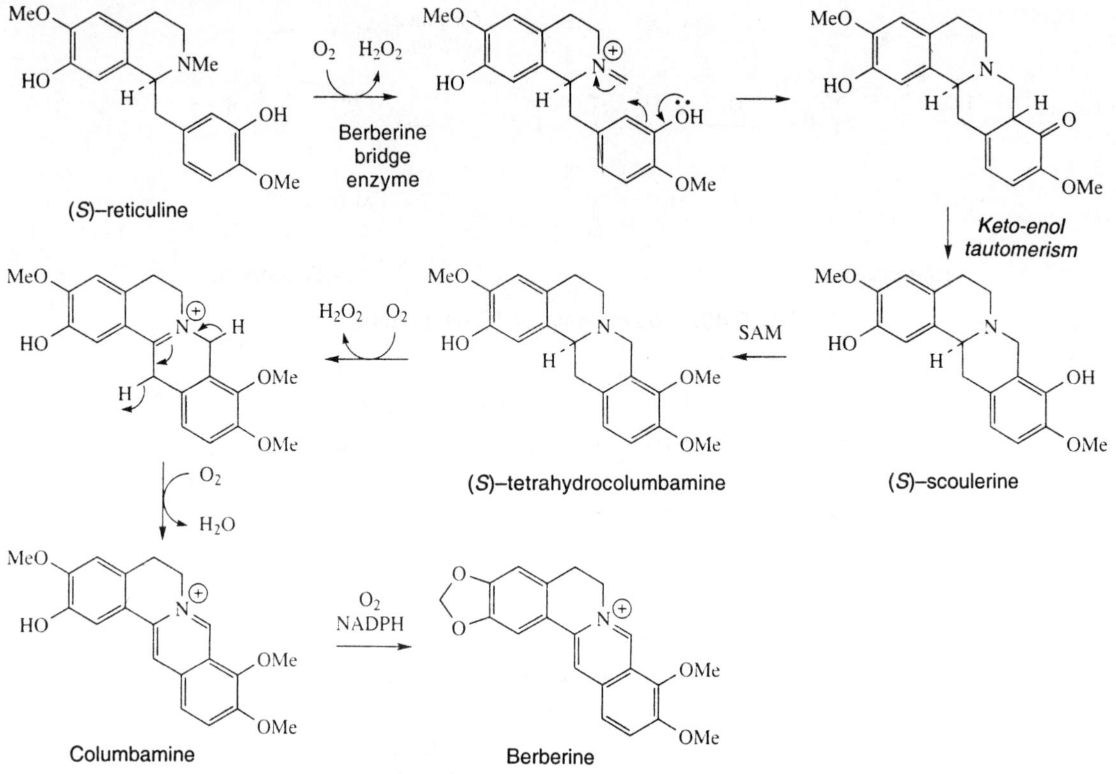

Fig. 9.52. Biosynthesis of berberine via scoulerine.

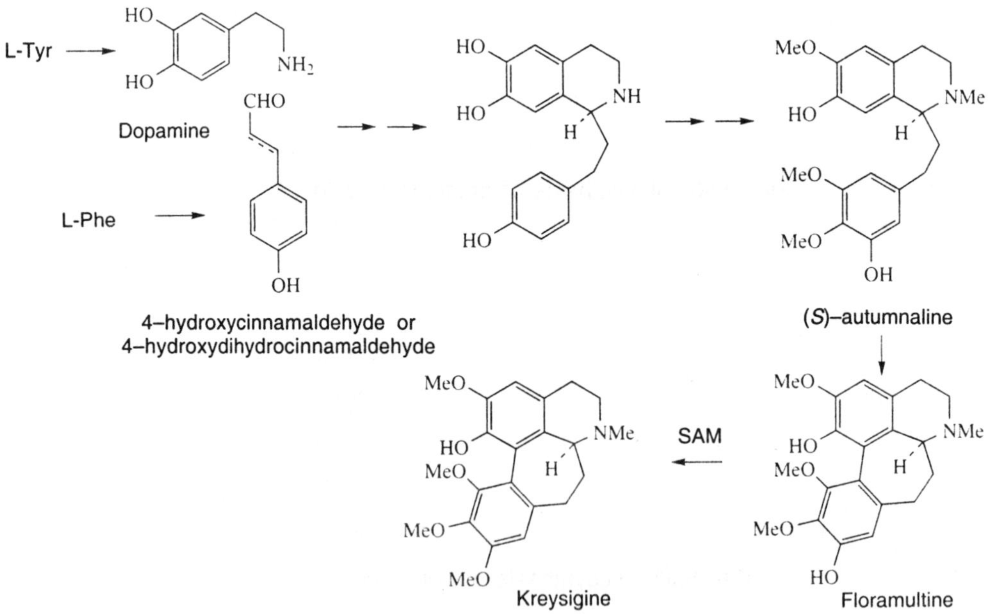

Fig. 9.53. Biosynthesis of phenethyl isoquinoline alkalods.

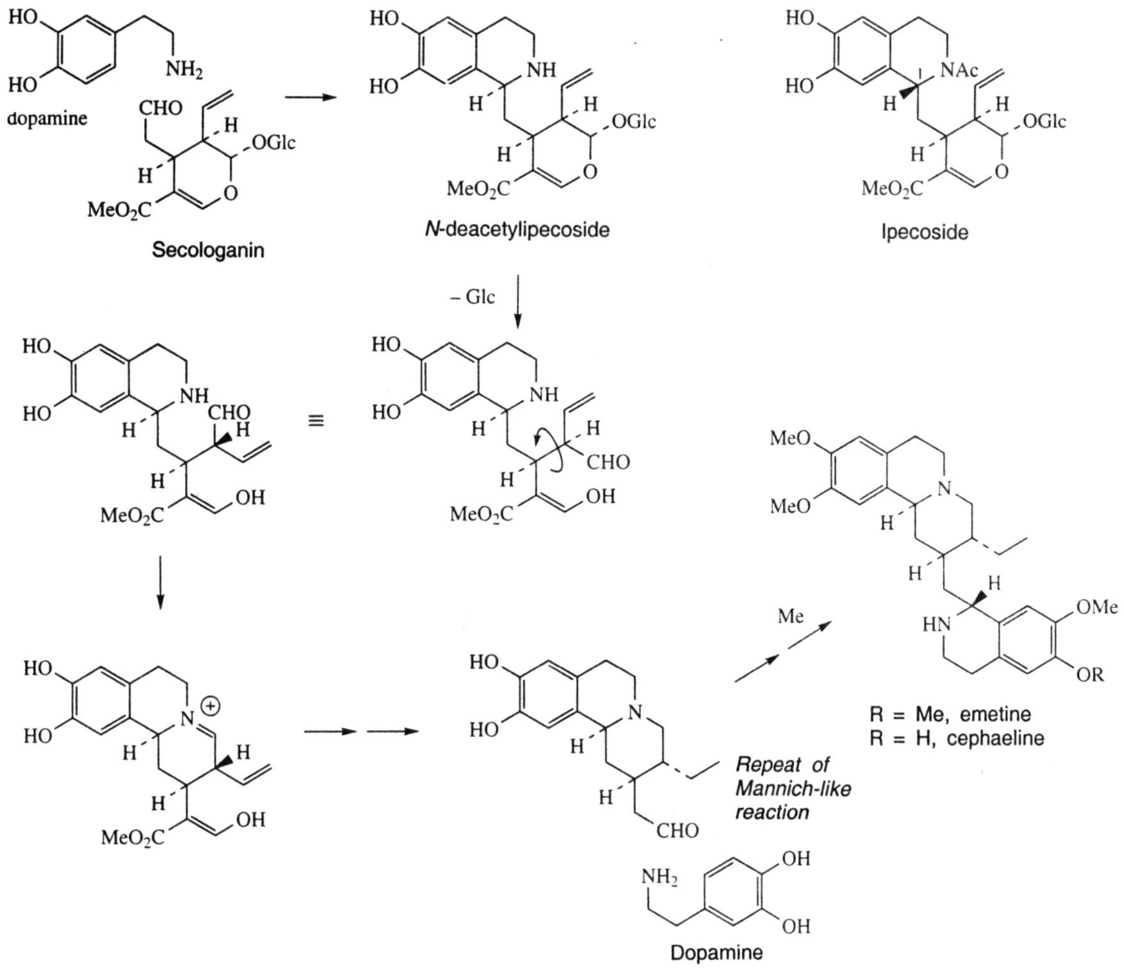

Fig. 9.54. Biosynthesis of ipecac alkaloids.

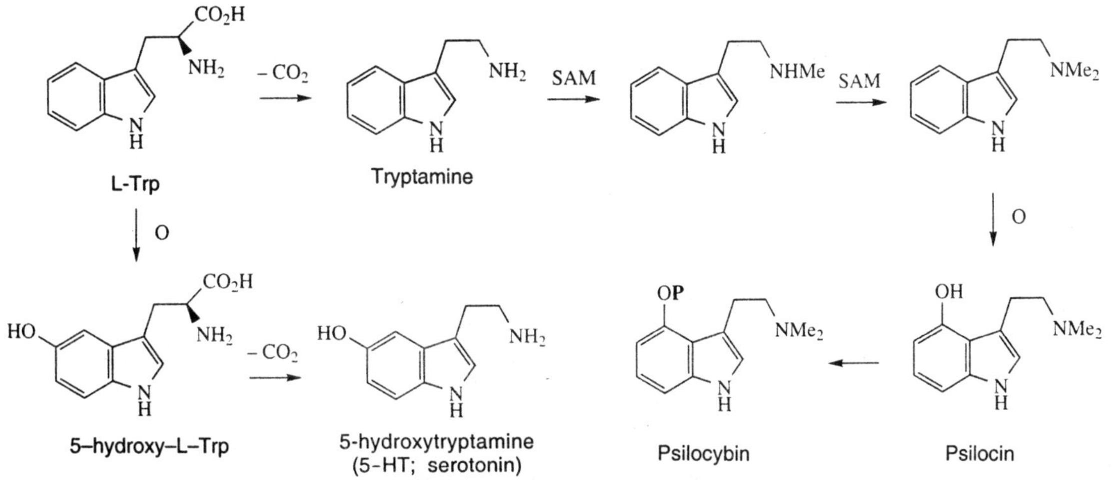

Fig. 9.55. Biosynthesis of psilocybin.

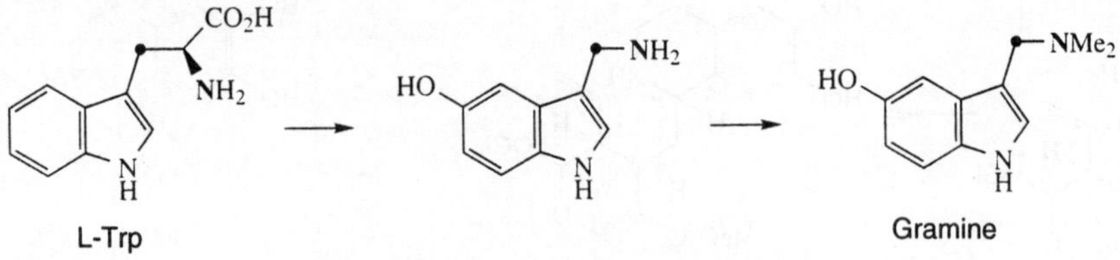

Fig. 9.56. Biosynthesis of gramine.

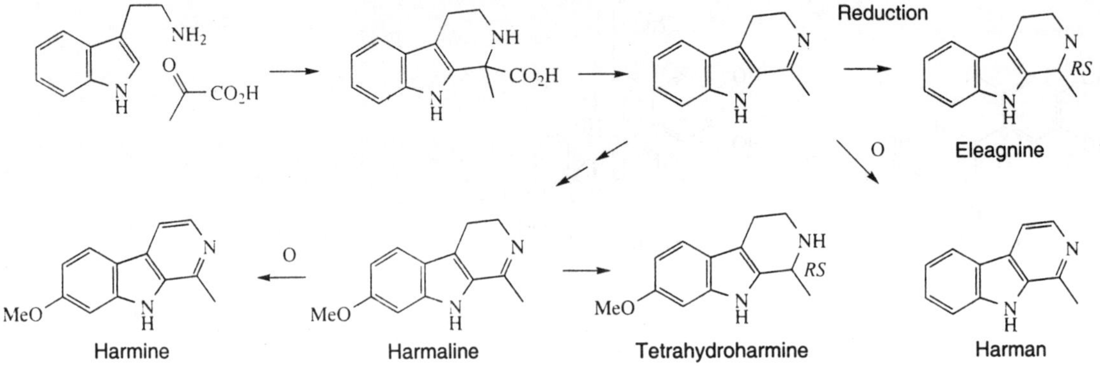

Fig. 9.57. Biosynthesis of harmine, harman and tetrahydroharmine.

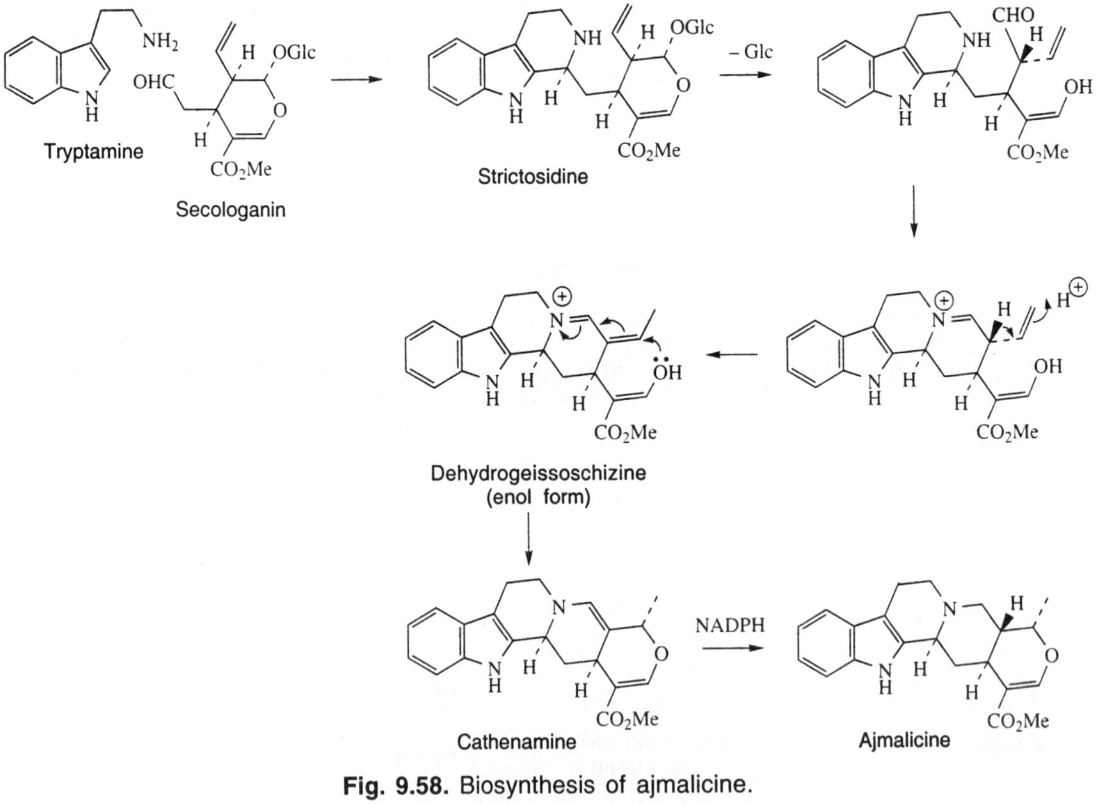

Fig. 9.58. Biosynthesis of ajmalicine.

Yohimbine is an indole alkaloid found in the bark of *Pausinystalia yohimbe* (Rubiaceae), Aspidosperma bark (*Aspidosperma* sp. *Apocyanaceae*) and Rauwolfia sp. (Apocyanaceae). Yohimbine arises from dehydrogeissoschizine (Fig. 9.59). Reserpine and deserpidine are trimethoxy benzoyl esters of Yohimbine like alkaloids, whilst rescinnamine is a trimethoxy cinnamoyl ester. All these alkaloids are present in *Rauwolfia* sp. (Fig. 9.60).

The structural changes involved in converting the corynanthe-type skeleton into those of the Aspidosperma and Iboga groups are quite complex, and are shown in Fig. 9.61. Preakuammicine arises from dehydrogeissoschizine, which inturn forms stemmadenine, catharthine, tebersonine and vindoline and the biosynthetic pathway is depicted in Fig. 9.61 The vinblastine and vincristine are anticancer indole alkaloids found in Madagascar periwinkle (*Catharanthus roseus*, Apocynaceae). The catharanthine is believed to involve firstly an oxidative reaction on catharanthine, catalyzed by a peroxidase, produces, vindoline which inturn reduction and hydroxylation forms vinblastine (Fig. 9.63).

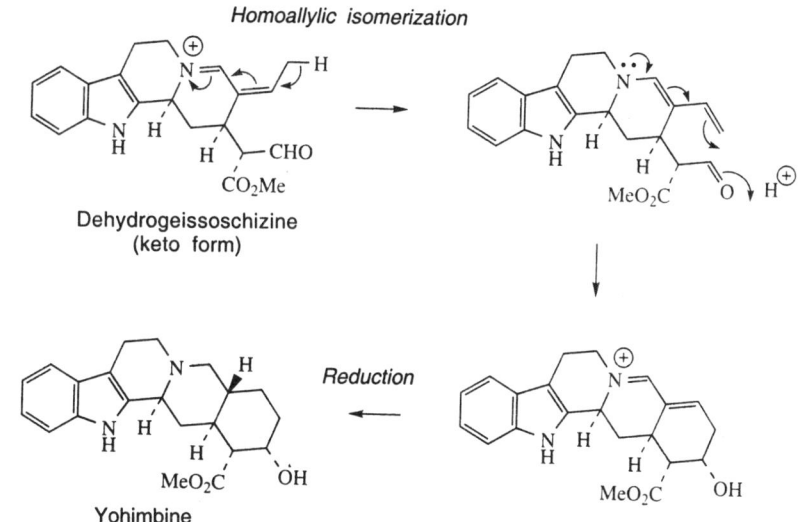

Fig. 9.59. Biosynthesis of yohimbine.

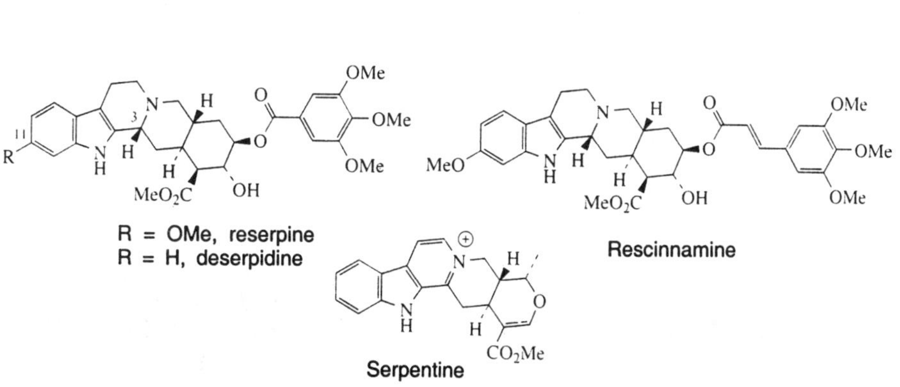

R = OMe, reserpine
R = H, deserpidine

Rescinnamine

Serpentine

Fig. 9.60. The chemical structures of Rauwolfia alkaloids.

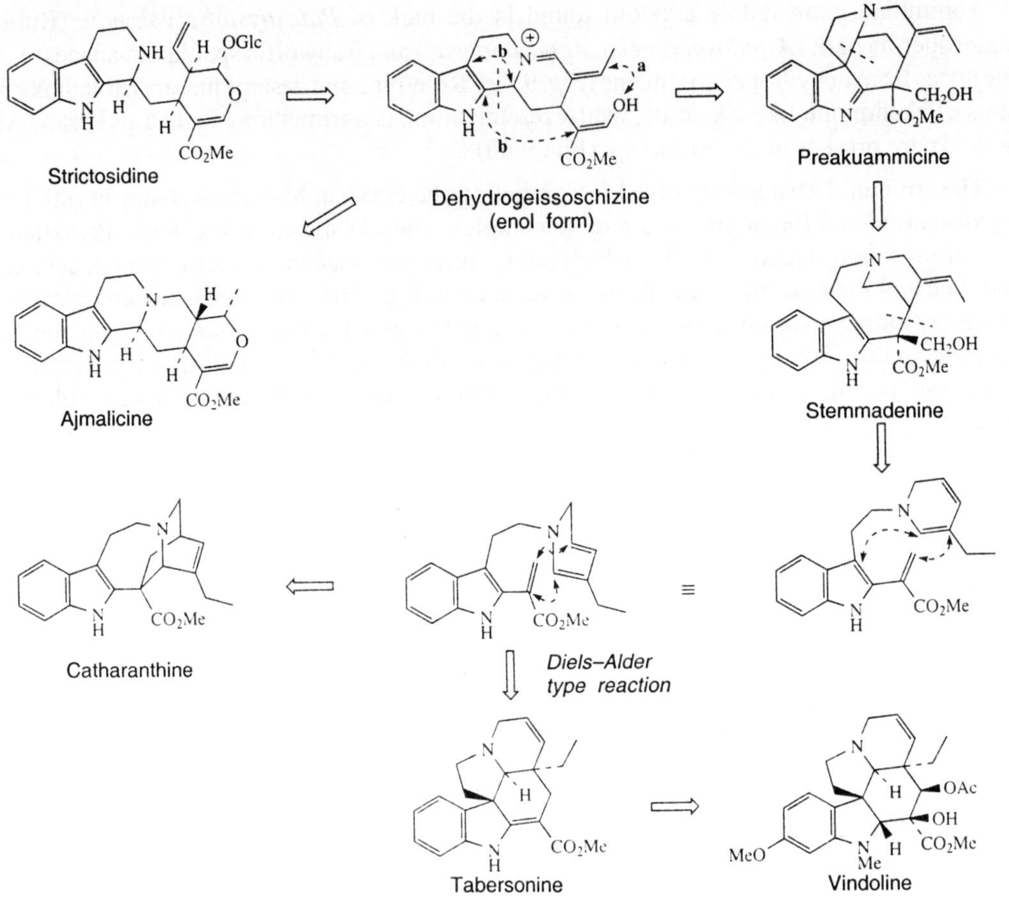

Fig. 9.61. Biosynthesis of vindoline, catharanthine, vindoline.

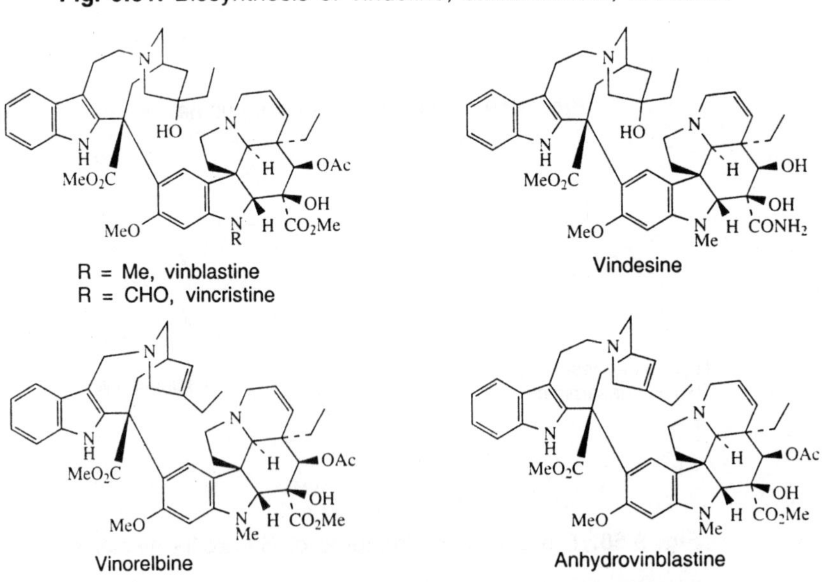

Fig. 9.62. The structures of Catharanthus alkaloids.

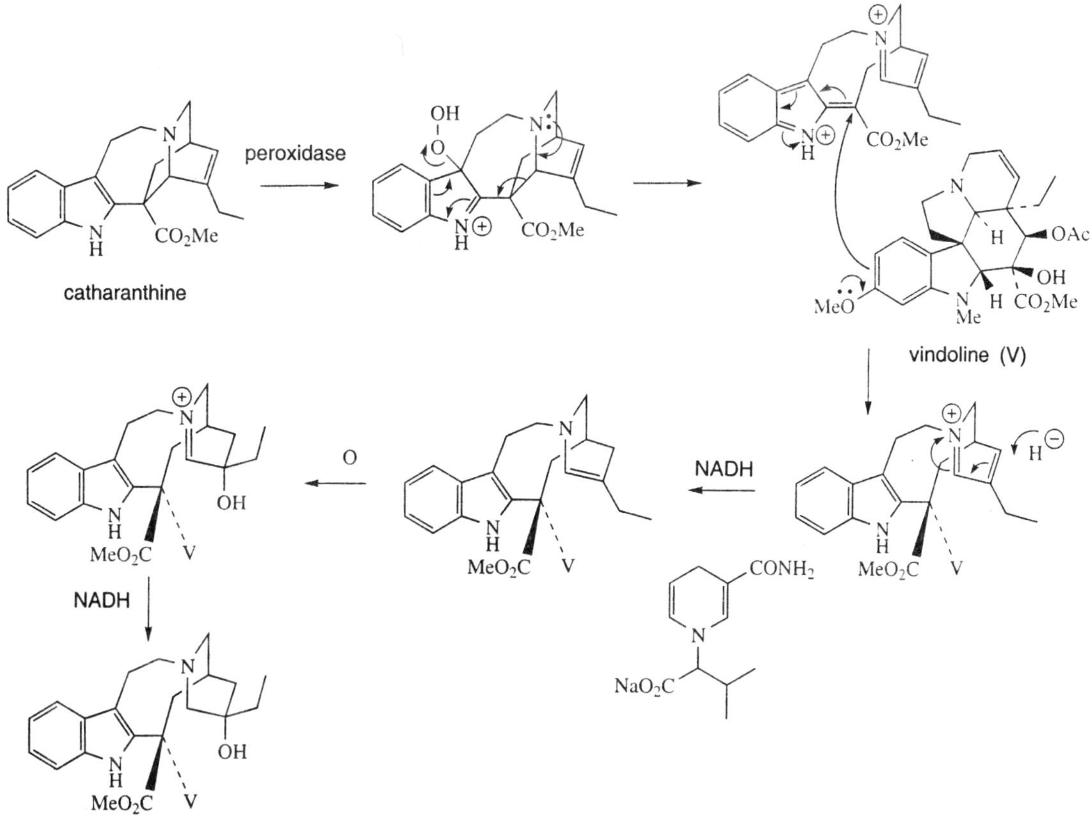

Fig. 9.63. Biosynthesis pathway of vinblastine.

The vincamine from *Vinca minor* and ajmaline from *Rauwolfia serpentina* are also biosynthesized from dehydrogeissoschizine (Fig 9.64). Ajmaline contains a C-9 corynanthe-type unit whereas vincamine C10 aspidosperma unit, and it orginates from tabersonine.

Nux-vomica consists of the dried ripe seeds of *Strychnos nux-vomica* (Loganiaceae), contains strychnine and its dimethoxy analogue brucine. The non-tryptamine portion of these compounds contains eleven carbons, and is derived from an iridoid-derived C9 unit, plus two carbons from acetate. The preakuammicine like structure upon hydrolysis/decarboxylation and addition of acetyl CoA form a hemiacetal in the weiland-gumlich aldehyde (Fig. 9.65).

9.7.3.4. *Pyrroloindole alkaloids*

Pyrroloindole type alkaloids are present in *Physostigma venenosum* (Leguminosae) Physostigmine (eserine), chimonanthine from *Chimonanthus fragrans* (Calycanthaceae) respectively. Physostigmine has anticholinesterase activity. A suggested pathway to physostigmine is by C-3 methylation of tryptamine, followed by ring formation involving attack of the primary amine function onto the minium ion. Further substitution produces the phytostigmine (eserine) (Fig. 9.66).

9.7.3.5. *Ergot Alkaloids*

Ergot is a fungal disease commonly found on many wild and cultivated grasses, and it caused by *Claviceps purpurea*. The disease is eventually characterized by the formation of hard, seed, like

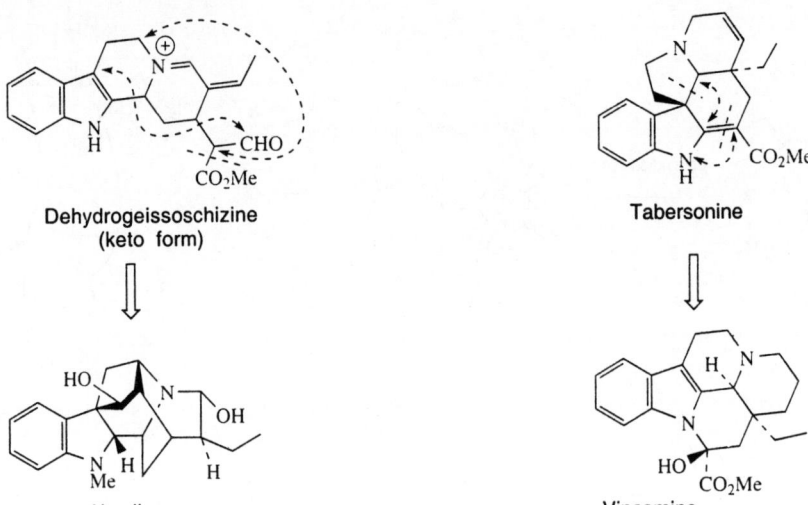

Dehydrogeissoschizine
(keto form)

Tabersonine

Ajmaline

Vincamine

Fig. 9.64. Biosynthesis of ajmaline and vincamine.

Preakuammicine

Acetyl–CoA

Wieland–Gumlich aldehyde

Wieland–Gumlich aldehyde
(hemiacetal form)

Strychnine

Brucine

Fig. 9.65. Biosynthesis of Nux-vomica alkaloids.

Table 10.4. Triterpenenoids

Substrates	Main products	Type of reaction	Plant species	Reference
18β-Glycyrrhetinic acid	3-O-(α-L-arabinopyrasyl-(1→2)-β-D-glucuronopyranosyl-24-hydroxyl-18β-glycyrrhetinic acid	Glucosylation hydroxylation	*Glycyrrhiza glabra*	Hayashi *et al.* (1990)
18β-Glycyrrhetinic acid	18β-Glycyrrhetinic acid 30β-glucopyranosyl ester 28-hydroxy-18β-glycyrrhetinic acid 30-β-glucopyranosyl ester	Glucosylation hydroxylation	*Eucalyptus perriniana*	Orihara and Furuya (1990)
18β-Glycyrrhetinic acid	18β-Glycyrrhetinic acid 30β-glucopyranosyl ester	Glucosylation	*Coffea arabica*	Orihara and Furuya (1990)
18β-Glycyrrhetinic acid	Glycosides, malonyl glycosides glycosyl ester	Glucosylation malonylation	Transformed *Panax ginseng*	Asada *et al.* (1993)

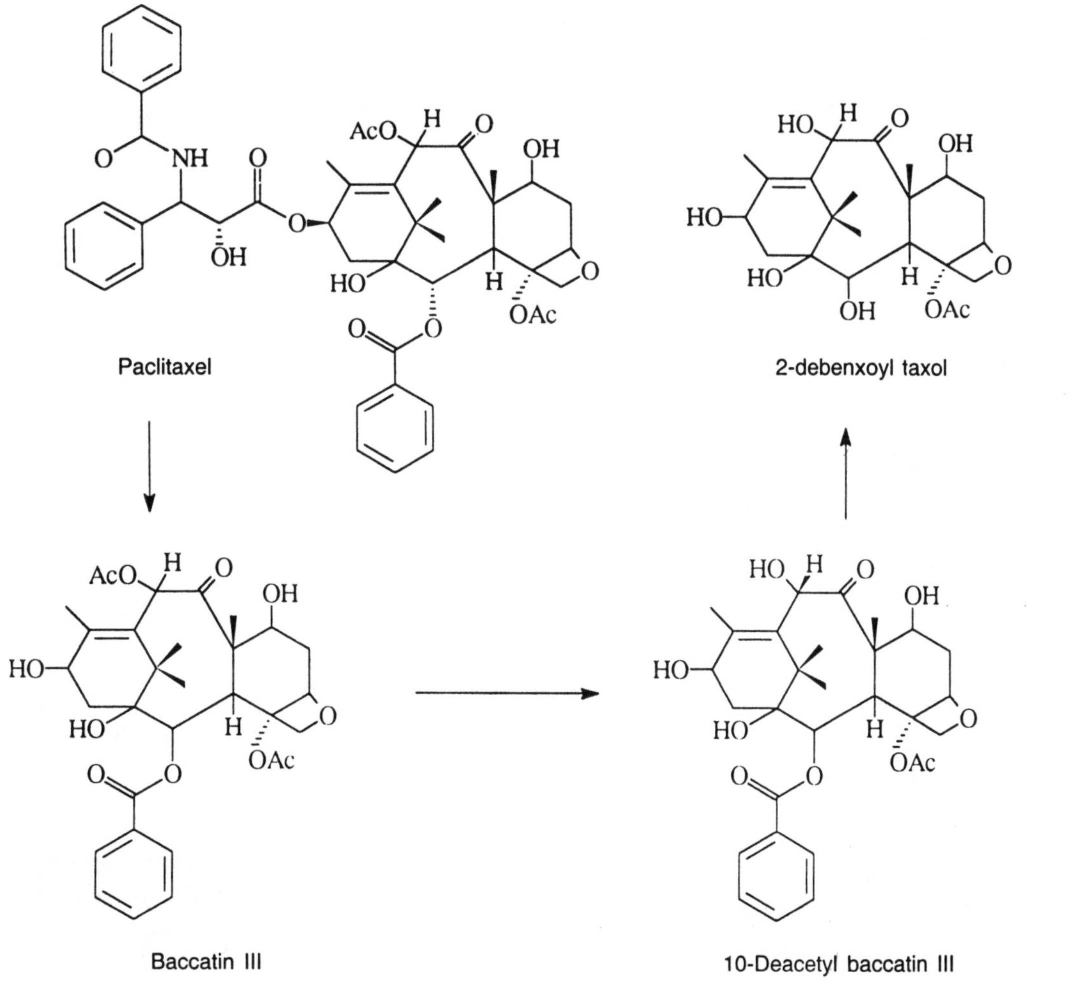

Fig. 10.7. Biotransformation of paclitaxel.

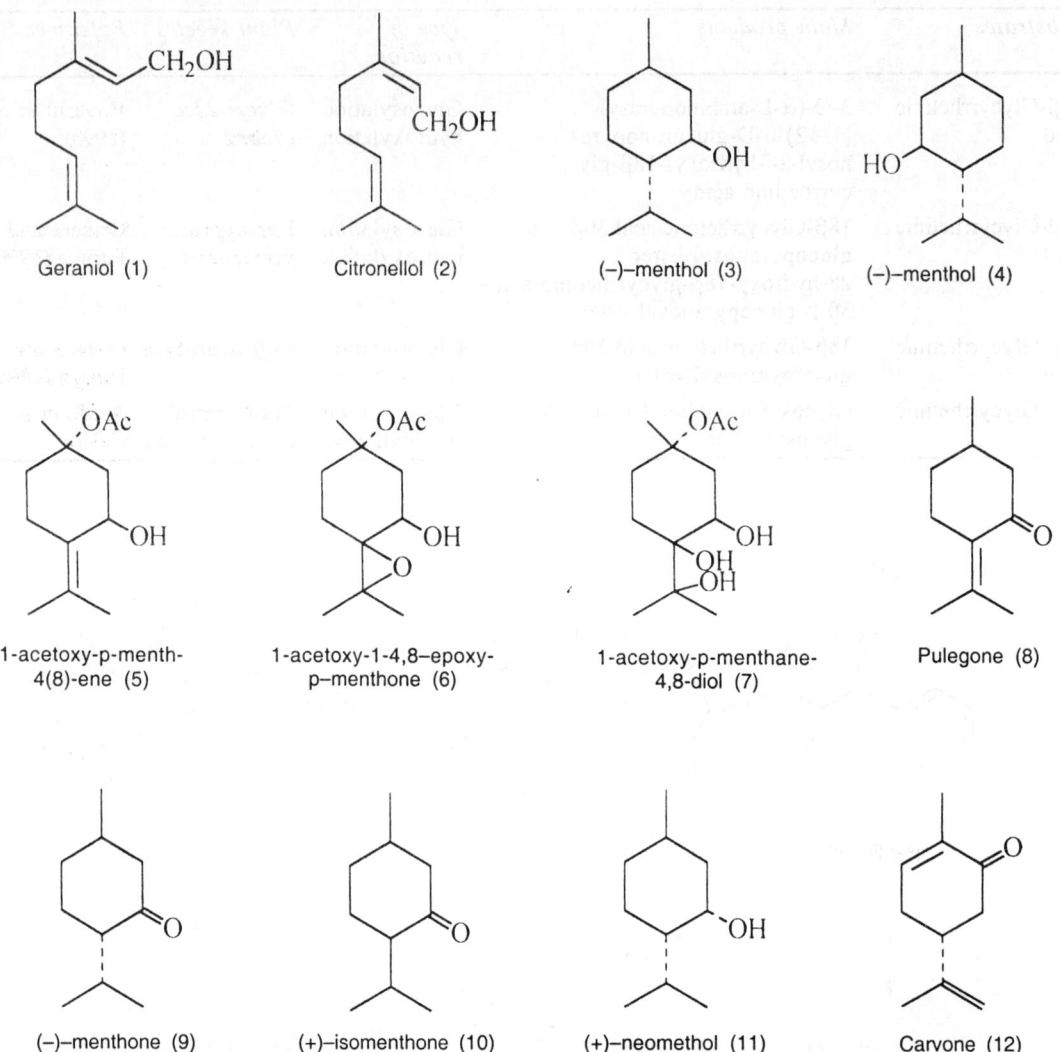

Fig. 10.8. Chemical structures relevant to Table 10.2.

Orihara and Furuya (1990) reported the biotransformation of 18 β-glycyrrhetinic acid by *E. perriniana* and *C. arabica* cell cultures. Both cell suspension cultures were able to biotransform 18 β-glycyrrhetinic acid to its glycosylester. *C. arabica* cell cultures yielded product 30-glucosyl ester and *E. perriniana* further biotransformed its glucosyl ester to the 28 hydroxy and the 23, 28 dihydroxy products.

Asada *et al.* (1993) described that the biotransformation of 18β-glycyrrhetinic acid by using hairy root cultures of *Panax ginseng*. Ginseng hairy roots glycosylated for 18β-glycyrrhetinic acid and produced 6 different glycosides including three new biotransformation products, but did not produce the hydroxylated product. 18β-glycyrrhetinic acid has been reported by microbial conversion

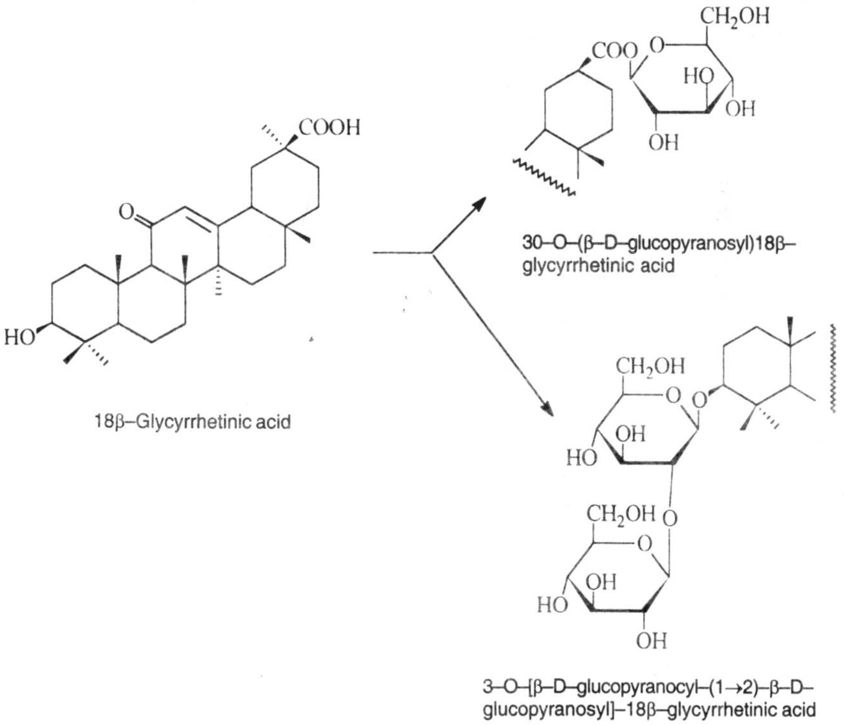

CH$_2$OH

COO

HO

OH

OH

30–O–(β–D–glucopyranosyl)18β–
glycyrrhetinic acid

COOH

O

HO

18β–Glycyrrhetinic acid

CH$_2$OH

O

OH

HO

CH$_2$OH O

O

OH

HO

OH

3–O–[β–D–glucopyranocyl–(1→2)–β–D–
glucopyranosyl]–18β–glycyrrhetinic acid

Fig. 10.9. Biotransformation of 18β-glycyrrhetinic acid.

e.g. hydroxylation by Streptomyces sp. (Sakano and Ohsima 1986) and ring cleavage by *Chainia antibiotica* (Sakano and Ohsima, 1986). No glycosides have been reported in the products of microbial biotransformation.

10.5.4. Steroids (Table 10.5)

Steroid hormones are of special pharmaceutical importance because they are widely used in medicine. Testosterone is one of the end products of steroid biosynthesis in the male human beings. Hamada *et al.* (1991a & b) studied the biotransformation of androstane-17, dione, testosterone, 1,4, androstadiene-3,17 dione and androsterone, by *Marchantia polymorpha* green cell suspension cultures. Testosterone was converted to 16 β-hydroxytestesterone in a yield of 18% (Fig. 10.10). The biotransformation of 1,4-androstadiene-3-17-dione and adrenosterone (4-androstene-3, 11,17-trione) is shown in Figure 10.10. 1,4-Androstadiene-3, 17-dione was converted to 17β-hydroxy-1, 4-androstadiene-3-one and 4- androstene-3-17-dione and testosterone and adrenosterone (4-androstene-3, 11,17-trione) was converted to 17β-hydroxy-4-androstene-, 11-dione as the sole product.

Thus it was found that the *M. polymorpha* green suspension cells stereoselectively reduce the carbonyl group at position 17 of 4-androstenes and regio- and stereo selectively hydroxylate the 6-position of testosterone. Such a hydroxylation has not previously been observed in the biotransformation of testosterone with plant cell suspension culture.

The feeding of testosterone to *N. tabacum* cell suspension cultures was able to metabolize testosterone through a range of reactions involving oxidation, reduction, glycosylation and esterification with fatty acid (palmitic acid). The same cells fail to metabolize related steroids such as cholesterol

Table 10.5. Steroids (Adopted from Dicosmo and Misawa, 1996)

Substrates	Main products	Type of reaction	Plant species	Reference
Testosterone	4-Androstan-3,17-dione 5α-androstan-17β-ol-3-one, testosterone glucoside, epiandrosterone epiandrosterone palmitate	Hydrogenation of C=C, reduction of C=O, glucosylation, palmityl esterification	Nicotiana tabacum	Hirotani and Furuya (1980)
Testosterone	Hydrolyzing product	Hydrolysis	Spirodela oligorrhiza	Tlomak et al. (1986)
Androsterone	Hydrolyzing product	Hydrolysis	S. oligorrhiza	Tlomak et al. (1986)
4-Androstene-3,17-dione	5α-Androstan-3β-ol-17-one, 5α-androstan-3β-17β-diol	Reduction of C=C and C=O	Dioscorea deltoidea	Stohs and Elolemy (1972)
Progesterone (pregn-4-en-3,20-dion)	5α-Pregnanolone and its palmitate palmityl esterification	Reduction of C=C, reduction of C=O,	N. tabacum Sophora angustifolia	Furuya et al. (1971)
Progesterone	5α-Pregnane-3,20-dione, 5α-pregnanolone, glucoside glycosylation	Reduction of C=C and C=O,	Digitalis purpurea	Graves and Smith (1967)
Progesterone 5α-pregnanolone	5α-Pregnane-3,20-dione, and C=O	Reduction of C=C	Hedera helix	Graves and Smith (1967)
Progesterone	Pregn-4-en-20-ol-3-one, 5α-pregnanolone	Reduction of C=O	Rosa sp.	Graves and Smith (1967)
Progesterone	5α-Pregnane-3,20-dione	Reduction of C=C	D. deltoidea, Cheriranthus cheiri (microsome)	Stohs (1969)
Progesterone	5α-Pregnane-3,20-dione 5α-pregnanolone	Reduction of C=C and C=O	Nicotiana rustica	Graves and Smith (1967)
Progesterone	5α-Pregnane-3,20-diol, 5α-Pregnanolone	Reduction of C=C and C=O	D.deltoidea	Stohs and Elolemy (1972)
Progesterone	Pregn-4-en-20α-ol-3-one	Reduction of C=O	Partheno-cissus sp.	Graves and Smith (1967)
Progesterone	5α-Pregnane-3,20-dione, 5α-pregnanolone	Reduction of C=C and C=O	Solanum tuberosum	Graves and Smith (1967)
Progesterone	5α-Pregnane-3,20-dione	Reduction of C=C	Digitalis lutea	Graves and Smith (1967)
Progesterone	5α-Pregnane-3,20-dione, 5α-pregnanolone	Reduction of C=C and C=O	Atropa belladonna	Graves and Smith (1967)
Progesterone	5α-Pregnane-3,20-dione, 5α-pregnanolone pregn-4-en-20á-ol-3-one	Reduction of C=C and C=O	N. tabacum	Graves and Smith (1967)
Pregnenolone	Progesterone, 5α-pregnane-3,20-dione	Oxidation of OH, reduction of C=C	D. lutea	Graves and Smith (1967)
Pregnenolone	Progesterone	Oxidation of OH	D. purpurea, N. tabacum	Graves and Smith (1967)
Pregnenolone	Pregnenolone palmitate, 5α-pregnanolone palmitate reduction of C=C	Palmityl esterification,	N. tabacum, Sophora angustifolia	Furuya et al. (1971)

(Contd.)

Substrates	Main products	Type of reaction	Plant species	Reference
5β-Pregnane-3,20-dione glucoside	Hydoxylating product, glucosylation	Reduction of C=O,	D. purpurea	Hirotani and Furuya (1975)
5βH-pregnane-3β-ol-20-one	5β-Pregnane-3,20-dione isomerization, glucosylation	Oxidation of OH,	Nerium oleander	Paper and Franz (1990)
Digitoxigenin	Digoxigenin Digitoxigenone	Hydroxylation, oxidation of OH	D. lanata	Alfermann et al. (1975)
Digitoxigenin	17βH-digitoxigenin, 3-epi-17βH-periplogenin	Epimerization, hydroxylation	Strophanthus gratus	Furuya et al. (1988)
Digitoxigenin	Periplogenin, 3-epidigitoxigenin glucosylation	Epimerization hydroxylation,	S. amboensis	Kawaguchi et al. (1988)
Digitoxigenin	Periplogenin, 3-eipi-17βH-digitoxigenin 3-epiditoxigenin	Epimerization hydroxylation, glucosylation	S.intermedius	Kawaguchi et al. (1989)
Digitoxigenin	Digitoxigenone,3-epidigitoxigenin, Glucosides	Glucosylation, epimerization, oxidation of OH	N. oleander	Paper and Franz (1990)
Digitoxigenin	Digitoxigenone, 3-epidigitoxigenin, Glucosides oxidation of OH	Glucosylation, epimerization	D. purpurea	Hirotani and Furuya (1980)
Digitoxigenin	3-Epidigitoxigenin-3β-D-glucoside, digitoxigenin-3β-D-glucoside, periplogenin-3β-D-glucoside, digitoxigenin sophoroside, -stearate, -palmitate, miristate	Glucosylation, esterification hydroxylation	Panax ginseng (hairy root)	Kawaguchi et al. (1990)
β-Methyldigitoxin (39)	βMethyldigoxin	Hydroxylation	D. lanata	Heins et al. (1978)
Digitoxin	Digoxin, deacetyllanatoside C	Hydroxylation, Glucosylation	D. lanata	Kreis and Reinhard (1986)
Digitoxin	Purpureaglucoside A	Glucosylation	D. lanata	Diettrich et al. (1987)
Gitoxigenin	5β-Hydroxygitoxigenin	Hydroxylation	Daucus carota	Jones and Veliky (1981)
Oleandrigenin	Gitoxigeninglucoside oleandrigeninglucoside	Glucosylation,	N. oleander	Paper and Franz (1990)

and sitosterol. This suggests the existence of a steric hindrance as a result of the β-carbon side chain in the latter substrates and its absence in testosterone.

Biotransformation of progesterone by plant cell cultures was first reported by Graves and Smith in 1967. Cultures of *Digitalis lanata* transformed the progesterone to 5α - pregnane – 3,20 dione. Stohs (1969) prepared microsomal fractions from undifferentiated suspension cultures of *Dioscroea deltoidea* and suspensions of plantlets of *Cheiranthus cheiri*. The microsomes were incubated with progesterone for 2hr. The reaction product found was 5α- Pregnane 3,20 dione (Fig. 10.11). The microsome fractions contained enzyme capable of only the reduction of the Δ^4 double bond of progesterone. Furuya et al. (1971) obtained after an incubation time of 4 weeks as a transformation product of progesterone 5β-pregnane-3β-01,20-one-palmitate and with *Sophora angustifolia* also

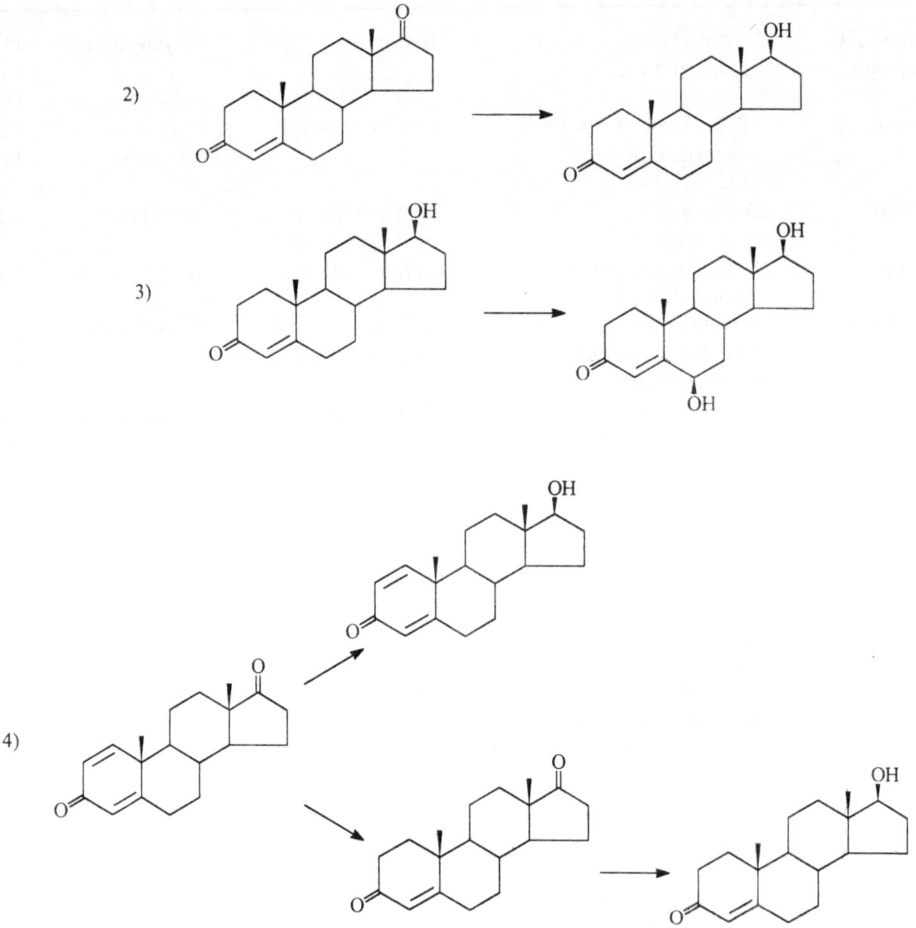

Fig. 10.10. Biotransformation of : 2) 4-androstene-3, 17-dione; 3) testosterone; 4) 1,4-androstadiene-3,17-dione.

5α-pregnane-3β-ol, 20- one. The results show that **progesterone** under goes stereospecific reduction of the Δ^4 double bond and keto group of position 3. Finally **esterification** with plamitic acid takes place.

Stohs and El- olemy (1972) reported that cell culture, of *D. deltoidea* are capable of metabolizing progesterone to 5α-pregnane-α3-ol, 20-one and 5α–pregnane–3β, 20β, -diol. Furuya *et al.* (1973) has examined the transformation of progesterone as a key intermediate in cardenolide biosynthesis by *D. purpurea* cell cultures (Fig. 10.12).

By these cultures progesterone is biotransformed into 5α-Pregnane-3β-o1, 20 one and its glycoside, 5α-pregnane-3β-ol, 20α-diol and its glycoside, 5α-pregnane-3β, 20β-diol and its glycoside, 4-pregnane α-ol-3-one its glucose,4-pregnen-20β-ol-3-one and its glycoside (Fig. 10.13).

Considering all the biotransformations of progesterone, so for studied one can state that plant cell suspension cultures are capable of the reduction of the Δ double bond always resulting in 5α-pregnane derivatives, the reduction of the keto group at position 3 always leading to the 3β-configuration the reduction of the keto group at position 20 leading either the c-20α or the 20β-configuration.

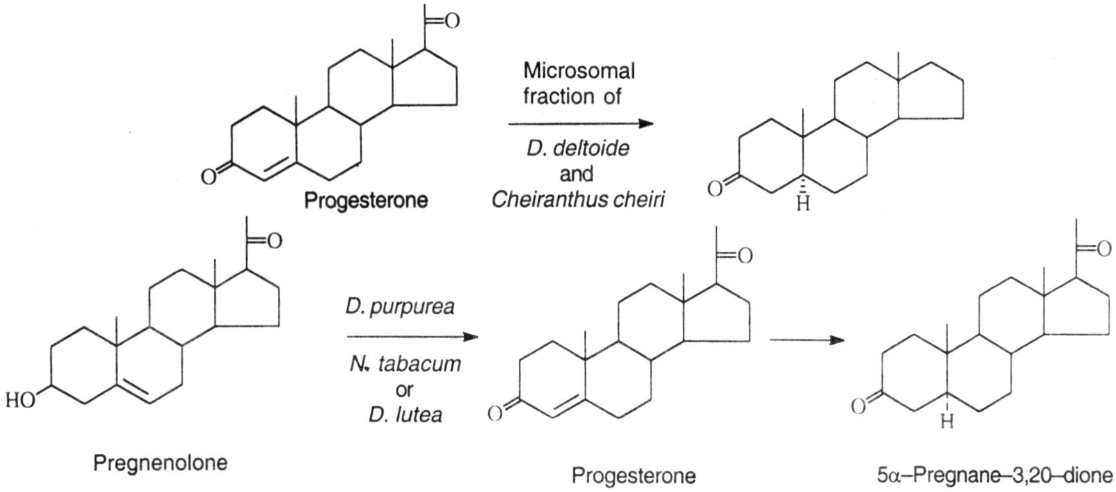

Fig-10.11. Biotransformation of Steroids.

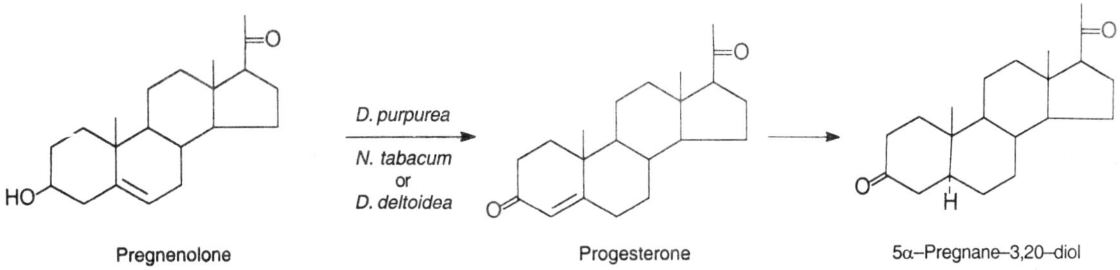

Fig. 10.12. Biotransformation of Pregnenolone.

Graves and Smith (1967) reported that pregnenolone is converted to progesterone by cell cultures of *D. purpurea*, *D. lutea* and *Nicotina tabcum*. This transformation involves oxidation of hydroxyl group at position 3 and the shift of the 5 double bond to the 4 position. Cultures of *D. lutea* then metabolize progesterone further to 5α-pregnane-3,20 dione (Fig. 10.13).

10.5.5. Cardenolides

There are some different opinions about transformation reactions by plant cell cultures involving cardenolides (Fig. 10.14). Here yields of transformation products, using plant cell cultures are high, compared to microbial biotransformation. The plant cells are genetically diverse compared to microbial cells and produce diversified products upon transformation reactions. Interest in the biotransformation of cardiac glycosides developed as a result of the need to obtain compounds with efficient cardiac activity and minimum undesirable side effect. The most of cardenolides are produced by certain plants e.g. *Digitalis purpurea* (Digoxin). The production of cardenolides by plant tissue culture, which are the same as the whole plant, is not reported by anyone. This makes it very important to study the transformation reaction of precursors by plant cell cultures and to pinpoint the differences between the cultures and the differentiated plant. Although cardiac glycosides are not produced by plant cell cultures, they very readily transform them if added, in some instances, to produce from compounds of less activity, compounds of high therapeutic value (Fig 10.15)

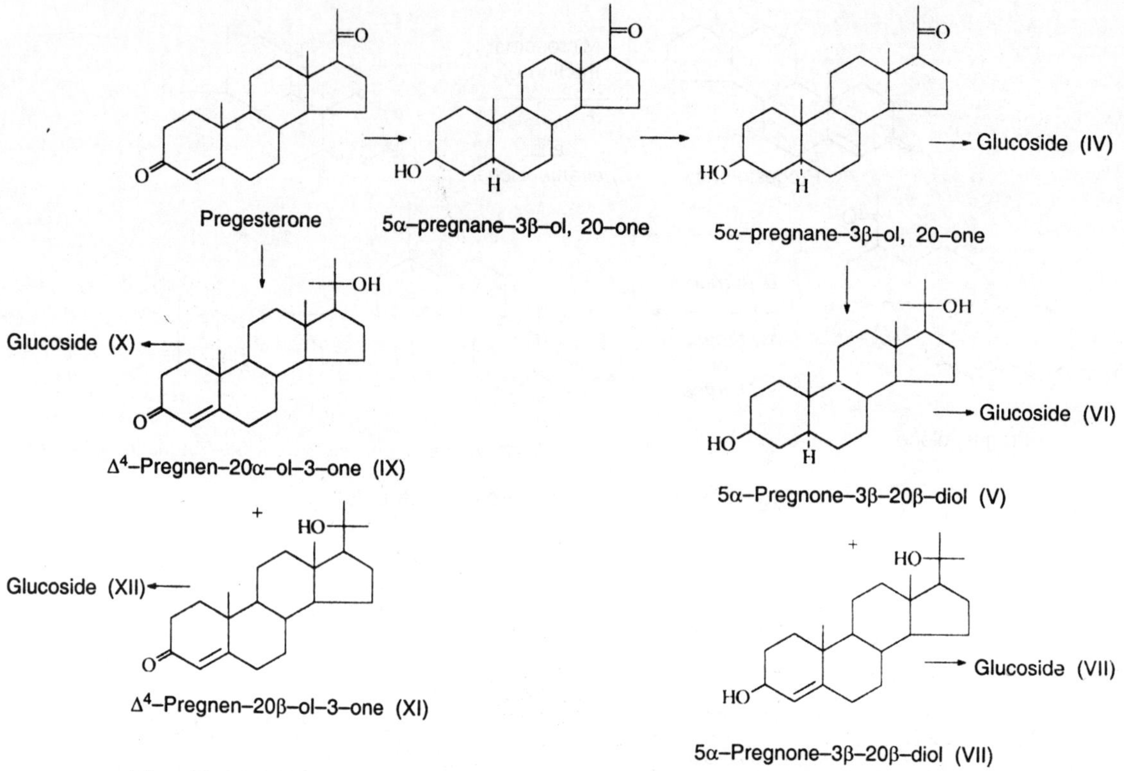

Pregesterone

5α–pregnane–3β–ol, 20–one

5α–pregnane–3β–ol, 20–one

→ Glucoside (IV)

Glucoside (X) ◄—

Δ⁴–Pregnen–20α–ol–3–one (IX)

→ Glucoside (VI)

5α–Pregnone–3β–20β–diol (V)

Glucoside (XII) ◄—

Δ⁴–Pregnen–20β–ol–3–one (XI)

→ Glucoside (VII)

5α–Pregnone–3β–20β–diol (VII)

Fig. 10.13. Biotransformation of progesterone by cell cultures of D. purpurea.

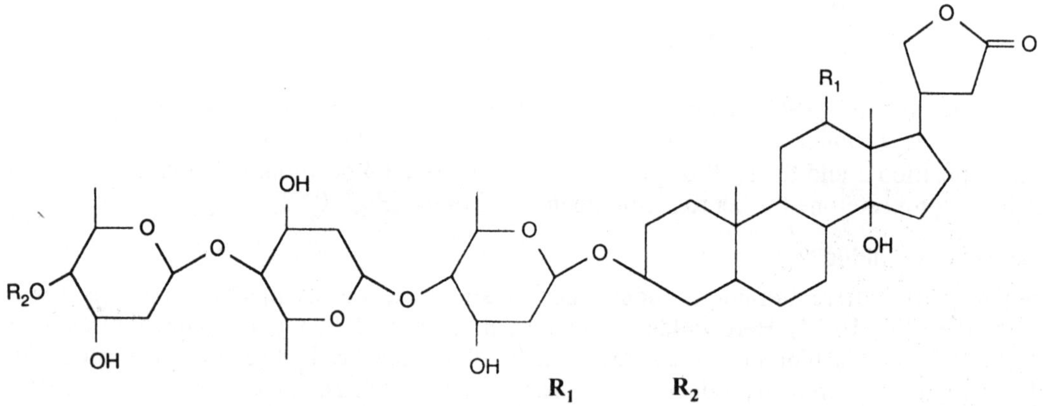

	R_1	R_2
β-methyldigitoxin	H	CH_3
β-methyldigoxin	CH	CH_3
Digitoxin	H	H
Digoxin	OH	H
Deacetyllanatoside C	OH	Glucose
Purpureaglucoside A	H	Glucose

Fig-10.14. Cardiac glycosides.

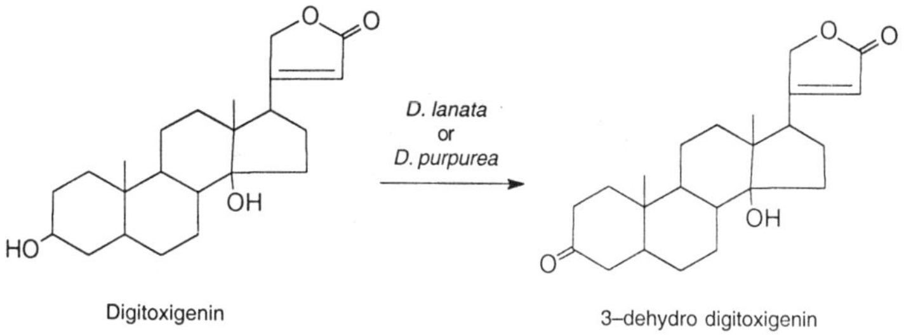

Digitoxigenin   3–dehydro digitoxigenin

Fig. 10.15. Biotransformation of Digitoxigenin.

More than 3 decades ago the first successful transformation involving cardiac glycosides was reported by Stohs and Staba (1964). They supplied culture of *D. lanata and D. purpurea* with digitoxigenin and after seven days isolated a product, which was later identified as 3-dehydro-digitoxigenin (Fig10.15). Biotransformation of digitoxigenin, a precursor of the cardiac glycosides has been studied with six plant species including two *digitalis* species (Alfermann *et al* 1977, Hirotoni and Furuya 1980). Three kinds of *Strophanthus* (Furuya *et al.* 1988, Kawaguchi *et al.* 1989), *Nerium oleander* (Paper and Franz 1990), *Panax ginseng* hairy root cultures (Kreis and Reinhard 1988), *Convallaria majalis* L. (Harkiss *et al.,* 1986). The general biotransformation routes for cardiac glycosides are depicted in Fig. 10.16. A common pathway observed in many species is the oxidation of the alcohol moiety (C-OH) to a carbonyl (C=O) and glycosylation at the C-3 position. In *N. oleander* (Paper and Franz 1990), 3-epidigitoxigenin 3β-D-glucoside and digitoxigenin –3β-D-glucoside were the main end products of the pathway. Also when the C19 steroid 5β-H pregnan-3β-Ol-20-one and oleandrigenin are the substrates. 3β-D-glucosides only were produced in *Nerium* cell cultures. Accordingly, *Nerium* cell cultures can be used as an appropriate tool for the production of labeled 5βH- steroid glucoside. The biotransformation of digitoxigenin by *Strophanthus gratus* is characterized by epimerization (17αH-17βH). *P. ginseng* hairy root cultures transformed digitoxigenin and yielding gentibioside and sophoriside and glycosides as well as stearate, Myristate and palmitate. The hairy root had shown excellent glycosylation reaction for digitoxigenin (Kawaguchi *et al.,* 1990).

Transformation of digitoxigenin to digoxigenin caused by 12-β- hydroxylation occur only in *D. lanata* cell culture. The cell cultures of some *D. lanata* cell lines are also well known to transform β-methyldigoxin from β-methydigitoxin. This is an important reaction of the plant cell mediated biotransformation because digoxin is the cardenolide that is used most extensively due to is pharmacological properties, although both digitoxin and digoxin are effective in the treatment of heart diseases. The digitoxin is used instead of β-methyldigitoxin as a substrate in *Digitalis* cell culture, most of the digitoxin is converted to purpurea glycoside A. Methylation to terminal digitoxose avoided the glucosylation to digitoxose. Kreis and Reinhard (1988, 1986) found that 8%glucose (or sucrose or fructose) solution was very effective for direct conversion of digitoxin to the 12β-hydroxylated products, digoxin and deacetyllanatoside C, which is the main product. In Murashige and Skoog medium (3%sucrose) most of the added digitoxin was glucosylated, leading to purpurea glycoside A, and accounted for 70% of the overall cardenolide content; 13% deacetyllanatoside C was produced and digoxin was detected in low yield. The method of using digitoxin instead of β-methyldigitoxin in the sugar rich solution has the following advantages: (1) digitoxin does not need to be modified chemically prior to biotransformation and the main product deacetyllanatoside C, can easily be deglucosylated to digoxin and (2) the substrate can be added to the culture medium as an aliquot, A two stage procedure in which *D. lanata* cells were grown in modified MS medium for 10 d and then transferred into 8% glucose medium has been developed for digitoxin biotransformation. Digitoxin was added to the cells

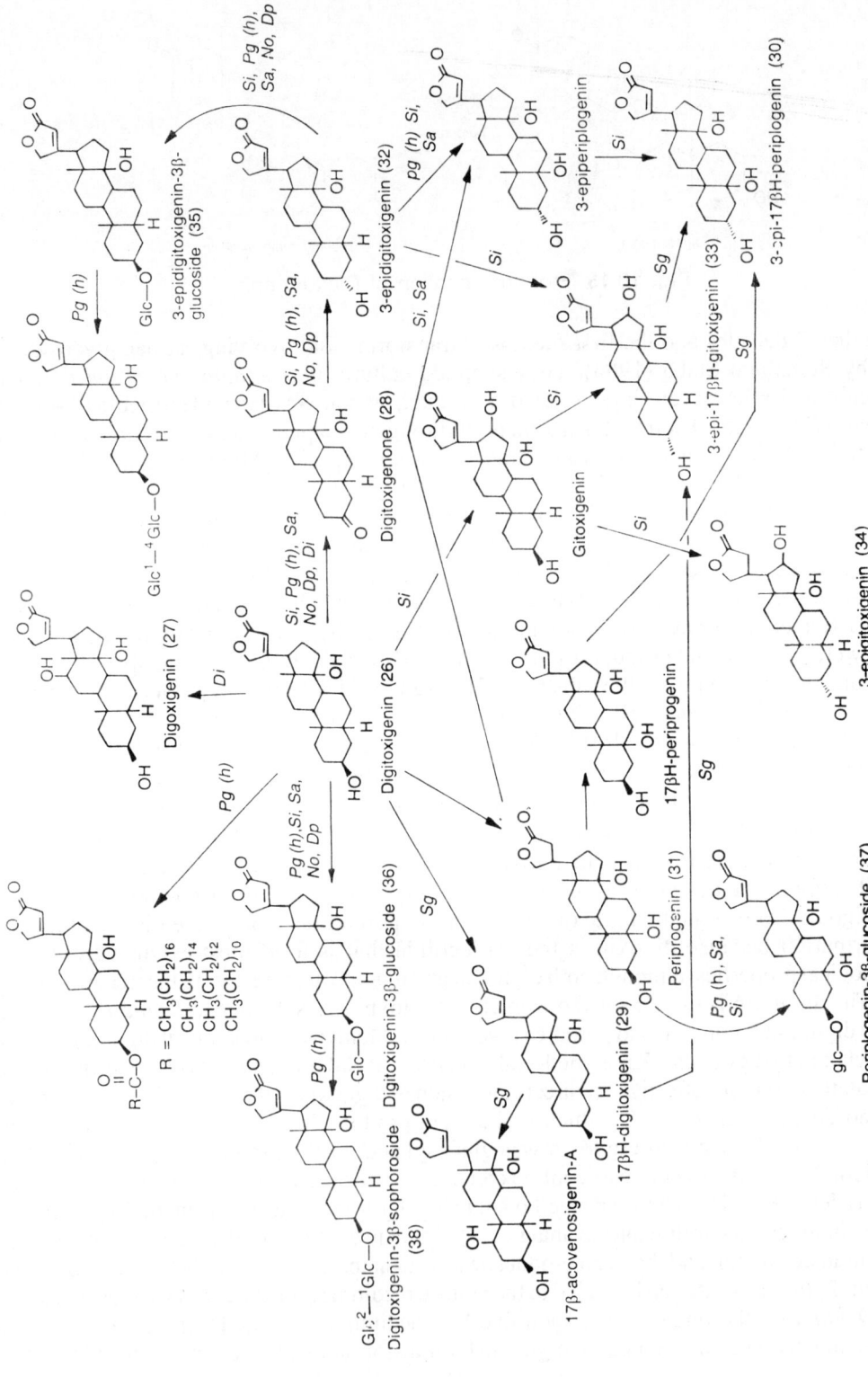

Fig. 10.16. Biotransformation of digitoxigenin by plant cell cultures (Adopted from Dicosmo and Misawa, 1996) Dp. Digitalis purpurea, De. Digitalis lanata, Sa. Strophanthus amboensis, Sg. Strophanthus gratus, St. Strophanthus remedius, No. Nerium oleander; Pg. (ts), Panax ginseng transformed roots.

3 d after transfer of cells to glucose containing medium. The digitoxin was biotransformed within 12d of incubation yielding 12β-hydroxylated products (digoxin and deacetyllanatoside C) and purpurea glycoside at 88 and12% yield respectively within the cells. The maximum yield of digoxin and deacetyllanatoside C was calculated to be about 0.5g/l. This is similar to the yield of β-methyldigoxin (Dicosmo and Misawa, 1996).

12β-Hydroxylase was isolated from suspension cultures of *D. lanata* (Petersen and Seitz, 1985). The enzyme was located in the microsomal fraction and had an optimum pH of 7.5. The Km value for β-methyldigitoxin was 7.1μM. Accetyldigitoxin, digitoxin and digitoxigenin were all hydroxylated. Reduced NAD was essential for the reaction; the optimal concentration was 1mM. The enzyme was suggested to be a cytochrome P-450-dependent monooxygenase because oxygen was critical and the inhibition by carbondioxide was reversible with 450-nm light. Furthermore, microsomes with 12β-hydroxylase were immobilized with alginate cross linked by $CaCl_2$. Seventy percent of the enzyme activity was retained after immobilization. The immobilized enzyme was identical to freely suspended microsomes with respect to the kinetics the optimum pH, and substrate concentration required for the optimum activity. Enzyme activity was observed for more than 20 hr using β-methylydigitoxin as a substrate.

In addition, UDP glucose digitoxin 16-O-glucosyltransferase, which forms purpurea glycoside A or deacetyllanatoside C from digitoxin or digoxin respectively, has been detected in the soluble fraction of *D. lanata* cultured cells. Enzyme activity was never found in either a purified vacuole preparation or lysed vacuoles. Ascorbate (10mM) increased the transferase activity as much as about fourfold. Other sugar nucleotides besides UDP-glucose failed to service as a glucose donor(Kreis et al 1986).

10.5.6. Phenolics

Phenolics are characterized by the strong tendency to be biotransformed to glucosides or its esters (Table 10.6-10.9).

Table 10.6. Phenolics (Adopted from Dicosmo and Misawa, 1996)

Substrates	Main Products	Type of reaction	Plant species	Reference
Phenol	Glucoside	Glucosylation	*Gardenia jasminoides*	Mizukami *et al.* (1987)
Catechol	Glucoside	Glucosylation	*G. jasminoides* *Datura innoxia*	Mizukami *et al.* (1987)
Resorcinol	Glucoside	Glucosylation	*G. jasminoides* *D. innoxia*	Mizukami *et al.* (1987)
Hydroquinone (1)	Arbutin (2)	Glucosylation	*G. jasminoides,* *D. innoxia,* *Catharanthus roseus,* *Rauwolfia serpentina*	Mizukami *et al.* (1987)
Hydroquinone	p-Hydroxyphenyl-O-primeveroside	Glucosylation	*R. serpentina*	Lutterbach *et al.* (1993)
o-Nitrophenol	Glucoside	Glucosylation	*G. jasminoides*	Mizukami *et al.* (1987)
m-Nitrophenol	Glucoside	Glucosylation	*G. jasminoides*	Mizukami *et al.* (1987)
p-Nitrophenol	Glucoside	Glucosylation	*G. jasminoides*	Mizukami *et al.* (1987)
Pentachlorophenol	Glucoside	Glucosylation	*Glycine max*	

(Contd.)

Substrates	Main Products	Type of reaction	Plant species	Reference
Benzoic acid	Glucosyl ester	Glucosylation	*Glycyrrhiza echinata*	Ushiyama *et al.* (1989)
Benzoic acid	Glucosyl ester, gentiobiolyl ester	Glucosylation	*Aconitum japonicum* *Coffea arabica*, *Dioscoreophyllum cumminsii*, *Nicotiana tabacum*	Ushiyama *et al.* (1989)
Benzylacetate	Phenol	Hydrolysis	*Spirodela oligorrhiza*	Pawtowicz *et al.* (1987)
Salicyl alcohol (3)	Salicin (4), isosalicin (5)	Glucosylation	*D. innoxia*, *Perilla frutescens*, *G. jasminoides*	Yokoyama (1991)
Salicylic acid (6)	Salicylic acid-O-glucoside	Glucosylation	*Mallotus japonicus*	Umetani *et al.* (1990)˙
Methyl salycilate	Glucoside, gentiobioside	Glucosylation	*Panax ginseng* (roots)	Ushiyama *et al.* (1989)
o-Hydroxybenzoic acid	Glucosyl conjugates	Glucosylation	*Perilla frutescens*, *C. roseus*, *M. japanicus*	Tabata *et al.* (1989)
m-Hydroxybenzoic acid	Glucosyl conjugates	Glucosylation	*D. innoxia*, *P. frutescens* *C. roseus*, *Lithospermum erythrorhizom*, *Buplerum falcatum*, *G. jasminoides*	Tabata *et al.* (1989)
2-Phenylpropionic acid(7)	3-(4-Hydroxyphenyl) propionic acid	Hydroxylation	*Dioscoreophyllum cumminsii*	Ushiyama *et al.* (1989)
2-(4-Hydroxyphenyl propionic acid	Glucosyl ester, glucosides	Glucosylation	*P. ginseng* (roots)	Ushiyama *et al.* (1989)
2-(4-Benzoylphenyl propionic acid	Glucosyl ester, glucosides	Glucosylation	*P. ginseng* (roots)	Ushiyama *et al.* (1989)
2-[2-(6-Methoxy-naphthyl] propionic acid	Glucosyl ester, glucosides	Glucosylation	*P. ginseng* (roots)	Ushiyama *et al.* (1989)
Tropic acid (8)	Glucosyl ester, glucosides	Glucosylation	*P. ginseng* (roots)	Ushiyama *et al.* (1989)
Tropic acid	Glucose ester	Glucosylation	*D. innoxia*, *N. tabacum*	Ushiyama *et al.* (1989)
Tropic acid	Glucose ester, glucoside	Glucosylation	*Eucalyptus perriniana*	Ushiyama *et al.* (1989)
Tropic acid	Glucose ester, sucrose ester Isotrehalose ester, glucoside	Glycosylation	*Coffea arabica*	Ushiyama *et al.* (1989)
Coniferyl alcohol (9)	Glucoside	Glucosylation	*P. ginseng* (roots)	
2-(p-Methoxy-phenyl) ethylamine	2-(p-Hydroxyphenyl) ethylamine	Demethylation	*C. roseus*	Hiraoka and Carew (1981)
2-(p-Methoxy-phenyl) ethylamine	2-(p-Methoxyphenyl) ethyl	Glucosylation	*Strobilanthes dyerianus*	Hiraoka and Carew (1981)
(4-Chloro-2-methyl phenoxy)-acetic acid	Hydroxylating product	Hydroxylation	*Phaseolus vulgaris*	Cole and Cough-mann (1984)
p-Hydroxybenzoic acid	Glucosyl conjugates	Glucosylation	*D. innoxia*, *P. frutescens* *B. falcatum*, *G. jasminoides* *M. japonicus*	Tabata *et al.* (1988)

(Contd.)

Substrates	Main Products	Type of reaction	Plant species	Reference
3,5-Dimethoxy-phenol gentiobioside	Glucoside, primeveroside,	Glucosylation	*P. ginseng* (roots)	Ushiyama *et al.* (1989)
1-Phenyl ethanol acetate	1-Phenyl ethanol	Hydrolysis	*Spirodela oligorrhiza*	Pawtowicz and Siewinski (1987)
Phenylacetic acid	Sucrose-bound products	Glycosylation	*C. arabica*	Furuya *et al.* (1988)
Phenylacetic acid	6-O-phenylacetyl-D-glucose, ethyl 6-O-phenylacetyl-α- D-glucopyranoside	Glucosylation	*A. japonicum*, *Glycyrrhiza echinata*	Ushiyama *et al.* (1989)
Phenylacetic acid	Phenethyl-α-D-glucopyranoside	Glucosylation	*D. cumminsii*	Ushiyama *et al.* (1989)
2-Phenylpropionic acid	Sucrose-bound products	Glycosylation	*C. arabica*	Furuya *et al.* (1988)
2-Phenylpropionic acid	Glucoside, xylopyranosyl, Glucose, myoinositol ester	Glycosylation	*P. ginseng* (roots)	Furuya *et al.* (1989)
2-Phenylpropionic acid	Glucose ester	Glucosylation	*A. japonicum*, *N. tabacum, D. cumminsii*	Furuya *et al.* (1987)
3-Phenylpropionic acid	Gentiobiosyl ester	Glycosylation	*A. japonicum*, *C. arabica*	Ushiyama *et al.* (1989)
Phenoxyacetic acid glucoside	4-Hydroxylating product, glucosylation	Hydroxylation,	*P. vulgaris*	Cole and Coughman (1984)
4-(p-Hydroxy-phenyl)-2-butanone	Rhododendrol, glucoside	Reduction of C=C, glucosylation	*Acer nikoense*	Fujita *et al.* (1991)
Vanillin (10)	Vanillin glucoside	Glucosylation	*C. arabica*	Kometani *et al.* (1993)
Vanillin	Vanillin glucoside	Glucosylation	*D. innoxia*	
Capsaicin	Capsaicin glucoside (11)	Glucosylation	*P. ginseng* (roots)	Kometani *et al.* (1993)
p-Hydroxyaceto-phenone	Glucoside	Glucosylation	*Daucus carota*,	Ushiyama *et al.* (1989)
Aromatic ketones (acetophenone, etc.)	Aromatic alcohols	Reduction of C=O,	*Daucus carota*, *N. tabacum, G. jasminoides*	Naoshima and Akakabe (1991)
2-(4-Methoxy-benzyl)-1-cyclo-hexanone (12)	Alcohol product, glucoside	Reduction of C=O, glucosylation	*Dioscorea deltoidea*	
2-(4-Methoxy-benzyl)-1-cyclo-hexanone demethylation	Alcohol product, glucoside, demethylating product	Reduction of C=O, glucosylation,	*Solanum aviculare*	Wimmer *et al.* (1987)
Cannabidiol, tetra-hydro cannabinol	Cannabielsoin, cannabicoumaronon	Hydrolysis	*Cannabis sativa*	Braemor and Paris (1987)
Demethoxyenceaclin	14-Hydroxyldeme-thoxyencecalin	Hydroxylation	*Ageratina adenophora*	Proksch *et al.* (1987)
Demethylencecallin	14-Hydroxyldemethyl-encecalin, Podo-phyllotoxin (13)	Hydroxylation Glucoside Glucosylation	*A. adenophora*	Proksch *et al.* (1987)

Table 10.7. Phenolics - Coumarins (Adopted from Dicosmo and Misawa, 1996)

Substrates	Products	Type of reaction	Plant species	Reference
Umbelliferone (1)	Skimmin (umbelliferone-7-glucoside)	Glucosylation	*Datura innoxia* *Perilla frutescens* *Catharanthus roseus* *Lithospermum erythrohizon,* *Bupleurum falcatum,* *Gardenia jasminoides*	Tabata *et al.* (1988)
Esculetin (2)	Esculin (3) chichoriin (4)	Glucosylation	*D.innoxia, P. frutescens* *C. roseus, L. erythrorhizon* *B. falcatum, G. jasminoides*	Tabata *et al.* (1988)
Esculetin	Esculin	Glucosylation	*L. erythrorhizon*	Tabata *et al.* (1984)
Daphnetin	O-glucoside	Glucosylation	*D. innoxia, P. frutescens* *C. roseus, L. erythrorhizon,* *B. falcatum, G. jasminoides*	Tabata *et al. (1988)*
Fraxetin	O-glucoside	Glucosylation	*D. innoxia, P. frutescens* *B. falcatum, G. jasminoides* *Mallotus japanicus*	Tabata *et al. (1988)*

Table 10.8. Phenolics - Flavonoids (Adopted from Dicosmo and Misawa, 1996)

Substrates	Products	Type of reaction	Plant species	Reference
Quercetin	3-O-glucoside, -digluco-side, Isorhamnetin 3-O-glucoside, -diglucoside	Glucosylation methylation	*Cannabis sativa*	Braemer *et al.* (1987)
Naringenin	Glucoside	Glucosylation	*Perilla frutescens* *Bulpleurum falcatum*	Tabata *et al.* (1988)
Naringenin	Naringenin 7-O-gluco-sides, -gentiobioside	Glucosylation	*Swertia jasminica*	Miura *et al.* (1986a)
Naringenin	Naringenin 7-O-gluco-sides,-gentiobioside	Glucosylation	*Duboisia myoporoides*	Miura *et al.* (1986b)
Naringenin	Naringenin 7-O-gluco-side (prunin) (66)	Glucosylation	*Citrus paradisi*	Lewinsohan *et al.* (1989)
Naringenin	Naringenin 7-O-gluco-sides, Narirutin	Glucosylation	*Citrus aurantium*	Lewinsohn *et al.* (1989)
Liquiritigenin	Glucoside	Glucosylation	*Datura innoxia,* *P. frutescens,* *Catharanthus roseus* *Lithospermum erythro-rhizon, B. falcatum,* *Gardenia jasminoides*	Tabata *et al.* (1988)

Table 10.9. Phenolics - Anthraquinones (Adopted from Dicosmo and Misawa, 1996)

Substrates	Products	Type of reaction	Plant species	Reference
Alizarin	Glucoside	Glucosylation	*Perilla frutescens* *Catharanthus roseus,* *Lithospermum erythrorhizon,* *Gardenia jasminoides*	Tabata *et al.* (1988)
Rhein	Glucoside	Glucosylation	*P. frutescens*	Tabata *et al.* (1988)
Emodin	Glucoside	Glucosylation	*P. frutescens*	Tabata *et al.* (1988)

Arbutin is one of the main active constituents in *Arctostaphylos uva-ursi* and is used as an urethral disinfectant. It is also proved to be a potent suppressor of the synthesis of melanin in human skin (Akiu *et al.*, 1988) without side effects (Itabashi et al 1988). The Shiseido, a cosmetic company (Japan) has developed this material as a skin depigmentation agent, and is also produced by a chemical process.

Tabata et al (1976) described that the *Datura innoxia* cell cultures were able to transform arbutin from hydroquinone. Yokayama and Yanagi (1992) developed a method of high level production of arbutin by employing cell cultures of *C. roseus*. The final yield in the biotransformation was more than 9 gm/l and was stable even when scaled up in a 20-l- bioreactor (Inomata *et al.*, 1991). Lutterbach *et al.* (1993) reported a yield of 18 g/l of arbutin using *Rauwolfia serpentina* cell cultures along with a side product of p-hydroxy phenyl-o-β-primeveroside (Lutterbach and Stockgit 1993).

2 phenyl propionic acid are converted to glucose or gentiobiose esters by *A. japonicum*, *Dioscoreophyllum cumminsii* or *Nicotiana tabacum* (Furuya *et al.*, 1987), cell cultures; however, *Coffea arabica* produced sucrose ester as the main compound (Furuya *et al.*, 1988).

Vanillin is the major component of vanilla flavor, which is extracted from the cured pods of the flowers of the vanilla used widely as the sweet flavor of a very wide range of foods, such as ice creams or confectioneries. Vanillin also has antimutagenic and antimicrobial activities (Kometani *et al.*, 1993). Vanillin was glycosylated by the cell cultures of *Medicago sativa*, *N. tabacum*, *C. arabica*, *Gardenia jasminoides*, *Theobroma* cacao and *Prunus amygdalus*. 80% of the vanillin added exogenously in (1 mm) was glucosylated within one day. Interestingly, the glucosyl vanillin had unchanged activities as an antimutagenic and retained antimicrobial activity. Glucosyl vanillin may be of great value as an edible compound with biological activities.

Capsaicins, the pungent, hot tasting component of the *Capsicum annuum* or *C. frutescens*. Capsaicin is reported to have carminative, counter-irritant properties and very widely used in food industry. The cell cultures of *C. arabica* produced capsaicin-β-D-glucopyronoside from capsaicin (Kometani *et al* 1993). Glucosylation of capsaicin decreased the pungent taste by more than 100 folds; the phenolic OH group is responsible for the pungent property. This glucoside can be used as a unique food component with biological activity.

2-(4-Methoxybenzyl)-1-cyclohexanone, an analogue of insect juvenile hormone, and used as an insect pest control agent. The cell cultures of *Solanum aviculare* converted this hormone into several products in which cis and trans alcohols were the main products formed through reduction of the carbonyl functionality (Wimmer *et al.*, 1987). However, the same product was converted into its glucoside through glucosylation of the alcohol moiety by *Dioscorea deltoidea* cell cultures (Vanek *et al* 1984). Plant tissue culture also biotransform synthetic compound. Bourgogne *et al.* (1989) reported that the synthetic 1,5 diphenyl sulfinyl-3-methyl–3-nitropentane that is devoid of toxicity to animal cells was fed to the cell cultures of *C. roseus* and biotransformed into 1-phenyl sulfonyl-5-phenyl-sulfinyl-3-methyl-3-nitorpentane through a regioselective oxidizing process. One of the earliest biotransformations studied involving phenols was undertaken with callus cultures of *Argostemma githago*, *Datura ferox*, and *Digitalis purpurea*.

Compounds with different hydroxylation patterns were supplied in the nutrient medium and the products were analyzed at the end of the growth cycle. Callus cultures of all three species were able to transform the exogenously supplied substrates with *D. purpurea* showing less potential than other two species. The substrates were catechol, resorcinol, vanillin, salicyl alcohol, and salicyl aldehyde, and the biotransformation reaction involved was glycosylation (Fig 10.17). Both salicyl alcohol and salicyl aldehyde were converted to isosalicin, with only trace of the expected product from the glycosylation of salicyl aldehyde, namely helicin, detected. These early results were supported by later work with cell suspension cultures of *Datura innoxia*. Although both resorcinol (*m-*

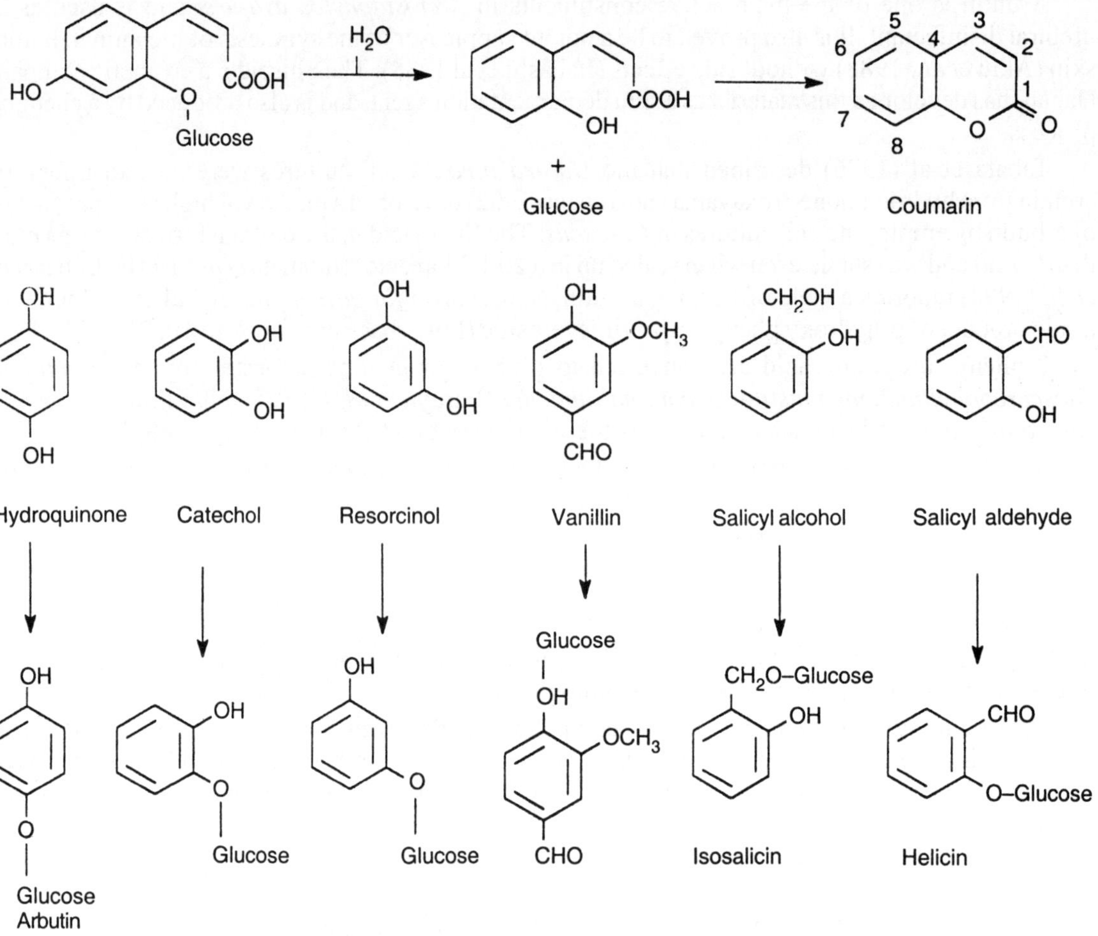

Fig. 10.17. Glycosylation of Phenols.

dihydroxybenzene and catechol (o-dihdroxybenzene) yielded their respective monoglucosides, the rate of hydroquinone biotransformation was quite remarkable. Within 10 days after the addition of hydroquinone to the cell cultures, it was totally converted to its glycoside, arbutin. The same cell culture converted both salicyl alcohol and salicyl aldehyde to isosalicin rather than to their respective glycoside salicin (Fig 10.18) and helicin. The story of the glucosylation pattern of salicyl alcohol is an illustrative example of how plant cell suspension cultures in general and biotransformation studies in

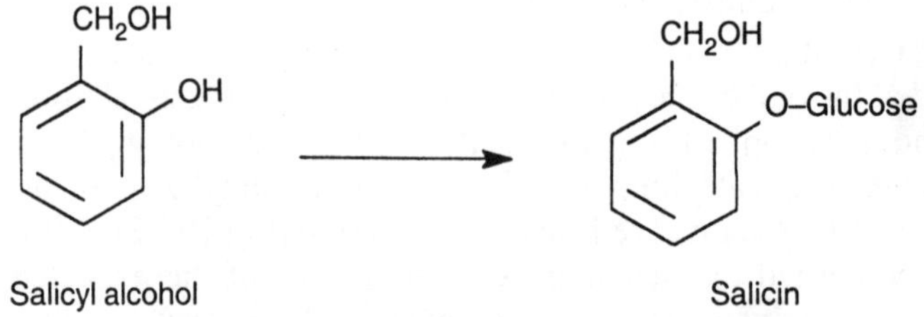

Fig. 10.18. Glycosylation of Salicyl alcohol.

particular, may contribute to the elucidation of biosynthetic pathways of secondary metabolites in plants.

The glucoside salicin (o-hydroxymethyl phenyl β-D-glucoside) was the first glycoside isolated from natural sources. Used externally, it is an analgesic for rheumatic pain. As such salicin was the precursor of aspirin (acetylsalicylic acid) with its analgesic and antipyretic properties. Salicin occurs in certain plants of the Rosaceae family, but salicyl alcohol is restricted to willows (Salix *sp.*). *In vivo* and *in vitro* studies with various plant materials (cultured cells and plant organs) in the 1960s and 1970s had shown that the glycoside salicin is derived from o-coumaric acid or benzoic acid via the intermediate salicyl aldehyde and its glucoside, helicin (Fig. 10.19). The same experiments also suggested that the glucosylation of salicyl alcohol involves the alcoholic rather than the phenolic group and consequently salicyl alcohol could not be a direct precursor of salicin. This pathway of salicin biosynthesis was widely accepted by the early 1980s. However, recently the question has come under review with salicyl alcohol reinstated as a precursor for salicin. A Japanese group led by H. Mizukami reported in 1983 that cell suspension cultures of *Lithospermum erythrorhizon* and *Gardenia jasminoides,* in addition to isosalicin production, were also able to glycosylate the phenolic group of salicyl alcohol to produce salicin. Later the same group isolated and partially characterized an enzyme catalyzing this glucosylation reaction in cultured *G. jasminoides* cells (Mizukami *et al* 1983). The enzyme, termed UDP-glucose; salicyl alcohol phenyl-glucosyltransferase, showed a clear specificity for the phenolic position of salicyl alcohol. The highest activity in the cells was during the exponential phase of growth when about 70% of the added substrate were glycosylated within four days. The growth stage of cells also appeared to influence the ratio of salicin/isosalicin formation, with the ratio decreasing as the cells reached the stationary phase. Furthermore, the 2,4-D and NAA suppressed salicin formation and increased concentrations of medium sucrose enhanced it. As an example the investigation of glycosylation of salicyl alcohol has illustrated the usefulness of biotransformation studies with cultured plant cells in elucidating the finer points of plant secondary metabolism pathways and their regulatory mechanisms.

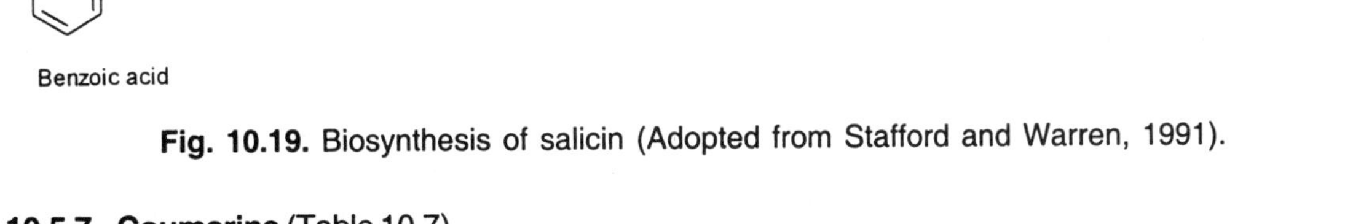

Fig. 10.19. Biosynthesis of salicin (Adopted from Stafford and Warren, 1991).

10.5.7. **Coumarins** (Table 10.7)

Of the several naturally occurring coumarins, the greatest majority is oxygenated at position 7. The simplest representative of this group is umbelliferone, which is considered to be the precursor of most coumarins p-hydroxylated at position 7. Cell cultures of *C. roseus* and *Conium maculatum* hydroxylated coumarin at position 7 to yield umbelliferone. The latter, when administered to cell cultures of *Ruta graveolens* is biotransformed into demethyl suberosin, which is considered to be the precursor of furanocoumarins. Warfarin, 3-(α-acetonyl benzyl)-4-hydroxy has been used for decades as an oral anticoagulant and rodenticide. Incubation of warfarin with *C. roseus* cell cultures reduces the carbonyl group of warfarin to the corresponding alcohol and regio selectively hydroxylated the 6 and 10 position of warfarin (Hamada and Furuya 1996) (Fig. 10.20).

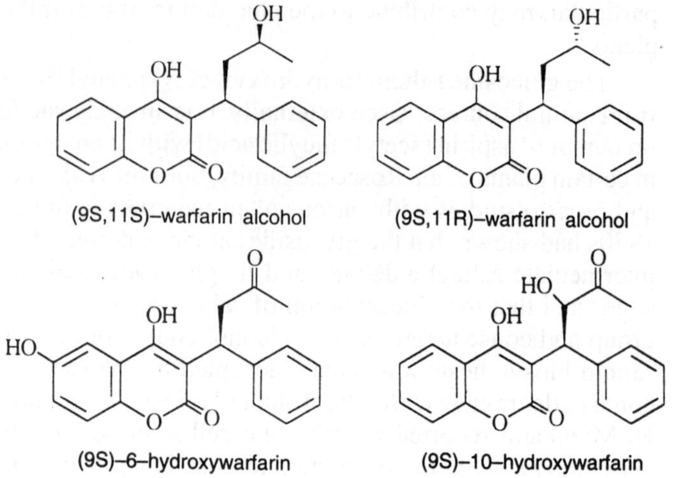

Tabata *et al.* (1988) studied the biotransformation of 4 kinds of hydroxyl coumarin derivatives i.e. umbelliferone, esculetin, daphnetin and fraxetin, by employing cell cultures of eight plant species. In general, umbelliferone was most efficiently converted into its glycoside and the conversion rate to glucosylation

(9S,11S)–warfarin alcohol   (9S,11R)–warfarin alcohol

(9S)–6–hydroxywarfarin   (9S)–10–hydroxywarfarin

Fig. 10.20. Biotransformation of Warfarin.

decreased with an increasing number of substituted groups on the chemical structure e.g. in the order of Umbelliferone, esuletin daphentin and fraxetin. *Perilla frutescens, L. erythro-rhizon, D. innoxia, C. roseus* had the best glucosylation ability two kinds of glycosides esculin and chichorin, *by D. innoxia* and *C.roseus* when *L. erythrorhizon* was used the product of the biotransformation was esculin (Fig. 10.21).

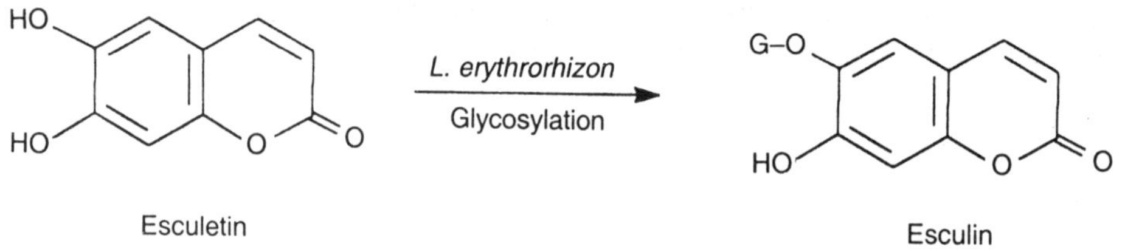

L. erythrorhizon
Glycosylation

Esculetin   Esculin

Fig. 10.21. Glycosylation of Esculetin - Esculin.

10.5.8. Flavonoids

Nearly all biotransformations with flavonoids as substrates result in one-step glycosylations. As such they have provided valuable information on the influence of the pattern of substitution in the flavonoid skeleton on the specificity of glycosyltransferase enzymes. For example when the three flavonoid liquiritigenin (7,4-dihydroxyflavanone), naringenin (5,7,4-trihydroxyflavanone) and baicalein (5,6,7-trihydro oxyflavone) (Fig. 10.22) with differing patterns and degrees of substitution were supplied to cell cultures of *Datura innoxia, Perilla frutescens, Catharanthus roseus,* and *Lithospermum erythrorhizon* on the highest yield of glycosylation was observed with the least substituted compound, namely liguiritigenin, Naringenin which has an additional hydroxyl group in position 5 was glycoslated only in *P. fructescens* and *Bupleurum falcatum* culture with substantially reduced yields. As for baicalein with its three hydroxy substitutions on adjoining positions no biotransformation was observed, most probably due to steric hindrance. The 2,3-dehydrogenation of naringenin gives rise to the corresponding flavone, apigenin. When administered to suspension cultures of *Cannabis sativa* the

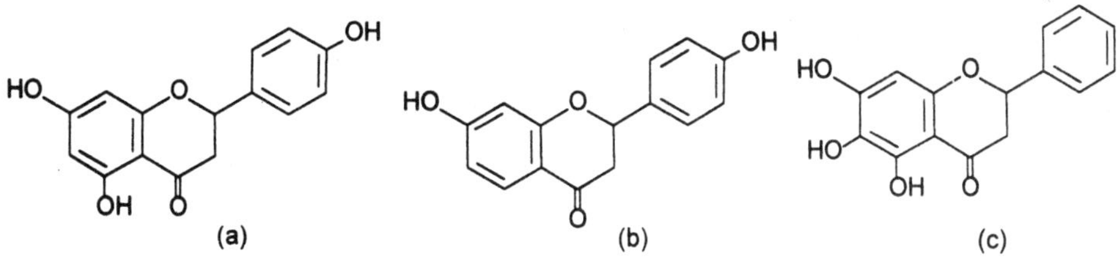

Fig. 10.22. Three flavonoids : (a) naringenin, (b) liquiritigenin, (c) baicalein.

latter is O-glycosylated at position 7 and C-glucosylated at position 8 to yield vitexin (Fig. 10.23). The O-glucosylation of flavones with free hydroxyl groups is very common in nature. The C-glycosides are much rare and are mainly apigenin derivatives. The exact mechanism of C-glycosylation in the intact plant is not fully understood, although experiments with labeled intermediates have suggested that it may occur prior to ring closure in flavones (Stafford and Warren, 1991). The bioconversion of apigenin with C-glycosylation was unknown until the report of Braemer *et al.* (1987) with cell suspensions of *C. sativa*. These cultures were also able to O-glycosylate the flavone quercetin in position 3 (Fig. 10.24). However they also exhibited a novel biosynthetic capacity for methylation of the hydroxyl group on C-3, as no O-methylation from whole plants has so far been reported this was yet another example of possibilities available with plant cell biotransformations.

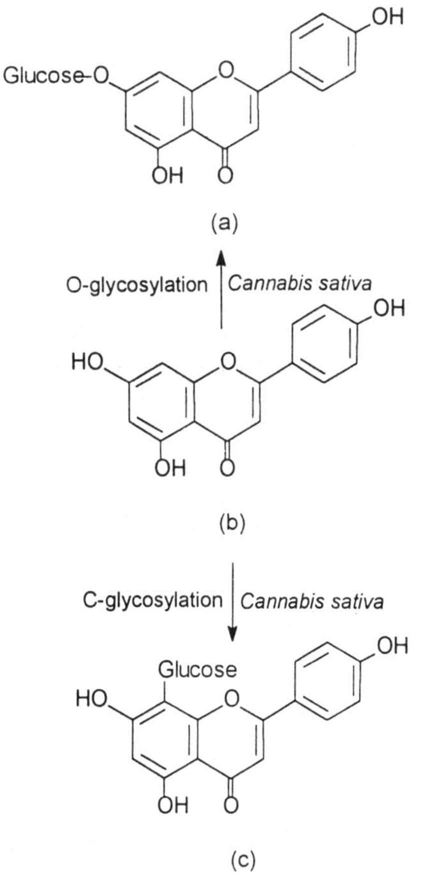

10.5.9. Anthraquinones (Table 10.9)

Although the biosynthesis of another class of coloured substances, namely anthraquinones (Fig. 10.25) has been studied in several plant species in culture, reports of their biotransformation are scarce. Of the naturally occurring quinones, anthraquinone pigments are the most numerous and most widely distributed in plants. Several anthraquinones have been used as natural dyes as well as drugs because of their cathartic properties. Most anthraquinones occur as glycosides in higher plants. Glycosylation is also the form of biotransformation of these pigments by plant cell cultures. Cells of *Perilla frutescens* glycosylate alizarin (a natural dye from the root of the madder, *Rubia tinctorum*), rhein

Fig. 10.23. (a) Apigenin-7-O-glucoside, (b) apigenin, (c) apigenin-8-C-glucoside (vitexin).

and emodin (a pigment with strong cathartic properties) (Fig.10.26) with decreasing efficiency in that order, probably because of increased substitution in the rings A and C. Rhein is also glycosylated, probably in position 8, by cell cultures of *Catharanthus roseus, Lithospermum erythrorhizon*, and *Gardenia jasminoides* with conversion rates ranging from 2.2 to 11.6% (Stafford & Warren, 1991).

10.5.10. Alkaloids (Table 10.10)

Among the great variety of secondary metabolites produced by the higher plants, alkaloids occupy a important place. Nearly 6000 different alkaloids are now known. Interesting biotransformation reactions

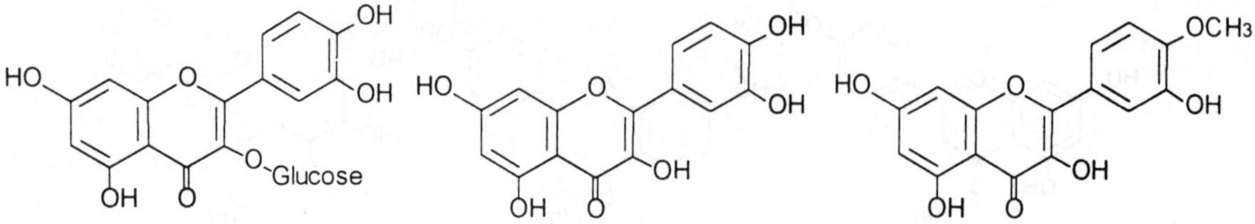

Fig. 10.24. (a) Quercitin-3-O-glucoside, (b) quercitin, (c) 2'-O-methylquercitin.

was done in benzylisoquinoline alkaloid papaverine an isoquinoline alkaloids, used as a smooth muscle relaxant and a cerebral vasodilator, was biotransformed mainly to 6 to 4 monodemethyl papaverine by *Silene alba* cell cultures. *Silene alba* does not produce any isoquinoline alkaloids. The reaction occurred was regioselective demethylation process (Verdeil *et al.,* 1986). This papaverine biotransformation is similar to that in animals, both 4 and 6-monodemethyl are major compounds in some animals including man, and further more 4 and 6 monodemethyl derivatives are noted

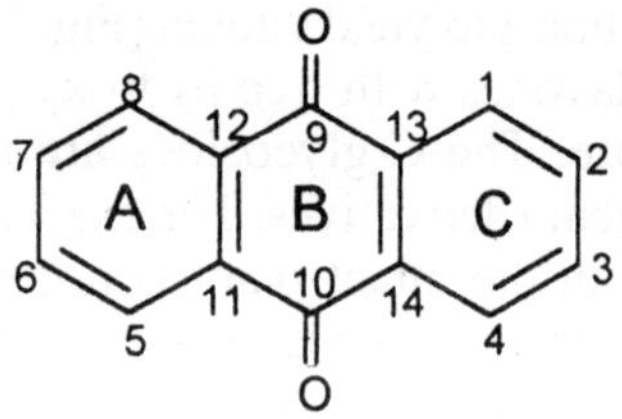

Fig. 10.25. Anthraquinone skeleton.

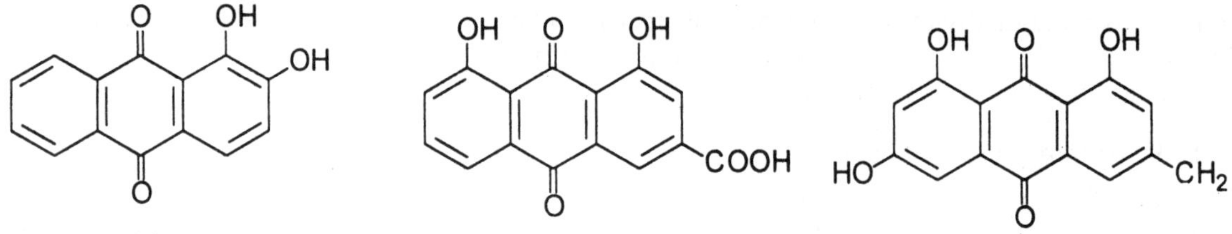

also in microorganisms. This reaction by plants, microorganisms and animals is a rare example of parallel biotransformation.

Fig. 10.26. (a) Alizarin, (b) Rhein, (c) emodin.

Glycyrrhiza glabra also converted papaverine to papaverinol (31.5%) papaveraldine (4%) and related products. However the papaverine was not biotransformed by the cell cultures of *Saponaria officinalis* (Dorisse *et al.,* 1988). *Papaver somniferum* L. is well known to contain morphinan type of alkaloids and the callus or cell cultures lack the ability to produce these alkaloids. Furuya *et al.* (1978) reported the pathway including (+) reticuline to (-)–codeinone, a precursor of morphine alkaloids. However (+)–(S)–reticuline is converted to (-)–(S)–scoulerine. Thebaine was converted to neopine by the cell cultures of *P. somniferum* (Tam *et al.,* 1982) but not by other strains of same species. Other morphine's, codeine, morphine or neopine were not metabolized (Fig. 10.27). The antitumour bisindole alkaloids, vinblastine and vincristine are important antitumour compounds. These alkaloids are obtained from aerial parts of *C. roseus*. However the yields are very low and no commercially viable chemical synthesis is reported. The cell cultures of *C. roseus* have failed to yield these complex alkaloids. However *C. roseus* cell culture has been shown to harbor the enzyme system that converts 3,4- anhydrovinblastine to leurosine, vinblastine, Catharine, vinamidine and hydroxy vinamidine, although 3,4, - anhydrovinblastine was not a direct precursor of vinblastine in the

Table 10.10. Alkaloids (Adopted from Dicosmo and Misawa, 1996)

Substrates	Main products	Type of reaction	Plant species	Reference
Tetrahydroberberine	Berberine	Dehydrogenesis	*Coptis japonica*	Yamada *et al.* (1985)
Papaverine (1)	6-Monodemethyl papaverine, 4'-monodemethyl papaverine	Demethylation	*Silene alba*	Verdeil et al (1986)
Papaverine	Papaverinol, papaveraldin, Demethylpapaverine	Oxidation, demethylation	*Glycyrrhiza glabra*	Dorisse et al (1988)
Isopapaverine	Isopapaveraldine, Isopapaverine N-oxide	Oxidation, demethylation	*S. alba*	Christinaki et al (1987)
(R, S)-reticuline	(-)-(s)-Scoulerine, (-)-(s)-cheilanthifoline	Demethylation	*Papaver somniferum*	Furuya et al (1978)
Codeinone (2)	Codeine (3)	Reduction	*P. somniferum*	Furuya et al (1978)
Thebaine	Neopine	Demethylation	*P. somniferum*	Tam et al (1982)
Salsolinol	O-methylated salslinol, N-methylated salsolinol	Methylation	*Corydalis pallida C. incisa*	Iwasa *et al.* (1992)
Anhydrovinblastine	Vinblastine	Hydroxylation Of C=C	*Catharanthus roseus*	Kutney *et al.* (1981, 1982, 1988)
Vinblastine	Vincristine	Oxidation to C=O	*C. roseus*	Hamada *et al.* (1991)
Tropane derivatives	Acetyltropane derivatives	Acetylation	*Datura innoxia*	Hiraoka & Tabata (1983)
Nicotine	Nomicotine	Demethylation	*Nicotiana tabacum*	Hobbs & Yeoman (1991)
Nicotine (shooty teratoma)	Nomicotine	Demethylation	*N. tabacum*	Saito *et al.* (1989)
Nornicotine, hydrastine	Hydrocotarnine Hydrohydrastine	Methylation	*Corydalis ochotensis*	Iwasa *et al.* (1987)

cell; the iminium ion was proven to be a real intermediate of end products such as vinblalstine (Fig 10.28) feeded 3,4, anhydrovinblastine is thought to be converted to the iminiumion. Hamada *et al.* (1991b) shown that exogenously added vinblastine could be biotransformed to vincristine in cell cultures of *C. roseus*. The same authors also reported that feeding of vindoline to the buffered cell cultures of *C. roseus* could transform into Deacetyl vindoline. Medicinally active members of tropane alkaloids include (-) hyoscyamine and hyoscine or scopolamine (Fig 10.29). The drug atropine is a racemic mixture of (+) and (-) –hyoscyamine and is used to dilate the pupil of the eye in ophthalmic practice. In the whole plant hyoscyamine is converted to hyoscine through the possible intermediates of 6,7-dehydrohyoscyamine and 6β-hydroxy hyoscyamine. The epoxide ring of hyoscine is formed when the β-hydrogen at positions C-6 and C-7 of the tropine ring are eliminated. Both hyoscyamine and hyoscine have been reported from cultured cells of the solanaceae species albeit at significantly reduced yields than in the whole plants. One of the species reported to produce these atropine alkaloids in culture is *Anisodus acutangulus*. These suspension cultures are also able to convert added hyoscyamine to hyoscine in a two-step biotransformation process involving 6-β hydroxyhyoscyamine as an intermediate.

10.5.11. Miscellaneous (Table 10.11)

Eucalyptus perriniana belong to the family Myrtaceae. These cells are capable of transforming caryophylleneoxide, the biotransformation reactions involved are epoxidation, reductive ring opening,

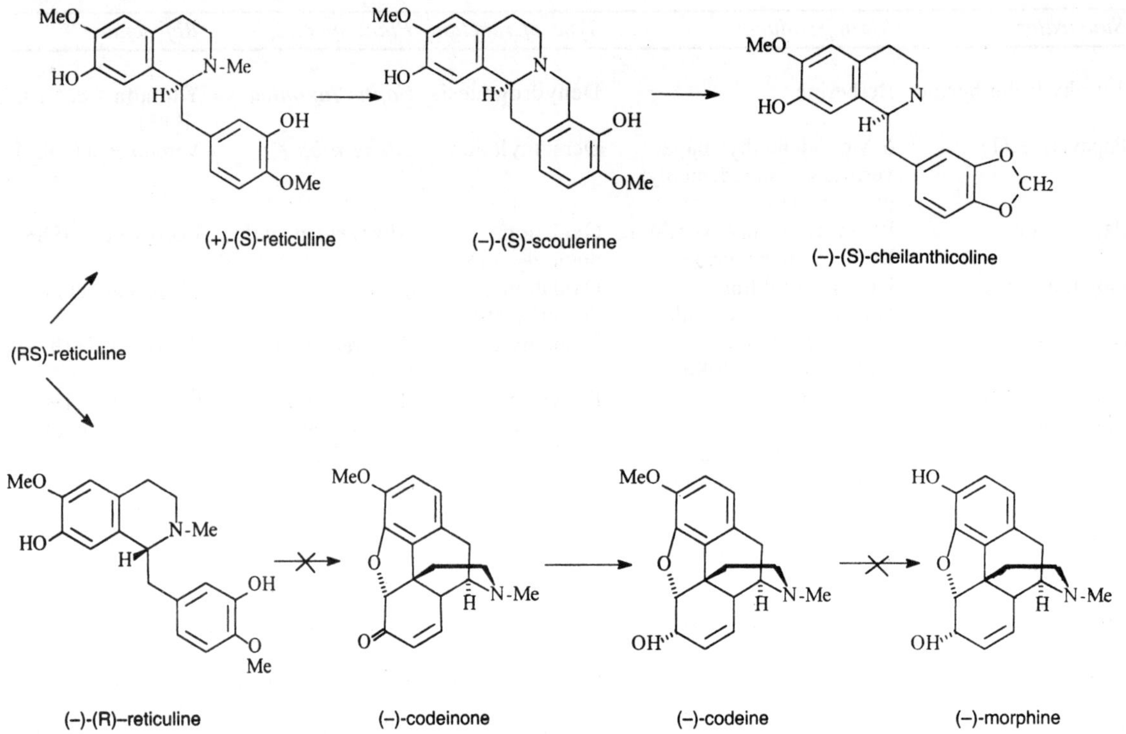

Fig. 10.27. Biotransformation of (RS)-reticuline and codeinone by cell culture of *Papaver somniferum.*

Table 10.11. Miscellaneous (Adopted from Dicosmo and Misawa, 1996)

Substrates	Products	Type of reaction	Plant species	Reference
Cyclohex-2-en-1-one	Cyclohexanone, Cyclohexanol	Reduction of C=C	*Medicago sativa*	Kergomard (1988)
2-,3-,4-Decanone	2-,3-,4-Decanol	Reduction of C=O	*Nicotiana tabacum*	Hamada *et al.* (1988)
2-Heptanone	Hectanol	Reduction of C=O	*N. tabacum*	Hamada *et al.* (1994)
2-Octanone	Octanol	Reduction of C=O	*N. tabacum*	Hamada *et al.* (1994)
3-Oxobutanoates	3-Hydroxybutanoates	Reduction of C=O	*N. tabacum*	Naoshima *et al.* (1989)
Cycloalkanol	Cycloalkanone	Reduction of C=O	*N. tabacum*	Suga *et al.* (1983)
Cycloalkanone	Cycloalkanol	Oxidation of OH	*N. tabacum*	Suga *et al.* (1983)
Tryptamine	Serotonin	Hydroxylation	*Peganum harmala*	Courtois *et al.* (1988)
Rhamnose	Glucose	Epimerization, Dehydroxylation	*Duboisia myoporoides*	Miura *et al.* (1985)
1,5-Diphenylsulphinyl, 3-methyl-3-nitropentane	1-Phenylsulphonyl-5-phenyl-3-memthyl, 3-nitropentane	Oxidation of S=O	*Catharanthus roseus*	Bourgogne *et al.* (1989)

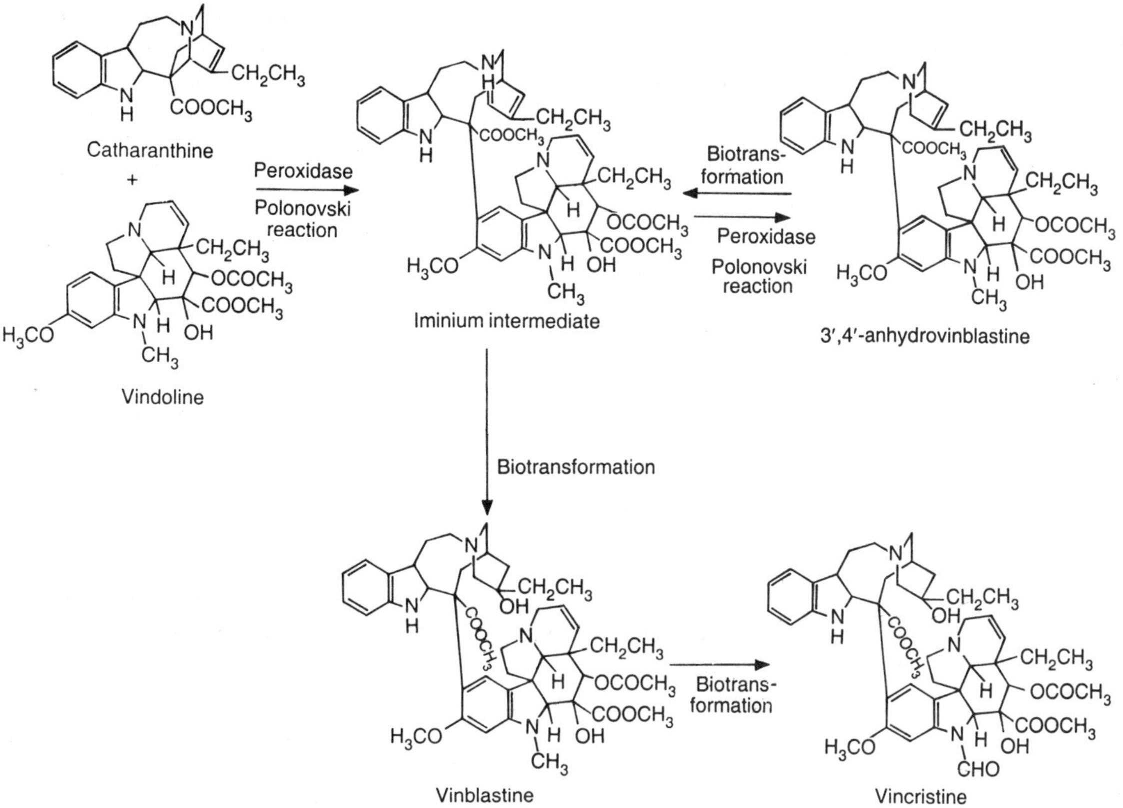

Fig. 10.28. Formation of vinblastine and vincristine by biotransformation.

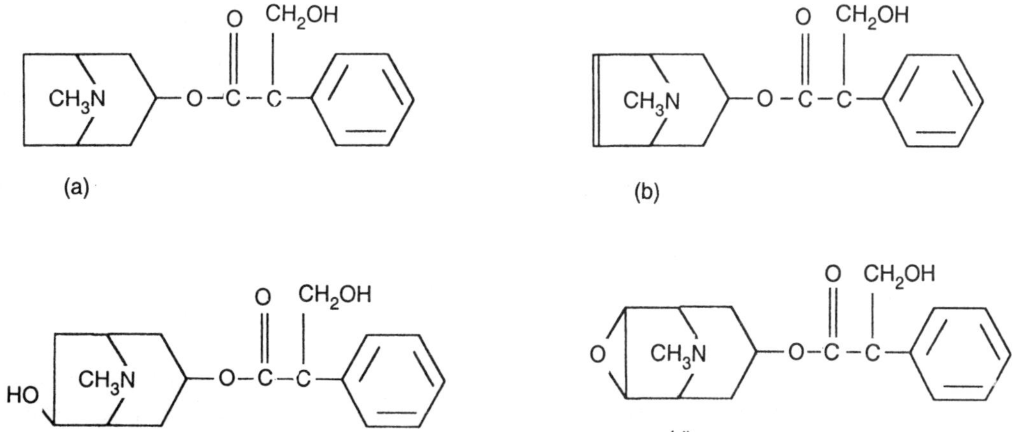

Fig. 10.29. (a) Hyoscyamine, (b) 6,7-dehydrohyoscyamine, (c) 6-hydroxyhyoscyamine, (d) hyoscine (scopolamine).

isomerization, hydroxylation and glucosylation. Seven new biotransformation products, caryophylla 3-ene-5, 14-O-β-gentiobioside, caryophylla-4-ol, 14-ene-5, 14 diol, 14-O-gentiobioside, 4,5-epoxycaryophyllane-14-ol, 14-O-D-glucopyranoside 4,5-caryophyllane-1,2-1,4-diol, 12-O-D-glucopyranoside,4,8,epoxy caryophyllane-5-14-diol, 14-O-D-glucopyranoside and 4, 11, 11-trimethyl tricyclododecane-5, 8-diol-5-O-D-glucopyranoside were isolatod from cell suspension culture of *Eucalyptus perriniana* following administration of caryophyllene oxide (Fig. 10.30) (Orihara *et al.,* 1994).

2-(14)-Methylamino-2', 4'dimethoxy-6-hydroxy benzophenone was synthesized and administered to *Ruta gravelons* (Rutaceae) cell suspension cultures (Fig 10.31). It was not incorporated into acridone alkaloids but glucosylated giving tecleanone-β-D-glucose. This reaction takes place also in cell suspension cultures of *Adhatoda vasica* and *Peganum harmala* not belonging to Rutaceae family (Baumert *et al.,* 1987).

Umetami *et al.* (1982) reported that the glucosylation of isomeric hydroxyl benzoic acids by cell suspension cultures of *Mallotus japonica*. Syahrani *et al.* (1999) reported some new biotrans-formation products, p-amino benzoic acid 7-O-β-D- glucopyranosyl ester, N-acetyl p-amino-benzoic acid 7-O-β-D-glucopyranosyl ester, o-amino benzoic acid 7-O-β-D- (β-1, 6-O-D-glucopyra-nosyl) glucopyranosyl ester and o-amino benzoic acid 7-O-β-D-glucopyranosyl ester were isolated from cell suspension cultures of *Solanum mammosum* following administration of p-amino benzoic acid, N-acetyl p-amino benzoic acid or o-aminobenzoic and respectively. N-acetyl p-amino benzoic acid and N-formyl p-aminobenzoic acids were also identified as cell suspension metabolites of p-amino benzoic acid (Fig. 10.31).

Thujaplicin (hinokitiol) is a naturally occurring aromatic seven-member tropolone compound found in the heartwood of several cupressaceous plants such as *Chamdecyparts obtuse var.* formosana (Taiwan-hinoki in Japanese*). Thuja plicata* (Westermn red cedar) and *Thujopsis dolabrata* var hondai (Hino-kisasunaro in Japanese). β-Thujaplicin has antibacterial and antifungal activities, and a regulatory action on ethylene production and respiration in fruits and vegetables. It is used as medicine, as food additive, as preservative and in cosmetics such as hair tonics. However the use of β-thujaplicin has been limited as a consequence of its water insolubility, sublimation, light decomposition and chelating with metal ions. Four new biotransformation product 4-isopropyltropolone 2-O-β-D-glucoside, 4-isopropyltropolone 2-O-β-D-gentiobioside, 6-isopropyltropolone 2-O-β-D-glucoside and 6-isopropyltropolone 2-O-β-D-gentiobioside were isolated from cell suspension culture of *Eucalyptus perriniana* following administration of β-thujaplicin. The chemical structures of all products were a determined by ^{1}H and ^{13}C NMR and MS spectroscopy. It also observed that the production ratio of biotransformation products was higher in iron deficient medium than in the medium containing iron ion (Furuya *et al.,* 1997).

Propio, acetophenone and other aromatic ketones have been reduced to their corresponding alcohols by immobilized cells of *Daucus carota* and *Gardenia jasminoides. D. carota* was able to reduce acetophenone and propiophenone to the corresponding (s)-alcohol with 54-70% yields and 89-94% enentiomeric excess (e.e.) respectively (Naoshima & Akakabe, 1991). Reducing properties of immobilized cells of *D. carota* and *Nicotiana tabacum* have also been reported for β-ketoesters, aromatic ketones and aromatic heterocyclic ketones, leading in almost all cases to the (s) enantiomer with enantiomeric excesses higher than 90% (Naoshima *et al.,* 1989; Naoshima & Akakabe, 1989; Akakabe *et al.,* 1995). The 18 species of plant cell could reduce 2-pentanone, acetophenone & ethyl acetoacetate to the s-alcohols with yields ranging from 20 to 100% (w/w) and optical purity ranging from 65 to 99% enantiomeric excess. Similar kind of glucosylation of p-amino benzoic acid by the cell cultures of *Eucalyptus perriniana* was also observed (Furuya *et al* 1998). Artemisinin a cadinane type sesquiterpene lactone containing an endoperoxide group has been established as an artemether component in the plant *A. annua* L. Artemisinin and its semisynthetic derivatives, Artemethers,

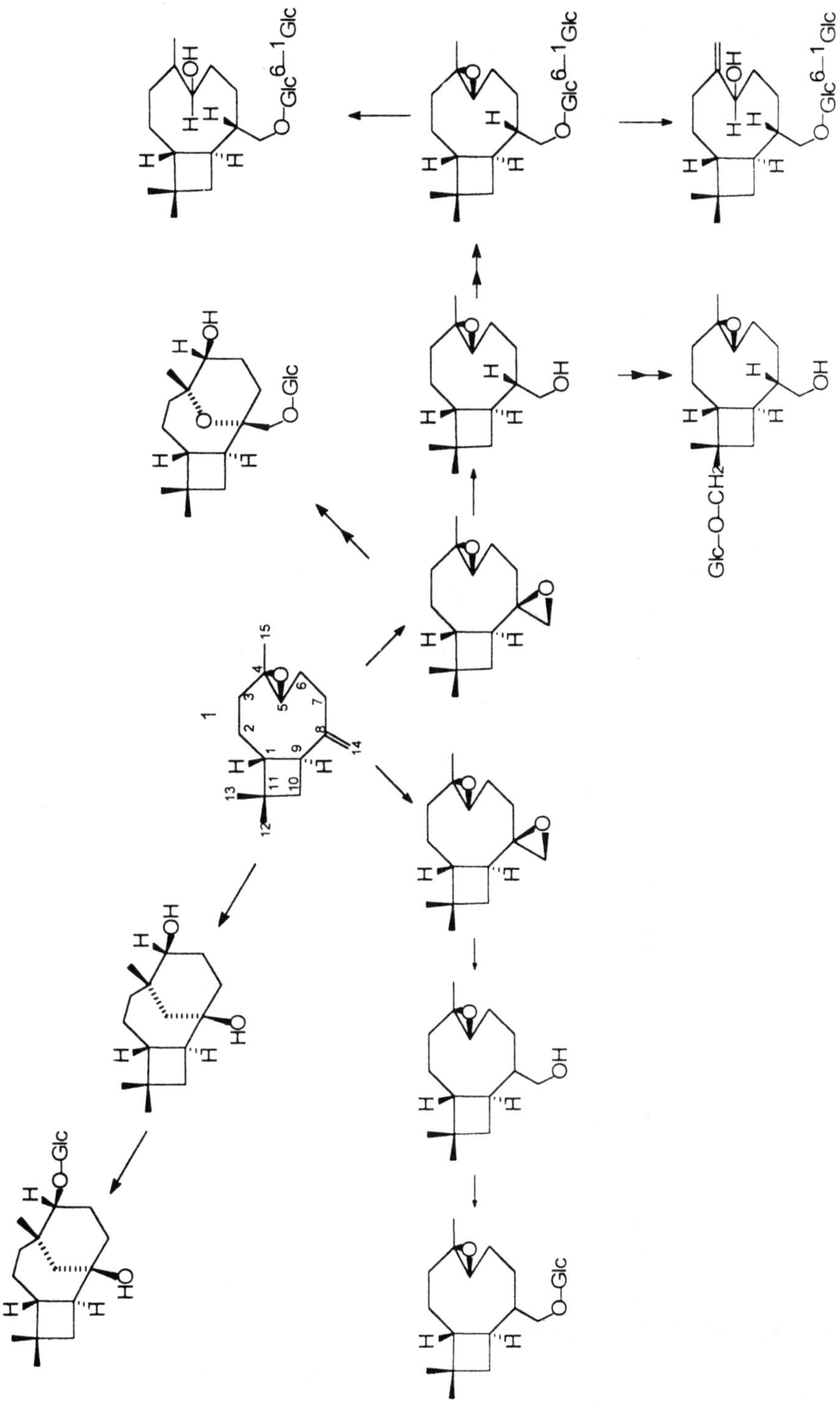

Fig. 10.30. Biotransformation of caryophyllene oxide (1) by cultured cells of *E. perriniana*.

artether and artesunate, have shown great promise against multidrug resistant strain of plasmodia (the malarial parasite). However, artemisinin remains expensive and is not available on a global scale total chemical synthesis of have been achieved, but they are too complicated to have commercial value. Therefore several groups have directed their investigations towards production of Artemisinin by tissue cultures of *A. annua*. However, little or none of this compound is synthesized de novo by undifferentiated cell cultures of *A. annua*. Artemisinic acid (AA) is a putative biogenetic precursor for the synthesis of artemisinin.Though AA has no antimalarial activity the utilization of AA as a starting material for the synthesis of Artemisinin has a practical importantce because it has a related chemical structure (cadinane-type sesquiterpene) to that of Artemisinin. Moreover, AA has been reported to be more abundant than Artemisinin in the leaves of *A. annua*. Methyl artemisinate was converted to methyl 3-oxoartemisintate by plant suspension cell culture of *Mentha piperita* L. (Kim and Kwan 1996).

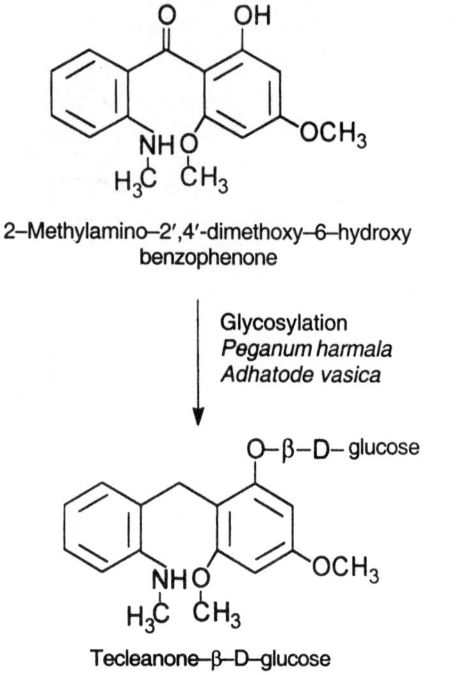

Three new biotransformation products, Artemisinic acid β-D-glucopyranosyl ester, 9-β-hydroxyartemisinic acid β-D-glucopyranosyl ester and 3-β-hydroxy-artemisinic acid β-D-glucopyranosl ester, were isolated

2–Methylamino–2′,4′–dimethoxy–6–hydroxy benzophenone

Glycosylation
Peganum harmala
Adhatode vasica

Tecleanone–β–D–glucose

Fig. 10.31. Biotransformation of 2 methyl amino 2'-4'-dimethoxy, 6-hydroxy benzophenone by plant cell culture.

from a cell culture of *Artemisia annua* following the administration of artemisinic acid (Fig. 10.32). Serotonin is presently extracted from coffee waxes and possibly useful as a drug for the central nervous system or may be useful for its anti-leprotic activity. *Peganum harmala* cell culture converts tryptamine to serotonin very efficiently (Sasse *et al* 1987). Furthermore, Courtois *et al* (1998) developed a method for high level production of serotonin in the same cell culture. They examined the effects of pH, temperature, and growth regulators on the yield of serotonin. Serotonin production increased in the range of pH 4.6 depending on the strength of alkali in a short-term experiment pH values above 7.2 resulted in the death of the cells. Serotonin production was independent of the growth stage; thus it was critical to increase the specific growth rate of the cells in order to generate biomass rapidly scale up to a 10-1 bioreactor allowed productivity of 2.5g secrotonin per liter in 20 d (20% of cell dry weight).

Miura *et al* (1985) noted an unexpected phenomenon that L-rhamnose when added to cell culture of *Duboisia myoporoides* was converted to D-glucose within the cell, not in the medium. If naringenin or phenol was added with L-rhamnose to the cell culture the conversion of L-rhamnose to D-glucose was stimulated, and naringenin glucosides or phenol glucosides were obtained from the cultured cells. Whether or not these reactions also occur with other sugars or plant species remains to be investigated.

10.6. INDUSTRIAL PROSPECTS

Plant cell cultures convert various substrates into diverse products with unique reactions, including regio and stereo specific biotransformations. However, the reactions except for glucosylation and hydroxylation, usually occur in concert with many other reactions, consequently reducing yield. The

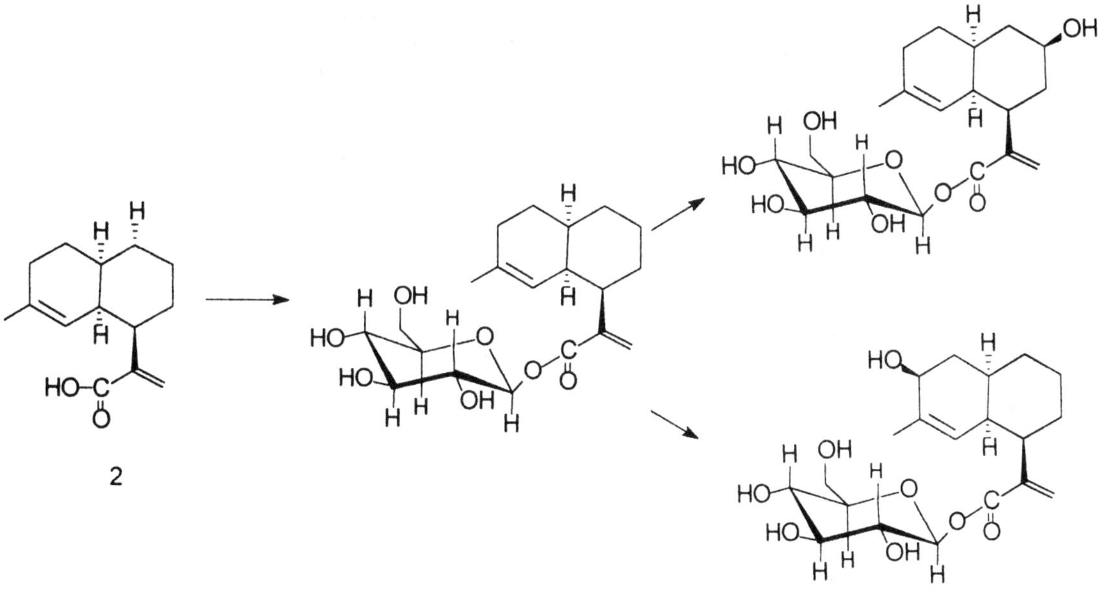

Fig. 10.32. Proposed scheme for the biotransformation of artemisinic acid (2) by cultured cells of *A. annua.*

reactions useful to industry must therefore arise through purified enzymes or from cells genetically engineered for a specific reaction.

On the other hand glucosylation and hydroxylation are relatively exclusive reactions in many cases, yielding high levels of product (Table 10.12); other reasons for glucosylation and hydroxylation on industrial scales have been presented. In general glucosylation of a precursor by chemical means is not a simple task; it requires several steps, including blocking the hydroxyl moieties of glucose via acetylating, conjugation of such OH-covered glucose to the precursor phenolic substance and saponification, such as deactivation. The conjugating reaction generally proceeds under high temperature. Therefore glycosides are relatively expensive chemical products.

Table 10.12. High-Yielding production by Biotransformation (Adopted from Dicosmo and Misawa, 1996)

Products	Substrate	Type of reaction	Plant species	Yield (g/l)	Reference
Arbutin	Hydroquinone	Glucosylation	*Rauwolfia serpentina*	18	Lutterback and Stockigt (1992)
P-Hydroxyphenyl *O*-primeveroside	Hydroquinone	Glucosylation	*R. serpentina*	9.2	Inomata (1991)
Serotonin	Tryptamine	Hydroxylation	*Peganum harmala*	2.5	Lutterbach and Stockigt (1993)
Skimmin	Umbelliferone	Glucosylation	*Datura innoxia*	1.6	Umetani *et al.* (1982)
Salicylic acid, *O*-glucoside	Salicylic acid	Glucosylation	*Mallotus japonicus*	1.1	Courtois *et al.* (1988)
Digoxin + deacetyll anatoside C	Digitoxin	Hydroxylation	*Digitalis lanata*	0.5	Kries and Reinhard (1986)
β-Methyldigoxin	β-Methyldigoxin	Hydroxylation	*D.lanata*	0.5	Heins *et al* (1978)

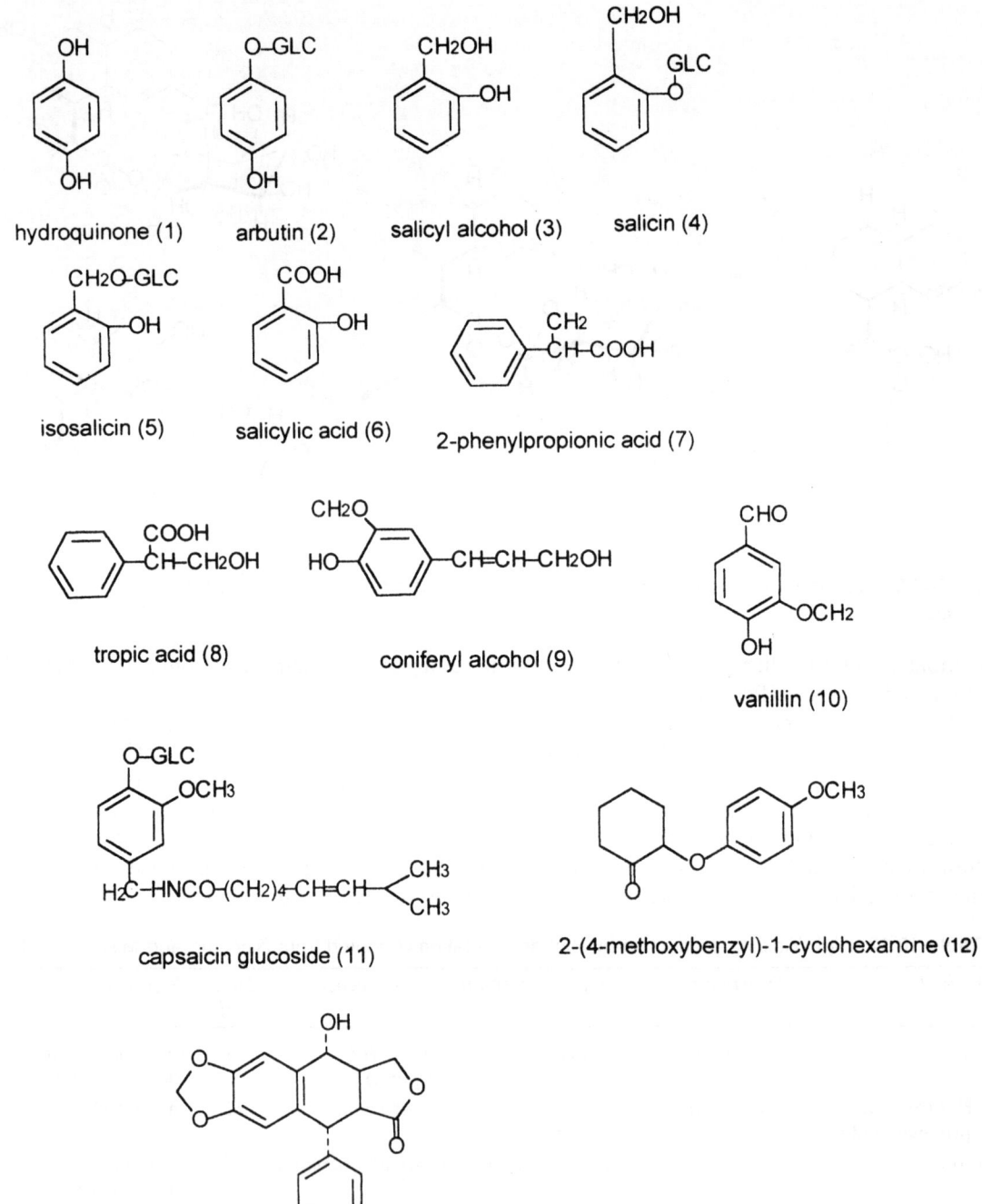

hydroquinone (1)   arbutin (2)   salicyl alcohol (3)   salicin (4)

isosalicin (5)   salicylic acid (6)   2-phenylpropionic acid (7)

tropic acid (8)   coniferyl alcohol (9)   vanillin (10)

capsaicin glucoside (11)   2-(4-methoxybenzyl)-1-cyclohexanone (12)

podophyllotoxin (13)

Fig. 10.33. Chemical structures related to the Table 10.6.

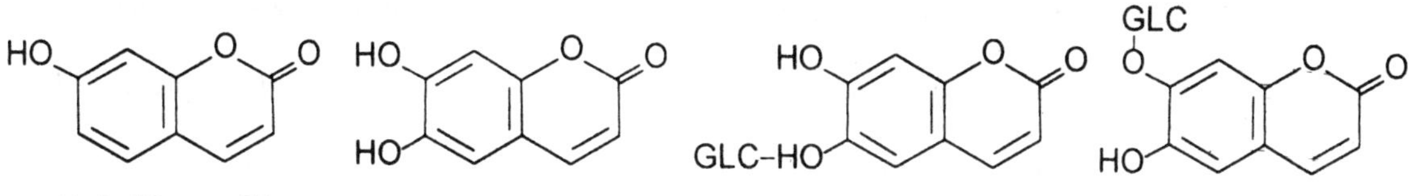

Umbelliferone (1) Esculetin (2) Esculin (3) Cichoriin (4)

Fig. 10.34. Chemical structures related to Table 10.7.

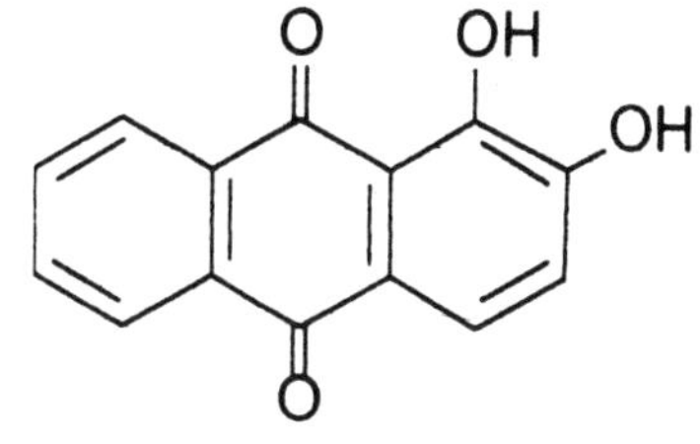

Alizarin

Fig. 10.35. Chemical structure of alizarin (Table 10.9).

The following considerations must be made when developing a cost analysis for biotransformation processes. Culture media for plant tissue culture is relatively inexpensive compared with the culture of microorganisms or animal cells. Power consumption is not a significant barrier to biotransformation processes. The major costs will arise from capital investment in facility construction and downstream processing (i.e., purification), as well as personnel expenses. Simple facilities using airlift type bioreactor are expected to be developed. Automation for downstream processing is also important in order to reduce personnel expenses. If the product yield is several grams per liter and if a new facility is not required, it should be possible to produce the material for under $300 to $400/kg product. When construction of general facilities is included, the cost will climb to approximately $1,000/kg product. Currently, a yield of several grams per liter can only be realized in the use of biotransformation for either glucosylation of hydroxylation reactions.

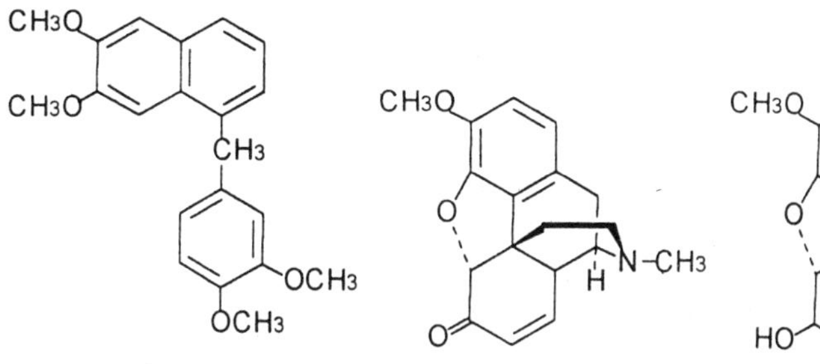

Papaverine (1) Codeinone (2) Codeine (3)

Fig. 10.36. Chemical structures related to Table 10.10.

REFERENCES

Akakabe, Y., Takahahi , M., *et al* (1995). *J. Chem. Perkin. Trans.*, 1. 1295-1298.

Akiu, S., Suzuki, Y., Fujinuma, Y., Asahara, T. and Fukuda, M. (1988). Inhibitory effect of arbutin on melanogenesis biochemical study in cultured B 16 melanoma cell and effect on the UV-induced pigmentation in human skin, *Proc. Spn. Soc. Invest. Dermatol.*, 12, 138.

Alfermann, A.W., Boy, H.M., Doller, P.C., Heins, M. and Wahl, J. (1977). Biotransformation of cardiac glycosides by plant cell cultures. In: *Proc. Int. Congr. Med. Plant Res.*, Uni. Munich. Germany. Barz. W., Reinhard. E., and Zenk. M. H., Eds. Springer Verlag, Berlin, 125.

Ambid, C., Moisseeff, M., and Fallot, J (1982). Biogenesis of monoterpenes: Bioconversion of citral by a cell suspension cultures of Muscat grapes, *Plant Cell Rep.*, 1, 91.

Asada, K and Kiso, K. (1973). Initiation of aerobic oxidation of sulfite by illuminated stomach chloroplasts. *Eur. J. Biochem.,* 33, 252.

Asada. Y., Saito. H., Yoshikawa. T., Sakamoto. K. and Furuya. T. (1993). Biotransformation of 18â–glycyrrhetinic acid by ginseng hairy root culture, *Phytochemistry*, 34, 1049.

Aviv, D., Dantes, A., Krochmal, E., and Galun, E (1983) Biotransfromation of monoterpenes by Mentha cell lines: conversion of pulegone-substituent and related unsaturated á-â ketones. *Planta medica*, 47,7.

Aviv, D., Dantes, A., Krochmal, E., and Galun, E (1981) Biotransfromation of monoterpenes by Mentha cell lines: conversion of menthone to neomenthol, *Planta medica*, 42, 236.

Balscvich, J., (1985) Biotransformation of 10-hydroxygeraniol and related compounds by cell suspension cultures of *Catharnathus roseus:* The fromation of reduced products, *Planta medica*, 51,128

Banthorpe. D.V., and Barrow. S. E. (1983) Monoterpene biosynthesis in extracts from cultures of *Rosa damascena*. *Phytochemistry*, 22. 2727.

Baumert, A., Rosza, ZS., Schliemann, W., Lewis, J.R., and Groger, D. (1987) Biotransformation of a ($^{14}CH_3$) –2-methylaminobenzophenone by plant cell cultures. *Planta Medicas*, 90-92.

Berger, R.G. and Drawert, F. (1988). Biotechnological preparation of glycosides using plant cell cultures (in German), German Patent, 3718340 (DEA1).

Bourgogne, V., Labidalle, S., Galons, H., Miocque, M., Foulquier, M., Jacquin-Dubreuil, A. and Cosson, L. (1989). Biotransformation of a synthetic compound, 1,5-diphenylsulphinyl-3-methyl-3-nitropentane, by cell suspensions of *Catharanthus roseus*. *Phytochemistry,* 28, 2345.

Braemer, R., Tsoutsias, Y., Huabielle, M. and Paris, M. (1987). Biotransformation of querctin and apigenin by a plant cell suspension cultures of *Cannabis sativa*. *Planta Medica,* 53, 225-226.

Braemer, R. and Paris, M. (1987). Biotransformation of cannabinoids by a cell suspension culture of *Cannabis sativa* L. *Plant Cell Rep*. 6, 150.

Brodellus, P. and Mosbach, K. (1982). Immobilised plant cells: general aspects. *J Chem. Tech. Biotechnol*, 32, 330.

Christinaki,H., Bister-Miel, F., Hammoumi,A., Guignard, J.L and Viel,C (1987) Biotransformation of isopapaverine by *silene alba* cell suspensions. *Phyotochemistry* 26,2991.

Cole, D and Loughman, B.C (1984) Transformation of phenoxyacetic acid herbicides by *Phaseolus vulgaris* L. callus cells, *Plant Cell Rep*.3,5.

Corbier, B. and Ehret. C., (1988) Biotransformation of monoterpenoids by cultured cells of *Rosa centifolis*. In: Proc. 10th Int. Congr. Essential Oils Fragrances Favors Sawrence, B. M., Mookherjee, B. D. and Willis. B. J. Eds., Elsevier Science Publishers B.V., Amsterdam. The Netherlands, 731.

Corchette, P. and Yeoman, M.M. (1989). Biotransformation of (-) codeinone to (-) codeine by *Papaver somniferum* cell immobilized in reticulate polyurethane foam, *Plant Cell Rep*. 8, 128.

Cormier, F. and Ambid, C. (1987). Extractive bioconversion of geraniol by *Vitis vinifera* cell suspension employing two phase system. *Plant Cell Rep.*, 6, 427.

Courtois, D., Yvernal, D., Flona, B. and Petiard, V. (1988). Conversion of tryptamine to serotonin by cell suspension cultures of *Peganum harmala*. *Phytochemistry*, 27, 3137.

Darise, M., Mizutani, K., Kasai, R., Tanaka, O. *et al*. (1984). Enzymic transglucosylation of rubusoside and the structure-sweetness relationship of steviol-bisglycosides. *Agric. Biol. Chem.,* 48, 2483.

Dicosmo, F. and Misawa, M. (1996). Plant Cell Culture, Secondary Metabolism towards Industrial Application CRC Press, BocaRaton, pp. 79-138.

Diettrich, B., Aster, U., Greidziak, N., Ross, W., and Lukner, M. (1987). Glucosylation of digitoxin and other cardenolides in cell cultures of *Digitalis lanata, Biochem.Physiol. Pflazen.*, 182, 245.

Dorisse, P., Glseye. J., Loise, P., Puig, P., Ed A.M. and Henry M. (1988). Papavenne biotransformation in plant cell suspension cultures. *J. Nat Prod.*, 51, 532.

Drawert, F., Berger, R.G. and Godelmann, R. (1984). Regioselective biotransformation of valencene in cell suspensin cultures of Citrus sp. *Plant Cell Rep.*, 3, 37.

Fukunaga, Y., Miyata, T., Nakayasu, N., Mizutani, K., Kasai, R. and Tanaka, O. (1989). Enzymic transglucosylation products of stevioside: separation and sweetness-evaluation. *Agric. Biol. Chem.*, 53, 1603.

Fulzzele, D., Kreis, W. and Reinhard, E. (1992). Cardenolide biotransformation by cultured *Digitalis lanata* cells; semi continuous cell growth and production of deacetyllanatoside C in a 40-1 stirred tank bioreactor, *Planta Med.*, 58, suppl, 1. A 601.

Furuya T., Nakano.M., and Yoshikawa. T. (1978). Biotransformation of (RS) –reticuline and morphinan alkaloids by cell cultures of *Papaver somniferum. Phytochemistry,* 17, 981.

Furuya T., Yoshikawa. T., Taira, M., (1984) Biotransformation of codeinone to codeine by immobilized cell of *Papaver somniferum, Phytochemistry*, 23, 999.

Furuya, T.,Orihara,Y., and Miyatake, H (1989) Biotransformation of (-) menthol by *Eucalyptus perriniana* cultured cells. *J.Chem.soc.Perkin Trans.* 1, 1711.

Furuya, T., Asada, Y., Mizobata, S., Matsuura, Y. and Hamada, H. (1998). Biotransformation of *p-amino* benzoic acid by cultured cells of *Eucalyptus perriniana. Phytochemistry,* 49(1), 109-111.

Furuya, T., Hirotoni, M and Kawaguchi, K (1971). Biotransformation of progesterone and pregnolone by plant suspension cultures. *Phytochemistry* 10,1013.

Furuya, T., Kawaguchi, K. and Hirotani, M. (1988). Biotronsformation of digitoxigenin by cell suspension cultures of *Strophanthus gratus, Phytochemistry*, 27, 2129.

Furuya, T., Ushiyama, M., Asada, Y., Yoshikawa, T. and Orihara, Y., (1988). Biotransformation of phenylacetic acid and 2-phenypropionic acid in suspension culture of *Coffea arabica. Phytochemistry*, 27, 803.

Furuya, T., Usiyama, M., Asada, Y. and Yoshikawa. T ., (1987) Glucosylation of 2 phenylpropionic acid and its ethyl ester in suspension cultures of *Nicotiana tabcum, Dioscoreophyllum cumminsii* and *A conitum japonicum, Phytochemistry*, 26, 2983.

Furuya,T., Asada,Y., Matsura, Mizobata, S., and Hamada H.(1997) Biotransformation of â-thujalplicin by cultured cells of *E. perriniana, Phytochemistry* 46(8), 1355 –1358 ,1997.

Furuya. T., Ushiyamna, M., Asada, Y. and Yoshikawa. T., (1989) Biotransformation of 2-phenylpropionic acid in root culture of *Panax ginseng., Phytochemistry*, 28. 483.

Galun, E., Aviv, D., Dantes, A and Freeman, A (1985) Biotransformation by division-arrested and immobilized plant cells: Bioconversion of monoterpenes by gamma-irradiated suspended and entrapped cells of Mentha and Nicotiana, *Planta Medica* 51, 511.

Gbolade, A.a. and Lockwood, G.B. (1989). Selective biotransformation of monoterpenoids by cell suspension cultures of *Petroselinum crispum. Z. Naturforsch.,* 44c, 1066-1068.

Graves, J.M.H. and Smith, W.K (1967) transformation of pregnenolone and progesterone by cultured plant cells.*Nature*, 214,1248.

Hamada, H., Umeda,N., Otusuka,N., and Kawabe,S (1988) The reductive transformation of decanones with immobilized cells of *Nicotiana tabcum. Plant Cell Rep.*,7, 493.

Hamada, H. and Kawabe. S. (1991a). Biotransformation of 4-androstene-3,17-dione by green cell suspension of *Marchantia polymorphia.* Stereolselective reduction at carbon 17. Life sciences 48: 613-615.

Hamada, H. and Furuya, T. (1996). Recent advances in plant biotransformation *Plant tissue culture and biotechnology,* 2(1), 52-60.

Hamada, H. and Nakazawa, K. (1991b). Biotransformation of vinblastane to vincristine by cell suspension cultures of *Catharanthus roseus. Biotechnology Lett.,* 13, 805.

Hamada, H and Furuya, T. Recent advances in plant biotransformation, *Plant tissue culture and Biotechnology,* 2(1), 52-60 (1996).

Hamada, H., Konishi, H., Williams, H.J., and Scoot, A.I. (1991b). Biotransformation of testosterone isomers by a green cell suspension cultures of *Machantia polymorpia. Phytochemistry,* 30: 2269-2270.

Hamada, H., Imura, M.,and Okano, T. (1994). Biotransformation of 1,1,1-trifluoroheptane-2-one and octanone-2-one by the cell suspension culture of *Nicotiana tabacum*-aceleration by fluorine substituion. *J. Biotechnology,* 32, 89.

Harkiss, K.J., Linley, P.A and Mesbah, M.K.M. Biotronsformation of (23-14 C)-Digitoxigenin and (23,14-C) Digitoxin in *Convallaria majalis* L. *Int. J. Crude Drug. Res.* 24(1) 1986, pp 25-30.

Hayashi, H., Fukui. H., and Tabata, M. (1990). Biotransformation of 18â glycyrrhetinic acid by cell suspenson cultures of *Glycyrrhiza glabra, Phytochemistry,* 29, 2149.

Hayashi, H., Yamada. K., Fukai. H. and Tabata, M. (1992). Metabolism of exogenous 18α glycyrrhetinic acid cultured cells. *Phytochemistry,* 31, 2729.

Heins, M.,Wahl, J., Lerch, H., Kaiser, F. and Reihard, E. (1978). Preparartion of â-methyldigitoxin by hydroxylation of â-methyldigitoxin in fermenter cultures of *Digitalis lanata, Planta Med,*33 , 57.

Hirata, T. Izumi, S., Ekida, T., and Suga, T., Hydroxylation of acetoxy-P-menthenes in cultured cells of *Nicotina tabacum.* Epoxidation of the Carbon Carbon double bond Bull Chem. Soc. Japan. 60, 289, (1987).

Hirata, T., Hamada. H., Aoki. T., a nd Suga. T. (1982) Stereoselectivity of the reduction of carvone and dihydrocarvone by suspenson cells of *Nicotiana tabacum. Phytochemistry* 21, 2209.

Hiraoka,N and Carew,D.P.(1981) Biotransfromation of 2-(p-methoxyphenyl)ethylamine by *Catharanthus roseus* and *Strobilanthes dyerianus cell lines, J.Nat.Products,* 44, 285.

Hiraoka, N and Tabata,M (1983) Acetylation o f t ropane d erivatives b y *D atura innoxia* c ell cultures, Phytochemistry,22, 409.

Hirotani, M. and Furuya, T. (1974). Biotransformation of testosterone and other androgens by suspension cultures of *Nicotiana tabacum "Bright yellow" Phytochemistry,*13, 2135.

Hirotani, M. and Furuya, T. (1975) Metabolism of 5α-pregnane-3,20-dione and 3β-hydroxy-5α-pregnan-20-one by Digitalis suspension cultures, 10, 2601.

Hirotani, M. and Furuya, T. (1980). Biotronsformation of digtoxigenin by cell suspension cultures of *Digitalis purpurea. Phytochemistry,* 19, 531.

Hobbs,M and Yeoman,M.M. (1991) Biotransformation of nicotine to nornicotine by cell suspensions of *Nicotiana tabcum* L. cv. Wisconsin-38, *New phytol.*119,477.

Hook,I.., Lecky, R. McKenna, B., and Sheridan,H (1990) Biotrnsformation of linalyl acetate by suspension cultures *Papaver bracteatum, Phytochemistry,* 29,2143.

Inomata. S, Yokayama, M., Seto, S. and Yanagi, M. (1991). High level production arbutin from hydroquinone in suspension cultures of catharanthus roseus plant cells. *Appl Microbiol Biotechnol.* 36, 315.

Itabashi, M., Aihara. Inoue, T., Yamate, J., Sanni, S., Tafima, M., Tanaka, C., and Wkisaka, Y. (1988). Reproduction study of arbutin in rats by subcutaneous administration, *Lyakuhin Kenkyu,* 19, 282.

Iwasa,K., Kamigauchi,M., and Takao,N., (1992) Transformation of(+)-salsinol into optically active o-and /orN-methylated derivatives by several papaveraceae plants and their tissue cultured cells, *J.Nat.Products.* 55,491

Iwasa,K., Kamigauchi,M., and Takao,N., (1987)Biotransformation of phthalideisoquinoline alkaloids by *Corydalis* tissue cultures, *Arch.Pharm.* 320,693.

Jones,A and Veliky,I.A (1981) Biotransformation of cardenolides by plant cell cultures.II. metabolism of gitoxigenin and its derivatives by suspension cultures of *Dacus carota, Planta Med,* 42,60.

Kawaguchi, K., Hirotani, M. and Furuya, T. (1988). Biotronsformation of digitoxigenin by cell suspension cultures of *Strophanthus amboenis. Phytochemistry,* 27,3475.

Kawaguchi, K., Hirotani, M. and Furuya, T. (1989). Biotronsformation of digitoxigenin by cell suspension cultures of *Strophanthus intermedius. Phytochemistry,* 28. 1093.

Kawaguchi, K., Hirotani, M., Yoshikawa, T. and Furuya, T. (1990). Biotransformation of digitoxigenin by ginseng hairy root cultures. *Phytochemistry,* 29, 837-843.

Kergomard, A., Renard, M.F., Veschambre, H., Courtois, D and Petiard, V. (1988). Reduction of α, β unsaturated ketones by plant suspension cultures, *Phytochemistry,* 27, 407.

Kim,. S.U and Kwon, H.J. (1996). *Planta Med.,* 359.

Kometani, T., Tanimota, H., Nishimura, T. and Okada, S. (1993). Glycosylation of vanilline by cultured plant cells. *Biosci. Biotech. Biochem.,* 57, 1290.

Kreis, W. and Reinhard E. (1988). 12â-hydroxylation of digitoxin by suspension –cultured *Digitalis lanata* cells. Production of deacetyl lanatoside C using a two-stage culture method, *Planta Med.,* 54, 143.

Kreis, W. and Reinhard, E. (1986). Highly efficient 12α-hydroxylation of digitoxin in *Digitalis lanata* cell suspension using two stage culture method. *Plant Med.,* 418.

Kreis, W., May,U., and Reinhard, E (1986). UDP-glucose: digitoxin 16' o-glycosyltransferase from suspension cultured *Digitalis lanata* cell. *Plant cell reports*, 5. 442.

Kurz, W.G.W. and Constabel, F. (1979). Plant Cell Cultures, a Potential Aource of Pharmaceuticals. *Advances in Applied Microbiology*, 25 : 209-240.

Kutney,J.P., Aweryn,B., Choi, L.S.L., and Kolodziejczyk, P (1981) Alkaloid production in *Catharanthus roseus* cell cultures,IX. Biotransformation studies with 3', 4'-dehydrovinblastine, *Heterocycle,* 16, 1169.

Kutney,J.P., Aweryn,B., Choi, L.S.L., and Kołodziejczyk, P (1982) Alkaloid production in *Catharanthus roseus* cell cultures,XI. Biotransformation studies with 3', 4'-dehydrovinblastine to other bisindole alkaloids, *Helv.Chin.acta,* 65, 1271.

Kutney,J.P., Botta,B.,Boulet,C.A., Buschi,C.A.,Choi,L.S.L., Golinski,J., Gumulka,M., et al (1988) Alkaloid production in *Catharanthus roseus* (L) Don. Cell cultures XVI. Biotransformation studies of , 3,' 4'-dehydrovinblastine with *Catharanthus roseus* cell cultures and enzyme systems, , *Heterocycle,* 27,629.

Langebartels, C. and Harms, H. (1984). Metabolism of pentacholrophenol in cell suspension cultures of soybean and wheat pentechlorophenol glucoside formation *Z. Pfanzsan Physiol,* 113, 201.

Lappin, G.J. Stride, J.D. and Tampion, J. (1987). Biotransformation of monoterpenoids by suspension cultures of *Lavandula angustifolia,* 26 (4), 995-997.

Lappin, G., Tampion, J.and Stride, J.D. (1986). The Biotransformation of monoterpenoids by plant tissue culture, In: Secondary Metabolism in plant cell cultures. Morris,P., Scragg, A.H., Stafford,A and Fowler, M.W. (eds) Cambridge Univ. Press. Cambridge, PP.113

Lee,Y.S., Hirata, T and Suga,T (19830 Biotransformation of 1-acetoxy-p-menthane-4(8)-ene with a suspension of cultured cells of *Nicotiana tabacum, J. Chem.Soc.Perkin trans.* 1, 2475.

Lewinsohn, E., Berman,E., Mazur,Y.,and Gressel, J., (1986) Glucosyaltion of exogenous flavones by grapefruit cell cultures,*Phytochemistry*, 25, 2531.

Lutterbach, R. and Stockigt, J. (1992). High-yield formation of arbutin from hydroquinone in suspension cultures of *Rauwolfia serpentina. Helv. Chim. Acta.,* 75, 2009.

Lutterbach, R., and Stockigt. J., and Kolshorn, H. (1993). P-Hydroxyophenyl-o-â-D-Primeveroside, a novel glycoside formed from hydroquinone by cell suspension cultures of *Rauwolfia serpentina. J. Nat. Prod.,* 56,1421.

Miura, H., Kitamura, Y. and Sugii, M. (1985). Conversion of L-rhamnose to D-glucose in *Duboisia myoporoides* cell suspension cultures. *Shoyakugaku Zasshi,* 39, 334.

Miura, H., Kitamura, Y. and Sugii, M. (1986). Studies on the tissue culture of swertia japonicum Makino.III glucosyaltion of naringenin in cultured cells. *Shoyakugaku Zasshi,* 40,40.

Miura, H., Kitamura, Y. and Sugii, M. (1986). Glucosylation of narigenin in *Duboisia myoporoides* cultured cells. *Shoyakugaku Zasshi,* 40, 113..

Mizukami, H., Terao, T., Miura, H., and Ohashi, H. (1983). Glycosylation of salicyl alcohol in cultured plant cells. *Phytochemistry,* 22, 679-680.

Mizukami, H.T., Ammano, A., and Ohashi H. (1986). Glucosylation of salicyl alcohol by *Gardenia jasminoides* cell cultures. *Plant Cell Physiol.,* 27, 645.

Mizukami, H., Hirano, A.,and Ohashi, H (1987) Effect of substituent groups on the glycosyl conjugation of xenobiotic phenols by cultured cells of *Gardenia jasminoides. Plant sci.* 48, 11.

Naoshima, Y., Akakabe, Y and Watanabe, F. (1989). Biotransformation of acetoacetic esters with reuse of immobilzed *Nicotiana tabacum* cells, *J Org. Chem.,* 54, 4237-4239.

Naoshima, Y and Akakabe, Y (1991) Biotransformation of aromatic ketones with cell cultures of carrot,tobacco and gardenia, *Phytochemistry,* 30, 3595-3597.

Oehar, Y., Saiki, K., and Furuya, T. (1991). Biotransformation of sterol by cultured cells of *Eucalyptus perriniana* and *Coffea arabica. Phytochemistry,* 30, 3989.

Ohtani, K., Aikawa, Y., Ishikawa, H., Kasai, R., Kithata, S., Mizutani, K. Doi, S., Nakaura, M. and Tanaka, O. (1991). Further study on the 1,4á-transglucosylation of rubusodide, a sweet steriol-bisglucoside from *Rubus suavissimus. Agric. Biol. Chem.,* 55, 449.

Orihara, Y. and Furuya, T. (1994). Biotransformation of (+) and (-) fenchone by cultured cells of *Eucalyptus perriniana. Phytochemistry,* 36, 55-59.

Orihara, Y., Miyatake, H. and Furuya, T. (1991). Triglucosylation of the biotransformation (+)-menthol by cultured cells of *Eucalyptus perriniana. Phytochemistry,* 30, 1843.

Orihara, Y., Saiki, K., and Furuysa T (1994) Biotransformation of carophyllene oxide by cultured cells of *Eucalyptus perriniana. Phytochemistry,* 35(3), 635-639.

Orihara, Y and Furuya. T. (1990). Biotranstormation of 18α- Glycyrretinic acid by cultured cells of *Eucalyptus perriniana* and *Coffea arabica, Phytochemistry,* 29, 3123.

Paper, D.H. and Franz, G. (1990). Biotronsformation of 5αH-Pregnan-3α-01-20-one and cardenolides in cell suspension cultures of *Nerium oleander L., Plant Cell Rep.,* 8, 651.

Pawtowicz, P and Siewinski, A., (1987) enatioselective hydrolysis of esters and the oxidation by *Spirodela oligorrhiza, Phytochemistry,*26,1001.

Petersen, M. and Seitz, H.U. (1985). Cytochrome dependent digitoxin 12β-hydroxylase from cell cultures of *Digitalis lanata. FEBS Lett.,* 188, 11.

Proksch, P., Witte, L., Wray, V. and Rahaus, L. (1987). Accumulation and biotransformation of chromenes and benzofurans in a cell suspension culture of *Ageratina adenophora, Planta Med.,* 53, 488.

Saito, K., Murakoshi, I., Inze, D., and Montagu, M. (1989). Biotransformation of nicotine alkaloids by tobacco shooty teratomas induced by T$_I$Plasmid mutant, *Plant Cell Rep*7, 607.

Sakano, K. and Ohsima, M. (1986). Structures of conversion products formation from 18β–glycyrrhetinic acid by Streptomyes sp. G.20. *Agric. Biol. Chem.* 50, 763.

Sakono, K., and Ohsima, M. (1986). Microbiol convertion of 18α–glycyrrhetinic acid and 22α–hydroxy-18β–glycyrrhetinic acid by *Chainia antibiotica. Agric. Boil. Chem.,* 50, 1239.

Sasse. F., Witte. L. and Berlin. J., (1987) Biotransformation of tryptamine to serotonin by ceil suspension culture *Peganum harmala, Planta Med.,*53, 354.

Stafford, A. and Warren, G. (1991). Plant Cell and Tissue Culture, Open University Press, Milton Keynes, U.K., p. 185.

Stohs, S.J. and Staba, E.J. (1964). Production of cardiac glycosides by plant tissue culture IV. Biotransformation of digitoxigenin and related substances. *Journal of Pharmaceutical Sciences,* 54, 56-64.

Stohs, S.J. (1969) Phytochemistry 8,1215

Stohs, S.J and El-olemy, M.M. (1972) Phytochemistry 11,1397.

Suga, T. (1992) Biotransformation of monoterpeniods by tobacco cultured cells, Presented at 3rd Plant Tissue Culture Colloquiura. The Japanee Association for Plant Tissue Culture, Nitgara July 22-23.

Suga, T. Hirata, T. and Futstsugi, M. (1984) The biotransformation of carvoxime and dihydrocarvoxime with cell suspension cultures of *Nicotiana tabacum,* Phytochemistry, 23, 1327-1328 .

Suga, T., Hirasa. T., Hamada, H., and Furassugi, M. (1983) Enantioselectivity in the biotransformation of bicyclic monoterpene alcohols with the cutltureed suspension cells of *Nicotiana tabaccum, Plant Cell Rep.* 2. 186.

Suga, T., Hamada, H., and Hirata, T(1987) Enantioselectivity in the biotransformation of bicyclo{3.1.1} heptanes with the cutltureed suspersion cells of *Nicotiana tabaccum, Chem.Lett.*471.

Suga, T., Hirata. T., Hamada. H., and Murakami, S. (1988). Biotransformation of 3-or-P – menthane derivatives by cultures cells of *Nicotiana tabbacum. Phytochemistry,* 27, 1041.

Suga. T. and Hirata. T. (1990). Biotransformation of exogenous substrates by plant cell cultures. *Phytochemistry,* 29, 2393.

Syahrani, A., Ratnasari, E., Indrayanto, G., and Wilkins, A. (1999). Biotransformation of – and P- amino benzoic acids and N –acetyl P-aminobenzoic acid by cell cultures of *Solanum mammosum, Phytochemistry* 51,615-620.

Tabata, M., Ikeda, F., Hiraoka, M. and Konoshima, M. (1976). Glucosylation of Phenolic compounds by *Datura innoxia* suspension cultures, *Phytochemistry*, 15, 1225.

Tabata, M., Umetani, Y., Ooya, M. and Tanks, S. (1988). Glucosylation of phenolic compounds by plant cell cultures. *Phytochemistry,* 27. 809.

Tlomak,E., Pawlowick,P., Czewinski,w., and Siewinski,A., (1986).Transformation of androstane derivatives by *Spirodela oligorrhiza*, *Phytochemistry*, 25, 61.

Tam, W, H. J., Kurz W. G. W., Constabel. F., and Chatson, K.B. (1982). Biotransformation of thebaine by cell suspension cultures of *Papaver somniferum* CV mananne. *Phytochemistry,* 21, 253.

Umetani, Y.,Tanaka,S and Tabata, M (1982) Glucosylation of extrinsic compounds by various plant cell cultures. In: Fujiwara, A., plant tissue culture , proc. 5th Intl. Congr. Plant. Tissue cell culture (pp 382-384). Tokyo, Maruzen.

Umetani, Y., Kodakari,E.,Yamamura, T., Tanaka,S.,and Tabata,M(1990) Glucosylaton of salicylic acid by cell suspension cultures of *Mallotus japonicus, Plant cell Rep.*9, 325.

Ushiyama,M., Kumagai,S., Furuya, T (1989) Biotransformation of phenylcarboxylic acids by plant cell cultures. *Phytochemistry* , 28,3335.

Ushiyama, M., and Furuya. T. (1989). Glucosylaton of phenolic compounds by root cultures of *Panax ginseng, Phytochemistry,* 28, 309.

Van Uden, W., Oeij. H., Woerdenbag H.J. and Pras. N. (1993) Glucosylaton of cyclodextrin complexed podophyllotoxin by cultures of *Linum flavum* L., *Plant Cell Tissue Organ Cult.*, 34, 169..

Vanek, T., Valterova, I. and Vaisar, T. (1999). Biotransformation of (S)-(-)- and (R)-(+)-limonene using *Solanum aviculare* and *Dioscorea deltoidea* plant cells. *Phytochemistry,* 50, 1347-1351.

Vanek. T., Umantseva, V. V., Wimmer, Z. and Macek. T., (1989) By *Dioscorea deltoidea* free and immobilized plant cells. Biotechnol. Lett.,11, 243..

Veeresham, C. and Kokate, C.K. (1997). Taxol: A novel anticancer drug from cell cultures of *Taxus* sp. *Indian Drugs,* 34(6), 354-359.

Veeresham, C., Rao, M.A., Babu, P.Ch., Mamatha, R. and Kokate, C.K. (2000). Biotransformation of paclitaxel (Taxol) by plant cell cultures. *Indian Drugs,* 37(2), 86-89.

Verdeil, J.L., Bisteri-Miel, F., Hammouni, A., Bury, M., Gugaard. J.L. and Viel, C. (1986). Papaverine biotransformation by *Silene alba* cell suspension. *Planta Med.,* 52, 1.

Wimmer, Z., Macek. T., Vaniek, T., Streinz, L., and Romamuk. M. (1987) Biotronsformation of 2-(4-Methoxybenzyl)- 1-Cyclohexanone by cell cultures of *Solanum aviculare. Biol. Plantar.,* 29, 88.

Woerdenbag, H J., Pras, N., Frijilink. H. W., Lerk,. C. F. and Malingre. T. M., (1990a) Cyclodextrin- facilitated bioconversion of 17β estradiol by a phenoloxidase from *Mucuna pruriens* cell cultures, *Phytochemistry*, 29, 1551.

Woerdenbag, H.J., van Uden, W., Frijilink, H. w., Lerk, C. F., Pras, N., and Malingre. T.M. (1990) Increased podophyllotoxin production in *podophyllum hexandrum* cell suspension cultures after feeding coniferyl alcohol as a α-cyclodextrin complex. *Plant Cell Rep.*, 9, 97.

Yokayama, M., Inomata, S., Yanagi, M., Wachi, Y. and Fukushima. S. (1993). Change of maximal celluar productivity of arbutin by biotransformation depending on the cultures stage of catharanthus roseus cells, presented at the 15th Int. Bot. Congr., Yokahama. Japan.

Yokoyama, M. (1991). Manufacture of phenolic glucosides. Japanese Patent No. 25909.

Yokoyama, M., Inomata, S., Seto, S. and Yanagi, M. (1990). Effects sugars gluosylation of exogenous hydroquinone by *C. roseus* cell in suspension culture, *Plant Cell Physiol*, 31, 555.

Yokoyama, M. and Yanagi, M. (1992). High level production of arbutin by biotransformation,. In: Plant cell culture in Japan.. Komamine A, Misawa. M., and Dicosmo F. (eds) CMC press. Tokyo.

Zacharis, R.M., and Kalan, E.B. (1984). Biotransformation of the potato stress metabolite, solavetivone, by cell suspension cultures of two solanaceous and three non-solanaceous species. *Plant Cell Rep.,* 3, 189.

Chapter 11

Immobilization of Plant Cells

INTRODUCTION

Plant cell culture has been for some time considered as an alternative method for the production of flavors, colors and pharmaceuticals to their extraction from plants. Characteristics of plant cell cultures, such as slow growth, the compounds produced should be of high value (\$ 500-1000kg^{-1}) and low volume. One of the major limiting factors in the development of a commercial production system using plant cell culture has been the production cost of phytopharmaceuticals. The use of high biomass levels for extended periods would be one method of increasing productivity and hence reducing the costs. This can be achieved by the immobilization of plant cells.

The immobilization of enzymes and cells has received increasing attention, and used to produce amino acids and carbohydrates. The immobilization of microbial cells is not new, it has known for sometime that cells will adhere to many surface in nature i.e. polymers.

Immobilization is the newest culture technology of plant cell, and considered as to be the most 'natural'. It has been defined as a technique, which confines to a catalytically active enzyme or to a cell within a reactor system and prevents its entry into the mobile phase, which carries the substrate and product. The first successful immobilization of plant cells was reported by Brodelius *et al*. (1979) and they entrapped *Catharanthus roseus* and *Daucus carota* cells in alginate beads. Following success with enzymatic and microbial process, immobilization has been suggested as a strategy to enhance the overall productivity of secondary metabolite in plant cell culture. The ability to immobilize plant cells has been reported for a large number of plant cells and protoplasts by using a variety of polymers (Table 11.1). Immobilization of plant cells has been used for a wide range of reactions, which can be divided into three groups. (1) Biotransformation or bioconversion, (2) Synthesis from precursor and (3) the de novo synthesis of compounds.

11.1. ADVANTAGES OF PLANT CELL IMMOBILIZATION

- Retention of biomass enables its continuous reutilization as a production system, a definite advantage with slow growing plant cells e.g. *Papaver somniferum* have remained stable and active for up to six months.
- High biomass levels: The immobilization of cells allows the use of a higher biomass level compared to cell suspension culture, because of the limitation of mass transfer and settling, e.g. Bead

Table 11.1. Some examples of the immobilization of plant cells

Species	Family	Method	References
Amaranthus tricolor	Amaranthaceae	Chitosan	Knorr and Teutonico (1986)
Apium graveolens	Umbelliferae	Alginate	Watts and Collin (1985)
Asclepias syriaca	Ascepiadaceae	Chitosan	Knorr et al. (1985)
Beta vulgaris	Chenopodiaceae	Polyurethane	Rhodes et al. (1985)
Brassica oleracea	Cruciferae	Alginate	Redenbaugh et al. (1986)
Brassica napa	Cruciferae	Agarose	Shillito et al. (1983)
Cannabis sativa	Cannabidaceae	Alginate	Jones and Veliky (1981)
Capsicum frutescens	Solanaceae	Polyurethane	Lindsey et al. (1983)
Capsicum annuum	Solanaceae	Alginate	Veeresham et al. (1991)
Catharanthus roseus	Apocynaceae	Alginate	Asada and Shuler (1989)
Cephaelis ipecacuanha	Rubiaceae	Alginate	Veeresham et al. (1994)
Cinchona pubescens	Rubiaceae	Polyurethane	Rhodes et al. (1985)
Commiphora wightii	Burseraceae	Alginate	Prashant et al. (1989)
Coffea arabica	Rubiaceae	Alginate	Haldimann and Brodelius (1986)
Crepis capillaris	Composite	Agarose	Shillito et al. (1983)
Datura innoxia	Solanaceae	Alginate	Gontier et al. (1994)
Daucus carota	Umbelliferae	Polyurethane	Lindsey et al. (1983)
Digitalis lanata	Scrophulariaceae	Alginate	Alfermann et al. (1980)
Glycine max	Leguminosae	Hollow fibres	Shuler (1981)
Glycyrrhiza echinata	Leguminosae	Alginate	Ayabe et al. (1986)
Haplopappus gracilis	Compositae	Alginate	Tramper (1985)
Humulus lupulus	Cannabidaceae	Polyurethane	Rhodes et al. (1985)
Hyocsyamus muticus	Solanaceae	Agarose/agar	Lorz et al. (1983)
Ipomoea sp.	Convolvulaceae	Alginate	Jones and Veliky (1981)
Jasminium sp.	Oleaceae	Polyurethane	Dainty et al. (1985)
Lavandula vera	Labiatae	Carrageenan/agar	Nakajima et al. (1985)
Lycopersicon peruvianum	Solanaceae	Agarose	Adams and Townsend (1983)
Mentha spp.	Labiatae	Polyacrylaminde	Galun et al. (1985)
Morinda citrifolia	Rubiaceae	Alginate	Brodelius et al. (1979)
Mucuna pruriens	Leguminosae	Alginate	Brodelius et al. (1979)
Nicotiana sylvestris	Solanaceae	Polyacrylaminde	Galun et al. (1985)
Eschscholtzia califorhia		Alginate	Villegas et al. (1999)
Papaver somniferum	Papaveraceae	Alginate	Furuya et al. (1984)
Petunia hybrida	Solanaceae	Agarose	Shillito et al. (1983)
Salvia miltiorrhiza	Labiatae	Alginate	Miyasaka et al. (1986)
Solanum aviculare	Solanaceae	Phenylene oxide	Jirku et al. (1981)
Solanum tuberosum	Solanaceae	Agarose	Adam and Townsend (1983)
Taxus cuspidata	Taxaceae	Glass fiber mats	Fett Neto et al. (1992)
Vicia faba	Leguminosae	Alginate	Schanable and Youngman (1985)

densities of 110 g dry weight/L have been obtained with calcium alginate entrapped cells whereas 30 g dry weight/L in suspension cultures. The high cell density allows a reduction in contact time in packed bed catalyst leading to an increased volumetric productivity.

- Separation of cells from medium: The immobilization separates cells from medium and the desired product is extra cellular, which will simplify down stream processing compared to extraction from tissue.

- Continuous process: Immobilization allows a continuous process, which increase volumetric productivity and allows the removal of metabolic inhibitors.
- Decoupling of growth and product formation: Immobilization is compatible with non-growth associated product formation.
- Reduces problems such as aggregate, growth and foaming: - The immobilization reduces some of the physical problems associated with the cultivation of plant cells such as the formation of aggregates, and susceptibility to mechanical damage (shear stress) are problems which do not affect immobilized system compared to cell culture.

11.2. DISADVANTAGES

- Secretion of secondary metabolites requires cellular transport or artificially altered membrane permeability.
- The efficiency of the production process depends on the rate of release of products rather than actual rate of biosynthesis.
- The immobilization process may reduce biosynthetic capacity.
- Products must be released from the cell into medium.
- Release of single cells from cell aggregate may make processing of the product more difficult.
- The microenvironment favouring optimal production can be unfavourable for released secondary metabolites and cause their degradation or metabolization.

    The prerequisites for a successful immobilization of plant cells are as follows :
    1. Non-growing cells must produce products.
    2. Products must be released from the cell into the medium.

11.3. NEED FOR IMMOBILIZATION

Plant cells are characterized by large size, sensitivity to shear and oxygen and need of a cell to cell contact for metabolite production. The secondary metabolites are triggered by short periods of stress in cultures. Immobilization can overcome many of the limiting factors of suspension cultures with the distinct advantages of easier operation of biocatalyst from the product and also being amenable for biotransformation of low value compounds to high value products.

11.4. DIFFERENT TYPES OF IMMOBILIZATION

The principle involved in immobilization of plant cell is depicted in Fig. 11.1. & Table 11.2
1. Direct intracellular binding due to natural affinity (adsorption, adhesion and agglutination).
2. Covalent coupling on otherwise inert matrices.
3. Intracellular connection via bi or poly functional reagent (cross-linking).
4. Mixing with suitable materials, changing their consistency with temperature (embedding).
5. Physical retention within the framework of diverse pore size and permeability (entrapment, micro encapsulation).

11.5. SELECTION OF IMMOBILIZATION SYSTEM

The choice of a suitable immobilization system is determined by the following requirements.
1. The polymer material (Table 11.4) used for immobilization must be available in large quantities; it must be inert, non-toxic and cheap.
2. It must be able to carry large quantities of biomass and its fixing potential must be high.

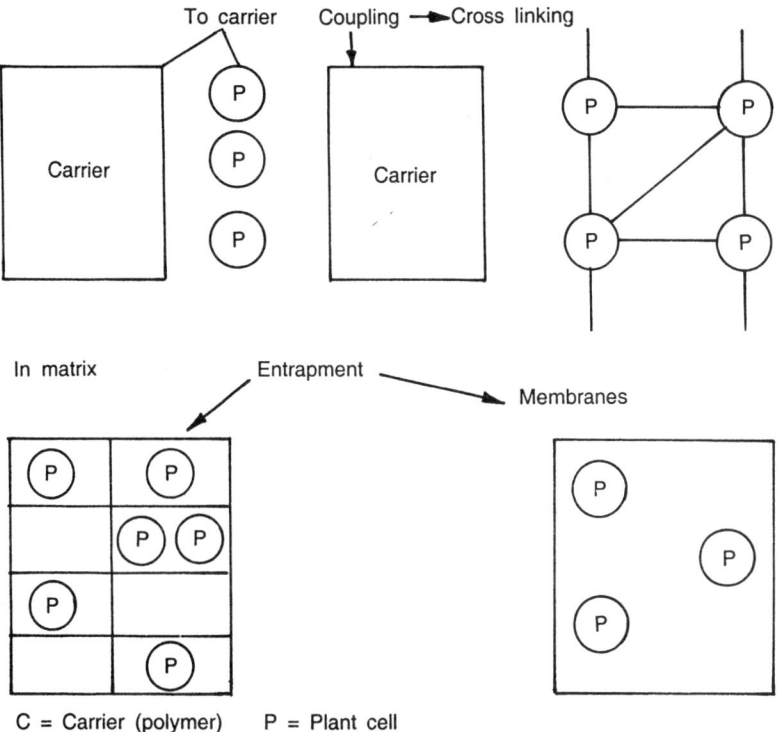

To carrier   Coupling ➝ Cross linking

Carrier

P

P

P

Carrier

P ─── P

P ─── P

In matrix   Entrapment

Membranes

P   P

P P

P

P

P

P

P

P

C = Carrier (polymer)   P = Plant cell

Fig. 11.1. Principles of plant cell immobilization.

3. The immobilization process must not diminish enzymatic activity of biological catalyst.
4. Manipulation of the biological catalyst must be as simple as possible (Hall *et al* 1988). (See Table 11.4)

11.6.1. Entrapment

11.6.1.1. Gel entrapment by polymerization

A monomer or a mixture of monomers is polymerized in the presence of a cell suspension, which is entrapped inside the lattice of the polymer. The most common example is polyacrylamide. The method is based on the free radical polymerization of acrylamide in an aqueous solution. As the linear polymers are soluble in water, they have to be insolubilized with bifunctional compounds such as N, N'-ethylene bisacrylamide. The free radical polymerization of acrylamide is conducted in an aqueous solution containing the cells and the cross-linking agent. Polymerization is commonly carried out in the absence of oxygen and at lower temperature ($10°C$) to avoid damage to the cell during the operation. An initiator N, N, N' N' - tetramethylethylene diamine (TEMED) is used (Table 11.5 and 11.6).

Both the initiation and the cross-linking agents are toxic to the cells and therefore, their viability can be lost e.g. *C. roseus* (Brodelius 1983) and *Silybum marianum* (Cabra *et al.,* 1984).

11.6.1.2. Gel entrapment by ionic net work formation (Table 11.8)

In this method, polymerization of polyelectrolytes is achieved by addition of multivalent ions. The most common method is the entrapment in calcium alginate. This is a non-toxic process in which

Table 11.2. Methods of Immobilization

- Adsorption
- Covalent attachment
- Entrapment
  - a) Natural polymer
     Alginate
     Agar
     Agarose
     κ-carrageenan
     Chitosan
  - b) Synthetic polymer
     Polyacrylamide
  - c) Porus structure
     Polyurethane foam
  - d) Membranes
     Hollow fiber
     Flat plate

Table 11.3. The factors influencing the stability of alginate beads.

- Chelating ion $Ca^{+2}, Ba^{+2}, Cu^{+2}, Sr^{+2}, Cd^{+2}, Mg^{+2}$ order of stability.
- Alginate concentration.
- Metal ion chelators (Phosphates, citrate, lactate).
- Composition of the alginate.
- Molecular weight of the alginate polymers.
- Duration of incubation in $CaCl_2$ solution.
- Density of the suspension culture.

Table 11.4. Polymers used for immobilization of plant cells

Polymer	Gelation mechanism
Agar	Gel formation in cold
Agarose	Ionotropic gel formation
Alginate	Ionotropic gel formation
κ-Carrageenan	Ionotropic gel formation
Chitosan	Ionotropic gel formation
Gelatin	Gel formation in the cold crosslinking
Gellan	Ionotropic gel formation, gel formation in the cold
Polyacrylamide	Polymerization
Polyacrylamide hydracide	Crosslinking
Polyphenylene oxide	Crosslinking
Agarose - gelatin	Gel formation in the cold, crosslinking
Alginate - gelatin	Ionotropic gel formation, crosslinking
Alginate/Chitosan	Iontropic-polyelectrolytic coacervate formation
κ-Carrageenan/Chitosan	Iontropic-polyelectrolytic coacervate formation
Pectin/Chitosan	Iontropic-polyelectrolytic coacervate formation

Table 11.5. Methods of Polymerization of polymers used in plant cell immobilization (Scragg 1991)

Polymer	Polymer Concentration %	Method used
Alginate	2.5	Ionic cross linking (0.1 M Ca^{+2})
Agar	2.1	Cooling from 50 - 20°C
Agarose	2.5	Cooling from 40 - 20°C
Alginate and gelatin	2.0	Ionic cross linking (0.1 M Ca^{+2}) and chemical cross linking (1 % glutaraldehyde)
Agarose and gelatin	2.1	Ionic cross linking (0.1 M Ca^{+2}) and chemical cross linking (1 % glutaraldehyde)
Carrageenan	1.5	Crosslinking from 50 to20°C ionic crosslinking with potassium

Table 11.6. Some polymers used for entrapment of cells

Polymer	Principle of Gel formation	Gel-forming procedure
Synthetic		
Polyacrylamide	Polymerization	Addition of initiators to a solution of monomer
Epoxy resin	Polycondensation	Epoxy and polyamine precursors are cured
Polyurethane	Polycondensation	Reaction of polysocyanates with water
Agar, agarose	Thermal gelation	Cooling of heated polymer solution
Kappa-carrageenan	Thermal gelation + Ionotropic stabilization	Cooling of heated polymer solution
Alginate	Ionotropic gelation	Dripping into a calcium chloride solution
Chitosan	Ionotropic gelation	Dripping into a polyphosphate solution
Proteins		
Collagen	Neutralization	Neutralization of an acidic solution
Gelatin	Thermal gelation	Cooling of heated polymer solution
Fibrin	Enzymatic gelation	Enzymatic conversion of fibrinogen to fibrin

Table 11.7. Gel entrapment by polymerization

Species	Immobilization method	Reference
C. roseus	Polysaccharide/Polyacrylamide	Lambe and Rosevear (1983)
Catharanthus roseus	Polyacrylamide	Brodelius (1983)
Lavandula vera	Polyvinylalcohol	Nakajima et al. (1985)
Mentha sp.	Prepolymerized/polyacrylamide	Galun et al. (1983)
Nicotiana sp.	Polysaccharide/polyacrylamide	Rosevear and Lambe (1985)
Silybum marianum	Polyacrylamide	Cabral et al. (1984)

sodium alginate solution containing the cell suspension is dropped into a mixture of counter ion solution such as calcium chloride. A uniform, spherical and highly microporous structure results, which retains the cells (Fig. 11.2).

The method may be inconvenient over long run in media containing calcium chelating agents, such as phosphates and certain cations such as Mg^{+2} that may accelerate disruption of gel by solubilizing bound Ca^{+2}. Carrageenan is also a suitable matrix for the immobilization of plant cells. The gels of calcium and potassium carrageenates are very stable at room temperature and at a pH above 4.5.

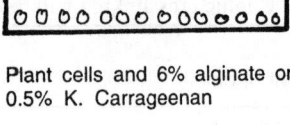
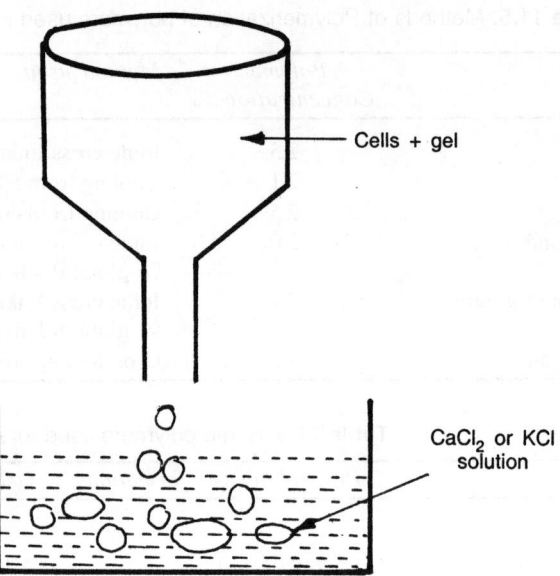

Plant cells and 6% alginate or
0.5% K. Carrageenan

Cells + gel

CaCl₂ or KCl
solution

Fig. 11.2. Protocol for gel immobilization of plant cells.

Table 11.8. Gel entrapment by ionic net work formation

Species	Gel	Reference
Catharanthus roseus	Alginate, Carrageenan	Brodelius *et al.* (1979)
	Carrageenen	Brodelius and Nilsson (1980)
	Alginate	Brodelius *et al.* (1981)
Daucus carota	Alginate	Jones and Veliky (1981)
Digitalis lanata	Alginate	Brodelius *et al.* (1979)
		Brodelius and Nilsson (1980)
		Alferman *et al.* (1983)
Lavendula vera	Alginate	Nakajima *et al.* (1985)
Morinda citrifolia	Alginate	Brodelius *et al.* (1981)
Papaver somniferum	Alginate	Furuya *et al.* (1984)

11.6.1.3. *Gel entrapment formation by precipitation* (Table 11.9)

Gels may be formed by precipitation of some natural and synthetic polymers by changing one or more parameters in the solution, such as temperature, salinity or pH of solvent. Several materials can be used for entrapment. The examples include methods involving thermal treatment. Some disruption of viability can occur naturally.

Table 11.9. Gel entrapment by precipitation

Species	Gel	Reference
Catharanthus roseus	Agarose	Felix *et al.* (1981)
	Agrose, agar	Brodelius and Nilsson (1980)
Silybum marianum	Agar	Cabral *et al.* (1983)

11.6.1.4. Entrapment in preformed structures

Hollow fiber reactors can be used to immobilize plant cells by entrapment. The cells are placed on the shell side of the reactor and nutrient medium is rapidly re-circulated through the fibers. This may have important applications in large-scale operation (Shuler, 1981).

In other examples, the cells are added to preformed polymerized structures such as polyurethane foam. When cells in suspension are mixed with these materials, they are rapidly incorporated into the net work and subsequently grown into the cavities of the mesh and are entrapped by physical restriction and attachment to the matrix material. The mechanism of this involvement is at first a mechanical entrapment and later, the fixation of the cells due to mechanisms of adsorption and adhesion or even due to their natural tendency for aggregation. These methods of immobilization have several advantages over other methods in that they are simple, cheap, gentle and rapid and maintain cellular functions (Table 11.10).

Table 11.10. Entrapment in preformed structures

Species	*Gel*	*Reference*
Capsicum frutescens	Preformed polyurethane foam	Lindsey *et al.* (1983)
		Mavituna and Park (1985)
Dioscorea deltoidea	Preformed polyurethane foam	Ishida (1988)
Glycine max	Hollow-fiber reactors	Shuler (1981)
Hop and Beet root	Nylon pan Scrubers	Rhodes and Kirsop (1982)
Lavendula vera	Urethane prepolymers	Tanaka *et al..* (1985)
Petunia hybrida	Hollow-fiber reactors	Jose *et al..* (1983)

11.6.2. Surface immobilization

Surface immobilization may occur on both natural and other matrices. Examples of natural matrices are deeper callus layers and cellulose, while synthetic one includes nets of steel and nylon. For e.g. Cells of *Solanum aviculare* were covalently linked to beads of polyphenylene oxide, which had been achieved using glutaraldehyde (Jirku *et al.*, 1981). Archambault *et al* (1986) described the spontaneous and rapid binding of *C. roseus* cells to a man made material.

11.6.3. Immobilization by embedding

The temperature dependent solubility of macromolecules like agarose, agar and carrageenan or the differing solubility of the sodium and calcium salts in the case of alginate are utilized to form polymeric gels or gel combination (Fig. 11.2). Insoluble are formed under cold conditions (agar) or in aqueous $CaCl_2$ solutions (alginate). Their structure is non-uniform, with differing pore diameters at the surface and in deeper layers. The size and form of the beads can be determined in part by stirring speed and using alginate, by the viscosity of the solution and dropping aperture.

11.7. STABILITY OF IMMOBILIZED BEADS

Independent of other factors (Table 11.3) alginate beads with a content of α-L-glutamic acid greater than 70% have the great mechanical strength. However, less viscous low molecular weight polymers are preferred because they can be sterilized by filtration.

11.8. VIABILITY TESTING OF IMMOBILIZED PLANT CELLS

In principle, the same standard methods can be employed to study the viability of immobilized plant cells as are used for suspended cells.

11.8.1. Viability staining

Various staining techniques may be employed to study the viability of cultivated plant cells. Fluorescein diacetate (FDA) is used to show viability of immobilized cells and protoplasts. The polymeric material (i.e. the gel) does not interfere when this dye is used.

Phenosafranine can also be used to study the viability of immobilized cells and protoplasts. An advantage of this dye is that a standard light microscope can be used. When this stain is used, cells appear red.

11.8.2. Plasmolysis

Plasmolysis offers a simple way to determine the integrity of plasma membrane. After the addition of a plasmolizing agent (e.g. glycerol or sorbitol), the immobilized cells are inspected under the microscope. To facilitate the inspection of cells, dye (e.g. phenosafranine) can be added to the hypertonic medium.

11.8.3. Respiration

Respiration is measured with an oxygen electrode of Clark type. The beads are suspended in nutrient medium, and the consumption of oxygen as a function of time is monitored.

11.8.4. Cell growth

The cell growth is of course a good indicator of viability. The dry weight increase of an immobilized cell preparation can be determined assuming constant dry weight of the polymer constituting the gel.

11.8.5. Cell division

The mitotic index is a measurement of how large a fraction of the cells is in mitosis. The immobilization does not affect the mitotic index for agarose-entrapped cells, as compared to freely suspended cells.

11.8.6. ^{31}P Nuclear Magnetic Resonance (NMR)

In *vivo*, ^{31}P NMR have recently emerged as a very useful method for the study of bioenergetics and metabolism of living systems. Vogel and Brodelius (1984) adopted this non-invasive technique to study both freely suspended and immobilized plant cells.

11.9. PHYSIOLOGICAL EFFECTS OF IMMOBILIZATION

Though immobilization was originally proposed for it's bioprocessing advantages, experimental evidence indicates that immobilization that can have an impact on cell physiology and production of secondary metabolites as well. Lindsey (1985) reported that the process of immobilization reduces the rate of cell division and protein synthesis and these effects were conducive to increase secondary metabolite yield. Hall *et al*. (1986) have reported that another important consequence of plant cell immobilization is to reduce the production of cell wall material which contains a substantial amount of bound phenolic compounds. This will cause increased availability of precursors for secondary metabolism. Some kind of differentiation is required to produce secondary metabolites in either plant or plant cell cultures (Veeresham, 1990). Differentiation refers to metabolic or morphological specialization of cells. "Biochemical differentiation" is defined as metabolic specialization, which controls the expression of specific enzymatic pathways for bioproduction of secondary metabolites. Structural differentiation is defined as the morphological differentiation. The process of immobilization causes the plant cells to feel as biochemical differentiation (Payne *et al.* 1992, Table 11.11) .

11.10. TYPES OF BIOREACTORS USED FOR IMMOBILIZED OF PLANT CELLS

The following types of reactors are generally used for immobilized plant cell.

11.10.1. Packed bed reactors

In this reactor, cells can be immobilized either on the surface or throughout the support, and fluid-containing S flows past the support particles. When the cells are immobilized through the support, the packed bed can accommodate a large number of cells per reactor volume (see Fig. 11.3).

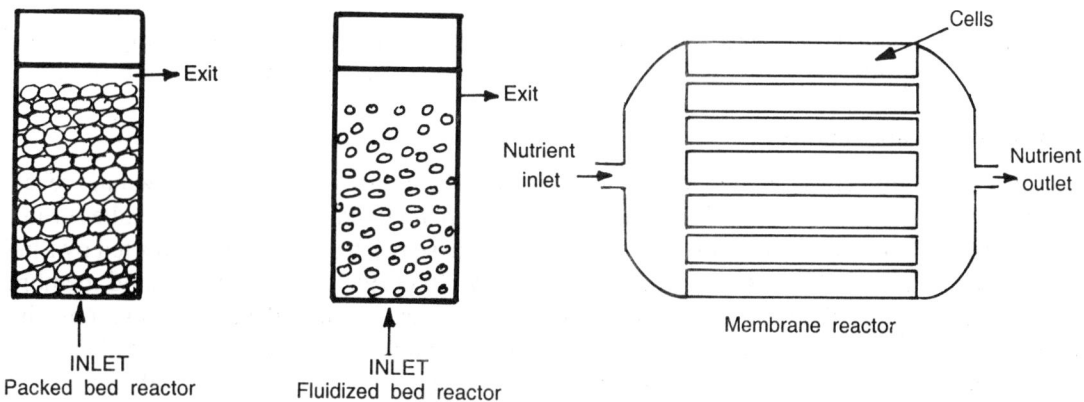

Fig. 11.3. Bioreactors used for plant cell immobilization.

Disadvantages

1. Low degree of mixing causes difficulties in transfers of oxygen, CO_2 etc., when it is necessary.
2. The pressure required to pump fluid through the packed bed is inversely proportional to the support particle size, and thus reduces the energy requirements for flow, large incompressible support particles are needed.
3. The packed bed reactors are having filters. When particulate matter are included in the media or when the support particles fragment during operation, they will be trapped in the column and block the pathways for fluid flow.

11.10.2. Well mixed reactors

The advantages of a well-mixed reactor for immobilized cells are that gaseous substrates can be directly sparged, the temperature and pH can be carefully controlled, and mass transfer rates can be improved over packed bed reactors. The suspension culture with complete cell recycle is essentially a well-mixed "immobilized" cell reactor. (see Fig. 11.3).

Disadvantages

The disadvantages of well-mixed reactor for immobilized cell systems are the possibility of fragmenting the support due to particle collision and shear. Using gentle pneumatic agitation can reduce this problem.

11.10.3. Fluidized bed reactors

Typical fluidized bed reactors utilize the energy of the flowing fluid (liquid and/or gas) to suspend the particles. Because the energy required for fluidization increases with increased particle size, small-immobilized particles are often employed. The mass transfer advantages of these small particles is one of the major benefits of the fluid must often be retained in the reactor for an extended period. This may be incompatible with the requirement that the fluid flow rate be sufficient to suspend the particles. To satisfy these two requirements, large gas volumes can be used to suspend the immobilized cells while maintaining low liquid flow rates, or the liquid can be rapidly recirculated through the bed. Both these conditions lead to a large degree of fluid mixing. Large recirculation rates were in the fluidized bed reactors (Prenosil and Pederson, 1983 and Morris et al., 1983).

Disadvantages

The disadvantages of the fluidized bed reactor are that shear and particle collisions may damage the immobilized supports and the complex fluid dynamics of such reactors make scale-up difficult.

11.10.4. Membrane reactors

The most commonly considered membrane reactors are the hollow fiber and spiral wound reactors shown in Fig. 11.3. In the hollow fiber reactor, the cells are retained either within the tubes or in the outer region. Preliminary studies to evaluate the use of these reactors have been reported by Shuler et al., (1983); Jose et al., (1983) and Prenosil and Pederson (1983). The spiral wound membrane reactor is essentially a flat plate reactor rolled into a cylindrical shape and although no reports have appeared on the use of spiral wound reactor. Hallsby and Shuler (1986) have studied plate reactor with tobacco cells. The studies have described that in addition to the membrane, the thickness of the cell layer is also important. The inner portions of thick cell layers are generally characterized by substrate deficiencies and product accumulation and these conditions may prove beneficial for the morphological and chemical differentiation of plant cells.

Advantages

The advantage of membrane reactors is the possibilities that these membranes can be reused. Thus, despite the high capital costs, the reusability may make this form of immobilization more economical. Also because plant cells are metabolically less active than microbes, thicker cell layers can be employed. Since reactor inoculation, or cell loading is easier with larger cell layer thicknesses it may be possible to develop plant cell reactors, which can be easily inoculated.

11.11. APPLICATION OF PLANT CELL IMMOBILIZATION

11.11.1. Biotransformations

Hydroxylation of cardiac glycosides has proved an interesting application of immobilized plant cells. Alfermann et al. (1982) used Digitalis lanata in alginate beads in a process aimed at converting β-methyldigitoxin into clinically more β-methyl-digoxin. In a batch process, the fluid bathing of the cells was changed every second day over a period of 59 days and resulted in a higher cell productivity than that of a free cell system. Brodelius (1981) successfully used a column of similar beads for upto 70 days. Using alginate immobilized cells of Daucus carota, Jones and Veliky (1981) achieved 5-hydroxylation of digitoxigenin into periplogenin over a period of 12 days, but noted a deterioration in biocatalytic activity due to accumulation of inhibitory levels of toxic substrate in the cells.

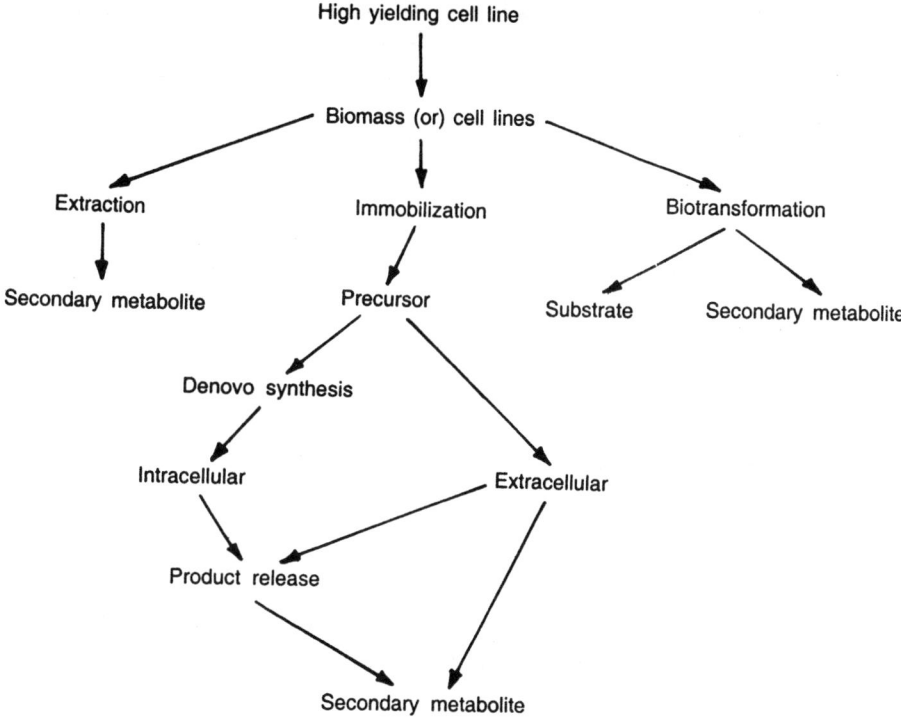

Fig. 11.4. Use of plant cell immobilization.

11.11.2. Biosynthesis

The biosynthetic capacity of immobilized viable plant cells has been demonstrated in a number of model studies, as summarized in Table 11.11. From all these model studies, it can be concluded that the immobilized cells can carry out the same biosynthetic reactions as the corresponding freely suspended cells. Infact, in certain instances, a higher productivity has been observed for the immobilized cells as shown in Table 11.11. It has been suggested that this increased synthesis by immobilized plant cells be seen because these cells are metabolically closer to cells immobilized in the whole plant than to the liquid suspended cells, from which they were derived.

Table 11.11. Effects of immobilization on secondary metabolite production in immobilized cultures compared to suspension culture (Adopted from Payne et al., 1992).

Species	Product	x-Fold change	Type of immo-bilization	References
Capsicum frutescens	Capsaicin	>100	Foam	Lindsey and Yoeman (1984a)
Capsicum annuum	Capsaicin	>100	Gel	Ravishankar et al. (1988)
Tagetes patula	Thiophenes	ca.$_{20}$	Natural aggregates	Hulst et al. (1989)
Coffea arabica	Methyl xanthines	13	Gel	Haldimann and Brodelius (1987)
Catharanthus roseus	Ajmalicine	3.5	Gel	Asada and Shuler (1989)
Mucuna pruriens	L-DOPA	*de novo* synthesis	Gel	Wichers et al. (1983)

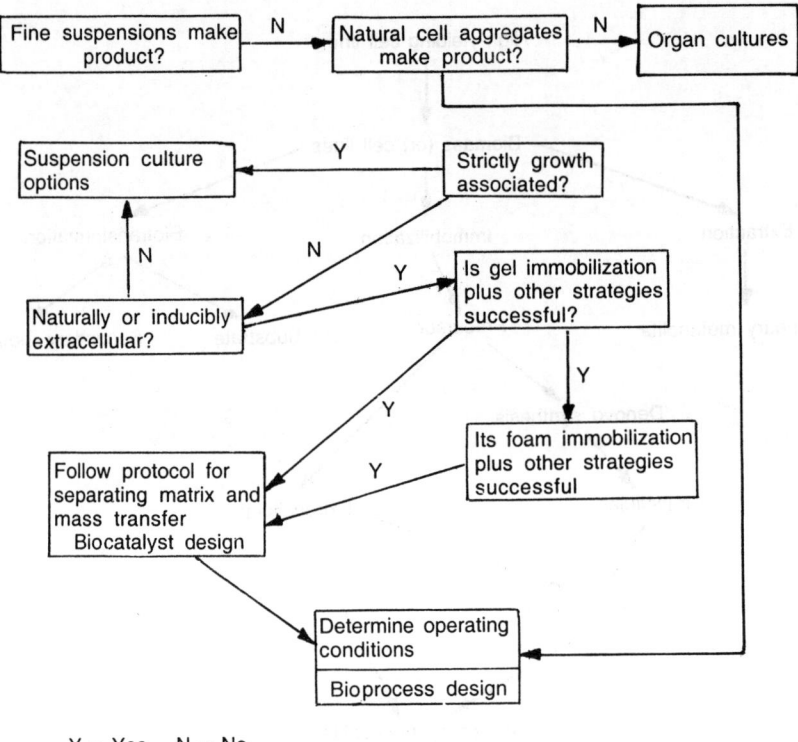

Y = Yes   N = No

Fig. 11.5. Immobilization - Decision making (Adopted from Payne et al., 1992).

Table 11.12. Comparison of the effects of various immobilization methods on *Catharanthus roseus* cells (Adopted from Brodelius 1980).

Methods of Immobilization	Plasmolysis	Respiration	Cells grown
Alginate	+	+	+
Carrageenan	+	+	+
Agar	+	+	+
Agarose	+	+	+
Gelatin	+	−	−
Alginate and gelatin	+	−	−
Agarose and gelatin	+	−	−
Polyacrylamide	−	−	−

Postive reference + ; Cells did not undergo −

REFERENCES

Adams, T.L., Townsend, J.A. (1983). A new procedure for increasing efficiency of protoplast plating and clone selection. *Plant Cell Rep.,* **2:**165-168.

Alfermann, A.W., Bergmann, W., Figur, C., Helmblod, U., Schwantag, D., Schuler, I., Reinhard, E. (1983). Biotransformation of α-methyldigitoxin to β-methyldigoxin by cell cultures of *Digitalis lanata.* In:Mantell SH, Smith H (eds) Plant Biotechnology. Univ. Press, Cambridge, pp 67-74.

Alfermann, A.W., Schuler, I., Reinhard, E. (1980). Biotransformation of cardiac glycosides by immobilized cells of *Digitalis lanata. Plant Medica,* **40:**218-223.

Archambault, J., Volseky, B., Kurz, W.G.W. (1986). Surface immobilization of plant cells. In:Somers DA, Gengenbach BG, Biesboer DD, Hacket WP, Green CE (eds) Proc VI Int Congr Plant tissue and cell culture, vol 1. IAPTC, Minneapolis, p 451.

Asada, M., Shuler, M.L. (1989). Stimulation of ajmalicine production and excretion from *Catharnthus roseus :* Effect of Adsorption In situ, Elicitors, and alginate Immobilization. *Appl. Microbiol.Biotechnol.* **30:** 475-481.

Ayabe, S., Iida, K., Furuya, T. (1986). Induction of stress metabolites in immobilized *Glycyrrhiza echinata* cultured cells. *Plant Cell Rep.,* **5:**186-189.

Brodelius, P. (1983). Production of biochemicals with immobilized plant cells. *Ann NY Acad Sci.,* **413:** 383-393.

Brodelius, P. (1985a). Immobilized plant cells. In:Laskin AI (ed) Enzymes and immobilized cells in biotechnology. Benjamin, Cummings, CA, pp 109-148.

Brodelius, P., Deus, B., Mosbach, K., Zenk, M.H. (1979). Immobilization of plant cells for the production and transformation of natural products. *FEBS Lett.,* **103:** 93-97.

Brodelius, P., Deus, B., Mosbach, K., Zenk, M.H. (1981). Catalysts for the production and transformation of natural products having their origin in higher plants. Process for production of the catalysts, and use there of. *European Patent Application 80850105.1.*

Brodelius, P., Nilsson, K. (1980). Entrapment of plant cells in different matrices. *FEBS Lett.,* **122:** 312-316.

Cabral J.M.S., Fevereiro, P., Novais, J.M., Pais, M.S. (1984). Growth of immobilized plant cells in reticulate polyurethane foam matrices. *Biotech. Lett.,* **7(9):** 637-640.

Cabral, J.M.S., Fevereiro, F., Novais, J.M., Pais, M.S.S. (1983). Comparison of immobilization methods for plant cells and protoplasts. Poster presented at Enzyme Engineering VII, White Haven, Pennsylvania, September, 25-20, 1983.

Dainty, A.L., Goulding, K.H., Robinson, P.K., Simpkins, I., Trevan, M.D. (1985). Effect of immobilization on plant cell physiology - real or imaginary? *Trends Biotechnol.,* **3:** 59-60.

Felix, H.R., Brodelius, P., Mosbach, K. (1981). Enzyme activities of the primary and secondary metabolism of simultaneously permeabilized and immobilized cells. *Analyt. Biochem.,* **116:** 462-470.

Fetto-Neto, A.G., Dicosmo, F., Reynolds, W.F., Sakata, K. (1992). Cell culture of Taxus as source of the antineoplastic drug taxol and related taxanes. *Biotechnology,* **10 :** 1572-1575.

Furuya, T., Yoshikawa, T., Taira, M. (1984). Biotransformation of codeinone to codeine by immobilized cells of *Papaver somniferum. Phytochemistry,* **23:** 999-1002.

Galun E, Aviv D, Dantes A, Freeman A (1985) Biotransformation by division-arrested and immobilized plant cells: Bioconversion of monoterpenes by Gamma- irradiated, suspended and entrapped cells of *Menta* and *Nicotiana. Planta Med.,* **51:** 511-514.

Galun, E., Aviv, D., Dantes, A., Freeman, A. (1983). Biotransformation by plant cells immobilized in crosslinked polyacrylamide-hydrazine. *Planta Med.,* **49:** 9-13.

Gontier, E., Sangwan, B.S., Barbotin, J.N. (1994). Effects of calcium, alginate and calcium alginate immobiliztion on growth and tropane alkaloid levels of stable suspension cell line of *Datura innoxia* mill. *Plant Cell Reports.,* **13:** 533-536.

Haldimann, D., Brodelius, P. (1987). Redirecting cellular metabolism by immobilization of cultured plant cells.A model study with *Coffea arabica. Phytochemistry,* **26 :** 1431-1434.

Hall R.D., Yeoman, M.M. (1986a) Temporal and spatial heterogeneity in the accumulation of anthocyanins in cell cultures of *Catharanthus roseus. J. Exp. Bot.,* **37:** 48-60

Hall, R.D., Holden, M.A., Yeoman, M.M. (1986a). Studies on the regulation of capsaicin biosynthesis in immobilized cell cultures and the developing fruit of the chilli pepper *Capsicum frutescens.* In : Morris P, Scragg aH, Stafford A, Fowler MW (eds) Secondary metabolism in plant cell cultures. Univ. Press, Cambridge, pp. 121-127.

Hall, R.D., Holden, M.A., Yeoman, M.M. (1986b). The accumulation of phenylpropanoid and capsaicinoid compounds in cell cultures and whole fruit of the chilli pepper *C. frutescens. Plant Cell Tiss. Org. Cult.,* **8** : 163-176.

Hall, R.D., Holden, M.A., Yeoman, M.M. (1988). Immobilization of higher plant cells. In:Bajaj YPS (ed) Medicinal and aromatic plants. Biotechnology in Agriculture and Forestry, Springer, Berlin, Heidelberg, New York, p. 136.

Hall, R.D., Yeoman, M.M. (1986b). Factors determining anthocyanin yield in cell cultures of *Catharanthus roseus. New Phytol.,* **103:** 33-43.

Hallsby, G.A., Shuler, M.L. (1986). Altering fluid flow patterns changes of cellular association in immobilized tobacco tissue cultured. *Biotechnol. Bioeng. Symp. Ser.,* **17** : 741-746.

Hulust, A.C., Meyer, M.M.T., Breteler H., Tramper, J. (1989). Effect of aggregate size in cell cultures of *Tagetes patula* on thiopene production and cell growth. *Appl. Microbiol. Biotechnol.* **30:** 18-25.

Ishida, B.K. (1988). Improved diosgenin production in *Dioscorea deltoidea* cell cultures by immobilization in polyurethane foam. *Plant Cell Reports.,* **7:** 270-273.

Jirku, V., Macek, T., Vank, T., Krumphanzl, V., Kubanek, V. (1981). Continuous production of steroid glycoalkaloids by immobilized plant cells. *Biotechnol. Lett.,* **3:** 447-450.

Jones, A., Veliky, A. (1981). Effect of medium constituents on the viability of immobilized plant cells. *Can. J. Bot.,* **59:** 2095-2101.

Jose, W., Pedersen, H., Chin, C.K. (1983). Immobilization of plant cells in a hollow-fibre reactor. *Ann. NY Acad. Sci.,* **413:** 409-412.

Jose, W., Pedersen, H., Chin, C.K. (1983). Immobilization of plant cell in a hollow fiber reactor. *Ann. NY Acad. Sci.,* **413:** 409-412.

Knorr, D., Miaza, S.M., Teutonico, R.A. (1985). Immobilization and permeabilization of cultured plant cells. *Food Technol.,* **39:**135-142.

Knorr, D., Teutonico, R.A. (1986). Chitosan immobilization and permeabilization of *Amaranthus tricolor. J. Agric. Food Chem.,* **34:** 96-97.

Lambe, C.A., Rosevear, A. (1983). Immobilized plant cells. *Proc. Biotech.,* **83:** 565-576.

Lindsey, K. (1985). Manipulation, by nutrient limitation, of the biosynthetic activity immobilized cells of *Capsicum frutescens* Mill. Ev. *Annuum. Planta.,* **165:** 126-133.

Lindsey, K., Yeoman, M.M. (1984a). The synthetic potential of immobilized cells of *Capsicum frutescens. Planta,* b162 :b 495-501.

Lindsey, K., Yeoman, M.M., Black, G.M., Mavituna, F. (1983). A novel method for the immobilization and culture of plant cells. *FEBS Lett.,* **155:** 143-149.

Lindsey, K., Yoeman, M.M. (1983b). The relationship between growth rate, differentiation and alkaloid accumulation in cell cultures. *J. Exp Bot.,* **34:** 1055-1065.

Lorz, H., Larkin, P.J., Thomson, J., Scoweroft, W.R. (1983). Improved protoplast culture and agarose media. *Plant Cell Tiss. Org. Cult.,* **2:** 217-226.

Mavituna, F. and Park, J.M. (1985). Growth of immobilized plant cells in reticulate poly urethane foam matrices. *Biotech. Lett.,* **7(9):** 637-640.

Morris, P., Smart, N.J., Fowler, M.W. (1983). A fluidised bed vessel for the culture of immobilized plant cells and its application for the continuous production of fine cell suspensions. *Plant Cell Tiss. Org. Cult.,* **2:** 207-216.

Myasaka, H., Nasu, M., Yamamoto, T., Endo, Y., Yoneda, K. (1986). Production of cryptotanshinone and ferruginol by immobilized cultured cells of *Salvia miltiorrhiza. Phytochemistry.,* **25:** 1621-1624.

Nakajima, H., Sonomoto, K., Usui, N., Sato, F., Yamada, Y., Tanaka, A., Fukui, S. (1985). Entrapment of *Lavandula* \vera cells and production of pigments by entrapped cells. *J. Biotechnol.,* **2:**107-117.

Payne, G.F., Bringi, V., Prince, C., Shuler, M.L. (1992). In; Plant cell and tissue cultures in liquid systems. Hanser publisher, Munnich, Vienna, Newyork, *Barcelona,* 181.

Prashant, P., Subramani, J., Bhatt, P.N., Mehta, A.R. (1989). Viability and gugulsteroid production in immobilized tissue cultured cells of *Commiphora weightii*. *Ind. J. Exp. Biology,* **27** : 338-340.

Prensoil, J.E., Pedersen, H. (1983). Immobilized plant cell reactors. *Enz. Microb. Technol.,* **5**: 323-331.

Ravishankar, G.A., Sarma, K.S, Venkarraman, L.V., Kadyan, A.K. (1988). Effect of Nutritional stress on *Capsaicin* production in immobilized cell cultures of *Capsicum annuum*. *Curr. Sci.* **57** : 381-383.

Redenbaugh, K., Paasch, B.D., Nichol, J.W., Kossler, M.E., Viss, P.R., Walker, K.A. (1986). Somatic seeds: encapsulation of asexual plant embros. *Biotechnology.,* **4:** 797-801.

Rhodes, M.J.C. (1985). Immobilized plant cell cultures. *Top Enz. Ferment. Biotechnol.,* **10:** 51-87.

Rhodes, M.J.C., Robins, R.J., Turner, R.J., Smith, J.I. (1985). Mucilaginous film production by plant cells immobilized in a polyurethane or nylon matrix. *Can. J. Bot.,* **63:** 2357-2363.

Rhodes, M.J.C., Kirsop, B.H. (1982). Plant cell cultures as sources of valuable secondary products. *Biologist.,* **29:** 134-140.

Rosevear, A. (1982). Improvements in or relating to composite materials. *European Patent Application 81304001.1.*

Rosevear, A., Lambe, C.A. (1985). Immobilized plant cells. *Adv. Biochem. Eng.,* **31** : 37-58.

Schaabl, H., Youngman, R.J. (1985). Immobilization of plant cell protoplasts inhibits enzymic lipid peroxidase. *Plant Sci.,* **40:** 65-69.

Scragg, A. (1991). The immobilization of plant cells. In: Stafford A. Warren G (eds) Plant cell and tissue culture, Open University press, Milton, Keynes p 205.

Shillito, R.D., Paszkowski, J., Potrykus, I. (1983). Agarose plating and a bead-type culture technique enable and stimulate development of protoplast-derived colonies in a number of plant species. *Plant Cell Rep.* **2:** 244-247.

Shuler, M.L. (1981). Production of secondary metabolites from plant tissue culture- problems and prospects. *Ann. NY Acad. Sci.* **369:** 65-79.

Shuler, M.L., Sahai, O.P., Hallsby, G.A. (1983). Entrapped plant cell cultures. *Ann.NY. Acad. Sci.,* **413** : 373-382.

Tanaka, H., Hirao, C., Semba, H., Tozawa, Y., Ohmomo, S. (1985). Release of intracellularly stored 5'-phosphodiesterase with preserved plant cell viability. *Biotechnol Bioeng.,* **27** : 890-892.

Tramper, J. (1985). Immobilizing biocatalysts for use in synthesis. *Trends Biotechnol.,* **3** : 45-50.

Veeresham, C. (1990) Studies on tissue cultures of higher plants. Ph.D thesis submitted to Kakatiya University, Warangal, India.

Veeresham, C., Kokate, C.K., Venkateshwarlu, V. (1994). Influence of precursors on production of isoquinoline alkaloids in tissue cultures of *Cephaelis ipecacuanha*. *Phytochemistry*, **35(4)** : 947-949.

Veeresham, C., Kokate, C.K., Venkateshwarlu, V., Apte, S.S. (1991). Enhanced capsaicin production in immobilized cell cultures of *Capsicum annum*. *Indian Drugs.,* **29(1)** : 12-14.

Villegas, M. Rosail, Brodelius, P.E. (1999). Effects of alginate and immobilization by entrapment in alginate on benzophenanthridine alkaloid production in cell suspension cultures of *Eschschottzia california, Biotechnology Letters,* **21:**49-55.

Vogel, H.J. and Brodelius, P. (1984). An *in vivo* ^{31}P NMR Comparison of freely suspended and immobilized *Catharanthus roseus* plant cells. *J. Biotechnology,* **1:** 159-170.

Watts, M.J., Collin, H.A. (1985). Growth and nutrient uptake by immobilized tissue culture cells of celery (*Apium graveolens*). *Plant Sci. Lett.,* **42** : 67-72.

Wichers, H.J., Malingre, T.M., Huizing, H.J. (1983). The effect of some environmental factors on the production of L-DOPA by alginate-entrapped cells of *Mucuna pruriens*. *Planta.,* **158:** 482-486.

Chapter 12

Elicitation

INTRODUCTION

Stimulation of particular facets of plant metabolism is achieved by treatment with biotic (microbes especially fungi and their extracts) and abiotic (certain chemicals like methyl jasmonate, arachidonic acid and some heavy metals like copper sulphate, sodium orthovandate and silver nitrate, etc.) compounds to enhance or increase the yield of desired secondary metabolite. Secondary compounds accumulated in response to microbial attack are referred as "Phytoalexins". Accumulation of phytoalexins, which results in chemical resistance, is an important factor in plant defense and has been demonstrated for a wide variety of species.

Phytoalexins are low molecular weight, antimicrobial compounds, synthesized by plants after microbial infection. These post infectional inhibitors are part of the defense reactions of higher plants against phytopathogens. Phytoalexins demonstrate a great diversity in chemical structure, which is characteristic of the genus or family of plant e.g., members of Solanaceae synthesize sesquiterpenoids and Papillionaceous plants produce isoflavones (Brooks *et al.*, 1986). There is an even increasing list of phytoalexins due to continued efforts in many laboratories. In addition to sesquiterpenes and isoflavones, various alkaloids, polyacetylic compounds, diterpenoids, etc. have been reported (Bailey and Mansfield, 1982; Brooks *et al.,* 1986).

The signals (compounds or preparations of fungal or bacterial cultures) triggering the formation of phytoalexins are called elicitors. In the widest sense, these are "molecules" inducing a reaction in plant cells assumed to be characteristic of its defensive responses.

12.1. ELICITOR-INDUCED EFFECTS IN PLANT CELLS

Elicitors induce different types of effects in plant cells which is reflected by the influenced cell metabolism viz.,

i) Ca^{2+} metabolism.

ii) Massive variations in membrane integrities, respiration, protein and phosphate metabolism, ethylene production and peroxidase activity.

iii) Differential gene expression, consequently forming enzymes concerned in the synthesis of polysaccharides as callose, hydroxyproline-rich glucoproteins. (HRGP) in cell walls *via* induction of proline hydroxylase, lignin and polyphenolics (deposited in cell walls), chitinases and protein inhibitors, specific proteins against pathogenic infections (PR), phytoalexins.

Hence, formation of phytoalexins is only one of several possible reactions.

12.2. MECHANISM OF ACTION

Removal of regulatory repressors, genetic manipulation of enzyme pathways, or the addition of specific metabolic inducers can increase secondary metabolism. It is suggested that the secondary biosynthetic capabilities of the plant cells are repressed in cell culture systems and need a stimulus for expression. Providing that stimulus to the culture is the basis of exploiting the biotechnological potential of plant cells. Elicitors have been used as a tool to understand the regulation of phenyl propanoid secondary metabolism in plants (DiCosmo and Misawa, 1985).

Dixon et al., (1990) proposed a model for the induction of plant defense responses. This model assumes that the elicitor binds to a specific receptor, probably located in the plant plasma membrane, and this binding indirectly leads to changes in the transcription activity of genes involved in the production of secondary metabolism by the host. A scheme of a hypothetical integrated system of the Ca^{2+}/c-AMP effect in plant cells (Grisebach, 1986; Kurosaki et al., 1987; Heim and Wagner, 1987; Kaus, 1987; Moesta and West 1985; Dhawan and Malik, 1979; Endress 1979; Brown and Newton, 1981; Endress, 1984), has been depicted in Fig. 12.1.

Ca^{2+} dependent enzymes in plant cells are:

(i) NAD kinase
(ii) Quinate – NAD^+ - oxidoreductase
(iii) Proteinkinase
(iv) Ca^{2+} transporting ATPases
(v) H^+ transporting ATPases
(vi) Membrane phospholipases

Evidence for the presence of specific high affinity elicitor binding sites in the soyabean plasma membrane has been shown (Dittrich et al., 1992; Schmidt and Ebel, 1987). The next step in elicitation is thought to be inhibition of plasma membrane ATP that reduces the proton electrochemical gradient across the membrane. The signal transduction chain between the elicitor receptor complex causes gene activation process (Cosio et al., 1990; Dittrich et al., 1992; Chappell and Hahlbrock, 1984; Brooks and Watson, 1991). It was suggested that these signals are either transported locally by diffusion through intercellular and extracellular fluids infection sites or systematically through the vascular system of the plants. Elicitors are believed to act by ionic communication, which can be brought about by changes in pH, electrolyte leakage, depolarization or inhibition of electronic ion pump (Kota and Stelzig. 1977; Katau et. al., 1982). For example, the biotic elicitor vanadate is known to inhibit cationic pumping of ATPases (O'Niell and Spanswick. 1984), which indirectly influence the intracellular Ca^{2+} level (Bhagyalaxmi and Bopanna, 1998)

12.3. SPECIAL FEATURES OF ELICITORS

1. The products, which accumulate in plant cell cultures due to elicitation, may be antimicrobial in nature, but they should not be confused with phytoalexins (a term used in plant pathology for pathogen-induced chemical defense in host plant), unless there is sufficient proof that the source plant responds to pathogens with rapid accumulation of the same product. Therefore, a new term that has been coined for those compounds, which in cell cultures are inducible by way of elicitation, is "Elicitation Product" or "Elicitation Metabolite".

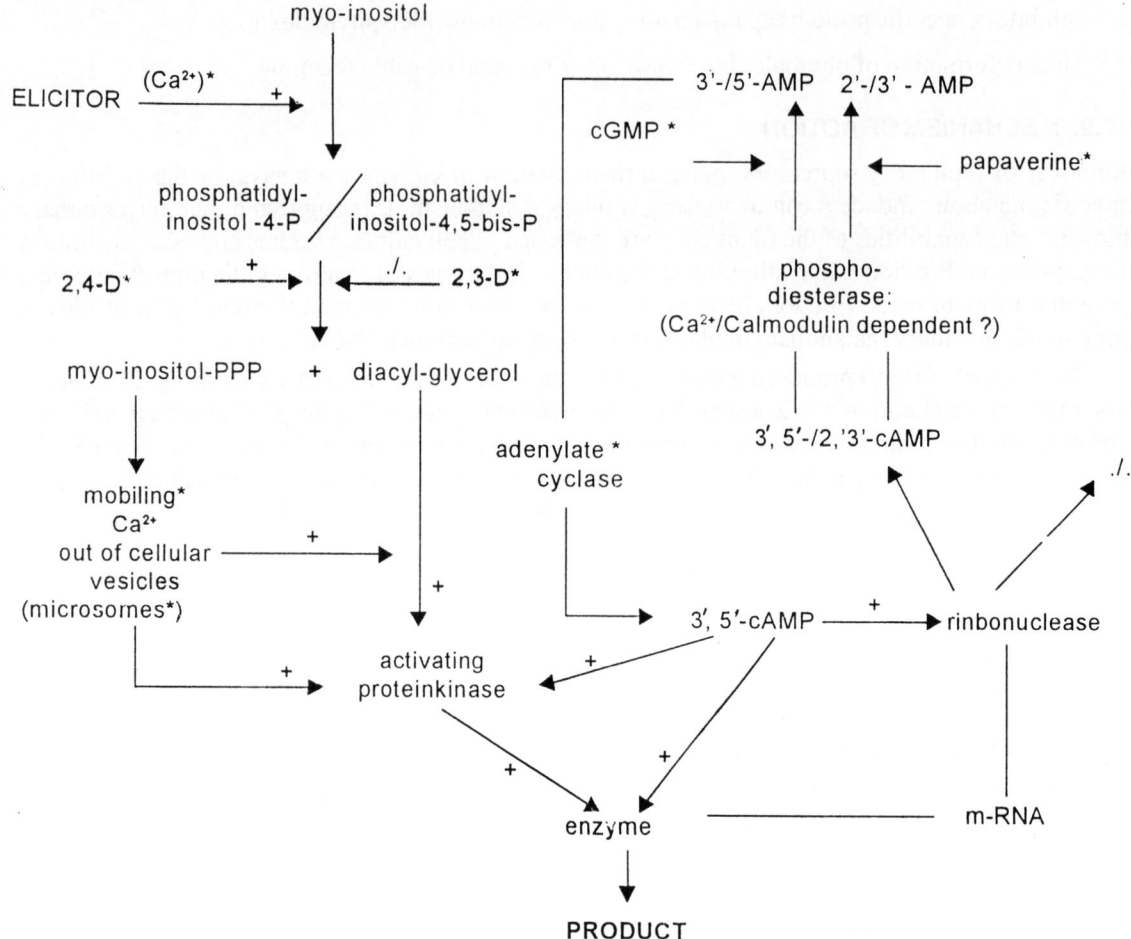

+   Stimulation

*   Effect demonstrated in different plant cells

./. Without influence

Fig. 12.1. Scheme of hypothetical integrated system of the Ca^{2+}/cAMP effect in plant cells (Adopted from Endress, 1979).

2.   Elicitors can be regarded as substitute of production media (that is optimum cultural conditions).

3.   Optimum employment of elicitors depends upon factors like :

    (a)   elicitor specificity,

    (b)   elicitor concentration,

    (c)   duration of elicitor contact,

    (d)   elicitor of cell line (clones),

    (e)   time course of elicitation,

(f) growth stage of culture,

(g) growth regulation, and

(h) nutrient composition.

4. Elicitation of cell in suspension culture may react in the following ways :

    (a) In a given cell line, different products may show highest levels of accumulation, at different times.

    (b) Product accumulation may be observed in cell lines, which are not known to synthesize.

    (c) Elicitation may not cause an additive effect when applied to cells in production media, but may shorten the culture period required for maximum product accumulation.

5. Product accumulation due to elicitation has also been observed in growth media. Such occurrence may be due to excretion or leakage caused by cell breakdown (Purohit and Mathur, 1996).

12.4. CLASSIFICATION OF ELICITORS

Elicitors formed inside or outside plant cells are distinguished as endogenous or exogenous elicitors respectively. Depending upon their nature, they are classified as biotic or abiotic (West 1981; Darvill and Albersheim, 1984).

[A] Origin as Classification Feature

12.4.1. Exogenous Elicitors

- Originated outside the cell, inducing the reaction immediately or *via* endogenous mediators
- Polysaccharides: Glucomannose, Glucans and Chitosan
- Peptides as:
  - Polycations e.g., Monilicolin, Poly-L-lysine, Polyamines and
  - Glycoproteins
  - Enzymes e.g., Polygalacturonase, Endo-polygalacturonic acid lyase and Cellulase
- Fatty acids e.g., Arachidonic acid, Eicosapentenoic acid
- Metallic ions
- UV light

12.4.2. Endogenous Elicitors

- Formed via secondary reactions induced by a signal of biotic or abiotic nature in the cell
- Dodeca- β-1, 4-D-galacturonide
- Hepta -β-glucosides

[B] Nature as Classification Feature

12.4.1. Biotic Elicitors

- Directly released by microorganisms and recognized by the plant cell (enzymes and cell wall fragments)
- Formed by the action of microorganisms on plant cell wall (fragments of pectins, etc.)
- Formed by the action of plant enzymes on microbial cell walls (chitosan and glucans)
- Compounds, endogenous and constitutive in nature, formed or released by the plant cell in response to various stimuli

12.4.2. Abiotic Elicitors

- Of physical or chemical nature working via endogenously formed biotic elicitors e.g., UV light, windfall, denatured proteins (RNase), freezing and thawing cycles, non-essential components of the media (agarose, tin, etc.) and heavy metals,

Chemicals

- With high affinity for DNA
- With membrane- destroying activities like
  Detergents: Xenobiochemicals
  Fungicides: Maneb, Butyl amine, Benomyl
  Herbicides: Acifluorofen

Biotic and Abiotic elicitors also differ in their dose-effect relationship. The type of elicitors usually has no effect on the type of product formed. Plant cells react to different stress factors by accumulating stress-specific compounds (Brodelius, 1988). However, not all enzymes of a biosynthetic pathway necessarily respond to every elicitor.

12.4.1. Biotic Elicitors

Originally, elicitors were called biotic if they formed during the plant defensive process against microbial infection. They mainly include conidia, enzymes that degrade cell walls, fragments of cell walls of both organisms and products experimentally produced therefrom (Halverson and Stacey, 1986), and the contents of culture filtrates (Eilert *et al.,* 1985). It is sufficient to dissolve 1% of polysaccharide rich in galacturonic acid from primary cell walls to release the elicitor concentration required to induce phytoalexin synthesis. However, their activity is considerably lower than that of exogenously applied hepta-β-glucosides (Table 12.1, Albersheim and Darvill, 1985; Ryan, 1987).

Table 12.1. Unpurified and purified biotic elicitors (Adopted from Endress, 1994)

Produced from	Biotic elicitors	
	Unpurified	*Purified*
Fungus cell wall by:		
β-1.3-endoglucanases of plants acid hydrolysis	Glucomannose Polysaccharide β-Glucan Glycoproteins	Hepta-β-glucoside (alditol) Glycosyl compounds Mannosyl compounds
Plant cell wall by Partial acid hydrolysis	α-1.4-bound galacturonosyl residues	Dodeca-α-1.4-D-galacturonides

Biotic elicitors are derived from microorganisms or they are endogenous compounds of the infected plant. The latter type of elicitors are often released by mechanical wounding or enzymatic hydrolysis of polymeric compounds in plants which are polysaccharides of plant cell walls (Dixon, 1986). These are of two types:

12.4.3.1. β-Linked Glucans

Apart from fungal elicitors, preparations from yeast and bacteria which were reported to enhance secondary metabolites are a beta-glucan, elicitor prepared by ethanolic precipitation of the yeast,

Saccharomyces cerevisiae was found to elicit alkaloid production in cultured cells of *Eschscholtzia california* (Brodelius *et al.,* 1989). A transient increase in rosmarinic acid (RA) content in cultured cells of *Lithospermum erythrorhizon* was observed after addition of yeast extract to the suspension culture, reaching a maximum at 24hr. (Mizukami *et al.,* 1992). Influence of several bacterial preparations was studied for their elicitor and antimicrobial activities in *Tagetes* cultures (Buitelaar *et al.,* 1992). Purified fractions from bacteria were also found to elicit diosgenin and capsaicin production (Rokem *et al.,* 1984).

12.4.3.2. Chitosan

Elicitation of capsaicin in *Capsicum frutescens* cultures was achieved by supplementing the culture medium with chitosan, curdlan and xanthan. Both curdlan and xanthan elicited two fold increase in the capsaicin synthesis when applied individually. Whereas when supplied together, 40-fold increase in the capsaicin production was observed, of which 75% leached into the medium (Johnson *et al.,* 1991). Treatment of culture of *Vanilla planifolia* with chitosan resulted in the induction of various enzymes of phenyl propanoid metabolism, while the amount of extractable phenolics decreased due to their rapid incorporation into polymeric ligneous material (Funk and Brodelius, 1990). Diterpene biosynthesis in rice is elicited by chitin (Ren and West, 1992). Enhanced production of anthraquinones and chitinases in *Morinda citrifolia* cell cultures is achieved by pectins, chitosans and especially by chitin-50 (Dornenburg and Knorr, 1994). In addition, Glycoproteins and proteins of fungal origin also serve as elicitors (Ebel, 1986).

Another groups of elicitors are enzymes with polygalactomarase activity, which release pectic fragments from plant cell walls. It has been observed that very small amounts of such water-soluble oligomers lead to rapid induction and accumulation of phytoalexin in cell cultures (Funk *et al.,* 1987). The production of sesquiterpenoids related to the phytoalexin debneyol from cell suspension cultures of *Nicotiana tabaccum,* was enhanced by cellulose (Whitehead *et al.,* 1988). A carbohydrate fraction isolated from yeast extract by ethanolic precipitation in cells of *Glycine max,* induced glyceolline synthesis and enhanced berberine synthesis upto four-fold in cells of *Thalictrum rugosum* (Funk *et al.,* 1987). The concentration of elicitor used is critical for maximum effect on cultures as it is frequently observed that product accumulation reaches a peak at a specific elicitor concentration and then declines at higher values. In *Thalictrum rugosum,* berberine accumulation is maximal at 0.3 mg yeast elicitor/g fwt. and declines with increase in elicitor concentration to 1.6 mg/g. fwt. (Funk *et al.,* 1987).

12.4.3.3. Fungal Elicitation

Fungal elicitor is normally one derived from a micro-organism (*Phytophthora, Botrytis, Verticilium, Alternaria, Fusarium,* etc.,) pathogenic to the plant species of interest (Grisebach and Ebel, 1978; Darvill and Albersheim, 1984; Ebel *et al.,* 1984) although, preparations derived from non-pathogenic (*Asperigillus, Micromucor, Rhodotorula,* etc.) microbes have been successfully employed for the elicitation purpose (Funk *et al.,* 1987).

For a successful elicitation process using microbial or fungal preparation, several factors have to be taken into consideration. Each one of the factors mentioned below is to be optimized independently for process of elicitation.

1. Sensitivity of cell culture strain
2. Elicitor specificity
3. Concentration of the elicitor
4. Timing and Duration of the elicitor exposure

The fungal elicitors stimulated accumulation of secondary compounds in cell cultures (*in-vitro*) is enlisted in Table 12.2. In last one decade numerous reports were recorded with respect to elicitation of phytoalexins (Grisebach and Ebel, 1978; Robbins *et al.,* 1985; Eilert *et al.,* 1986; Buitelaar *et al.,* 1992). The stimulation of metabolites includes alkaloids, codeine and morphine by fungal spores in *Papaver somniferum* cultures (Heinstein, 1985). Autoclaved fungal mycelia induced the accumulation of diosgenin in *Dioscorea deltoidea* culture (Rokem *et al.,* 1984). The production of berberine and shikonin was enhanced in cultured cells treated with fungal extracts (Funk *et al.,* 1987; Kim and Chang, 1990). A successful large-scale application of elicitation method is in the cell culture of *Papaver somniferum* cell line 2009, the opium poppy with *Botrytis* preparation, which resulted in the production of sanguinarine, dihydrosanguinarine a benzophenanthridine alkaloid usually not found in the normal poppy plant (Eilert *et al.,* 1986). Twelve hours after the elicitor addition, 40-60 % of the total alkaloids leached into the medium (Constabel, 1990). It is also reported that upon elicitation of 2 days old suspension culture of *Capsicum annuum,* with sterile spores of *Gliocladium deliquescens,* a significant increase in incorporation of ^{14}C phenylalanine into capsaicin is observed (Holden *et al.,* 1987). Some secondary products that accumulate in cultured plant cells in response to contact with fungal elicitors are given in Fig. 12.2.

Table 12.2. Fungal elicitor stimulated accumulation of secondary metabolites in cell cultures (in vitro)

S.No.	Elicitor used	Plant cell culture	Secondary metabolite produced	Reference
1.	*Phytopthora megasperma*	*Glycine max*	Glyceollin	Keen and Horsch (1972)
2.	Fungal glycoprotein	*Phaseolus vulgaris*	Phenylpropanoids (Kervitone, Phaseollin)	Dixon and Fuller (1977)
3.	*Botrytis cinerea*	*Phaseolus vulgaris*	Phaseollin	Dixon and Bendall (1978)
4.	Ribonuclease(denatured)	*Phaseolus vulgaris*	Krevitone, Phaseollin	Dixon and Bendall (1978)
5.	Fungal polysaccharide	*Phaseolus vulgaris*	Krevitone, phaseollin	Dixon *et al.,* (1981)
6.	Fungal polysaccharides (*Botrytis allii*)	*Ruta graveolens*	Alkaloid (Rutacridone epoxide)	Wolters and Eilert (1982)
7.	Hemicellulase	*Brugmansia candida*	Hyosujamine	Sandra *et al.* (1988)
8.	Fungal polysaccharides	*Petroselinum hortense*	Coumarins (Psoralen, Xanthotoxin, Graveolone)	Tietjen *et al.,* (1982)
9.	Fungal glucan	*Glycine max*	Phenylpropanoid (Glyceollin)	Hille *et al.,* (1982)
10.	Fungal culture filtrate	*Bidens pilosa*	Polyacetylene (Phenylheptatriyn)	DiCosmo and Misawa (1985)
11.	Agaropectin	*Lithospermum erythrorhizon*	Naphthoquinone (Shikonin)	Fukui *et al.,* (1983)
12.	*Alternaria carthami*	*Petroselinum hortense*	Furanocoumarines	Tietjen *et al.,* (1982)
13.	Fungal polysaccharide	*Carthamus tinctorius*	Polyacetylenes (unspecified)	Tietjen *et al.,* (1982)
14.	Chitosan	*Ruta graveolens*	Alkaloid (Rutacridone epoxide)	Eilert *et al.,* (1984)

(Contd.)

S.No.	Elicitor used	Plant cell culture	Secondary metabolite produced	Reference
15.	Chitosan, Poly-L-lysine	*Glycine max*	Phenylpropanoid (Glyceollin)	Kohle *et al.,* (1984)
16.	Fungal mycelia	*Dioscorea deltoidea*	Steroid (Diosgenin)	Rokem *et al.,* (1984)
17.	Several biotic	*Capsicum annuum*	Capsaicin	Lindsey and Yeoman, (1984)
18.	Fungal mycelia Extracts *(Botrytis* sp.)	*Papaver somniferum*	Alkaloid (Sanguinarine)	Eilert *et al.,* (1985)
19.	*Rhodotorula rubra*	*Papaver somniferum*	Alkaloid (Sanguinarine)	Eilert *et al.,* (1985)
20.	Fungal spores	*Papaver somniferum*	Alkaloids, (Codeine, Morphine)	Heinstein, (1985)
21.	Fungal cell wall material (*Colletotrichum lindemuthianum*)	*Phaseolus vulgaris*	Phenylpropanoids (Krevitone, Phaseollin)	Robbins *et al.,* (1985)
22.	*Botrytis* sp.	*Papaver somniferum*	Dihydro-sanguinarine	Eilert *et al.,* (1986a)
23.	*Pythium aphanidermatum*	*Catharanthus roseus*	N-acetyl-tryptamine	Eilert *et al.,* (1986b)
24.	*Ascochyta rabiei*	*Cicer arientinum*	Medicarpin, Maackiain	Barz *et al.,* (1988)
25.	Fungal spores	*Datura stramonium*	Lubimin	Whitehead *et al.,* (1990)
26.	Fungal filtrate	*Ricinus communis*	Casbene	Sitton & West (1975)
27.	*Phytophthora cactorum*	*C. roseus*	Ajmalicine	Asada & Shuler (1989)
28.	Cellulase	*C. annuum*	Capsidol	Patrica *et al.* (1996)
29.	*Vertcillium dahliae*	*Cephalotaxus harringtonia*	Alkaloid	Heinstein (1985)
30.	*Vertcillium dahliae*	*Papaver somniferum*	Morphine, codeine	Heinstein (1985)

The preparations (cells extracts and culture filtrates) from *Cytospora abietis* and *Pencillium minioluteum* induced the production of taxol from *Taxus brevifolia* cell suspension cultures (Christen *et al.,* 1991). Strobel *et al.,* (1992), reported that the semi purified fungal (void glucan) preparation stimulated ^{14}C-taxol biosynthesis in *T. brevifolia* bark. The preparations from *Vertcilium dahliae* (Cline *et al.,* 1993) and *Botrytis* sp. culture (Eilert and Constabel, 1986a) have been effective elicitors with other plant species. Ciddi *et al.,* (1995) demonstrated that the addition of cell extracts and cell filtrates of *Vertcillium dahliae* and *Gliocladium deliquescens* on the 10th day after transferring *Taxus* sp. cell suspensions into an induction medium, further improved the production of taxol and total taxanes.

Upon addition of fungal cell wall fractious, i.e., a skleroglucan or an elicitor from either *Phytophthora megasperma* or *Alternaria carthami,* the cell suspension cultures of *Ammi majus* excreted large amounts of umbelliferone (Daria. *et al.,* 1990). Large amounts of sesquiterpenoid capsidiol accumulated in the media of *Nicotiana tabaccum* cell suspension cultures upon addition of fungal elicitor (Chappell and Noble, 1987). The same fungal elicitor (*Vertcillium dahliae*) enhanced benzophenanthridines content in cell suspension cultures of *Sanguinaria canadensis* and *Papaver bracteatum,* which are taxonomically related, but did not elicit the accumulation of the same alkaloid in the two different plant cultures (Cline *et al.* 1993). Addition of cell wall fragments of *Phytophthora megasperma* to *Datura stramonium* cell cultures, enhanced the final tropane alkaloid yield by five fold compared with control cultures (Ballica *et al.,* 1993). Elicitor preparations from either homogenized

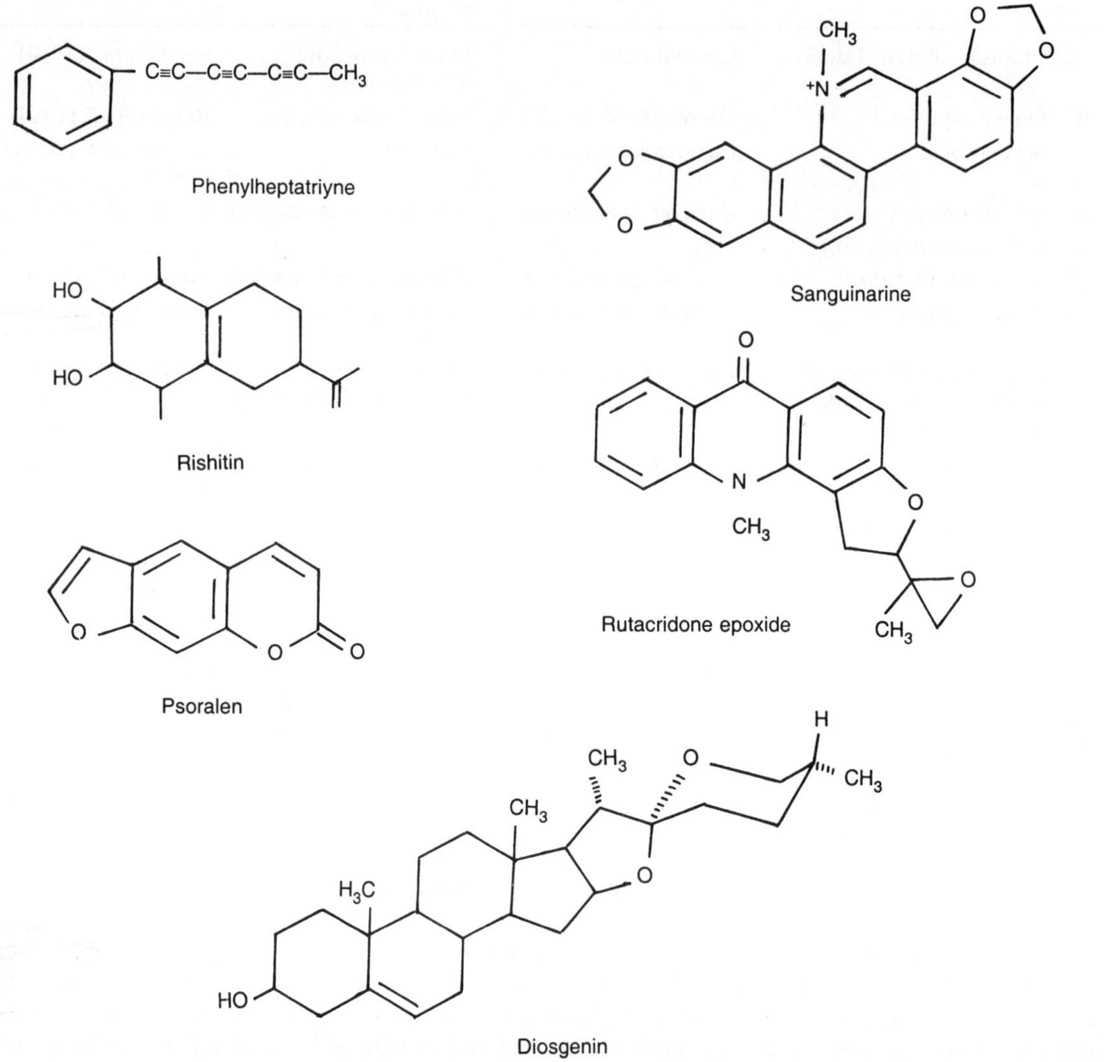

Fig. 12.2. Some secondary metabolites that accumulate in cultured plant cells in response to contact with fungal elicitors (Adopted from DiCosmo and Misawa, 1985).

mycelia of *Dendryphion pencillatum*, a specific pathogen of *Papaver* species, or conidia of *Vertcillium dahliae*, general pathogens were added to 14 day old suspension cultures of *Papaver bracteatum*. *Dendryphion* extracts elicited an accumulation of the benzophenanthridine alkaloid, sanguinarine whereas, *Vertcillium* elicited the cultures accumulated sanguinarine in an elicitor dose dependent manner only under conditions of hormonal deprivation, resulting in an elevation of sanguinarine levels 5 to 500 fold greater than controls (2-10% dry weight) (Cline and Coscia, 1988).

Monoterpene indole alkaloid formation in suspension cultures of *Catharanthus roseus* was elicited with homogenates of various fungi, namely *Alternaria zinnae, Pythium aphanidermatum, Vertcillium dahliae* and *Rhodotorula rubra* (Eilert *et al.,* 1986).

Catharanthine synthesis had been elicited and the production of major indole alkaloids was stimulated by *Catharanthus roseus* cells under non-growth altering treatment with *Pythium vexans* extracts (Nef *et al.*, 1991). Treatment of cell suspension cultures of *Tripterygium wilfordii* with autoclaved *Botrytis* sp. homogenate rapidly increased the synthesis of a family of anti-inflammatory oleanane and triedelane triterpenes (Kutney *et al.*, 1993).

A glucan elicitor from the cell wall of the fungus *Phytophthora megasperma* sp. *Glycinea* caused increase in the activities of the phytoalexin biosynthetic enzymes phenylalanineammonialyase and chalcone synthase, and induced the production of the phytoalexin, glyceollin in soybean (*Glycine max*) cell suspension cultures (Stab and Ebel, 1987). Suspension cultured carrot cell (*Daucus carota*) and their protoplasts respond to a fungal elicitor prepared from culture medium of *Pythium aphanidermatum* by accumulating 4-hydroxy benzoic acid (4-HBA) (Bach *et al.*, 1993). The carrot (*Daucus carota* L.) cells respond to treatment with fungal elicitors from *Pythium aphanidermatum* by synthesizing cell wall bound p-hydroxy benzoic acid (p-HBA) (Schnitzler *et al.*, 1992). Treatment of suspension cultured cells of carrot with yeast glucon elicitor induced generation of ethylene followed by accumulation of 6-methoxymellin (Guo and Ohta, 1994). Treatment of *Pinus sylvestris* (Scots pine) cell suspension cultures with an elicitor preparation from the pine needle pathogen *Lophodermium seditiosum* resulted in a several hundred to thousand fold accumulation of the stilbenes, pinosylvin and pinosylvin 3-O-methyl ether in methanolic cell extracts (Lange *et al.*, 1994).

Changes in lipid composition occurred when tobacco cells (*Nicotiana tabaccum* var. *Xanthi*) were treated with cryptogein, a proteinaceous elicitor from *Phytophthora cryptogea*. The most striking change was an increase in acylated steryl glycoside and steryl ester levels, certainly resulting from the glycosylation and/or esterification of free sterols (Tavernier *et al.*, 1995). A sonicate of *Phytophthora infestans* mycelium and a spore suspension of *Penicillium chrysogenum* elicited the formation of the sesquiterpenoid phytoalexins, lubimin, 3-hydroxylubimin and rishitin in fruit cavities of *Datura stramonium* (Whitehead *et al.*, 1990).

12.4.4. Abiotic Elicitors

All factors, which cannot be regarded as a natural component of the environment of a plant cell, are considered as abiotic elicitors (Table 12.3). As indicated earlier, abiotic elicitors are of non-biological origin mainly the metal ions. Threfall and Whitehead (1988) tested aluminum, lithium, magnesium, calcium, chromium, manganese, iron, cobalt, nickel, copper, zinc, mercury, cadmium, lead and bismuth on the elicitation of sesquiterpenoid phytoalexins of *Datura stramonium* where, only aluminum, chromium, cobalt, nickel, copper, zinc, cadmium and lead elicited the alkaloid to various degrees.

Abiotic elicitors are of physical or chemical nature working via endogenously formed biotic elicitors. e.g., UV light, windfall, denatured proteins (RNase), freezing and thawing cycles, non-essential components of the media (agarose, tin, etc.,), heavy metals and chemicals (detergents, fungicides and herbicides). Colchicine treated suspension cultures of *Valeriana wallichii* produced higher amounts of valepotriates than did the respective untreated cultures (Becker and Chavadej, 1985).

Apart from the above chemicals, the chemicals like arachidonic acid, eicosapentenoic acid, methyl jasmonate, cinnamic acid, salicylic acid, salicin, copper sulphate, sodium orthovandate, etc., can also be employed for eliciting the plant cell cultures.

The use of metal ions as elicitors offers many advantages over their biotic counterparts, these include:

- their ready availability
- relative cheapness
- ease of use
- they are chemically defined

Table 12.3. Influence of abiotic elicitors on production of secondary metabolites in several cell culture systems

S.No.	Elicitors Used	Plant Cell Culture	Secondary Metabolite Produced	Reference
1.	Activated Carbon	*Lithospermum erythrorhizon*	Benzoquinone (Echinofuran)	Fukui *et. al.,* (1983)
2.	Na-alginate	*Glycyrrhiza echinata*	Echinatin	Ayabe *et al.,* (1986)
3.	Curdlan, Xanthan	*Capsicum frutescens*	Capsaicin	Johnson *et al.,* (1991)
4.	Agaropectin	*Lithospermum erythrorhizon*	Naphthoquinone (Shikonin)	Fukui *et al.,* (1983)
5.	Ribonuclease (denatured)	*Phaseolus vulgaris*	Phenylpropanoids (Krevitone, Phaseollin)	Dixon and Bendall, (1978)
6.	Diethyl amino ethyl dichloro phenyl ether	*Catharanthus roseus*	Alkaloids (Ajmalicine, Catharanthine)	Lee *et al.,* (1998)
7.	Colchicine	*Valeriana wallichii*	Valepotriates	Becker and Chavadej (1985)
8.	Arachidonic Acid	i) *Capsicum annuum*	Capsidiol, Rishitin	Hoshino *et al.,* (1994)
		ii) *Taxus* sp.	Taxol	Ciddi *et al.,* (1995)
		iii) *Datura stramonium*	Sesquiterpenoids (Lubimin, 3-hydroxy lubimin, Rishitin)	Whitehead *et al.,* (1990)
		iv) *Taxus wallichiana*	Taxanes	Prasad Babu (2000)
		v) *Coleus forskohlii*	Forskolin	Prasad Babu (2000)
9.	Methyl Jasmonate	i) *Lithospermum erythrorhizon*	Rosmarinic acid	Mizukami *et al.,* (1993)
		ii) *Taxus xmedia* var. *Hatfieldii*	Paclitaxel, Cephalomannine	Furmanowa *et al.,* (1997)
		iii) *Taxus* sp.	Taxane diterpenes	Yukimune *et al.,* 1995
		iv) *Taxus baccata*	Paclitaxel, Cephalomannine	Furmanowa *et al.* (1995), & Yukimune *et al.,* (1996)
		v) *Vaccinium pahlae* (Ohelo)	Anthocyanins	Fang *et al.,* (1999)
		vi) *Hyoscyamus muticus*	Sesquiterpenes	
		vii) *Glycine max*	Vegetative storage protein	Singh *et al.,* (1998), Anderson, (1991)
		viii) *Oryza sativa* leaves (rice)	Putrescine	Chen *et al.,* (1994)
		ix) *Solanum tuberosum*	Polyamines (free and conjugated)	Mader (1999)
		x) *Catharanthus roseus* and *Cinchona ledgeriana* (seedlings)	Monomeric alkaloids	Aerts *et al.,* (1994, 1996)
		xi) *Coleus blumei*	Rosmarinic acid	Szabo *et al.,* (1999)
		xii) *Hyoscyamus albus* (hairy roots)	Phytoalexins with vetispyrane skeleton	Kuroyanagi *et al.,* (1998)
		xiii) *T.cuspidata*	Taxol	Mirjalili and Linden, (1996)

(Contd.)

S.No.	Elicitors Used	Plant Cell Culture	Secondary Metabolite Produced	Reference
		xiv) *Rauwolfia serpentina* X *Rhazya stricta* (Somatic hybrid cells)	Total alkaloidal content	Sheludko *et al.,* (1999)
		xv) *T. baccata* var. *pendula*	Taxanes	Moon *et al.,* (1998)
		xvi) *T. canadensis* and *T. cuspidata*	Taxoids	Ketchum *et al.,* (1999)
		xvii) *T.canadensis*	Taxanes	Phisalaphong *et al.,* (1999)
		xix) *T. wallichiana*	Taxanes	Prasad Babu. (2000)
		xx) *C. forskohilii*	Forskolin	Prasad Babu (2000)
10.	Salicylic acid	*Daucus carota* (carrot)	Chitinase	Muller *et al.,* (1994)
		C. forskolii	Forskolin	Prasad Babu(2000)
11.	Metal ions			
	i) Al^{3+}, Cr^{3+}, Co^{2+}, Ni^{2+}, Cu^{2+}, Zn^{2+}, Cd^{2+}, Au^{3+}, Pb^{2+}	*Datura stramonium*	Sesquiterpenoids (Lubimin, 3-hydroxy lubimin, Rishitin)	Threfall and Whitehead, (1998)
	ii) Cu^{2+}, Mn^{2+}, Hg^{2+}	*Cicer arientinum*	Pterocarpenes	Threfall and Whitehead, (1998)
	iii) Copper Sulphate	*Lithospermum erythrorhizon*	Shikonin	Fujita *et al.,* (1982)
	iv) Vanadium Sulphate	*Catharanthus roseus*	Catharanthine, Ajmalicine	Smith *et al.,* (1987)
	v) Ca^{2+}, Mn^{2+}, Zn^{2+}, Co^{2+}, Fe^{2+}, V	*Daucus carota*	Anthocyanins	Suvarnalatha *et al.,* (1994)
	vi) Cu^{2+}, Cd^{2+}	*Atropa belladonna*	Tropane alkaloids	Lee *et al.,* (1998)
	vii) Copper Sulphate	*Hyoscyamus albus* (hairy roots)	Phytoalexins with vetispyrane skeleton	Mader, (1999)
	viii) Silver Nitrate	*Solanum tuberosum* (roots and shoots)	Free and conjugated polyamines	
12.	Calcium chloride	*C. forskohlii*	Forskolin	Prasad Babu (2000)

Darvill and Albersheim (1985) proposed that the abiotic elicitors function by activating biotic elicitors present in an inactive form. Even though, it is very difficult to explain the mechanisms, the fact that they achieve the desired result is worth considering.

Biotic and abiotic elicitors also differ in their dose effect relationship. In *Cicer arientinum* suspension cultures, the effect of biotic elicitors from *Ascochyta rabiei* and accumulation of the same pterocarpanes induced by heavy metals, namely Cu^{2+}, Mn^{2+}, Hg^{2+} are characterized (Threfall and Whitehead, 1988). The type of elicitor usually has no effect on the type of product formed (Table 12.3). Plant cells react to different stress factors by accumulating stress specific compounds (Brodelius, 1988). However, not all enzymes of a biosynthetic pathway necessarily respond to every elicitor.

12.4.4.1. Arachidonic Acid

Arachidonic acid is a fatty acid present in the lipids of plant pathogenic Oomycete fungi and a potent elicitor of sesquiterpenoid phytoalexins and suppressor of steriod glycoalkaloid synthesis in solanaceous species (Choi and Bostock, 1994).

Addition of arachidonic acid as an elicitor, to green pepper (*Capsicum annuum*) cell suspension cultures induced the extracellular accumulation of capsidiol and rishitin (Hoshino *et al.*, 1994). It was reported that arachidonic acid (1mg/l) addition at the time of inoculation increased taxol production by 150% in *Taxus* species cell cultures. A higher concentration of arachidonic acid (5 mg/l) caused browning and cell death (Ciddi *et al.* 1995). Arachidonic acid elicited the formation of the sesquiterpenoid phytoalexins lubimin, 3-hydroxy lubimin and rishitin in fruit cavities of *Datura stramonium* (Whitehead *et al.*, 1990). The advantage of working with arachidonic acid is that while it mimics the effects of other fungal elicitors, it does not require the preparation time like fungal elicitors since it is a pure compound and can be applied reproducibly. The site of action of arachidonic acid is unknown at this time.

12.4.4.2. Methyl Jasmonate

Methyl Jasmonate (methyl ester of jasmonic acid), a linoleic acid derived molecule, can act as a volatile signal that induces the accumulation of proteinase inhibitor proteins to even higher levels than can be induced by wounding (Farmer *et al.*, 1992). The concentrations of jasmonic acid or methyl jasmonate in different tissues and species are typical of plant hormones, which range from 0.01 to 3 µg/gram of fresh weight (Staswick, 1992).

Thirty-six plant species tested in cell suspension culture could be elicited by exogenously supplied methyl jasmonate (MJ) to accumulate secondary metabolites. The induced secondary metabolites were accompanied by an increase in PAL poly (A) + RNA that was followed by an increase in phenylalanine ammonialyase (PAL) activity to maximal values 25 and 38 hours after elicitation, respectively. It was suggested that addition of MJ initiates *de novo* transcription of PAL genes (Gundlach *et al.*, 1992). This enzyme is a key regulator of the phenylpropanoid pathway that yields a diverse range of phenolics with defense related functions. The elicitation activity of MJ in suspension cultures of *Lithospermum erythrorhizon* was much higher than that of yeast extract in terms of the induction of PAL and (4-hydroxyphenyl) pyruvate reductase activities and rosmarinic acid accumulation (Mizukami *et al.*, 1993).

It was observed that ethylene and MJ have a synergistic induction of plant defense genes; ethylene/MJ resulted in osmotin protein accumulation to levels similar to those induced by osmotic stress. Particular signal combinations may synergistically hyperinduce plant defense genes and are more specifically related to gene function than any single inductive signal. The responsive sequences of the osmotin promoter to combinations of ethylene and MJ are present on the same DNA fragment, where responsiveness to other signals has been mapped. It was suggested that the binding of ethylene to its receptor on the plasma membrane might sensitize methyl jasmonate receptors on the membrane (Xu *et al.*, 1994). MJ conditioning causes parsley cells to become more sensitive to low concentrations of fungal elicitor and suggests an improved cellular signal perception/transduction system (Kauss *et al.*, 1992).

Different elicitor signals induce different parts of the defense response (Grosskope *et al.*, 1991). Particular combinations of signal molecules produce a specific "signature", set of inducers, which initiate a specific type of environmental protection. The occurrence of "cross-talk" between signaling molecules can be explained by an interactive signal transduction system (Xu *et al.*, 1994). Ethylene and MJ may represent "cross-talk" signaling in such plant culture system.

Mirjalili and Linden (1996) also observed that the methyl jasmonate enhanced the production of paclitaxel by 15 fold in cell cultures of *T.cuspidata*. Moon *et al.*, (1998), reported elicitation kinetics of taxane production in suspension cultures of *T.baccata* var. *pendula*. Methyl jasmonate increased taxane production in suspension cultures of *T.baccata* var. *pendula*.

It was reported that, on White-Rangasamy medium, with MJ (100 µM), the paclitaxel content increased from 2.37 µg/g to 90 µg/g and cephalomannine from 5.14 µg/g to 29.14 µg/g (dry weight), whereas growth of the cultures ceased in callus culture of *Taxus xmedia* var. *Hatfieldii* (Furmanowa *et al.*, 1997). Jasmonates showed various morphological and physiological activities when exogenously applied to plants. The jasmonates are possibly signal compounds in the elicitation process leading to *de novo* transcription and translation and ultimately, to the biosynthesis of secondary metabolites in plant cell cultures. They induced transcriptional activation of genes involved in the formation of secondary metabolites (Yukimune *et al.*, 1996).

Jasmonic acid was used for the production of taxane diterpenes in tissue cultures of many *Taxus* species and the cytostatic agents can be produced in high yield on an industrial scale (Yukimune *et al.*, 1995). At the same time, it was also found that MJ stimulated the paclitaxel and cephalomannine production in callus and suspension cultures of *Taxus baccata* (Furmanowa *et al.*, 1995). The highest paclitaxel accumulation (104.2 mg/l) was observed when 10μM of MJ was added to the medium (Yukimune *et al.*, 1996).

A hypothetical mechanism of action was proposed for MJ (Gundlach *et al.*, 1992): An elicitor receptor complex activates a lipase releasing α-linolenic acid, which is then transformed by constitutive enzymes to jasmonic acid and MJ activates in different plant systems, "Jasmonate-induced" proteins. A multitude of species specific genes involved in the function of high and low molecular weight compounds are expressed in response to these signal transducer molecules (Mueller *et al.*, 1998). It was suggested that the cell wall fragments are involved in the rapid induction of the defence genes, whereas MJ acts later in this process (Creelman *et al.*, 1992). Other evidence that MJ acts at the level of transcription was also reported (Tamari *et al.*, 1995), (Fig. 12.3).

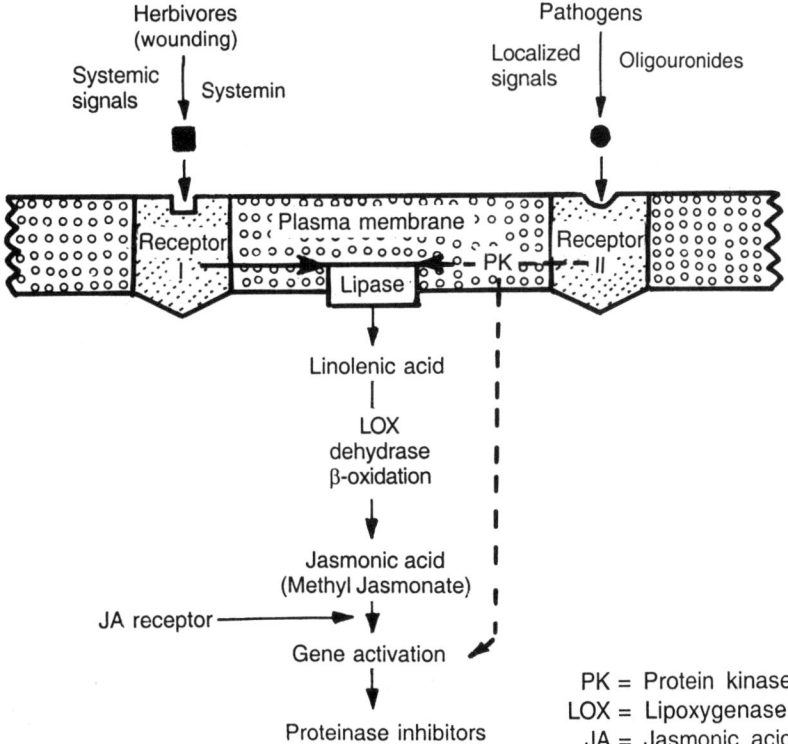

Fig. 12.3. Proposed model for the signaling that leads to the expression of wound-inducible proteinase genes in tomato leaves (Edward and Ryan, 1992).

A concentration dependent response was exhibited by *Vaccinium pahalae* (Ohelo) cell cultures treated with exogenous methyl jasmonate. The addition of 0.5 μM MJ alone provoked a 2-3 fold increase in anthocyanin production over that of the control (Fang *et al.*, 1999). Although, MJ alone could not induce sesquiterpene in unwounded root cultures of *Hyoscyamus muticus,* MJ added in the presence of wounding displayed a dose-dependent response saturating at 50 μM (Singh *et al.*, 1998). High concentrations of methyl jasmonate (2-4.5 μM per disk surface) suppressed hmgl mRNA and SGA accumulation but did not affect hmg2 mRNA abundance or induce phytoalexins (Choi and Bostock, 1994). Addition of jasmonic acid to suspension cultures of soybean specifically increased the level of soybean vegetative storage protein (Anderson, 1991). Methyl jasmonate induced the accumulation of putrescine, but not proline in detached rice leaves (Chen *et al.*, 1994). Distribution and accumulation of free and conjugated polyamines in roots and shoots of *Solanum tuberosum in vitro,* enhanced by jasmonic acid (Mader, 1999). The synthesis of monomeric alkaloids in *Catharanthus roseus* and *Cinchona ledgeriana* seedlings, enhanced by methyl jasmonate vapour (Aerts *et al.*, 1994, 1996). Suspension cultures of *Coleus blumei* treated with methyl jasmonate enhanced the accumulation of rosmarinic acid approximately threefold (Szabo *et al.*, 1999). The treatment of hairy roots of *Hyoscyamus albus* with methyl jasmonate produced several phytoalexins having the vetispyrane skeleton (Kuroyanagi *et al.*, 1998). The treatment of *Rauwolfia serpentina x Rhazya stricta* somatic hybrid cell suspension culture with 100 μM of methyl jasmonate led to a general increase in indole alkaloid content and to qualitative changes in the alkaloid pattern (Sheludko *et al.*, 1999). Ketchum *et al.*, (1999) reported the kinetics and distribution of taxoids elicited with methyl jasmonate in cell and suspension cultures of *T. canadensis* and *T. cuspidata.* Phisalaphong and Linden (1999) reported suspension cultures of *T. canadensis* were elicited with methyl jasmonate under defined headspace ethylene concentrations.

12.4.4.3. Metal Ions

In addition to biotic elicitors, metal ions are effective at inducing the formation of sesquiterpenoid phytoalexins (lubimin, 3- hydroxy lubimin and rishitin) in cell suspension cultures of thorn apple (*Datura stramonium*), copper (II) ions (Cu SO_4) were optimal for inducing the accumulation of high levels of these compounds in fruit cavities whilst, in cell suspension culture the highest levels of product were formed in response to 1 mM copper (II) ions (Whitehead *et al.*, 1990). The results obtained with gold, silver and lead were perhaps the most surprising particularly, since 0.1 mM gold (III) induced four times more material to accumulate than 1 mM lead (II). This means gold (III), on a molar basis, is effective forty times more active than lead (II) and thus, in this particular system, gold (III) can be considered to be a super-elicitor.

The addition of metal ions to the cell suspension culture produced several directly observable changes. When lead nitrate, copper sulphate or vanadium sulphate was added to the cultures at 1 mM, they caused both a drop in the p^H of the culture media (by at least 1 unit) and cellular aggregation. In addition lead (II) ions gave a precipitate in the culture medium (probably the hydroxide or sulphate salt). However, since both lead (II) and copper (II) ions were active as elicitors but vanadyl [VO (II)] ions were not, it appears that none of the above effects are involved in the elicitation response. The metal ions could be one of the followings- (a) the metal ions cause stress by indiscriminate binding to biological constituents of the cells or (b) the metal ions act through a common mechanism in a particular system. These two possibilities probably, represent the extremes between specificity and non-specificity and are not necessarily mutually exclusive of each other.

Epperlein *et al.*, (1986) provided evidence that some abiotic elicitors of the phytoalexin response (including some metal salts) may act by causing the formation of the hydroxyl radical -OH which, through mechanisms thought to include membrane damage (e.g., lipid peroxidation), leads to phytoalexin formation. Their evidence is based primarily on the results of experiments with OH scavengers like mannitol and benzoate. 10 mM mannitol completely inhibited by the formation of phytolexins in thornapple cell suspension cultures elicited with 1 mM Pb(II) ions. This effect was not due to chelation of metal ions by mannitol and although certain metal ion species (i.e., Cu (II) and Fe (II)) in the presence of hydrogen peroxide readily form -OH, metals like lead and cadmium do not. Thus, although -OH may be involved in the elicitation sequence at some stage (thus explaining the inhibitory effect of mannitol), the formation of such radical species would probably occur subsequently to the specific complex formation envisaged above and not directly from the metal ions themselves. This again suggests that complex formation may be the primary event in metal-ion-induced phytoalexin accumulation (Threfall and Whitehead, 1988).

Fujita *et al.*, (1982) reported that the addition of copper sulphate to the cell suspension cultures of *Lithospermum erythrorhizon,* enhanced the production of shikonin. Smith *et al.*, (1987), reported that the addition vanadium (II) sulphate (0.3mM) to cell suspension cultures of *Catharanthus roseus* resulted in the enhanced accumulation of two medicinally important indole alkaloids, ajmalicine and catharanthine.

The abiotic stress agents Ca, Mn, Zn, Co, Fe and V enhanced anthocyanin production in cell cultures of *Daucus carota* (Suvarnalatha *et al.*, 1994). The addition of Cu^{+2} and Cd^{+2} (5mM) to transformed root cultures of *Atropa belladonna,* improved the excretion of tropane alkaloids into the medium. However, their addition led to lysis of transformed roots (Lee *et al.*, 1998). The treatment of hairy roots of *Hyoscyamus albus* with copper sulphate (Cu^{2+}) produced several phytoalexins having the vetispyrane skeleton (Masanori *et al.*, 1998). The distribution and accumlation of free and conjugated polyamines in roots and shoots of *Solanum tuberosum in vitro,* were enhanced by silver nitrate (Mader, 1999).

12.4.4.4. Salicylic Acid

Salicylic acid (Fig.12.4) affects numerous processes in plants (Raskin, 1992). In recent years, the importance of this phenolic acid in plant pathogen interaction has become apparent. Significant insight has been gained in the past few years into the roles of salicylic acid (SA) in plant-pathogen interactions. The ability to accumulate salicylic acid has been shown to be essential for systemic acquired resistance in tobacco plants and have also been shown that SA is apparently not a systemic vascular mobile signal, but rather is required for signal transduction at the local level. Its mode of action may include inhibition of catalase activity, leading to increased levels of hydrogen peroxide (Vernooij *et al.*, 1994).

The signal perception mechanisms in plants for eukaryotic cell were found (Jones, 1994). We, now, know something about how plants sense blue light. The salicylic acid, ethylene and class of growth promoting hormones designated as auxins, of indole 3-acetic acid are reported to be the representative members of this mechanism (Fig. 12.4).

Chen and co-worker (1993) reported the identity of a receptor for salicylic acid, the endogenous signal required for the systemic acquired resistance (SAR) response of plants. SAR is part of a defence response that is induced locally by pathogen or pest attack, but that spreads systemically to protect the entire plant. Salicylic acid likely mediates this response. Application of salicylic acid or certain derivatives such as aspirin (acetylsalicylic acid) induces the rapid expression of the pathogenesis- related genes, which serve as molecular markers for the SAR response (Ward *et al.*, 1991; Raskin, 1992). Moreover, if the timing is right, salicylic acid levels increases throughout the plant after only one part of the plant is attacked by a pathogen and this increase occurs before the expression of pathogen related genes (Fig. 12.5; Malamy *et al.*, 1990; Klessig and Malamy, 1994).

Transgenic tobacco and *Arabidopsis thaliana* expressing the bacterial enzyme salicylate hydroxylase cannot accumulate salicylic acid (SA). This defect not only makes the plants unable to

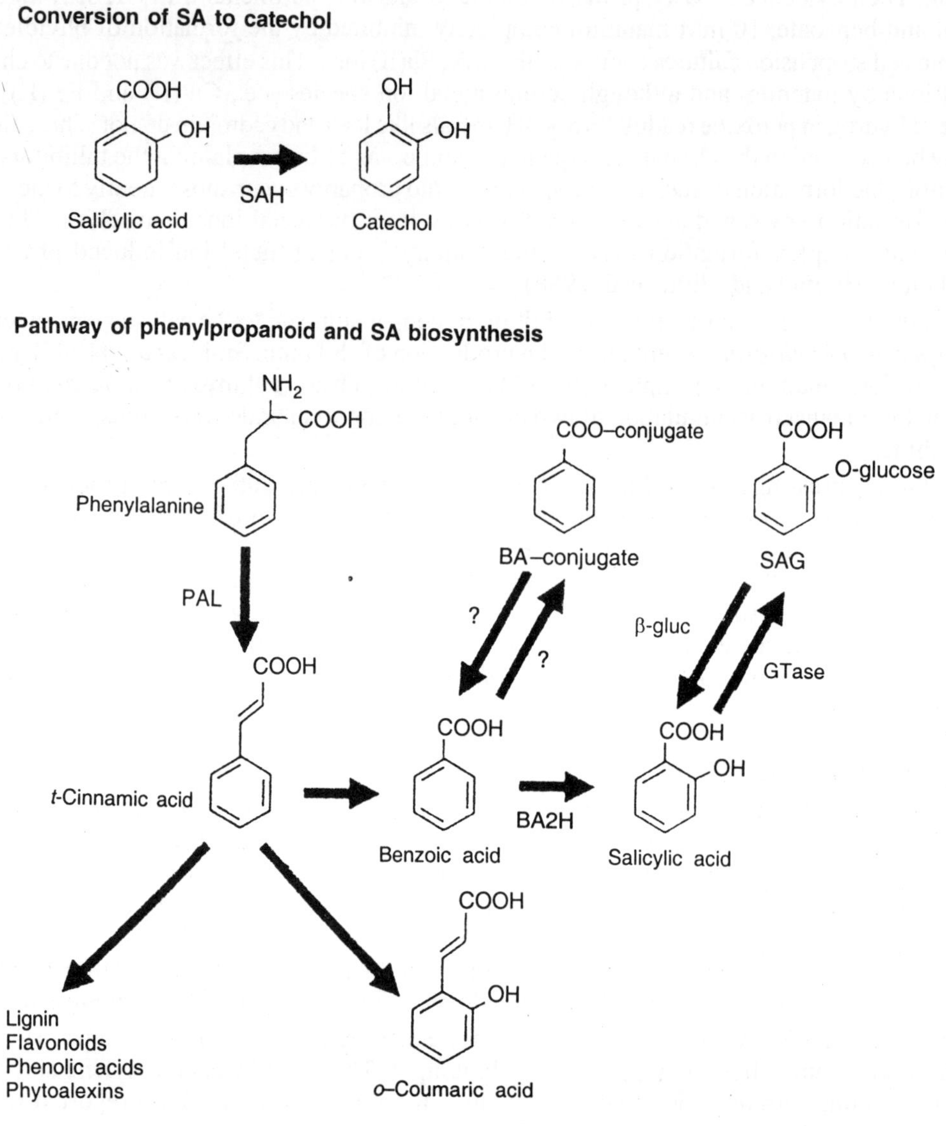

Conversion of SA to catechol

Salicylic acid → Catechol (SAH)

Pathway of phenylpropanoid and SA biosynthesis

PAL   phenylalanine ammonia lyase; Gtase   UPP-glucose; SA 3-O-glucosyltransferase;
β-gluc = β-glucosidase; SAG = salicylic acid-glucoside; ? = possible reactions

Fig. 12.4. Biochemical pathways involving salicylic acid (Adopted from Vernooij et al., 1994).

induce systemic acquired resistance, but also leads to increased susceptibility to viral, fungal and bacterial pathogens. The enhanced susceptibility extends even to host pathogen combinations that would normally result in genetic resistance. Therefore, SA accumulation is essential for expression of multiple modes of plant disease resistance (Delaney *et al.*, 1994).

Treatment of carrot (*Daucus carota*) suspension culture with a fungal cell wall preparation induced the production of chitinase. This treatment also caused an accumulation of salicylic acid,

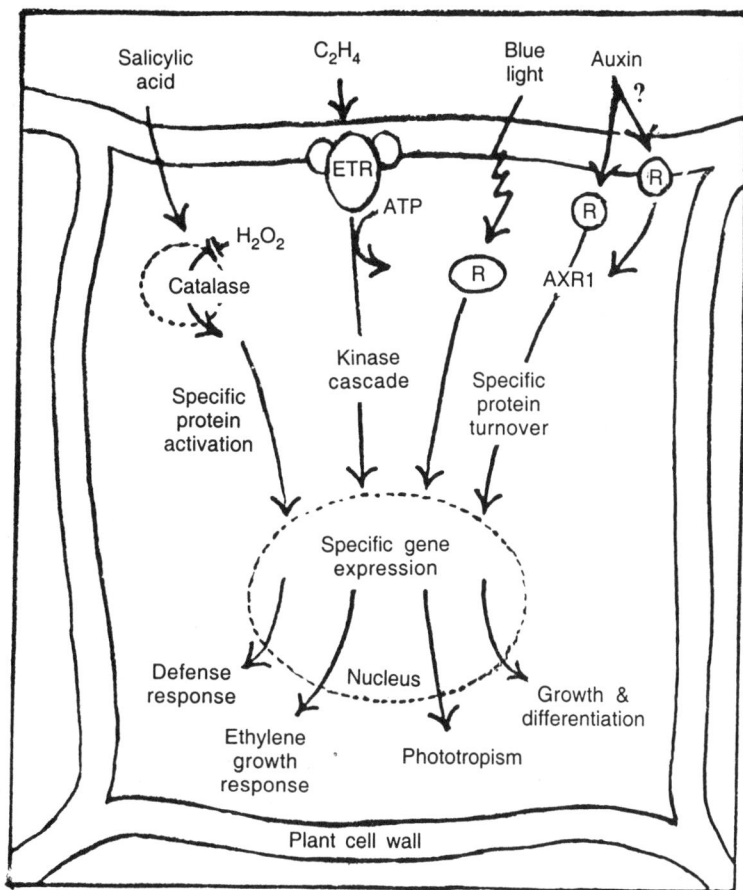

Fig. 12.5. Surprising signals : Discovered mechanisms by which plants sense their environments (Adopted from Jones, 1994).

both in the cells and in the extracellular medium. Culture medium of induced cells, which contained elevated levels of salicylic acid, was able to induce chitinase activity in a non-induced suspension culture (Muller *et al.*, 1994).

REFERENCES

Aerts, J.R., Gisi, D., Carolis, D.E., Luca, D.V. and Baumann, W.T. (1994), Methyl jasmonate vapor increases the developmentally controlled synthesis of alkaloids in *Catharanthus* and *Cinchona* seedlings, *The Plant J.*, **5(5)**: 635-643.

Aerts, J.R., Schaefer, A., Hesse, M., Baumann, W.T. and Slusarenko, A. (1996), Signaling molecules and the synthesis of alkaloids in *Catharanthus roseus* seedlings, *Phytochemistry*, **42(2)**: 417-422.

Albersheim, P. and Darvill, A.G. (1985), Oligosaccharins, *Sci. Am.*, **253(3)**: 44.

Anderson, M.J., (1991), Jasmonic acid-dependent increase in vegetative storage protein in soybean tissue cultures, *J. Plant Growth Regul.*, **10**: 5-10.

Asada, M. and Shuler, M.L. (1989), Stimulation of ajmalicine production and excretion from *C. roseus*: effects of adsorption *in situ* elicitors and alginate immobilization. *Appl. Microbiol. Biotechnol.*, **30** : 475-381.

Ayabe S., Lida, K. and Furuya, T. (1986), Stress induced formation of echinatin and a metabolite 5-prenyl licodione in cultured *Glycyrrhiza echinata* cells. *Phytochemistry,* **25:** 2803- 2806.

Bach, M., Schnitzler, P.J. and Seitz, U.H. (1993), Elicitor. Induced Chages in Ca^{2+} Influx, K^+ Efflux and 4- hydroxy benzoic acid synthesis in protoplasts of *Daucus carota* L., *Plant Physiol.*, **103:** 407-412.

Bailey, J.A. and Mansfield, J.A. (1982). In: *Phytoalexins*, Manipulating secondary metabolism in culture. Robins, R.J. and Rhodes, M.J.C. (eds) Cambridge University Press, Cambridge, New York, **57**.

Ballica, R., Ryu, Y.D.D. and Kado, I.C. (1993), Tropane alkaloid production in *Datura stramonium* suspension cultures: Elicitor and Precursor effects, *Biotechnol. Bioeng.* **41:** 1075-1081.

Barz, W., Daniel, S.I., Hinderer, W., Jaques, U., Kebmann, H., Koster, J. and Tieinann, K. (1988), In: *Application of Plant Cell and Tissue Culture*, PP. 179,Yamada,Y.,(eds) CIBA Foundations Symposium No.**137:** Wiely Chichester.

Becker, H. and Chavadej, S. (1985). Valepotriate production of normal and colchicine treated cell suspension cultures of *Valeriana wallichii. J. Nat. Prod.,* **48(1) :** 17-21.

Bhagyalakshmi, N. and Bopanna, K. (1998), Elicitation and immobilization of cell cultures for enhanced synthesis of pharmaceutical compounds, In: Irfan A. Khan and Atiya Khanum (eds) *Role of Biotechnology in Medical and Aromatic plants.* Ukaaz publications, Hyderabad, A.P., India **I:** 279.

Brodelius, P. (1988), Permeabilization of plant cells for release of intracellularly stored products: Viability studies, *Appl. Microbiol. Biotechnol.* **27:** 561-566.

Brodelius, P., (1988), Stress induced secondary metabolism in plant cell cultures In: Pais, M., Mavituna, F. and Novais, J.M (eds) *Plant Cell Biotechnology. Proc.* NATO advanced study Institute on plant cell Biotechnology, Series H: *Cell Biology,* **18:** Springer Berlin, Heidelberg, New York, 195.

Brodelius, P., Collinge, M.A., Funk, C., Gugler, K. and Marque, E. (1989), In: *Primary and Secondary Metabolism of Plant Cell Cultures* (Kurz, W.G.W., ed.), Springer-Verlag, Berlin, 191.

Brooks, C. and Watson, D. (1991), Terpenoid phytoalexins, *Nat. Prod. Rep.*, **8:** 367.

Brooks, C.J.W., Watson, D.G. and Freer, I.M. (1986), Elicitation of capsidiol accumulation in suspended callus cultures of *Capsicum annuum, Phytochemistry*, **25:** 1089-1092.

Brown, E.G. and Newton, R.P. (1981), Cyclic AMP and higher plants, *Phytochemistry*, **20:** 2453.

Buitelaar, R.M., Cesario, M.T. and Tramper, J. (1992), Elicitation of thiophene production by hairy roots of *Tagetes patula, Enzyme Microb. Technol.*, **14:** 2-7.

Chappell, J. and Hahlbrock, K. (1984), Transcription of plant defense gene in response to UV light or fungal elicitor, *Nature* (London), **311:** 76.

Chappell, J. and Noble, R. (1987), Induction of sesquiterpenoid biosynthesis in Tobacco cell suspension cultures by fungal elicitor, *Plant Physiol.*, **85:** 469-473.

Chen, T.C., Chou, M.C. and Kao, H.C. (1994), Methyl jasmonate induces the accumulation of putrescine but not proline in detached rice leaves, *J. Plant Physiol.*, **143:** 119-121.

Chen, Z., Silva, H. and Klessig, D.F. (1993). Involvement of reactive oxygen species in the induction of systemic acquired resistance by salicylic acid in plants. *Science,* **262 :** 1883-1886.

Choi, D. and Bostock, R.M. (1994), Involvement of *de novo* protein synthesis protein kinase, extracellular Ca^{2+} and lipoxygenase in arachidonic acid induction of 3-hydroxy-3-mehtyl glutaryl-coenzyne-A reductase genes and isoprenoid accumulation in potato (*Solanum tuberosum* L.), *Plant Physiol.*, **104:** 1237-1244.

Christen, A.A., Gibson, D.M. and Bland, T. (1991), Production of Taxol or taxol – like compounds in *Taxus brevifolia* callus culture, *US Pat.* **No. 5,019,504**.

Ciddi Veeresham., Srinivasan, V. and Shuler, M.L. (1995), Elicitation of *Taxus* sp. Cell cultures for production of Taxol, *Biotechnology Letters*, **17(2):** 1343-1346.

Cline, D.S. and Coscia, J.C. (1988). Stimulation of Sanguinarine production by combined fungal elicitation and hormonal deprivation in cell suspension cultures of *Papaver bracteatum. Plant Physiol.,* **56 :** 0161-0165.

Cline, S.D., McHale, R.J. and Coscia, C.J. (1993), Differential Enhancement of Benzophenanthridine alkaloid content in cell suspension cultures of *Sanguinaria canadensis* under conditions of combined hormonal deprivation and Fungal elicitation, *Journal of Natural Products*, **56(8):** 1219-1228.

Constabel, F. (1990). Medicinal plant biotechnology. *Planta Medica,* **56 :** 421-425.

Cosio, E., Frey, T., Verduyn, R., Van, B.J. and Ebel, J. (1990), High affinity binding of a synthetic heptaglucoside and fungal glucan phytoalexin elicitors soybean membranes, *FEBS Lett.*, **271:** 223.

Creelman, R.A., Tierney, M.L. and Mullet, J.E. (1992), Jasmonic acid/methyl jasmonate accumulate in wounded soybean hypocotyls and modulate wound gene expression, *Proc. Natl. Acad. Sci., USA*, **89:** 4938-4941.

Daria, H., Ross, C.B., Richard, E.K., Ulrich, M. and Karl, H. (1990), Accumulation of coumarins in elicitor – treated cell suspension cultures of *Ammi majus*, *Phytochemistry*, **29(4):** 1137-1142.

Darvill, A.G. and Albersheim, P. (1984). Phytolexins and their elicitors – a defense against microbial infection in plants. *Ann. Rev. Plant. Physiol.,* **35 :** 243-275.

Delaney, D.T., Uknes, S., Vernooij, B., Friedrich, L., Weymann, K., Negrotto, D., Gaffney, T., Rella, G.M., Kessmann, H., Ward, E., Ryals, J. (1994), A central role of salicylic acid in plant disease resistance, *Science*, **266:** 1247-1250.

Dhawan, A.K. and Malik, C.P. (1974). Cyclic AMP control of some oxidoreductases during pine pollen germination and tube growth. *Phytochemistry,* **18 :** 2015.

DiCosmo, F. and Misawa, M. (1985), Eliciting secondary metabolism in plant cell cultures, *Trends Biotechnol.* **3:** 318 – 322.

Dittrich, H., Kutchan, T. and Zenk, M. (1992), The jasmonate precursor, 12-oxo-phytodienoic acid, induces phytoalexin synthesis in *Petroselinum crispum* cell cultures, *FEBS Lett.* **309:** 33.

Dixon, R., Choudhary, A., Edwards, R., Harrison, M., Lamb, C., Lawton, M., Mavandad, M., Stermer, B. and Yu, L. (1990), Elicitors and defense gene activation in cultured cells, In: *Signal Perception and Transduction in higher plants*, (Ranjeva, R. and Boudet, A. eds.), Springer-Verlag, Germany, 283.

Dixon, R.A. (1986), The phytoalexin response: elicitation, signalling and host gene expression, *Biol. Rev.*, **61:** 239.

Dixon, R.A. and Bendall, D.S. (1978), Changes in phenolic compounds associated with phaseollin production in cell suspension culture of *Phaseolus vulgaris*, *Physiol. Plant Pathol.* **13 :** 283.

Dixon, R.A. and Fuller, K.W. (1977), Characterization of components from culture filtrates of *Botrytis cinerea* which stimulate phaseollin biosynthesis in *Phaseolus vulgaris* cell suspension cultures, *Physiol. Plant Pathol.* **11:** 287.

Dixon, R.A., Dey, P.M., Murphy, D.L., and Whitehead, I.M. (1981), Dose responses for *Colletotrichum lindemuthianum* elicitor – mediated enzyme induction in French bean cell suspension cultures, *Planta*, **151:** 272.

Dornenburg, H., and Knorr, D. (1994), Elicitation of chitinases and anthraquinones in *Morinda citrifolia* cell cultures, *Food Biotechnology*, **8(1):** 57-65.

Ebel, J. (1986), Phytoalexin synthesis; biochemical analysis of the induction process, *Annu. Rev. Phytopathol.* **24:** 235 -264.

Ebel, J., Schmidt, W.E. and Loyal, R. (1984), Phytoalexin synthesis in soybean cells: Elicitor induction of Phenylalanine ammonia-lyase and chalcone synthase mRNA and correlation with phytoalexin accumulation, *Arch. Biochem. Biophys.* **232:** 240-248.

Edward, E. F. and Ryan, C. A. (1992), Octadecanoid precursors of jasmonic acid activate the synthesis of wound-inducible proteinase inhibitors, *The Plant Cell*, **4:** 129-134]

Eilert, U. and Constable, F. (1986a), Elicitation of sanguinarine accmulation in *Papaver somniferum* cells by Fungal homogenates – An Induction Process, *J. Plant Physiol.* **125:** 167-172.

Eilert, U., Constable, F., and Kurz, W.G.W. (1986b), Elicitor-stimulation of monoterpene indole alkaloid formation in suspension cultures of *Catharanthus roseus*, *J. Plant Physiol.* **126:** 11-12.

Eilert, U., Ehmke, A., and Wolters, B. (1984), Elicitor-induced accumulations of acridone alkaloid epoxides in *Ruta graveolens* suspension cultures, *Planta Med.*, **50:** 507.

Eilert, U., Kurz, W.G.W. and Constabel, F. (1985), Stimulation of Sanguinarine Accumulation in *Papaver somniferum* cell cultures by Fungal Elicitors; *J. Plant Physiol.* **119:** 65-76.

Eilert, U., Kurz, W.G.W. and Constabel, F. (1986), Elicitor induction of sanguinarine formation in *Papaver somniferum* cell cultures and semicontinuous sanguinarine production by re-elicitation, *Planta Med.*, **5:** 417-418.

Endress, R. (1979), Allosteric regulation of phosphodiesterase from *Portulaca* callus by c-GMP and Papaverine, *Phytochemistry*, **18:** 15.

Endress, R. (1994). Plant Cell Biotechnology, Springer-Verlag, Berlin, p. 222.

Endress, R., Jager, A. and Kreis, W. (1984), Catecholamine biosynthesis dependent on the dark in betacyanin forming *Portulaca* callus, *J. Plant Physiol.*, **115:** 291.

Epperlein, M.M., Noranha, D.A.A. and Strange, R.N. (1986), Involvement of the hydroxyl radical in the abiotic elicitation of phytoalexins in legumes, *Physiol. Mol. Plant Pathol.* **28:** 67-77.

Fang, Y., Smith, L.A.M. and Pepin, F.M. (1999), Effects of exogenous methyl jasmonate in elicited anthocyanin producing cell cultures of Ohelo (*Vaccinium pahalae*), *In Vitro Cell. Dev. Biol. Plant*, **35:** 106-113.

Farmer, E.E., Johnson, R.R. and Ryan, A.C. (1992). Regulation of expression of proteinase inhibitor gene by methyl jasmonate and jasmonic acid. *Plant Physiol.*, **98 :** 995-1002.

Fujita, Y., Tabata, M., Nishi, A. and Yamada, Y. (1982), In: *Plant tissue Culture 1982*, ed., Fujiwara, A., pp.399-400. Japanese Association for Plant Tissue Culture, Tokyo.

Fukui, H., Yoshikawa, N. and Tabata, M. (1983), Induction of shikonin formation by agar in *Lithospermum erythrorhizon* cell suspension cultures, *Phytochemistry,* **22(11):** 2451.

Funk, C. and Brodelius, P. (1990), Influence of growth regulators and an elicitor on phenylpropanoid metabolism in suspension cultures of *Vanilla planifolia*, *Phytochemistry*, **29(3):** 845-848.

Funk, C., Gugler, K. and Brodelius, P. (1987), Increased secondary product formation in plant cell suspension cultures after treatment with a yeast carbohydrate preparation (Elicitor), *Phytochemistry,* **26(2):** 401-405.

Furmanowa, M., Glowniak, K., Baranek, K.S., Zgorka, G. and Jozefczyk, A. (1997), Effect of picloram and methyl jasmonate on growth and taxane accumulation in callus culture of *Taxus x media* var. *Hatfieldii*, *Plant Cell, Tissue and Organ Culture*, **49:** 75-79.

Furmanowa, M., Gowniak, K., Zobel, A., Guzewska, J., Zgorka, G., Rapczewska, L. and Jozefczyk, A. (1995), Taxol in *Taxus baccata* L. var. *elegantissima* organs and in tissue culture, *Med. Fac. Landbouww. Univ. Gent.* **60:** 2115-2118.

Grisebach, H. (1986), Regulation of secondary metabolite formation, In: *Phytoalexin Accumulation*, Kleinkauf, H., von Dohren, H., Dorauer, H. and Nesemann, G. (eds.), Woksh Conf. Hoechst **16**, VCH, Weinheim, 355.

Grisebach, H. and Ebel, J. (1978), Phytoalexins: Chemical defense substances of higher plants, *Angewandte Chemie International Edn. in English*, 17: 635-647.

Grosskope, D., Felix, G. and Boller, T. (1991), A yeast derived glycopeptide elicitor and chitosan or digitonin differentially induce ethylene biosynthesis, phenyl alanine ammonia lyase and callose formation in suspension cultured tomato cells, *J. Plant Physiol.*, **138:** 741.

Gundlach, H., Muller, J.M., Kutchan, M.T. and Zenk, H.M. (1992), Jasmonic acid is a signal transducer in elicitor incduced plant cell cultures, *Proc. Natl. Acad. Sci., USA.* **89:** 2389-2393.

Guo, Z.J. and Ohta, Y. (1994), Effect of ethylene biosynthesis on the accumulation of 6-methoxymellein induced by elicitors in carrot cell. *J. Plant. Physiology,* **144 :** 700-704.

Halverson, L.J. and Stacey, G. (1986), Signal-exchange in plant-microbe interactions, *Microbiol. Rev.*, **50(2):** 193.

Heim, S. and Wagner, K.G. (1987), The Phosphatidyl inositol species of suspension cultured plant cells, *Z.Naturforsch*, **42C:** 1003.

Heinstein, F.P. (1985), Future approaches to the formation of secondary natural products in plant cell suspension cultures, *Journal of Natural Products*, **48(1):** 1-9.

Hille, A., Purwin, C., and Ebel, J. (1982), Induction of enzymes of phytoalexin synthesis in cultured soybean cells by an elicitor from *Phytophthora megasperma* var. *glycinea*, *Plant Cell Rep.*, **1:** 123.

Holden, M.A., Hall, R.D., Lindsey, K. and Yeoman, M.M. (1987), Capsaicin biosynthesis in cell cultures of *Capsicum frutescens*. In Plant and Animal Cells/Process Possibilities, Eds. C.Webb and F.Mavituna, 45-63, Ellis Harwood Books.

Hoshino, T., Chida, M., Yamaura T., Yoshizawa, Y. and Mizuatani, J. (1994), Phytoalexin induction in green pepper cell cultures treated with arachidonic acid, *Phytochemistry*, **36(6)**: 1417-1419.

Johnson, T.S., Ravishankar, G.A. and Venkataraman, L.V. (1991), Elicitation of capsaicin production in freely suspended cells and immobilized cell cultures of *Capsicum frutescens, Food Biotechnol*. **5**: 197-205.

Jones, M. A. (1994), Surprising signals in plant cells, *Science*, **263**: 183-184.

Katau, K., Tomiyama, K. and Okamoto, H. (1982), Effects of hyphal wall components of *Phytophthora infestans* on membrane potential of potato tuber cells, *Physiol. Plant Pathol*. **21**: 311-317.

Kaus, H. (1987), Some aspects of calcium dependent regulation in plant metabolism, *Annul. Rev. Plant Physiol.*, **28**: 47.

Kauss, H., Krause, K. and Jeblick, W. (1992), Methyl jasmonte conditions parsley suspension cells for increased elicitation of phenylpropanoid defense response, *Biochemical and Biophysical Research Communications*, **189(1)**: 304-308.

Keen, N.T. and Horsch, R. (1972), Hydroxyl phaseollin production by various soybean tissues: a warning against use of 'unnatural' host-parasite systems, *Phytopathology*, **62**: 439-442.

Ketchum, R.E.B., Gibson, D.M., Crouteau, R.B. and Shuler, M.L. (1999), The kinetics of taxoid accumulation in cell suspension cultures of *Taxus*, following elicitation with methyl jasmonate, *Biotech. and bioengg.* **62(1)** :97-105.

Kim, D.J. and Chang, H.N. (1990), Increased shikonin production in *Lithospermum erythrorhizon* suspension cultures with *in situ* extraction and fungal cell treatment (elicitor), *Biotech. Lett*. **12**: 443-446.

Klessig. F.D. and Malamy. J. (1994). The salicylic acid signal in plants, *Plant Molecular Biology*, **26**: 1439-1458.

Kohle, H., Young, D.H. and Kauss, H. (1984), Physiological changes in suspension cultured soybean cells elicited by treatment with chitosan, *Plant Sci. Lett*. **33**: 221.

Kota, D. and Stelzig, D.A. (1977), Electrophysiology as a means of studying the role of elicitors in plant disease resistance, *Proc. Amer. Phytopathal. Soc.*, **4**: 216-217.

Kutoyanagi,M., Arakava,T., Mikami,Y., Yoshida,K., Kawahar,N.,Hayashi,T., and Ishimaru,H (1998) Phytoalexins from hairy root cultures of *Hyoscyamus albus* treated with methyl jasmonate. *J.Nat.Products*. 61, 1516-1519.

Kurosaki, F., Tsurusawa, Y., Nishi, A. (1987), Breakdown of phosphatidyl inositol during the elicitation of phytoalexin production in cultured carrot cells, *Plant Physiol.,* **85**: 601.

Kutney, J.P., Samija, D.M., Hewitt, M.G., Bugante, C.E. and Gu, H., (1993), Anti-inflammatory oleanane triterpenes from *Tripterygium wilfordii* cell suspension cultures by fungal elicitation, *Plant Cell Reports*, **12**: 356-359.

Lange, M.B., Trost, M., Heller, W., Langebartels, C. and Sandermann, H.Jr. (1994), Elicitor- induced formation of free and cell wall-bound stilbenes in cell suspension cultures of Scots pine (*Pinus sylvestris* L.), *Planta*, **194**: 143-148.

Lee, T.K., Yamakawa, T., Kodama, T. and Shimomura, K. (1998), Effects of chemicals on alkaloid production by transformed roots of *Belladonna*, *Phytochemistry*. **49(8)**: 2343-2347.

Lindsey, K. and Yeoman, M.M. (1984), The synthetic potential of immobilized cells of *Capsicum frutescens* Mill. *cv. Annum. Planta* **162**: 495-501.

Mader, C.J. (1999), Effects of jasmonic acid, silver nitrate and L-AOPP on the distribution of free and conjugated polyamines in roots and shoots of *Solanum tuberosum, in vitro, J. Plant Physiol.*, **154**: 79-88.

Malamy, J., Carr, P.J., Klessig, D.F., Raskin, I. (1990), Salicylic acid: a likely endogenous signal in the resistance response of Tobacco to viral infection, *Science*, **250**: 1002-1004.

Masanori, K., Arakawa, T., Mikami, Y., Yoshida, K. Kawahar, N., Hayashi, T. and Ishimaru, H. (1998), Phytoalexins from Hairy roots of *Hyoscyamus albus* treated with Methyl Jasmonate and copper sulphate, *J. Nat. Prod.* **61**: 1516-1519.

Mirjalili, N. and Linden, J. C. (1996), Methyl Jasmonate induced production of taxol in suspension cultures of *Taxus cuspidata*; ethylene interaction and induction models, *Biotech. Progress*, **12**: 110.

Mizukami, H., Ogawa, T., Ohashi, H. and Ellis, E.B. (1992), Induction of rosmarinic acid biosynthesis in *Lithospermum erythrorhizon* cell suspension cultures by yeast extract, *Plant Cell Reports*, **11**: 480-483.

Moesta, P. and West, C.A. (1985). Casbene synthetase : regulation of phytoalexin biosynthesis in *Ricinus communis* L. seedlings, *Arch. Biochem. Biophys.*, **238** : 325.

Moon, W.J., Yoo, B.S., Kim, D. and Byun, S.Y. (1998), Elicitation kinetics of taxane production in suspension cultures of *Taxus baccata* pendula, *Biotechnology Technique*, **12(1)**: 79-81.

Mueller, U.I., Parthier, B. and Nover, L. (1998), Jasmonate- induced alteration of gene expression in barley leaf segments analyzed by *in vivo* and *in vitro* protein synthesis, *Planta*, **176**: 241.

Muller, S.S., Kurosaki, F. and Nishi, A. (1994). Role of salicylic acid and intracellular Ca^{2+} in the induction of chitinase activity in carrot suspension culture, *Physiological and Molecular Plant Pathology*, **45**: 101-109.

Nef, C., Rio, B. and Cherstin, H. (1991), Induction of Catharanthine synthesis and stimulation of major indole alkaloids production by *Catharanthus roseus* cells under non-growth-altering treatment with *Pythium vexans* extracts, *Plant Cell Reports*, **10**: 26-29.

O' Niell, S.D. and Spanswick, R.M. (1984), Solubilization and reconstitution of a vanadate-sensitive hydrogen ATPase from the plasma membrane of *Beta vulgaris*, *J. Membr. Biol.*, **79**: 231-243.

Patrica, M. Moctezuma, L. and Gloria, L.E. (1996), Biosynthesis of sesquti terpene phytoalexin capsidol in elicited root cultures of chilli peppers (*C. annuum*). *Plant Cell Reports*, **15** : 360-366.

Phisalaphong. M. and Linden, J.C.(1999), Kinetic studies of paclitaxel production by *T. canadensis* cultures in batch and semi-continuous with total cell recycle., *Biotechnol.Prog.*, **15** : 1072-1077.

Prasad Babu, Ch. (2000). Elicitation of tissue cultures for production of Biomedicines, Ph.D. thesis, Kakatiya University, Warangal, A.P., India.

Purohit, S.S. a nd Mathur, S .K. (1996), I n: Biotechnology: p roduction o f se condary plant m etabolites; Biotechnology: Fundamentals and Applications, *Agrobotanical Gardens*, Bikaner, India, 366-393.

Raskin, I. (1992), Role of salicylic acid in plants. *Annu. Rev. Plant Physiol. Plant Mol. Biol.*, **43**: 439-463.

Ren, Y.Y. and West, A.C. (1992), Elicitation of Diterpene Biosynthesis in rice (*Oryza sativa* L.) by chitin, *Plant Physiol.*, **99**: 1169-1178.

Robbins, M.P., Bolwell, G.P., and Dixon, R.A. (1985), Metabolic changes in elicitor-treated bean cells; Selectivity of enzyme induction in relation to phytoalexin accumulation, *Eur .J. Biochem.* **148**: 463-569.

Rokem, J.S., Schwarzberg, J. and Goldberg, I. (1984), Autoclaved fungal mycelia increase diosgenin production in cell suspension cultures of *Dioscorea deltoidea*, *Plant Cell Rep.*, **3**: 159-160.

Ryan, C.A. (1987), Oligosaccharide signalling in plants. *Annu. Rev. Cell Biol.*, **3**: 295.

Sandra, I., Alvarez, P. and Giulietti, A.M. (1988). Novel biotechnological approaches to obtain scopolamine and tryosumine : the influence of biotic elicitors and stress agents on cultures of transformed rats of *Byugmansia candida*, *Phytotherapy Research*, **12** : 518-520.

Schmidt, W. and Ebel, J. (1987), Specific binding of a fungal glucan phytoalexin elicitor to membrane fractions from soybean *Glycine max*, *Proc. Natl. Acad. Sci., USA*. **84**: 4117.

Schnitzler, P.J., Madlung, J., Rose, A. and Seitz, U.H. (1992), Biosynthesis of p-hydroxy benzoic acid in elicitor-treated carrot cell cultures, *Planta*, **192**: 594-600.

Sheludko, Y., Gerasimenko, I., Unger, M., Kostenyuk, I., Stoeckigt, J. (1999), Induction of alkaloid diversity in hybrid plant cell cultures, *Plant Cell Reports*, **18**: 911-918.

Singh, G., Gavrieli, J., Oakey S.J., Curitis, R.W. (1998), Interaction of methyl jasmonate, wounding and fungal elicitation during sesquiterpene induction in *Hyoscyamus muticus* in root cultures, *Plant Cell Reports*, **17**: 391-395.

Smith, J.I., Smart, N.J., Misawa, M., Kurz, W.G.W., Tallevi, S.G. and DiCosmo, F. (1987), Increased accumulation of indole alkaloids by some cell lines of *Catharanthus roseus* in response to addition of vanadyl sulphate. *Plant Cell Reports*, **6**: 142-145.

Stab, R.M. and Ebel, J. (1987), Effects of Ca²⁺ on Phytoalexin Induction by Fungal Elicitor in Soybean Cells; *Archives of Biochemistry and Biophysics*, **257(2):** 416-423.

Staswick, P. (1992), Jasmonate, genes and fragrant signals, *Plant Physiol*. **99:** 804.

Strobel, G.A., Stierle, A. and van Kujik, F.J.G.M. (1992), Factors influencing the *in vitro* production of radio-labelled taxol by Pacific Yew, *Taxus brevifolia*, *Plant Science*, **84:** 65-74.

Suvarnalatha, G., Rajendran, L., Ravishanker, G.A. and Venkatraman, C. (1994), Elicitation of anthocyanin production in cell cultures of carrot (*Daucus carota* L.) by using elicitors and abiotic stress, *Biotechnology Letters*, **16(12):** 1275-1280.

Szabo, E., Thelen, A., Petersen, M. (1999), Fungal elicitor preparations and methyl jasmonate enhance rosmarinic acid accumulation in suspension cultures of *Coleus blumei*, *Plant Cell Reports*, **18:** 485-489.

Tamari, G., Borochov, A., Atzorn, R. and Weiss, D. (1995), Methyl jasmonate induces pigmentation and flavonoid gene expression in *Petunia corollas*: A possible role in wound response, *Physiol. Plant*, **94:** 45.

Tavernier, E., Stallaert, V., Blein, P.J. and Pugin, A. (1995), Changes in lipid composition in tobacco cells treated with Cryptogein, an elicitor from *Phytophthora cryptogea*, *Plant Science*, **104:** 117-125.

Threfall, D.R. and Whitehead, I.M. (1988), The use of biotic and abiotic elicitors to induce the formation of secondary plant products in cell suspension cultures of Solanaceous plants. *Biochem. Soc. Trans.*, **16:** 71-75.

Tietjen, K.G. Hunkler D., and Matern, U. (1982), Differential response of cultured parsley cells to elicitors from two non-pathogenic strains of fungi-I, Identification of induced products as coumarine derivatives, *Eur. J. Biochem.* **131:** 401-413.

Vernooij, B., Uknes, S., Ward, E. and Ryals, J. (1994), Salicylic acid as a signal molecule in plant-pathogen interactions, *Current Opinion in Cell Biology*, **6:** 275-279.

Ward, E.R., Uknes, S.J., Williams, S.C., Dincher, S.S., Wiederhold, D.L. Alexander, D.C., Ahl, G.P., Metraux, J.P. Ryals, J.A., (1991), Coordinate gene activity in response to agents that induce systemic acquired resistance, *Plant Cell*, **3:** 1085-1094.

West, C.A. (1981), Fungal elicitors of the phytoalexin response of higher plants, *Naturswissenschaften*, **68:** 447-457.

White, P.R. (1934), Potentially unlimited growth of excised tomato root tips in a liquid medium, *Plant Physiol.*, **9:** 585-600.

Whitehead, M.I., Atkinson, L.A. and Threlfall, R.D. (1990), Studies on the biosynthesis and metabolism of the phytoalexin lubimin and related compounds in *Datura stramonium* L., *Planta*, **182:** 81-88.

Whitehead, M.I., Ewing, F.D. and Threlfall, R.D. (1988), Sesquiterpenoids related to the phytoalexin debneyol from elicited cell suspension cultures of *Nicotiana tabaccum*, *Phytochemistry*, **27(5):** 1365-1370.

Wijnsma, R., van Weeden, I.N., Verpoorte, R., Harkes, P.A.A., Lugt, Ch.B., Scheffer, J.J.C. and Baerheim Svendsen, A. (1986), An anthraquinones in *Cinchona ledgeriana* bark infected with *P. cinnamomi. Planta Med.*, **52:** 211-12.

Wolters, B. and Eilert, U. (1982), Accumulation of acridone-epoxides in callus cultures of *Ruta graveolens* incresed by mixed culture with fungi, *Z.Naturforsch.* **37:** 575.

Xu, Y., Chang, L.F.P., Liu, D., Narasimhan, L.M., Raghothama, G.K., Hasegawa, M.P. and Bressan, A.R. (1994), Plant defense genes are synergistically induced by ethylene and methyl jasmonate, *The Plant Cell*, **6:** 1077-1085.

Yukimune, Y., Hara, Y., Higashi, Y., Oshnishi, N., Tabata, H., Suga, C., and Matsubara, K. (1995), Process for producing taxane diterpene and method for harvesting cultured cell capable of producing taxane diterpene in high yield, *WO Patent* Nº: 95/1410.3.

Yukimune, Y., Tabata, H., Higashi, Y. and Hara, Y. (1996). Methyl jasmonate-induced over production of paclitaxel and baccatin-III in *Taxus* cell suspension cultures. *Nature Biotechnology*, **14 :** 1129-1132.

Zenk, M.H., El-Shagi, H. and Schulte, U., (1975), Anthraquinone production by cell suspension culture of *Morinda citrifolia*, *Planta Med. Suppl.* **79:**101.

Chapter 13

Organogenesis and Regeneration in vitro

INTRODUCTION

Potentiality of a plant cell to regenerate the entire plant is termed as 'totipotency'. Since the potential lies mainly in cellular differentiation, this indicates that all genes responsible for differentiation are present within individual cells and many of them, remain inactive in differentiated tissues or organs are able to express only under adequate culture conditions. Development of an adult organism from a single cell or zygote is the result of the integration of cell division and cell differentiation.

Isolated cells from differentiated tissues are generally non dividing and quiescent; to express totipotency the differentiated cell first undergoes **dedifferentiation** and then **redifferentiation**. The phenomenona of a mature cell reverting to a meristematic state and forming undifferentiated callus tissue is termed dedifferentiation, whereas the ability of a dedifferentiated cell to form a whole plant or plant organs is termed redifferentiation.

13.1. ORGANOGENESIS

Organogenesis refers to the process whereby explants, tissues or cells can be induced to form root/ and or shoot and even whole plantlets (Fig. 13.1). In other words, the formation of organs is called organogenesis. It may be categorized as rhizogenesis and caulogenesis. The process of root formation is called rhizogenesis and the process of shoot initiation is known as caulogenesis.

The earliest reports on controlled organogenesis *in vitro* were by White (1939) who obtained shoots on callus of a *Nicotiana* hybrid and by Nobecourt (1937), who observed root formation in carrot callus. Skoog (1944), showed that auxin could stimulate rooting and inhibit shoot formation. Further studies of Skoog and Miller (1957) established that a balanced combination of auxin and cytokinin controls the root and shoot formation. They also reported that a high ratio of auxin to cytokinin in the medium favored root formation, a reverse ratio having high cytokinin favors shoot formation and that intermediate ratio promotes callus formation. Thus root and shoot differentiation is a function of quantitative interaction between an auxin and cytokinin. Casein hydrolysate or tyrosine also induces kinetin type bud formation even in the presence of higher levels of IAA in the medium.

Light intensity plays an important role in organogenesis. High light intensity has been shown to be inhibitory for shoot bud formation in tobacco. The quality of light also influences organogenesis. Blue light promotes shoot bud differentiation in tobacco callus while red light stimulates rooting. In general,

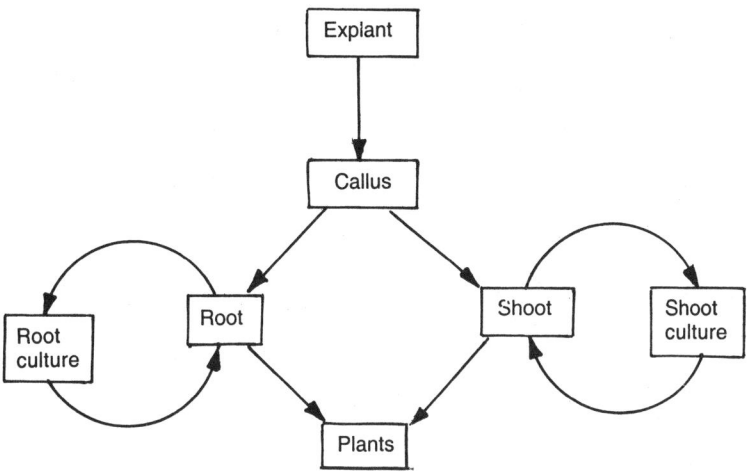

Fig. 13.1. The various steps in organogenesis.

maintenance of callus under alternating light and dark period (15-16h) may prove satisfactory for differentiation of shoots. Temperature also affect differentiation. Increase in temperature upto 33°C may be associated with increased growth of callus in tobacco while low temperature favor shoot bud differentiation (18°C is optimal). A medium solidified with agar favors bud formation although there are some reports about the development of leafy shoot buds on cultures grown as suspension. As an outcome of this approach, several hundred-plant species have been reported to form shoot and/or root *in vitro* .

The process of embryo development is called embryogenesis. Somatic emrbyogenesis is defined as "a non-sexual developmental process which produces a bipolar embryo from somatic tissue". The earliest report on controlled somatic embryogenesis *in vitro* was with carrot reported by Reinert (1958). Somatic embryo can be formed from callus, cell cultures, protoplasts or organized structure such as stem segments or zygotic embryo.

13.2. REGENERATION

Regeneration is defined as the tendency shown by a developing organisms to restore any part of it which has been removed or physiologically isolated and thus to produce a complete individual. The regeneration through plant biotechnological methods using organogenesis or embryogenesis has several advantages compared to conventional method of propagation. These are 1. Efficiency of process (reduction in labour cost and time, the formation of plantlet is fewer steps), 2. The potential for the production of much higher number of plantlets and the morphological and cytological uniformity of the plantlets. The first report on plantlet formation is an orchid through the tissue culture was reported by Morel in (1960). They are several plants produced through tissue culture on commercial scale, which includes food crops, vegetables, spices, fruits, medicinal and aromatic plants. There are four ways by which micropropagation can be achieved (Fig. 13.2). The detailed chapter is devoted on micropropagation (See Chapter 16)

13.2.1. Factors affecting regeneration

Regeneration via *de novo* organogenesis is a complex multi-stage phenomenon. The factors affecting the process of regeneration are:

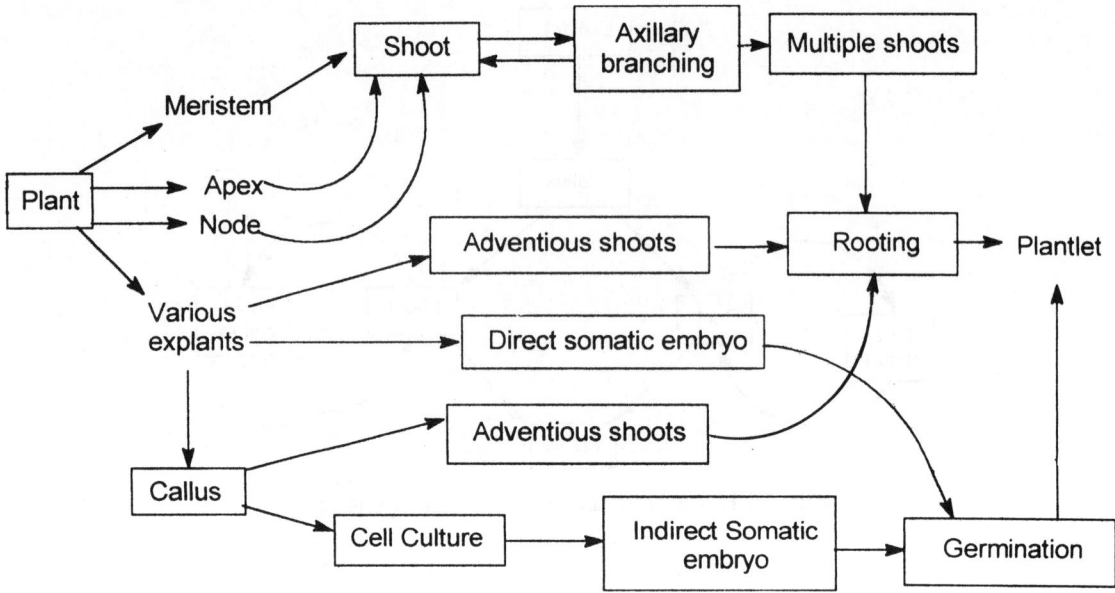

Fig. 13.2. Various methods of plant regeneration from explant and callus/cell cultures.

13.2.1.1. *Source of explant*

The factors which influence the response of the inoculum in the culture are :

(a) Size of the explant

(b) The season in which the explant is procured.

(c) The physiological age of the organ

(d) The organ that is to be served as tissue source

(e) The overall quality of the plant from which explants are taken.

Regeneration from any plant part can be achieved proper combination of the factors. The explants like stem segments and apices, leaf pieces, flower petals, hypocotyls, root segments and seed embryo etc., can give rise to organs and embryo directly or indirectly via callus. Explants consisting or mersitamatic actively dividing cells have been useful in initiating the cultures and subsequent regeneration. Young explants are generally more responsive in producing somatic embryogenesis.

13.2.1.2. *Nutrient media and constituents*

The following medium constituents can influence the regeneration

(a) **Inorganic salts :** MS medium is the most widely used salt formation compared to White, Heller, B5 and SH medium. The major difference in composition of these media is in the amount and form of nitrogen, and level of calcium.

(b) **Organic substances :** Sucrose is the best carbon source for regeneration media (2-4%) although sometimes glucose or fructose can also be used. The most commonly used vitamin are inositol, pyridoxine, thiamine and nicotinic acid. All these media contain a balanced combination of an auxin and cytokinin.

13.2.1.3. Culture environment

There are many aspects of culture environment that can influence growth and organized development. Those include 1) the pH of the medium, 2) the light quality and quantity, 3) temperature, 4) humidity, 5) presence or absence of agar.

13.3. REGULATION OF REGENERATION

By employing a suitable explant and proper medium, callus or explants can be induced to form organs or embryos. The organogenesis and embryogenesis may occur in the same cultures. The induction of shoot or root is formed depending upon the growth regulators.

13.3.1. Organogenesis

Evan *et al.* (1984) found that for 75% of the species forming shoots, either kinetin or BA (0.05-46 μM). Auxins such as NAA/IAA (0.06-27 μM). Usually a high cytokinin concentration favors shoot formation. 2,4-D promotes cell proliferation and supress cellular and organ differentiation in dicots.

13.3.2. Embryogenesis

The process of somatic embryogenesis has two stages (1) induction of somatic embryogenic competence (achieved through high concentration of auxin) (2) Development of embryogenic cells into embryos (achieved through auxins). 2,4-D is the commonly used auxin in somatic embryogenesis, the others such as 2,4,5-T, Dicamba and Picloram etc. The commonly used media in somatic embryogenesis is MS medium. Increased concentration of carbohydrates (from 2 to 6% w/v) in the medium caused osmatic stress and it has been observed that it enhances somatic embryogenesis in several species. It has also been observed that a high nitrate to low ammonium nitrogen favors somatic embryogenesis.

REFERENCES

Evans, D.A., Sharp W.R. and Medina-Filho, H.P. (1984). Somaclonal and gametoclonal variation. *Am. J. Bot.,* 71: 759-774.

Flick, C.E. (1983). Isolation of mutants from cell cultures. *In :* D. Evans *et al.* (eds.) Handbook of *Plant cell culture,* vol. 1, *Techniques for Propagation and Breeding.* Macmillan Publishing Co., New York, pp. 393-441.

Gengenbach, B.G., Green C.E. and. Donovam, C.M (1977). Inheritance of selected pathotoxin resistance in maize plants regenerated from cell cultures. *Proc. Natl. Acad. Sci., U.S.A.,* 74 : 5113-5117.

Larkin, P.J. and Scowcroft, W.R. (1981). Somaclonal variation – a novel source of variability from cell cultures for plant improvement. *Theor. Appl. Genet.,* 60 : 197-214.

Morel, G., (1960). Producing virus freecymbidium, *Am. Orchid Soc, Bull.,* 29: 495-497.

Nobecourt,P. (1937). Cultures en serie de tissus vegetaux sur milieu artificiel. *C.R seane. Soc.Biol.,* 205: 521-523.

Razdan, M.K. (1993). *An Introduction to Plant Tissue Culture,* Oxford & IBH Publ. Co., New Delhi.

Reinert, J. (1958). Morphogenesis und ihre Kontrolle an Gewebckuluren aux carotten. *Naturwissenschaften,* 45: 344-345.

Skoog,F. (1944).Growth and organ formation in tobacco tissue cultures.*Am. J.Bot.,* 31: 19-24.

Skoog, F. and Miller, C.O. (1957). Chemical regulation of growth and organ formation in plant tissues culture in vitro. *Symp. Soc. Exp. Biol.* 11: 118-131.

Sun, Z.X., Zhao, C.Z.. Zheng, K.L. Qi, X.F and Fu, Y.P. (1983). Somaclonal genetics of rice. *Oryza Sativa* l. *Theor. Appl. Genet.,* 67: 67-73.

White, P.R. (1939). Potentially unlimited growth of excised plant callus in artificial nutrient medium. *Am. J. Bot.,* 26 : 59-64.

Chapter 14

Somaclonal Variation and Genetic Instability in Plant Cell Cultures

INTRODUCTION

The growth of plant cells *in vitro* and their regeneration into whole plants is an asexual process, involving mitotic division of the cell. Therefore, it is expected that the process will produce genetically uniform plants or clonal multiplication is possible through callus regeneration. This expectation has formed the basis for micropropagation work and provided a basis for genetic manipulation in plants using callus. The occurrence of uncontrolled variation during the callus regeneration was unexpected and undesirable. This variation is known as somaclonal variation and is defined as the variation found in plants regenerated from cell cultures. Larkin and Scowcroft (1981) coined the term somaclonal variation. The cause of variation is attributed to changes in the chromosome number and structure. Recent investigations have revealed that cell or tissue cultures undergo frequent genetic changes (aneuploidy, polyploidy, translocation, gene amplification, and mutation) and these are also expressed at biochemical or molecular levels. Plant cell and tissue cultures provide increased genetic variability relatively rapidly and without applying a sophisticated technology. Genetic variability in cultures is expressed in the form of variant traits in regenerated plants, which are then transmitted to the progeny through sexual or asexual propagation.

14.1. SOURCE MATERIAL AND CULTURE CONDITIONS

Attempts directed towards the evolution of new crop varieties through somaclonal variation are influenced by genotype, explant source, duration of culture, and growth hormone effect. The genotype influences both the frequency of regeneration and the frequency of somaclonal variation. Sun *et al*. (1983) reported that the frequency of polyploidy regenerates in 18 varieties of rice under identical culture conditions and recovered multiploids only in the indica variety but not in the japonica. The source of explant has been considered to be a critical variable for somaclonal variation. It is likely that different selective pressure would be exerted against different explant resources in culture, resulting in the spectrum of somaclonal variation in regenerating plants.

High proportions of phytohormones effect karyotypic alterations in cultured cells. BAP, 2,4-D and others have been shown to induce chromosomal variability in cultured plant cells. Growth hormones are essential for induction of shoot differentiation or organogenesis. The careful monitoring of phytohormones is required for *in vitro* propagation of somaclonal variants.

14.2. MOLECULAR BASIS OF VARIATION

The variants in tissue cultures may arise as a result of more subtle changes due to single gene mutations in cultured cells. Changes in the cytoplasmic genome have also been observed in somaclones. Gengenbach *et al.* (1977) observed that maize genotypes having male sterile cytoplasm were also sensitive to the toxin secreted by the *Drechslera maydis* race T. In an attempt to combine resistance to toxin with the Texas male sterile cytoplasm trait in maize plant, they succeeded in selecting resistant plants associated with concomitant reversion to male fertility. Another aspect of single gene mutation responsible for somaclonal variation relates to transposable elements. Variations have been reported as a result of insertion of plasmid like DNA in the mitochondrial genome of cell cultures of tobacco maize, wheat and alfalfa. The somaclonal variation may also be due to molecular changes caused by mitotic crossing over in regenerated plants. This could be either symmetric or asymmetric variation. Single gene mutations by mitotic crossing over may constitute a unique mechanism of including new genetic variations. Recent studies revealed that changes in the organelle, DNA, isoenzyme and protein profile correlate with the occurrence of somaclonal variation in plants.

14.3. ISOLATION OF VARIANTS

Somaclonal variants can be isolated either selection pressure or without selection pressure. In case of without selection pressure method, callus or cells are grown in cultures for various periods on a medium that contains a selective agent; such calli are induced to differentiate whole plants. The regenerated plants are ultimately transferred to a field and screened for somaclonal variation. e.g. potato, wheat, maize and sugarcane etc. In the selection pressure method, variant cell lines are screened by their ability to survive in the presence of a substance in the medium that may be inhibitory or toxic or may survive under the conditions of environmental stress. The variants may be produced by direct or indirect selection. In the case of direct selection the new variant cell types have a selective advantage over the rest of the population because of their tolerance to a specific toxic compound. As for indirect selection, the wild type cells are selectively killed and only mutant recovered by transferring the surviving cells to an enriched medium.

Amino acid analogues either act as false feed back inhibitors of amino acid biosynthesis or may be incorporated into proteins, causing the loss of enzyme activity. Growth inhibition can be reversed by the addition of corresponding amino acid. The change in the feed back sensitivity results in the evolution of cell lines resistant to the analogue and resulting in over production or accumulation of free amino acid. The accumulated amino acid may be metabolized or excreted into the culture medium. The cell lines resistant to analogues of proline, lysine, phenylalanine, tryptophan and methionine have been isolated (Table 14.1). Exposure of cell and protoplast cultures to various herbicides induces mutation that may yield tolerant cell lines, which ultimately regenerate into plants. Many herbicide-resistant, salt-tolerant and pathotoxin resistant plants have been regenerated (Table 14.2). Various auxotrophic mutants and variant lines selected in cell cultures are shown Table 14.3. Cell lines resistance to the antibiotics like lincomycin, kanamycin, chloramphenicol and streptomycin has been developed from various plant species (Table 14.4).

14.4. CAN GENETIC INSTABILITY BE CONTROLLED IN PLANT TISSUE CULTURES?

Introduction

The somaclonal variation is useful for crop improvement by selecting variants for desired character such as disease resistant crops (Table 14.5).

Table 14.1. Amino acid and amino acid analogue-resistant cell lines selected in cultures (Adopted from Flick, 1983, Ignacimuthu, 1998)

Selective agent	Species	Mutagen	Plant regeneration
p-Fluorophenyl alanine	A. pseudoplatanus	None	No
	D. carota	None	No
	D. carota	None	No
	D. carota	None	No
	D. innoxia	None	Yes
	N. tabacum	UV or EMS	No
	N. tabacum	None	No
	N. tabacum	None	No
	N. tabacum	None	No
	N. tabacum	None	No
	N. tabacum	None	Yes
6-Fluorotryptophan	Petunia hybrida	NG	No
5-Fluorouracil	D. carota	None	Yes
Threonine	N. tabacum	None	Yes
Valine	N. tabacum (haploid)	UV	Yes
	N. tabacum (halpoid)	NG or irradiation	No
5-Methyl tryptophan	Catharanthus roseus	None	No
	D. carota	None	No
	N. tabacum	UV or EMS	No
	N. tabacum	None	No
	N. tabacum	None	Yes
	N. sylvestris	None	No
Methione sulfoximine	Z. mays	None	Yes
	N. tabacum	None	Yes
Lysine plus threonine	Z. mays	Azide	Yes
Hydroxy proline	D. carota	EMS	No
	H. vulgare	Azide	Yes
Hydroxy Lysine	N. tabacum	EMS	No
	N. tabacum	UV or EMS	No
Glycine hydroxamate	N. tabacum	None	Yes
S-(aminoethyl)-L-cysteine	Arabidopsis thaliana	EMS	Yes
	Daucus carota	None	No
	D. carota	None	No
	Hordeum vulgare	Azide	Yes
Aminopterin	Oryza sativa	EMS	No
	Datura innoxia (haploid)	None	Yes
Azaguanine	Acer pseudoplatanus	NTG	No
	Glycine max	EMS	No
	Haplopappus gracilis	EMS	No
	Medicago sativa	EMS	No

(Contd.)

Selective agent	Species	Mutagen	Plant regeneration
Azauracil	*H. gracilis*	EMS	No
	Zea mays	EMS	No
Azetidine-2-carboxylic acid	*D. carota*	None	No
	D. carota	None	No
Bromodeoxy-uridine	*G. max*	NG	No
	M. sativa	EMS	No
	N. tabacum	None	Yes
Ethionine	*D. carota*	EMS	No
	D. carota	None	No
	M. sativa	EMS	No
Hydroxyurea	*N. tabacum*	None	Yes
Seleno-amino acids	*N. tabacum*	None	No

NG = N-methyl-N'-nitro-N-nitrosoguanidine

Table 14.2. Agriculturally useful mutant cell lines obtained in vitro (Adopted from Flick, 1983, Ignacimuthu, 1998)

Selective agent	Species	Mutagen	Plant regeneration
Salt tolerance	*Capsicum annuum*	None	No
	Datura innoxia	None	Yes
	Kickxia ramosissima	None	Yes
	Medicago sativa	None	No
	Oryza sativa	None	No
	N. sylvestris	None	No
	N. tabacum	EMS	Yes
Pathotoxin resistance			
Fusarium oxysporium	*Solanum tuberosum*	None	Yes
Helminthosporium maydis	*Zea mays T-cms*	None	Yes
Phytophthora infestans	*S. tuberosum*	None	Yes
Herbicide resistance			
Asulam	*Apium gravaeolens*	None	No
Bentazone	*Nicotiana tabacum* (Haploid)	γ-ray	Yes
2,4-D	*Lotus comiculatus*	None	Yes
2,4-D; 2,4,5-T;			
Trifolium repens	None	No	
Isopropyl-N-phenyl carbamate	*N. tabacum*	EMS	Yes
Paraquat	*N. tabacum*	X-ray	Yes
Phenmedifarm	*N. tabacum* (haploid)	γ-ray	Yes
Picloram	*N. tabacum*	None	Yes
Cold Tolerance	*Daucus corota*	EMS	Yes
N. sylvestris	None	Yes	

Table 14.3. Auxotrophic mutants and various other variant cell lines selected in culture (Adopted from Flick, 1983, Ignacimuthu, 1998)

Nutrient requirement/ phenotype	Species	Mutagen	Plant regeneration
1. Autotrophs			
Lysine	N. tabacum (haploid)	EMS	Yes
Isoleucine	N. plumbaginifolia (haploid)	γ-rays	Yes
Histidine	Hyoscyamus muticus	NTG	Yes
Hypoxanthine	N. tabacum (haploid)	EMS	Yes
Arginine	N. tabacum (haploid)	EMS	Yes
Biotin	N. tabacum	EMS	Yes
Proline	N. tabacum (haploid)	EMS	No
p-aminobenzoic acid	Nicotiana tabacum (haploid)	EMS	Yes
Adenine	Datura innoxia (hapaloid)	EMS	No
Nicotinamide	H. muticus (haploid)	NG	Yes
Pantothenate	D. innoxia	EMS	No
2. Miscellaneous			
a) Carbohydrate metabolism			
Glycerol ultilization	N. tabacum	None	Yes
Maltose utilization	Glycine max	None	No
b) Plant growth regulator			
Abscisic acid	N. tabacum	EMS	No
Auxin heterotrophic and auxin resistant	N. sylvestris gall	γ-ray	No
Carboxin resistant	N. tabacum	None	Yes
Cell wall	D. innoxia	Azide	No
UV-resistant	Rosa damascena	None	No

Table 14.4. Selection of cells for antibiotic resistance (Adopted from M.K. Razdan, 1993)

Streptomycin	N. tabacum	None	Yes
	N. sylvestris	None	Yes
	N. sylvestris	None	Yes
Nystatin	N. tabacum	EMS	No
Kanamycin	N. tabacum	None	Yes
	N. sylvestris (haploid)	None	Yes
Cycloheximide	Daucus carota	None	Yes
	N. tabacum	EMS	Yes
Colchicine	Acer pseudoplatanus	None	No
Chloramphenicol	N. sylvestris (haploid)	None	Yes
Amphotericin B	Nicotiana tabacum	EMS	No

Table 14.5. Disease resistant crop plants obtained by in vitro selection (Adopted from Chawla, 2000)

Crop	Pathogen	Selection agent
Food crops		
Wheat	*Helminthosporium sativum, Fusarium graminearum, Pseudomonas syringae*	CT
Maize	*Helminthosporium maydis*	Hm T Toxin
Barley	*Helminthosproium sativum, Fusarium* sp.	CT
Rice	*Helminthosprium oryzae, Xanthomonas oryzae*	Bacterial cells, CT
Oats	*Helminthosporium victoriae*	Victorin
Sugarcane	*Helminthosporium sacchari*	Toxin
2. Horticulture and other crops		
Potato	*Phytophthora infestants, Fusarium oxysporum, Erwinia carotovora*	CF Pathogen
Tobacco	*Pseudomonas syringae* pv. *tabaci*	Methionine sulfoxime
	Alternaria alternata; P. syringae pv *tabaci*	Toxin
	Tobacco mosaic virus,	Virus
	Fusarium oxysporum f.sp. *nicotianae*	CF
Tomato	Tobacco mosaic virus X	Virus
	Pseudomonas solanacearum	CF
Alfalfa	*Foxysporum* f.sp. *medicagnis*	CF
Eggplant	*Vertcillium dahlie*	CF
	Little leaf disease	CF
Peach	*Xanthomonas campestris* pv. *prumi*	CF
Hop	*Vertcillium albo-alrum*	CF
Celery	*Septoria apiicola*	CF

CF = Culture filtrate, CT = Crude toxin

Instability in culture is also a problem, however, in genetic manipulation systems based on regeneration, such as protoplast fusion, and genetic transformation, where specific genetic changes are desired in otherwise unaltered genomes. To satisfy both the needs for stability and the exploitation of somaclonal variation, it would be clearly beneficial if we would control the level of stability. If genetic stability in culture to be controlled it is necessary to establish.

14.4.1. What is the nature of variation?

Efforts to control instability in culture should take into account the nature of the variation in terms of its phenotypic and genotypic characteristics. From numerous studies in a wide range of plant species, a number of different and important characteristics have emerged.

14.4.1.1. The variation is widespread

Somaclonal variation is widespread amongst plant species and genera. It has been described in inbreeders, in vegetatively propagated and seed propagated plants and in cultivated and non-cultivated species. It is, however, only associated with regeneration through a callus or cell suspension phase and not with true micropropagation or meristem-tip culture.

14.4.1.2. Changes can occur at high frequencies

The frequencies at which changes arise are difficult to estimate and vary from one somaclonal population to another. Moreover, where experiments have been repeated, the results are not always consistent (Creissen and Karp, 1985; Fish and Karp, 1986). Nevertheless, it is clear that variation can occur at high frequencies. In regenerants from leaf explants of the tropical legume *Stylosanthes guianensis* (Aubl) SW., for example, 39.9% of SC families (SC = regenerants, were found to be variant (Godwin *et al.,* 1987). Similarly in one cultivar of *Begoniax haemalis* , 43% of regenerated plants were variants (Roest *et al.,* 1981). High incidences of variant plants are particularly prevalent in somaclones derived from protoplasts, where every single clone may differ (Shepard *et al.,* 1981).

14.4.1.3. Variation occurs in quantitative and qualitative traits

14.4.1.4. Changes can occur in homozygous form

In mutation breeding homozygotes expressing a mutant phenotype can only be obtained in the sexual progeny of treated plants. In contrast, many variants observed in tissue culture derived plants are already homozygous. The explanation for this is not clear, although gene conversion and mitotic recombination events have been postulated (Evans *et al.,* 1984). Homozygous mutants can appear amongst somaclones at quite high frequencies, as illustrated in *Brassica juncea*, where 7 out of 85 regenerants were homozygous for the recessive trait of yellow seed (George and Rao, 1982).

14.4.1.5. Changes can occur as whole population shifts

In some cases, changes amongst somaclonal populations have been seen in the form of population shifts.

14.4.1.6. Not all the variation is stable

A number of somaclonal variants express phenotypes that are not inherited. Variation of this kind has been termed 'epigenetic' and is loosely attributed to 'carry-over' effects of the culture conditions. To determine accurately the extent of somaclonal variation, sexual progeny should therefore be examined. Although many of these changes become recognisable to the experienced eye, recent evidence suggest that this may be a more serious aspect of somaclonal variation than originally assumed. Oono (1985) isolated dwarf mutants amongst rice somaclones, which appeared stable on selfing but, when crossed with control plants, the dwarf character never reappeared amongst segregating progenies. The explanation of this is not known. Other unstable changes arising in somaclones may be due to the activation of transposable elements. A possible candidate for this is an unstable flower mutation in alfalfa (Grose and Bingham, 1986), the genetic basis of which is currently under study. Activation of transposable elements during the culture phase has been demonstrated in maize (Peschke *et al.,* 1986). In addition, some unstable changes may have their basis in changes in DNA methylation. It has been shown in maize, that changes in DNA methylation do occur during culture (Brown and Lorz, 1986) but how these relate to changes in phenotype is not yet known.

14.4.1.7. Some changes are peculiar to somaclonal variation

Although much of the variation from *in vitro* culture has been described before, some features do appear to be peculiar to somaclonal populations. Included in these are the occurrence of homozygous changes at high frequency and whole population shifts. Gavazi and colleagues (1987) compared the types of mutants present in the progeny of tomato somaclones with those present in progeny from chemically mutagenized plants. Their results indicated that *in vitro* culture resulted in a higher frequency of some mutations and a different spectrum of variation. Furthermore, some mutants were found exclusively in the somaclonal population.

14.4.1.8. Somaclonal variation involves the whole range of genetic variation

The whole gamut of genetic changes from point mutations and amplification or deletion of DNA sequences to gross changes in chromosome number and structure occurs during *in vitro* culture.

Moreover, changes occur in nuclear, mitochondrial and chloroplast genomes (Karp and Bright, 1985 and Larkin, 1987).

14.4.2. What factors affect the nature and frequency of somaclonal variation

Having established the characteristics of somaclonal variation, the next issue to address is what factors affect the nature and frequency of the instability? To date, this is not known for all the genetic changes described above. In terms of changes at the chromosomal level, however, at least five factors have been clearly identified. Although these will be dealt separately, in nature they interact in ways that have yet to be determined. Some of the factors have also been shown to influence the frequencies of other genetic changes.

14.4.2.1. Genotype

Evidence is accumulating that there is a genotypic component to instability in culture. Some cultivars, for example, have been shown to give rise to more somaclonal variation than others. Larkin (1987) has reported greater variability amongst regenerants of CV Yaqui 50E than for cvs Millewa and Warigal. Similarly, differences in somaclonal variation have been described amongst regenerated plants of winter wheat varieties (Galiba *et al.,* 1985) and different cultivars of *Medicago* (Nagarajan and Walton, 1987), cocoyams (Gupta, 1985) and rye (Linacero and Vazquez, 1986). In addition, although Roest *et al.* (1981) found 43% of regenerants were variant in one variant of *Begonia haemalis,* in another variety, only 7% were variant.

14.4.2.2. Ploidy

In addition to genotype, the ploidy of the source material influences the frequency and nature of the variation obtained. Although, in general, polyploids exhibit greater somaclonal variation than diploids, the influence of ploidy manifests in a variety of ways. In protoplast regeneration of potato, polyploids show a greater range of chromosomes instability. On average, 50-60% retain their tetraploid complement of 48 chromosomes whilst 10-15% are aneuploid at the tetraploid level and 20-30 % are octaploid or aneuploid at the octaploid level (Fish and Karp, 1986). In contrast, regeneration from diploid protoplasts of *Solanum brevidens* yielded a higher incidence of instability, with only 25% of plants remaining diploid. However, the range of variation was much smaller, 75% of plants having undergone chromosome doubling to give tetraploids (50%) or aneuploids at the tetraploid level (25%). A similar pattern can be seen in regeneration from cultured leaf pieces, where chromosome doubling is far more frequent from monohaploid leaves (almost 100% in most lines) than from dihaploid leaves (50-60% depending on the line) (Jacobsen1981; Karp *et al,* 1984; Sree Ramulu *et al,* 1985), and regeneration from tetraploid leaf pieces gives rise to very little chromosome doubling (Wheeler *et al,* 1985).

14.4.2.3. Tissue Culture Method

Protoplast regeneration is associated with more somaclonal variation than regeneration from cultured immature embryos and explants. A qualification of this is that regeneration from protoplasts gives rise to more aneuploidy and polyploidy, and therefore more gross abnormalities, as it has yet to be determined whether changes at a finer level, such as point mutations, differ in frequency in protoplast Vs explant populations. Either way, the choice of regeneration system is one factor that affects the degree of variability observed. This may be a reflection of the increased complexity, (in terms of time in culture and hormones used) of the protoplasts are subjected to during de-differentiation and the onset of division.

In general, the longer the culture phase, the greater the instability, at least in terms of chromosomes (Orton, 1980; McCoy *et al,* 1982; Cassells and Morrish, 1987; Armstrong and Phillips 1988). In addition , there is some evidence indicating that the length of time between subculturing is important, with greater stability being obtained by frequent subculturing (Evan and Gamborg, 1982; Cassells and Morrish, 1987).

A more complex issue is the importance of the mode of regeneration. It has been suggested that regeneration via somatic embryogenesis largely avoid the problem of instability. In wheat, however, Karp and Maddock (1984) found variation in plants regenerated from both embryogenic and organogenic cultures (37.2% vs 24.2%), although the frequency of chimeric plant was much lower in regenerates from somatic embryoids. Similarly, in rye (Bebeli *et al*. 1988) found that one of a number of sister lines showed superior response in terms of somatic embryogenesis. Studies of somaclonal variation in regenerated plants, however, indicated that the 'superior' line was amongst the most variable . Clearly, any relationship between somatic embryogenesis and stability is not an exact one, and other factors such as genotype may be overriding in some lines. Nevertheless, somatic embryogenesis does have advantages as a mode of regeneration since the embryos derive largely from single cells (Haccius, 1978; Vasil and Vasil, 1982) thereby limiting the occurrence of mosaics and chimeras.

14.4.2.4. Tissue source

Early studies on chromosome stability in callus derived from different plant tissues indicated that tissue source is an influential factor. Murashige and Nakano (1967) studied ploidy levels in tobacco pith and its derivative callus, for stem sections at increasing distances from the apex. They found that older stem sections contained higher frequencies of polyploid cells, in addition to some aneuploid cells, and that callus cultures derived from these older sections contained more chromosome variation.

Plant regenerated from different tissues may also show differences in somaclonal variation. In potato plants derived from different cultured explants, Wheeler *et al.* (1985) noted higher frequencies of aneuploidy in plants regenerated from tuber pieces, although the sample size was small. In more recent studies, higher frequencies of octaploids and aneuploids at the octoploid level were found in potato regenerants from tuber protoplasts compared with those from leaf mesophyll protoplasts. In explant derived *Chrysanthemum* differences in somaclonal variation were observed between petal and pedicel-derived plants (De Jong and Custers, 1986). Similarly, Bush *et al.* (1976) compared plants regenerated from petal cultures with those from shoot-tip calluses, and found a much higher frequency of variants in regenerants from petals. Kunakh and coworkers (1984) also found differences in stability between different pea tissues, stem callus being more stable than root callus.

Tissue age appears to be important too. In a study of chromosome doubling of anther-culture derived haploids of tobacco, Kasperbauer and Collins (1972) found that plants regenerated from the midvein region of young (less than fully expanded) leaves were all haploids whilst regeneration from midvein regions of aged leaves (3-4 weeks after full expansion) included diploids. They also found that plants regenerated from stem pith callus were nearly all-aneuploid and showed a high degree of leaf abnormalities such as variegation and/or irregular veination.

14.4.2.5. Media composition

There are a number of reports describing the influence of hormones on chromosome variation in tissue cultures (Karp and Bright, 1985). The problem lies in the interpretation of these results with respect to the compilation of general rules. The auxin 2,4-D has been advocated as a possible generator of instability and it is generally considered good practice to limit 2,4-D levels in tissue cultures. Where experiments on the influence of hormones on somatic mutations have been carried out, however, there has been no clear evidence that 2,4-D or any other hormones are mutagenic (Dolezal and Novak, 1984). On the other hand, changes in ploidy level during culture have been shown to be a function of endogenous growth substances (Ghosh and Gadil, 1979). The possibility remains, therefore, that hormones influence chromosome variation by promoting cell division in selected ploidy levels.

14.4.3. Where does the variation come from?

14.4.3.1. Somaclonal variation arises from variation already present in plant tissues

Spontaneous mutation rates in plants have been estimated as between 10^{-4} to 10^{-7} per locus (Bhatia *et al*. 1985). Together with the fact that endopolyploidy and polyteny occur during differentiation of some plant tissues, this indicates that some variation is already present in the source tissue.

It must be accepted, therefore, that some of the variation observed in somaclones will have arisen from variation already present, whether this is in the tissues or amongst the donor plants. Although the extent of this contribution is difficult to estimate, it has strong implications on experimental design. To account for the possibility of this origin, all experiments in somaclonal variation should have adequate controls and the origin of each regenerant should be clearly known. If a number of plants, which can all were traced to the same source, show the same mutation, the interpretation will be obvious.

14.4.3.2. Somaclonal variation is induced by the peculiar conditions of the culture phase

A callus derived from a single protoplast can give rise to a range of phenotypic and genotypic variation which indicates that not all of somaclonal variation can be accounted for mutations in source material. The importance of the callus phase is indicated by the observations that tissue culture systems, which avoid it, are largely free of somaclonal variation and the variation increases with duration in culture. Another possible origin of somaclonal variation is, therefore, that it arises from the special conditions associated with a callus or *in vitro* culture phase.

It could be argued that somaclonal variation originates through lack of constraints during the callus phase (Bhatia *et al.*, 1985). In normal plant development, any mutations arising during DNA replication will be subjected to selective constraints during growth or reproduction. Cells in cultures, however, are separated by many cycles of cell division from the original source plant. Each cycle of DNA replication will not only propagate any variation inherited from the ancestral cell but also generate new aberrations, which will accumulate if selective constraints are removed.

A corollary to this explanation is that the mutations generated would be similar to those observed to occur spontaneously or in mutation experiments. As discussed earlier, where comparisons have been made (Gavazzi *et al.*, 1987) the data suggest that there are differences in both the range and frequency observed. However, in these experiments only a limited range of mutagens was tested and more data is needed before general conclusions can be drawn.

An alternative explanation for the role of the tissue culture phase is that it imposes stress. This stress could then be viewed as enhancing the spontaneous muations rate disproportionately from the rate of DNA repair, or as inducing latent mechanisms of instability such as chromosome breakage and DNA transposition. The evidence described earlier which indicates a genotypic component to instability finds context in this explanation, as different genotypes may respond in different ways depending on the sequences present in their genomes.

There is not enough data to discriminate between stress induction and lack of constraints, most likely both are involved, but there is some evidence indicating that particular DNA sequences have a role in generating instability in culture. In maize and oats, chromosome breaks induced by tissue culture have been related to the presence of heterochromatin (Lee and Phillips, 1987; Johnson *et al.*, 1987). Similarly, in wheat-rye hybrids the majority of translocation break-points in tissue culture derived plants were found to be at or near heterochromatic regions (Lapitan *et al.*, 1984). It seems likely that repetitive sequences in general may have a role in somaclonal variation, but more experiments are required before the nature of this role can be fully understood.

14.4.4. How predictable is somaclonal variation?

Given that the characteristics of somaclonal variation are now definable and that factors affecting the variation can be identified, it would seem that somaclonal variation should be predictable in outcome. Unfortunately, this is not the case for at least three reasons: (1) it is not yet understood how the factors interact and whether the influence of some are stronger than others, (2) the basis of the genotypic component to instability is not fully known and consequently conditions resulting in variability for one genotype may not for another and (3) somaclonal variation originates from chance events and, even when conditions are identical, repeat experiments may give inconsistent results.

14.4.5. Can guidelines of control be identified?

Although considerable progress has been made in the understanding of the nature and cause of somaclonal variation, it is difficult to outline guidelines of control because the variation is still not predictable. It is in this area that effort is needed now. Until then, tentative guidelines can be listed which it is hoped will stimulate further work in addition to providing points of focus (Table 14.6). (The asterixes are meant to represent the level of confidence with which the recommendation can be made based on the existing evidence, high confidence being indicated by three***, medium confidence by two** and inconclusive one*).

Table 14.6. Guidelines for control of somaclonal variation (Karp, 1989)

Recommendation	e.g. to reduce variability
1. Pay attention to callus phase and regeneration system	Limit the duration*** Choose speedy and efficient regeneration*** Choose somatic embryogenesis* Avoid protoplasts***
2. Select plant material carefully	Select stable genotypes** Use diploids* Avoid genomes with large amounts of heterochromatin*
3. Select tissue source carefully	Test different tissue sources** Avoid old or very specialized tissues (eg. Storage or nutritive tissues)*
4. Pay attention to the media	Avoid high levels of hormones*** Limit 2,4-D*
5. Subculture is important	Subculture at frequent intervals** Change media compositions during culture*

REFERENCES

Armstrong, C.L., and Phillips,R.L. (1988) Genetic and cytogenetic variation in plants regenerated from organogenic and friable embryogenic tissue of maize, *Crop.Sci.* 28, 363-369.

Bebeli,P., Karp,A., and Kaltsikes,P.J., (1988) Plant regeneartion and somaclonal variation from cultured immature embryos of sister lines of rye and triticate differing in their content of heterochromatin 1. Morphogenetic response. *Theor.Appl.Genet.* 75, 929-936.

Bhatia,C.R., Joshua,D.,and Mathews, H., (1985) Somaclonal variation: A genetic interpretation based on the rates of spontaneous chromosomal aberrations and mutations. *Trends.Pl.Res*, 317-326.

Brown,P.T.H. and Lorz,H., (1986) Molecular changes and possible origins of somaclonal variation.In: Somaclonal variation and crop improvement. Semal,J., (eds) Martinus Nijhoff, Dordrecht, PP148-159.

Bush, S.R., Earle, E.D. and Langhans, R.W. (1976). Plantlets from petal segments, petal epidermis and shoot tips of the periclinal chimera *Chrysanthemum morifolium* 'Indianapolls'. *Am. J. Bot.,* 63 : 729-737.

Cassells, A.C. and Morrish, F.M. (1987). Variation in adventitious regenerants of Begonia rex Putz. 'Lucille Closon' as a consequence of cell ontogeny, callus ageing and frequency of callus sub-culture. *Scientia Horticulturea,* 32 : 135-143.

Cassells, A.C., Deadman, M.L., Coleman, M.C. and Brown, C. (1989). Randomand non-rand variation in potato somaclones. In : Genetic Manipulation in Plant Breeding. *Proc. Eucarpia Congress.*

Chawla, H.S. (2000). Introduction to plant biotechnology, Oxford & IBH Publishing Co. Pvt. Ltd., New Delhi, p. 129.

Creissen, G.P. and Karp, A. (1985). Karyotypic changes in potato plants regenerated from protoplasts. *Plant Cell, Tissue Organ Culture,* 4 : 171-182.

De Jong, J., and Custers, J.B.M., (1986) Induced changes in growth and flowering of Chrysanthemum after irradiation and invitro culture of pedicels and petal epidermis, *Euphytica,* 35, 137-148.

Dolezal,J., and Novak,F.J., (1984) Effect of plant tissue culture media on the frequency of somatic mutants in *Tradescantia* stamen hairs. *Z.Pflazenphysiol,* 144, 51-58.

Evans, D.A. and Gamborg, O.L. (1982). Chromosome stability of cell suspension cultures of *Nicotiana* spp. *Plant Cell Reports,* 1 : 104-107.

Evans, D.A., Sharp, W.R. and Medina-Filho, H.P. (1984). Somaclonal and gametoclonal variation. *Am. J. Bot.,* 71: 759-774.

Fish, N. and Karp, A. (1986). Improvements in regeneration from protoplasts of potato and studies on chromosome sability 1. The effect of initial culture media. *Theor. Appl. Genet.,* 72 : 405-412.

Flick, C.E. (1983) Isolation of mutants from cell culture, In: Hand book of plant cell culture. Evans, D *et al.* (eds) 1, Techniques for propagation and Breeding, Mac Millan Publishing Co, NY, USA, PP 393-441.

Galiba, G., Kertesz, Z., Sulka, J. and Sagi, L. (1985). Differences in somaclonal variation in three winter wheat *Triticum aestivum* varieties. *Cereal Res. Commun.,* 13 : 343-350.

Gavazzi, G., Toneeli, C., Todesco, G., Arreghini, E., Raffalid, F., Vecchio, F., Barbuzzi, G., Biasini, M.G. and Sala, F. (1987). Somaclonal variation versus chemically induced mutagenesis in tomato (*Lycopersicon esculentum* L.) *Theor. Appl. Genet.,* 74 : 733-738.

George, L. and Rao, P.S. (1982). Yellow-seeded variants in *in vitro* regenerants of mustard (*Brassica juncea*) Cross var. Rai-5. *Plant Sci. Lett.,* 30 : 327-330.

Gengenbach, B.G., Green, C.E., and Donovam, C.M.,(1977) Inheritance of selected pathotoxin resistance in maize plants regenerated from cell cultures. *Proc. Natl. Acad. Sci. USA,* 74 : 5113-5117.

Ghosh, A. and Gadgil, V.N. (1979). Shift in ploidy of callus tissues. A function of growth substances. *Indian J. Exp. Biol.,* 17 : 562-564.

Godwin, I.A., Gordon, G.H. and Cameron, D.F. (1987). Callus culture-derived somaclonal variaton in the tropical pasture legume *Stylosanthes guianensis* (Aubl.) SW. *Plant Breeding,* 98 : 220-227.

Grosse, R.W. and Bingham, E.T. (1986). An unstable anthocyanin mutation recovered from tissue culture of alfalfa *Medicago sativa* 1. High frequency of reversion upon reculture. *Plant Cell Rep.,* 5 : 104-107.

Gupta, P.P. (1985). Plant regeneration and variabilities from tissue cultures of cocoyams (*Xanthosoma sagittifolium* and *X. violaceum*). *Plant Cell Reports,* 4 : 88-91.

Haccius, B. (1978). Question of unicellular origin of nonzygotic embryos in callus cultures. *Phytomorphology,* 28 : 74-81.

Ignacimuthu, S.J.S. (1997). Plant biotechnology, Oxford & IBH Publishing Co. Pvt. Ltd., New Delhi, pp. 251-255.

Jacobsen, E. (1981). Polyploidization in leaf callus tissue and in regenerated plants of dihaploid potato. *Plant Cell, Tissue and Organ Cult.,* 1 : 77-84.

Johnson, S.S., Phillips, R.L. and Rines, H.W. (1987). Possible role of heterochromatin in chromosome breakage induced by tissue culture in oats (*Avena sativa* L.). *Genome,* 29 : 439-446.

Karp, A., Risiott, R., Jones, M.G.K. and Bright, S.W.J. (1984). Chromosome doublng in monohaploid and dihaploid potatoes by regeneration from cultured leaf explants. *Plant Cell, Tissue and Organ Culture,* 3 : 363-373.

Karp, A. and Maddock, S.E. (1984). Chromosome variation in wheat plants regenerated from cultured immature emrbyos. *Theor. Appl. Genet.,* 67 : 240-255.

Karp, A. and Bright, S.W.J. (1985). On the causes and origins of somaclonal variation. In : Oxford Surveys of Plant Molecular and Cell Biology, vol. 2 (ed. B.J. Millin) Oxford University Press, Oxford, pp. 199-234.

Karp, A. (1989). Can genetic instability be controlled in plant tissue cultures? *Newsletter*, IAPTC, 58 : 2-11.

Kasperbauer, M.J. and Collins, G.B. (1972). Reconsititution of diploids from leaf tissues of anther-derived haploids in tobacco. *Crop. Sci.,* 12 : 98-101.

Kunakh, V.A., Alkhimova, E.G. and Voityuk, L.I. (1984). Variability of the chromosome number in pea cullus tissues and regenerates. *Tsitol. Genet.,* 18 : 20-25.

Lapitan, N.L.V., Sears, R.G. and Gill, B.S. (1984). Translocations and other karyotypic changes in wheat x rye hybrids regenerated from tissue culture. *Theor. Appl. Genet.,* 68 : 547-554.

Larkin, P.J. (1987). Somaclonal variation, history, method and meaning. *Lowa State Journal of Research*, 61 : 393-434.

Larkin, P.J. and Scwcroft, W.R. (1981).Somaclonal variation-a novel source of variability from cell cultures for plant improvement. *Theor. Appl. Genet.* 60: 197-214.

Lee, M. and Phillips, R.L. (1987). Genomic rearrangements in maize induced by tissue culture. *Genome,* 29 : 122-128.

Linacero, R. and Vazquez, A.M. (1986). Somaclonal variation in plants regenerated from embryo calluses in rye *Secale cereate* L. in : Genetic manipulation in plant breeding (eds. Horn, W., Jensen, C.J., Odenbach, W. and Schieder, O.) W. de Gruyter, Berlin, pp. 479-481.

McCoy, T.J. and Phillips, R.L. (1982). Chromosome stability in maize (*Zea mays*) tissues cultures and sectoring in some regenerated plants. *Can. J. Genet.,* 24 : 559-565.

Murashige, T. and Nakano, R. (1967). Chromosome complement as a determinant of the morphogenetic potential of tobacco cells. *Am. J. Bot.,* 54 : 963-970.

Nagarajan, P. and Walton, P.D. (1987). A comparison of somatic chromosomal instability in tissue culture regenerants from *medicago media* Pers. *Plant Cell Rep.,* 6 : 109-113.

Oono, K. (1985). Putative homozygous mutations in regenerated plants of rice. *Mol. Gen. Genet.,* 198 : 377-385.

Orton, T.J. (1980). Chromosomal variability in tissue cultures and regenerated plants of *Hordeum. Theor. Appl. Genet.,* 56 : 101-112.

Peschke, V.M., Phillips, R.L. and Gengenbach, B.G. (1986). Transposable element activity in progeny of regenerated maize plants. In : *Proc. 6th Intern. Conbg. Plant Tissue and Cell Culture.* IAPTC Minneapolis.

Razdan, M.K. (1993) An Introduction to plant tissue culture, Oxford & IBH Publishing Co, New Delhi.

Roest, S., Van Berkel, M., Bokelmann, G.S. and Broertjes, C. (1981). The use of an *in vitro* adventitious bud technique for mutation breeding of *Begonia x hiemalis. Euphytica,* 30 : 381-388.

Shepard, J.F., Bidney, D. and Shahin, E. (1980). Potato protoplasts in crop improvement. *Science,* 208 : 17-24.

Sree Ramulu, K., Dijkhuis, P., Bokelmann, G.S. and Roest, S. (1985). Genetic instability *in vitro* of monohaploid, dihaploid and tetraploid genotypes of potato. In : *Proc. Symp. Somaclonal variations and Crop Improvement* (ed. J. Sernal) Martinus Nijhoff, Dordrecht, pp. 213-218.

Sun, Z.X., Zhao, C.Z., Zheng, K.L., Qi, X.F. and Fu, Y.P., (1983) Somaclonal genetics of rice, *Orzya sativa* L. *Theor. Appl. Genet.* 67 : 67-73.

Wheeler, V.A., Evans, N.E., Foulger, D., Webb, K.J., Karp, A., Franklin, J. and Bright, S.W.J. (1985). Shoot formation from explant cultures of fourteen potato cultivars and studies of the cytology and morphology of regenerated plants. *Ann. Bot.,* 5 : 309-320.

Vasil, V. and Vasil, I.K. (1982). The ontogeny of somatic embryos of *Pennisteum americanum* (L.) K Scheme 1 in cultured immature embryos. *Bot. Gaz.,* 143 : 454-465.

Chapter 15

Isolation, Culture and Regeneration of Protoplasts

INTRODUCTION

One of the most striking characteristics of plant cells is the presence of a thick and relatively rigid cell wall. This wall functions as the mechanical support of the cell and in defence against physical damage and attack by pathogens. It may also play more subtle roles in cellular communication. Protoplasts are plant cells with a plasma membrane but without cell wall. Because of this, the protoplasts provide the starting point for many of the techniques of genetic manipulation of plants, in particular the induction of somaclonal variation, somatic hybridization and genetic transformation. E.C. Cocking in 1960 from University of Nottingham, U.K. found that cellulose-degrading enzymes, isolated from wood-rotting fungi, were able to dissolve away the cell wall and yield the fragile but still viable protoplast, a cell surrounded only by its membrane.

15.1. APPLICATIONS OF PROTOPLAST ISOLATION AND CULTURE

Plant cells are normally connected to each via many plasmodesmata in multicellular tissues. Plant protoplast culture provides an excellent system to obtain single cells or higher organism and to study its functioning in a controlled environment. Protoplast isolation provides millions of cells, which is comparable to microbial system and can be used many ways.

1. To develop novel hybrid plant through protoplast fusion.
2. Isolation of mutants, developed spontaneously or through mutagens, is easier in single cell derived colony.
3. Single cell cloning can be easily performed with protoplasts.
4. The protoplast in culture can be regenerated into an entire plant.
5. Genetic transformation through DNA or organelles uptake can be achieved with protoplasts.

15.2. ISOLATION OF PROTOPLASTS

Protoplasts are the living material of the cell where as an isolated protoplast is the cell from which the cell wall is removed. Protoplasts can be isolated from almost all plant parts i.e., roots, leaves, fruits, tubers, root nodules, endosperm, pollen mother cells, pollens, callus, and suspension culture. Protoplasts are isolated from cells by two methods.

(i) Mechanical

(ii) Enzymatic

15.2.1. Mechanical Method

The cells were kept in a suitable plasmolyticum and cut with a fine knife. In this process some of the plasmolysed cells were cut only through the cell wall, releasing intact protoplast.

This method of isolation are suitable for limited variety of higher plant tissues, such as leaf, bulb scale, fruit epidermis, or storage tissues. Disadvantages of the technique is that it yields a very small number of protoplast after a rather tedious procedure.

15.2.2. Enzymatic Method

A) Isolation of Protoplasts from leaves

This involves four steps :

 i) Sterilization of leaves.

 ii) Peeling off the lower epidermis.

 iii) Incubation in enzyme solution and

 iv) Isolation and cleaning of the protoplasts.

Healthy leaves are obtained from plants growing under green house or wild source and are surface sterilized by using alcohol (70% v/v) for 1 min. then treating them with sodium hypochlorite (2%) solution for 15-20 min. Then leaves are rinsed 3 times with sterile distilled water and these operations are done under laminar flow. The lower epidermis of the sterilized leaves is carefully peeled off and the stripped leaves are cut into small pieces. From these peeled leaf segments mesophyll protoplasts can be obtained. The protoplasts can be isolated using any of the following methods.

 i) *Direct (One Step Method)* : In this method, treatment with macerozyme (or pectinase) and cellulase is done simultaneously. The leaf segments are incubated overnight (14-18 hrs) with enzyme mixture (0.5% macerozyme + 2% cellulase in 13% sorbitol or manitol at pH 5.4) at 25°C and treated gently to liberate the protoplasts. The mixture is filtered with wire gauze to remove leaf debris, transferred to screwcap tubes (13-x 1000 mm) and centrifuged for 1 min. The protoplasts form a pellet and supernatant removed. The process is repeated for 3

Table 15.1. The Optimal condition for the isolation of protoplasts in two species

Parameter	Cultured cells of tobacco (Uchimiya and Murashige, 1974)	Mesophyll cells of cereals (Scott et al., 1978)
Plant material	4-5 day-old subculture	5-6-day-old seedlings
Cellulase	1% Onozuka R-10	2% Cellulysin
Macerozyme	0.1-0.2% Onozuka R-10	0.5%
Hemi-cellulase	--	1%
pH of enzyme solution	4.7 or 5.7	4.6 – 5.4
Volume of enzyme solution / fresh weight of tissue	10 mlg^1	10 mlg^1
Incubation period	2-3 h	2h
Incubation temperature	22-37°C	20-25°C
Rate of agitation	50 rpm	80 rpm
Osmoticum	300-800 mmol^{-1} mannitol	600 mmol^{-1} sorbitol

times and protoplasts are washed with 13% sorbitol solution, which is later replaced by sucrose (20%) solution and centrifuged for 1min, the cleaned protoplast, which will float can be pipetted out.

ii) *Sequential (Two Step) Method* : The leaf segments with mixture A (0.5% macerozyme + 0.3% potassium dextran sulphate in 13% mannitol at pH 5.8) are vacuum filtered for 5-7 min., transferred to a water bath at 25°C and subjected to slow shaking. After 20 min, the enzyme mixture is replaced by fresh 'enzyme mixture A' and leaf segments incubated for another one hour. The mixture is filtered using nylon mesh, centrifuged (100 g) for 1 min. and washed 3 times with 13% mannitol to get a pure sample of isolated cells. These cells are then, incubated with 'enzyme mixture B' (2% cellulase in a 13% solution of mannitol at pH 5.4) for about 90 min at 30°C. After incubation, the mixture is centrifuged at 100g for 1 min so that protoplasts form a pellet, which is cleaned 3 times to get a cleaned protoplast.

Table 15.2. Enzyme mixtures and conditions for protoplast isolation

Species	Tissue	Enzyme mixture	Comments
Arabidopsis thaliana	Leaves	1% (w/v) Cellulase R10, 0.25% (w/v) Macerozyme R10, 8mM $CaCl_2$, 0.4M mannitol, pH 5.5	Cut leaves into 1 mm² pieces, plasmolyse for 60 min in 0.5M mannitol; digest for 14-18h, 25°C, dark 16h
Brassica napus	Hypocotyl, stem and petiole	1% (w/v) Cellulysin, 0.1% (w/v) Macerozyme R10, in K3 solution (22) containing 136.9 g/litre sucrose, pH 5.6	
Dimorphotheca aurantiaca	Leaves, seeding cotyledons	2% (w/v) Meicelase, 2% (w/v) Rhozyme HP 150, 0.03% (w/v) Macerozyme R10, CPW8M, antibiotics, pH 5.8	Plasmolyse for 1h, 5-6 h static incubation for leaves, 10-12h for cotyledons
Glycine canescens	Cotyledons from 10-day-old seedlings	1% (w/v) Rhozyme HP 150, 2% (w/v) Cellulase R10, 0.1% (w/v) Pectolyase Y23, CPW9M, H 5.6	Plasmolyse for 20 min, incubate for 16h in the dark.
Glycine max	Hypocotyls from 5- to 7-day -old seedings	As for *Dimorphotheca* leaves, but dissolved in CPW13M solution	Plasmolyse 30 min, incubate with enzyme for 16 h with shaking (10 rpm) in the light (2.5 iE.m⁻²s⁻¹)
Nicotiana tabacum cv Xanthi nc	Expanded leaves	1.5% (w/v) Meicelase, 0.05% (w/v) Macerozyme R10, CPW13M, antibiotics, pH 5.8	Lower epidermis removed by peeling; 16h static incubation
Oryza sativa v. Taipei 309	Embryogenic cell suspension	0.33% (w/v) Cellulase RS, 0.033% (w/v) Pectolyase Y23, 325.4 mg/litre MES, CPW13M, pH 5.6	
Rudbeckia hirta	Leaves, cotyledons, ray florets	0.8% (w/v) Cellulase R10, 0.4% (w/v) Driselase, 0.8% (w/v) Rhozyme HP 150, 0.2% (w/v) Macerozyme R10, CPW8M antibiotics[a], (CPW13M for ray florets), pH 5.8	Plasmolyse for 1h, then 3-4 enzyme incubation with shaking[b], followed by 8-10h stationary incubation in the dark (23°C)

[a]Antibiotic mixture : 400 mg/litre ampicillin, 10 mg/litre gentamycin, 10 mg/litre tetracycline
[b]Shaker : 30 rpm, 2 cm throw.

Table 15.3. Commercially available enzymes commonly used for protoplast isolation (Adopted from M.K. Razdan, 1993)

Enzyme	Source	Supplier
Cellulases		
Cellulase Onozuka R-10	*Trichoderma viride*	Kinki Yokult Mfg, Co. Ltd. 8-12, Shingikancho, Nishinomiya, Japan
Cellulase Onozuka RS	*T. viride*	Yokult Honsha Col., Tokyo Japan
Cellulase YC	*T. viride*	Seishin Pharma Col. Ltd., 9-500-1, Nagareyama Nagareyama-shi Chiba-ken, Japan
Cellulase CEL	*T. viride*	Copper Biomedica inc. Malvern, PA, USA
Cellulysin	*T. viride*	Calbiochem. San Diego, CA, USA
Driselase	*Irpex lacteus*	Kyowa Hakko Kogyo Co. Ltd., Tokyo, Japan
Meicelase P-1	*T. viride*	Meiji Seiki Kaisha Ltd., No. 8, 2-Chome, Kyobashi, Chou-Ku, Japan
Hemicellulases		
Helicase	*Helix pomatia*	Industrie Biologique, Francaise, Gannevilliers, France
Hemicellulase	*Aspergillus niger*	Sigma Chemical Co., St. Louis, MO, USA
Hemicellulase H-2125	*Rhizopus* sp.	Sigma Chemical Co., Munieh, Germany
Rhozyme HP 150	*Aspergillus niger*	Genencor Inc., San Francisco, CA, USA
Pectinases		
Macerase	*Rhizopus arrhizus*	Calbiochem, San Diego, CA, USA
Macerozyme R-10	*R. arrhizus*	Yakult Hosha Co., Tokyo, Japan.
PATE	*Bacillus polymyxa*	Farbwerke-Hoechst AG Frankfurt, Germany
Pectinol	*Aspergillus* sp.	Rohm and Haas Co. Independence Hall West, Philadelphia, PA 19105, USA.
Pectolyase Y23	*A. japonicus*	Seishin Pharma Co. Ltd., Tokyo, Japan
Zymolyase	*Arthrobacter luteus*	Sigma Chemical Co. Ltd., St. Louis, MO, USA.

15.3. ISOLATION OF PROTOPLASTS FROM CULTURE CELLS

Actively dividing cells in suspension cultures are proved as most suitable material for isolation of non-germ protoplast in large amount. Protoplasts have been isolated from suspension culture of many plant species like, Rose, Atropa sp., Carrot, Sugarcane etc. Protoplasts can be easily isolated from such cultures by treating the filtered suspension with 2-4% Onozuka cellulase in 0.6M mannitol, for 4-6 hrs at 32°C in a gently shaking water bath.

15.3.1. Factors Affecting yield and viability of Protoplasts

15.3.1.1. Source of Material

The most convenient source of plant protoplasts is the leaf because it allows the isolation of large number of uniform cells without the necessity of killing the plants. Since the mesophyll cells are loosely arranged, the enzymes have an easy access to the cell wall. When the protoplasts are prepared from leaves the age of the plant and the conditions under which it has been grown may be critical. To achieve maximum control on the growth conditions of source plants, several researchers have used *in vitro* growing shoots (Binding 1975, Schieder 1978). Most research, however, use glass house or green house grown plants. In these, the plants grown under low light intensity (1000 μcm^{-2}), short day conditions, temperature in the range 20-25°C, relative humidity (60-80%) and a good supply of nitrogen

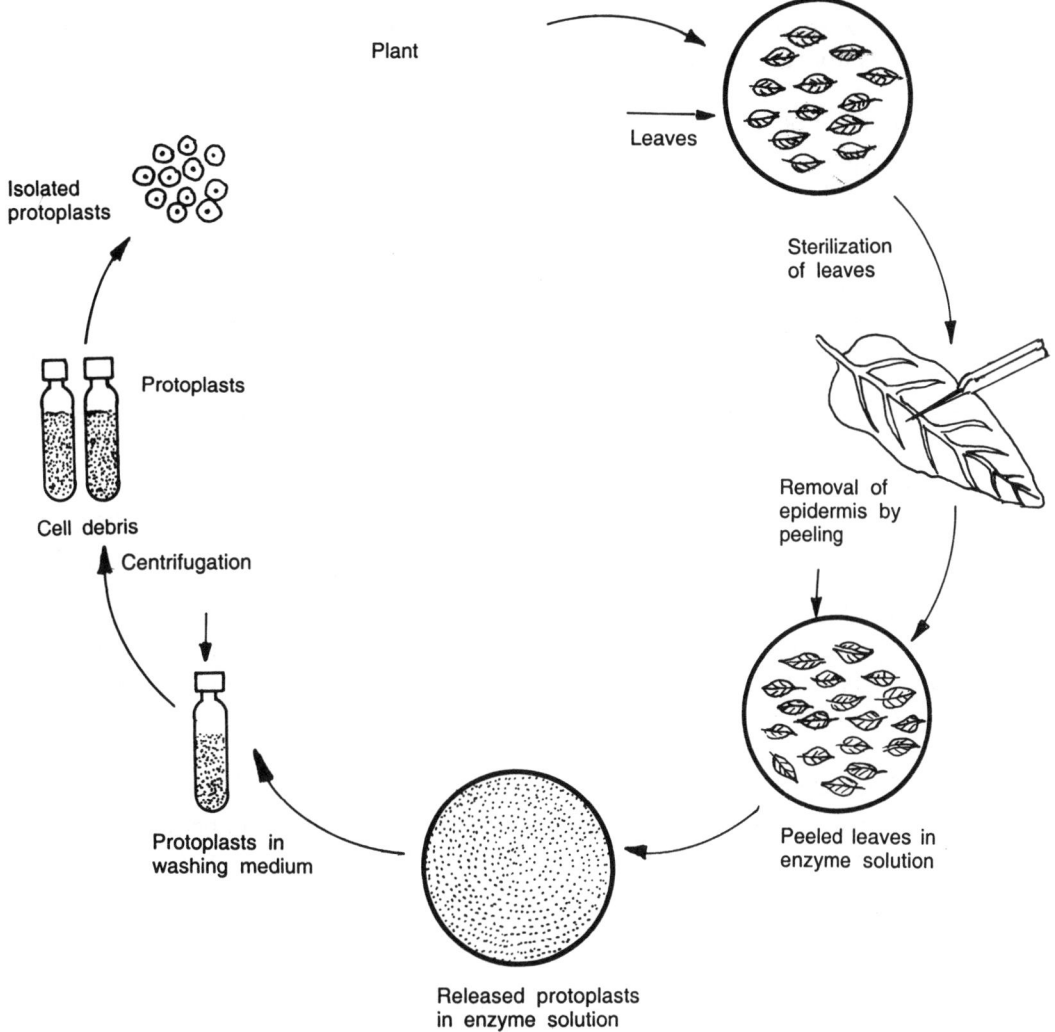

Fig. 15.1. Isolation of mesophyll protoplasts using leaves.

fertilizer have gave better results (Eriksson, 1977). It is very difficult to isolate protoplasts from leaf cells of cereals and other species. The alternative is the cultured cells. The yield of protoplasts from culture cells depends on the growth rate and growth phase of the cells. Frequently subcultured (3 days) suspension cultures, and cells taken from early log phase are most suitable (Kao *et al.* 1971; Uchimiya and Murashige, 1974).

15.3.1.2. Pre-enzyme Treatments

With tissues taken from plants from wild source the first step is to surface sterilize them. Generally, the methods adopted for this purpose are the same as those described in earlier chapter 2 and 5 for initiation of callus. Scott *et al.* (1978) observed that the most effective and efficient ways to surface sterilize cereal leaves was to rinse them in Zephiran (alkyl dimethylbenzyl ammonium chloride) (0.1%)-ethanol (10% solution) for 5 min. To facilitate the penetration of enzyme solution into the intracellular spaces of leaf is to peel the lower epidermis and float the stripped pieces of leaf on the enzyme solution. Agitation of the incubation mixture during the enzyme treatment improves the yields of protoplasts from culture cells.

15.3.1.3. Enzyme Treatment

The release of protoplasts is depends on the nature and concentration of the enzymes used. The two essential enzymes used for isolation of protoplasts are cellulase and pectinase. The latter degrades the middle lamella and the former is required to digest the cellulytic cell wall. The commercially available enzymes are shown in Table 15.3. Driselase have a number of zymolytic activities, such as pectinase, laminarinase, cellulase and xylanase (Kao *et al.* 1974), these are especially useful for the isolation of protoplasts from cultured cells. Pectolyase Y-23, a highly powerful macerozyme, in combination with cellulase released protoplasts from mesophyll cells of pea within ½ hr (Nagata and Ishii, 1979). The optimal temperature for the activity of these enzymes is 40-50°C, which happens to be high for the cells. Generally 25-30°C is adequate for day as well as overnight isolation of protoplasts.

15.3.1.4. Osmoticum

A fundamental property of isolated protoplasts is their osmotic fragility and hence, the need for a suitable osmotic stabilizer in the enzyme solution, the protoplast-washing medium, and the protoplasts culture. Most commonly used osmotica are sorbital and mannitol in the range of 450-800 mmol^{-1}. Uchimiya and Murashige (1974) reported that for isolating protoplasts from tobacco suspension cultures several soluble carbohydrates, including glucose, fructose, galactose, sorbital & mannitol, were equally effective. When non-ionic substances are $CaCl_2$ (50-100 Mmol l^{-1}), KCl (335 Mmol l^{-1}) and $MgSO_4.7H_2O$ (40 Mmol l^{-1}) and they improve the stability of the plasma membrane.

15.4. PURIFICATION OF PROTOPLASTS

After the material has been incubated in enzyme solution for an adequate period the incubation vessel is gently swirled or the leaf pieces are gently squeezed to release the protoplasts held in the original tissue. The digestion mixture at this stage would consist of sub-cellular debris, especially chloroplasts, vascular elements, undigested cells and broken protoplasts. So the purification of protoplasts is required to remove all the cell debris to get only the released protoplasts. The purification is done by the following methods.

15.4.1. Sedimentation and Washing

In this method, the crude protoplasts suspension is centrifuged at low speed (50-100 g/5 min) and the supernatant discarded. The pellet is gently resuspended in fresh culture media plus mannitol and rewashed. This process is repeated for 3 times to get clean protoplast preparation.

Table 15.4. Interspecific hybrids produced through protoplast fusion

Some parent species are			
1. *N. tabacum*	X	*N. glauca* -	Evans *et al.* (1980)
2. *N. tabacum*	X	*N. glutinosa* -	Nagao (1979)
3. *Dacus carota*	X	*D. capillifolius* -	Dudits *et al.* (1977)
4. *Datura innoxia*	X	*D. stramonium* -	Schieder (1978)
5. *D. innoxia*	X	*D. canadida* -	Schieder (1980)
6. *Solanum tuberosum*	X	*S. chacoense* -	Butenko & Kuchko (1980)

15.4.2. Flotation

Protoplasts being lighter (low density) than other cell debris, gradients may be used, which will allow the protoplasts to float and the cell debris to sediment. A solution of mannitol, sorbitol and sucrose (0.3–0.6 m) can be used as a gradient and crude protoplast suspension may be centrifuged in this gradient at an appropriate speed. Protoplasts can be pipetted off from the top of the tube after centrifugation. This method causes less loss or damage than the sedimentation and washing method.

15.5. VIABILITY OF THE PROTOPLASTS

Viability of the freshly prepared protoplasts can be checked by a number of methods.

(a) Observation of cyclosis or cytoplasmic streaming as indication of active metabolism (Pelcher *et al.,* 1974);

(b) Oxygen uptake measured by an oxygen electrode which indicate respiratory metabolism (Taiz and Jones, 1971);

(c) Photosynthetic activity (Kanai and Edwards, 1973);

(d) Exclusion of Evans blue dye by intact membrane (Glimelius *et al.,* 1974) and

(e) Staining with fluorescein diacetate.

15.6. PROTOPLAST CULTURE

The culture methods and the culture requirements of isolated protoplasts are often similar to those of single cells. Isolated protoplasts are usually cultured in either liquid or semisolid agar media plates. They require somatic protection in the culture medium until they generate a strong cell wall. Osmolarity in the medium is adjusted to the same level as in the enzyme and washing solution. Prolonged culture at high osmotic pressure can result in browning of the cultures and inhibition of callus growth. The osmolarity of the medium is gradually decreased with cell wall formation and cell division. Abrupt decrease in osmolarity may cause bursting of protoplasts and cells and also affect the growth of cells.

15.6.1. Nutritional Requirements

Generally the basic constituents in protoplasts culture medium are similar to those used for culturing of cells and tissues e.g. MS or B5 medium. However, since the protoplasts are devoid of a cell wall, they tend to be very efficient in uptake of nutrients from the medium. Usually carbon, nitrogen, vitamins and phytohormones are suitably modified. Kao *et al* (1973) demonstrated that the addition of 1 Mmol calcium chloride to B5 medium enhanced the percentage of dividing cells in protoplast cultures of *Vicia hajastana* and *Bromus inermis*. However, supplementing the medium with 20 millimol ammonium nitrate reduced the frequency of dividing cells.

The vitamins used for protoplast culture are the same as those used in standard tissue culture media. Phytohormones, particularly auxins and cytokinins are always required. The most commonly used auxin is 2,4-D, but Uchimiya and Murashige (1976) reported that NAA was superior to 2,4-D or IAA for the culture of protoplasts from cell suspension of Nicotiana. The commonly used cytokinins are BAP, kinetin, 2-ip or zeatin.

15.6.2. Environmental factors

High light intensities often inhibit protoplast growth when applied from the beginning of culture. It is better to initiate protoplast culturing in darkness or dim light for few days and later transfer the cultures to light of about 2000-5000 lux. Protoplast cultures are generally cultured at temperature ranging between 20-28°C. A pH in the range of 5.5-5.9 is recommended for protoplast culture media.

15.7. METHODS OF CULTURE

In general, protoplast of most species prefer to be embedded at a density between 5.0×10^2 and 1.0×10^6/ml in a medium made semi-solid with agar or agarose, rather than being cultured in liquid medium. This is presumably because of the enhanced support provided by agar or agarose, which encourages cell wall development. However, liquid media allow faster diffusion of nutrients and waste products, as well as facilitating reduction of the osmotic pressure as the protoplasts grow and resynthesize cell walls.

15.7.1. Agar Embeded culture

Protoplasts are suspended at the required plating density in culture medium containing molten (40°C) agarose (1.2% w/v). The suspension is dispensed into Petri dishes allowed to cool and solidfy. The dishes are sealed with Parafilm and incubated in culture room, either in the dark or under low intensity illumination with a suitable photoperiod. The agar or agarose layer containing the embedded protoplasts may be cut into sections and the latter transferred to larger petri dishes containing liquid culture medium of the same composition. Alternatively, the molten agarose medium containing the suspended protoplasts is dispensed as droplets (50-150 μl) in the bottom of petri dishes. After solidification, the droplets are bathed in liquid medium of the same composition.

15.7.2. Liquid-over-agar or agarose

A thin layer of agarose solidified medium is formed in the bottom of a Perti dish and liquid medium containing the protoplasts at twice the required plating density is poured over the agarose layer.

15.7.3. Hanging drop culture

This permits culture of small numbers of protoplasts, often at high density, in liquid medium. Small droplets, 20-40 μl in volume, containing the protoplasts are dispensed into the lids of Petri dishes. Culture medium or sterile distilled water is placed in the base of the petri dish. The lid with the droplets of protoplasts is inverted and placed over the base before the petridish is sealed with Parafilm.

15.7.4. Micro culture chamber

The microculture chamber technique was first developed by Jones *et al.* (1960) and used by Vasil and Hildebrandt (1965) after some modifications. This method consists of culturing 30-50 μl of medium containing one or more protoplasts on a microscope slide of the drop. The cultures are sealed with sterile paraffin oil and incubated in light at 23-25°C.

15.7.5. Multidrop array technique

Potrykus *et al.* (1977) developed the multiple drop array (MDA) technique for systematically screening a large number of multiple combinations of media constituents for protoplast culture. The MDA screening technique uses hanging droplets of 40 μl as the experimental unit. Each droplet represents one combination of factors to be tested. The droplets are arranged in regular array of 7×7 drops on the lid of a 9 cm petri dish. To test seven different auxins in combination with four different cytokinins in the medium, each of these factors is used in seven different concentrations. The whole experiment includes 4×7 petri dishes. Since each petri dish contains 49 droplets this results in a total of $4 \times 7 \times 49 = 1372$ two-factor combinations. This experiment can be performed by one person within 5-6 h in addition to the time required for media preparation, protoplast isolation, and culture evaluation.

15.7.6. Feeder layer technique

In this technique a feeder cell layer is prepared by exposing protoplasts to irradiation with X-ray which inhibits division of cells but allows them to remain metabolically active. The irradiated protoplasts are then plated in soft agar medium and serves as the feeder layer for the non-irradiated protoplasts, which are placed above this.

15.8. REGENERATION FROM PROTOPLASTS

The process of cell wall formation begins within few hours after isolation of protoplasts and may be completed in 2-7 days. Soon after the formation of a wall around the protoplasts the reconstituted cells show considerable increase in size and first divisions generally occur between 2-7 days. Subsequent divisions gives rise to small cell colonies. Subsequent sustained divisions give rise to macroscopic colonies, which upon transfer to an osmotic free medium will develop a callus. The callus may be induced to undergo organogenetic differentiation or whole plant regeneration (Fig.15.2) e.g. *Oryza sativa*, *Zea mays*, *Asparagus officinalis*, *Capsicum annuum*, *Rosa* Sp. and *Chrysanthemum* sp.

15.9. PROTOPLAST FUSION AND SOMATIC HYBRIDIZATION

Introduction

Plant protoplats represent the finest single cell system and offer exciting possibilities in the fields of somatic cell genetics and crop improvement. The protoplast fusion allows us to bring any desirable plant traits (for e.g. disease resistance, salt tolerance and high yielding varieties) in combination that are not possible by sexual means. The technique of hybrid production through the fusion of isolated somatic protoplats under *in vitro* conditions and subsequent development of their product to a hybrid plant is called as somatic hybridization. It provides a means to circumvent sexual barrier to plant breeding. Somatic hybridization involves four discrete stages, which are protoplast isolation, fusion of protoplasts, selection of hybrid cells and identification of hybrid plants. Somatic hybridization is significant in improvement of crops such as banana, potato, sugarcane, yam in which sexual reproduction either weak or absent.

15.9.1. Mechanism of Fusion

Protoplast fusion consists of 3 main phases.
1. Agglutination, during which the plasma membrane of 2 or more protoplasts are brought into close proximity (A & B).
2. Membranes of protoplasts agglutinated by fusogen get fused at the point of adhesion. This results in the formation of cytoplasmic bridges between the protoplasts.
3. Rounding off of the fused protoplasts due to the expansion of the cytoplasmic bridges forming spherical hetero or homokaryons (A-B) Binucleate heterokaryons.

   The fusion of the two nuclei results in a tetraploid hybrid cell or synkaryocyte. If one of the nuclei degenerate, a cybrid or heteroplast is produced (Fig. 15.3).

15.9.2. Techniques of Protoplast Fusion

The protoplast fusion may be of three kinds.

15.9.2.1. Spontaneous Fusion

During isolation of protoplasts for culture, when enzymatic degradation of cell wall is affected, some of the adjacent protoplast fuses together forming homokaryons or homokaryocytes, each with 2-40

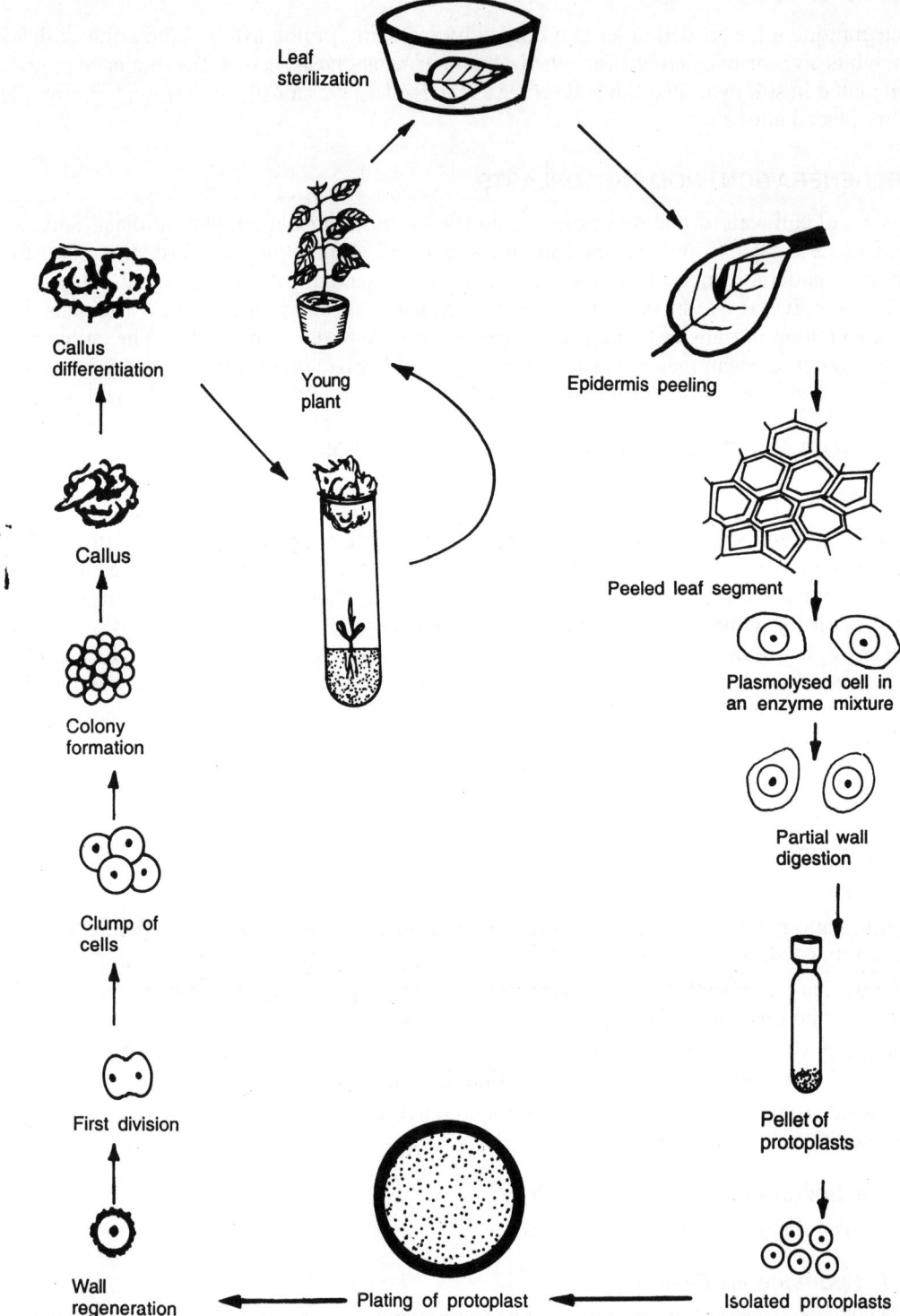

Fig. 15.2. The technique used for isolation, culture and regeneration of plants from leaf protoplasts.

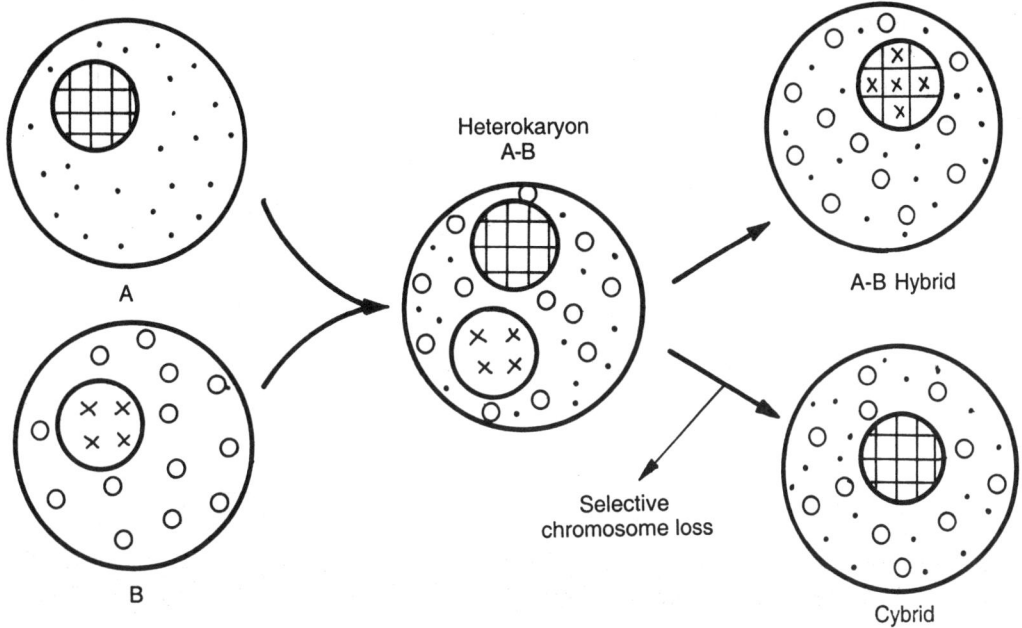

Fig. 15.3. Mechanism of fusion of protoplasts.

nuclei. The occurrence of multinucleate fusion bodies is more frequent, when protoplasts are prepared from actively dividing cells. This spontaneous fusion, however, is strictly intraspecific. Young leaves are more likely to undergo this fusion.

15.9.2.2. Mechanical Fusion

The giant protoplasts of Acetabularia have been fused mechanically. This kind of fusion is not dependent upon the presence of fusion-inducing agent. However, in this method protoplasts are likely to get injury.

15.9.2.3. Induced Fusion

So far as somatic hybridization is concerned spontaneous fusion is of no value; it requires the fusion of protoplasts of different origin. To achieve the induce fusion a suitable agent (fusogen) is necessary. The various fusogens used for induced fusion are $NaNO_3$ (Power et al., 1970), artificial sea water (Eriksson, 1971), Lysozyme (Potrykus, 1973), mechanically induced adhesion (Schenk and Hildebrandt 1970, Ito, 1973), virus (Withers, 1973), gelatin (Kameya, 1973), high pH and high Ca^{2+} (Keller and Melchers, 1973), polyethylene glycol (Kao and Michayluk, 1975; Wallin et al., 1974), antibodies (Hartmann et al., 1973; Burgess and Fleming, 1974), plant lectin Concanavalin A (Glimelius et al., 1974), Polyvinyl alcohol (Nagata, 1978) and electric stimulation (Senda et al 1979, Zimmermann and Scheurich, 1981).

15.9.2.3.1. NaNO₃ Treatment

The technique suffers from a low frequency of heterokaryon formation especially when highly vacuolated mesophyll protoplasts are involved. The method involves the following stages.

(i) The isolated protoplasts are suspended in fusion inducing mixture (5.5% sodium nitrate in 10% sucrose solution) and causes fusion on incubation on water bath at 35°C. In order to obtain a high frequency of fused protoplasts, mixture may be centrifuged and the pellet resuspended and incubated for one or more additional cycles.

(ii) Finally the mixture is replaced by a liquid medium and the protoplasts in this mixture are incubated again; the cycle may be repeated twice before plating the protoplasts on a solid medium. e.g. interspecific somatic hybrids in the genus *Nicotiana* (Fig.15.4).

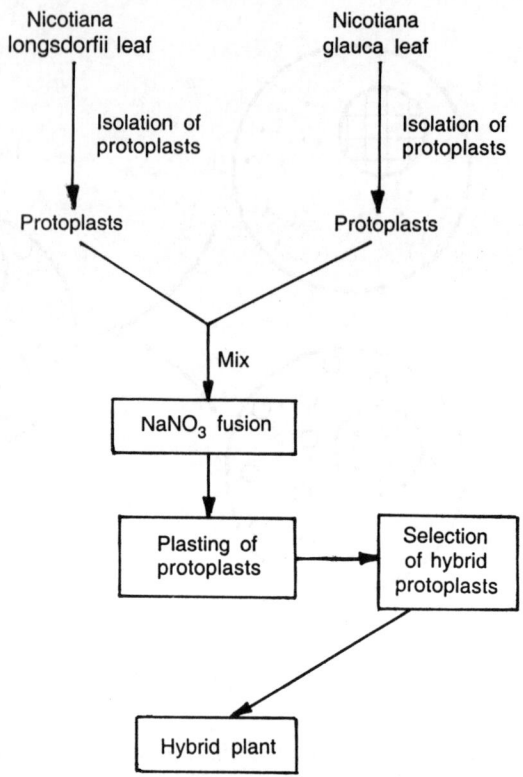

Fig. 15.4. Production of interspecific somatic hybrid plant of *Nicotiana* sp.

15.9.2.3.2. Treatment with high pH and high Ca²⁺

The method involves centrifugation of the protoplasts in a fusion inducing solution (0.05 M calcium chloride in 0.4M mannitol at pH 10.5) for 30 min at 50g at 37°C. This leads to fusion of 20-50% of the protoplasts. Using this method, Melchers and Labib (1974) and Melchers (1977) produced intra and interspecific somatic hybrids of Nicotiana. The major disadvantage of this method is, the high pH may be toxic to the protoplasts (Kao and Wetter, 1977).

15.9.2.3.3. Polyethylene Glycol (PEG) Treatment

PEG has achieved widespread acceptance as a fusogen of plant protoplasts because of the reproducible high frequency heterokaryon formation with low cytotoxicity. Another advantage of PEG-induced fusion is that the formation of a high proportion of binucleate heterokaryons (Wallin *et al.* 1974; Kao 1977). PEG-induced fusion is non-specific, e.g. soyabean-tobacco, soyabean-maize, soyabean-barley (Kao *et al.,* 1974), animal cells with yeast (Ahkong *et al.,* 1975).

The freshly isolated protoplasts from the two selected parents are mixed in appropriate proportions and treated with 28-50% PEG (1500-1600 molecular weight) solution for 15-30 min. followed by gradual washing of protoplasts with the culture medium. Kao *et al.* (1974) reported that eluting PEG with a highly alkaline solution (pH 9-10) containing a high Ca²⁺ ion (50 mMol calcium chloride) led to higher frequency of fusion than washing with the culture medium. Some modifications are suggested to the PEG method (Kao and Michayluk, 1974) is the addition of Concanavallin A to the PEG solution for the purpose of increasing the incidence of fusion.

Factors Affecting Protoplast Fusion by PEG

(a) PEG of molecular weight higher than 1000 induces tight adhesion and high frequency fusion of protoplasts.

(b) High temperature (35-37°C) promotes fusion frequencies while low temperature (15°C) promotes protoplast adhesion.

(c) Excessive dilution of the enzyme solution leads to poor fusion.

(d) Protoplast from young leaves and fast growing calli give better fusions.

(e) Prolonged incubation in PEG solution reduces heterokaryon formations.

15.9.2.3.4. Electrical Fusion

If protoplasts are placed into a small culture cell containing electrodes and a potential difference is applied, then the protoplasts will line up between the electrodes. It now an extremely short, square wave electric shock is applied, protoplasts can be induced to fuse.

15.9.3. Selection of Fused Protoplasts

After the fusion treatment as above mentioned methods, the protoplast population consists of a mixture of parentral types, homokaryons and heterokaryans, of which heterokaryons often make only 0.5%-10%. The following methods have been used for selection of fused protoplasts.

The hybrids *Nicotiana gluca x N. longsdorfii* could only grown on Nagata and Takebe medium but not of the parents.

The protoplasts of two parents may be labelled by different fluorescent compounds, which will then enable for the selection of hybrids.

REFERENCES

Ahkong, Q.F., Howell, J.I., Lucy, J.A., Safwat, F., Davey, M.R. and Cocking, E.C. (1975). Fusion of hen erythrocytes with yeast protoplasts induced by polyethylene glycol. *Nature* (London), 255 : 66-67.

Binding, H. (1975). Reproducibly high plating efficiencies of isolated mesophyll protoplasts from shoot cultures of tobacco. *Physiol. Plant.,* 35 : 225-227.

Burgess, J. and Fleming, E.N. (1974). Ultrastructural studies of the aggregation and fusion of plant protoplasts. *Planta,* 118 : 183-193.

Butenko, R. and Kuchko, A.A. (1980). Physiological aspects of procurement, cultivation, and hybridization of isolated potato protoplasts. Sov. *Plant Physiol.* (English transl.), 26 : 901-909.

Cocking, E.C. (1960). A method for the isolation of plant protoplasts and vacuoles. *Nature* (London), 187 : 927-929.

Dudits, D., Kao, K.N., Costabel, F. and Gamborg, O.L. (1977). Fusion of carrot and barley protoplasts and division of heterokaryocytes. *Can. J. Genet. Cytol.,* 18 : 263-269.

Eriksson, T. (1971). Isolation and fusion of plant protoplasts. In : Les Cultures de Tissues de Plantes. *Colloq. Int. C.N.R.S., Paris, No.* 93, pp. 297-302.

Eriksson, T. (1977). Technical advances in protoplast isolation and cultivation. In : Barz,W. et al (Eds) plant tissue culture and its biotechnological application, Springer-Verlag, Berlin, pp. 313-322.

Evans, D.A., Wetter, L.R. and Gamborg, O.L. (1980). Somatic hybrid plants of *Nicotiana glauca* and *Nicotiana tabacum* obtained by protoplast fusion. *Physiol. Plant.,* 48 : 225-230.

Glimelius, K., Wallin, A. and Eriksson, T. (1974). Agglutinating effects of concanavalin, A. on isolated protoplasts of *Daucus carota. Physiol. Plant,* 31 : 225-230.

Hartmann, J.X., Kao, K.N., Gamborg, O.L. and Miller, R.A. (1973). Immunological methods for the agglutination of protoplasts from cell suspension cultures of different genera. *Planta,* 112 : 45-56.

Ito, M. (1973). Studies on the behaviour of meiotic protoplasts. II. Induction of a high fusion frequency in protoplasts from liliaceous plants. *Plant Cell Physiol.,* Tokyo, 14 : 865-872.

Jones, L.E., Hildebrandt, A.C., Riker, A.J. and Wu, J.H. (1960). Growth of somatic tobacco cells in microculture. *Am. J. Bot.,* 47 : 468-475.

Kameya, T. (1973). The effects of gelatin on aggregation of protoplasts from higher plants. *Planta,* 115 : 77-82.

Kanai, R. and Edwards, G.E. (1973). Purification of enzymatically isolated mesophyll protoplasts from C_3, C_4 and crassulacean acid metabolism plants using an aqueous dextran-polyethylene glycol two phase system. *Plant Physiol.,* 52 : 484-490.

Kao, K.N. (1977). Chromosomal behaviour in somatic hybrids of soybean – *Nicotiana glauca. Mol. Gen. Genet.,* 150 : 225-230.

Kao, K.N. and Michayluk, M.R. (1974). A method for high-frequency intergeneric fusion of plant protoplasts. *Planta,* 115 : 355-367.

Kao, K.N. and Michayluk, M.R. (1975). Nutritional requirements for growth of *Vicia hajastana* cells and protoplasts at a very low population density in liquid medium. *Planta,* 126 : 105-110.

Kao, K.N. and Wetter, L.R. (1977). Advances in techniques of plant protoplast fusion and culture of heterokaryocytes. In : B.R. Brinkley and K.R. Porter (Editors). International Cell Biology 1976-1977. The Rockefeller University Press, Boston, MA, pp. 216-224.

Kao, K.N., Gamborg, O.L., Miller, R.A. and Keller, W.A. (1971). Cell division in cells regenerated from protoplasts of soybean and *Haplopappus gracilis. Nature* (London), *New Biol.,* 232 : 124.

Kao, K.N., Gamborg, O.L., Michayluk, M.R., Keller, W.A. and Miller, R. (1973). The effects of sugars and inorganic salts on cell regeneration and sustained division in plant protoplasts. In : Les cultures de Tissus de Plantes. Colloq. Int. C.N.R.S., Paris, No. 212, pp. 207-213.

Kao, K.N., Constabel, F., Michayluk, M.R. and Gamborg, O.L. (1974). Plant protoplast fusion and growth of intergeneric hybrid cells. *Planta,* 120 : 215-227.

Keller, W.A. and Melchers, G. (1973). The effects of high pH and calcium on tobcco leaf protoplast fusion. *Z. Naturforsch.,* 28 : 737-741.

Melchers, G. (1977). Microbial techniques in somatic hybridization by fusion of protoplasts. In : B.R. Brinkley and K.R. Porter (editors), International Cell Biology 1976-1977. The Rockefeller University Press, Boston, MA, pp. 207-215.

Melchers, G. and Labib, G. (1974). Somatic hybridization of plants by fusion of protoplasts I. Selection of light resistant hybrids of haploid light sensitive varieties of tobacco. *Mol. Gen. Genet.,* 135 : 277-294.

Morel, G. (1965). Clonal propagation of orchids by meristem culture. *Cymbidium Soc. News,* 20 : 3-11.

Nagata,T and Ishii,S. (1979) A rapid method for isolation of mesophyll protoplasts. *Can.J.Bot.* 57: 1820-1823.

Nagao, T. (1979). Somatic hybridization by fusion of protoplasts. II. The combination of *Nicotiana tabacum* and *N. glutinosa* and of *N. tabacum* and *N. alata. Jpn. J. Crop Sci.,* 48 : 385-392.

Nagata, T. (1978). A novel cell-fusion method of protoplasts by polyvinyl alcohol. *Z. Naturwissenschaften,* 65: 263-264.

Pelcher, L.E., Gamborg, O.L. and Kao, K.N. (1974). Bean mesophyll protoplasts, production, culture and callus formation. *Plant Sci. Lett.,* 3 : 107-111.

Potrykus, I. (1973). Isolation, fusion and culture of protoplasts of petunia. In : J.R. Villanueva *et al.* (Editors), Yeast, Mould and Plant Protoplasts. Academic Press, London, pp. 319-322.

Potrykus, I., Harms, C.T., Lorz, H. and Thomas, E. (1977). Callus formation from stem protoplasts of corn. *Mol. Gen. Genet.,* 156 : 347-350.

Power, J.B., Cummins, S.E. and Cocking, E.C. (1970). Fusion of isolated plant protoplasts. *Nature* (London), 225 : 1016-1018.

Razdan,M.K.(1993) An introduction to plant tissue culture. Oxford and IBH Publishing and Co. Newdelhi

Schenk, R.U. and Hildebrandt, A..C. (1970). Production, manipulation and fusion of plant cell protoplasts as steps towards somatic hybridization. In : Les Culture de Tissue de Plantes. Colloq. Int. C.N.R.S., Paris, No. 193, pp. 319-331.

Schieder, O. (1978). Somatic hybrids of *Datura innoxia* Mill + *Datura discolor* Benth and of *Datura innoxia* Mill + *Datura stramonium* L. var. *tatula* L. *Mol. Gen. Genet.,* 162 : 113-119.

Schieder, O. (1980). Somatic hybrids between a herbaceous and two tree *Datura* species. *Z. Pflanzenphysiol.,* 98 : 119-127.

Scott, K.J., Chin, J.C. and Wood, C.J. (1978). Isolation and culture of cereal protoplasts. In : Proceedings of the Symposium on Plant Tissue Culture. Science Press, Peking, pp. 298-315.

Senda, M., Takeda, J., Abe, S. and Nakamura, T. (1979). Induction of cell fusion of plant protoplasts by electrical stimulation. *Plant Cell Physiol.,* Tokyo, 20 : 1441-1443.

Skoog, F. (1954). Growth and organ formation in tobacco tissue culture. *Am. J. Bot.,* 31 : 19-24.

Taiz, L. and Jones, R.L. (1971). The isolation of barley aleurone protoplasts. *Planta,* 101 : 95-100.

Uchimiya, H. and Murashige, T. (1974). Evaluation of parameters in the isolation of viable protoplasts from cultured tobacco cells. *Plant Physiol.,* 54 : 939-944.

Uchimiya, H. and Murashige, T. (1976). Influence of the nutrient medium on the recovery of dividing cells from tobacco protoplasts. *Plant Physiol.,* 57 : 424-429.

Vasil, V. and Hildebrandt, A.C. (1965). Differentiation of tobacco plants from single, isolated cells in microculture. *Science,* N.Y., 150 : 889-892.

Wallin, A.,Glimelius, K. and Eriksson, T. (1974). The induction of aggregation and fusion of *Daucus carota* protoplasts by polyethylene glycol. *Z. Pflanzenphysiol.,* 74 : 64-80.

Withers, L.A. (1973). Plant protoplast fusion : methods and mechanism. In : Les Cultures de Tissus de Plantes. Colloq. Int. C.N.R.S., Paris, No. 212, pp. 215-241.

Zimmermann, U. and Scheurich, P. (1981). High frequency fusion of plant protoplasts by electric fields. *Planta,* 151 : 26-32.

Chapter 16

Micropropagation

INTRODUCTION

Micropropagation is a field dealing with the ability to regenerate plants directly from explants. It is defined as "True-to-type propagation of selected genotypes using *in vitro* culture techniques". Unlike animals where differentiation is generally irreversible, in plants (due to an intact membrane system and a variable nucleus) even highly mature and differentiated cells retain the ability to regress to a meristematic state. The phenomenon of a mature cell reverting to the meristematic state and forming undifferentiated callus tissue is termed 'dedifferentiation'. The degree of regression a cell can undergo would depend on the cytological and physiological state it had reached *in situ*. The ability of plant cells to be cultured indefinitely on fully defined medium and their capacity to regenerate (cellular totipotency) into the whole plant *via* organogenesis or embryogenesis (Redifferentiation) have helped to study some of the problems haunting plant scientists (Fig. 16.1).

The vegetative method of propagating plants is termed as micropropagation or Cloning Tissue culture or growing *in vitro*. Morel (1965) used the technique for the first time for orchids. This propagation is now commonly accepted and adopted widely with lot of significance in commercial horticulture. It has important agroeconomic applications in medicinal plant preservation, floriculture and forestry.

Hundreds of commercial laboratories in different parts of the world are currently involved in micropropagation (Atkinson *et al.* 1996) some of them producing over 20 million plants a year.

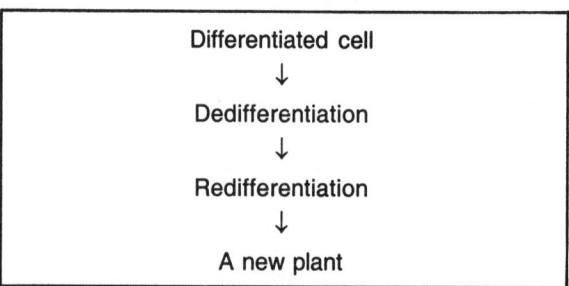

Fig. 16.1. In-vitro culture techniques of propagation

16.1. ADVANTAGES

The various advantages of micropropagation are :

1. Useful for plants that are difficult to propagate by conventional methods (e.g. plants producing little viable seeds) (Schoner and Reinhard 1986, Diettrich et al. 1982).
2. In relatively short time (due to increased multiplication rate) and space, a large number of plants that is
   (a) Genetically stable and true-to-type progeny [by rapid clonal propagation] (Hu and Wang 1988),
   (b) virus-free (Hollings 1965, Vine 1968, Jones and Vine 1968, Vine and Jones 1969, Wilmar and Hellendoorn 1968),
   (c) insect-resistant, disease-resistant, herbicide-resistant, salt-resistant (Maliga et al. 1973, Dix and Street 1975, Ebel et al. 1976, Bojwani and Razdan 1983) and
   (d) With special phenotypic character (i.e.) changed genotypes (tetraploids, haploids, hybrids) can be produced from single individual.
3. Conservation of genetic resources of species and threatened medicinal plants [by axillary bud proliferation] (Pierik 1987, Chu 1992).
4. Plant improvement by regeneration technique in conjunction with *in vitro* cell manipulation and
5. To solve some theoretical problems connected with the pathway of biogenesis of chemical compounds in plants and the relation between organogenesis and production of metabolites.

16.2. DISADVANTAGES

Though, there are several advantages, there are also certain disadvantages with the system. These are :

1. Micropropagation methods through use of tissue culture involve capital intensive expensive materials like autoclave, laminar airflow bench, controlled culture rooms etc.
2. This is a technically skilled work, knowledge about material, techniques and decisions making (during subculture and multiplication of propagule) are required in the personnel.
3. Contamination is a serious threat and cause severe damage to material and add substantively lot to the cost of production, affects time schedule delivery of the material.
4. Specific conditions of micropropagation, rooting and hardening may be required. Therefore, each material requires separate research methods.
5. Small delicate plantlets are produced, which take longer initial time to grow.
6. Genetic stability is doubtful in certain methods.
7. It is a capital-intensive industry, if plants are produced in small number, they cost too much, otherwise also cost is a major factor for the production and sale of tissue culture raised plants.

   Several investigators in the case of many medicinal and aromatic plants through micropropagation (Table 16.1) have achieved an appreciable level of success.

16.3. FACTORS THAT INFLUENCE MICROPROPAGATION

The factors that play important roles in the degree of success achieved in a given micropropagation or plant regeneration system include

1. The genotype of the donor,
2. The physiological conditions of the donor material,

3. The explant source,
4. The orientation and size of explant in culture,
5. The culture medium composition(s),
6. Interactions of endogenous hormones with exogenously supplied growth regulators,
7. The incubation conditions (including light quality and intensity, temperature, relative humidity and air quality) and
8. The timing of the subculture interval / changes in medium / incubation treatment.

Table 16.1. Micropropagation of Medicinal and Aromatic Plants

Name	Reference
Aconitum napellus	Watad *et al.* 1995
Allium sativum	Ayabe and Sumi 1998
Ancistrocladus abbreviatus	Bringmann *et al.* 1999
Arnica chammisonis	Cassels *et al.* 1999
Bacopa moniera	Ali *et al.* 1996
Bergenia crassifolia	Furmanowa and Rapczewska 1993
Catharanthus roseus	Suk *et al.* 1994
Centella asiatica	Tiwari *et al.* 2000
Cephaelis ipecacuanha	Yoshimatsu and Shimomura 1993
Cleistanthus collinus	Quraishi *et al.* 1996
Coleus forskohlii	Sen and Sharma 1991
Commiphora weightii	Barve and Mehta 1993
Eucalyptus grandis	Warrag *et al.* 1990
Eucalyptus sideroxylon	Burger 1987
Eucalyptus sp.	LeRoux and VanStaden 1991
Glycyrrhiza glabra	Thengane *et al.* 1998
Gymnema sylvestre	Reddy *et al.* 1998
	Komalavalli and Rao 2000
	Balakumar and Selvakumar 2000
Heracleum candicans	Wakhlu and Sharma 1998
Hemidesmus indicus	Sreekumar *et al* 2000
Hypericum perforatum	Sulochana rani *et al.* 2001
Melia azedarach	Thakur *et al.* 1998
Ocimum basilicum	Sahoo *et al.* 1997
Picrorhiza kurroa	Lal *et al* 1988
Pothomorphe umbellata	Ana Maria *et al.* 2000
Rauwolfia serpentina	Roja and Heble 1996
Scilla natalensis	McCartan and VanStanden 1998
Stevia rebaudiana	Ales and Tomas 1998
Swertia chirata	Wawrosch *et al.* 1999
Trichopus zeylanicus	Krishnan *et al.* 1995
Valeriana jatamansi	Kaur *et al.* 1999
Vitex negundo	Sahoo and Chand 1998
Withania somnifera	Anjali Kulkarni *et al.* 1996

Cells must be physiologically receptive to the hormone induction signals before they can be induced to regenerate a plant or organ. Thus, the timing of the application of induction signals can be critical. This physiological receptiveness to be induced is termed 'competence' (Graham and Wareing 1984, Christianson and Warnick 1987). Komalavalli and Rao (2000) reported that the nature of the explant, seedling age, medium type, plant growth regulators, complex extracts (casein hydrolysate, coconut milk, malt extract and yeast extract) and antioxidants (activated charcoal, ascorbic acid, citric acid and polyvinylpyrrolidone) markedly influence *in vitro* propagation of *Gymnema sylvestre*.

The other problems commonly encountered in the micropropagation industry are

* *Vitrification*: In repeated cycles of *in-vitro* shoot multiplication a percentage of culture show water-soaked (almost translucent) leaves and may eventually die.
* *High price of propagated plants:* Plant propagation by tissue culture technique is encumbered by the intensive labour requirement for the multiplication process; thus, scaling-up systems and automation of unit operations (i.e.) increased mechanisation, especially of the multiplication step or by placing them in 'low wage' countries are necessary to cut down the production costs (Aitken-Chiristie 1991, Vasil 1991).
* *Somaclonal variation:* It is defined, as the variation, which occurs in cultures of cells and tissues that may be either genetic or epigenetic. By selecting somaclonal variants, disease resistant as well as herbicide resistant plants could be achieved. This phenomenon has not yet been used to improve medicinal plants but several examples of changes in the colour of flowers of ornamental plants indicate that secondary metabolism can also be affected (Evans et al. 1987). The factors that increase the frequency of somaclonal variation among regenerated plants are
* Length of time in culture (Reisch 1983, Evans and Sharp 1986, Lee and Phillips 1988).
* Cultural stresses (improper media components or mutagens, certain growth regulator treatments, delayed subculture intervals leading to nutrient stress, or exposure to extreme or highly variable incubation conditions) and
* Explant source (non-meristematic explants that do not orderly mitoses, used in adventitious or de novo regeneration systems).

It is known from histological studies that the regenerated shoot organ or somatic embryo may arise either from a single cell or from a cluster of cells acting in a co-ordinate manner, depending upon the species (Thorpe 1980, Ammirato 1987). Regenerants derived from multiple-celled origins are more likely to be chimeric (cells consisting of two or more different genotypes) due to the accumulation of genetic variation during culture, compared to regenerants derived from single celled origins. Variegated cultivars usually are chimeric and therefore the location and type of origin of a regeneration event in such plant material will determine the phenotype of the regenerant (Evanari 1989). Induction of somatic embryogenesis has a strong dependence on cultivar and genotype. For self-pollinated plant species, somatic embryogenic differences also may be important between individuals within cultivars. Cultivars identified by screening or developed by selective breeding to have the capability for high expression of plant regeneration have been reported (Bingham et al. 1975, Brown and Atanassov 1985). The further detail see Chapter 14.

16.4. GENERAL TECHNIQUES OF MICROPROPAGATION

The various stages of the micropropagation are shown in Fig. 16.2.

16.4.1. Stage I

In stage I, suitable media (usually MS / modified MS media), plant growth regulator levels and their combinations are selected in order to promote explant establishment and shoot growth. Physiological

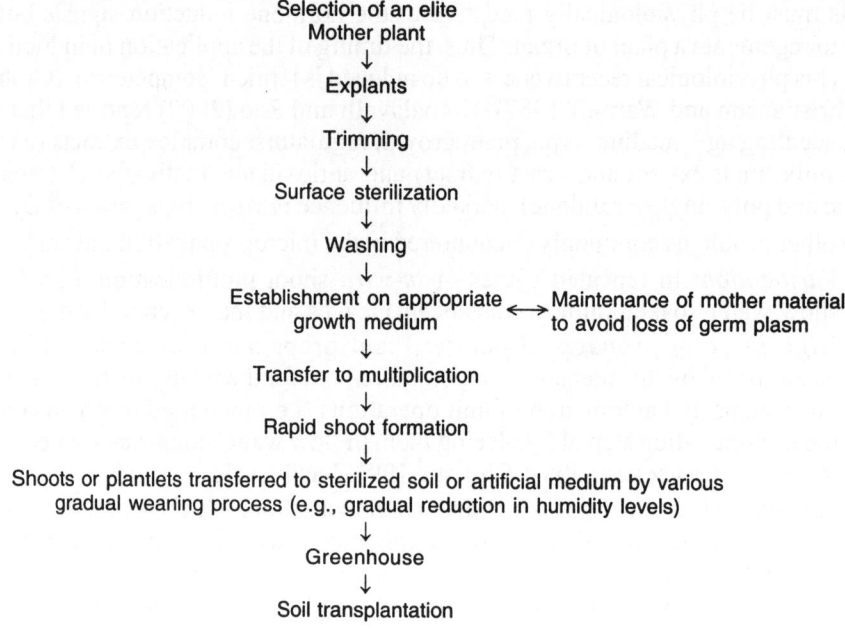

Fig. 16.2. Various stages of micropropagation of plants

stabilization may require (3-24) months and (4-6) sub cultures on stage I medium. Failure to do so before transfer to stage II medium containing higher cytokinin levels (to disrupt apical dominance of shoot tip) may result in diminished shoot multiplication rates or production of undesirable basal callus and adventitious shoots. In many commercial labs, stabilized cultures verified as having specific pathogen tested and free of cultivable contaminants, are often maintained on media that limit shoot production to maintain genetic stability. These cultures called "Mother blocks" serve as sources of shoot tips or nodal segments for initiation of new stage II cultures. The following factors may affect successful stage I establishment of meristem explants:

* Explantation time: (Beginning of growing seasons generally gives best results),
* Position of explant on the stem,
* Explant size and
* Polyphenol oxidation (tissue/medium browning). Excision of explant promotes release of polyphenol, which stimulates the activity of polyphenol oxidase. Tissue browning can be reduced by the use of liquid medium with frequent transfer, the addition of antioxidants (ascorbic acid, PVP etc.,) or culturing in reduced light intensity/darkness.

Cytokinins and/or auxins are most frequently added to stage I media to enhance explant survival and shoot development. BA, 2 ip, NAA and IBA are most widely used plant growth regulators. 2,4-D is more effective in somatic embryogenesis.

16.4.2. Stage II

This is characterized by repeated enhanced formation of axillary shoots from shoot tips or lateral buds cultured on medium supplemented with a relatively higher cytokinin level to disrupt apical dominance of the shoot tip. Subcultures inoculated with explants that had been shoot apices in the previous subculture often exhibit higher multiplication rates than lateral bud explants. Inverting shoot explants in the medium can double/triple the number of axillary shoots produced on vertically oriented explants per culture period in some species. Selecting only terminal shoots of axillary origin for subculture, instead of shoot bases, decreases the frequency of off-types including the preclinal chimeras. Addition of auxin, often mitigates the inhibitory effect of cytokinin on shoot elongation, (but may form callus) thus increases the number of usable shoots of sufficient length for rooting. Selection of Stage II cytokinin type and concentration is based on shoot multiplication rate, shoot length, frequency of genetic variation and possibility of adverse carryover effects on the survivability and rooting of plantlets in stage IV. Using liquid shake cultures has the special advantage that the shoots broke apart as they multiplied and the manual cutting of shoot cultures is not required. Media with reduced salt levels are used if necessary. The nature of organogenic differentiation is determined by the relative concentration of auxins and cytokinins (Skoog-Miller hypothesis).

16.4.3. Stage III

This may involve elongation of shoots prior to rooting (GA3 may be added), rooting of individual shoots / shoot clumps, fulfilling dormancy requirements of storage organs by cold treatment and pre-hardening cultures to increase survival. Where possible, commercial labs have developed procedures to transfer stage II micro cuttings to soil, thus by-passing stage III rooting. A low salt medium with optimum auxin concentration is determined based upon percentage of rooting, root number and root length (upto max. 5 mm, approx. 15 days) to prevent root damage during transplanting.

16.4.4. Stage IV

This involves acclimatizing or hardening of plantlets to conditions of significantly lower humidities and higher light intensities. Micropropagated plants are difficult to transplant for two primary reasons: A heterotrophic mode of nutrition and poor control of water loss. Before senescencing, the older leaves function as "life boats" by supplying stored carbohydrate to the developing and photosynthetically competent new leaves. However, this is not the rule for all plants. Plantlets are transplanted to a well-drained "sterile" growing medium and maintained initially at high relative humidity (using humidity tents or single tray propagation domes), reduced light intensity (160 (mol/m2/sec ~ 1000-5000 lux) with diurnal regime of 16 h day and 8 h night at 20-27°C. Transplants are acclimatized by gradually lowering the relative humidity over a (1-4) week period. Plants are gradually moved to lower relative humidity and higher light intensities (3000-10,000 lux) and increased sucrose in the medium to promote vigorous growth. Usually the most difficult step during micropropagation is the recovery of plants from the culture vessels into the soil. The plantlets were propagated *in vitro* under conditions of 100% relative humidity with little or no need to perform photosynthesis or to control respiration. The plantlets transferred to soil conditions must now perform all of their own photosynthesis, and adapt to lower relative humidity by developing a waxy cuticle and regulating stomatal function. The hardening-of process during the transition from petri dish to greenhouse conditions must be gradual. Often plantelets go through this process more readily when they are larger and better developed. Establishment of delicate, herbaceous plants also requires a fine balance of water relations: too little water leads to permanent wilt, while too much water leads to rot.

6.5. METHODS OF MICROPROPAGATION

The various methods of micropropagation are :
1. Adventitious shoot proliferation
2. Micropropagation by proliferation of axillary buds
3. Plant regeneration by organogenesis from callus and cell suspension cultures (Further details see Chapter 13).

4. Plant regeneration from callus and cell suspension cultures by somatic embryogenesis.

5. Direct (adventitious) somatic embryogenesis (Non-zygotic embryo-genesis).

6. Artificial seeds (see Chapter 17).

6.5.1. Adventitious shoot proliferation (shoot culture)

It is the method of choice for commercial production of plants through micropropagation and relies on the stimulation of axillary shoot growth from lateral buds following disruption of apical dominance.

Compared with other methods, shoot cultures :

• Provide reliable rates and consistency of multiplication following culture stabilization.

• Are less susceptible to genetic variation (due to the highly organized structure of shoot tip) and

• May provide for clonal propagation of preclinal chimeras.

Meristem culture is the culture of apical meristematic dome alone whereas meristem tip culture is the culture of (0.2-0.5 mm) long meristem tip (apical meristem with one or two leaf primordia) explants that have undergone thermo/chemotherapy. Meristem culture is rarely used as it exhibits both low survival rates and increased genetic variability following callus formation and indirect shoot organogenesis e.g. Begonia, *Ficus lyrafa*, etc.

6.5.2. Micropropagation by proliferation of Axillary buds

This technique is the simplest type that exploits the normal ontogenetic route for branch development by lateral axillary meristems. The axillary buds are treated with hormones to break dormancy and produce shoot branches. The shoots are then separated and either rooted to produce plants or used as propagules for further propagation. Many ornamental plants are woody species and are propagated commercially by axillary bud proliferation (Mantell *et al.* 1985, Pierik, 1987, Chu, 1992, Huetteman and Preece, 1993). Axillary bud proliferation typically results in an average tenfold increase in shoot number per monthly culture passage. In a period of 6 months, it is feasible to obtain as many as 10,00,000 propagules or plants, starting from a single explant. Adventitious bud formation is better than callusing approach (from cytologically abnormal plants), as they form uniformly diploid cells. However, they involve the risk of splitting genetic chimeras leading to pure type plant.

REFERENCES

Aitken-Christie, J. (1991). In: Micropropagation Automation. Eds. Debergh, PC and Zimmerman RH. 342-354 *Kluwer Acad. Publ*, Dordrecht, Boston, London.

Ales, N. and Tomas, V. (1998). In vitro propagation of *Stevia rebaudiana* plants using multiple shoot culture. *Planta Med.,* 64: 775-776.

Ali, G. *et al.* (1996). A rapid protocol for micropropagation of *Bacopa monniera* (L), Wettst-An important medicinal plant. *Plant Tiss. Cult. Biotechnol.,* 2(4): 208-211.

Ammirato, P.V. (1987). Organizational events during somatic embryogenesis. In: Plant tissue and cell culture. Eds. Green CE, Somers DA, Hackett WP and Biesboer DD. 57-81 Alan R Liss, New York.

Ana Maria, S.P., Binaca, W.B., Beatriz Appezzato-da-Gloria, Alba, R.B.A., Ana Helena, J., Miriam, V.L., Suzelei, C.F. (2000). Micropropagation of *pothomorphe umbellta* via direct organogenesis from leaf explants. *Plant Cell Tiss. Org. Cult.,* 60: 47-53.

Anjali Kulkarni, A., Thengane, S.R., Krishnamurthy, K.V. (1996). Direct in vitro regeneration of leaf explants of *Withania somnifera* (L) Dunal. *Plant Sci.,* 119: 163-168.

Atkinson, P.J., Fay, M.F., Walter, K.S. (1996). Development of a new micropropagation protocols database at the Royal Gardens, Kew. *Plant Tissue Culture and Biotechnology,* 2(3) : 154.2.

Ayabe, M., Sumi, S. (1998). Establishment of a novel tissue culture method, stem-disc culture and its practical application to micro-propagation of garlic (*Allium sativum* L). *Plant Cell Rep.,* 17: 773- 776.

Balakumar, T., Selvakumar, V. (2000). Certain preliminary observations on the micropropagation of the antidiabetic herb *Gymnema sylvestre* RBr. In: Recent trends in species and medicinal plants research. Ed. De AD. 92-95 Assoc. press, New Delhi.

Barve, D.M., Mehta, A.R. (1993). Clonal propagation of mature elite trees of *Commiphora wightii. Plant Cell Tiss. Org. Cult.,* 35: 237-244.

Bingham, E.T., Hurley, L.V., Kaatz, D.M., Saunders, J.W. (1975). Breeding alfalfa, which regenerates from callus tissue in culture. *Crop Sci.,* 15: 719-721.

Bojwani, S.S., Razdan, M.K. (1983). Development in Crop Sciences 5. In: Plant Tissue Culture: Theory and Practice. Ed. Bojwani, S.S. 37-45 Elsevier Amsterdam, Oxford, New York.

Bringmann, G., Rischer, H., Schlauer, J. Ake Assi, L. (1999). *In vitro* propagation of *Ancistrocladus abbreviatus* Airy Shaw (*Ancistrocladaceae*). *Plant Cell Tiss. Org. Cult.,* 57: 71-73.

Brown, D.C.W. Atanassov, A. (1985). Role of genetic background in somatic embryogenesis in Medicago. *Plant Cell Tiss. Org. Cult.,* 4: 111-122.

Burger, D.W. (1987). *In vitro* micropropagation of *Eucalyptus sideroxylon. Hort. Sci.,* **22(3):** 496-501.

Cassels, A.C., Walsh, C., Belin, M., Cambornac, M., Robin, J.R., Lubrano, C. (1999). Establishment of a plantation from micropropagated Arnica chamissonis, pharmaceuticals substitute for the endangered A. montana. *Plant Cell Tiss. Org. Cult.,* 56: 139-144.

Christianson, M.L., Warnick, D.A. (1987). Physiological genetics of organogenesis invitro. In: Genetic manipulation of woody plants. Eds. Hanover JW and Keathly DE. 101-115 Plenum Press, New York.

Chu, I.Y.E. (1992). Perspectives of micropropagation industry. In: Transplant production systems. Eds. Kurata K, Kozai T. 137-150 Kluwer *Acad. Publ*, Amsterdam.

Diettrich, B., Neumann, D., Luckner, M. (1982). Clonation of protoplast derived cells of *Digitalis lanata* suspension cultures. *Biochem. Physiol. Pflanzen* 177:176-431.

Dix, P.J., Street, H.E. (1975). Sodium chloride resistant cultured cell lines from *Nicotiana sylvestris* and *Capsicum annuum. Plant. Sci. Lett.,* 5: 231-237.

Ebel, J., Ayers, A.R., Albersheim, P. (1976). Host-pathogen interactions: XII, Response of suspension cultured soybean cells to the elicitor isolated from Phytophthora megasperma varsojae, a fungal pathogen of soybeans. *Plant Physiol.,* 57: 775- 779.

Evanari,M (1989). The history of research on white-green variegated plants. *Bot. Rev.,* 55: 106-133.

Evans, D.A., Sharp, W.R. (1986). Somaclonal and gameto clonal variation. In: Handbook of plant cell culture. Eds. Evans DA, Sharp WR and Ammirato PV. 4 : 97-132 Macmillan, New York.

Evans, D.A., Sharp, W.R., Bravo, J.E. (1987). Plant somaclonal variation and mutagenesis. In: Neste Research News 1986/87. Ed. Nestec Ltd., 63-73 Vevey: Nestec Ltd.

Furmanowa, M., Rapczewska, L. (1993). Bergenia crassifolia (L) Fritsch (Bergenia): micropropagation and arbutin contents. In: Medicinal and aromatic plants IV, Biotechnology in agriculture and forestry. Ed. Bajaj, YPS 21:18-33 Springer-Verlag, Berlin, W. Germany.

Graham, C.F. Wareing, P.F. (1984). Development control in animals and plants. 2nd edn , 73-88 Blackwell Scientific, Boston.

Hollings, M. (1965). Disease Control through virus-free stock. *Ann. Rev. Phytopathol.,* 3: 367- 396.

Hu, C.Y., Wang, P.J. (1988). Meristem, shoot tip and bud culture. In: Handbook of plant cell culture. Eds. Evans DA, Sharp WR, Ammirato PV and Yamada Y. 1: 11-48 New York, Macmillan, 177-227.

Huetteman, C.A., Preece, J.E. (1993). Thidiazuron: a potent cytokinin for woody plant tissue culture. *Plant Cell Tiss. Org. Cult.,* 33: 105-119.

Jones, O.P. and Vine, S.J. (1968). The culture of gooseberry shoot tips for eliminating viruses. *J. Hort. Sci.,* 43: 289-292.

Kaur, R., Sood, M., Chander, S., Mahajan, R., Kumar, V., Sharma, D.R. (1999). *In vitro* propagation of *Valeriana jatamansi*. *Plant Cell Tiss. Org. Cult.,* 59: 227-229.

Komalavalli, N., Rao, M.V. (2000). *In vitro* micropropagation of *Gymnema sylvestre* - A multipurpose medicinal plant. *Plant Cell Tiss. Org. Cult.,* **61:** 97-105.

Krishnan, P.N., Sudha, C.G., Seeni, S. (1995). Rapid propagation through shoot tip culture of *Trichopus zeylanicus* Gaertn, a rare ethnomedicinal plant. *Plant Cell Rep.,* **14:** 708.

Lal, N. Ahuja, P.S. Kukreja, A.K. Pandey, B. (1988) Clonal propagation of *Picrorhiza kurroa* Royale ex benth. By shoot tip culture. Plant cell reports,7:202-205.

Le Roux, J.J., Van Staden, J. (1991). Micro-propagation and tissue culture of *Eucalyptus*- a review. *Tree Physiol.,* 435-441

Lee, M., Phillips, R.L. (1988). The chromosomal basis of somaclonal variation. *Ann. Rev. Plant Physiol. Mol. Biol.,* **39:** 413-437

Maliga, P., Breznovita, S.A., Marton, L. (1973). Streptomycin-resistant plants from callus culture of haploid tobacco. *Nature New Biol.,* **244:** 29- 30

Mantell, S.H., Mathews, J.A., McKee, R.A. (1985). Principles of plant biotechnology. 130-157 Blackwell Scientific, Boston

McCartan, S.A., van Staden, J. (1998). Micropropagation of the medicinal plant, *Scilla natalensis* Planch. *Plant Gro. Reg.,* **25:** 177-180.

Morel,G.(1965). Clonal propagation of orchids by meristem culture. cymbidium Soc. News, 20:3-11.

Pierik, R.L.M. (1987). *In vitro* culture of higher plants. Ed. Nijhoff M. 183-230 Kluwer Acad. Publ, Dordrecht, Boston, London.

Quraishi, A., Koche, V., Mishra, S.K. (1996). *In vitro* micropropagation from nodal segments of *Cleistanthus collinis*. *Plant Cell Tiss. Org. Cult.,* **45:** 87-91

Reddy, S.R.P., Rama Gopal, G., Sita, G.L. (1998). *In vitro* multiplication of *Gymnema sylvestre* an important medicinal plant. *Curr. Sci.,* **75 (8)**: 843

Reisch, B. (1983). Genetic variability in regenerated plants. In: Handbook of Plant Cell Culture. Eds. Evans DA, Sharp WR, Ammirato PV and Yamada Y. 1: 748-769 Macmillan, New York

Roja, G., Heble, M.R. (1996). Indole alkaloids in clonal propagules of *Rauwolfia serpentina* benthe ex kurz. *Plant Cell Tiss. Org. Cult.,* **44:** 111-115.

Sahoo, Y., Chand, P.K. (1998). Micropropagation of *Vitex negundo*, a woody aromatic medicinal shrub, through - frequency axillary shoot proliferation. Plant Cell Rep 18: 301

Sahoo, Y., Pattnaik, S.K., Chand, P.K. (1997). *In vitro* clonal propagation of an aromatic medicinal herb *Ocimum basilicum* L (Sweet Basil) by axillary shoot proliferation. *In vitro Cell Dev. Biol. Plant,* **33:** 293-296.

Schoner, S., Reinhard, E. (1986) Long term cultivation of *Digitalis lanata* clones propagated *in vitro*: Cardinolide content of regenerated plants. *Planta med.,* **6:** 478-481

Sen, J., Sharma, A.K. (1991). *Invitro* propagation of *Coleus forskohlii* Briq for forskolin synthesis. *Plant Cell Rep.,* 696-701.

Sreekumar, S. Seeni, S. Pushpagandan, P. (20000. Micropropagation of *Hemidesmus indicus* for cultivation and production of 2-hydroxy 4- methoxy benzaldehyde. *Plant cell tissue and organ culture*, 62: 211-218.

Sulochana Rani, N., Balaji, K., Veeresham, C. (2001). Micropropagation of *Hypericum perforatum* by tissue culture. *Ind. J. Nat. Products* (in press).

Suk, W.K. Nam, H.S. Kyung, H.J. Sang, S.K., Jang, R.L.(1994) High frequency plant regeneration from anther derived cell suspension cultures via somatic embryogenesis in *Catharanthus roseus*. *Plant cell reports.* 13:319-322.

Thakur, R., Rao, P.S., Bapat, V.A. (1998). In vitro plant regeneration in Melia azedarach L. *Plant Cell Rep.,* **18:** 127-131.

Thengane,S.R., Kulkarni,D.K., Krishnamurthy, K.V., (1998) Micropropagation of Liquorice (*Glycyrrhiza glabra*) through shoot tip and nodal cultures. *In-vitro cell Dev. Biol.Plant,* **34**, 331-334.

Thorpe, T.A. (1980). Organogenesis in vitro: structural, physiological and biochemical aspects. *Int. Rev. Cytol. Suppl.,* **11A:** 71-112.

Tiwari, K.N., Sharma, N.C., Tiwari, V., Singh, B.D. (2000). Micropropagation of *Centella asiatica* (L), a valuable medicinal herb. *Plant Cell Tiss. Org. Cult.,* 63: 179-185.

Vasil, I.K. (1991). Rationale for the scale-up and automation of plant propagation. In: Cell cultures and somatic cell genetics of plants. Ed. Vasil IK. 8:1-12 Acad. Press, San Diego

Vine, S., Jones, O.P. (1969). The culture of shoot tips of hop (*Humulus lupulus* L) to eliminate virus. *J. Hort. Sci.,* 44: 281-284.

Vine, S.J. (1968). Improved culture of apical tissues for production of virus-free strawberries. *J. Hort. Sci.,* 43: 292-297.

Wakhlu, A.K., Sharma, R.K. (1998). Micropropagation of *Heracleum candicans* Wall: A rare medicinal herb. *In vitro Cell Dev. Biol. Plant,* 35: 79-81.

Warrag, E.I., Lesney, M.S., Rockwood, L. (1990). Micro-propagation of field-tested superior *Eucalyptus grandis* hybrids. *New forests,* 4: 67-71.

Watad, A.A., Kochba, M., Nissim, A., Gaba, V. (1995). Improvement of *Aconitum napellus* micropropagation by liquid culture on floating membrane rafts. *Plant Cell Rep.,* 18: 345.

Wawrosch, C., Maskay, N., Kopp, B. (1999). Micropropagation of the threatened nepalese medicinal plant Swertia chirata Buch - Ham ex Wall. *Plant Cell Rep.,* 18: 997-1001.

Wilmar, C., Hellendoorn, M. (1968). Growth and morphogenesis of Asparagus cells cultures in vitro. *Nature* (London) 217: 369-370.

Yoshimatsu, K., Shimomura, K(1993). Cephaelis ipecacuanha A. Richard: Micropropagation and the production of emetine and cephaeline. In: Medicinal and aromatic plants IV. Biotechnology in Agriculture and Forestry (ed) Bajaj, Y.P.S. 21: 87-103. Springer-verlag Berlin

Chapter 17

Synthetic Seeds

The synthetic seed (also referred as artificial or somatic seed or synserts or synseeds) technology is one of the most rapidly developing fields of plant science in the last few decades. Synthetic seeds are defined as artificially encapsulated somatic embryos, shoot buds, cell aggregates, or any other tissue that can be used for sowing a seed and that possess the ability to convert into a plant under *in vitro* or *ex vitro* conditions and that retain this potential also after storage (Capuano *et al.*, 1998).

Synthetic seeds offer a potential technology towards use of somatic embryogenesis for large-scale propagation of plants through automation. Synthetic seeds are generally prepared by encapsulating the somatic embryos obtained from tissue culture in a protective jelly capsule, which is usually prepared with sodium alginate, or by desiccating the somatic embryos with or without coating.

The micropropagation industry started with Orchids and Ferns and is still largely restricted to horticulture and ornamental plants. During the past few decades' production of bulb and tuber plants, such as lily, potato, gladiolus and cassava has become quite common. Large-scale propagation requires planting material in suitable age, shape and of germination percentage. Somatic embryos are suitable materials for this purpose as large number of somatic embryos of same age can be produced in a bioreactor. Technology is developed to encapsulate them to produce artificial or synthetic seeds. Encapsulated embryos are protected against desiccation and mechanical injury by the gel.

The term "Artificial seed" was first coined by Murashige in 1977. In simple terms it means a somatic embryo entrapped in a biodegradable synthetic polymer coating that acts as an artificial seed coat. The coating should not adversely affect the embryo and allow embryo to plant conversion to occur without introducing variation inducing components. The first construction of synthetic somatic embryos was achieved by Kitto and Janick (1982) and single embryo encapsulation of hydrated alfa alfa (*Medicago sativa*) somatic embryos was obtained by Redenbaugh *et al.* (1984). However, it was Kamada (1985) who broadened the scope of the technique by suggesting a new definition of somatic propagules i.e., Artificial seeds. According to him "an artificial seed comprises a capsule prepared by coating a cultured matter, like a tissue piece or an organ, which can grow into complete plant along with nutrients and artificial film/covering" (Fig. 17.1).

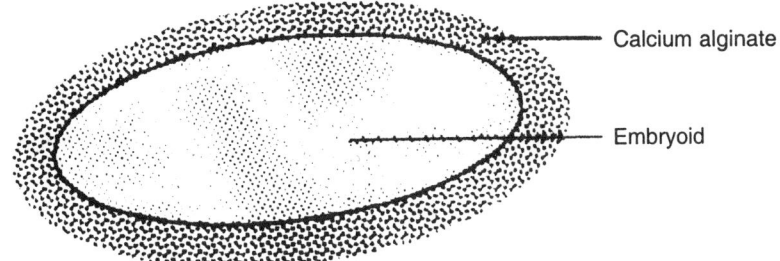

Fig. 17.1. Synthetic seed.

Capsule gel with hydrophilic membrane, the :

A. Artificial seed coat

B. Somatic embryo

C. Artificial endosperm.

Today, the artificial seed technology has been extended to a large number of plant species belonging to both dicot and monocot groups (Table 17.1).

Table 17.1. List of plant species where artificial seed production was reported

In vitro propagules used for encapsulation	Plant	Reference
1. Somatic embryos	Apium graveolens	Kim & Janick, 1987
	Daucus carota (Carrot)	Kitto & Janiek, 1985
	Gossypium hirsutum	Redenbaugh et al., 1988
	Lactuca sativa	Redenbaugh et al., 1988
	Medicago sativa (Alfa alfa)	Redenbaugh et al., 1984
	Loblolly pine	Gupta & Durzan, 1987
	Dactvlis glomerata	Gray et al., 1987
	Santalum album	Bapat & Rao, 1988
	Celery	Redenbaugh et al., 1986
	Brinjal	Rao & Singh, 1991
	Brassica	Redenbaugh et al., 1987
	Lettuce	Redenbaugh et al., 1987
	Oryza sativa (rice)	Suprasanna et al., 1995
	Arachis hypogea (Ground nut)	Padmaja et al., 1995
	Asparagus cooperi	Ghosh and Sen 1994
	Brassica compestris (Mustard)	Arya et al., 1998
	Camellia japonica	Janeiro et al., 1997
	Eucalyptus citridora (Eucalyptus)	Muralidharan and Mascarenhas, 1995
	Mangifera indica (Mango cv. Amrapali)	Ara et al., 1999
	Picea abies (Norway spruce)	
	Picea glauca (White spruce)	Gupta et al., 1987
	Picea glauca Engelmanii (Interior spruce)	Attree et al., 1994
	Pistacia vera (Pistachio)	Gupta et al., 1987

(Contd.)

In vitro propagules used for encapsulation	Plant	Reference
	Psidium guajava (Guava)	Onay *et al.*, 1996
	Solanum melongena (Eggplant)	Lakshmana Rao & Singh, 1991
	Vitis vinifera (Grape)	Gray and Purohit, 1991
2. Shoot buds (or) shoot tips	Banana	Ganapathi *et al.*, 1992
	Cardamon	Ganapathi *et al.*, 1994
	Carum carvi	Furamanova *et al.*, 1991
	Picrorhiza kurroa	Ahuja *et al.*, 1989
	Valeriana wallichii	Mathur *et al.*, 1989
	Actinidia deliciosa (Kiwifruit)	Piccioni & Standardi, 1995
	Betula pendula (Birch)	Piccioni & Standardi, 1995
	Crataegus oxyacantha (Hawthorn)	Piccioni & Standardi, 1995
	Malus pumila (Apple rootstock M. 26)	Capuano *et al.*, 1998
	Morus indica (Mulberry)	Picioni, 1997
	Rubus idaeus (Raspberry)	Bapat, 1993
	Rubus (Blackberry cv. Jumbo veten)	Piccioni & Standardi, 1995
	Zingiber officinale (Ginger)	Piccioni & Standardi, 1995
3. Axillary buds	Eucalyptus	Huang & Cheng, 1990
	Morus indica	Bapat *et al.*, 1987
	Valeriana wallichii	Mathur *et al.*, 1989
	Dioscorea floribunda	Ahuja *et al.*, 1989
	Picrorhiza kurroa	Ahuja *et al.*, 1989
	Salvia sclarea	Ahuja *et al.*, 1989
	Pogostemon cablin	Ahuja *et al.*, 1989
	Syringa vulgaris	Refouvelet *et al.*, 1998
4. Protocorm like bodies	*Cymbidium giganteum* (orchid)	Sharma *et al.*, 1992
	Geodorum densiflorum (orchid)	Datta *et al.*, 1999
	Phains tankervillae (orchid)	Malemngaba *et al.*, 1996
	Spathaglotis plicata (orchid)	Singh, 1991

17.1. SYNTHETIC SEED TECHNOLOGY

Synthetic or artificial seeds are the living seed-like structure derived from somatic embryoids *in vitro* cultures after encapsulation by a hydrogel. The preserved embryoids are termed as synthetic seeds. Somatic embryoids are identical with zygotic embryos but they lack important accessory tissues, i.e. endosperm and protective coatings, which make them inconvenient to store and handle. Furthermore, they are generally regarded to lack a quiescent resting phase and to be incapable of undergoing dehydration. Therefore, the synthetic seed technology is primarily aimed to produce somatic embryos that resemble more closely the seed embryos in storage and handling characteristics so that they can be utilized as a unit for clonal plant propagation and germplasm conservation. In achieving such a goal the technology of encapsulation has evolved as the first major step for production of synthetic seeds. Encapsulated synthetic seed should also contain growth nutrients, plant growth promoting microorganisms (e.g. micorrhizae), and/or other biological components necessary for optimal embryo to plant development. The success of the synthetic seed technology is constrained due to scarcity and undesirable qualities of somatic embryos making it difficult for their development into plants. The choice of coating material for making synthetic seeds is also an important aspect for synthetic seed production.

17.2. CLASSIFICATION OF SYNTHETIC SEEDS

Based on technology established so far, two types of synthetic seeds are known: desiccated and hydrated.

These two are again classified into encapsulated and uncoated.

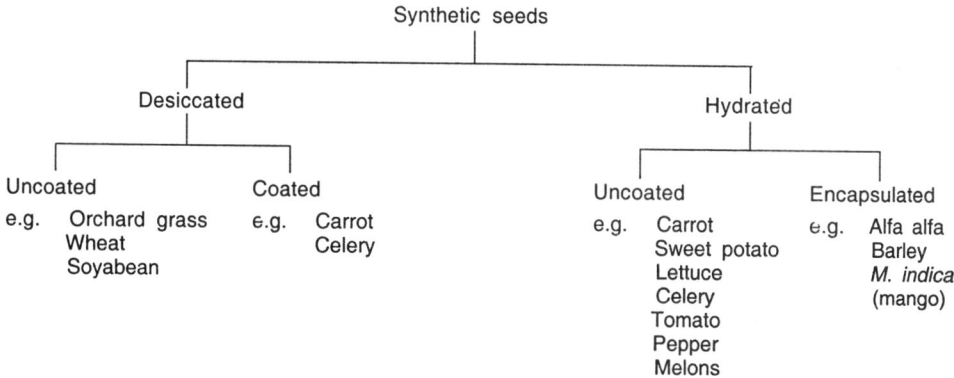

The necessary components of synthetic seeds depend on whether they are hydrated or desiccated. Various characteristics of hydrated and desiccated somatic embryos used for synthetic seed formation are given in Table 17.2.

Table 17.2. Synthetic seed components (Ramawat, 2000)

Component	Hydrated	Desiccated
High quality somatic embryos	Essential	Essential
Matured somatic embryos	Essential	Essential
Dormant or quiescent somatic embryos	Not required	Essential
Protective coating	Beneficial	Beneficial
Germination arresting reagents	Essential	Not required

The desiccated synthetic seeds are produced from somatic embryos either naked or encapsulated in polyoxy ethylene glycol (Polyox) followed by their desiccation. Desiccation can be achieved either slowly over a period of one or two weeks sequentially using chambers of decreasing relative humidity, or rapidly by unsealing the petri-dishes and leaving them on the bench overnight to dry. Such types of artificial seeds are produced only in plant species whose somatic embryos are desiccation-tolerant. The polyox coating provides embryo protection by preventing the lethal embryo desiccation. The encapsulated embryos, like true seeds must retain a critical level of moisture in order to remain viable.

On the other hand, hydrated synthetic seeds are produced in those plant species where the somatic embryos are recalcitrant and sensitive to desiccation. Hydrated synthetic seeds are produced by encapsulating the somatic embryos or somatic propagules in hydrogel capsules.

17.3. ENCAPSULATION OF SYNTHETIC SEEDS

In vitro somatic propagules (embryoid or shoot buds) develop from callus tissue and their induction is initiated by supplementing the medium with auxin and cytokinins in proper ratio. Such seeds are contaminated with microbes and desiccate quickly, when they are subjected to field conditions. Therefore, to get rid from this problem, they are encapsulated by a protective gel like PEG or calcium alginate. These encapsulated embryoids can resist unfavorable field conditions without desiccation. These seeds so developed behave like a true seed and are used as a substitute of natural seeds. They can also be sown directly in the greenhouse or in fields.

Kitto and Janick (1982, 1985) reported that the polyoxyethylene is the best coating material to produce encapsulated desiccated syn seeds. Because it is readily soluble in water and dries to form a thin film does not support the growth of microorganisms and is non-toxic to the embryo.

Redenbaugh *et al.* (1984) developed a technique for hydrogel encapsulation of individual somatic embryos of alfaalfa to produce encapsulated hydrated syn seeds. Since then encapsulation in hydrogel remain to be the most studied method of artificial seed production. A number of substances like potassium alginate, sodium alginate, carrageenan, agar, gelrite ™ are used. The main features that make sodium alginate a choice hydrogel material include (a) Easy encapsulation with $CaCl_2$ through an ion-exchange reaction, (b) Biologically non-damaging (c) Bio-degradable, (d) Easy, universal and availability and (e) low price.

A variety of hydrogels that have been employed in the hydrated syn seed production are given in Table 17.3.

Table 17.3. Commonly used hydrogels for artificial seed production

Hydrogel	Working concentration (w/v %)	Complexing agent	Working concentration (mM)
Sodium alginate	0.5–5.0	Calcium chloride	30–100
Sodium alginate +	2.0–4.0	Calcium chloride	30–100
gelatin	5.0–6.0		
Carrageenan +	0.2–0.8	Potassium chloride or Ammonium chloride	500
Locust bean gum	0.4–1.0		
Gelrite™	0.25–0.40	Lowering of temperature	

Two methods have been used to coat artificial seeds. (A) Gel complexation via dropping technique and (B) Molding.

In the former method, the somatic embryos are served from the suspension cultures and then mixed with sodium alginate gel in amounts appropriate for the maturation and plantlet formation worked out for the species in question. Alternatively, axillary (apical) or adventitious buds are excised and trimmed to smallest possible size and mixed with sodium alginate. The suspension of somatic propagule and alginate is then dropped into calcium chloride or calcium nitrate solution (containing the growth supplements) where ion-exchange reaction occurs and sodium ions are replaced by calcium ions forming calcium alginate beeds or capsules surrounding the somatic propagule. The size of the capsule is controlled by varying the inner diameter of the pipette nozzle. Hardening of the calcium alginate is modulated with the concentrations of sodium alginate and calcium chloride as well as the duration of complexing. The beads are then allowed to stand in calcium chloride solution for 30-60 minutes, sieved, washed with water and plated on suitable substratum.

Alternatively, the propagule can be mixed in a temperature dependent hydrogel like gelrite, placed in the well of a microtiter plate and gelled to form capsules by lowering the temperature.

However, because of the rapid drying and the stickiness of the alginate capsules, a hydrophobic coating is required for mechanical handling. Moreover, these hydrated capsules are more difficult to store because of the requirement of embryo respiration. Coating the capsules with Elvax 4260 copolymer (ethylene vinyl acetate acrylic tropolymer, Du pont, USA), which is suitable for producing a slow-drying, non-tacky coating which allows embryo conversion can offset these problems.

17.3. MECHANIZATION OF THE PROCESS OF SYNSEED PRODUCTION

For horticultural and agronomic crops millions to billions of somatic propagules are required for large-scale commercialization. This can be met only with mechanization of the procedure of synseed production. Mechanization can be done in two stages.

A. Bioreactor development for somatic embryos production

B. Mechanization or Automation of encapsulation process.

Bioreactor technology has to be improved as somatic embryos on agar plates are found better with better conversion frequencies than in liquid cultures. This technology has been used in alfa alfa, celery, carrot and lilies. The air lift column / bioreactor with cell immobilization is a superior cultivation system identified as good for embryo culture.

Automate encapsulation process is the recent and quick method for artificial seed production. An encapsulation machine can be used successfully to encapsulate somatic embryos, e.g., for alfa alfa. However more efforts and improvement is required for mechanical encapsulation to accommodate various types of somatic embryos. A method to encapsulate somatic propagules is given in simplified and schematic manner in Fig. 17.2 and 17.3. Somatic embryos move in a vibrating bowl, fall one by one and get encapsulated in sodium alginate gel. At various points sensors are incorporated to monitor the embryo movement, air flow and gel formation.

The artificial seed systems coupled with artificial intelligence and microcomputer systems like the most advanced robots which can mimic the motions and functions of a living being (i.e., automated encapsulation) would tremendously increase the efficiency of encapsulation and production of artificial seeds, and revolutionize the plant propagation method in the years ahead (Fig 17.4).

17.4. ADVANTAGES OF ARTIFICIAL SEEDS

Artificial seed technology offers following advantages :

1. Synseeds offer rapid and large-scale multiplication with economy of space, nutrients, labor, which lead to better cost-benefit ratio.

2. It is an effective delivery system that provides viable alternative to presently employed high cost incurring vegetative propagation techniques.

3. Direct sowing in the fields of seed-size propagule would by-pass the acclimatization step require during transplantation in a conventional micropropagation procedure.

4. When compared to large plantlet (few cm) the small size (few mm) of synseeds in very convenient for transfer of propagules from laboratory to land, for storage, shipping and planting in fields.

5. The encapsulation coating of synseeds acts as a carrier for beneficial supplements such as plant growth regulators, microorganisms, pesticides, fertilizers and nutrients.

6. Artificial seeds have particular relevance in the propagation of hand-pollinated hybrids, elite germplasm and genetically engineered hybrids with sterility or meiotically unstable genotype combinations.

7. Synseeds are breed true propagules.

8. Synseeds can be produced within a short time where as the natural seed production by plant is a

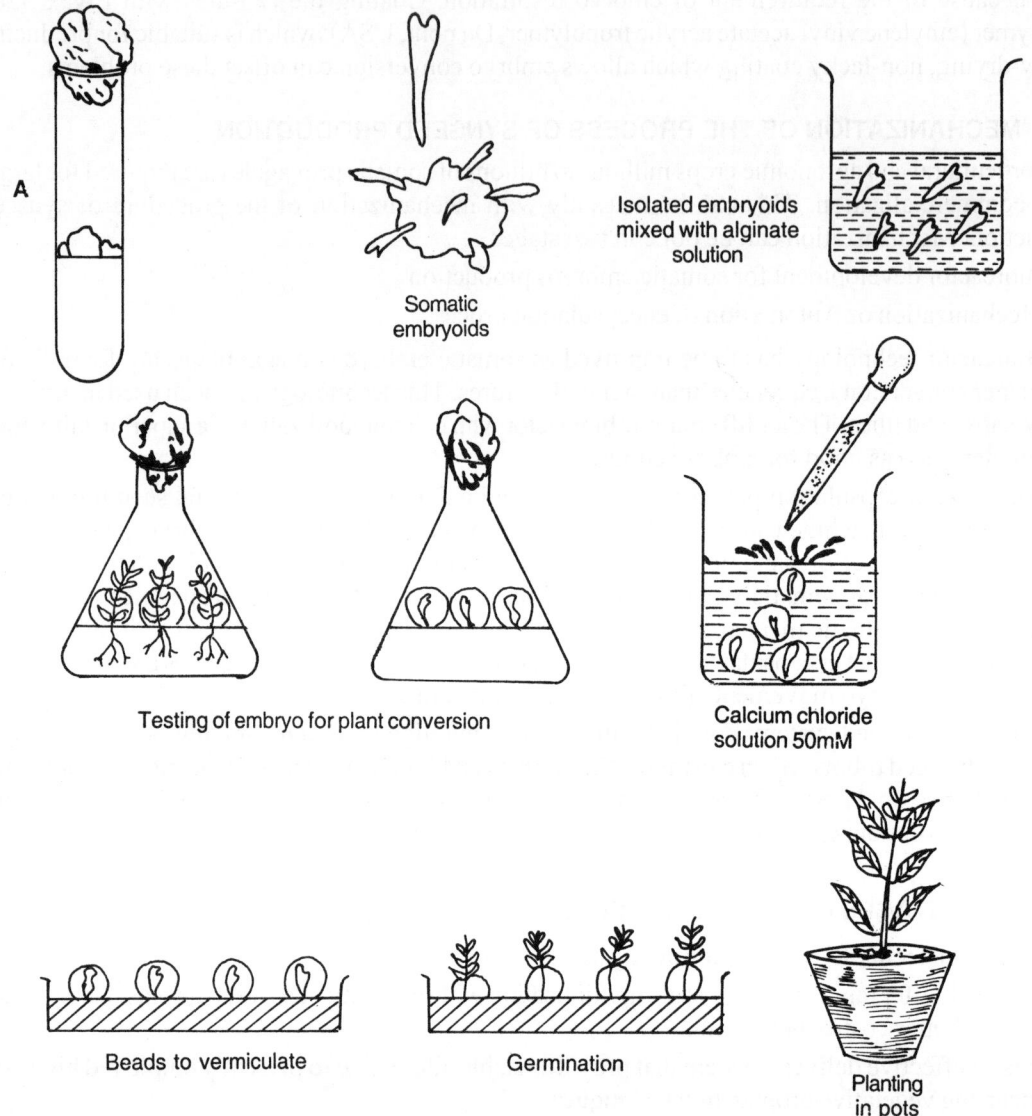

Fig. 17.2. Diagrammatic presentation of procedure of synthetic seed production and plant conversion. A-Callus.

time consuming and complex process and breeders have to wait for a long time for development of new varieties.

9. Artificial seeds can be produced at any time and in any season of a year.
10. By means of synthetic seeds the dormancy period can be reduced to a great extent, there by shortening the life cycle of a plant.
11. They are useful in preserving germplasm.
12. As an academic tool, artificial seeds offer comparative aid for better understanding of zygotic embryogony, role of endosperm during embryo conversion and seed coat formation.

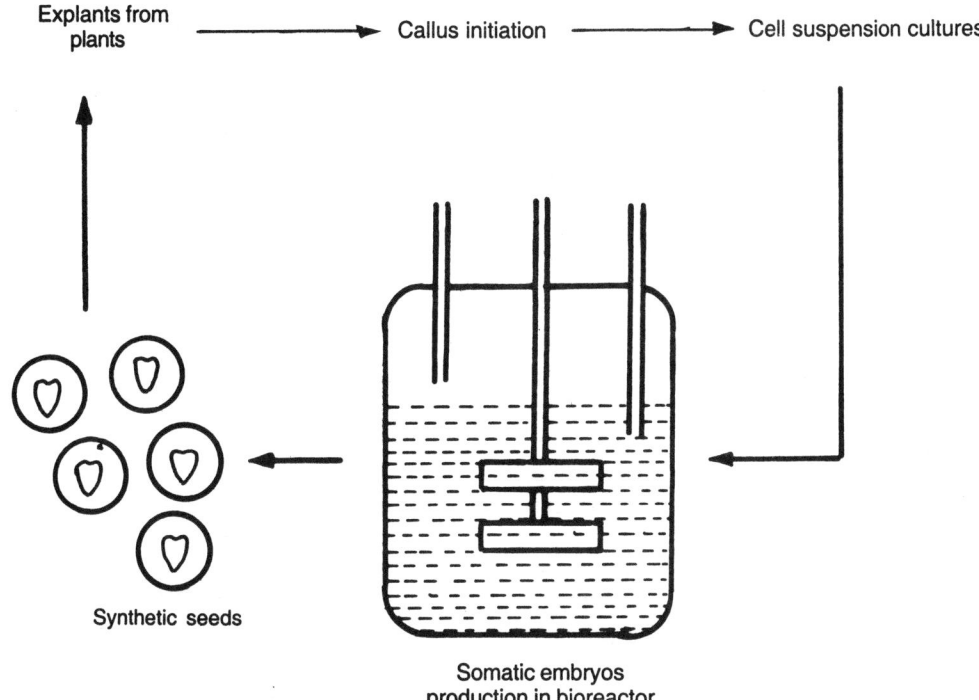

Explants from plants ⟶ Callus initiation ⟶ Cell suspension cultures

Synthetic seeds

Somatic embryos
production in bioreactor

Fig. 17.3. Synthetic seeds production using a bioreactor on large scale.

17.5. LIMITATIONS

Although synthetic seeds research is being carried out in many laboratories and seems promising for propagating a number of plant species, practical implementation of the technology is constrained due to the following reasons.

1. The principle limitation for commercialization of synseeds has been the quality and quantity of somatic embryogenesis, limited production of viable micropropagules useful in syn seed production.

2. Anomalous and asynchronous development of somatic embryos that makes the artificial seeds inefficient for germination and conversion into normal plants, is one of the major bottlenecks for synseed production.

3. Lack of dormancy and stress tolerance in somatic embryos that limit the storage of synseeds.

4. Poor conversion of even apparently normally matured somatic embryos and other micropropagules into plantlets that limit the value of the synthetic seeds and ultimately the technology itself.

17.6. APPLICATIONS OF SYNTHETIC SEEDS

1. Micropropagation through artificial seeds may be commercially exploited on a large scale by generating millions of plants in a few days and this may become a profitable multibillion rupees industry in near future. Most likely, the technology will first be used with hybrid vegetable crops such as celery or with high-value flower and ornamental species.

2. Artificial seed technology can be very useful for the propagation of a variety of crop plants, especially crops for which true seeds are not used or not readily available for multiplication (e.g. potato) or the true seeds are expensive (e.g. Cucumber and Geraniums), hybrid plants (e.g.

Chapter 11

Immobilization of Plant Cells

INTRODUCTION

Plant cell culture has been for some time considered as an alternative method for the production of flavors, colors and pharmaceuticals to their extraction from plants. Characteristics of plant cell cultures, such as slow growth, the compounds produced should be of high value (\$ 500-1000kg $^{-1}$) and low volume. One of the major limiting factors in the development of a commercial production system using plant cell culture has been the production cost of phytopharmaceuticals. The use of high biomass levels for extended periods would be one method of increasing productivity and hence reducing the costs. This can be achieved by the immobilization of plant cells.

The immobilization of enzymes and cells has received increasing attention, and used to produce amino acids and carbohydrates. The immobilization of microbial cells is not new, it has known for sometime that cells will adhere to many surface in nature i.e. polymers.

Immobilization is the newest culture technology of plant cell, and considered as to be the most 'natural'. It has been defined as a technique, which confines to a catalytically active enzyme or to a cell within a reactor system and prevents its entry into the mobile phase, which carries the substrate and product. The first successful immobilization of plant cells was reported by Brodelius *et al.* (1979) and they entrapped *Catharanthus roseus* and *Daucus carota* cells in alginate beads. Following success with enzymatic and microbial process, immobilization has been suggested as a strategy to enhance the overall productivity of secondary metabolite in plant cell culture. The ability to immobilize plant cells has been reported for a large number of plant cells and protoplasts by using a variety of polymers (Table 11.1). Immobilization of plant cells has been used for a wide range of reactions, which can be divided into three groups. (1) Biotransformation or bioconversion, (2) Synthesis from precursor and (3) the de novo synthesis of compounds.

11.1. ADVANTAGES OF PLANT CELL IMMOBILIZATION

- Retention of biomass enables its continuous reutilization as a production system, a definite advantage with slow growing plant cells e.g. *Papaver somniferum* have remained stable and active for up to six months.
- High biomass levels: The immobilization of cells allows the use of a higher biomass level compared to cell suspension culture, because of the limitation of mass transfer and settling, e.g. Bead

hybrid rice) and many vegetatively propagated plants which are more prone to infections e.g. Day lily, Garlic, Potato, Sugarcane, Sweet potato, Grape and Mango.

3. The artificial production of seeds has already been obtained successfully in *Zea mays, Apium graveolens, Daucus carota, Lactuca sativa, Medicago sativa, Brassica* spp, *Gossypium hirsutum, Valerina* sp., *Santalum* sp. etc.

4. For development of plants for breeding purposes, e.g. alfa alfa.

5. In forestry for transplanting improved/ selected material reduce the cost of breeding, e.g. used in eastern white pine, Black spruce, white spruce, Red spruce, Norway spruce, Japanex larch, European larch, and hybrid larch.

6. Germplasm conservation of endangered species through cryopreservation of synseeds is another potential application of artifical seeds in near future.

Synthetic seed technology is gradually moving towards the commercial propagation of high value crops. However, there is a great need for refinement of this technology by the tackling of certain technical problems such as the need to produce high-quality and high fidelity somatic embryos, and to avoid the genetic instability and variability of tissue culture derived plant. Further, the understanding about the storage, transport, handling, growth habit and harvest index of artificial seeds is essential. Similarly, the efforts to increase the output of plants/ gram of callus tissue input and conversion frequency of artificial seeds are also needed. If all these problems are rectified with technical progress, no doubt this novel method can become valuable tool in agriculture to propagate crop species. The two important factors in the production of synseeds are :

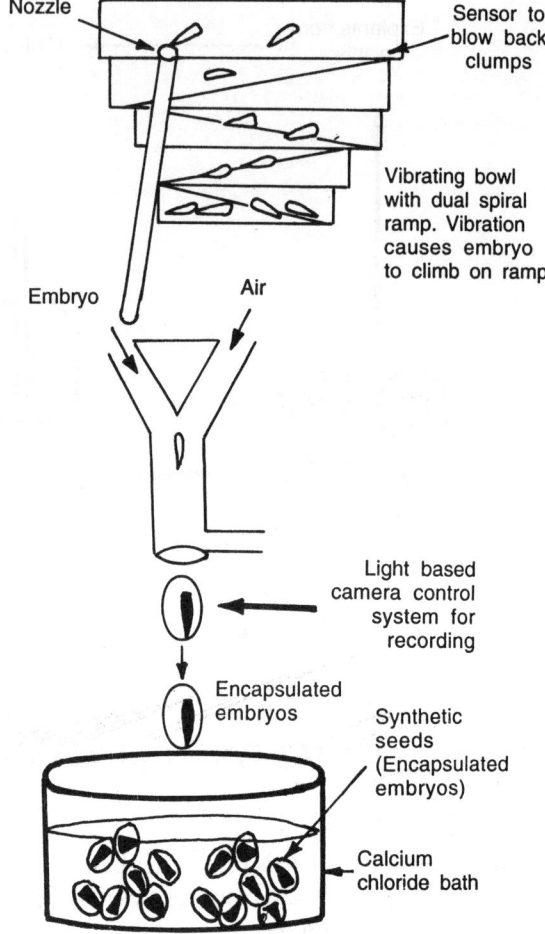

Fig. 17.4. Schematic presentation of automation unit for the production of synseeds using sodium alginate (after Plant Genetics Inc. USA).

A) Embryo to Plantlet Conversion

Embryo to plantlet conversion frequency is defined as percentage of the somatic embryos that produces complete plant with normal phenotype 'Conversion' is the term used instead of 'Germination' in context to artificial seeds. Green house conversion of somatic embryos depends on quality of somatic embryo and watering system. Watering should be perfect and adequate, neither high, nor dry. Good quality somatic embryos are characterized by a dipolar entity with well-marked root and shoot axis, fully developed cotyledons and lack of callusing. Such embryos undergo following sequence of events during plantlet conversion.

A. Radicle elongation

B. Development of vigorous root system with secondary and tertiary branching

C. Elongation of shoot meristem

D. Expansion of cotyledons and appearance of atleast two true leaves without hypocotyl swelling or callusing

E. Production of a plant with normal growth and phenotype.

For commercial applications, somatic embryos must germinate rapidly and should be able to develop into plant atleast at rates and frequencies more are less similar if not superior to true seeds. To achieve conversion of somatic embryos into plantlets and to overcome deleterious effects of recurrent somatic embryogenesis it is necessary to provide optimum nutritive and environmental conditions.

Maltose has been found valuable for improving alfa alfa somatic embryo conversion (Redenbaugh and Walker, 1990). Addition of sucrose in the medium is necessary for viability of somatic embryos and their subsequent development, maturation and germination in many plant species.

In many plant species the somatic embryos have been found to be sensitive to desiccation. Nevertheless, desiccation and subsequent rehydration have been found useful in inducing a high frequency conversion of somatic embryos into plantlet in some species.

Gradual drying of alfa alfa somatic embryos with progressive and linear loss of water gave better response and improved the quality of embryos in comparison to uncontrolled drying. Desiccation improved the germination frequency in soybean also.

B) Embryo Hardening

Another important aspect in synthetic seed technology is embryo 'hardening'. Hardening imposes biochange on precocious germination and increase embryo survival upon desiccation. This is an important consideration for increasing the shelf life of these somatic embryos. High inoculum density, elevated sugar level, chilling and abscisic acid (ABA) have been identified to be effective embryo hardening agent.

Abscisic acid was shown to improve desiccation tolerance of encapsulated embryos of celery and soyabean (Rangaswamy, 1986). Kim and Janick (1987) reported that positive influence of ABA on embryo hardening in celery was further improved when used along with 10 mM protein. Desiccation tolerance in alfa alfa somatic embryos can be induced by external stimuli such as ABA, exposure to cold, heat, water and osmotic stress and sub-lethal levels or increasing the sucrose content in the medium (Senaratna et al., 1995). The spruce somatic embryos matured in presence of PEG and ABA were very tolerant to low moisture levels (Fowke and Attree, 1996). The moisture content of such seeds was reduced to 10% by desiccation. These embryos were stored at –20°C for a year and there after successfully germinated following inhibition with no loss in viability.

The beneficial effects of high sucrose concentration are attributed to its osmotic action. Osmotic potential imposed by 12% sucrose (0.35 M) bring about an increase in endogenous abscisic acid level, resulting into buffer embryo hardening.

Chilling treatment also favors abscisic acid accumulation. Besides this, low temperatures improves embryo hardness by enhancing carbohydrate accumulation (Fujii et al., 1990)

The main aim of all the above pre-treatments is to impart "Quiescence" to the encapsulated embryos (Smith and Drew, 1990).

17.7. THE VARIOUS STEPS OF COMMERCIAL ARTIFICIAL SEED PRODUCTION

The following steps are needed for commercial synseed production.

1. Production of large-scale embryogenic tissue from transformed cells or tissue.
2. Large scale production of synchronous somatic embryos.
3. Maturation of somatic embryos.
4. Non-toxic encapsulation or coating process.
5. Artificial endosperm/megagametophyte, depending on species.
6. Storage capability of artificial seeds.
7. High frequency, direct green house/nursery field conversion, depending on production requirements.
8. Low genetic and epigenetic variation.
9. Appropriate expression of engineered trait.

REFERENCES

Ahuja, P.S., Mathur J., Lal N., Mathur A., Mathur A.K. and Kukreja A.K. (1989). Towards developing artifical seeds by shoot bud encapsulation. In tissue culture and biotechnology of medicinal and aromatic plants (Kukreja A.K., Mathur A.K., Ahuja P.S. and Thakur P.S. (eds.)). Paramount Publishing House pp. 22-28.

Akhtar, N (1997). Ph.D. thesis, Banaras Hindu University, Varanasi.

Ara, H., Jaiswal U. and Jaiswal V.S. (1999). *Plant Cell Rep.,* 19, 166-170.

Ara H., Jaiswal U. and Jaiswal V.S. (2000). Synthetic seed : Prospects and limitation. *Current Science,* 78(12), 1438-1444.

Arya K.R., Beg M.V. and Kukreja A.K. (1998). *Indian J. Exp. Biol.,* 36, 1161-1164.

Attre S.M., Pomeroy M.K. and Fowke L.C. (1994). *Plant Cell Rep.,* 13, 601-606.

Bapat V.A. (1993). In : synseeds (Rendebaugh K (ed)), CRC Press, Boca Raton, pp. 381-407.

Bapat V.A. and Rao P.S.(1988). Sandalwood plantlets from synthetic seeds, *Plant Cell Rept.,* 7, 434-436.

Bapat V.A., Mhatre M. and Rao P.S. (1987). *Plant Cell Rep.,* 6, 393-395.

Capuano G., Piccioni, E. and Standardi A., (1998). *J. Hortic. Sci. Biotechnol.,* 73, 299-305.

Datta K.B., Kanjilal B. and Sarker D. (1999). *Curr Sci.,* 76, 1142-1145.

Fowke L. and Attree S. (1996). *Plant Tissue Cell Biotechnol.,* 2, 124-130.

Fujii J.A.A., Slade D., Olsen R., Ruzin S.E. and Redenbaugh K., (1990). *Plant Sci.,* 72, 93-100.

Furamova M., Sowinaska D. and Pietrosiuk A. (1991). Carum Carvi L. (Caraway) : *In vitro* culture embryo genesis and the production fo aromatic compounds. In : Bajan YPS (ed.) Biotechnology in Agriculture & Forestry, Vol. 15, Medicinal and Aromatic Plants III, Springer-Verlag, New York, Heidelberg, pp. 176-192.

Ganapathi T.R., Bapat V.A. and Rao P.S. (1994). *Biotech Techniques,* 8(4), 239-244.

Ganapathi T.R., Suprasanna P., Bapat V.A. and Rao P.S. (1992). Propagation of banana through encapsulated shoot tips. *Plant Cell Rep.,* 11, 571-575.

Ghosh, B. and Sen S. (1994). *Plant Cell Rep.,* 13, 381-385.

Gray D .J., Conger B.V. and Songstad D.D. (1987). Desciccated quiscent somatic embryos of orchard grass for use as synthetic seds. *In Vitro Cell Dev. Biol.,* 23, 29-33.

Gray D.J. and Purohit A. (1991). *Crit. Rev. Plant. Sci.,* 10, 33-61.

Gupta P.K. and Durzan D.J. (1987). Biotechnology of somatic polyembryogenesis and plantlet regeneration in lablolly pine, *Biotechnology,* 4, 147-151.

Gupta R.K., Shaw D. and Durzan D.J. (1987). In: Cell and Tissue Cultrue in Forestry (Bonga J.M. and Durzan D.J. (eds.)) Martinus Mijhoff Publishers, Dordrecht, pp. 101-108.

Huang M.J. and Cheng Z. (1990). Preliminary studies on artifical seeds of eucalyptus. In : Studies on artifical seeds of plant (Li Q-Q (ed.)) Peking Univ. Press, Peking, China, pp. 139.

Janeiro, L.V. Ballester A. and Vieitez A.M. (1997). *Plant Cell Tissue Org. Cult.,* 51, 119-125.

Kamada H., (1985), Artificial seed. In Practical Technology on the Mass Production of Clonal Plants (Tanaka R. (ed)) CMC Publisher, Tokyo, Japan, 48.

Kim Y.H. and Janick J. (1987). Production of synthetic seeds of celery. *Hort. Sci.,* 22, 89.

Kim Y.H. and Janick J. (1990). Synseed Technology : Improving desiccation tolerance of somatic embryos of celery. *Acta. Hort.,* 280, 23-28.

Kitto S.K. and Janick, J. (1982). *J. Hort. Sci.,* 17, 488.

Kitto S.L. and Janick J. (1985). Production of synthetic seeds by encapsulating asexual embryos of carrot. *J. Amer. Soc. Hort. Sci.,* 110, 277-82.

Kumar, U. (2000) Synthetic seeds for commercial crop production, studies in Biotechnology series No. 3. Agro Botanica Publishers & Distributros, Bikaner.

Lakshmana Rao P.V. and Singh B. (1991). *Plant Cell Rep.,* 10, 7-11.

Malemngaba H., Roy B.K., Bhattacharya S. and Deka P.C. (1996). *Indian J. Exp. Biol.,* 34, 801-805.

Mathur A.K. and Ahuja P.S. Artificial seeds : Some emrging Trends, CIMAP Publication No. 1080.

Mathur J, Ahuja P.S., Lal N. and Mathur A.K. (1989). *Plant Sci.,* 60, 111-116.

Muralidharan E.M. and Mascarenhas A.F. (1995). In : somatic embryogenesis in woody plants (Jain S, Gupta P and Newton R (eds.)) Kulwer Academic Publishers, Dordrecht, pp. 101-108.

Murashige T., (1977). Plant cell and organ culture as horticultural practices. *Acta. Hort.,* 78, 17-30.

Onay A, Jeffree C.E. and Yeoman M.M. (1996). *Plant Cell Rep.,* 15, 723-726.

Padmaja G., Reddy L.R. and Reddy G.M. (1995). *Indian J. Exp. Biol.,* 33, 967-971.

Piccioni, E. (1997). *Plant Cell Tissue Org. Cult.,* 47, 255-260.

Picioni E. and Standardi A. (1995). *Plant Cell Tissue Org. Cult.,* 42, 221-226.

Ramawat K.G. (2000). Synthetic seed technology In: Plant Biotechnology, S. Chand & Co. Ltd., pp. 154-159.

Rangaswamy N.S. (1986). Somatic embryogenesis in angiosperm cell, tissue and organ cultures. *Proc. Ind Acad. Sci.,* 96, 247-271.

Rao, P.V. and Singh B. (1991) Plantlet regeneration from encapsulated somatic embryos of hybrid *Solanum melongena. Plant Cell Rep.,* 10, 7-11.

Redenabugh K., Nichol J., Kossler M.E. and Paasch B. (1984). *In vitro,* 20, 256-257.

Redenbaugh K, Fujii J. and Slade D. (1988). Encapsulated plant embryos. In : Advances in Biotechnological Process Vol. 9 (Mizrahi A. (ed.)) Alan R., Liss Inc., New York, pp. 225-2248.

Redenbaugh K. and Walker K. (1990). In: Plant Tissue Culture : Applications and Limitations (Bhojwanis (ed.)). Elsevier, Amsterdam, pp. 102-135.

Redenbaugh K., Paasch B., Nichol J., Kossler M., Viss P. and Walker K. (1986). Somatic seeds : Encapsulation of asexual plant embryos. *Biotechnology,* 4, 797-801.

Redenbaugh K., Viss P., Slade D. and Fujii J. (1987) Scale-up : Artificial seeds, In: plant tissue and cell cultures (Green C. Somers D, Hacxett W. and Biesboer D. (eds.)). Alan R. Liss Inc. New York, pp. 473-493.

Refouvelet E., Le Nours S., Tallor C. and Daguin F. (1998). *Sci. Hortic.,* 74, 233-241.

Senaratna T., Sexena P.K., Rao, M.V. and Afela J. (1995). *J. Plant Cell. Rep.,* 14, 375-379.

Sharma A, Tandon P. and Kumar A. (1992). *Indian J. Exp. Biol.,* 30, 747-748.

Sharma T.R., Singh B.M. and Chauhan R.S. (1994). *Plant Cell Rep.,* 13, 300-302.

Singh F. (1991). *Lindleyana,* 6, 61-64.

Smith M.K. and Drew R.A. (1990). *Aust. Pl. Physiol.,* 17, 267-286.

Suprasanna P., Ganapathi T.R. and Rao P.S. (1995). *Journal of Genetics & Breed,* 49, 9-14.

Kabuchi H. (1992) Artificial seeds. In Seed of Techniques, eds Rita Y.U. (Ed) Seed of Food Plants Theory (eds) CMC Publishers, Tokyo, Japan 16.

Kim T. and Janick (1989) under two of endosperm nutrient effect. J Amar Soc 114: 1165.

Kong H. and Yeoul J. (1991) sense of embryolous and germination frequency of somatic embryos of ....

Amer Soc J Short 7: (1989) ....

Kruse SA. and Janic J. (1991) carrot ....
....

....
....

....
....

Chapter 18

Transgenic Plants

INTRODUCTION

Conventionally, the genetic variation necessary for crop improvement is generated through hybridization, mutagenesis and polyploidy. More recently, biotechnological approaches have become available for creating genetic variation because plant cells *in vitro* generates a considerable amount of genetic variation, called somaclonal variation (Chapter 14), from which useful variants can be isolated; some of which have been released as commercial varieties. The somatic hybridization (Chapter 15) which often yields hybrid plants between sexually incompatible species. The micropropagation will be useful for large scale production of phenotype plants (Chapter 16).

However, the most potent biotechnological approach for the transfer of specifically constructed gene assemblies through various transformation methods, constitutes the genetic engineering. The plants obtained through genetic engineering contain a gene usually from an unrelated organism, such genes are called transgenes, and the plants containing transgenes are called as transgenic plants. Genetic transformation can be defined as the transfer of foreign genes isolated from plants, viruses, bacteria into a new genetic background. The first step in gene transfer technology is to select cells that are capable of giving rise to whole transformed plants. Transformation without regeneration and regeneration without transformation are of no use or limited use. The targetted cells for gene transformation are cultured cells or protoplasts, meristem cells from embryos, pollens, zygote and cells from immature embryos, shoot and flowers.

18.1. GENERAL FEATURES OF GENE CONSTRUCTS OF TRANSFORMATION

Gene construct is the assembly of various DNA sequences designed for easy identification of the construct and its efficient expression in transgenic individuals. It also contain a reporter gene for an easy identification or selection of the transgenic individual.

18.1.1. Plant Gene

A typical plant gene has the following regions beginning with the 5' end (i) protomer (transcription initiation) (ii) Enhancer/silencer (regulation), (iii) transcriptional start, (iv) leader sequence, (v) initiation codon, (vi) exans (vii) introns, (viii) stop codon(s), (ix) untranslated region, (x) poly (A) tail.

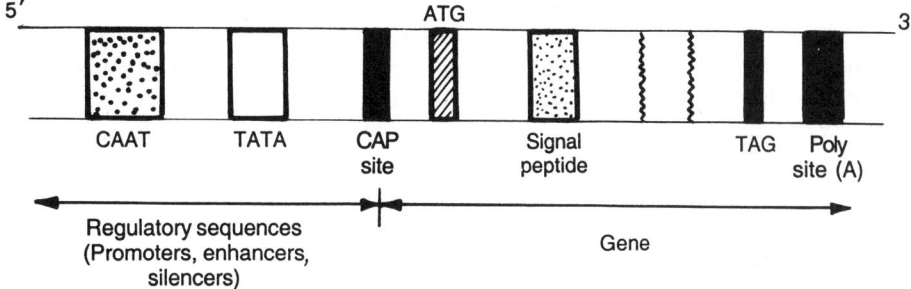

Fig. 18.1. A Typical Plant Gene.

The promoter provides the site for binding of RNA polymerase, and is involved in transcription initiation. The enhancer/silencer sequences, make the regulation of gene action. A typical gene organization is shown in Fig. 18.1.

18.1.2. Promoters/Enhancers

It is needless to emphasize that for an efficient expression in plant cells, foreign genes must have an appropriate promoters, 5' leader and 3' terminator sequences. A suitable enhancer sequence will also be needed if the gene is required to be expressed either in a specific tissue, during a specific developmental stage or in response to a specific stimulus. A list of promoter sequences have been used to drive genes in plant cells rare shown in Table 18.1. Among all the promoters CaMV 35S promoter is very commonly empolyed in transformation experiments among monocots.

Table 18.1. Some of the promoters used for driving the expression of transgenes (Adopted from Singh, 1998)

Promoter	Source	Remarks
35 S	CAMV 35S RNA gene	High activity; most commonly used
35S + Adh1-I 1	35S Promoter + first intron of maize Adh I gene	Enhanced promoter activity; constitutive
35S + sh 1-I 1	35S promoter + first intron of maize shrunken 1 gene	Better than 35S + Adh-1-I 1 in monocots; constitutive
19 S	CaMV 19S RNA gene	Constitutive; moderate activity
nos	*Agrobacterium* nopaline synthase gene	Moderate activity
ocs	*Agrobacterium* octopine synthase gene	Moderate activity
mas	*Agrobacterium* mannopine synthase gene	Moderate activity
Gene VI	CaMV gene VI encoding the matrix protein	Constitutive expression
Adh 1	Promoter of alcohol dehydrogenase gene of maize	Moderate activity in cereals; anaerobic expression
Emu	Modified from Adh1 promoter and its first intron	Moderate activity in cereals; anaerobic expression
Ubi1 + Ubi1-I 1 (or 16)	Maize ubiquitin 1 gene promoter + its first (or sixth) intron	High activity in cereals; constitutive
Vicilin promotoer	Pea vicilin storage protein gene	Seed-specific promoter
PHA-L	*Phytohaemagglutinin* gene of *P. vulgaris*	Strong, seed-specific activity

Expression of many genes is confined to specific tissues and/or induced by specific stimuli and such genes are known as tissue specific or stimulus-responsive genes respectively. This expression of genes is due to certain DNA sequences, called enhancers/silencers. An enhancer is defined as a DNA sequence which enhances the activity of promoter of a gene, while DNA sequences suppressing promoter activity are known as silencers. They only activate/inhibit the promoter activity but themselves do not have any activity e.g. Adh 1 and shrunken 1 genes of maize act as enhancer of 35 S promoter of CaMV, the Adh 1 enhancer sequence suppress Adh 1 expression under aerobic condition.

18.1.3. Reporter genes

The genes need to be tagged with another gene (usually bacterial gene, plant/animal genes) are called as reporter genes, whose expression is easily detected either through enzyme assays (scorable reporter genes) or through expression of resistance to a toxin (selectable reporter genes e.g. nos (nopaline synthase from Agrobacterium) etc. (Table 18.2).

Table 18.2. Selectable marker genes for plant transformation (adopted from Yoder and Goldsbrough, 1994)

Selectable gene	Gene product	Source	Selection
nptll	neomycin phosphotransferase	Tn5	Kanamycin, G418 paromomycin, Neomycin
ble	Bleomycin resistance	Tn5 and Streptoalloteichus hindustanus	Bleomycin Phleomycin
dhfr	Dihydrofolate reductase	Plasmid R67	Methotrexate
cat	Chloramphenicol acetyl transferase	Phage p1 Cm	Chloramphenicol
aphIV	Hygromycin phosphotransferase	E. coli	Hygromycin B
ept	Streptomycin phosphotransferase	Tn5	Streptomycin
aacC3	Gentamycin-3-N	Serratia	Gentamicin
aacC4	Acetyltransferase	Marcescens; Klebsiella pneumoniae	
bar	Phosphinothricin acetyl transferase	Streptomyces hygroscopicus	Phosphinothricin bialophos
Epsp	5-Enolpyruvylshikimate-3-phosphate synthase	Petunia hybrida	Glyphosate
bxn	Bromoxynil specific nitrilase	Klebsiella ozaenae	Bromoxynil
PsbA	Q_a protein	Amaranthus hybridus	Atrazine
FfdA	2, 4-D monooxygenase	Alcaligenes	2, 4-dichlorophenoxy acetic acid
dhps	Dihydrodipicolinate synthase	E. coli	S-aminoethyl, L-cysteine
a k	Aspartate kinase	E. coli	High concentrations of lysine and threonine
sul	Dihydropteroate synthase	Plasmid R46	Sulfonamide
Csr1-1	Acetolactate synthase	Arabidopsis thaliana herbicides	Sulfonylurea
Tdc	Tryptophan decarboxylase	Catharanthus roseus	4-Methyl tryptophan

18.2. GENE TRANSFER METHODS

Plasmids have a wide range of structures, they may be composed of RNA or DNA of single or double stranded nature in circular or linear. The size of plasmid varies from 1.5 kilobase to 1500 Kb.

They are defined as, self replicating double stranded, extra chromosomal DNAs maintained as independent molecule by most of the bacterial genera. The presence or absence of plasmid from a bacterial cell makes no difference to the bacterial cell. Plasmids are of two types.

a) Conjugative or transmistible plasmids eg. $_p$BR 322

b) Non-conjugative or nontransmistible plasmids e.g. Col EI.

Transgenic plants can be produced by two methods.

1. Vector mediated gene transfer (Indirect gene transfer)
   a) Bacterial vectors e.g. Agrobacterium
   b) Viral vectors e.g. *E. coli*
2. Direct gene transfer
   a) Direct gene transfer to protoplasts
      i) Chemical treatment (PEG)
      ii) Electrical treatment (Electroportion)
      iii) DNA delivery (liposomes)
   b) Direct gene transfer to cells and tissues
      i) Microprojectile bombardment (Biolistic or DNA particle gun)
      ii) Microinjection of DNA into cells and protoplasts.
   c) Macro injection of DNA into plants
   d) DNA uptake into imbibing zygotic embryos
   e) Transmission of pollen
   f) Fibre-mediated DNA delivery into plant cells

18.2.1. Indirect gene transfer (Vector mediated gene transfer)

Vectors are the DNA carriers into which 'foreign' DNA or genes of interest are inserted to make a recombinant DNA. Vectors along with this 'foreign' DNAs are then introduced into appropriate host cells. Vectors are divided into two categories.

a) cloning vectors (used for making millions of copies of DNA segment)

b) expression vectors (used for expression of cloned gene to produce the product).

Most vectors carry a reporter gene which allow recognition e.g. antibiotic resistance, restriction site, origin of replication.

18.2.1.1. *Ti and Ri plasmids of Agrobacterium*

Crown gall disease of higher plants is caused by infection of *Agrobacterium tumefaciens*, and is a neoplastic growth. This bacteria utilizes opines produced by host (tumour) cells as carbon and nitrogen sources. *A. tumefaciens* possess a large plasmid ($90 - 150 \times 10^6$ daltons) known as tumour inducing (Ti) plasmid. Transformation is associated with and accomplished by transfer of a stable, replicating portion of Ti plasmid into the plant cell.

(i) Structure of Ti plasmid (Fig 18.2)

18.2.2. Structure of T-DNA (Fig 18.3)

The T-DNA is defined, delimited by two 25 bp directly repeats its ends, the T-DNA borders any DNA and only DNA between these borders is transported to the plant cell. The right border is critical, T-DNA transfer must start at this ends. Transfer-DNA (T-DNA) is transferred and integrated into host genome during the infection. It brings about physiological and morphological changes due to expression of genes located on T-DNA. The different regions of T-DNA are :

i) An **onc** region, which consists of three genes (tms 1 and tms 2 representing shooty locus and tmr representing rooty locus) is responsible for the biosynthesis of phytohormones IAA (an auxin) and isopentyladenosine 5' monophosphate (a cytokinin).

ii) An OS region, which is responsible for synthesis of an unusual amino acid or sugar derivatives, collectively called as opines. These are low molecular weight nitrogen containing compounds e.g. octopine, nopaline, lysopine, histopine etc.

iii) 25 bp direct repeat sequence, which is essential for T-DNA transfer acting only in cis-orientation .

iv) Vir region - It is essential for transformation of cis or trans orientation. The vir region (35 kbp) is organised into six operons, namely Vir A, Vir B, Vir C, Vir D, Vir E and Vir G (Fig 18.4)

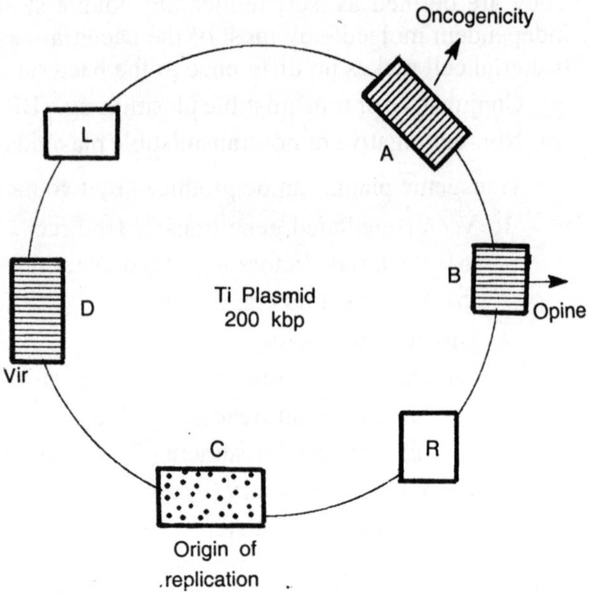

Fig. 18.2. Structure of Ti Plasmid. A - T-DNA responsible for tumor formation; B - responsible for replication; C - responsible for conjugation; D - responsible for virulence; L = left, R = right.

Vir genes are required for T-DNA movement. Vir-A is a product specific inner membrane protein that recognizes and is responsive to the plant phenolic compounds. Vir-A transduces information most likely by protein phosphorlyation to the product of Vir G.

Vir G : It can act as a transcriptional activator of it self and other loci of Vir region.

Vir C and *Vir D* : The products of Vir C and Vir D are involved in the generation and processing of T-DNA

Vir-B and Vir-E : The products of Vir-B and Vir-E are involved in forming most of the structure component that facilitates T-DNA movement.

Vir-H : Products of Vir H may allow the bacteria to survive in the presence of bactericidal or bacteriostatic plant compounds during the infection.

18.2.3. T-DNA transfer process

An early event in the T-DNA transfer process is the nicking of Ti plasmid between 3rd and 4th base of the bottom strand of each 25 bp repeat. The Vir D operon encodes an endonuclease that produces nicks in the border sequences. Then the initiation of DNA synthesis in 5'-3' direction (Fig18.5). The involvement of bacterial genome causes the synthesis and secretion of glucose, cellulose, fibrils and cell surface proteins. This is common physiological response in all soil bacteria and is involved with pathogenic characters.

The generation of the T-strand is the complex process of Agrobacterium mediated plant cell transformation (Fig. 18.5) this DNA must transverse the bacterial cell membrane, the bacterial cell wall, the plant cell wall and plant cell and nuclear membranes. Once inside the nucleus, the T-DNA, finally integrate stably into the plant cell genome. During this entire transit process, the T-DNA

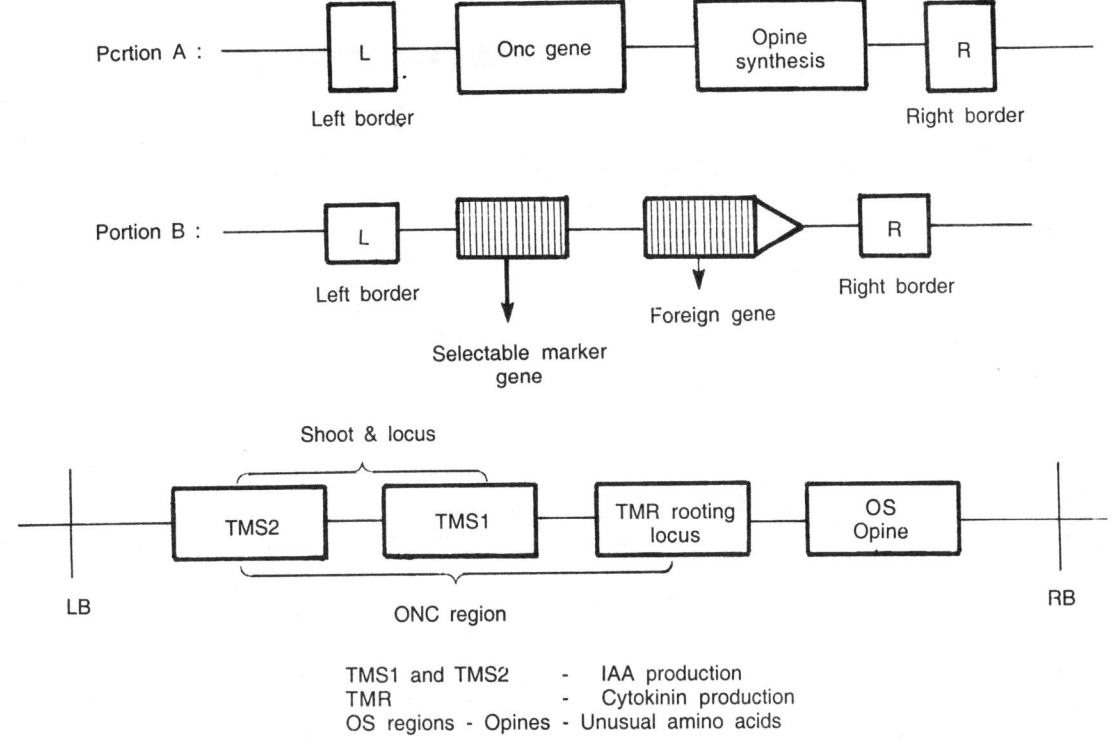

TMS1 and TMS2   -   IAA production
TMR   -   Cytokinin production
OS regions - Opines - Unusual amino acids

Fig. 18.3. Structure of T-DNA.

strand also must avoid degradation by nuclease. The T-DNA exists as a DNA protein complex. The T-DNA complex protects it and mediates its travel. The final step in the genetic transformation of plant cell is integration of T-DNA copy, presumably T-strand, into plant cell DNA. It has been argued that the T-strand might be converted to a double standard (ds) DNA prior to integration (Fig 18.6).

Thus by placing foreign genes into T-DNA region of Ti-plasmid, it is possible to clone (make copies) the introduced genes with the multiplication of plasmid residing inside the bacteria which is grown on a medium and with the multiplication of bacterial population, residing plasmid is also multiplied by this method. It is possible to exploit the natural ability of *Agrobacterium* to transfer new DNA into the plant genome.

18.2.4. Vectors based on Ti and Ri plasmid

The Ti plasmid or Ri plasmid cannot be used directly, because of the following properties.

  i) large size
 ii) Absence of unique restriction enzyme site
iii) Tumour induction

The tumour cells cannot develop into normal shoots, disarming of Ti plasmids or removal of tumour induction property is an essential step in designing useful vectors. This can be achieved by

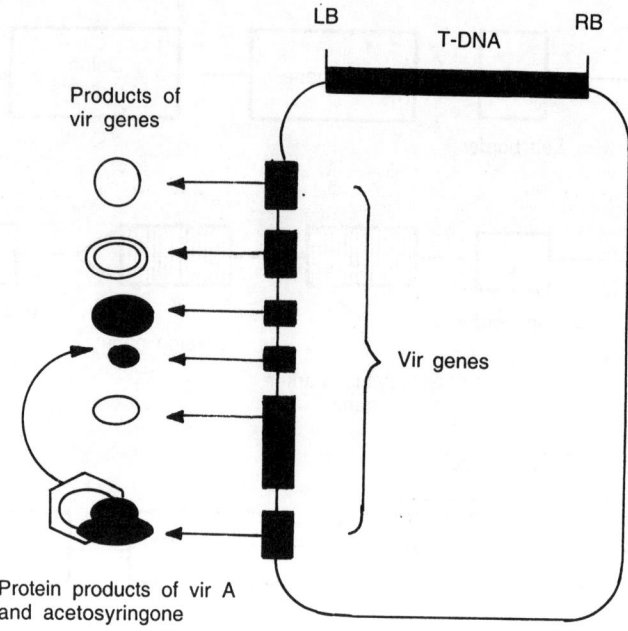

Fig. 18.4. Vir genes and products of vir genes.

replacement of tumour inducing genes in transfer DNA by selectable markers (eg. npt-II), providing resistance against antibiotics like kanamycin.

Promoters and polyadenylation signal isolated from nopaline or octapine synthase genes were used for expression of selectable markers. The other promoters such as CaMV35S and CaMV 19S can be used.

There are two types of Agrobacterium vectors that are currently used.

(i) Cointegrative vectors

Cointegrative vectors recombine, via DNA homology with an intermediate cloning vectors, which is used for manipulation and cloning of the gene in *E. coli*. *Agrobacterium* containing cointegrative vector and *E. coli* containing intermediate cloning vector are allowed to undergo conjugation, but the intermediate vector cannot replicate in *Agrobacterium*. So it has to transfer the marker genes as well as the DNA segment to the resident Ti plasmid (cointegrative vector) through recombination in the region of DNA homology. e.g. $_p$GV 3850 from a nopaline-type Ti plasmid, where almost all T-DNA has been replaced by $_p$BR 322, a small *E. coli* cloning vector.

(ii) Binary vector

These are based on the principle that vir genes may be located on a 'helper' Ti plasmid having the whole of T-DNA deleted (eg. $_p$AL 4404), because 'vir' genes can function even in trans configuration. In this case, T-DNA is found on a separate vector (binary vector) designed to replicate in both *E. coli* and *Agrobacterium* and capable of conjugal transfer between these two bacterial species.

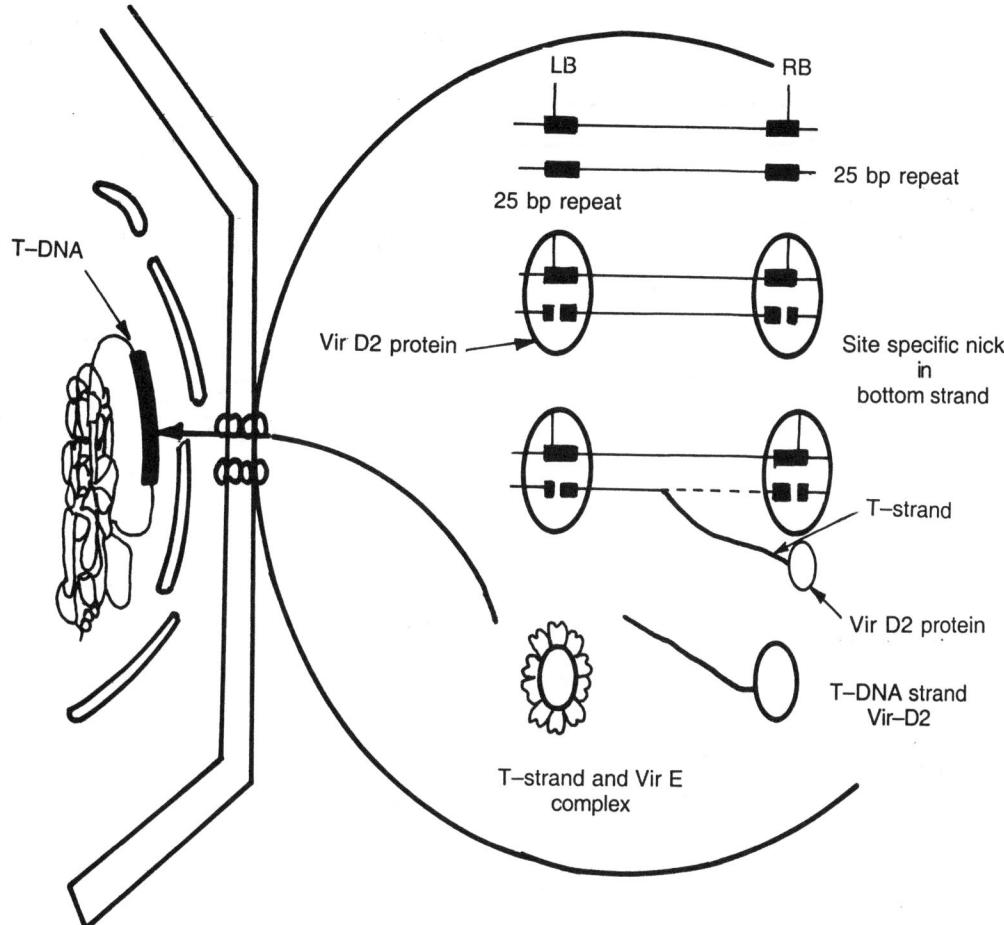

Fig. 18.5. T-DNA transfer process (adopted from plant genetic engineering, Grierson, D. 1991).

18.3. TRANSFORMATION TECHNIQUES

The uptake of foreign DNA or transgene by plant cells is called transformation. The techniques used can be placed into 2 categories (i) *Agrobacterium* mediated (ii) Direct gene transfer.

18.3.1. *Agrobacterium* mediated gene transfer

Agrobacterium mediated gene transfer is achieved by two ways :

(a) co-culture with tissue explants

(b) in planta transformation

The pre-requisites for *Agrobacterium* mediated gene transfer in higher plants include the following

   i) The plant explants must produce acetosyringone or other active compounds in order to induce 'vir' genes or virulence; alternatively *Agrobacterium* may be pre-induced with synthetic acetosyringone.

   ii) The induced Agrobacteria should have access to cells that are competent for transformation.

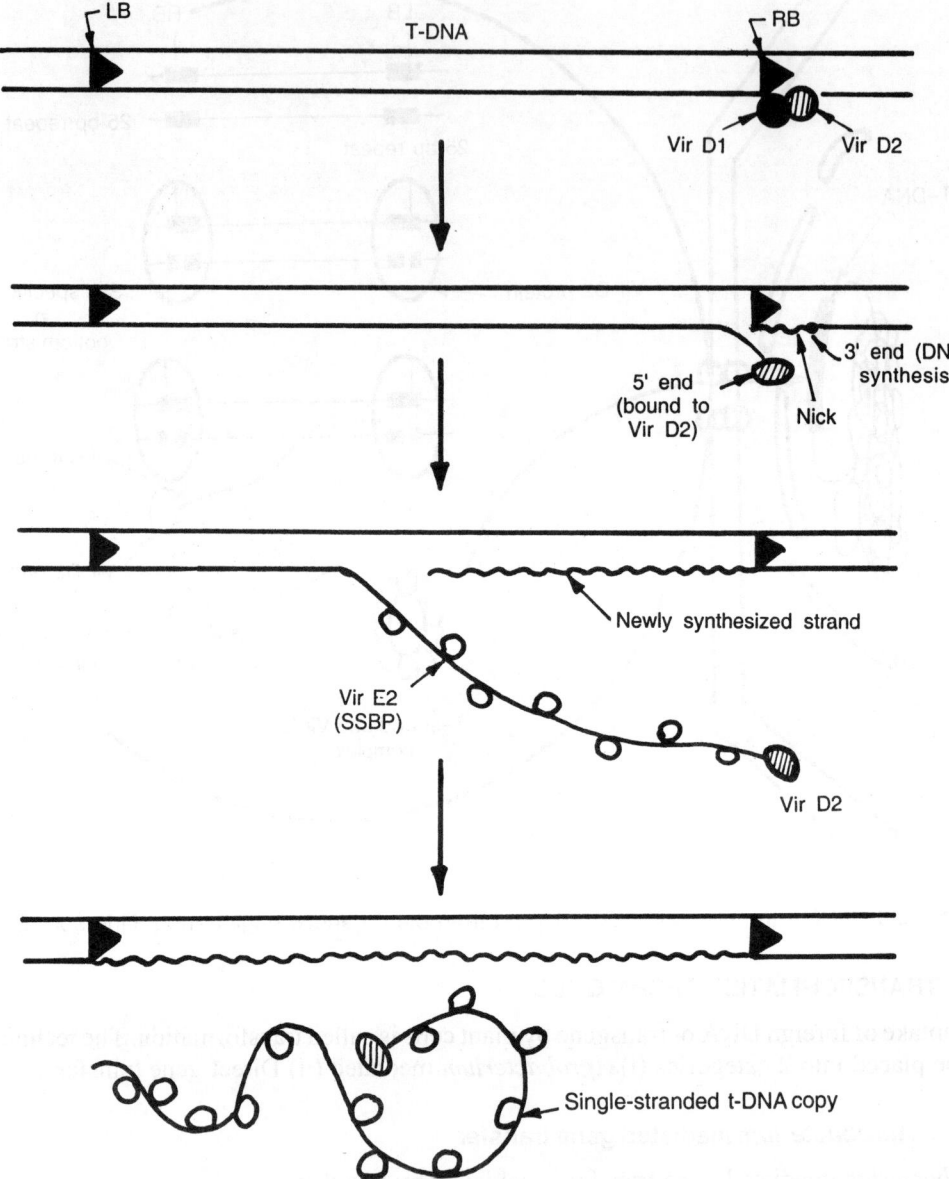

Fig. 18.6. The production of t-DNA copy (single-stranded) for transfer into plant cells (Adopted from Singh, 1998).

iii) Often transformed tissues or explants do not regenerate and it is diffcult to combine transformation competence with totipotency.

18.3.2. Explants used for transformation

The explants used for inoculation or co-cultivation with *Agrobacterium* carrying the vector, include callus cell clumps, suspension cultured cells, protoplasts, tissue slices, whole organ sections etc.

18.3.3. Marker genes for selection of transformed cells/callus or shoots

After explants are inoculated with *Agrobacterium* carrying the requisite vector having the gene of interest, we need to select the transformed tissues. This is facilitated by the presence of selectable marker genes available in the vector (Table 18.2). The selectable marker genes enable the transformed cells to survive in media containing toxic levels of the selection agent, which is usually an antibiotic or a herbicide. Following successful selection, the next obstacle is the regeneration as shoots from transformed calli. Since the explants may be heterogenous and non-transformed tissue may escape initial selection, several measures are used to ensure that very few non-transformed shoots escape the selection procedure.

Screening technique is an alternative in which no selection pressure is imposed on cells or shoots developing from inoculated explants. The samples of tissue are taken from all regenerated shoots and tested for the expression of a marker gene. A number of marker genes, described reporter genes or scoreable genes or screenable genes are shown in Table 18.3. The commonly used reporter genes are GUS, Cat, lux, npt II etc.

Table 18.3. Reporter genes used as scoreable markers, and their enzyme assays (Adopted from Gupta, 1996)

Reporter gene for	Substrate and assay	Identification
β-glucuronidase (GUS)	Glucuronides (PNPG, x-GLUC, NAG, REG)	Fluorescence detection; colourimetric and histochemical
β-galactosidase (lac Z)	β-galactoside (X-gal)	Colour of the colony
Octopine synthase (OCS)	Ariginine + pyruvate + NADH	Electrophoresis
Nopaline synthase (NOS)	Arginine + ketoglutaric acid + NADH	Electrophoresis
Neomycin phosphotransferase (NPT II)	Kanamycin + (^{32}P) ATP (in situ assay)	Radioactivity dection
Chloramphenicol acetyltransferase (CAT)	(^{14}C) chloramphenicol + acetyl CoA; TLC separation	Detection of acetyl chloramphenicol with autoradiography
Luciferase	(a) Decanal and FMNH$_2$ (b) ATP + O$_2$ + luciferin	Bioluminiscence (exposure of X-rays film)

PNPG = p-Nitrophenyl glucuronide;
X-GLUC = 5-bromo, 4-chloro, 3-indolyl glucuronide;
NAG = naphtol AS-B1 glucuronide;
REG = resorufin glucuronide

18.3.4. Co-culture with tissue explants

The appropriate gene construct is inserted within the T-region of a disarmed Ti plasmid; either cointegrative or a binary vector is used. The recombinant vector is placed in *Agrobacterium*, which is co-cultured with the plant cells or tissues to be transformed for about 2 days. In case of many plant species, small leaf discs are excised from surface sterilized leaves and used for co-cultivation. e.g. tobacco, petunia, tomato etc. In general, the transgene construct includes a selectable marker gene e.g. bacterial neo gene.

During the leaf disc-*Agrobacterium* co-culture, acetosyringone released by plant cells includes the vir genes which brings about the transfer of recombinant T-DNA into many of the plant cells. The T-DNA would become integrated into the plant genomes, and the transgene would be expressed. As a result, the transformed plant cells would become resistant to kanamycin . After 2 days, the leaf discs are transferred onto a regeneration medium containing appropriate concentrations of kanamycin

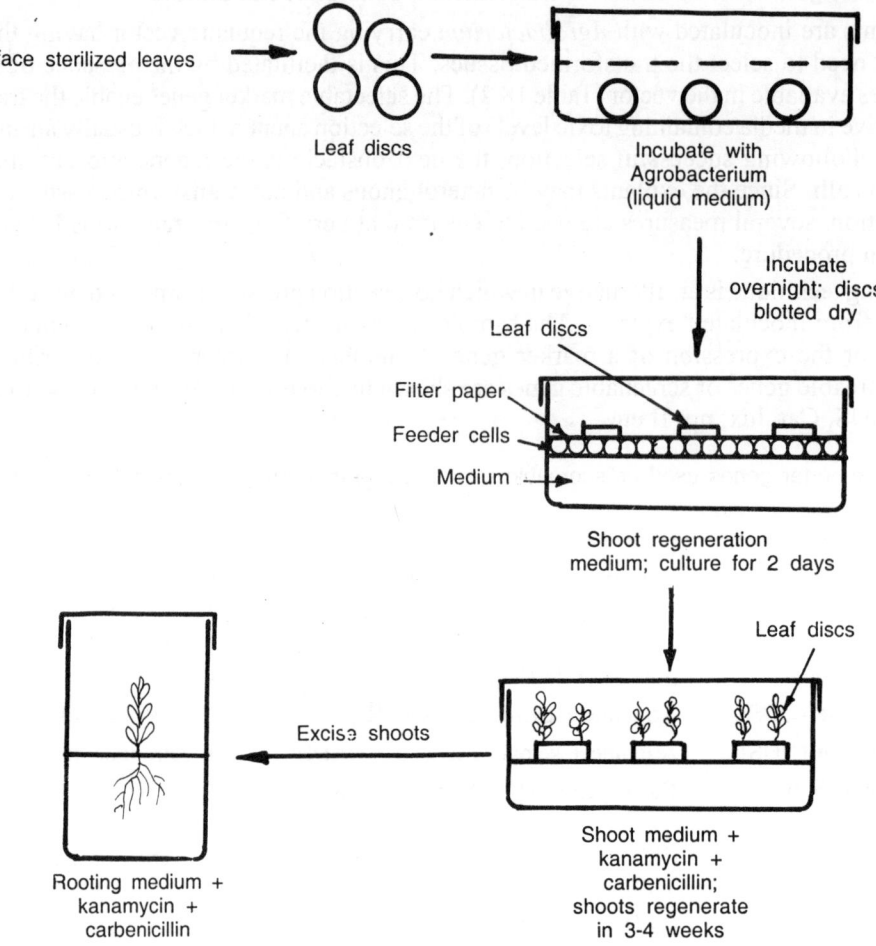

Fig. 18.7. Procedure of Agrobacterium mediated gene transfer (Adopted from Singh, 1998).

and carbencillin (Fig. 18.7). Kanamycin allow only trans-formed plant cells to divide and regenerate shoots in about 3-4 weeks, while carbencillin kills *Agrobacterium* cells. The shoots are separated, rooted and finally transferred into soil.

Agrobacterium does infect some monocot plant species and forms crown galls e.g. chlorophytum, Naricuss and *Allium cepa*.

18.3.5. In Planta transformation

Imbibition of *Arabidopsis* seeds in fresh cultures of *Agrobacterium* leads to stable integration of T-DNA in the *Arabidopsis* genome. It appears that *Agrobacterium* cells which enter the seedlings during germination, are retained within the plants, and when flowers develop they transform either the zygotes or the cells that give to zygote.

Alternatively, arabidopsis plants about to flower are immersd in a fresh culture of *Agrobacterium*, and partial vaccum is created to faciliate entry of bacterial cells into the plants. The plants are grown, and progeny screened for identification of transformation.

18.4. DIRECT GENE TRANSFER OR PHYSICAL DELIVERY METHODS OR DNA MEDIATED GENE TRANSFER

Introduction of DNA into plant cells without the involvement of a biological agent e.g. *Agrobacterium* and leading to stable transformation is known as direct gene transfer. The spontanenous uptake of DNA by plant cells is quite low. In many crops including cereal and legumes, tissue culture techniques for regeneration are not very successful. These limitations forced to look for alternative methods of gene transfer. The various methods of direct gene transfer are as follows.

18.4.1. Electroporation

Introduction of DNA into cells by exposing them for very short period to high voltage electrical pulses which is thought to induce transient pores in the plasma lemma is called *electroporation*. There are two types of systems of electroporation : (1) low voltage-long pulses method and (2) high voltage-short pulses approach. For tobacco mesophyll protoplasts, for low voltage-long pulses method, 300-400 V cm^{-1} for 10-50 ms (milliseconds; exponential decay), and for the high voltage-short pulses approach, 1000-1500 V cm^{-1} for μs (microseconds; square wave pulse generators).

Protoplasts are suspended in a suitable ionic solution containing recombinant plasmid DNA . The electroporation mixture is then exposed to the chosen voltage-pulses combination for the desired number of cycles. Protoplasts are then cultured to obtain cell colonies and plants. The optimal voltage and time will depend on the plant species, the source of protoplasts and the resistance of medium. Ordinarily a doubling of the protoplast diameter would reduce the optimal voltage of one-half.

Generally, low voltage-long pulses produce high rates of transient transformation while high voltage-short pulses give high rates of stable transformation. In most studies electroporation conditions giving about 50% protoplasts survival also give the highest rates of stable transformation. Transformation frequencies are increased several-fold by a heat-stock to protoplasts just prior to electroporation and the presence of low concentration (about 8%) of PEG during electroporation. For most species PEG-induced gene transfer is considered to be more reliable and efficient than electroporation. But some plant species are sensitive to PEG for which electroporation may be the method of choice. It has been used to produce stable transformed cell lines and/or plants in several plant species, e.g., maize, rice, petunia, tobacco, wheat etc. Electroporation has been used to delieer DNA into intact plant cells.

18.4.2. Chemical methods

Certain chemicals e.g. calcium phosphate, PEG (polyethylene glycol) and polyvinyl alcohol, enhance the uptake of DNA by protoplasts. PEG mediated transformation involves mixing of isolated protoplasts with DNA and immediately adding PEG (15-20%) dissolved in a buffer containing divalent cations (such as Ca^{+2}). This mixture is incubated for 30 min, protoplasts are washed and then plated in petriplate for culture and growth. The optimization of transformation frequencies by this method will depends on many factors such as (1) PEG concentration in the mixture (2) the pH of the solution (3) composition and concentration of salts used (4) concentration of the foreign DNA and (5) culture and selection techniques used for protoplasts. PEG stimulated the uptake of liposomes and improves the efficiency of electroporation e.g. petunia 40% transformed calli derived from mesophyll protoplasts could be induced to form fertile plants.

18.4.3. Lipo infection

Introduction of DNA into cells via liposomes is known as lipoinfection. Liposomes are small lipid vesicles, in which large number of plasmids are enclosed. They can be induced to fuse with protoplast using devices like PEG, and therefore have been used for gene transfer. There are several advantages for the use of this technique. (i) protection of DNA/RNA from nuclease digestion (ii) low cell toxicity (iii) stability and storage of nucleic acids due to encapsulation in liposomes (iv) high degree of reproducibility and enters the protoplasts due to endocytosis of liposomes (Fig. 18.8). This method

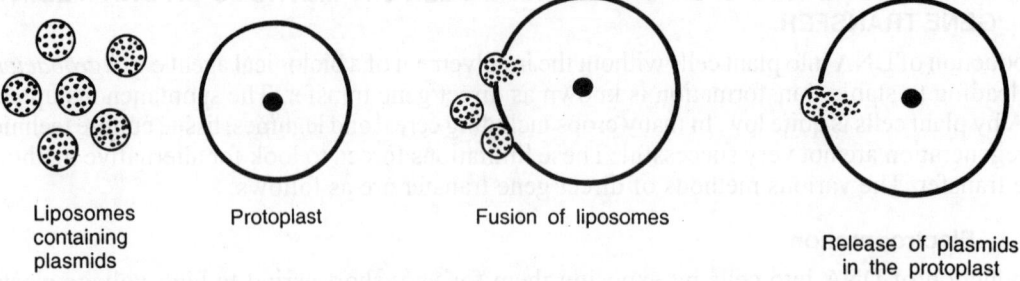

Fig. 18.8. Liposome mediated gene transfer (Adopted from Ramawat, 2000).

can also be used in the gene transfer for the production of transgenic animal. The technique has been successfully used to deliever DNA into the protoplasts of a number of plant species e.g. carrot, petunia and tobacco.

18.4.4. Calcium phosphate precipitation method

Foreign DNA can also be carried with the Ca^{++} ions, to be released inside the cell due to the precipitation of calcium in the form of calcium phosphate.

18.4.5. Microinjection

Plant regeneration from transformed protoplasts, still remains a problem. Therefore cultured tissues, that encourage the continued development of immature structures, provide alternative cellular targets for transformation. The immature structures includes germinating pollen, isolated ovules, immature embryos, meristems etc.

The DNA solution is injected directly inside the cell using capillary glass micropipettes with the help of micromanipulators of a microinjection assembly. It is easier to use protoplasts than cells since cell wall interferes with the process of microinjection. The protoplasts are usually immobilized in agarose or on glass coated with polylysine or by holding them under suction by a micropipette (Fig. 18.9). The process of microinjection is technically demanding and time-consuming; a maximum of

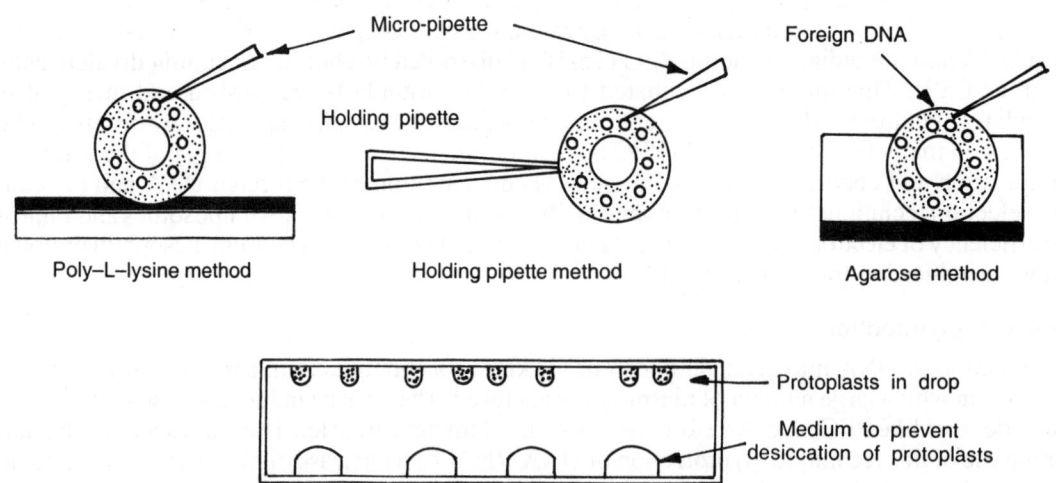

Fig. 18.9. Methods for the Microinjection and culture of plant protoplasts (Adopted from Gupta, 1996).

40-50 protoplasts can be microinjected in one hour. Successful transformation by microinjection of cells/protoplasts have been achieved with tobacco, alfalfa, *Brassica* sp. etc., the transformation frequencies ranging between 14 and 66%.

It is necessary to introduce the DNA into the nucleus or the cytoplasm of the cell for high transformation rates. Therefore, it is the most successful with densely cytoplasmic non-vacuolated embryonic cells. Large vacuolated cells show very low transformation rates possibly because, the DNA gets delivered into the vacuole, and as a consequence is degraded.

18.4.6. Fibre-mediated DNA delivery

In this method DNA is delivered into the cell cytoplasm and nucleus by silicon carbide fibres of 0.6 μm diameter and 10-80 μm length. This method is similar to that of microinjection and successfully used for maize and tobacco with cell cultures.

18.4.7. Laser-induced DNA delivery

Laser has been successfully used for high frequency (10^{-3}) transfection of animal cells, Lasers puncture transient holes in the cell membrane through which DNA may enter into the cell cytoplasm. Lasers have been used to deliver DNA into plant cells, but there is no information on transient expression or stable integration.

18.4.8. Incubation of dry seeds, embryos, tissues or cells in DNA

Incubation of dry seeds, embryos, tissues or cell in a known DNA has been tried in many cases and expression of defined genes has been witnessed, However, there is no case of proof of integrative transformation could be available.

18.4.9. Transformation by ultrasonication

In wheat, tobacco and sugarbeet, explants after being cultured for a few days, were sonicated with plasmid DNA (carrying marker genes like cat). When sonicated calli were transfered to selective medium, shoots were obtained, although all control calli (not sonicated with plasmid DNA) died. In tobacco transgenic plants were obtained at a frequency of 22 per cent.

18.4.10. Macro injection

The injection of plasmid DNA into the lumen of developing inflorescence using a hypodermic syringe is called macroinjection. In *secale cereale*, a marker gene was macroinjected into the stem below the immature floral meristem, so as to reach the sporogenous tissue leading to successful production of transgenic plants (De lapena *et al* 1987).

18.4.11. Microprojectiles (biolistics or particle-gun) for gene transfer

In this method, 1-2 μm tungsten or gold particles, coated with the DNA to be used for transformation, are accelerated to velocities which enable their entry into plant cells/nuclei. Particle acceleration is achieved by using a device which varies considerably in design and function. The most successful device accelerates particles in one of the two ways : (1) by using pressurised helium gas or (2) by the electrostatic energy released by a droplet of water exposed to a high voltage. This method is also called as biotistic or ballistic or particle gene gun method.

The components of a helium pressure device are: gas acceleration tube, rupture disc, stopping screen, macrocarrier carrying particles coated with DNA, and target cells (Fig. 18.10). These components are enclosed in a chamber to enable for the creation of partial vacuum which facilitates particle acceleration and reduces damage to plant cells. After creation of partial vacuum sufficiently pressurised helium gas is released in the acceleration tube to break the rapture disc. This generates

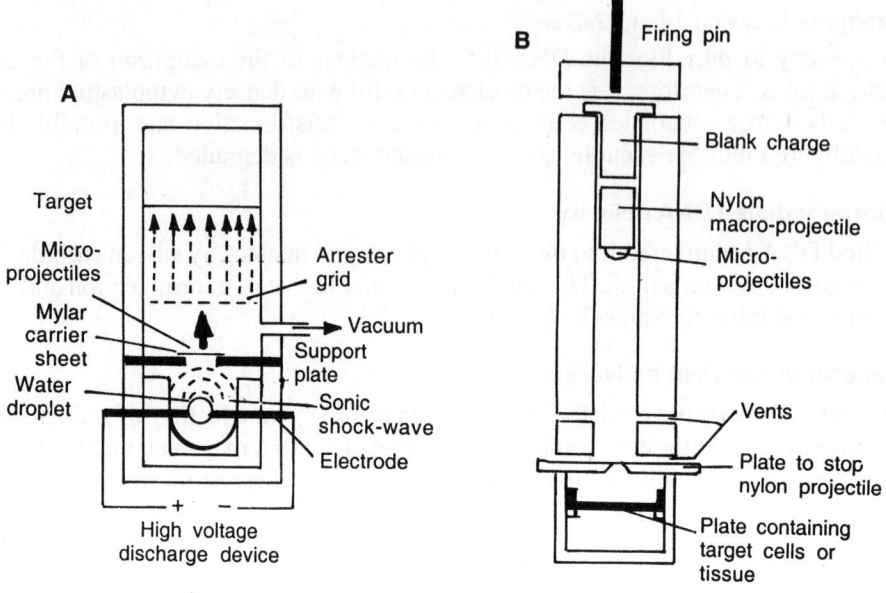

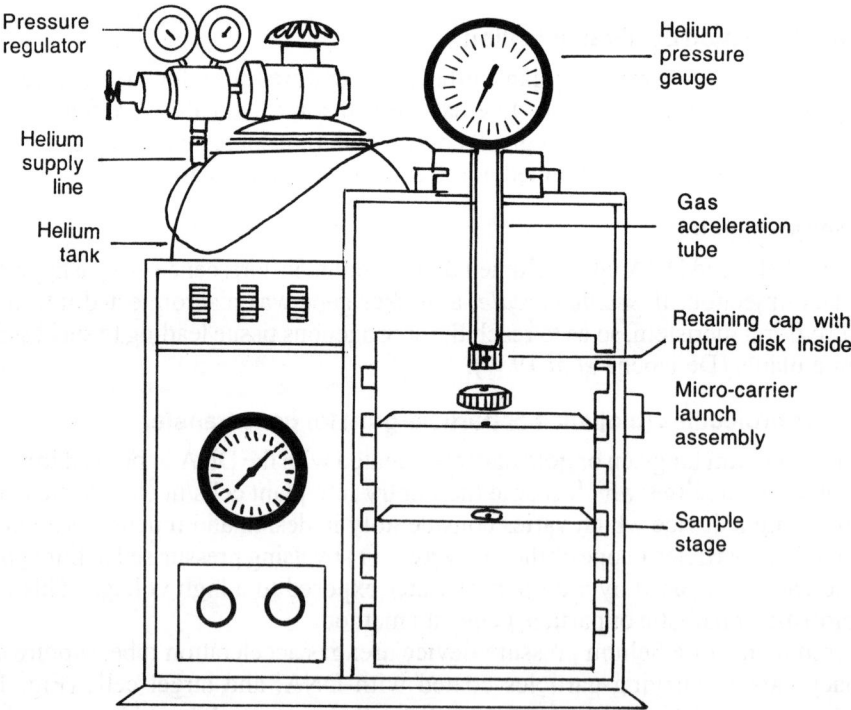

Fig. 18.10. Helium-powered biolistic particle gene gun (The PDA-1000/He is manufactured by Biorad, Life sciences division, Hercules, California, USA).

helium shock waves which accelerates the macroprojectile to which DNA coated microprojectiles are attached. The microprojectile is stopped by a stopping screen and pass through this screen and become embedded in the cells kept about 10 mm below the stopping screen microprojectiles. Helium is preferable to air since it is lighter and offers certain advantages. Generally a 1000 psi of helium pressure is used for acceleration (Singh, 1998).

The macroprojectile is a 2.5 cm diameter 0.06 mm thick plastic membrane which is used only once. The light mass of macroprojectile offers certain advantages, including rapid acceleration. The microprojectiles, microparticles or microcarriers vary in diameter from 0.5 to 2.0 µm; the average size 1.0 µm is commonly used. Tungsten particles are cheaper, but are of irregular shape and size, may be toxic to certain cell types and show surface oxidation which may lead to precipitation of DNA. In addition, they tend to form aggregates after addition of DNA which reduces particle dispersion. In comparison gold particles are more uniform in size(1-3 µm) and shape and show much lower toxiciy but they are much costlier and show variable coating with DNA.

The most critical factor concerns the coating of microparticles with DNA by precipitation. One approach of DNA coating is to mix 1.25-18 mg microparticles with 0.5-70 µg of plasmid DNA in a $CaCl_2$ (0.25-2.5 M) and spermidine (0.1M) solution. This mixture is continously vortexed to ensure uniform coating. After DNA precipitation, the microparticles are transferred onto the macrocarrier membranes, allowed to dry and immediately used (Singh, 1998).

Biolistic technique has been used to produce stable gene transfers in papaya, wheat, tobacco, sugarcane, soybean, sorghum, rice, maize, cotton etc.

This gene transfer method is very attractive because DNA can be delivered into cells of shoot meristems of shoot tips/embryos which makes gene transfer independent of regeneration ability of the species. This makes the technique applicable to virtually all plant species and even to animal cells. But its chief limitation is the need for a costly specialized acceleration device (particle gun).

18.5. VALUABLE PRODUCTS FROM TRANSGENIC PLANTS

The valuable products that can be manufactured by transgenic plants are antigen and antibodies, high value pharmaceutical oligopeptides, apart from crop improvement and production to specific starch through transgenic potato. There are two main strategies for the manufacture of high value products by transgenic plants (Fig. 18.11).

In one, the expression of the heterologous gene is transient. Organs (usally leaves) that express this latter gene are then harvested and the respective product is purified. In the other strategy, the heterologous gene is integrated into the genome of the transgenic plants. In the latter case the transgene is expressed usually in storage organs (e.g. seeds, tubers) from which the product is extracted. In the first strategy mediated by plant viruses, each generation of plants must be infected *de novo* by the respective viral vector, while in the second strategy the genetic transformation is performed only once and the sexual progeny maintains the expression of the heterologous gene.

18.5.1. Transient Expression

The strategy of using plant virus coat proteins as carriers of antigenic epitopes was already suggested by Haynes *et al.* (1986). These investigators used the TMV coat protein gene with a C-terminal extention that contained a poliovirus type 3 antigenic epitope. The fused gene was then expressed in bacteria (*E.coli*), where the respective fused protein polymersied.

TMV and related viruses have several features that render them good candidates as carriers of heterologous genes to infected plants. The latter viruses have a wide host range, replicate quickly and spread not only from cell to cell but also, via the pholem, throughout the infected plants. Their genomic RNA can be manipulated in *vitro* at the level of cDNA, and then used to infect plants in

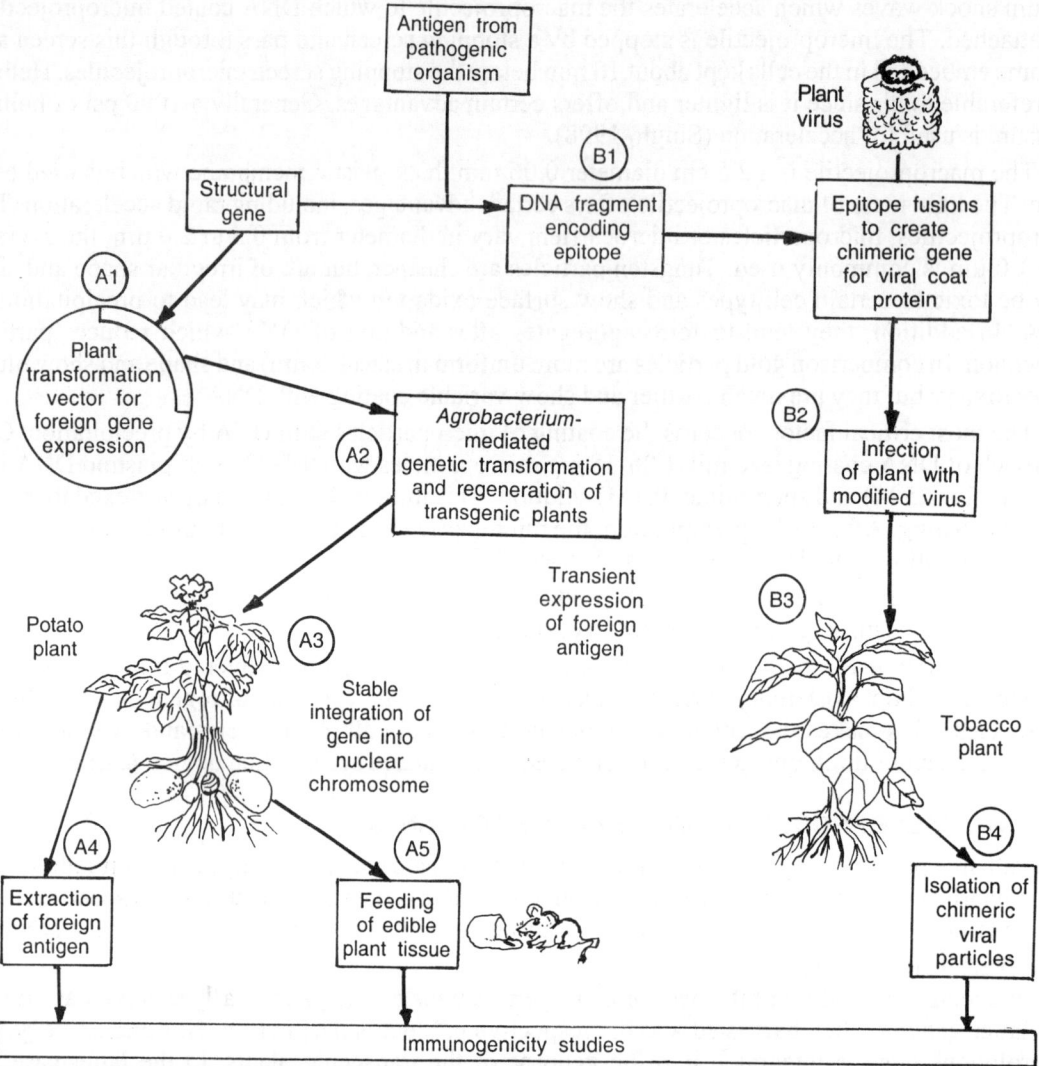

Fig. 18.11. Strategies for the production of candidate vaccine antigens in plant tissues. Genes encoding antigens from pathogenic organisms (viruses, bacteria or parasites) that have been characterised and for which antibodies are available, can be handled in two ways. In one case, the entire structural gene is inserted into a plant transformation vector between 5' and 3' regulatory elements (A1); this will allow transcription and accumulation of the coding sequence in all, or selected, plant tissues. This vector is then used for the Agrobacterium-mediated transformation of plant cells (A2), or for stable integration of the expression cassette by other means, and regeneration of transgenic plants. The resulting plants contain the expression cassette stably integrated into the nuclear chromosomal DNA (A3), and can be used either for extraction and partial purification of the foreign antigen (A4), or for direct feeding of plant tissues (A5; in this case, a potato tuber) for assessment of immunogenicity. Alternatively, if epitopes within the antigen are identified, DNA fragments encoding these can be used (B1) to construct chimeric genes by fusion with a coat protein gene from a plant virus, or in the case of TMV and CPMV, even the viral RNA made in vitro from the plasmid clone, is then used to infect established plants (B2). Virus replication and systemic spread allow high-level transient expression of the chimeric coat protein in most plant tissues (B3). The viral particles, expressing the foreign epitope on their surfaces, are then purified (B4) and used for immunogenicity studies (Adopted from Mason and Arntzen, 1995).

confined areas (e.g. greenhouse) and the resulting virus can then serve to mass-inoculate plants at a large scale.

However, simply adding an additional coding sequence with a homologous promoter to the existing coding RNA was found to be problematic. Such an addition has to be performed in an appropriate manner that will cause the additional RNA to stably replicate with the whole coding RNA and also maintain the spread of the virus in the infected host plant. Due manipulations finally resulted in a working system. The hybrid viral RNA and termed it TB2, contained sequences from two tobamoviruses : TMV-U1 and odontoglossum ring-spot virus. Donson et al (1991) inserted either of two bacterial selectable marker genes into TB2: the gene of neomycin phosphotransferase (npt II) or the gene for dihydrofolate reductase (dhfr). The TMV-U1 and the bacterial-gene sequence were encapsulated by the odontoglossum ring-spot virus coat protein. Infected plants (*Nicotiana benthamiana*) had less severe symptoms than TMV-infected plants. The nptII gene was expressed in the infected plants but at a lower rate than expected. Nevertheless this research established the possibility that a viral vector can provide a rapid means of expressing heterologous genes and gene variants in plants.

The latter system was thus utilised by a team of dozen investigators (Kumagai *et al.*, 1993) to produce an anti-human-immunodeficiency-virus (HIV) drug in transfected plants. The drug, termed α-trichosanthin, from *Trichosanthes kirilowii* is an eukaryotic ribosome-inactivating protein, which in its mature form inhibits protein synthesis by affecting the ability of the 60S ribosomal subunit to interact with elongation factors. By a mechanism, which was not revealed completely, this drug was reported to have an anti-HIV effect. The coding sequence (0.88 kbp) for α-trichosanthin was placed under the control of the TMV-U1 coat protein promoter and by due manipulation an infectious RNA was transcribed *in vitro*. The transcibed RNA was used to infect *N. benthamiana* plants. The hybrid virus is spread in the infected plants. The virion had the expected shape and reached 2.9-8.6 mg/g of leaf fresh weight. In the upper leaves of the infected cells the heterologous protein reached 2 per cent of the total soluble protein. The α-trichosanthin produced in the inoculated plants was verified by immunoblotting. Moreover the recombinant α-trichosanthin caused *in vitro*, a concentration-depedent inhibition of protein synthesis. This indicated that two weeks after inoculation a high level of the heterologous drug could be obtained by this procedure.

Kumagai *et al.* (1995) again used the system of expressing heterologous genes by hybrid tobamovirus infection. *Nicotiana benthamicana* was used for viral infection. The expression vector was modified. It did contain the TMV-U1 genome but this was fused (at the DNA level) with part of a tomato mosaic virus open reading frame that codes for the coat protein. Hamamoto *et al.* (1993) also used a TMV-based vector to transiently express a heterologous gene in inoculated plants. They inserted six-base sequence after the stop codon for 130 k of TMV and then inserted a coding sequence for an angiotensin-converting enzyme-inhibitor peptide (ACE 1). It is a 12-amino acid peptide, found in milk-casein hydrolysate that has an antihypertensive effect. When tobacco protoplasts were transfected with this transcript the cells produced both the CP-ACEI fused protein and CP, at a ratio of 1:200. The inoculation of tobacco plants with this transcript resulted after seven days in the spread of the virus from inoculated leaves to other leaves. A single substitution in the stop codon of 130k prevented this spread.

The genome of CPMV consists of two separately encapsidated positive-RNA strands (RNA1, 5880 nucleotides and RNA2, 3481 nucleotides). Both strands contain a single open-reading-frame and the resulting polyprotein is subsequently processed. Both RNAs are required for infecting plants by RNA 1 by itself can infect protoplasts. RNA 2 codes for the two CPs and cannot replicate in protoplasts in the absence of RNA1. Galun and Breiman (1997) constructed chimeras (at the DNA level) in a manner that the *in vitro* synthesised recombinant transcript will replicate in plants and that the heterologous protein (antigen) will be exposed on the outside of the viral particle. They found that a specific sequence of the small CP protein, which is rather variable among CPMVs, is a suitable site for adding the alien epitope. As epitope they choose the "FMDV loop" from the foot-and-mouth disease virus (FMDV), which encompasses 20-25 amino acids. They thus produced a modified cDNA for RNA 2 in which the code for the "FMDV-loop" was inserted. The insertion was performed either as a replacement or as an addition to the wild type small CP. When the viral vector with the addition was used the virus could replicate in protoplasts and also in cowpea (*Vigna unginiculata*) plants. But when the vector with replacement was used the virus could replicate in protoplasts but not in the plants.

Sugiyama *et al.* (1995) reported on the extension of their TMV-based procedure (Hamamoto *et al.*, 1993). They constructed viral vectors that led to the formation of TMV particles that carried three different epitopes. Two of them from influenza virus hemagglutinin (HA) and one from HIV-1 envelope protein. The *in vitro* transcribed RNAs were encapsidated with CP, *in vitro* and then used to inoculate tobacco plants. Mosaic symptoms were revealed 2-3 weeks after inoculation in the upper, non-inoculated, leaves. This indicated that infective virus particles were formed and were able to spread systematically in the plants. Analysis of leaves from inoculated plants indicated that the level of modified TMV particles reached about 1:1000 of the fresh weight and that the expected epitopes were formed. They suggested a model for the location of these epitopes on the CP of the modified TMV.

18.5.2. STABLE EXPRESSION

18.5.2.1. Production of antigens (Table 18.4)

The attempts to express a heptatitis B surface antigen (HBsAg) in transgenic plants was made by Arntzen and associates (Boyces Thompson Institute for Plant Research, Cornell University, Ithaca, New York, USA) (Mason *et al.*, 1992). Hepatitis B virus infection is a widespread human disease that can cause acute and chronic hepatitis that may lead to hepatocellular carcinoma. The infectious viral particle contains a 3.2 kb DNA genome that is encapsulated in a core particle and surrounded by a viral envelope. The latter contains phospholipids and the major surface antigens, HBsAg. HBsAg for vaccines can be produced from the serum of infected patients and by other means such as recombinant DNA expressed in yeast followed by disulphide linkages. However, such vaccine production procedures are quite expensive.

Table 18.4. Important recent examples of antigens produced in transgenic plants (Adopted from Herbers and Sonnewald, 1999)

Antigen	Source	Plant	Reference
CT-B	*Vibrio cholerae*	Potato	Arakawa *et al.*, 1997, 1998
LT-B	ETEC	Tobacco / potato	Haq *et al.*, 1995
LT-B	ETEC	Potato	Tacket *et al.*, 1998
LT-B	ETEC	Potato	Masson *et al.*, 1998
Insulin linked to CT-B	Human / ETEC	Potato	Arakawa *et al.*, 1998
Capsid protein	Norwalk virus	Tobacco / potato	Masson *et al.*, 1996
Surface antigen	Hepatitis B virus	Tobacco	Thanavala *et al.*, 1995
Malaria epitope	*Plasmodium* spp.	Tobacco	Turpen *et al.*, 1995
V3 loop of HIV-1	HIV	Tobacco	Yusibov *et al.*, 1997
Rabies virus Drg 24 antigene	Rabies virus	Tobacco	Yusibov *et al.*, 1997
Rabies virus Drg 24 antigene	Rabies virus	Spinach	Modelska *et al.*, 1998

Mason *et al*. (1992) therefore intended to produce effective HBsAg in transgenic plants with the ultimate goal of developing an oral vaccine for the "developing world". They thus constructed a binary plasmid that contained part of the coding sequence for HBsAg with the CaMV 35S promoter and the NoS terminator. Two such plasmids were constructed. One of them contained a 35S dual enhancer as well as the untranslated leader sequence from tobacco etch virus (pHB102). The other plasmid (pHB101) contained the regular 35S promoter and no leader sequence. The plasmids also contained the npt II gene for kanamycin resistance with due cis regulatory sequences. These plasmids were moved into *Agrobacterium* strains and the latter were used for genetic transformation of tobacco leaves.

Shoots and subsequently rooted shoots were selected in the appropriate media and transgenic plants were obtained. The pHB102 was found to be much more effective than pHB101. Transgenic plants which harboured the former plasmid, produced HBsAg that reached 25-65 ng protein per mg soluble protein in the leaves. The HBsAg from human serum and from plasmid-transformed yeasts occurs as spherical particles (~22 nm). The transgenic-tobacco produced spherical HBsAg particles of a similar form. The latter were also analysed with monoclonal antibodies and found to be immunologically similar to human HBsAg. Thus this endeavour looked promising for future manufacture of antibodies in transgenic plants but the yield was rather : about 0.01 per cent of soluble leaf protein (Galun and Breiman, 1997).

It should be noted that there was an earlier claim for the production of edible antibodies - a surface protein from a *Streptococcus* mutants, but it was patented and not published in a reviewed journal (see : Mason and Arntzen, 1995). Haq *et al.,* 1995 reported oral immunisation with a recombinant bacterial antigen that was produced in transgenic plants. The enterotoxigenic *Escherichia coli* (ETEC) causes an acute watery diarrhoea by colonising the small intestine and producing the heat-labile-enterotoxin (LT) as well as other toxins. The former toxin contains a pentamer of 11.6 kDB (binding) subunits (LT-B). Antibodies interfere with the binding of the B subunits to the intestine cells. These investigators thus constructed transformation vectors to produce anti-LT-B antibodies in transgenic plants.

They contained a 35S CaMV promoter followed by a TEV leader sequence after which a sequence coding for LT-B was inserted. After this coding sequence they added an endoplasmic-reticulum retention signal and then a termination sequence; or the latter signal was omitted. It was found that the retention signal was effective; transgenic tobacco plants that were obtained after due agrobacterial transformation that harboured the vector with this signal produced more antibodies than transgenic tobacco that harboured the vector without this signal. In addition to tobacco, potato plants were also genetically transformed.

Mice were fed with transgenic potato tubers (5g that were calculated to contain 20 μg of the antigen). The mice that consumed these tuber samples developed serum IgG and mucosal IgA that bind specifically to LT-B.

Cholera toxin (ctx) and its subunits have the capability to stimulate oral immunity. Ctx thus functions as an antigen as well as an adjuvant in mucosal immune response : high titer, specific antibody responses were observed in serum and in mucosa after feeding the ctx or its B subunit.

The A chain of ctx was used because it enhances the antigen and adjuvant characteristics of ctx on oral administration. Since the natural ctA chain by itself has a diarrhoegenic effect these investigator made changes in the coding sequence (e.g. Agr to Lys change). The modified ctA chain was expressed in transgenic tobacco but its level was lower than the natural ctA chain. In other transgenic plants the ctB chain was expressed (Arakawa 1997, 1998).

18.5.2.2. Production of Antibodies

Genetic transformation is now a routine method in many crop plants. The expression of antibodies can be directed to storage organs (e.g. seeds, tubers) and accumulate there. Once stably integrated into the plant genome under *cis* regulatory sequences that will induce ample expression, the transgenic plants can be propagated sexually and thus the production of antibodies can be moved to commercial fields.

The advantage of transgenic plants as producers of antibodies is that plant cells translate, glycosylate and process recombinant proteins faithfully and the correct assembly of complicated, multimeric proteins was demonstrated with full-length antibodies.

Furthermore crossing two transgenic plants each producing either the light-chain or the heavy-chain, can lead to the correct assembly of the full-length antibodies in the endoplasmic reticulum (ER) of transgenic plants.

The lymphoid cells, immunoglobulin light-chains and immunoglobulin heavy-chains are synthesized as precursor proteins. They contain a signal peptide that leads the nascent chains into the lumen of the ER. Inside the ER, a heavy-chain binding-protein (BiP) interacts with the immunoglobulin light and heavy chains, and another stress-protein (GRP94) - both acting as molecular chaperones - thus producing functional antibodies. Similar chaperonic proteins exist in plant ER.

These studies demonstrated the principle that the co-expression of two recombinant gene products can lead in plants to correct folding and assembling of antibody components into a product that is functionally identical to its mammalian counterpart (Fig. 18.11). This functional identity is interested because albeit plant-produced antibodies are glycosylated in a similar manner as mammalian antibodies, there is a difference in the complex glycans; these are probably more heterogeneous and smaller in plants than in mammalian complex glycans. There is also a difference in the terminal sugar residues. The secretion of processed antibodies is an important consideration. Such a secretion, outside of the cells, occurs in mammals. The accumulation of plant-produced antibodies in the apoplasm could protect them from hydrolytic destruction (Table 18.5).

Table 18.5. Antibodies and antibody fragments produced in transgenic plants[a] (Adopted from Ma and Hein, 1996).

Antibody Form	Valency	Antigen	Reference
Single domain (dAb)	1	Substance P (neuropeptide)	Benvenuto *et al.*, 1992
Single chain Fv	1	Phytochrome	Firek *et al.*, 1993
Single chain Fv	1	Artichoke mottled crinkle virus coat protein	Tavladoraki *et al.*, 1993
IgM (lambda)	2	NP (4-hydroxy-3-nitro-phenyl)acetyl hapten	During *et al.*, 1990
Fab; IgG (Kappa)	2	Human creatine kinase	De Neve *et al.*, 1993
IgG (kappa)	2	Transition state analog	Hiatt *et al.*, 1989
IgG (kappa)	2	Fungal cutinase	Van Engelen *et al.*, 1994
IgG (kappa) and IgG/A hybrids	2	*S. mutans* adhesin	Ma *et al.*, 1994
SIgA/G	4	*S. mutans* adhesin	Ma *et al.*, 1995)

[a] In all cases, the antibody molecules have been expressed in *Nicotiana*, except for FaB and IgG to human creatine kinase, which has also been expressed in Arabidopsis.

18.5.2.3. Oligopeptides and Proteins

One of the early efforts to manufacture pharmaceutically-valuable products in transgenic plants was reported by investigators of the PGS company and the university in Gent, Belgium (Vandekerckhove *et al.,* 1989). These investigators intended to produce leuenkephalin in seeds of transgenic plants. Leuenkephalin is a neuropeptide comprising five amino acids (Try-Gly-Gly-Phe-Leu) that displays opiate activity. The strategy was to insert the pentapeptide in the 2S albumins of seeds. The insertion was intended to be bordered by amino acid sequences that should facilitate excision and purification. These albumins were chosen because they can be solubilised (in low salt solution) and they are abundant among storage proteins in seeds. It was also important to choose a correct site of insertion so that the correct folding, in protein-bodies, shall not be impaired.

Agrobacterium-mediated transformation resulted in transgenic *Arabidopsis* and rape seed (*Brassica napus*). The seeds were harvested, the albumin was extracted, leaved (trypsin) and following further purification the expected leuenkephalin was detected in the seeds of transgenic *Arabidopsis* and rape seed; the yield was about 200 n 1ol/g seed and 10-50 nmol/g seed, respectively.

Investigators of the Mogen International Company (Leiden, The Netherlands) explored the production of human serum album (HSA) in transgenic potato and tobacco plants (Sijmons *et al.,* 1990). In man HSA is synthesized in the liver as a prepro-albumin. It is released from the ER after removal of a 18-amino-acid aminoterminal-prepeptide, resulting in the pro-albumin. The latter is further processed in the Golgi complex by cleavage of another six amino acids and the mature 585-amino-acid HSA is secreted. Targeting and processing in plant cells have similarities with those of metazoa.

Galuan and Breiman (1997) attempted to express the human gene that encodes HSA in transgenic plants and possibly modify the gene to cause the formation of the mature HSA in such plants. They thus constructed several plant expression vectors. All of the latter had a *npt* II selectable gene, a modified 35S promoter and an alfalfa mosaic virus leader, 3' of the promoter. The vectors differed in the codes for the signal peptides. These codes were either from the natural HSA or from the presequence of the tobacco extracellular PR-S protein. In all cases the code for the four amino acids - Asp-Ala-His-Lys - that are an essential component of the signal, were retained. Potato tuber discs were used for *Agrobacterium*-mediated transformation and transgenic potato plants were obtained after selection in kanamycin.

The levels of HSA were evaluated by immunoblotting. Many transgenic plants with HSA were obtained. Leaves contained upto 0.02 per cent of HSA in their total soluble protein. Similar genetic transformation was also performed with tobacco and they obtained transgenic cell suspension of tobacco. The HSA was also evaluated in extracellular extracts. It was found that the processing of the precursor protein was dependent on the type of signal sequence. While the expression of the natural prepro HSA led only to partial processing and secretion, fusion genes that encoded the plant PR-S presequence were correctly processed, resulting in authentic HSA.

18.5.2.4. Sugar Oligomers and Polymers

Transgenic potato tubers were used for the manufacture of another commercially valuable product : cyclodextrins (CDs). CDs are cyclic oligosaccharides, or several α, β and γ1, 4-linked glucopyranose units. Because they have an a polar cavity with a hydrophilic exterior they form inclusion complexes with hydrophobic substances rendering the latter more stable and water-soluble. CDs have therefore pharmaceutical and other uses.

CDs can be produced in some bacteria but this production is rather expensive. A team of researchers of the Calgene company in Davis, CA (USA), explored the production of CD in transgenic potato tubers. They took the gene for cyclodextrin glycosyl-transferase from the bacterium *Klebsiella pneumoniae* and fused it with a tuber-specific patatin promoter. To further target, the expression of the transgene in the plastids (i.e. amyloplasts) of the tubers they used the coding sequence for a transit peptide of the small subunit of ribulose bisphosphate carboxylase. A termination signal (NOS) as well as other required components were added to obtain an expression vector for *Agrobacterium*-mediated transformation. After due genetic transformation, they obtained transgenic potato tubers. The final yield of CD in the tubers could not be evaluated precisely but it was rather low; moreover the mRNA for the respective enzyme could not be detected (Galun and Breiman, 1997).

Fructans of a low degree of polymerisation (DP) also have a sweet taste and can serve as food sweeteners. While several plants store fructans. These plants are either not edible or there are other reasons that render these plants an unattractive source of fructans.

Several bacterial species have the capability to produce high DP fructans (Levans). *Bacillus subtilis* produces an extracellular enzyme, fructosyltransferase, that operates in the coversion of sucrose to a high DP fructan. The gene encoding this enzyme, sacB, was used by Smeekens and associates (see Galun and Breiman, 1997, p. 239) to launch a project to produce fructans in transgenic plants. In their earlier work they constructed a fused gene in which the sacB gene was fused with sequences that encoded signals for leading the enzyme into the Er and then to the vacuoles. This fused gene was then moved into an appropriate transformation vector and *Agrobacterium*-mediated transformation resulted in the respective transgenic tobacco plants. Such plants did accumulate high-DP fructan but at a relatively low level (3-8 per cent of dry weight) and the level of the sacB mRNA was below detection.

18.5.2.5. Alkaloids and phenolics

The tropane alkaloids hyoscyamine (its racemic form, being atropine) and scopolamine was used as anticholinergic agents that act on the parasympathetic nerve system. Considering the metabolic pathway from hyoscyamine to scopolamine, Yamada *et al.* (1990) focused on the enzyme hyoscyamine-6β-hydroxylase (H6H). H6H converts hyoscyamine to 6-β-hydroxylhyoscya-mine, and also the epoxidation of the latter compound to scopolamine. Their strategy was therefore to isolate the cDNA encoding H6H from *Hyoscyamun niger* and express it in transgenic *A. belladonna*. Consequently they engineered a fusion-gene in which the H6H cDNA was inserted between the 35S promoter and the NOS terminator. The fusion-gene was moved to an appropriate *Agrobacterium*-transformation vector that contained the *npt*" gene.

Genetic transformation and selection in the kanamycin-containing medium have resulted in transgenic *A. belladonna* plants. The latter were screened by western blotting to focus on plant and obtained the T_1 progeny. It should be noted that it is assumed that the alkaloids are synthesised in the roots and transferred, in part to the upper parts. The T_1 plants had a high scopolamine level in their leaves (but no hyoscyamine). One of them reached 1.2 per cent of scopolamine in the leaves dry weight. Control *A. belladonna* contain about 0.3 percent alkaloids and these are mostly hyoscyamine. Four transgenic tobacco plants were obtained; all had normal morphology. These plants contained that ubiC gene and they were self-pollinated. The progeny segregated (3:10) and those with antibiotic resistance were further analysed for CPL activity and 4HB dervatives. Three plants among the later progeny showed high per gram dry weight of leaves. A further study with ^{13}C-labelled shikimic acid, which was added to a cell suspension derived from the transgenic plants, verified that the transgenic plants could perform the direct conversion of chromismate to 4HB (refer Chapter 9 for Biosynthetic Pathway).

18.5.2.6. Degradable polymers

The use of plant-derived raw materials to manufacture industrial polymers ("plastics") was made possible several decades ago. Poly β-hydroxybutyrate (PHB) -an aliphatic polyester -and other polymers of the group of polyhydroxalkanoates (PHAs) are candidates for the production

One of these bacteria is *Alvaligenes eutrophos*. This bacterium has an exceptionally efficient path way for the synthesis of PHB, which accumulates to 8.0 per cent of dry weight in the form of small (ca 0.2-0.5 μm) inclusions of polymers of 10^3-10^4 monomers. Moreover, when cultured in media in which glucose is supplemented by other carbon sources (e.g, propionic acid, valeric acid,), *A. eutrophus* will produce random copolymers (e.g of 3-hydroxybutrate and 3-hydroxyvalerate). There are three key enzymes that are involved, in bacteria, with the synthesis of PHB : (1) 3-ketothiolase (catalyses the reversible condensation of two acetyl-CoA moieties, to form acetoacetyl-CoA); (2) acetoacetyl-CoA reductase (reduces acetoacetyl-CoA to D-(-) 3 hydroxybutyryl-CoA); (3) PHB synthase (polymerises D-(-) 3-hydroxybutyryl CoA to PHB). The genes encoding these enzymes were cloned and sequenced. Of these only the gene that codes for 3-ketothiolase also exists in plant genomes.

The gene encoding acetoacetyl-CoA reductase (phbB) was engineered into an Agrobacterium plasmid and used for genetic transformation of *Arabidopsis thaliana* plants. In parallel the gene that encodes PHB synthase (pbhC) was introduced into another plasmid and the respective transgenic *A. thaliana* plants were produced. The plants of the two transformations were selfed to obtain homozygous transgenic plants. These were first analysed for the respective expressions of mRNAs and enzymes. Transgenic *A. thaliana* plants with integrated phbB expressed the respective mRNA and also had a high level of acetoacetyl-CoA reductase activity. On the other hand, those with integrated phbC expressed the respective mRNA for PHB synthase but they did not translate it to the respective enzyme. It appears that without the substrate (D-(-)3-hydroxybutyric-CoA) no accumulation of PHB synthase takes place (Fidler and Dennis, 1992; Galun and Breiman, 1997).

18.6. ENZYMES FOR MAN AND OTHER ANIMALS (Table 18.6)

Monogastric animals lack the ability to utilise the phosphorous-containing phylate of crop seeds. Thus, they require either the addition of phosphorus of fungal phytase to their feed. Pen *et al.* (1993) reported on a study that aimed to engineer *Aspergillus niger* phytase in tobacco seeds. The aim was to explore the possibility of adding phytase-containing crop-seeds to the feed of monogastric animals in order to supply sufficient phosphorus in the diet. They thus fused the sequence that encodes the signal peptide of tobacco PR-S protein, upstream of the coding sequence of *A. niger* phytase. The chimeric gene was then moved into an *Agrobacterium* (binary) vector and due genetic transformation

Table 18.6. Technical enzymes produced in stably transformed plants (Herbers and Sonnewald, 1999)

Enzyme	Origin of gene	Plant	Reference
α-amylase	B. licheniformis	Tobacco	Penetal, 1992
Phytase	A. niger	Tobacco	Verwoerd et al., 1995
Phytase	A. niger	Tobacco	Pen et al., 1993
Phytase	A. niger	Soybean	Denbow et al., 1998
β(1, 3-1, 4)glucanase	R. flavefaciens	Tobacco	Herbers et al., 1996
β(1, 3-1, 4)glucanase	B. amyloliquefaciens/B. macerans	Barley	Jensen et al., 1996
β(1, 4)xylanase	C. thermocellum	Tobacco	Herbers et al., 1995
β(1, 4)xylanase	R. flavefaciens	Tobacco	Herbers et al., 1996

resulted in transgenic tobacco plants. Seeds of such plants contained up to 1 per cent phytase in their soluble protein. By simulating poultry feeding these investigators concluded that their approach was feasible. Moreover adding transgenic seeds that contained phytase to poultry-feed could substitute the addition of phosphorus or fungal phytase. The phytase was stable in the transgenic seeds for at least one year.

Many therapeutic strategies require human bioreactive proteins. There are several limitations to produce such human proteins in human (or other mammalian) tissue cultures. There are also limitations in the production of such proteins in bacteria, due to the complex processing required for human proteins and for glycoproteins of pharmaceutical importance.

There are three required characteristics that are of prime consideration in many products of genetic engineering; medical safety, purity and identity to the natural protein, and low cost of production. These characteristics are especially important when dealing with heterologous expression of human proteins for medical treatment. The hPC is one of several highly processed vitamin-K-dependent proteases and clotting factors in blood, which are critical in the coagulation/anticoagulation cascade. In order to perform anticoagulation function hPC is converted to a compound with serine-protease activity. This activation of hPC is exerted by thrombin bound to trombomodulin - in the presence of Ca^{2+}. After activation hPC functions to inactivate clotting factors (Va and VIII) by limited proteolysis. Before it is secreted from the liver cells into the blood, the hPC zymogen undergoes extensive co- and post-translational modifications such as proteolytic cleavages, glycosylation of aspargine residues, formation of disulphide bonds, β-hydroxylation of aspartic acid and γ-carboxylation of glutamyl residues.

Thus either the cDNA must be modified extensively or/and additional human genes should be expressed in the transgenic plants to result in the final product (or a product that can be easily further codified *in vitro*). Cramer *et al.* (1996) reported that they produced a fused gene in which the coding sequence for the full-length hPC gene was linked to the 35S CaMV promoter. The fused gene was used to obtain transgenic tobacco plants by routine *Agrobacterium*-mediated transformation. The yield of the heterologous human protein was very low-evaluation by the respective antibodies resulted in an estimation of hPC content that reached 0.002 per cent of soluble protein in the transgenic tobacco seedlings. Tobacco cells performed some, but not all, of the processing that is performed in man, e.g. there was a deficiency of cleavages in tobacco. On the other hand, the transgenic plants induced unnecessary modifications suggested to retain tobacco but to use it for heterologus expression of human genes, to manufacture medically valuable proteins.

Deficiencies in hGC cause the Gaucher disease - a severe lysosomal storage disease. In Gaucher patients there is pathologic accumulation of glucosylceramides, primarily in reticuloendothelial cells of bone-marrow, spleen and liver. Enzyme replacement was undertaken as a therapeutic treatment. This involves regular and frequent intravenous administration of a modified human glucocerebrosidase. "Ceredase" is a very expensive drug; the price of "Ceredase" (Brand name of Genzyme Corporation, USA) required for one year treatment may reach US$300, 000 per patient. The production of hGC in tobacco is therefore very attractive, commercially.

Consequently, the cDNA for hGC was engineered into an *Agrobacterium*-transformation cassette; but rather than the 35S promoter these investigators used a "Proprietary inducible plant promoter" (MeGA promoter of Crop Tech). Transgenic plants that were obtained expressed a high level of the hGC transcription upon induction. Moreover the induced transgenic tobacco plants produced a glycoprotein of appropriate size (~68 kD) that cross reacted with anti-hGC monoclonal antibodies and is apparently enzymatically active. This indicated that at least part of the heterologous protein is correctly glycosylated and folded.

18.7. AGRICULTURAL/HORITCULTURAL/FORESTRY APPLICATIONS

18.7.1. Herbicide resistance

Herbicides normally affect processes like photosynthesis or biosynthesis of essential amino acids (Table 18.7).

Table 18.7. Examples of herbicides resistance gene transfer

Active principle of herbicide	Inhibited pathway inhibitor	Use	Basis of resistance
1. Amino acid biosynthesis inhibitor sulphonyl urea and imidazolinones	Branched chain amino acids	Selected crops	Mutant ALS gene
Phosphinothricin	Glutamine biosynthesis	Broad spectrum	Gene amplication, bar gene, detoxification
2. Photosynthesis inhibitory Albbazine	Photosystem II	Selected crop	Mutant psba gene, GST gene, detoxification

18.7.2. Insect resistance

18.7.2.1. Insecticides

Use of pesticides and insecticides are very common for the control of insect pests.

Bt-toxins produced by *Bacillus thuringiensis* is used as biological insecticides. Insecticidal property of bacteria is due to a protein delta endotoxin synthesized during sporulation. This is very specific toxin having high cost of production. This crystal protein is instable in the field.

Bt-2 gene from *B. thuringiensis* is transferred to cotton, tomato, tobacco, by plasmid mediated transformation. Transgenic plants produce toxin protein for which Bt gene is encoded. The plants become resistant to *Manducta sextra*, a pest of tobacco. Feeding of larvae of this insect shows 75-100% mortality. Transgenic cotton containing Bt gene is being been introduced in India by Maharashtra Hybrid Seeds Co. (MAHYCO), Jalna.

18.7.2.2. Protease inhibitors

Vigna anguiculata (Cow pea) trypsin inhibitor (CpTi) level was responsible for resistance to storage pests (Bruchid beetle). CpTi was shown to be toxic to a variety of insects but cow pea seeds with high level of CpTi are not toxic to human beings. Using a number of binary vectors, CpTi was joined with CaMV 35S promoter and one or more marker gene. CpTi gene has been transferred in tobacco without 'yield penalty' (No decrease in yield).

18.7.2.3. Resistance against viral infection

i) Cross protection - previous infection by a wild strain gives protection by a virulent strain. Following systems have been used :
   Citrus - Citrus tristeza virus
   Potato - Potato spindle tuber viriod
   Tomato - Tomato mosaic virus

ii) Gene of virus coat of capsid protein - Coat protein gene from tobacco mosaic virus (TMV) was transferred to tobacco and stable expression in tobacco was recorded.

iii) Gene for nucleocapsid (N) protein from tomato spotted wild virus (TSWV) has been isolated.

Negative (–) strand virus - TSWV is pathogen for *Chrysanthemum, Impatiens,* groundnut, lettuce, pepper, and tomato. RNS is tightly associated with nucleocapsid protein. The plants transformed with N gene showed expression of N gene and resistance to TSWV.

iv) Resistance against bacterial and fungal pathogens has been achieved by transferring the appropriate gene (Table 18.8).

Table 18.8. Transgenic plants for bacterial and fungal pathogens

S.No. Pathogen	Disease	Res. gene	Source of gene	Transgenic crop
1. *Alternaria longipes*	Brown spot	Chitinase gene	Soil bacteria	Tobacco
2. *Phytophthora infestans*	Late Bright	Osmotin gene	Potato	Potato
3. *Pseudomonas syringae*	Wild fire	Acetyl transferase gene	--	Tobacco
4. *Rhizoctonia solani*	--	Chitinase gene	Bean	Tobacco

18.7.2.4. Examples of Transgenic plants

Tobacco, *Petunia hybrida*, tomato, potato, egg plant, *Arabidopsis thaliana,* lettuce, *Apium graveolens* (celary), sunflower, flax, rape oil seed, cauliflower, cabbage, cotton, soyabean, pea, chicory, carrot, licorice, sweet potato, kiwi, papaya, grape, rose, *Chrysanthemum, Populus, Malus, Pyrus communus, Azadirachta indica, Asparagus, Secale cereale, Oryza sativa, Triticum aestivum, Zea mays, Avena sativa, Picea glauca.*

REFERENCES

Arakawa, T., Chong, D.K.X., Langridge, W.H.R. (1998). Efficacy of a food plant-based oral cholera toxin B subunit vaccine. *Nat. Biotechnol.,* 16 : 292-297.

Arakawa, T., Chong, D.K.X., Lawrence, M., Langridge, W.H.R. (1997). Expression of cholera toxin B subunit oligomers in transgenic potato plants. *Transgenic Res.,* 6 : 403-413.

Arakawa, T., Yu, J., Chong, D.K.X., Hough, J., Engen, P.C., Langridge, W.H.R. (1998). A plant-based cholera toxin B subunit-Insulin fusion protein protects against the development of autoimmune diabetes. *Nat. Biotechnol.,* 16 : 934-938.

Benvenuto, E., Ordas, R.J., Tavazza, R., Ancora, G., Biocca, S., Cattaneo, A. and Galeffi, P. (1991). 'Phytoantibodies': A general vector for the expression of immunoglobulin domains in transgenic plants. *Plant Mol. Biol.,* 17: 865-874.

Cramer, L.L., Weissenborn, D.L., Oishi, K.K., Grabau, E.A., Bennett, S., Ponce, E., Grabowski, G.A., Radin, D.N. (1996). Bioproduction of human enzymes in transgenic tobacco. *Ann. N.Y. Acad. Sci.,* 792 : 62-71.

De Lapena *et al.* (1987). Transgenic plants obtained by injecting DNA into young floral tillers. *Nature,* 325 : 274-76.

De Neve, M., De Loose, M., Jacobs, A., Van Houdt, H., Kaluza, B., Weidle, U., Van Montagu, M., Depicker, A. (1993). Assembly of an antibody and its derived antibody fragment in *Nicotiana* and *Arabidopsis. Transgenic Res.,* 2 : 227-237.

Denbow, D.M., Grabau, E.A., Lacy, G.H., Kornegay, E.T., Russell, D.R., Umbeck, P.F. (1998). Soybeans transformed with a fungal phytase gene improve phosphorus availability for broilers. *Poultry Sci.,* 77 : 878-881.

Donson, J., Kearney, C.W., Hilf, M.E., and Dawson, O.W., (1991) *Proc.Natl.Acad.Sci. USA,* 88, 7204.

Draper, J. and Scott, R (1991) Gene transfer to plants In: Plant genetic engineering, Plant Biotechnology series, (Eds) Grierson, D. 1, Blackie, Glasgow and London, pp. 39 - 76.

During, K., Hippe, S., Kreuzaler, F., Schell, J. (1990). Synthesis and self assembly of a functional antibody in transgenic *Nicotiana tabacum*. *Plant Mol. Biol.,* 15 : 281-293.

Firek, S., Draper, J., Owen, M.R.L., Gandecha, A., Cockburn, B., Whitelam, G.C. (1993). Secretion of a functional single-chain Fv protein in transgenic tobacco plants and cell suspension cultures. *Plant Mol. Biol.,* 23 : 861-870.

Galun.E and Brieman, A., (1997) Transgenic Plants, Imperial college press, Lodon.PP 214-240.

Grierson, D., (1991) Plant Genetic Engineering, Blackie, Glasgow and London, Plantbiotechnology, 1, PP 1-31.

Gupta, P.K. (1996). Elements of Biotechnology, Rastogi and Company, Meerut, India, p. 359-362.

Hamamoto, H., Sugiyama, Y., Nakagawa, N., Hashida, E., Matsunaga, Y., Takemoto, S., Watanabe, Y., Okade, Y. (1993). A new tobacco mosaic virus vector and its use for the systemic production of angiotensin-1-converting enzyme inhibitor in transgenic tobacco and tomato. *Bio/Technology,* 11 : 930-932.

Haq, T.A., Mason, H.S., Clements, J.D., Arntzen, C.J. (1995). Oral immunization with a recombinant bacterial antigen produced in transgenic plants. *Science,* 268 : 714-716.

Haynes, J.R., Cunningham, J., Vonseefried, A., Lennick, M., Garvin, R.T., and Shen, S., (1986) Development of a genetically-engineered, candidate polio vaccine employing the self asembling properties of the tobacco mosiac virus coat protein. *Bio/technology*, 4, 637-641.

Herbers, K. and Sonnewald, U. (1999). Production of new / modified proteins in transgenic plants. *Current Opinion in Biotechnology*, 10(2) : 163-168.

Herbers, K., Flint, H.J., Sonnewald, U. (1996). Apoplastic expression of the xylanase and β(1-3, 1-4) glucanase domains of the *xyn*D gene from *Ruminococcus flavefaciens* leads to functional polypeptides in transgenic plants. *Mol. Breeding,* 2 : 81-87.

Herbers, K., Wilke, I., Sonnewald, U. (1995). A thermostable xylanase from *Clostridium thermocellum* expressed at high levels in the apoplast of transgenic tobacco has no detrimental effects and is easily purified. *Bio/Technology,* 13 : 63-66.

Hiatt, A., Cafferkey, R., Bowdisk, K. (1989). Production of antibodies in transgenic plants. *Nature,* 342 : 76-78.

Jensen, L.G., Olsen, O., Kops, O., Wolf, N., Thomsen, K.K., von Wettstein, D. Transgenic barley expressing a protein-engineering, thermostable (1-3, 1-4)-β-glucanase during germination. *Proc. Natl. Acad. Sci. USA,* 93:3487-3491.

Kumagai, M.K., Donson, J., Della-Cioppa, G., Harvey, D., Hanley, K., Grill, L.K. (1995). *Proc. Natl. Acad. Sci. USA,* 92 : 1679.

Kumagai, M.K., Turpen, T.H., Weinzettl, N., Della-Cioppa, G., Turpen, A.M., Donson, J., Hilf, M.E., Grantham, G.L., Dawson, W.O., Chow, T.P., Piatak Jr., M., Grill, L.K. (1993). *Proc. Natl. Acad. Sci. USA,* 90 : 427.

Ma, J.K.C., Hiqtt, A., Hein, M., Vine, N.D., Wang, F., Stabila, P., Van Dolleweerd, D., Mostov, K. and Lehner, T. (1995). Generation and assembly of secretory antibodies in plants. *Science,* 268 : 716-719.

Ma, J.K.-C., Lehner, T., Stabila, P., Fux, C.I., Hiatt, A. (1994). Assembly of monoclonal antibodies with IgGI and IgΛ heavy chain domains in transgenic tobacco plants. *Eur. J. Immunol.,* 24 : 131-138.

Ma, K.C.J. and Hein, M.B. (1996). Antibody production and engineering in plants. Annals. The New York Academy of Sciences, New York, 792 : 72-80.

Mason, H.S. and Arntzen, C.J. (1995). Transgenic plants as vaccine production systems. *Trends in Biotechnology,* 13(9) : 388-392.

Mason, H.S., Ball, J.M., Shi, J.J., Jiang, X., Estes, M.K., Arntzen, C.J. (1996). Expression of Norwalk virus capsid protein in transgenic tobacco and potato and its oral immunogenicity in mice. *Proc. Natl. Acad. Sci. USA,* 93: 5335-5340.

Mason, H.S., Haq, T.A., Clements, J.D., Arntzen, C.J. (1998). Edible vaccine protects mice against *E. coli* heat-labile enterotoxin (LT) : potatoes expressing a synthetic LT-B gene. *Vaccine,* 16 : 1336-1343.

Mason, H.S., Lam, D.M-K. Arntzen, C.L. (1992). Expression of hepatitis B surface antigen in transgenic plants. *Proc. Natl. Acad. Sci. USA,* 89 : 11745-11749.

Modelska, A., Dietzschold, B., Sleysh, N., Fu, Z.F., Steplewski, K., Hooper, D.C., Koprowski, H., Yusibov, V. (1998). Immunization against rabies with plant-derived antigen. *Proc. Natl. Acad. Sci. USA,* 95 : 2481-2485.

Pen, J., Verwoerd, T.C., VanParidon, P.A., Beudaker, R.F., VandenElzen, P.J.M., Geerse, K., Vander Klis, J.K., Verteegh, H.A.J., Van Ooxen, A.J.J. and Hoekema, A. (1993). Phytase-containing transgenic seeds as a novel feed additive for improved phosphorus utilization. *Bio/Technology,* 11 : 811-814.

Ramawat, K.G. (2000). Plant Biotechnology, S. Chand & Company Ltd., New Delhi, India, p. 199.

Sijmons, P., Dekker, B.M.M., Schrammeijer, B., Versoerd, T.C., Van Den Elzen, P.J.M., Hoekema, A. (1990). Production of correctly processed human serum albumin in transgenic plants. *Bio/Technology,* 8 : 217-221.

Singh, B.D. (1998). Biotechnology, Kalyani Publishers, Ludhiana, India, pp. 299-325.

Sugiyama, y., Hamamoto, H., Takemoto, S., Watanabe, Y., and Okada, Y., (1995) Systemic production of foreign peptides on the particle surface of tobacco. *FEBS Lett.* 359, 247-250.

Tacket, C.O., Mason, H.S., Losonsky, G., Clements, J.D., Levine, M.M., Arntzen, C.J. (1998). Immunogenicity in humans of a recombinant bacterial antigen delivered in a transgenic potato. *Nat. Med.,* 4 : 607-609.

Tavladoraki, P., Benvenuto, E., Trinca, S., De Martinis, D., Cattaneo, A., Galeffi, P. (1993). Transgenic plants expressing a functional single-chain Fv antibody are specifically protected from virus attack. *Nature,* 366 : 469-472.

Tavladoraki, P., Benvenuto, E., Trinca, S., De Martins, D., Cattaned, A., Galeffi, P. (1993). Transgenic plants expressing a functional single-chain Fv antibody are specifically protected from virus attack. *Nature,* 366 : 469-472.

Thanavala, Y., Yang, Y.F., Lyons, P., Mason, H.S., Arntzen, C. (1995). Immunogenicity of transgenic plant-derived hepatitis B surface antigen. *Proc. Natl. Acad. Sci. USA,* 92 : 3358-3361.

Turpen, T.H., Reini, S.J., Charoenvit, Y., Hoffman, S., Fallarme, V., Grill, L.K. (1995). Malarial epitopes expressed on the surface of recombinant tobacco mosaic virus. *Bio/Technology,* 13 : 53-57.

Vanderkerckhove, J., Van damme, J., Van Lisebettens, M., Botterman, J., et al (1989) Enkephalins produced in transgenic plants using modified 2S seed stroage protein, *Biotechnology*, 7, 929-932.

Van Engelen, F.A., Schouten, A., Molthoff, J.W., Roosien, J., Salinas, J., Dirkse, W.G., Schots, A., Bakker, J., Gommers, F.J., Jongsma, M.A. *et al.* (1994). Coordinate expression of antibody subunit genes yields high levels of functional antibodies in roots of transgenic tobacco. *Plant Mol. Biol.,* 26 : 1701-1710.

Verwoerd, T.C., van Pandon, P.A., van Ooyen, A.J.J., van Lent, J.W.M., Hoekema, A., Pen, A. (1995). Stable accumulation of *Aspergillus niger* phytase in transgenic tobacco leaves. *Plant Physiol.,* 109 : 1199-1205.

Waugh, R.and Brown, J.W.S. (1991) Plant gene structure and expression, In: Plant genetic engineering, (eds) Grierson, D. 1. Blackie, Glasgow and London. PP 1-31.

Yamada, Y., Okabe, S., Hashimoto, T. (1990). Homogenous hyoscyamine 6â-hydroxylase from cultured roots of *Hyoscyamus niger*. *Proc. Japan Acad. Ser. B,* 66 : 73-76.

Yoder, J.I., Goldbrough, A.P. (1994). Transformation systems for generating marker for transgenic plants. *Bio/Technology,* 12 : 263-67.

Yusibov, V., Modelska, A., Steplewski, K., Agadjanyan, M., Weiner, D., Hooper, C.D., Koprowski, H. (1997). Antigens produced in plants by infection with chimeric plant viruses immunize against rabies virus and HIV-1. *Proc. Natl. Acad. Sci. USA,* 94 : 5784-5788.

Chapter 19

Hairy Root Cultures

INTRODUCTION

Plants remain a major source of pharmaceuticals and fine chemicals. Many of these compounds are derived from tropical species and their quality, availability and cost, make it attractive to consider alternative sources of supply. Generally their structural complexity makes their chemical synthesis uneconomic. Thus, the biosynthesis of these so called 'secondary' metabolites using plant cells in culture has long been identified as a worthwhile objective. Despite considerable efforts, only a few commercial processes have been achieved using cell cultures (e.g. paclitaxel, shikonin and berberine). The major constraint with cell cultures is that they are genetically unstable and cultured cells tend to produce low yields of secondary metabolites. The attempts at enhancing product formation in rapidly growing suspension cultures showed that some biosynthetic pathways are spontaneously well expressed under such conditions. Poorly expressed pathways, however, do not respond well to manipulations aimed at enhancing the productivity. There are some secondary metabolites formed at low levels in undifferentiated cell cultures, but after induction of some morphological differentiation, however, enhanced biosynthesis of such secondary metabolites. The organ cultures of plant cells such as root and shoot cultures are known for quite some time. However, from a biotechnological point of view, these cultures were of low interest, because they are difficult to establish, they often grow slowly and are some time very unstable. A new route for enhancing secondary metabolite production is by transformation using the natural vector system, *Agrobacterium rhizogenes*, and the causative agent of hairy root disease in plants. *A. rhizogenes* is a soil dwelling gram negative bacterium which infects a wide range of plant species and causes the nepotistic plant disease syndrome known as "Hairy root" disease. This typical disease syndrome is characterized by numerous fast growing, highly branched adventitious roots at the site of infection, which continue to grow *in-vitro* in hormone free culture medium. These fast growing hairy roots can be used as a continuous source for the production of valuable secondary metabolites. Moreover, transformed roots are able to regenerate into whole viable plants and maintain their genetic stability during further sub culturing and plant regeneration. The biosynthetic capacity of the hairy root cultures is equivalent or more (sometimes) to the corresponding plant roots, and they can be grown in bioreactors.

The history of using hairy roots grown *in-vitro* as an experimental tissue is a comparatively short one. The first report on the initiation of hairy root cultures for production purpose was reported by Flores and Filner (1985) on *Hyoscyamus* species for the production of tropane alkaloids (0.5%).

This was followed by huge number of reports on the initiation of metabolite producing hairy root cultures (Table 19.1).

Table 19.1. Secondary metabolite production from hairy root cultures (Adopted from Giri and Lakshmi Narasu, 2000)

Plant	Secondary Metabolite	References
Aconitum heterophyllum	Aconite	Giri *et al.,* 1997
Ajuga replans var. *atropurpurea*	Phytoecdysteroids	Matsumoto and Tanaka, 1991
Ambrosia artemisifolia	Thiarubine AXB	Flores et al 1988
Amsonia elliptica	Indole alkaloids	Yoshikawa and Furuya 1987
Anisodus luridus	Tropane alkaloids	Flores *et al.,* 1991
Armoracia laphthifolia	Peroxidase, Isoperoxidase, Fusicoccin	Taya *et al.,* 1989; Bahakov *et al.,* 1985
Artemisia absinthum	Essential oils	Nin *et al.,* 1997
Artemisia annua	Artemisinin	Jaziri *et al.,* 1995; Teoh *et al.,* 1996; Weathers *et al.,* 1994; Rao *et al.,* 1998
Astragalus mongholicus	Cycloartane saponin	Ionkava *et al.,* 1997
Atropa belladonna	Atropine	Christen 1999; Lee *et al.,* 1998
Azadirachta indica A. Juss	Azadirachtin	Sowjanya *et al.,* 1998
Antirrhinum manus	Tannins	Carron *et al.,* 1994
Beta vulgaris	Betalain pigments	Taya *et al.,* 1992
Bidens sp.	Polyacetylenes and thiophenes	Flores *et al.,* 1988
Brugamansia candida	Tropane alkaloids	Giulietti *et al.,* 1993
Calystegia sepium	Cuscohygrine	Jung and Tepfer, 1987
Campanula medium	Polyacetylenes	Tada *et al.,* 1996
Carthamus	Thiophenes	Flores *et al.,* 1988
Cassia obtusifolia	Anthraquinone Polypeptide pigments	Asamizu *et al.,* 1988; Ko *et al.,* 1995
Catharanthus roseus	Indole alkaloids, Ajmalicine	Parr *et al.,* 1988; Toivonen *et al.,* 1990, 1993
Catharanthus tricophyllus	Indole alkaloids	Davioud *et al.,* 1989
Centranthus ruber	Valepotriates	Christen, 1999; Granicher *et al.,* 1995
Cephaelis ipecacuanha	Emetic alkaloids	Jha *et al.,* 1991
Chaenatis douglasis	Thiarubrins	Constabel and Towers, 1988
Cinchona ledgeriana	Quinine	Hamill *et al.,* 1989
Coleus forskohlii	Forskolin	Sasaki *et al.,* 1998
Coreopsis Sp.	Polyacetylene	Marchant, 1988
Datura candida	Scopolamine, Hyoscyamine	Christen *et al.,* 1989
Datura ferox	Hyoscyamine, scopolamine	Parr *et al.,* 1990
Datura innoxia	Hyoscyamine, Scopolamine	Flores *et al.,* 1988
Datura metel	Hyoscyamine, scopolamine	Parr *et al.,* 1990
Datura stramonium	Hyoscyamine, Sesquiterpene	Payne *et al.,* 1987; Hilton and Rhodes, 1993

(Contd.)

Plant	Secondary Metabolite	References
Daucus carota	Flavonoids, Anthocyanin	Bel-Rhlid *et al.*, 1993; Kim *et al.*, 1994
Digitalis purpurea	Cardioactive glycosides	Saito *et al.*, 1990
Duboisia myoporoides	Scopolamine	Deno *et al.*, 1987
Duboisia leichhardtii	Scopolamine	Muranaka *et al.*, 1992
Echinacea purpurea	Alkamides	Mukundan and Hjortso, 1990: Trypsteen et al 1991
Echinops pappii	Thiophenes	Abegaz 1991
Ephedra gerardiana	Ephedrine	Dowd and Richardson 1994
Ephedra minima	Ephedrine	Dowd and Richardson 1994
Ephedra minima hybrid	Ephedrine	Dowd and Richardson 1994
Ephodra saxatilis	Ephedrine	Dowd and Richardson 1994
Fagra zanthoxyloids Lam.	Benzophenanthridine Furoquinoline alanine	Couilterot *et al.*, 1999
Fagopyrum esculentum	Flavanol	Trotin *et al.*, 1993
Fragaria Sp.	Polyphenol	Motomari *et al.*, 1995
Geranium thubergee	Tannins	Ishimaru and Shimomura, 1991
Glycyrrhiza glabra	Flavonoids	Asada *et al.*, 1998
Gynostemma pentaphyllum	Saponin	Fei *et al.*, 1993
Hyoscyamus albus	Tropane alkaloids, Phytoalexins	Kuroyanagi *et al.*, 1998
Hyoscyamus muticus	Tropane alkaloids Hyoscyamine, Proline	Vanhala *et al.*, 1995; Halperin and Flores, 1997
Hyoscyamus canariensis	Hyoscyamine, Scopolamine	Hashimoto *et al.,* 1986
Hyoscyamus niger	Hyoscyamine	Jaziri *et al.*, 1988
Hyoscyamus pusillus	Hyoscyamine, scopolamine	Hashimoto *et al.,* 1986
Hyssopus officinalis	Rosmarinic acid	Murakami *et al.,* 1998
Lactuca virosa	Sesquiterpene lactones	Kisiel *et al.*, 1995
Leontopodium alpinum	Anthocyanins & Essential oil	Hook 1994
Linum flavum	Lignans (5-methoxy podophyllotoxins)	Oostdam *et al.*, 1993
Lippia dulcis	Sesquiterpenes, (hernandulcin)	Sauerwein *et al.*, 1991
Lithospermum erythrorhizon	Shikonin, benzoquinone	Shimomura *et al.*, 1981; Fukui *et al.*, 1998
Lobelia cardinalis	Polyacetylene glucosides	Yamanaka *et al.*, 1996
Lobelia inflata	Lobeline, Polyacetylene	Yonemitsu *et al.*, 1990
Lobelia sessilifolia	Lobetyolin	Ishimaru *et al.*, 1994
Lotus corniculatus	Condensed tannins	Carron *et al.*, 1994
Nicandra physaloides	Hygrine	Parr, 1992
Nicotiana africana	Nicotine	Parr and Hamill, 1987
Nicotiana cavicola	Nicotine	Parr and Hamill, 1987
Nicotiana glauca	Anabasine, nicotine	Fecker *et al.*, 1992
Nicotiana hesperis	Nicotiane	Walton and BelShaw 1988
Nicotiana rustica	Nicotine, N'-ethyl-S-nornicotine	Rhodes *et al.,* 1987
Nicotiana tabacum	Nicotine, cadaverine, anabasine	Fecker *et al.,* 1993
Nicotiana umbratica	Nicotine	Parr and Hamill, 1987

(Contd.)

Plant	Secondary Metabolite	References
Nicotiana velutina	Nicotine	Parr and Hamill, 1987
Nicotiana hesperis	Nicotine, Anabasine	Parr and Hamill, 1987
Nicotiana rustica	Nicotine, Anabasine	Hamill *et al.*, 1986
Nicotiana tabacum	Nicotine, Anabasine	Flores and Filner, 1985
Ocimum basilicum	Rosmarinic acid	Tada *et al.*, 1996
Panax ginseng	Saponins	Yoshimatsu *et al.*, 1996; Yoshikawa and Furuya, 1987
Panax hybrid (*P. ginseng* × *P. quinquefolium*)	Ginsenosides	Washida *et al.*, 1998
Papaver somniferum	Codeine	Williams and Ellis, 1993; Yoshimatsu and Shimomura, 1992
Perezia cuernavacana	Sesquiterpene quinone	Arellano *et al.*, 1996
Pimpinella anisum	Essential oils	Santos *et al.*, 1998
Platycodon grandiflorum	Polyacetylene elkucosides	Tada *et al.*, 1995; Ahn *et al.*, 1996
Psoralea sp.	Diadzein and related flavonoids	Bourgaud *et al.*, 1999
Rauwolfia serpentina	Reserpine	Sato *et al.*, 1991; Benjamin *et al.*, 1994
Rubia peregrina	Anthraquinones	Lodhi *et al.*, 1996
Rubia tinctorum	Anthraquinone	Sato *et al.*, 1991
Rudbeckia spp.	Polyacetylenes and thiophenes	Flores *et al.*, 1988
Salvia miltiorrhiza	Diterpenoid	Hu and Alfermann, 1993
Sanguisorba officinalis	Tannins	Ishimaru *et al.*, 1990
Scopolia carniolica	Hyoscyamine, Scopolamine	Knoop *et al.*, 1988
Scopolia stramonifolia	Hyoscyamine, Scopolamine	Parr *et al.*, 1990
Scopolia tangutica	Hyoscyamine, Scopolamine	Shimomura *et al.*, 1991
Scopolia japonica	Hyoscyamine	Mano *et al.*, 1987
Scutellaria baicalensis	Flavonoids and phenylethanoids	Zhou *et al.*, 1997
Serratula tinctoria	Ecdysteroid	Delbeque *et al.*, 1995
Sesamum indicum	Naphthaquinone	Ogasawara *et al.*, 1993
Solanum aculeatissi	Steroidal saponins	Ikenaga *et al.*, 1995
Solanum laciniatum	Steroidal alkaloids	Hamill *et al.*, 1987
Solanum aviculare	Steroidal alkaloids	Yu *et al.*, 1996
Swainsona galegifolia	Swainsonine	Ermayanti *et al.*, 1994
Swertia japonica	Xanthons	Ishimaru *et al.*, 1990
Tagetus patula	Thiophenes	Arroo *et al.*, 1995
Tanacetum parthenium	Sesquiterpene coumarin ether	Kisiel and Stojakowska, 1997
Trichosanthes kirilowii maxim var *japonicum*	Defense related proteins	Savary and Flores, 1994
Trigonella foenum graecum	Diosgenin	Merkli *et al.*, 1997
Tropaeolum majus	Glucotrapaeolin and Myrosinase	Weilanek and Urbanek, 1999
Vitis vinifera	Reservatol	Guellec *et al.*, 1990
Valeriana officinalis L.	Valepotriates	Granicher, 1994
Vinca minor	Indole alkaloids (vincamine)	Tanaka *et al.*, 1994
Withania somnifera	Withanolides	Banerjee *et al.*, 1994

Invasion of dicotyledonous plant tissues by *A. rhizogenes* soil bacteria usually occur at a wounded site possibly caused by insect or mechanical damage. Wounded site produces phenolics (e.g., acetosyringone, 3', 5'-dimethoxy-4'-hydroxy acetophenone) which attracts *A. rhizogenes* by chemotaxis. Other phenolics include hydroxy acetosyringone, coniferyl alcohol, and sinapinic acid. All these stimulate efficient vir gene expression. It causes roots to proliferate rapidly at the infection site causing hairy root disease. The distinctive symptoms of hairy root disease are the formation of a mass of roots. A large number of small roots protrude as fine as hairs directly from the infection site (Banerjee et al., 1995). Riker et al. (1928) named the organism as *Phytomonas rhizogenes,* which was later renamed as *Agrobacterium rhizogenes*. This phenotypic response results from the insertion into the plant genome of T-DNA (transfer DNA), carried on the bacterial Ri-plasmid (Root inducing plasmid), coding for auxin synthesis and other rhizogenic functions. The transformed segment T-DNA contains genes for opine biosynthesis and sensitivity. Products of virulence (Vir) genes located on non transferred segment of the Ri plasmid are responsible for excision of the T-DNA for transfer into the plant cell, and possibly for chromosomal integration in the nucleus for the recipient cell. Transferred genes often act as dominant. Mendelian alleles, with about 75% of the R_1 or selfed progeny carrying the transgenes.

The roots can be removed from the parent tissue and cultured indefinitely in simple defined media free of plant growth hormones. In this respect they differ from untransformed root cultures which are often dependent upon added auxin and which generally have low growth rates.

Hairy root cultures are potentially applicable to the production of all root derived secondary products from dicotyledonous plants. Hairy root cultures have been developed as an alternate source for the production of root biomass and to obtain root derived compounds.

A number of roots synthesized compounds, including the tropane alkaloids, atropine and hyoscyamine, steroidal precursors such as solasodine, and *Catharanthus* alkaloids, are of sufficiently high value to justify the exploitation of hairy root cultures for their commercial production.

19.1. ESTABLISHMENT AND NATURE OF HAIRY ROOT CULTURES

19.1.1. Establishment of hairy root cultures

In the production of hairy root cultures, the explant material is inoculated with a suspension of *A. rhizogenes* generated by growing bacteria in YMB medium for 2 days at 25°C with gyratory shaking, pelleting by centrifugation (5×10 rpm; 20 min) and resuspending the bacteria in YMB medium to form a thick suspension (approx. 10^{10} viable bacteria/ml). Transformation may be induced on aseptic plants grown from seed, of on detached leaves, leaf-discs, petioles of stem segments from greenhouse plants following sterilization of the excised tissue with 10% (v/v) Domestos (Lever Bros.) for 20 mins. Scratching the midrib of a leaf or the stem of a plantlet, with the needle of a hypodermic syringe containing the thick bacterial suspension allows inoculation with small (about 5-10 µl) droplets containing *A. rhizogenes*.

In some species, a profusion of roots may appear directly at the site of inoculation, but in others a callus will form initially and roots emerge subsequently from it. In either case, hairy roots normally appear within 1-4 weeks. Although the susceptibility of species to infection is variable. Infection of particularly resistant species might be enhanced by including 10 µm acetosyringone in the medium in which the bacteria are resuspended. This compound produced during the wounding response of *Agrobacterium* aiding plasmid T-DNA transfer (Stachel et al., 1985).

An alternative approach to transformation is to use co-cultivation of plant protoplasts with *A. rhizogenes*. Wei et al. (1986) obtained transformation of *Solanum nigrum* at a frequency of abou' 3×10^{-3} and found calli cultured on hormone-free media.

For many species, establishment of hairy root cultures in suspension is achieved simply by excising the transformed roots from the explant and placing them in growth medium (Murashige and Skoog 1962; Gamborg et al., 1968) with sucrose (30 g/l) as the energy source. However, some species, are more difficult to establish in independent culture. Various techniques, including suspending roots on the explant until they are established, buffering the pH of the medium and/or using half-or quarter-strength, high phosphate or nitrate, or the addition of exogenous hormones, have been found to cause extensive callus formation. Cultures may be cleared of bacteria by several passages in media containing 500 mg/l ampicillin or 500 mg/l ampicillin + 200 mg/1 cephalosporin C.

Alternatively, allowing root tips to grow through medium containing high agar concentration or treatment with elevated temperatures have been used to free transformed roots of bacteria. Chilton et al (1982) established clonal experimental cultures by placing single root tips in 50-ml aliquots of the appropriate culture media. Individual transformed root tips have been reported to be of single-cell origin (Tepfer 1984).

19.2. GENETICS OF TRANSFORMATION (FIG. 19.1)

To utilize *Agrobacterium* for the genetic transformation, it is essential to have T-DNA vectors into which target gene can be inserted, preferably adjacent to selectable marker genes such as antibiotic resistance and usually with the phytohormone genes deleted.

Two types of vectors are available, cointegrative and binary. Both types of vectors are used for plant transformation. In the more common binary vector,

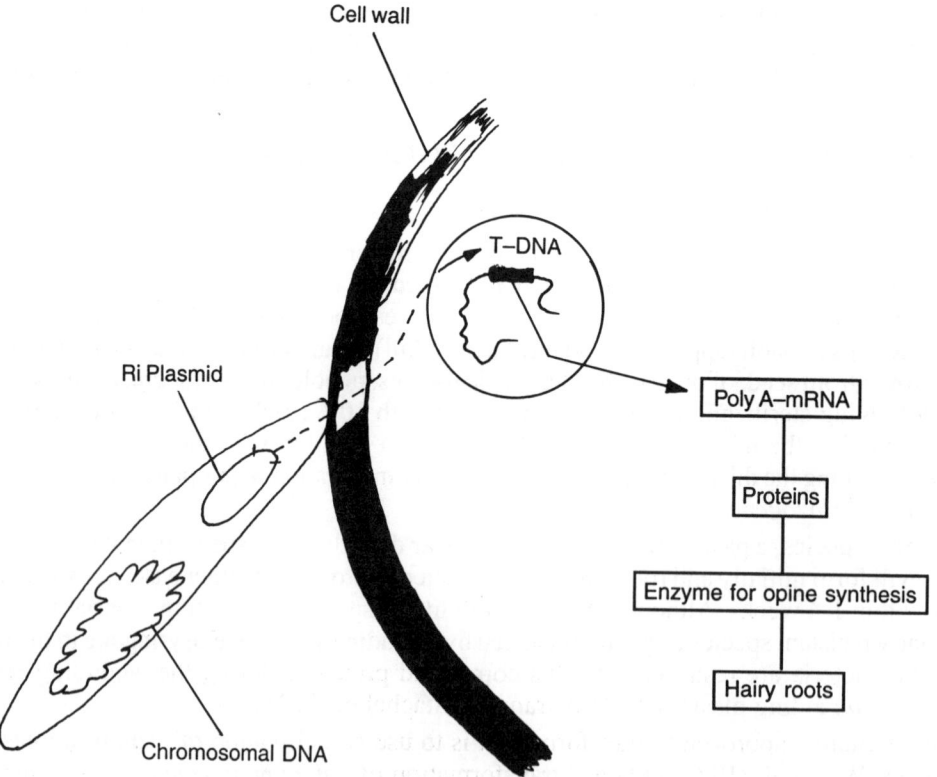

Fig. 19.1. Genetics of transformation.

A) T-DNA region with opines and phytohormone genes deleted and cloning sites inserted are engineered onto small, autonomously replicating, and wide-host-range plasmid. A second plasmid usually the native Ri plasmid, carries the virulent genes necessary for transformation. In the cointegrative vector, both the T-DNA and the virulent genes are located on the same plasmid.

B) Recipient cells capable of accepting and integrating the T-DNA to express the transformed and

C) Transformed cells capable of regeneration into plants, which express the transgenes.

The infection of plants with the bacterium causes one or both of two pieces of T-DNA (T_L and T_R) to be inserted into plant genome.

T-DNA region of Ri plasmid contains different genes, functions of which are given below. The exact functions may vary from species to species in the plants.

T_L region:

1. rol A-Development of hairy root morphology,
2. rol B-Induction of hairy roots by hydrolysing bound auxins such as indole-3-acetyl-β-D-glucosides leading to an increase in the intracellular levels of indole-3-acetic acid.
3. rol C-Reduction in the levels of isopentenyl adenosine (iPA) and increase in the levels of GA_{19}

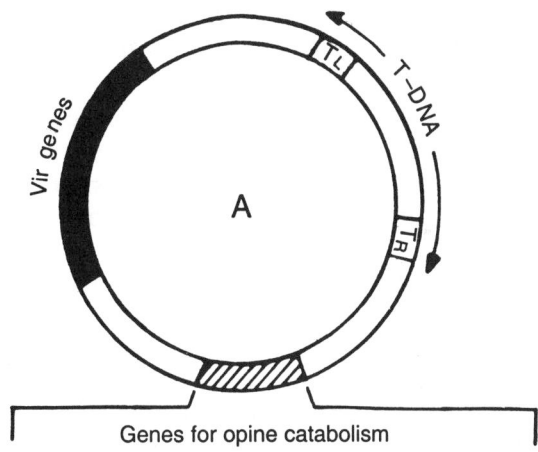

Genes for opine catabolism

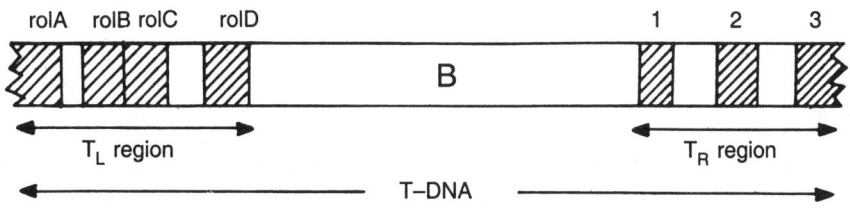

Fig. 19.2. Structure of part of T-DNA.

T_R region: Genes 1 and 2 code for auxin biosynthesis by coding for tryptophan-2-monooxygenase and indole-3- acetamide hydrolase, Genes coding for agropine synthase (Rhodes *et al.,* 1989).

Integration, although apparently random with to both copy number and location alters the auxin metabolism of transformed tissue in such a way that the hairy root phenotype is expressed and amino acid metabolism is modified such that specific metabolites, opines are produced (Rhodes *et al.,* 1989).

A. rhizogenes strains and Ri plasmids are classified according to the opines formed in transformed roots and degraded by plasmid encoded catabolic pathways. So far four strain types have been described. They are,

1. *Agropine type (A₄ 1855, 15834, TR105, HRI)* : Agropine mannopine, mannopinic acid, agropinc acid and agrocinopine A.
2. *Mannopine type (8196, TR107)* : Mannopine, mannopinic acid, agropinic acid and agrocinopine C
3. *Cucumopine type (NCPPB2657, NCPPB2659)* which corresponds to a newly identified opine. Cucumopine. 4-carboxy-4 (2-carboxy ethyl) spinacine, a condensation product of histidine and 2-Ketoglutaric acid, and
4. *Mikimopine type (NIAES 1724, MAFE 301724)* Mikimopine

Agropine and Mannopine type strains belong to *Agrobacterium* biovar. (biotype) 2, where as Cucumopine type strains belongs to biovar.1. Ri plasmids are conjugative and they can be transferred to both biovar.1 and 2 strains. Only one T-DNA region has been identified on Mannopine type or Cucumopine type, whereas Agropine type Ri plasmids contain 2 T-DNA regions, respectively called T_L-DNA and T_R-DNA.

The T_R-DNA region carries genes encoding auxin and agropine synthesis. The other has homology to the unique T-region of the two other strain types.

The strains of the bacterium most commonly used for transformation contain the plasmids p[RiA4], p[RiHR1], p[Ri1855] and p[Ri8196], the first three of which are almost identical. These strains all show good virulence characteristics. The bacterial strain LBA9402, which contains the plasmid p[Ri1855] and induces production of the agropine, has been used to induce roots on a wide range of dicotyledonous plants.

Homology has been reported between part of T_R-DNA of agropine type strains and the auxin loci (tms) of *A. tumefaciens*. Furthermore, two auxin genes of *A. rhizogenes* are capable of complementing Ti (tumor inducing) mutants, indicating that the auxin locus of *A. rhizogenes* contains at least 2 genes (Ri aux-1 and Ri aux-2) which determine functions similar or identical to those of the Ti aux-1 and aux-2 genes.

19.3. FACTORS INFLUENCING THE TRANSFORMATION

19.3.1. Virulence of A. rhizogenes strains

Strains with agropine type Ri plasmid have an extended host range by comparison to mannopine or cucumopine type strains. Here host range does not only concern the list of plant species that can be successfully inoculated, but also, for a given species, the response to inoculation at a particular site.

19.3.2. Medium

The growth medium has a significant effect on hairy root induction. High salt media such as LS or MS favors hairy root formation in some plants. Low salt media such as B5 favor excessive bacterial multiplication in the medium and therefore the explant needs to be transferred several times to fresh antibiotic containing medium before incubation. The bacterial concentration also plays an important role for the production of transformed roots, suboptimal concentrations may result in lower availability of bacteria for transforming the plant cells while high concentrations may decrease it by competitive inhibition (Mukundan *et al.*, 1998). Christen *et al.* (1992) reported that the effects of various culture media containing 3 or 5% sucrose on growth and alkaloid content was examined after 19 days in hairy root cultures of *Hyoscyamus albus*. The media tested were half strength MS (1/2MS), MS, B5 (Gamborg *et al.*, 1968) and Woody Plant medium. The five alkaloids were detected in all the hairy root cultures analysed. Hyoscyamine was the most abundant compound throughout. The highest concentration of hyoscyamine was observed in B5 medium containing 3% sucrose, while in 1/2 MS supplemented with 3% sucrose higher content of 6β-hydroxy hyoscyamine and scopolamine were obtained. The transformed roots cultured in media with 3% sucrose showed a higher alkaloid content. In WP media (3 and 5% sucrose), the hairy roots showed the fastest growth but the alkaloid concentration was relatively low. 7β-Hydroxyhyoscyamine content remains very low in the eight culture media tested, while the concentration of littorine was found to be remarkably stable (0.08-0.09% dry wt) except in the MS media (0.04% dry wt).

Uozumi *et al.* (1996) reported that fructose is an excellent carbon source for the growth of carrot hairy roots. The hairy root cultures of *Hyoscyamus muticus* (Oksman Caldentey et al., 1994) *Armoralia rustica* and *Datura stramonium* (Payne *et al.*, 1987) were unable to grow on medium containing glucose as sole carbon sucrose. Of all the carbon sources sucrose was found to be ideal for hairy root cultures.

There have been many studies on the relationship between carbon source and culture growth. High sucrose concentration (2-3%) increases the scpolamine production in one strain of *Datura stramonium* hairy roots from 0.015 to 0.15% (Jaziri *et al.*, 1988).

Sauerwein and Shimomura (1991) reported that nitrate concentration (1-50mM) remarkably affected growth and alkaloid yield in hairy roots of *Hyoscyamus albus* transformed with *A. rhizogenes* MAFF 03-01724. The highest concentration of hyoscyamine was obtained with WP medium 3% sucrose with additional 15 mM nitrate. The optimal concentration for *D. stramonium* hairy roots was 20 mM (Payne *et al.*, 1987), where as the cell cultures of *Holarrhena antidysenterica* required 60 mM for the optimal accumulation of conessine (Panda *et al.*, 1992).

The additional amounts of metal ions Fe^{+2}, Ca^{+2} or Zn^{+2} had no effect on growth and production of tropane alkaloids in hairy root cultures of *H. albus*. However, the addition of Cu^{+2} (0.5 and 1.0 μm) had effect on growth and production of alkaloids in hairy root cultures of *H. albus*. Similar kind of results were observed with the cell suspension cultures of *Lithospermum erythrorhizon* (Fujita *et al* 1981). However, the light had no effect on growth and production of alkaloids from hairy root cultures of *H. albus*.

The effects of exogenously supplicd auxin/cytokinins on growth of transformed root cultures to vary between species. In some species such as Tagetes (Croes *et al.*, 1989) and Panax (Yoshikawa and Furuya 1987) auxins either alone or in combination with cytokinins stimulate growth by increase of branching. In others such as *Beta vulgaris* are without effect. However, in some solanaceous species these hormones cause significant effect on root growth and indeed stimulate disorganisation of the root structure, e.g. *Nicotiana rustica*. Gibberellic acid (0.1 and 10 μM) stimulated growth of transformed root cultures of *Brugmansia candia*. This stimulation in growth is characterised by a small effect on the rate of elongation (upto 30%) but a much more substantial effect on the rate of branching (Rhodes *et al.*, 1994).

19.3.3. Age and nature of explant

The age and nature of explant also has influence on the hairy root induction. Hairy root development can be described as a two-stage process.

a) Primary root meristem initiation, and

b) Root growth and secondary meristem formation.

Of these two, only the first one requires high auxin levels. Diffusion of auxin in the inoculated area, from cells with T_R-DNA, is sufficient to trigger the development of roots from those cells that have received only the T_L-DNA.

T_L-DNA and T_R-DNA both have rhizogenic functions but in most species the T_L-DNA appears to be more important in determining the hairy root phenotype. However, cooperation between the two T-DNA regions is probably necessary for vigorous root initiation, root growth being maintained by T_L-DNA alone. The ineffectivity of *A. rhizogenes* may also depend on the hormonal balance of the plant tissue or species, which is being infected.

19.4. CONFIRMATION OF TRANSFORMATIONS

Following are the ways to get confirmation of transformations :

19.4.1. Biochemical markers

a) Opines (sugar + amino acid groups) and

b) Mannopines

19.4.2. Genetic markers

a) Southern hybridization, and

b) Polymerase Chain Reaction (PCR).

19.4.3. Morphological characters

By using specified morphological changes we can get the result of confirmation of transformation.

19.4.1. Biochemical markers

Since the opine synthesis in *A. rhizogenes* infected plants is encoded by T-DNA of PRi plasmid (White *et al.,*1982, 1985); opines serve as an effective biochemical marker in elucidating transformed nature of the cultured root tissue (Tepfer, 1984). The detection of opines is a firm indication of transformed nature of hairy roots (Petit *et al.,* 1983). However, expression of the opine genes may be unstable with time (Kamada *et al.,* 1986).

There are several examples for which the transformation of hairy root culture were confirmed by the detection of opines and mannopines through paper electrophoresis.

Table 19.2. Confirmation of the hairy roots by paper electrophoresis

S.No.	Name of the culture	Amino acids	Reference
1.	*Rauwolfia serpentina*	Agropine and Mannopine	Benjamin *et al.,* 1993
2.	*Lawsonia inermis*	Mannopine	Bakkali *et al.,* 1997
3.	*Datura stramonium*	Agropine, Mannopine agrocinopine	Sikuli and Demeyer 1997
4.	*Atropa belladona*	Agropine, Mannopine	Kamada *et al.,* 1986
5.	*Papaver somniferum*	Mikimopine	Yoshimatsu and Shimomura, 1992
6.	*Solanum nigrum*	Agropine and Mannopine	Wei *et al.,* 1986
7.	*Valeriana officinalis*	Agropine and Mannopine	Granicher *et al.,* 1992
8.	*Swertia japonica*	Agropine and Mannopine	Ishimaru *et al.,* 1990
9.	*Trichosanthes kirilowii*	Agropine and Mannopine	Savary and Flores, 1994
10.	*Coleus forskohlii*	Mikimopine	Sasaki *et al.,* 1998
11.	*Cichorium intybus*	Agropine and Mannopine	Bais *et al.,* 2001

19.4.2. Genetic marker

Although synthesis of the opines is a firm indication that hairy roots are indeed transformed, the expression of the opine genes in hairy root tissue may be unstable with time (Kamada *et al.*,1986).

A localization in host plant genome serves as a reliable genetic marker to confirm transformation. There are a number of techniques available to demonstrate and locate T-DNA incorporation in the host plant chromosomal DNA. These include localization of T-DNA by southern hybridization as described by White *et al.*, (1982); Verification of transformed nature of a tissue by screening for the presence of foreign gene sequence by DNA 'blot dotting' (Draper and Scott; 1988); localization of T-DNA in plant chromosome tissue by *in situ* hybridization (Ambros *et al.*, 1986), and Dong *et al.* (1992). Among these techniques, southern hybridization remains one of the early used method and also widely employed till today (Flores *et al.*, 1987; Rhodes *et al.*, 1987b; Berthmieu and Jouanin, 1992; Han and Dougall, 1992).

Rhodes *et al.* (1994) have observed *Nicotiana rustica* root cultures which did not synthesize opines but which possessed the T_R- DNA on which the opine genes were reported to reside. Evidence from Southern analysis of transformed root cultures of several different species had confirmed. However, this analysis also indicates variations between root lines in the copy number and locus of insertion of the T-DNA and evidence for internal rearrangement of the T-DNA during transformation. The transformation of hairy root cultures of *Artemisia annua* was confirmed by southern hybridization and T_L-DNA was also detected. The transformation of hairy root culture of *Daucus carrota, (*David *et al.,* 1988*), Artemisia annua* (Chen *et al.*, 1999), *Cinchona ledgeriana* (Hamill *et al.*,1989) was also confirmed by southern hybridization.

19.4.2.1. Polymerase Chain Reaction (PCR)

PCR offers the fastest and accurate tool for verifying transgenes as well as determination of changes in a particular gene sequence resulting from tissue culture. PCR in combination with **RAPD** (Random Amplified Polymorphic DNA) technology has been used to confirm the presence of transforming DNA in single cells and protoplasts, and to follow them as recipient cells undergo mitotic division. A modification of PCR technology known as inverse PCR has been used to determine the T-DNA copy number in transgenic plants generated by *Agrobacterium* mediated transformation. Reverse transcriptase - PCR (RT- PCR) used to detect antisense transcripts in transgenic plants. The expression of foreign DNA in transformation experiments can be determined by RT-PCR. Southern analysis is still required to confirm the stable integration and inheritance of the gene. The integration of farnesyl diphosphate synthase (gene) was confirmed by PCR and southern blot analysis of transformed hairy root cultures of *Artemisia annua* (Chen *et al.*, 1999). Giulietti *et al.*, (1993) used PCR with primers for the rol B gene to amplify a 700bp fragment from a total DNA preparation from transformed roots of hairy root cultures of *Brugmansia candida* for the confirmation of transformation.

Reporter genes, such as the gene encoding β-glucuronidase (GUS) enzyme activity, may be used in T-DNA vectors to identify cells expressing foreign DNA or to optimise the transformation procedures. β-Glucuronidase enzyme activity (GUS-assay) was used to determine the genetic transformation of hairy root cultures of *Gentiana* species (Momcilovic *et al* 1997).

19.4.3. Morphological marker

It refers to confirming transformation by transformed root morphology exhibited by hairy root cultures and their transformed regenerants. The hairy roots have altered phenotype and these roots show high degree of lateral branching, profusion of root hairs and lack of geotropism (Tepfer and Tempe, 1981; Tepfer, 1982, 1983). Also the transformed regenerants of hairy roots inherit an aberrant phenotype in having wrinkled leaves, shortened internode and several other morphologically distinct features compared to their normal counterparts (Chilton *et al.,* 1982; Ooms *et al.,* 1985; Guerche *et al.,* 1987). Thus Ri T-DNA of *A.rhizogenes* can be said to provide two kinds of markers that function in root organ culture. The first is the selectable, rapid two growth. The second is the morphological, increased branching and plagiotropism.

19.5. TRANSFORMATION AND ESCAPE FREQUENCIES

Escapes arise in several ways, including the loss of T-DNA, loss of gene expression, protection by transformed cells, and somaclonal variation during tissue culture. Transformation efficiency, escape frequency and transformation frequency are the terms useful to evaluate the success of an experiment and the feasibility for gene introduction into crop plants.

19.5.1. Transformation efficiency (TE)

It is defined by the equation :

TE = Number of confirmed transformed shoots × 100/Total number of explants treated.

19.5.2. Escape frequency (EF)

It is defined by the equation :

EF = Number of non transformed shoots after selection × 100 / Total number of shoots after selection.

19.5.3. Transformation frequency (TF)

It is defined by the equation :

TF = Number of transformed shoots after selection × 100 / Total number of shoots after selection, e.g., Tobacco, TE is 25-50% and EF is 10-30%

19.6. SCREENING OF TRANSFORMANTS

The following assays are generally used initially to identify potential transformants.

19.6.1. Rooting assay

The explants are transferred into *Agrobacterium* suspension and they are fully submerged for several seconds. The sections are blotted on a filter paper and cultured on cocultivation media in perti dishes and incubated them for 2-3 days. Transformed roots and shoot will remain green and healthy on selective media, whereas non-transformed shoots do not root and turn bleached in appearance. There is a possibility of root formation by non-transformed shoots regenerated on selective medium by escapes. Therefore, additional means of verification are required for transformation, including molecular evidences. Molecular verification is recommended to verify the integration of introduced DNA into the plant genome. e.g., Screening of tobacco transformants.

19.6.2. Rooting and bleaching assay

Shoots are cut off from the explants and place them cut end down in the medium. Positive transformants will continue to grow without bleached explants and will root in the medium. The roots must come from the cut surface of the stem in the medium. Roots growing from the stem above the medium and then down into the medium or running along the medium do not indicate the presence of a transformed plant e.g., Screening for *Chrysanthemum* transformants.

19.6.3. Leaf callus assay

This is for leaf drugs. Leaf callus medium is prepared (BNCK 50[1] or BNCK 100[2]). Cut two leaves from each of the shoots to be tested. Wound the leaves slightly by running a scalpel over the epidermis of the leaf. Include a non-transformed control leaf for visual comparison when evaluating callus formation on BNCK 50 or BNCK 100. After the specified period (12-15 days for Chrysanthemum transformants) the assay can be scored. Transformed leaves will have formed callus, where as escapes will not.

19.6.4. GUS assay

This method is used when the GUS reporter gene is used in the vector. Place the plant tissue pieces in different wells of a multiwell plate. Add GUS stain, 200 microlitres in each well, for 24 well plate or 50 microlitres in each well, for 96 well plate. Incubate at 37°C or at room temparature for several hours. Examine the tissue under a dissection microscope. A blue colour develops when β-glucuronidase enzyme reacts with the GUS stain. Deep blue stained areas represent transformed tissues.

19.7. PROPERTIES OF HAIRY ROOTS

The hairy roots show certain distinct advantageous, growth characteristics over both normal root cultures and cell suspension cultures. The differences between hairy root, tap roots and adventitious roots are shown in Table 19.3.

Table 19.3. Differences between hairy roots, tap roots and adventitious roots (Banerjee *et al.,* 1995)

Sl. No.	Taproot	Normal adventitious roots	Transformed adventitious roots (Hairy roots)
1	*Occurrence* Only in dicotyledons	Occur in both monocotyledons and dicotyledons	Occur both in monocotyledons and dicotyledons
2	*Origin* From the root meristem of embryo irrespective of hypogeal or epigeal germination	They arise from large roots, or different parts of shoots from primordia. They are known to occur on hypocotyl, stem and rarely on leaves	A transformed cell can develop into hairy roots
3	*Vascular supply* Normal xylem and phloem cells are present in the vascular system and have connection to the main vasculature	Normal xylem and phloem cells are present in the vascular system and have connection to the main vasculature of the plant	Xylem and phloem kinds of cells are seen in the vascular system; whether or not they are connected to the normal vascular system of the plant is not yet established
4	*Tropism* Directed toward earth	Directed toward earth	Any direction
5	*Involvement of* DNA rearrangement Not known	Not known	Root induction is associated with random insertion of T-DNA from *Agrobacterium rhizogenes* into the host plant chromosomal DNA

BNCK 50[1] contains BNCK media MS major and minor salts, B_5 vitamins, sucrose (3% w/v), 2,4D (0.1 mg/l), NAA (0.1 mg/l), BA (1 mg/l), Gelrite (0.2% w/v) + Carbenicillin (500 mg/l) + Kanamycin (50 mg/l). BNCK 100[2] contains BNCK 50[1] media + Kanamycin (50 mg/l).

19.7.1. Hairy roots

Hairy roots are characterized by a high degree of lateral branching, profusion of root hairs and an absence of geotropism (Tepfer, 1984). They have high growth rates in culture, due to their extensive branching, resulting in the presence of many meristerms. Measurement of root growth in terms of increase in dry weight of biomass reflects largely involvement of process of cell enlargement rather than an increase in cell number (Banerjee-chatopadhyaya et al.,1985). The increase in the number of branches is approximately logarithmic during the early stages of growth and thus overall pattern of growth is very similar to cell suspension cultures. (Flores and Filner, 1985; Flores, and Flores, 1992, 1986). However, hairy root cultures do not require conditioning of the medium. Rhodes et al., (1987a) successfully inoculated fermentors with roots without apparent detriment to the subsequent growth rate.

Both the growth rate and extent of branching, highly inter-related characteristics vary between species and with the culture conditions. With *D. stramonium,* the density of branching per unit length of primary root is markedly affected by the ionic strength of the medium.

The growth of cultures of *Nicotiana rustica* (Robins et al 1987) and *Datura stramonium* (Payne et al., 1987) is influenced by the initial pH of the culture medium. Raising the initial pH from 5.8 to 7.0 increased the length of the lag phase before growth commenced and, particularly in *N. rustica,* affected the subsequent rate of growth. This contrasts with the findings of Mano et al (1986) in *Scopolia japonica,* in which raising the pH stimulated growth. It is anticipated, however, that the development of improved growth media and the use of different explant material and transforming bacterial strains may lead to cultures of increased growth rate.

19.7.2. Hairy roots are genetically stable

Hairy roots are genetically stable (Aird et al., 1988 a,b) and consequently they exhibit biochemical stablity that leads to stable, and high-level production of secondary metabolites (Kamada et al., 1986). Although the productivity of normal root cultures may be similarly exploited, their establishment and maintenance, with all but few species, is difficult and the auxin supplements needed for optimal growth can depress productivity. This is not the case with hairy root cultures. The cell and callus cultures are well known to give rise to mixed populations consists of polyploid and aneuploid cell (D'Amato 1985). Hairy root cultures apparently retain diploidy in all species so far studied (Banerjee-chatopadhyaya et al., 1985). The stable production of hairy root cultures dependent on the maintenance of organized state. The factors which promote disorganization and callus formation depress secondary metabolite production.

The productivity of hairy root cultures is stable over many generations in contrast to disorganised cell cultures. This stability is reflected in both the growth rate and the level and pattern of secondary metabolite production. Alkaloids from *H. muticus* and *N. rustica,* which have remained constant for one year. The cell division shows genetic stability. Integration of Ri T-DNA into plant chromosomes is stable which accounts for genetic stability of transformed root cultures (Banerjee et al., 1995) major variations between different hairy root clones are usually evident because probably T-DNA does not insert preferentially into any particular chromosomal location and may be present in multiple copies. The growth of disorganised cultures is associated with high level of genetic abnormalities leading to excessive polyploidy or aneuploidy. This somaclonal variation is valuable for screening. Sevon et al., (1998) obtained substantial somaclonal variation in growth rate, morphology and alkaloid production of *H. muticus* hairy root clones showed improved alkaloid production in relation to the parent clone. A major disadvantage is variation in chromosome number in some suspension cultures derived from various plants (Rhodes et al., 1987). The various advantages of hairy root cultures are given in Table 19.4.

Table 19.4. The Various Advantages Of Hairy Root Cultures over Cell Culture

Parameter	Hairy root culture	Cell culture
Biomass doubling time	2-7 days	0.7-14 days
Media	Simple, no hormones and vitamins	Complex
Genetic stability	More stable	Less stable
Product accumulation	Same as in parent	Less stable
Product release	More often released	Altered
Maximum biomass density	About 30g.d.w./1	200g.d.w./1
Inoculum size	Size independent	Dependent
Shear sensitivity	highly sensitive	Sensitive
Biomass handling	difficult	easy

Both hairy root cultures and normal root cultures show levels and patterns of secondary products similar to those in roots of intact plants. Parr and Hamill (1987) compared hairy root cultures of a range of *Nicotiana* species and the plants from which they were formed. The species were chosen on the basis of their different alkaloid compositions. The hairy root cultures derived from them reflected closely these differences in alkaloids content. The biosynthesis of certain compounds present in the roots of plants occur in the aerial parts and these compounds are deposited after back-transport. The accumulation in root cultures of compounds normally transported to the shoot in intact plants might occur. For e.g. *D. stramonium*, which has a very active transport system for translocating alkaloids in the plant, there is an increased hyoscyamine content in hairy roots relative to roots of an intact plant.

The effect of state of development of the root systems also affects the secondary metabolite production. Unusual tropane alkaloid levels are reported, relative to normal roots, in *H. niger* (Flores and Filner, 1985) and some hairy root lines of *Scopolia japonica*. Levels of tropane alkaloids in many root cultures are known to vary with age, and hence with root maturity. In both *Beta vulgaris* and *D. stramonium,* the concentration of the product increases in culture during stationary phase.

The accumulation of secondary metabolites approximately growth associated, the product content per unit biomass stays roughly constant. This may indicate the product is synthesized only in a segment of root of precise physiological age, the quantity of tissue of this type depend on the growth rate and branching rate of the cultures. Growth-independent production is desired for low concentration products. For e.g. betalain synthesis in *Beta vulgaris* and hyoscyamine synthesis in *D. stramonium* continue after growth of the culture ceases.

Some transformed root cultures release products into the growth medium. For example, nicotine is released by *N. rustica* (Hamill *et al.*, 1986), shikonin by *L. erythrorhizon* (Shimomura *et al.*, 1986), and thiophene by *Tagetes patula* (Ketel *et al.,* 1986). Release of nicotine into the culture medium could involve either diffusion from the surface of the root or by operation in culture of the normal mechanism for long distance transport. The culture environment can be used to enhance product release. The released products can be recovered by using flow-through systems or by using adsorbents in the medium. Nicotine can be recovered by flow-through systems (Table 19.5).

An increase in the productivity of the hairy root cultures is easy to achieve by manipulating the culture conditions (Nabeshima *et al.*, 1986; Parr and Hamill, 1987; Westcott, 1988; Protacio *et al.*, 1990; Mukundan and Hjortso, 1991; Sauerwein, 1991a,b). Secondary metabolite production has been successfully enhanced by the addition of precursors and metabolic intermediates to the growth medium or by the addition of biotic or abiotic elicitors (Eilert, 1987; Mukundan and Hjortso, 1990).

Table 19.5. Levels of secondary metabolites released into media

Plant	Compound	Percent	Reference
N. rustica	Nicotine	10-15	Hamill et al (1986)
N. tabacum	Nicotine	5-30	Wilson et al (1987)
L. erythrorhizon	Shikonin	25	Shimomura et al (1991)
C. roseus	Indole alkaloids	3-5	Rhodes et al (1989)
D. stramonium	Hyoscyamine	5-15	Rhodes et al(1989)

The genetic and biochemical stability of hairy roots facilitates the process designing and scale up of hairy roots at a commercially feasible level (Flores and Curtis, 1992). The large scale cultivation of hairy roots in fermenters without affecting the productivity from the cultures has been documented in fermenters (Hilton and Rhodes, 1990; Wilson *et al.,* 1990) and also in bioreactors (Hamill *et al.,* 1987; Scheidegger, 1990; Toivonen, 1993).

19.8. APPLICATIONS OF HAIRY ROOT CULTURES

19.8.1. Secondary metabolite production

Hairy root cultures are characterized by a high growth rate and are able to synthesize root derived secondary metabolites. Normally, root cultures need an exogenous phytohormone supply and grow very slowly, resulting in poor or negligible secondary metabolite synthesis. However, the use of hairy root cultures has revolutionized the role of plant tissue culture for secondary metabolite synthesis. These hairy roots are unique in their genetic and biosynthetic stability. Their fast growth, low doubling time, ease of maintenance, and ability to synthesize a range of chemical compounds offers an additional advantage as a continuous source for the production of valuable secondary metabolites. To obtain a high-density culture of roots, the culture conditions should be maintained at the optimum level. Hairy root cultures follow a definite growth pattern, however, the metabolite production may not be growth related. Hairy roots also offer a valuable source of root derived phytochemicals that are useful as pharmaceuticals, cosmetics, and food additives. These roots can also synthesize more than a single metabolite and therefore prove economical for commercial production purposes. Transformed roots of many plant species have been widely studied for the *in vitro* production of secondary metabolites (Table 19.1). Transformed root lines can be a promising source for the constant and standardized production of secondary metabolites. Hairy root cultures produce secondary metabolites over successive generations without losing genetic or biosynthetic stability. This property can be utilized by genetic manipulations to increase biosynthetic capacity. Sevon *et al.* (1997) characterized transgenic plants derived from hairy root cultures of *Hyoscyamus muticus* and concluded that a single hairy root that arises from the explant tissue is a clone.

19.8.2. Enhanced production of secondary metabolites from transformed root cultures

Agrobacterium mediated hairy roots of medicinal plants, in certain cases the production of secondary metabolites was higher in transformed root culture than in the roots of field grown plant (Table 19.6).

The contents of atropine and scopolamine in hairy roots of *Atropa belladonna* were examined (Kamada *et al.,* 1986). Though roots of *in vitro* developed plantlets contained these alkaloids, their concentration was lower than those of the normal plants grown in field. In contrast, hairy roots contained higher scopolamine and atropine content (0.024 and 0.371 % d.wt. basis) as compared to one-year-old field grown plant containing 0.008 and 0.34% d.wt. scopolamine and atropine, respectively (Table 19.7).

Table 19.6. Hairy root in which specific secondary metabolites are synthesized at higher levels than in the corresponding normal roots (Banerjee et al., 1995)

Sl. No.	Plant species	The strains of Agrobacterium rhizogenes (Ar) that induced hairy root	Secondary metabolite examined	Degree of hyper-production of metabolite by hairy roots $= \dfrac{\textit{Amout in hairy root}}{\textit{Amount in normal root}}$	Remarks	References
1.	Atropa belladonna	Ar ATCC 15834 carrying T-DNA the 35-S promoter H6H gene	Scopolamine	5		Hashimoto et al., 1993
2.	A. belladonna	Ar A$_4$	Atropine and hyoscyamine	2		Jung and Tepfer, 1987
3.	Cinchona ledgeriana	Ar LBA 9402	Quinine, quinidine and cinchonidine	3	Normal B$_5$ medium proved superior over others	Hamill et al., 1989
4.	Datura innoxia	Ar ATCC 15834	Hyoscyamine and scopolamine	6	Normal MS medium	Shimomura et al., 1991a
5.	D. quercifolia	Ar LBA 9402	Scopolamine and hyoscyamine	20		Dupraz et al., 1994
6.	D. candida	Ar ATCC 15834	Scopolamine and hyoscyamine	2		Christen et al., 1991
7.	Duboisia leichhardtii	Ar ATCC 15834 and Ar A$_4$	Scopolamine	2	Heller's HF medium proved best	Mano et al., 1989
8.	Fagopyrum esculentum	Ar ATCC 15834	(+) Catechin (-) epicatechin-3-O-gallate procyanidin B$_2$-3'-O-gallate	8		Trotin et al., 1993
9.	Hyoscyamus niger	Ar ATCC 15834	Hyoscyamine and scopolamine	2		Shimomura et al., 1991
10.	Rubia tinctoria	Ar ATCC 15834	Anthraquinone	19	Normal MS medium supplemented with 5 ìM IAA proved best	Saito et al., 1991
11.	Tagetes patula	Ar LBA 9365	Thiophene	25		Croes et al., 1989
12.	Valeriana officinalis	Ar R 1601	Valpotriates	7	Half-strength B$_5$ sallts containing 20% sucrose	Granicher et al., 1992

The scopolamine and hysocyamine content in hairy roots of *Datura candida* hybrid was quantified using GLC and HPLC and compared with the non-transformed plants (Christen *et al.*, 1991). The alkaloid yield (0.68% d.wt.) obtained with hairy roots was 1.6 and 2.6 times the amount found in the aerial parts and in the roots of the parent plants, respectively. Scopolamine being the principal alkaloid, scopolamine/hyoscyamine ratio 5:1 makes these hairy root cultures as potential source of scopolamine.

Table 19.7. Growth and Production of Secondary Metabolites by Transformed Roots As Compared To Those Of Non Transformed Roots.

Plant species	Biomass increase (dry wt.)		Producer Yield (%dry wt)	
	Normal	Hairy root	Normal	Hairy root
N. tabacum	0.03	0.2	2	3
A. belladonna	0.10	0.4	0.8	1.32
P. ginseng	0.31	0.4	0.91	0.36
H.muticus	0.05	0.5	0.61	0.52
H.niger	0.05	0.5	0.52	0.55

Hairy root cultures of *Valeriana officinalis* var *sambucifolia* (Granicher *et. al.*, 1992) were established using *Agrobacterium* strain R 1601. The valepotriate content in the hairy roots cultured on 40 days in 10 different media containing 2-7% sucrose was examined and compared with 9 months old non-transformed plants. In each case, the valepotriate concentration in the transformed root cultures was significantly increased in comparison with the roots of non transformed plants.

Transformed roots of *Catharanthus roseus* produce indole alkaloids (Davioud *et al.*, 1989). Transformed roots of *C. roseus* were obtained following infection of detached leaves with *A. rhizogenes* strain LBA 9402 (Parr *et al.*, 1990). About 20-fold increase in root mass production was observed within 28 days of culture and showed a high level of alkaloid production at all stages of growth cycle.

A. rhizogenes mediated hairy root cultures of *Lithospermum erythrorhizon* (Shimomura *et al.*, 1991) produced a large amount of red pigment shikonin, which leached into the solid or liquid medium. The addition of adsorbents such as Amberlite XAD-2, to liquid medium not only resulted in 3-fold increase of shikonin but also increased growth of hairy roots.

19.8.3. Production of Novel secondary metabolites from hairy root cultures

Hairy roots of *Swertia japonica* (Ishimaru *et al.*, 1990) an important bitter stomachic and effective in the treatment of hepatitis, produce bellidifolin, methylbellidifolin, amarogentin, and amaroswerin. These bitter compounds were identical with those of the authentic isolated from mother plants. A new xanthone derivative 8-O-primeverosyl-bellidifolin not found in mother plant, was also reported in hairy roots of *S. japonica* (Table 19.8).

Table 19.8. Hairy roots that have been observed to sythesize novel secondary metabolites and not reported to be present in the control tissues (Banerjee et al., 1995)

Sl. No.	Plant species	Agrobacterium rhizogenes strain Which induced Hairy root	Class of secondary metabolites/name of compound was observed	Culture condition(s) under which synthesis	Reference
1	Astragalus membranaceous	Ar pRi 15834, PGs Gluc	Agroastragaloside I saponins and poly-saccharides	Gamborg B_5 medium 3% sucrose at 25°C in dark at 60 rpm	Hirotani et al., 1994
2	Lobelia inflata	Ar ATCC 15834	Polyacetylene/Catechin, Robetyolinin	Normal MS medium	Ishimaru et al., 1991
3	Rauwolfia serpentina	Ar A_4	12-hydroxyajmaline	Normal B5 liquid medium	Faskenbagen et al., 1993
4	Swertia japonica	Ar ATCC 15834	8-O-primeverosyl bellidifolin	Normal root culture medium	Ishimaru et al., 1990

Panax ginseng, which has been used in Chinese traditional medicine since ancient times, is the main source of ginsenosides, the pharmacologically active compounds. Thirty-four strains of fast growing roots were selected and maintained in liquid normal LS medium. A strain R 52 was selected amongst the 34 strians based on the growth rate, ginsenoside content and stability. The total ginsenoside content 1.7-2.2% (which is higher than values in the control study) was obtained when the media was repeated exchanged at 7 days interval in shake flask or in bioreactor. The ginsenoside content in hairy roots was equal to 5 years old field grown plants. Strain R52 grew fast with profused lateral branching, stably producing ginsenoside in hormone free medium for five years. This high and stable ginsenoside production by strain R52 and the achievement of bioreactor cultivation imply that there is a good potential for commercial application of R 52 (Inomata *et al* 1993).

Hairy root cultures of *Fagopyrum esculentum* transformed with *A. rhizogenes* strain 15834 are a useful tissue for the production of catechins and procyanidin with a higher productivity than in plant roots. The values for (+) catechin were 8-fold higher; (-) epicatechin-3-0-gallate, 4 fold higher and procyanidin B 23-0 gallate 8-fold higher, whereas the production of other flavonals remained about the same in both types of cultures.

19.8.4. Improving secondary product formation in hairy root cultures

It should be beneficial to initiate cultures from individual plants of high biosynthetic capacity. Screening natural plant populations shows wide variation to exist in their biosynthetic capacity (Weiler and Westekemper, 1979), while in some genera (e.g. *Datura*) high-yielding varieties have been generated in interspecific hybridization and selective breeding programs. With disorganized cultures it is disputed whether initiating cultures from high-grade explants is advantageous (Roller, 1978), but, as already discussed, hairy roots have productive capacities similar to the plants from which they were derived. Inspite of this, some variations between hairy root clones in the level and relative proportions of products does occur. For instance, Mano *et al*. (1986) selected from a population of 1500 hairy roots of *Scopolia,* one line relatively high in hyoscyamine accumulation and one in scopolamine.

Hairy root cultures are amenable to screening in a manner analogous to that used for dispersed systems. In the presence of high levels of exogenous auxins, hairy root cultures spontaneously form callus, from which suspension cultures can be obtained. If, however, such transformed suspension cultures are returned to hormone-free media, roots generate. Flores (1986) has demonstrated that those roots, which generate after passage through a callus stage, regain the capacity for alkaloid production lost in the disorganized state. Rhodes *et al*. (1987) shown those hairy roots of *N. rustica* can be generated from individual protoplasts prepared from suspension cultures, which have been initiated from hairy root tissue. In this case, considerable variation in both the growth rate and morphology and in nicotine accumulation was observed between different regenerants. The phenotype of the regenerated roots appears to be stable, in line with the return to an organized state. Thus, the variation induced by disorganization of the transformed culture in the presence of auxins can be exploited with the added advantage that selected cell lines can be stabilized as hairy roots.

It is also possible to select positively lines with the desired properties at the whole root, dispersed culture or protoplast level. Robins *et al*. (1987) have shown that transformed roots of *N. rustica* are unable to grow at levels of nicotinic acid above abut 2.5 mM, yet are insensitive to nicotine at 4-fold this level. As nicotinic acid is the direct precursor of nicotine, cells or roots with a high synthetic capacity should grow in the presence of nicotinic acid since the formation of nicotine may be one of the mechanisms by which it will be detoxified.

This selection pressure, applied at the single-cell derived colony stage, can potentially generate clonal roots of elevated stable synthetic capacity. In *Datura stramonium*, tropic acid, the precursor of hyoscyamine, behaves similarly (Robins *et al*., 1987) and may be applied in a similar way. Indeed the same approach might be applicable to any system where an acidic compound is proximal precursor of the desired product.

Hairy root cultures can be used to obtain selected products. In case of *Valeriana officinalis*, the main constituents of the essential oil of non transformed roots were bornyl acetate and valerenal. The main constituents of the oil from hairy roots are kessyl alcohol and kessyl acetate. A new iridoid ester, validate, was isolated from hairy roots, which is not detected in non-transformed roots.

Since plasmid transfer is involved in the formation of hairy roots, the system is highly amenable to manipulation at the genetic level. Foreign genes have also been inserted into plants during transformation with *A. rhizogenes* by either insertion into the Ri plasmid T-DNA (Comai *et al.*, 1985) or in binary co-transformations using disarmed vectors derived from *A. tumefaciens* (e.g. pBIN19). In the latter events, T-DNA from both plasmids is inserted into the plant genome, infection being driven by the DNA of the *vir* region on the Ri plasmid.

Table 19.9. Hairy root culture growth indices and duplication times in days(*) (Adopted from Vargas and Ham, 1995)

Family	Genus and species	Growth (times/days)	Reference
Apocynaceae	*Amsonia elliptica*	4	Sauerwein *et al.*, 1991b
	Catharanthus roseus	28/17	Toivonen *et al.*, 1989
		52/40	Bhadra *et al.*, 1993
		2.8*	Ciau-Uitz *et al.*, 1994
	Catharanthus trichophyllus	2045, 25/66	Davioud *et al.*, 1989
	Rauwolfia serpentina	6.2/35	Benjamin *et al.*, 1994
Araliaceae	*Panax ginseng*	3.07/21	Yoshikawa and Furuya, 1987
Asteraceae	*Tagetes patula*	81/14	Kyo *et al.*, 1990
Campanulaceae	*Lobelia inflata*	20-60.28	Yonemitsu *et al.*, 1990
Gentianaceae	*Swertia japonica*	150.56	Ishimaru *et al.*, 1990
Labiatae (Lamiaceae)	*Ajuga reptans*	230.45	Matsumoto and Tanaka, 1991
Linaceae	*Linum flavum*	9*	Oostdam *et al.*, 1993
Pedaliaceae	*Sesamum indicum*	33.3*	Ogasawara *et al.*, 1993
Polygonaceae	*Fagopyrum esculentum*	70/21	Trotin *et al.*, 1993
	Cinchona ledgeriana	6-8/28	Hamill *et al.*, 1989
Solanaceae	*Atropa belladonna*	60/28	Kamada *et al.*, 1986
	Datura candida X	20.28	Christen *et al.*, 1989
	Datura stramonium	0.95*	Mendoza *et al.*, 1992
		55/28	Payne *et al.*, 1987
	Duboisia leichhardtii	64/28	Marn *et al.*, 1989
	Hyoscyamus albus	366/21	Shimomura *et al.*, 1991
	Hyoscyamus niger	3.86/22	Jaziri *et al.*, 1988
	Nicandra physaloides	28.26	Parr, 1992
Valerianaceae	*Valeriana officinalis*	112/50	Granicher *et al.*, 1992

An approach yet to be tested with hairy root tissue is the uptake of the naked DNA of engineered plasmids into protoplasts, which has proved a successful method of transformation (Lorz *et al.*, 1985). Using the binary system, hairy roots of several species containing the neomycin in phosphotransferase gene (from Tn5) have been generated which are resistant to kanamycin (Simpson *et al.*, 1986; Shahin *et al.*, 1986). As the gene copy number may be influenced by the level of kanamycin used to select the roots initially, such a system might be used to insert into the roots varying levels of genes relevant to the pathway under consideration. For example, the amount of a rate-limiting enzyme could be increased by having a higher copy number of level of expression.

Resistance to the herbicide glyphosate was demonstrated in *Petunia* by introduction of a gene for 5-enolpyruvylshikimate-3-phosphate (EPSP) synthase under the control of a high expression viral promoter, which led to an elevated level of EPSP synthase and conferred tolerance to otherwise toxic levels of herbicide. Alternatively genes coding for an enzyme of altered regulatory properties can be inserted, as again demonstrated with EPSP synthase and glyphosate-resistance.

The advantage of these methods involving specific manipulations, is that the genome of the host plant may be augmented with foreign genes, or genes identified as playing key roles in the pathway can be returned to the host with different regulatory properties. In both cases the inherent advantages of the growth and productivity of the hairy root tissue are retained. Thus, the manipulation of secondary metabolism by altering the expression of key genes in the pathways is now a real possibility. Studies aimed at identifying key regulatory enzymes in the pathways of secondary metabolite formation are a prerequisite to genetic manipulation but the necessary information is at present only available in a few cases.

Polyploidy may be induced sometimes to get increased yield of secondary metabolites. Ermayanti *et al.* (1994) attempted enhancement of swainsonine production by induction of polyploidy by colchicine in hairy root cultures of *Swainsona galegifolia.*

By using precursors, the productivity of hairy root cultures can be improved e.g., alkaloids from *N. rustica*. Hairy root cultures can be used in anabasine biosynthesis from cadaverine in transformed root cultures of *Nicotiana* sp. Cadaverine fed to root cultures of *Nicotiana* transformed with *A. rhizogenes* stimulates the production of anabasine.

By using elicitors, the productivity of hairy root cultures can be improved. e. g., phytoalexins from *H. muticus* by fungal elicitors. Vazquez-flota *et al.,* (1994) used fungal elicitors viz., Aspargillus sp., *Trichoderma viride, Trichoderma reseii, Trichoderma longibrachiatum, Rhodotorula marina* for the production of ajmalicine and catharanthine in hairy roots of *C. roseus.* Shimomura *et al* (1991) used adsorption by polymeric resins for the improvement in production of shikonin in hairy root cultures of *L. erythrrorrhizon.* Sugars may be used sometimes to get increased yield of secondary metabolites. Ermayanti *et al.,* (1994) attempted enhancement of swainsonine production in hairy roots of *Swainsonia galegifoila* by treatment with sugars like sucrose, glucose and fructose.

19.8.5. Plant regeneration

Transformed roots are able to regenerate whole plants; hairy roots as well as the plants regenerated from hairy roots are genetically stable. However, in some instances transgenic plants have shown an altered phenotype compared to controls. Plants regenerated from Ri transformed roots display 'hairy root' syndrome. Combined expression of the *rol A B C* loci of the Ri plasmid is responsible for this expression (Fig. 19.2). Each locus is responsible for a typical phenotypic alteration; that is *rolA* is associated with internode shortening and leaf wrinkling; *rolB* is resonsible for protruding stigmas and reduced length of stamens; *rolC* causes internode shortening and reduced apical dominance (Cardarelli *et al.*, 1987; Palazon *et al.*, 1998).

Plants can be regenerated from hairy root cultures either spontaneously (directly from roots) or by transferring roots to hormone-containing medium. The advantage of Ri plasmid-based gene transfer is that spontaneous shoot regeneration is obtained avoiding the callus phase and somaclonal variations. Ri plasmid-based gene transfer also has a higher rate of transformation and regeneration of transgenic plants; transgenic plants can be obtained without a selection agent thereby avoiding the use of chemicals

Table 19.10. Transgenic plants obtained by Agrobacterium rhizogenes mediated transformation (Adopted from Giri and Lakshmi Narasu, 2000)

Plant	Gene introduced	References
Ajuga s pp	GUS	Uozumi *et al.* 1986
Anthyllis vulneraia	NPTT II, ipt	Stiller *et al.* 1992
Atropa b elladonna	Bar, 6 βH	Saito *et al.* 1992
Brassica n apus	GUS, NPT II,ALS	Christey and Sinclair 1992
B. napus	NPT II	Christey *et al.* 1997
B. campestris	GUS, NPT II,ALS	Christey *et al.* 1997
B. o leracea	NPT II GUS	Downs *et al.* 1994
B. o leracea	GUS, NPT II,ALS	Christey *et al.*1997
B. campestris	NPT II	Trulson *et al.* 1986
B. napus	GS	Rech *et al.* 1989
Brassica sps	NPT, Bt, GUS,35 EFE5 '7' gene	Kumar *et al.* 1991
Cucumis satives	NPT II	Otani *et al.* 1993
G. canescens	NPT II	Morgan *et al.* 1987
Glycine a rgyea	NPT II	Shin *et al.* 1994
Ipomoea b atatus	NPT II GUS	Forde *et al.* 1989
L. peruvianum	NPT II	Shanin *et al.*1986
Larix decidua	NPT II, aro A, BT	Thomas *et al.* 1992
Lotus c orniculatus	GS from *Phaseolus vulgaris*	Damini and Arcionic, 1991
Lycopersicon e sculentum	NPT II	Davery *et al.* 1987
Medicago t runcatula	NPT II	Hatamoto *et al.* 1990
M. arborea	HPT II	Hamill *et al.*1990
Nicotiana debneyi	NPT II	Palazon *et al.* 1998
Nicotiana p lumaginifolia,		
N. t abacum	NPT II	Pyrhould *et al.* 1987
N. rustica	ODS	Pyrhould *et al.* 1987
Nicotiana spp	ROL	Davey *et al.* 1987
Populus triocarpa ×		
P. d eltoides	NPT II	Visser *et al.* 1989
Robina pseudoacasis	NPT II	Lodhi *et al* 1996
S. nigrum	NPT II	Lodhi *et al.* 1996
S.tuberosum	NPT II, GUS	Mciness *et al.* 1991
Solanum d ulcamara	NPT II rol	Manner and Hay1989
Stylosanthes h umilis	NPT II	Manner and Hay 1989
Verticordia g randis	NPT II, GUS	Stummer *et al.* 1995
Vinca minor	NPT II,GUS	Tanaka *et al.* 1994
Vitis vinifera	NPT II, GUS	Flores *et al.* 1988

that inhibit shoot regeneration; high rate of co-transfer of genes on binary vector can occur without selection (Table 19.10). Further, *Agrobacterium tumefaciens* mediated transformation consistently yields only transformed cells that can be obtained after several cycles of root tip cultures. These hairy roots can be maintained as organ cultures for a long time and subsequent shoot regeneration can be obtained without any cytological abnormality. Rapid growth of hairy roots on hormone-free medium and high plantlet regeneration frequency allows clonal propagation of elite plants. In *in-vitro* cultures, the hairy root regenerants show rapid growth, increased lateral bud formation, and rapid leaf development, these regenerants are useful for micropropagation of plants that are difficult to multiply (Perez-Molphe *et al.*, 1998; Hoshino and Mii, 1998; Gutierrez-Pesce *et al.*, 1998). Altered phenotypes are produced from hairy root regenerants and some of these have proven to be useful in plant breeding programs. Morphological traits with ornamental value are abundant in adventitious root formation, reduced apical domainance, and altered leaf and flower morphology. Dwarfing, altered flowering, wrinkled leaves, or increased branching may also be useful for ornamentals. Dwarf phenotype is an important characteristic for flower crops such as *Eustoma grandiflorum* and *Dianthus* (Giovanni *et al.*, 1997). Higher levels of some target metabolites have been found in the levels of plants regenerated from hairy roots, so plant regeneration is an important aspect for production of these chemicals. Pellegrineschi *et al.* (1994) improved the ornamental quality of scented *Pelargonium* sp. This plant has pleasant odor but its long internodes and ungainly growth makes it unattractive, and hairy root regenerants are of shorter stature. In snapdragon, the flower number was increased upon transformation (Handa *et al* 1995). Some perennial forage legumes turned annual after transformation (Damiani and Aricioni, 1991).

19.8.5.1. Tree improvement

A major limitation of tree improvement programs is their generation cycle. Conventional breeding programs in trees are slow and tedious and it is difficult to introduce specific genes for genetic manipulation by crossing parental lines. *Agrobacterium rhizogenes* mediated transformation can be a useful alternative, as a rapid and direct route for introduction and expression of specific traits (Huang *et al.*, 1991). Transformation of trees and subsequent regeneration of transgenic plants has been reported for only a few genera (Zhan *et al.*, 1988, Mc Granahan *et al.*, 1988, 1993, Cabrera-Ponce *et al.*, 1996.). The ability to manipulate tree species at cellular and molecular level shows great potential and *in-vitro* transformation and regeneration from hairy roots facilitates application of plant biotechnology to tree species. This significantly reduces the time necessary for tree improvement and gives rise to new gene combinations that cannot be obtained using conventional breeding methods. In some tree species root initiation limits vegetative propagation; by using *A.rhizogenes* rooting of cuttings from recalcitrant woody species have been improved. Roy (1989) demonstrated this for some fruit trees such as peach, apple, cherry and olive. McAfee *et al.* (1993) reported it for *pinus* and *Larix* sp. Rugini and Mariotti (1991) demonstrated successful rooting of some tree species. These methods have the potential to increase the efficiency of plant propagation in crops where propagation is difficult *A. rhizogenes* mediated transformation had the potential to introduce foreign genes specifically into root systems.

19.8.5.2. Genetic manipulation

Transformed roots provide a promising alternative for the biotechnological exploitation of plant cells. *A. rhizogenes* mediated transformation of plants may be used in a manner analogous to the well-known procedures employing *A.tumefaciens*. *A. rhizogenes* mediated transformation has also been used to produce transgenic hairy root cultures and plantlets have been regenerated. With the exception of the border sequences, none of the other T-DNA sequences are required for the transfer. The rest of the T-DNA can be replaced with the foreign DNA and introduced into cells from which whole plants can be regenerated. These foreign DNA sequences are stably inherited in a mandelian manner (Stougaard *et al.*,1987, Zambryski *et al.*, 1989). The *A. rhizogenes* mediated transformation has the advantage that any foreign gene of interest placed in binary vector can be transferred to the transformed hairy root clone.

It is also possible to selectively alter some plant secondary metabolites or to cause them to be secreted by introducing genes encoding enzymes that catalyze certain hydroxylation, methylation and glycosylation reactions. An example of a gene of interest with regard to secondary metabolism that was introduced into hairy roots is the 6-β-hydroxylase gene of *H. muticus,* which was introduced to hyoscyamine rich *A. belladonna* by binary vector systems using *A. rhizogenes.* In another instance, engineered roots showed an increased amount of enzyme activity and a five-fold higher concentration of scopolamine (Hashimoto *et al.,*1993). Hairy root cultures of *Nicotiana rustica* with ornithine decarboxylase gene from yeast (Hamill *et al.,*1990) and *Peganum harmala* with tryptophan decarboxylase gene from *C. roseus* (Berlin *et al.,*1993) have been shown to produce increased amounts of the secondary metabolites nicotine and serotonin when expressing transgenes from yeast. Transgenic plants produced either by binary or co-integrate vectors are summarized in Table 19.10.

In 12 Brassica cultivars transgenic plants with genes from binary vectors have been obtained and the plant showed hairy root phenotype to varying degrees and were fertile. Segregation analysis confirmed the transmission of traits to the progeny (Christey *et al.,* 1997). Due to independent insertion of the Ri T-DNA and binary vector T-DNA in subsequent generations, phenotypically normal transgenic plants were produced in tobacco (Hatamoto *et al.,*1990) and in *Brassica napus* (Boulter *et al.,*1990). Downs *et al.* (1994) reported transgenic hairy roots in *Brassica napus* containing a glutamine synthase gene from soybean showed a three-fold increase in enzyme activity. When a bacterial isochorismate synthase gene was cloned in a binary vector and then mobilized into *A. rhizogenes,* the transgenic hairy root *Rubis peregrina* cultures containing this gene expressed twice as much isochorismate synthase activity as the roots of control plants and accumulated 20% higher levels of total anthraquinones (Lodhi *et al.,* 1996). Recently, there has been considerable attention given to the specific induction of secondary metabolite in transgenic plant cell cultures using inducible promoters (Sommer *et al.*, 1998). This approach can be extrapolated to hairy root cultures for yield enhancement. In addition new secondary metabolites can be induced in transgenic hairy roots by introducing anthocyanin transactivators (Damiani *et al.,* 1998). In the near future, this approach may, be a reality for the commercial production of pharmaceutically important compounds using transgenic hairy root culture system. Recently a number of genes including tryptophan decarboxylase, strictosidine synthase, tropinone reductase, berberine bridge enzyme, and berbamunine synthase have been isolated and used for the metabolic engineering of secondary metabolic pathways.

An additional advantage of hairy root cultures is for enzymological studies. Abundant quantities of sterile, rapidly growing tissue can be generated. In hairy roots the proportion of meristematic tissue is high and phenolic contents are lower than in normal plant roots, leading to an increased level of enzyme activity (McLauchlan *et al.,* 1993; Walton *et al.,* 1994).

Artificial seeds have been developed by encapsulating root segments and shoot primordial (Nakashimada *et al.,* 1995). Root tips of hairy roots of *Panax ginseng* (Yoshimatsu *et al.,* 1996) and shoot tips of hairy roots regenerants have been cryopreserved in horseradish (Phunchindwan *et al.,* 1997). These can be regenerated and cultured when needed. Hairy roots in the form of transformed plant organs provide a promising means for the biotechnological exploitation of plant cells. Artificial seeds are a reliable delivery system for clonal propagation of elite plants with genetic unifomity, high yield, and low cost of production. Plant cells used for artificial seed production must have a good ability to regenerate. Micropropagation can be done from hairy roots using artificial seeds (Uozumi *et al.,* 1996). Artificial seeds using hairy roots have further potential for mass propagation, and modifications in bioreactor design, image analysis with computers and robotics can improve the process.

19.8.6. Production of proteins

Production of industrial and therapeutic proteins in hairy roots is receiving a lot of attention. In the past few years, hairy root cultures of *Trichosanthes kirlowii* (Shih *et al.,* 1998) and *Luffa cylindrica* (Poma *et al.,* 1997) produced ribosome inactivating protein. Hairy roots, besides being a source of important proteins and enzymes can also be used as a source of heterologus proteins. Following the recent advances in plant genetic engineering, plant cells have been used for expression of a range of foreign proteins of direct commercial value (Hogue *et al.,* 1990, Gao *et al.,* 1991, Gao and Lee 1992, Domansky *et al.,* 1995). Of all these proteins antibodies and antibody fragments have received maximim attention. Monoclonal antibodies were secreted by hairy roors of *N. tabacum* (Wongsamuth and Doran 1997). Monoclonal antibodies are produced in bulk using hybridoma cells obtained by fusion between lymphocytes and myeloma cells. Recently, several heterologous systems such as bacteria, yeast, insect cells and non-lymphoid mammalian cells have been tested for antibody production (Better *et al.,* 1988; Horowitz *et al.,* 1988; Weidle *et al.,* 1987). Plant cells can serve as an alternative to animal cell culture for production of monoclonal antibodies and hairy roots fit in as the material of choice (Wongsmuth and Doran, 1997; Doran 2000). Transgenic tobacco plants expressing Guy's 13 antibody - a monoclonal antibody which binds to a surface protein of *Streptococcus mutans,* a causative agent of dental caries were used for induction of hairy roots. These produced the antibody, which was secreted into the medium and was higher than reported in other heterologus systems (Wongsamuth and Doran 1997) Secretion of antibody into the medium is beneficial for downstream processing. Antibody degradation appears to occur to a greater extent in hairy roots than in hybridoma system and further work is required to stabilize the "plantibody" production in hairy root cultures (Wongsamuth and Doran 1997).

19.8.7. Use of Hairy root cultures in rhizosphere research

The rhizosphere is a site of convergence of several biological factors, which are important to the life of the plant (Tepfer 1989). The root is surrounded by a variety of organisms - fungi, bacteria, animals and roots of other plants and the complexity and inaccessibility hamper the study of rhizosphere. One solution is to use transformed hairy root cultures for studying the interactions between various organisms in the rhizosphere. The hairy roots were used to study the interaction between aphids and plant roots (Wu *et al.,* 1999) and *Bradyrhizobium* and peanut hairy root (Akasaka *et al.,* 1998). Transformed roots provide a source of axenic material for modeling plant microbe interaction in the rhizosphere.

Plant roots exude substances, which act as attractants, stimulants and repressors of the activity of other organisms. Some of the secondary metabolite molecules may act as chemoattractants and give signals to the bacterium. It may be possible to target a soil bacterium to a species. The hairy root culture thus provides an ideal experimental mode system for unraveling the mysteries of plant-microbe interactions in the rhizosphere.

19.8.8. Hairy Roots for Culture of Obligate Parasites

Hairy root culture can be used to establish *in-vitro* cultures of soil organisms, which require roots for growth. Obligate root parasites including *Plasmodiophora brassicae* and *Polymyxa betae* have been cultivated on transformed root cultures of sugarbeet (Mugnier, 1987). Infections with vesicular-arbusculaar mycorrhizal fungi, *Glomus mosseae* Gerdemann and Trappe and *Gigaspore margarita* Backer and Hall have been obtained on hairy root cultures of *Convolvulus sepium* L. (Mugnier and Musse, 1987). Besides, the root nematode *Meliodogyne javonica* was successfully cultured on transformed root cultures of potato, tomato (Verdejo and Jaffee, 1988). Soybean cyst nematode (Cho *et al.,*2000) was successfully propagated in soybean hairy roots. $Hs_1 Pro_1$ gene that confers resistance to beet cyst nematode (*Heterodera schachtii* Schmidt), a major pest in the cultivation of sugarbeet, was cloned and expressed in susceptible sugarbeet hairy roots where it conferred resistance to infection by the beet cyst nematode (Cai *et al.,*1997).

Besides study of host-parasite interaction, screening of agrochemical for fungicide, herbicide, nematicide activity and research into crop disease resistance have been enhanced through the study of hairy root cultures. •

19.8.9. Hairy Roots for Phytoremediation of elemental and Organic Pollutants

Hairy roots can be used for studies on phytoremediation of elemental or organic pollutants. Elemental pollutants include toxic heavy metals and radionuclides such as arsenic, cadmium, cesium, chromium, lead, mercury, strontium, technetium, uranium, and organic pollutants include polychlorinated biphenyls (PCBs), polycyclic aromatic hydrocarbons (PAHs) and linear and halogenized hydrocarbons. Phytoremediation of elemental pollutants may occur by adsorption, translocaton and storage in vacuoles. Rhizofiltration is the use of plants such as Indian mustard and *Chenopodium amaranticolor* were shown to take up uranium from a solution of up to 500 mM within a short period of incubation (Eapen *et al.,* 2000). Hairy root cultures of *Solanum nigrum* and *Solanum aviculare* were shown to take up cadmium (Macek *et al.,*1994,). Large-scale culture of hairy roots from hyperaccumulators and use in rhizofiltration of elemental pollutants will help in containing the dangerous pollutants.

Hairy roots can also be used for degradation of organic pollutants. Removal of polychlorinated biphenyl was studied using hairy roots of *Solanum nigrum* (Kas *et al.,* 1997, Kucerova *et al.,* 2000) *Armoracea rusticana, Solanum aviculare* and *Atropa belladonna* (Mackova *et al.,* 1997). Hairy roots of *Myrophyllum aquaticm* (Hughes *et al.,*1997) and *Catharanthus roseus* (Bhadra *et al.,* 1999) were used to study the fate of nitroaromatic explosive TNT.

The availability of genes for heavy metal tolerance and degradation of organic pollutants and introduction into hairy roots will help developing high potential hairy root cultures for rhizofiltration. Some of the genes could be introduced into model hairy roots to study the mechanism of uptake and transport of heavy metals. Plants can be regenerated from hairy roots, where these genes have been introduced. Genetic manipulation to enhance bioaccumulation using hairy roots will be one of the prospective areas of research.

19.8.10. Hairy Roots for Modifying Plant Architecture

When plants are regenerated from hairy roots induced by *A. rhizogenes,* the phenotypes are stably altered which though are regarded as undesirable, yet these can have potential applications in modifying the plant architecture. The dwarf ornamental potted plants with increased branching and number of flowers having more appeal to the customers may probably be the first commercial products from this genetic transformation. Pellegrineschi *et al.,* (1994) improved the ornamental quality of scented *pelargonium* by producing short-statured, highly-branched plants architecture modification has been reported in *Rosa hybrida (*Handa *et al.,*1994) *Begonia tuberhybrida* (Kiyokawa *et al.,* 1996). In *Osteospermum ecklonis,* different ornamental characters could be altered by introduction of 'rol' A,B,C gene were shown to break the stem apical dominance resulting in increased branching growth rate, production index and delayed winter dormancy (Tzfira *et al.,*1999). In trifoliate orange, 'rol'C gene produced shorter plants with enhanced rooting (Kaneyoshi and Kobayashi 1999). 'Rol genes were introduced into rice plants through particle gun bombarment, which increased tillering capacity (Nayak *et al.,* 1999). Increased tillering will result in more biomass in grasses, which are used as fodder.

The 'rol' genes also are known to have effect on the life cycle of plants changing biennials to annuals as in the case of *Cichorium intybus* (Sun *et al.,* 1991) and *Daucus carota* (Limami *et al.,*1998). Damiani and Arcioni (1991) noted that in forage legumes too, change from perenial to annual and delayed flowering inhibition of flowering as in *Ajuga reptans* (Tanaka and Matsumoto 1993) and *Pelargonium* (Pelligrineschi *et al.,* 1994). However, in genitan, flowering was accelerated in transformed plants (Suginuma and Akihama 1995). 'Rol' C gene was shown to induce male-sterility (Schmulling *et al.* 1993), which may have specific implication in development of male sterile lines in crop plants. In potato, plants having the rol C gene had an increased tiller number, improved root growth and altered tuber morphology (Romanov *et al.,* 1998). The demonstration of specific functions associated with specific 'rol' genes and combinations provide useful information to predict the altered phenotype.

19.8.11. Enhancing Rooting in Cuttings

In vegetative propagation using cuttings, root induction is a problem at times. Since the root induced by *A.rhizogenes* is highly branched, it can be a feasible method for developing extensive root system. Therefore an increase in rooting of several crops such as peach, olive, apple, cherry, Magnolia (Roy 1989), *Pinus monticola, P. banksiana* and *Larex larcina* (Mc Afee *et al.,* 1993), Eucalyptus (Macrae and Van Staden 1993), Hazelnut (Bassil *et al.,*1991), Almond (Damiano *et al.,*1995) and Sequoia (Mihaljevic *et al.,* 1999) has been observed on treatment with *A. rhizogenes*. Introduction of 'rol' genes such as 'rol' A, 'rol' B and 'rol' C via *A. tumefaciens* showed the characteristic hairy root phenotype (Rugini and Mariotti, 1991). Hence *A. rhizogenes* has tremendous potential to increase the rooting efficiency of cuttings. Production of increased root mass may help in increased water and nutrient uptake enable better adaptation to the environment.

19.9. PROBLEMS AND THEIR SOLUTIONS

Following are the problems associated with hairy root cultures.

19.9.1. Excess bacterial growth

Excess bacterial growth on explants during co-cultivation is common and may stress the plant tissues. To reduce this one has to wash the explants after co-cultivation in water or liquid medium plus antibiotics or one has to use dilute inoculum for co-cultivation. The explants are transferred to regeneration medium with carbencillin or cefatoxime to kill the bacteria, and kanamycin to prevent proliferation of non- - transformed cells. Frequent subculturing is necessary because the antibiotics gradually become ineffective.

19.9.2. Slow regeneration

Some times regeneration of shoots from transformed cells may be slow or infrequent (e.g., tobacco, chrysanthemum), because regeneration is highly dependent on genotype. Some cultivars may be more suitable than others. Binary vectors may be combined with various helper plasmids and *Agrobacterium* strains to optimize infection and transformation of a particular cultivar. Use of a co-integrative vector system may be more effective in specific cases. Some antibiotics used in counter selection may stimulate or reduce shoot regeneration.

19.9.3. Decreased cell division and transformation due to stress

Stress can be reduced by pre-conditioning of the explants prior to co-cultivation by culturing on hormone containing media, or modification of the culture on hormone containing media, or modification of the culture medium such as reducing basal salts, or stimulation of *Agrobacterium* virulent genes as by addition of phenolic compouds, such as acetosyringone or by lowering the pH to 5.5.

19.9.4. No transformation

No transformation will be observed if unsuitable promoters are used for the selectable marker gene; or if the plasmids carrying the virulent genes and/or the T-DNA are lost from the Agrobacterium, which often happens when the proper antibiotic selection is not made during bacterial culture.

19.9.5. Loss of gene effects

Loss of gene effects may occur, due to the result of position effects rather than the loss of the T-DNA itself.

19.9.6. Chimeric Shoots

If shoots are derived from a mixture of transformed and non-transformed cells, chimeric shoots composed of antibiotic resistant and antibiotic sensitive or of GUS positive and GUS negative sectors may develop. In chimeric transformants, the frequency of transmission of the foreign DNA to progeny is reduced. Excision and subculture of such shoots on fresh antibiotic containing medium may help. Reinduction of shoot morphogenesis may also help.

19.10. MEASURMENT OF GROWTH PARAMETERS IN HAIRY ROOT CULTURES

The growth can be determined by recording fresh and dry weights, on respective days from the mean of 3-4 replicates. The fresh weight can be measured after removing culture medium by filtration through a preweighed whatman #1 filter paper under vacuum. The dry weights estimated after the sample is dried at 60^0 C for 48 hr.

Conductometric estimation of the medium filtrate can be done by using a conductivity meter and expressed in microsiemen units. Osmolarity can be measured by using an automatic Cryoscopic Osmometer and expressed in m Osmol/kg units (see Chapter 5).

19.11. DETERMINATION OF PRODUCTS IN HAIRY ROOT CULTURES

For determining the products formed, the hairy roots are to be extracted and isolated for active constituents. The hairy roots are lyophilized, powdered and extracted at room temperature with eluent solvents in the order of increasing polarities viz., petroleum ether, ethyl acetate, chloroform, acetone, water. The extracts are filtered and the filtrate is subjected to different quantitative analytical methods.

The different methods used for determination of products in hairy root cultures are.

19.11.1. High performance Liquid Chromatography (HPLC)

HPLC analysis of culture of genetically transformed hairy roots derived from anthraquinone producing European madder plant, R. tinctorum reveals the range of anthraquinones resembles that of intact roots and rhizomes. HPLC determination of polyacetylenes (lobetyol, lobetyolin and lobetyolinin) in *Platycodon grandiflorum* hairy root cultures (Tada *et al* 1995)

George *et al.* (1999) obtained HPLC profile of esculin in hairy root cultures of *Cichorium intybus* eluted in petroleum ether with 5% ethyl acetate. Ionkova *et al.,* (1989) quantitatively analysed hyoscyamine and scopolamine in hairy root cultures, lobeline in hairy root cultures of *Lobelia inflata* by HPLC. Kyo *et al.,*(1990) analysed nematicidal compounds in hairy root cultures of *Tageta patula* by HPLC. Kamada *et al.* (1986) examined the presence of atropine and scopalamine in hairy root cultures of *Atropa belladonna* by HPLC. Zheng *et al.,* (1998) determined 6-isoflavonoids in the hairy roots of *Astragalus membranaceous* by reversed phase HPLC.

19.11.2. Gas Chromarography

Ionkova et al., (1989) identified the complete spectrum of alkaloids in transformed root cultures of *D. innoxia* using GC/MS. Chirsten *et al.,* (1989) applied capillary GC and MS for rapid screening of the alkaloid extract of hairy roots of *Datura candida* hybrid and identified 18 alkaloids in which hyoscine and hyoscyamine are the main contituents. Kamada *et al.,* (1986) analysed the contents of atropine and scopolamine in hairy root cultures of *A. belladonna* by GLC.

19.11.3. Nuclaer Magnetic Resonance (NMR)

NMR analysis may be used to analyse secondary metabolism *in vivo* in transformed root cultures. Ford *et al.,* (1996) studied *in vivo* NMR analysis of tropane alkaloid metabolism in hairy root cultures of *D. stramonium.* The metabolism of [^{15}N] tropinone in transformed root cultures was studied. Labelled metabolites were readily detected within 2-4 weeks of application of [^{15}N] tropinone.

NMR *in vivo* study can be used to detect phytohormone induced GABA production in hairy root cultures of *Datura stramonium* (Ford *et al.,*1996).

19.11.4. Infrared Spectroscopy

The production of perezone by transformed root cultures of *Perezia cuernavancana* was evidenced by IR spectroscopy (Arellano *et al.,*1996.)

The scaling up for large-scale cultivation of hairy root using bioreactors is discussed in Chapter 20.

Table 19.11. Nutrient media employed for the production of various secondary metabolites from hairy root cultures

Plant spp.	Induction medium	Production medium	Reference
C.roseus	Monnier's salts vitamins of Morel	Same	Brillanceau *et al.,*(1989
H. albus	MS	WP	Shimomura *et al.,* (1991)
F. esculentum	Helle's	B5	Trotin *et al.,*(1993)
N. tabacum	MS	MS	Whitney (1992)
Panax sp.	MS	MS	Yoshikawa *et al.,*(1993)

REFERENCES

Abegaz, B.M. (1991). Polyacetylenic thiophenes and terpenoids from the roots of *Echinops pappii*. *Phytochemistry,* 30 : 879-881.

Aird, E.L.H., Hamill, J.D. and Rhodes, M.J.C. (1988a). Cytogenetic analysis of hairy root cultures from a number of plant species transformed by *Agrobacterium rhizogenes. Pl. Cell Tiss. Org. Cult.,* 15 : 47-57.

Aird, E.L.H., Hamill, J.D., Robins, R.J. and Rhodes, M.J.c. (1988b). Chromosome stability in transformed root culture and the properties of variant lines of *Nicotiana rustica* hairy root. Cambridge University Press, Cambridge, UK, pp. 137-144.

Akasaka, V., Mii, M. and Daimon, H. (1998). Morphological alterations and root nodule formation *Agrobacterium rhizogenes* hairy roots of peanut (*Arachis hypogaea* L.). *Annal. Bot.,* 81 : 355-362.

Ambros, P.F., Matizke, A.J.M. and Matzke, M.A. (1986). Localization of *Agrobacterium rhizogenes* T-DNA in plant chromosomes by *in situ* hybridization. *EMBO,* 5 : 2073-2077.

Arellano, J., Vazquez, F., Villegas, T. and Hernandez, G. (1996) Establishment of transformed root cultures of *Perezia cuernavacana* producing the sesquiterpene quinone perezone. *Plant Cell Reports,* 15 : 455-8.

Arroo, R.R.J., Develi, A., Meijers, H., Van de Westerlo, E., Kemp, A.K., Croes, A.F. and Williems, G.J. (1995) Effect of exogenous auxin on root morphology and secondary metbaolism in *Tagetes patula* hairy root cultures. *Physiol. Plant,* 93 : 233-40.

Asada, Y., Li, W., Yoshikawa, T. (1998). Isoprenylated flavonoids from hairy root cultures of *Glycyrrhiza glabra. Phytochemistry,* 47 : 389-92.

Asamizu, T., Abiyam, K. and Yasuda, I. (1988). Anthraquinones production by hairy root culture in *Cassia obtusifolia. Yakagaku zasshi,* 108 : 1215-8.

Bahakov, A.V., Bartova, L.M., Dridze, I.L., Maisuryan, A.N., Margulis, G.U., Oganian, R.R., Voblikova, V.D. and Muromeisev, G.S. (1995). Culture of transformed horse radish roots as source of Fusicoccin-like ligands. *J. Plant Growth Reg.,* 14 : 163-7.

Bais, H.P., Venkatesh, R.T., Chandrashekar, A. and Ravi Shankar, G.A. (2001). *Agrobacterium rhizogenes* mediated transformation of witloof chicory – *In vitro* shoot regeneration and induction of flowering. *Current Science,* 80(1): 83-87.

Bakkali, A.T., Jazin, M., Foriers, A., Vanderheyden, Y., Vanhaelen, N. and Homes, J. (1997). Lawsone accumulation in normal and transformed cultures of *Lawsonia inermis. Plant Cell Tiss. Org. Cult.,* 51 : 83-87.

Banerjee, S., Naqui, A.A., Mandal, S. and Ahuja, P.S.(1994) Transformation of *Withania somnifera* (L.) Dunal by *Agrobacterium rhizogenes* : Infectivity and phytochemical studies. *Phytotherapy Res.,* 8 : 452-5.

Banerjee, S., Zehra, M., Kukreja, A.K. and Sushil Kumar (1995). Hairy root in medicinal plants *Current Research on Medicinal and Aromatic Plants,* 17 : 348-378.

Banerjee-Chattopadhyaya, S., Schwemmin, A.M. and Schwemmin, D.J. (1985). A study of karyotypes and their alterations in cultured and agrobacterium transformed roots of *Lycopersicon peruvianum. Mill. Theor Appl. Genet.,* 71 : 258-262.

Bassil, N.V., Proebstring, W.M., Moore, L.W. and Lightfoot, D.A. (1991). Propagation of hazelnut stem cuttings using *Agrobacterium rhizogenes. Plant Cell Rep.,* 11 : 334-338.

Bel-Rhlid, R., Chabot, S., Piche, Y. and Chenevert, T. (1993). Isolation and identification of flavonoids from Ri T-DNA transformed roots (*Daucus carota*) and their significance in vesicular-arbuscular Mycorrhiza. *Phytochemistry,* 35 : 381-3.

Benjamin, B.D., Roja, G. and Heble, M.R. (1993). *Agrobacterium rhizogenes* mediated transformation of *Rauwolfia serpentina*. Regeneration and alkaloid synthesis. *Plant cell tissue org cult,* 35 : 253-257.

Benjamin, B.D., Roja, G. and Heble, M.R. (1994). Alkaloid synthesis by root cultures of *Rauwolfia serpentina* transformed by *Agrobacterium rhizogenes. Phytochemistry,* 35 : 381-383.

Berlin, J., Ruegenhagen, C., Dietze, P., Fecker, L.F., Goddijn, O.J.M. and Hoge, J.H.C. (1993). Increased production of serotonin by suspension and root cultures of *Peganum harmala* transformed with a tryptophan decarboxylase cDNA clone from *Catharanthus roseus. Transgenic Res.,* 2 : 336-344.

Berthomieu, P. and Jouanin, L. (1992). Transformation of rapid cycling cabbage (*Brassica oleracea* var. capitata) with *Agrobacterium rhizogenes. Plant Cell Rep.,* 11 : 334-338.

Better, M., Chang, C.P., Robinson, r.R. and Horwitz, A.H. (1988). *Escherichia coli* secretion of an active chimeric antibody fragment. *Science,* 24(1) : 1041-1043.

Bevan, M. (1984). *Nucleic Acid Res.,* 12 : 8711-8712.

Bhadra, R., Vani, S. and Shanks, J.V. (1993). Production of indole alkaloids by selected hairy root lines of *Catharanthus roseus. Biotechnol. Bioeng.,* 41 : 581-592.

Bhadra, R., Wayment, D.G., Hughes, J.B. and Shanks, J.V. (1999). Confirmation of conjugation process during T.N.T. metabolism by axenic plant roots; *Environ. Sci. Technol.,* 33 : 446-452.

Boulter, M.E., Croy, E., Simpson, P., Shields, R., Croy, R.R.D., Shirsat, A.H. (1990). Transformation of *Brassica napus* L. (oil seed rape) using *Agrobacterium tumefaciens* and *Agrobacterium rhizogenes* : A comparison. *Plant Sci.,* 70 : 91-99

Bourgaud, F., Bouque, V. and Guckert, A. (1999). Production of flavanoids by psoralea hairy root culture. *Plant Cell Tissue and Organ Culture,* 56 : 97-104.

Brillanceau, M.H., David, C. and Tempe, J. (1989). Genetic transformation of *Catharanthus roseus* G. Don. By *Agrobacterium rhizogenes*. *Plant Cell Rep.* 8 : 63-66.

Cabrera-Ponce, J.C., Vegas-Garcia, A.and Herrera-Estrella, L. (1996). Regeneration of transgenic papaya plants via somatic embryogenesis induced by *Agrobacterium rhizogenes. In vitro Cell Dev. Biol. Plant*, 32 : 86-90.

Cai, D., Kleine, M., Kifle, S., Hardoff, H.J., Marcker, K.A. *et al.* (1997). Positional cloning of a gene for nematode resistance in sugar beet. *Science*, 275 : 832-834.

Cardarelli, M., Mariotti, D., Pomponi, M., Spano, L., Capone, I. and Constantino, P. (1987). *Agrobacterium rhizogenes* T-DNA genes capable of inducing hairy root phenotype. *Mol. Gen. Genet.,* 209 : 475-480.

Carron, T.R., Robbins, M.P. and Morris, P. (1994). Genetic modification of condensed tannin biosynthesis in *Lotus corniculatus*. I. Heterologous and antisense dihydroflavonol reductase down-regulate tannin accumulation in "hairy root" cultures. *Theor. Appl. Genet.,* 87 : 1006-15.

Chen, D.H., Liu, C.J., Ye, H.C., Li, F.G., Liu, B.L., Meng, L.X. and Chen, X.Y. (1999). Ri-mediated transformation of *Artemisia annua* with a recombinant farnesyl diphosphate synthase gene for artemisinin production. *Plant Cell Tissue and Organ Culture,* 57 : 157-162.

Chilton, M.D., Tepfer, D.A, Petit, A., David, C., Casse-Delbart, F. and Tempe, J. (1982). *Agrobacterium rhizogenes* inserts T-DNA into the genomes of the host plant root cells. *Nature,* 295 : 432-434.

Cho, H.J., Ferrand, S.K., Noel, G.R. and Widholm, J.M. (2000). High efficiency induction of soybean cysnematode. *Planta,* 210 : 195-204.

Christen, P. (1999). *Centrathus* species : *In vitro* culture and the production of valeopotriates and other secondary metabolites. In : Bajaj, Y.P.S., editor. Biotechnology in Agriculture and Forestry, Medicinal and Aromatic Plants XI, vol. 43, Berlin : Spinger-Verlag, 42-56.

Christen, P., Aoki, T. and Shimomura, K. (1992). Characteristics of growth and tropane alkaloid production in *Hyoscyamus albus* hairy roots transformed with *Agrobacterium rhizogenes* A$_4$. *Plant. Cell Rep.,* 11 : 597-600.

Christen, P., Roberts, M.E., Phillipson, J.D. and Evans, W.C. (1989). High-yield production of tropane alkaloids by hairy-root cultures of a *Datura candida* hybrid. *Plant Cell Rep.,* 8 : 75-77.

Christen, P., Roberts, M.F., Phillipson, J.D. and Evans, W.C. (1991). Alkaloids of hairy root cultures of a *Datura candida* hybrid. *Plant Cell Rep.,* 9 : 101-104.

Christey MC Sinclair BK, Braun RH, Wyke L. (1997). Regeneration of transgenic vegetable *Brassicas (brssica aleracea and B. campesrtris)* via Ri-medicated transformation *Plant Cell Reports,* 16 587-93.

Christey, M.C. and Sinclair, B.K.(1992) Regeneration of transgenic kale (*Brassica oleracea* var. *acephal*), rap (*B. napus*) and turnip (*B. campestris* var *rapifera*) plants via *Agrobacterium rhizogenes* mediated transformation. *Plant Sci.,* 87 : 161-9.

Ciau-Uitz, R., Miranda-Ham, M.L., coello-Coello, J., Chi, B., Pacheco, L.M. and Loyola-vargas, V.M. (1994). Indole alkaloid production by transformed non-transformed roo cultures of *Catharanthus moseus. In vitro Cell Dev. Biol.,* 30 : 84-88.

Col,D., Kleine, M., Kifle, S., Harloff, H.J., Sandal, N.N., Marcker, K.A., Klein-Lankhorst, R.M., Salentijn, E.M., Lange, W., Stickema, W.J., Wyss, U., Grundler, F.M.W. and Jung, C. (1997). Positional cloning of a gene for nematode resistance in sugar beet. *Science,* 275832-834.

Comai, L., Facciotti, D., Hiatt, W.R., Thompson, G., Rose, R.E. and Stalker, D.M. (1985). *Nature,* 317 : 741-744.

Constabel, C.P. and Towers, G.H.N (1988). Thiarubrine accumulation in hairy root cultures of *Chaenactis dauglasei. J. Plant Physiol.,* 133 : 67-72.

Couilterot, E., Caron, C., Trentesaux, C., Chenieux, J.C. and Audran, J.C. (1999). *Fagra zanthoxyloides* Lam. (Rutaceae) : *In vitro* culture and the production of benzophenanthridine and furoquinoline alanine. In : Bajaj YPS, editor. Biotechnology in Agriculture and Forestry, Medicinal and Aromatic Plants XI, vol. 43, Berlin : Spinger-Verlag, pp. 136-56.

Croes, A.F., Vander Berg, A.J.R., Bosveld, M., Breteler, H. and Wullems, G.J. (1989). Thiophene accumulation in relation to morphology in roots of *Tagetes patula*. Effects of auxin and transformation by *Agrobacterium*. *Planta Med.,* 179 : 43-50.

D'Amato, F. (1985). Cytgenetics of plant cell and tissue culture and their regenerants. *CRC Critical Rev. Plant Sciences*, 3 : 73-112.

Damiani, F. and Arcioni, S. (1991). Transformation of *Medicago arborea* L. with an *Agrobacterium rhizogenes* binary vector carrying the hygromycin resistance gene. *Plant Cell Rep.*, 10 : 300-303.

Damiani, F., Paoloci, F., Consonni, G., Crea, F., Tonelli, C., Aricioni, S. (1998). A maize anthocyanin transactivator induces pigmentation in hairy roots of dicotyledonous species. *Plant Cell Reports*, 17 : 339-44.

Damiano, C., Archilletti, T., Caboni, E., Lauri, P., Falasca, G., Mariotti, D. and Ferraiolo, G. (1995). *Agrobacterium mediated* transformation of almond : *in vitro* rooting through localized infection of *A. rhizogenes* w.t. *Acta. Hort.*, 392 : 161-169.

Davey, M.R., Mulligan, B.J., Gartland, K.M.A., Peel, E., Sargent, A.W. and Morgan, A.J. (1987). Transformation of *Solanum* and *Nicotiana* species using a Ri plasmid vector. *J. Exp. Bot.*, 38 : 1507-16.

David, C., Petit, A. and Tempe, J. (1988). T-DNA length variability in mannopine hairy root : more than 50 kilobase pairs of P^{Ri} T-DNA can integrate in plant cells. *Plant Cell Reports*, 7 : 92-95.

Davioud, E., Kan, C., Hamon, J., Tempe, J. and Husson, H.P. (1989). Production of indole alkaloids by *in vitro* root cultures from *Catharanthus trichophyllus*. *Phytochemistry*, 28 : 2675-80.

Delbeque, J.P., Beydon, P. and Chapuis, L. (1995). *In vitro* incorporation of radio labelled cholesterol and mevalonic acid into ecdysteron by hairy root cultures of a plant *Serratula tinctoria*. *Eur. J. Entomol.*, 92 : 301-7.

Deno, H., Yamagata, T., Emoto, T., Yoshioka, T., Yamada, Y. and Fujita, Y. (1987). Scopolamine production by root cultures of *Duboisia myoporoides* II. Establishment of a hairy root culture by infection with *Agrobacterium rhizogenes*. *J. Plant Physiol.*, 131 : 315-23.

Domansky, N., Ehsani, P., Salmanian, A.H. and Medvedeva, T. (1995). Organ specific expression of hepatits B surface antigen in potato. *Biotechnol. Lett.*, 17 : 863-866.

Dong, L.C., Sun, W., Thies, K.L., Luthe, D.S. and Graves, C.H. (1992). Use of polymerase chain reaction to detect pathogenic strains of *Agrobacterium*. *Phytopath.*, 82 : 434-439.

Doran, P.M. (2000). Foreign protein production in plant tissue cultures. *Curr. Opinion Biotechnol.*, 11 : 199-204.

Downs, C.G., Christey, M.C., Davies, K.M., King, G.A., Seelye, J.F., Sinclair, B.K., Stevenson, D.G. (1994). Hairy roots of *Brassica napus* : II glutamine synthase over expression alters ammonia assimilation and the response to phosphinothricin. *Plant Cell Reports*, 14 : 41-46.

Draper, J and Scott, R (1991) Gene transfer to plants. In: Plant genetic engineering (Eds) Grierson, D. PP38-72. Blackie Glasgow and London

Draper, J. and Scott, R. (1988). The isolation of plant nucleic acids. In : *Plant Genetic Transformation and Gene Expression – A Laboratory Manual* (eds. J. Draper, R. Scott, P. Armitage and R. Walden), pp. 192-236, Blackwell Scientific Publications.,London

Dupraz, J.M., Christen, P. and Kapetanidis, I. (1994). Tropane alkaloids in transformed roots of *Datura quercifolia*. *Pl. Med.*, 60 : 158-162.

Eapen, S. Sueelan, K.N. Tivarekars, S. Kotwal and Mitra, R (2000). Development of hairy root cultures of *Brassica Juncea* and chemopidum *amarantticolor* for rhizofiltration of Uranium in DAE, BRNs sympossium on the use of nuclear and molecular techniques in crop important, BARIC, mumbao, Dec 6-8, PP 303-314.

Eilert, U. (1987). Elicitation : Methodology and aspects of application. In : *Cell Culture and Somatic Cell Genetics of Plants* (Eds. I.K. Vasil and F. Constabel), 4 : 153-197, Academic Press, New York.

Ermayanti, T.M., McComb, J.A. and O'Brien, P.A. (1994). Stimulation of synthesis and release of swainsonine from transformed roots of *Swainsona galegifolia*. *Phytochemistry*, 36 : 313-317.

Faskenhagen, H., Stockigt, J., Kuzovkina, I.N., Alterman, I.E. and Kolshorn, H. (1993). A novel indole alkaloid from the hairy root of *Rauwolfia serpentina*. *Can. J. Chem.*, 71 : 2201-2203.

Fecker, L.F., Hildebrandt, S., Rugenhagen, C., Herminghaus, S., Landsmann, J. and Berlin, J. (1992). Metabolic effects of a bacterial lysine decarboxylase gene expressed in hairy root cultures of *Nicotiana glauca*. *Biotechnol. Lett.*, 14 : 1035-1040.

Fecker, L.F., Rugenhagen, C. and Berlin, J. (1993). Increased production of cadaverine and anabasine in hairy root cultures of *Nicotiana tabacum* expressing a bacterial lysine decarboxylase gene. *Plant Mol. Biol.,* 23: 11-21.

Fei, H.M., Mei, K.F., Shen, X., Ye, Y.M., Lin, Z.P. and Peng, L.H. (1993). Transformation of *Gynostemma pentaphyllum* by *Agrobacterium rhizogenes.* Saponin production in hairy root cultures. *Acta. Bot. Sincia,* 35: 626-31.

Flores, H. (1992). Plant roots as chemical factories. *Chemistry and Industry,* 18 : 374-377.

Flores, H. and Filner, P. (1985). Metabolic relationships of putrescine GABA and alkaloids in cell and root cultures of Solanaceae. In : Neumann K., Barz, W., Reinhard, E. editors. Primary and Secondary Metabolism of Plant Cell Cultures, Berlin : Springer-Verlag, 174-85.

Flores, H.E. (1986). Use of plant cell and organ culture in the production of biological chemicals. In : Applications of biotechnology to agricultural chemistry. *Amer. Chem. Soc. Symp. Series,* pp. 96.

Flores, H.E. and W.R. Curtis (1992). Approaches to understanding and manipulating the biosynthetic potential of plant roots. In : *Biochemical Engineering VII* (Eds. H. Pederson, R.D. Mutharasan and D. Biasio), 665 : 188-209, The New York Academy of Science, New York.

Flores, H.E., Hoy, M.W. and Pickard, J.J. (1987). Secondary metabolites from root cultures. *TIBTECH,* 5 : 64-69.

Flores, H.E., Pickard, J.J. and Hoy, M.W. (1988). Production of polyacetylenes and thiophenes in heterotrophic and photosynthetic root cultures of *Asteraceae.* In : Lam J., Breheler, H., Arnason, T., Hansen, L. editors Chemistry and Biology of Naturally occurring Acetylenes and Related Compounds (NOARC). *Bioactive Molecules,* 7 : 233-54.

Ford, Y.Y., Ratcliffe, R.G., Robins, R.J. (1996). *In vivo* NMR analysis of tropane alkaloid metabolism in transformed root and de-differentiated cultures of *Datura stramonium* L. *Phytochemistry,* 43(1) : 115-120.

Forde, B.G., Day, H.M., Turton, J.F., Shen, W.J., Cullimore, V. and Oliver, J.E. (1989). Two glutamine synthase genes from *Phaseolus vulgaris* L. display contrasting developmental and spatial patterns of expression in transgenic *Lotus corniculatus* plants. *Plant Cell,* 1 : 391-401.

Fujita. Y. Hara, Y. Suga , C. and Morimoto, T (1981). Production of Shikonin derivatives by cells suspension cultures of *Lithospermum erythrohizon.* A new medium for the production of shikonin derivates. *Plant Cell Rep* 1 : 61-63.

Fukui, H., Feroj Hasan, A.F.M., Veoka, T. and Kyo, M. (1998). Formation and secretion of a new brown benzoquinone by hairy root cultures of *Lithospermum erythrorhizon. Phytochemistry,* 47 : 1037-1039.

Furze, J.M., Rhodes, M.J.C., Parr, A.J., Robins, R.J., Whitehead, I.M. and Threlfall, D.R. (1991). Abiotic factors elicit sesquiter-penoid phytoalexin production but not alkaloid production in transformed root cultures of *Datura stramonium. Plant Cell Reports,* 10 : 111-4.

Gamborg, L., Miller, R.A. and Ojima, K. (1968). Nutrient requirements of suspension culture of soybean root cells. *Exp. Cell Res.,* 50 : 155-158.

Gao, J. and Lee, J.M. (1992). Effect of oxygen supply on the suspension cultures of genetically modified plant cells. *Plant Cell Rep.,* 10 : 533-536.

Gao, J., Lee, J.M. and An, G. (1991). The stability of foreign protein production in genetically modified plant cells. *Plant Cell Rep.,* 10 : 533-536.

George, J., Bais, H., Ravishankar, G.A. (1999). Production of esculin by hairy root cultures of chicory (*Cichorium intybus* L. cv. Lucknow Local). *Indian J. Exp. Biol.,* 37(3) : 269-273.

Giovanni, A., Pecchioni, N., Rabaglio, M. and Allavena, A. (1997). Characterization of ornamental *Datura* plants transformed by *Agrobacterium rhizogenes. In vitro Cell Dev. Biol. Plant,* 33 : 101-106.

Giovannini, A., Zottini, M., Morreale, G., Spena, A. and Allavena, A. (1999). Ornamental trait modification by rol genes in *Osteospermum ecklonis* transformed with *Agrobacterium tumefaciens. In vitro Cell Dev. Biol.,* 35: 70-75.

Giri, A., Banerjee, S., Ahuja, P.S. and Giri, C.C. (1997). Production of hairy roots in *Aconitum heterophyllum* wall. Using *Agrobacterium rhizogenes. In vitro Cell Dev. Biol-Plant,* 33 : 280-284.

Giri, A. and Lakshmi Narasu, M. (2000). Transgenic hairy roots : recent trends and applications, Biotechnology Advances, 18 : 1-22.

Giulietti, A.M., Parr, A.J. and Rhodes, M.J.C. (1993). Tropane alkaloids production in transformed root cultures of *Brugmansia candida*. *Planta Medica*, 59 : 428-31.

Granicher, F., Christen, P. and Kapetandis, I. (1992). High yield production of valepotriates by hairy root cultures of *Valeriana officinals* L. var. *sambucifolia* Mikan. *Pl. Cell Rep.*, 11 : 339-342.

Granicher, F., Christen, P. and Vuagnat, P. (1994). Rapid high performance liquid chromatographic qualification of *Valeriana officinalis* L. var. *sambutifolia* hairy roots. *Phytochemistry*, 38 : 103-5.

Granicher, F., Cristen, P. and Kaptanidis, I. (1995). Production of valepotriates by hair root cultures of *Centranthus ruber* DC. *Plant Cell Reports*, 14 : 294-8.

Guellec, V. David, V. Branchard, M. and TempeJ (1990). Agrobacterium rhizogenes mediated transformation of grapevine (*Vitis vinifera*), *Plant Cell, Tissue and Organ Culture*, 20 211-215.

Guerche, P., Jouanin, L., Tepfer, D. and Pelletier, G. (1987). Genetic transformation of oil seed rape (*Brassica napus*) by the Ri T-DNA of *Agrobacterium rhizogenes* and analysis of inheritance of the transformed phenotype. *MGG.*, 206 : 382-386.

Gutierrez-Pesce, P., Taylor, K., Muleo, R. and Rugini, E. (1998). Somatic embryogenesis and shoot regeneration transgenic roots of cherry rootstock Colt (*Prunus avium X P. pseudocerasus*) mediated by pRi 1855 T-DNA of *Agrobacterium rhizogenes*. *Plant Cell Reports*, 17 : 581-585.

Halperin, S.J. and Flores, H.E.(1997) Hyoscyamine and proline accumulation in water stressed *Hyoscyamus muticus* hairy root culture.s *In vitro Cell Dev. Biol. Plant*, 33 : 240-4.

Hamill, J.D., Parr, A.J., Rhodes, M.J.C., Robins, R.J. and Walton, N.J. (1987). New routes to plant secondary products. *Biotechnol.*, 5 : 800-4.

Hamill, J.D., Parr, A.J., Robins, R.J. and Rhodes, M.J.C. (1986). Secondary product formation by cultures of *Beta vulgaris* and *Nicotiana rustica* transformed with *Agrobacterium rhizogenes*. *Plant Cell Reports*, 5 : 111-114.

Hamill, J.D., Robins, R.J. and Rhodes, M.J.C. (1989). Alkaloid production by transformed root cultures of *Cinchona ledgeriana*. *Planta. Med.*, 55 : 354-357.

Hamill, J.D., Robins, R.J., Parr, A.J., Evans, P.M., Furze, J.D. and Rhodes, M.J.C. (1990). Over expressing a yeast ornithine decarboxylase gene in transgenic roots of *Nicotiana rustica* can lead to enhanced nicotine accumulation. *Plant Mol. Biol.*, 15 : 27-38.

Han, A. and Dougall, D.K. (1992). The effect of growth retardants on anthocyanin production in carrot cell suspension cultures. *Plant Cell Reports*, 11 : 304-309.

Han, K.H., Keathley, D.E., Davis, J.M. and Gordon, M.P. (1993). Regeneration of a transgenic woody legume *Robinia pseudoacacia* L. (Black locust) and morphological alterations induced by *Agrobacterium rhizogenes* mediated transformation. *Plant Sci.*, 88 : 149-57.

Handa, T., Sujimura, T., Kato, E., Kamada, H. and Takayanagi, K. (1994). Genetic transformation of *Eustoma grandiflorum* with rol genes. *Acta Hort.*, 392 : 209-218.

Hashimoto, T., Yukimune, Y. and Yamada, Y. (1986). Tropane alkaloid production in *Hyoscyamus* root cultures. *J. Plant Physiol.*, 124 : 61-75.

Hashimoto, T., Yun, D.J. and Yamada, Y. (1993). Production of tropane alkaloids in genetically engineered root cultures. *Phytochemistry*, 32(3) : 713-718.

Hatamoto, H Boulter, M.E, Shirsat, AH. Croy E.J, Ellis, J.R. (1990). Recovery of morphologically normal transgenic tobacco from hairy roots co –transformed with *Agrobacterium rhizogenes* and a binary vector plasmid. *Plant Cell Reports*, 9 : 88-92.

Hilton, M.G. and Rhodes, M.J.C. (1990). Growth and hyoscyamine production by hairy root cultures of *Datura stramonium* in a modified stirred tank reactor. *Applied Microb. Biotech.*, 33 : 132-138.

Hilton, M.G. and Rhodes, M.J.C. (1993). Factors affecting the growth and hyosyamine production during batch culture of transformed roots of *Datura stramonium*. *Planta Medica*, 59 : 340-4.

Hirotani, M., Zhou, Y., Hekai, L. and Furuya, T. (1994). Astraglaosides from hairy root cultures of *Astragalus membranaceous. Phytochem.,* 36 : 665-670.

Hogue, R.S., Lee, J.M. and An, G. (1990). Production of a foreign protein product with genetically modified plant cells. *Enzyme Microb. Technol.,* 12 : 533-538.

Hook, I. (1994). Secondary metabolites in hairy root cultures of *Leontopodium alpinum* cass (edelweiss). *Plant Cell Tissue and Organ Culture,* 38 : 321-6.

Horwitz, A.H., Chang, C.P., Better, M., Hellstrom and Robinson, R.R. (1988). Secretion of a functional antibody and Fab fragment from yeast cells. *Proc. Nat. Acad. Sci. USA,* 85 : 8678-8682.

Hoshino, Y. and Mii, M. (1998). Bialaphos stimulates shoot regeneration from hairy roots of snapdragon (*Antirrhinum majus* L.) transformed by *Agrobacterium rhizogenes. Plant Cell Reports,* 17 : 256-261.

Hu, Z.B. and Alfermann, A.W. (1993). Diterpenoid production in hairy root cultures of *Salvia miltiorrhiza. Phytochemistry,* 32 : 699-703.

Huang, Y., Diner, A. and Karnosky, F. (1991). *Agrobacterium rhizogenes* mediated genetic transformation of a conifer : *Larix decidua. In vitro Cell Dev. Biol. Plant,* 27 : 201-207.

Hughes, J.B., Shanks, J., Vanderford, M., Lauritzen, J. and Bhadra, R. (1997). Transformation of TNT by aquatic plants and plant tissue culture. *Env. Sci. Technol.,* 51 : 266-271.

IIshimaru, K., Sudo, H., Satake, M. and Shimomura, K. (1990). Phenyl glucosides from a hairy root cultures of *Swertia japonica. Phytochem.,* 29 : 3823-3825.

Ikenaga, T., Oyama, T. and Muranaka, T(1995). Growth and steroidal saponin production in hairy root cultures of *Solanum aculeatissi. Plant Cell Reports,* 14 : 413-7.

Inomata, S., Yokoyama, M.Y., Goza Shimizu, T. and Yanagi, M. (1993). Growth pattern and ginsenoside production of *Agrobacterium* transformed *Panax ginseng* roots. *Pl. Cell Rep.,* 12 : 681-686.

Ionkava, I., Kartnig, T. and Alfermann, W. (1997). Cycloartane saponin production in hairy root cultures of *Astragalus mongholicus. Phytochemistry,* 45(8) : 1597-1600.

Ionkova, I., Witte, L. and Alfermann, A.W. (1989). Production of alkaloids by transformed root cultures of *Datura innoxia. Planta Medica.,* 55 : 229-230.

Ishimaru, K. and Shimomura, K. (1991). Tannin production in hairy root cultures of *Geranium thunmbergii. Phytochemistry,* 30 : 825-8.

Ishimaru, K., Agakawa, H., Yamanaka, M., Shimomura, K. (1994). Polyacetylenes in *Lobelia sessilifolia* hairy roots. *Phytochemistry,* 35 : 365-369.

Ishimaru, K., Sudo, H., Satake, M., matsugana, Y., Hasegawa, Y., Takemoto, S. and Shimomura (1990). Amarogentin, amaroswertin and four xanthones from hairy root cultures of *Swertia japonica. Phytochemistry,* 29 : 1563-1565.

Ishimaru, K., Yonemitsu, H. and Shimomura, K. (1991a). Lobetyolin and lobetyol from hairy root culture of *Lobelia inflata. Phytochemistry.,* 30 : 2255-2257.

Jajzri, M., Legros, M., Homes, J. and Vanhaelen, M. (1988). Tropane alkaloids production by hairy root cultures of *Datura stramonium* and *Hyoscyamus niger. Phytochem.,* 27 : 419-420.

Jaziri, M., Shimomura, K., Yoshimatsu, K., Fauconnier, M.L., Marlier, M. and Homes, J. (1995). Establishment of normal and transformed root cultures of *Artemisia annua* L. for artermisinin production. *J. Plant Physiol.,* 145 : 175-7.

Jaziri, M., Yoshimatsue, K., Homes, J. and Shimomura, K. (1994). Traits of transgenic *Atropa belladona* doubly transformed with different *Agrobacterium rhizogenes* strains. *PCTOC,* 38 : 257-262.

Jha, S., Sahu, N.P., Sen, J., Jha, T.H. and Mahato, S.B. (1991). Production of emetine and cephaeline from cell suspension and excised root cultures of *Cephaelis ipecacuanha. Phytochemistry,* 30 : 3999-4003.

Jung, G. and Tepfer, D. (1987). Use of genetic transformation by the Ri T-DNA of *Agrobacterium rhizogenes* to stimulate biomass and tropane alkaloid production in *Atropa belladonna* and *Calystegia sepium* roots grown *in vitro. Plant Sci.,* 50 : 145-51.

Kamada, H., Okamura, N., Satake, M., Harada, H. and Shimomura, K. (1986). Alkaloid production by hairy root cultures of *Atropa belladona*. *Plant Cell Reports,* 5 : 239-242.

Kaneyoshi J and Kobayashi S (1999 Characteristics of transgenic trigolliate orange (*Porcirus trifoliate* Raf) possessing the rol 'C' gene of *Agrobacterium rhizogenes* Ri plasmid. *Journal of Jap. Soc. Hort. Sci.,* 68 : 734-738.

Kas, J., Burkhard, J., Demnerova, K., Kostal, J., Macek, T., Mackovq, M. and Pazalrova, J. (1997). Prospectives in biodegradation of alkanes and PCBs. *Pune Appl. Chem.,* 69 : 2357-2369.

Ketel, D.H., Breteler, H. and de Groot, B. (1986). Poster presented at 'Process Possibilities for Plant and Animal Cell Cultures. *Inst. Chem. Engin. Symp.,* 25/26th March 1986, UMIST, Manchester, U.K.

Kim, C.H., Lee, S.W., Chung, I.S. (1994). Hairy root cultures of *Ducus carota* for anthocyanin production in fluidized bed bioreactor. *Agric. Chem. Biotechnol.,* 37 : 237-42.

Kisiel, W., Stojakowska, A. (1997). A sesquiterpene coumarin either from transformed roots of *Tanacetum parthenium*. *Phytochemistry,* 46(3) : 5515-6.

Kisiel, W., Stojakowska, A., Malarz, J. and Kohlmunzer, S. (1995). Sequiterpene lactones in *Agrobacterium rhizogenes* transformed hairy root cultures of *Lactuca virosa*. *Phytochemistry,* 40 : 1139-40.

Kiyokawa, S., Kikuchi, Y. and Kamada, H. (1996). Genetic transformation of Beagonia tuberhybrida by R$_1$ rolgenes. *Plant Cell Rep.,* 15 : 606-609.

Knopp, E., Strauss, A. and Wehrli, W. (1988). Root induction on several Solanaceae species by *Agrobacterium rhizogenes* and the determination of root tropane alkaloid content. *Plant Cell rep.,* 7 : 590-593.

Knopp, E., Strauss, A. and Welhrli, W. (1988). Root induction on several *Solanaceous* species by *Agrobacterium rhizogenes* and the determination of root tropane alkaloid content. *Plant Cell Reports,* 7 : 590-593.

Ko, K.S., Ebizuka, Y., Noguchi, H. and Sankawa, U.('1995) Production of polypeptide pigments in hairy root cultures of *Cassia plants*. *Chem. Pharm. Bull.* (Tokyo), 43 : 274-8.

Kucerova P,Mackova, M, Chroma L, Burkhard j, Trisja J, Demnerova K and Macek T(2000) Metabolism of polychlorinated triphenyls by *Solanum nigrun* , hairy root clone SNC-90 and analysis of transformation products, *Plant and Soil* 225 109-115

Kuroyanagi, M., Arakava, T., Mikami, Y., Yoshida, K., Kawahar, N., Hayashi, T. and Ishimaru, H. (1998). Phytoalexins from hairy roots cultures of *Hyoscayamus albus* treated with methyl jasmonate. *J. Nat. Prod.,* 61 : 1516-9.

Lee, K.T., Yamakawa, T., Kodama, T. and Shimomura, K. (1998). Effects of chemicals on alkaloid production by transformed roots of *Belladonna*. *Phytochemistry,* 49 : 2343-7.

Limami, M.A, Sun L.Y, Douatc, C, Helgeson, J and Tepfer, D. Nautral genetic transformation by *Agrobacterium rhizogenes, Plant physiolog,* 118,543-550

Lodhi, A.H., Bongaerts, R.J.M., Verpoorte, R., Coomber, S.A., Charlwood, B.V. (1996). Expression of bacterial isochrismate synthase (EC 5.4.99.6) in transgenic root cultures of *Rubia peregrina*. *Plant Cell Reports,* 16: 54-57.

Lorz, H., Baker, B. and Schell, J. (1985). *Mol. Gen. Genet.,* 199 : 178-182.

Macek, T.E, Kotraba P, Suchova M, Skacel, F, Demnerova, M. and Rumi, T. (1994). Accumulation of Cadmium by hairy root cultures of *Solanum nigram*. *Biotechnol. Lett.,* 16 : 621-624.

Mackova, M., Macek, T., Oceanaskova, J., Burkhard, J., Demneova, K. and Pazlarova, J. (1997). Biodegradation of polychlorinated biphenyles by plant cells. *Int. Biodeterioration Biodegrad.,* 39 : 317-325.

Macrae, S. and Van Staden, J. (1993). *Agrobacterium rhizogenes* mediated transformation to improve rooting ability of *Eucalyptus*. *Treee Physiol.,* 12: 411-418.

Manners, J.M. and Way, H. (1989). Efficient transformation with regeneration of the tropical pasture legume *Stylosanthes humilis* using *Agrobacterium rhizogenes* and a Ti plasmid-binary vector system. *Plant Cell reports,* 8 : 341-5.

Mano, Y., Nabeshima, S., Matsui, C. and Ohkawa, H. (1986). Production of tropane alkaloids by hairy root cultures of *Scopolia japonica*. *Agric. Biol. Chem.,* 50 : 2715-2722.

Mano, Y., Ohkawa, H. and Yamada, Y. (1989). Production of tropane alkaloids by hair root cultures of *Duboisia leichhardtii* transformed by *Agrobacterium rhizogenes*. *Plant Sci.,* 59 : 191-201.

Mano, Y., Sabeshima, S., Matsui, C., Ohkawa, H. (1987). Production of tropane alkaloids by hairy-root cultures of *Scopolia japonica. Agric. Biol. Chem.,* 50 : 2715-2722.

Marchant, Y.Y. (1988). *Agrobacterium rhizogenes*-transformed root cultures for the study of polyacetylene metabolism and biosynthesis. In : Lam, J., Brecheler, H., Arnason, T., Hansen, L. editors. Chemistry and Biology of Naturally occurring Acetylenes and Related Compounds (NOARC). *Bioactive Molecule,* 7 : 217-31.

Matsumoto, T. and Tanaka, N. (1991). Production of phytoecdysteroids by hairy root cultures of *Ajuga reptans* var Atropurpurea. *Agric. Biol. Chem.,* 55 : 1019-1025.

McAfee, B.J., White, E.E., Peicher, L.E. and Lapp, M.S. (1993). Root induction in pine (*Pinus*) and larch (*Larix*) spp using *Agrobacterium rhizogenes. Plant Cell Tiss. Org. Cult.,* 34 : 53-62.

McGranahan, G.H., Leslie, C.A. and Dandekar, A.M. (1993). Transformation of pecan and regeneration of transgenic plants. *Plant Cell Reports,* 12 : 634-638.

McGranahan, G.H., Leslie, C.A. and Uratsu, S.L. (1988). *Agrobacterium*-mediated transformation of walnut somatic embryos and regeneration of transgenic plants. *Bio/Technol,* 6 : 800-804.

McInnes, E., Morgan, A.J., Mulligan, B.J. and Davey, M.R. (1991). Phenotypic effects of isolated pRiA4 TL-DNA rol genes in the presence of intact TR-DNA in transgenic plants of *Solanum dulcamara* L. *J. Exp. Bot.,* 42 : 1279-86.

McLauchlan, W.R., McKee, R.A. and Evans, D.A. (1993). The purification and immunocharacterization of N-methyl putrescine oxidase from transformed root cultures of *Nicotiana tabacum* L. cv Sc 58. *Planta,* 191 : 440-5.

Merkli, A., Christen, P. and Kapetanidis, I. (1997). Production of diosgenin by hair root cultures of *Trigonella foenum-graecum* L. *Plant Cell Reports,* 16 : 632-6.

Mihaljevic, S., Katavic, V. and Jelaska, S. (1999). Root formation in micropropagated shoots of *Sequoia sempervirens* using *Agrobacterium. Plant Sci.,* 141 : 73-80.

Momcilovic, I., Grubisic, D., Kojic, M. and Nejkoric, M. (1997). *Agrobacterium rhizogenes*-mediated transformation and plant regeneration of four gentian species. *Plant Cell Tissue and Organ Culture,* 50 : 1-6.

Morgan, A.J., Cox, P.N., Turner, D.A. Peel, E., Davey, M.R., Gartland, K.M.A. and Mulligan, B.J. (1987). Transformation of tomato using an Ri plasmid vector. *Plant Sci.,* 49 : 37-49.

Motomari, Y., Shimomura, K., Mori, K., Kunitake, H., Nakashima, T., Tanaka, M., Miyazaki, S. and Ishimaru, K (1995). Polyphenol production in hairy root cultures of *Fragaria X ananassa. Phytochemistry,* 40 : 1425-8.

Mugnier, J. (1987). Infection by *Polymyxa betae* and *Plasmodiphora brassicae* of roots containing root-inducing transferred DNA of *Agrobacterium rhizogenes. Phytopathology,* 77 : 539-542.

Mugnier, J. and Musse, B. (1987). Vesicular-arbuscular mycorrhizal infection in transformed root-inducing T-DNA roots grown axenically. *Phytopathology,* 77 : 1045-1050.

Mukundan, U. and Hjortso, M.A. (1990). Effect of fungal elicitor on thiophene production in hairy root cultures of *Tagetes patula. App. Microb. Biotech.,* 33 : 145-147.

Mukundan, U., Rai, A., Dawda, H., Ratnaparkhi, S. and Bhinde, V. (1998). Secondary metabolites in *Agrobacterium rhizogenes* mediated transformed root cultures. In : Srivastava, P.S., (eds). Plant Tissue Culture and Molecular Biology Applications and Prospects. New Delhi : Narosa Publishing House, pp. 302-331.

Mukundan, Y., Hjortso, MA. (1990). Growth and thiophene accumulation by hair root cultures of *Tagetus patula* in media of varying initial pH. *Plant Cell reports,* b9 :b 627-30.

Murakami, Y. Omoto, T. Asai, I. Shimomura. K. Yashihira K and Ishomaru, K. (1998). Rosmarinic acid and related phendics in tranfirmed root cultures of *Hyssopus officinalis. Plant Cell, Tissue and Organ Culture,* 53: 75-78

Muranaka, T., Ohakawa, H. and Yamada, Y. (1992). Scopolamine release into media by *Duboisia leichhardtti* hairy root clones. *Appl. Microbiol. Biotechnol.,* 37 : 554-9.

Murashige, T. and Skoog, F. (1962). A revised medium for rapid growth and bioassays with tobacco tissue cultures. *Physiol. Plant,* 15 : 473-493.

Nabeshima, S., Y. Mano and H. Ohkawa (1986). Production of tropane alkaloids by hair root cultures of *Scopolia japonica. Symbiosis,* 2 : 11-18.

Nakano, M., Hoshino, Y. and Mii, M. (1994). Regeneration of transgenic plants of grape vine (*Vitis vinifera* L.) via *Agrobacterium rhizogenes* mediated transformation of embryogenic calli. *J. Exp. Bot.,* 45(274) : 649-56.

Nakashimada, Y., Uozemi, N. and Kobayashi, T. (1995). Production of plantlets for use as artifical seeds from horseradish hairy roots fragmented in a blender. *J. Ferment. Bioeng.,* 79 : 458-64.

Nayak, P., Mishra, S., Banerjee, N.S. and Sen, S.K. (1999). Transgenic indica rice lines with increased tillering capacity; in *National Symposium on Role of Plant Tissue Culture in Biodiversity Conservation and Economic Development,* 7-9 June, G.B. Pant Inst. Himalayan Environment and Development, Kosi-Katarmal, Almora, Abstract pg. 51.

Nin, S., Bennei, A., Ronelli, G., Mariotti, D., Schiff, S. and Magherini, R. (1997). *Agrobacterium* mediated transformation of *Artemisia absinthum* L. (worm wood) and production of secondary metabolites. *Plant Cell Reports,* 16 : 725-30.

O'Dowd, N.A. and Richardson, D.H.S. (1994). Production of tumours and roots by *Ephedra* following *Agrobacterium rhizogenes* infection. *Can. J. Bot.,* 72 : 203-207.

Ogasawara, T., Chiba, K. and Tada, M. (1993). Production in high-yield of a naphthaquinone by a hairy root culture of *Sesamum indicum. Phytochemistry,* 33 : 1095-1098.

Oksman-Caldentey, K.M., Sevon, N., Vanhala, L. and Hiltunen, R. (1994). Effect of nitrogen and sucrose on the primary and secondary metabolism of transformed root cultures of *Hyoscyamus muticus. Plant Cell Tissue and Org. Cult.,* 38 : 263-272.

Ooms, G., Karp, A., Burrell, M.M., Twell, D. and Roberts, J. (1985). Genetic modification of potato development using Ri-T-DNA. *Theor. App. Gen.,* 70 : 440-446.

Oostdam, A., Mol, J.N.M. and Vanderplas, L.H.W. (1993). Establishment of hairy root cultures of *Linum flavum* producing the lignan 5-methoxypodophyllotoxin. *Plant Cell Reports,* 12 : 474-7.

Otani, M., Mu, M., Handa, T., Kamada, H. and Shimada, T. (1993). Transformation of sweet potato (*Ipomoea batatus* (L.) Lam.) plants by *Agrobacterium rhizogenes. Plant Sci.,* 94 : 151-9.

Palazon, J., Cusido, R.M., Roig, C. and Pinol, M.T. (1998). Expression of the rol gene and nicotine production in transgenic hairy roots and their regenerated plants. *Plant Cell Reports,* 17 : 384-390.

Panda, A.K., Bisaria, V.S. and Mishra, S. (1992). Alkaloid production by plant cell cultures of *Holarrhena antidysentrica* II. Effect of precursor feeding and cultivation in stirred tank bioreactor, *Biotechnology Bioeng.,* 39 : 1052.

Parr, A.J. (1992). Alternative metabolic fates of hygrine in transformed root cultures of Nicandra physaloides. *Plant Cell Rep.,* 11 : 270-273.

Parr, A.J. and Hamill, J.D. (1987). Relationship between *Agrobacterium rhizogenes* transformed hairy roots and intact uninfected *Nicotiana* plants. *Phytochemistry,* 26 : 3241-5.

Parr, A.J., Payne, J., Eagles, J., Chapmau, B.T., Robins, R.J. and Rhodes, M.J.C. (1990). Variation in tropane alkaloid accumulation with in the Solonaceae and strategies for its exploitation. *Phytochem.,* 29 : 2545-2550.

Parr, A.J., Peerles, A.C.J., Hamill, J.D., Walton, N.J., Robbins, R.J. and Rhodes, M.J.C. (1988). Alkaloid production by transformed root cultures of *Cathoranthus roseus. Plant Cell Reports,* 7 : 309-312.

Payne, J., Hamill, J.D., Robins, R.J. and Rhodes, M.J.C. (1987). Production of hyoscyamine by hairy root cultures of *Datura stramonium. Pl. Med.,* 53 : 474-478.

Pellegrineschi, A., Damon, J.P. Valtorta, N., Paillard, N., Tepfer, D. (1994). Improvement of ornamental characters and fragrance production in lemon scented geranium through genetic transformation by *Agrobacterium rhizogenes. Bio/Technol.,* 12 : 64-68.

Perez-Molphe, E., Ochoa-Alejo, N. (1998). Regeneration of transgenic plants of Mexican lime from *Agrobacterium rhizogenes* transformed tissues. *Plant Cell reports,* 17 : 591-596.

Petit, A., David, C., Dahl, G.A., Ellis, J.G. and Guyon, P. (1983). Further extension of the opine concept : Plasmids in *Agrobacterium rhizogenes* cooperate for opine degradation. *MGG.,* 190 : 204-214.

Phunchindwan, M., Hirata, K., Sakai, A. and Miyamoto, K. (1997). Cryopreservation of encapsulated shoot primordia induced in horseradish (*Armoracia rusticana*) hairy root cultrues. *Plant Cell Reports,* 16 : 469-473.

Poma, A., Galeota, K., Miranda, M. and Spanol (1997). A ribosome inactivating protein principle from hairy roots and seeds of *Luffa cylindrica* (L) Roem and its cytotoxicity on melanotic and amelanotic melanoma cell lines. *Int. J. Pharmacognosy,* 35 : 212-214.

Production of flavonoids by psoralea hairy root cultures, plant cell, tissue and organ culture 56 : 97-104.

Protacio, C.M., Dai, Y.R. and Flores, H.E. (1990). Effect of dopamine on growth and morphogenesis in thin cell layers and hairy root cultures. In : *Abstracts of the VII International IUPAC Congress on Plant Tissue and Cell Culture,* pp. 294. Amsterdam, June 24-29, Kluwer Academic Publishers, Dordrecht.

Pythoud, F., Sinbar, V.P., Nester, E.W. and Gordon, M.P. (1987). Increased virulence of *Agrobacterium rhizogenes* conferred by the *vir* region of pti B0542 : Application to genetic engineering of *Poplar. Bio/Technol.,* 5 : 1323-7.

Rech, E.L., Golds, T.J., Husnain, T., Vainstein, M.H., Jones, B., Hammat, N., Mulligan, B.J. and Davey, M.R. (1989). Expression of a chimaeric kanamycin resistance gene introduced into the wild soybean *Glycine canescens* using a co-integrate Ri plasmid vector. *Plant Cell Reports,* 8 : 33-6.

Rhodes, M.J.C., Hilton, M., Parr, A.J., Hamil, J.D. and Robins, R.J. (1987a). Nicotine production by hairy root cultures of *Nicotiana rustica.* Fermentation and product recovery. *Biotech. Letters,* 8 : 415-420.

Rhodes, M.J.C., Robins, R.J., Hamill, J.D., Parr, A.J. and Walton, N.J. (1987b). Secondary product formation using *Agrobacterium rhizogenes* transformed hairy root cultures. *IAPTC Newsletter,* 53 : 2-15.

Rhodes, M.J.C., Robins, R.J., Lindsay, E., Aird, M., Payne, J., Parr, A.J. and Walton, N.J. (1989). Primary and secondary metabolism of plant cell cultures. Springer-Verlag, New York, pp. 58-72, 73.

Rhodes, M.J.C., Parr, A.J., Giulietti, A. and Aird, E.L.H. (1994). Influence of exogenous hormones on the growth and secondary metabolite formation in transformed root cultures. *Plant Cell Tiss. and Org. Cult.,* 38 : 143-151.

Riker, A.J., Banfield, W.M., Wright, W.H. and Keitt, G.W. (1928). The relation of certain bacteria to the development of roots. *Sci.,* 68 : 357-359.

Robins, R.J., Hamill, J.D., Parr, A.J., Smith, K., Walton, N.J. and Rhodes, M.J.C. (1987). Potential for use of nicotinic acid as a selective agent for isolation of high nicotine producing lines of *Nicotiana rustica* hairy root cultures. *Pl. Cell Rep.,* 6 : 122-126.

Roller, U. (1978). In : Production of Natural compounds by Cell Culture Methods (Alfermann, A.W. and Reinhard, E., eds.), 95-108, Gesellschaft fur Strahlenund Umweltforschung mbh, Munich.

Romanov, G.A., Konstantinova, T.N., Sergeeva, L.T., Golyanovaskaya, S.A., Kossmann, J., Willmitzer, L., Schmulling, T. and Aksenova, N.P. (1998). Morphology and tuber formation of *in vitro* grown potato plants harboring the year invertase gene and / or the *rol* C gene. *Plant Cell Rep.,* 18 : 318-324.

Roy, M.C. (1989). Plant growth response to *Agrobacterium rhizogenes* M. Appl. Sc. Thesis, Lincoln College, University of Canterbury, New Zealand.

Rugini, E. and Mariotti, D. (1991). *Agrobacterium rhizogenes* T-DNA genes and rooting in woody species. *Acta. Hort.,* 300 : 301-308.

Saito, K., Yoshimatsu, K. and Murakoshi, T.(1990) Genetic transformation of fox glove (*Digitalis purpurea*) by chimeric foreign genes and production of cardioactive glycosides. *Plant Cell Reports,* 9 : 121-4.

Saito, K., Yamazaki, T., Okuyama, E., Yoshihira, K. and Shimomura, K. (1991). Anthraquinone production by transformed root cultures of *Rubia tinctorium.* Influence of phytohormones and sucrose concentration. *Phytochem.,* 30 : 1507-1509.

Saito, K., Yamazaki, M., Anzai, H., Yoneyama, K. and Murakoshi, I. (1992). Transgenic herbicide-resistant *Atropa belladonna* using an Ri plasmid vector and inheritance of the transgenic trait. *Plant Cell Reports,* 11: 219-224.

Santos, P.M., Figueiredo, A.C., Olivera, M.M., Barroso, J.G., Pedro, L.G., Deans, S.G., Younus, A.K.M. and Scheffer, J.J.C.(1998) Essential oils from hairy root cultures and from fruits and roots of *Pimpinella anisum*. *Phytochemistry,* 48 : 455-60.

Sasaki, K., Udagava, A., Ishimaru, H., Hayashi, T., Alfermann, A.W., Nakanishi, F. and Shimomura, K. (1998). High forskolin production in hairy roots of *Coleus forskohlii*. *Plant Cell Reports,* 17 : 457-9.

Sato, K., Yamazaki, T., Okuyama, E., Yoshihira, K. and Shimomura, K. (1991). Anthraquinone production by transformed root cultures of *Rubia ticntorum* : Influence of phytohormone and sucrose concentration. *Phytochemistry,* 30 : 2977-8.

Sauerwein, M. and Shimomura, K. (1991). Alkaloid production in hairy roots of *Hyoscyamus albus* transformed with *Agrobacterium rhizogenes*. *Phytochemistry,* 30(10) : 3277-3280.

Sauerwein, M., Yamazaki, T. and Shimomura, K. (1991a). Hernandulcin in hairy root cultures of *Lippia dulcis*. *Plant Cell Reports,* 9 : 579-581.

Sauerwein, M., Ishimaru, K. and Shimomura, K. (1991b). Indole alkaloids in hairy roots of *Amsonia elliptica*. *Phytochemistry,* 30 : 1153-1155.

Sauerwein, M., Wink, M. and Shimomura, K. (1992). Influence of light and phytohormones on alkaloid production in transformed root cultures of *Hyoscyamus albus*. *J. Plant Physiol.,* 140 : 147-152.

Savary, B.J. and Flores, H.E. (1994). Biosynthesis of defense related proteins in transformed root cultures of *Tricosanthes kirilowii* Maxim. Var. *Japonium* (Kitam). *Plant Physiol.,* 106 : 1195-204.

Scheidegger, A. (1990). Plant Biotechnology goes commercial in Japan. *TIBTECH,* 8 : 197-198.

Schmulling, T., Rohrig, H., Pilz, S., Walden, R. and Schell, J. (1993). Restoration of fertility by anti-sense RNA in genetically engineering male sterile tobacco plants. *Mol. Gen. Genet.,* 237 : 385-394.

Schmulling, T., Schell, J. and Spena, A. (1998). Single genes from *Agrobacterium rhizogenes* influence plant development. *EMBO. J.,* 7 : 2621-2619.

Sevon, N., drager, B., Hiltunen, R., Oksman-Caldentey, K.M. (1997). Characterization of transgenic plants derived from hairy roots of *Hyoscyamus muticus*. *Plant Cell reports,* 16 : 605-611.

Shah, D.M., Horsch, R.B., Klee, H.J., Kishore, G.M., Winter, J.A., Tumer, N.E., Hironaka, C.M., Sanders, P.R., Gasser, C.S., Aykent, S., Siegel, N.R., Rogers, S.G. and Fraley, R.T. (1986). *Science,* 233 : 478-481.

Shahin, E.A., Sukhainda, K., Simpson, R.B. and Spivey, R. (1986). Transformation of cultivated tomato by a binary vector in *Agrobacterium rhizogenes* : Transgenic plants with normal phenotypes harbor binary vector T-DNA, but no Ri-plasmid T-DNA. *Theor. Appl. Genet.,* 72 : 770-777.

Shih, N.J.R., McDonald, K.A., Dandekar, A.M., Girbes, T., Iglesias, R. and Jackman, A.D. (1998). A novel type-1 ribosome inactivating protein isolated from supernatant of transformed suspension cultures of *Trichosanthes kirlowii*. *Plant Cell Rep.,* 17 : 531-537.

Shimomura, K., Sataka, M. and Kamada, H. (1986). Production of useful secondary metabolites by hairy roots transformed with Ri plasmid. In : International Congress of Plant Tissue and Cell Culture, University of Minnesota, Minneapolis, pp. 155.

Shimomura, K., Sauerwein, M. and Ishimaru, K. (1991a). Tropane alkaloids in the adventitious and hairy root cultures of Solanaceous plants. *Phytochemistry,* 30 : 2275-2278.

Shimomura, K., Sudo, H., Saga, H. and Kamada H. (1991). Shikonin production and secretion by hairy root cultures of *Lithospermum erythrorhizon*. *Plant Cell Reports,* 10 : 282-285.

Shin, D.I., Podila, G.K., Hung, Y. and Karnosky, D.F. (1994). Transgenic larch expressing genes for herbicide and insect resistance. *Can. J. For. Res.,* 4 : 2059-67.

Sikuli, N.N. and Demeyer, K. (1997). Influence of the ion-competition of the medium on alkaloid production by 'hairy roots" of *Datura stramonium*. *Plant Cell Tissue Organ Culture,* 47 : 261-267.

Simpson, R.B., Spielmann, A., Margossian, L. and McKnight, T.D. (1986). *Plant Mol. Biol.,* 6 : 403-415.

Sommer, S., Siebert, M., Bechthold, A., Heide, L. (1998). Specific induction of secondary product formation in transgenic plant cell cultures using a inducible promoter. *Plant Cell Reports,* 17 : 891-9.

Sowjanya, D. Veeresham,C. Kokate, C.K and Apte, S.S. (1998) Production of *Azadirachtin* from hairy root cultures of *Azadirachta indica*. Indian drugs 35 (8), 485-487

Stachel, S.M., Messens, E., Van Montagu, M. and Zambryski, P. (1985). Identification of the signal produced by wounded plant cells that activate T-DNA transfer in *Agrobacterium tumefaciens*. *Nature,* 318 : 624-629.

Stiller, J., Nasinec, V., Svoboda, S., Nemcova, B. and Machackova, T. (1992). Effects of agrobacterial oncogenes in kidney vetch (*Anthyllis vulneraria* L.). *Plant Cell Reports,* 11 : 363-7.

Stougaard, J., Abildsten, D. and Marcher, K.A. (1987). The *Agrobacterium rhizogenes* pRi T$_L$-DNA segment as a gene vector system for transformation of plants. *Mol. Gen. Genet.,* 207 : 251-255.

Stummer, B.E., Smith, S.E. and Langridge, P. (1995). Genetic transformation of *Verticordia grandis* (Myrtaceae) using wild type *Agrobacterium rhizogenes* and binary *Agrobacterium* vectors. *Plant Sci.,* 111 : 51-62.

Suginuma, C. and Akihama (1995). Transformation of gentian with *Agrobacterium-rhizogenes*. *Acta. Hort.,* 392 : 153-160.

Sun, L.V., Touraud, G., charbonnier, C. and Tepfer, D. (1991). Modification of phenotype of Belgian endive (*Cichorium intybus*) through genetic transformation by *Agrobacterium rhizogenes* conversion from biennials to annual flowering. *Transgenic Res.,* 1 : 14-22.

Tada, H. Murakami, Y. Omoto, T. Shimomura., K and Ishimaru,K (1995) phenolics in Hairy root cultures of *Ocimum Basilicum*. Phytochemistry, 42(2)**,** 431-434.

Tada, H., Shimomura, K. and Ishimaru, K. (1995). Polyacetylenes in *Platycodon grandiflorum* hairy roots of *Platycodon grandiflorum*. *Phytochemistry,* 42(1) : 69-72.

Tada, H., Nakashima, T., Kuntake, H., Mori, K., Tanaka, M. and Ishimaru, K. (1996). Polyacetylenes in hairy root cultures of *Campanula medium* L. *J. Plant Physiol.,* 147 : 617-9.

Tanaka, M. and Matsumoto, T. (1993). Regenerants from *Ajuga* hairy roots with high productivity of 20-hydroxyecdysone. *Plant Cell Rep.,* 13 : 87-90.

Tanaka, N., Takao, M. and Matsumoto, T. (1994). *Agrobacterium rhizogenes* mediated transformation and regeneration of *Vinca minor* L. *Plant Tiss. Cult. Lett.,* 11 : 191-8.

Taya, M., Yoyama, A., Nomura, R., Kondo, O., Matsui, C., Kobayashi, T. (1989). Production of peroxidase with horse radish hairy root cells in a two step culture system. *J. Ferment. Bioeng.,* 67 : 31-4.

Taya, M., Mine, K., Kinoka, M., Tone, S. and Ichi, T. (1992). Production and release of pigments by cultures of transformed hairy roots of red beet. *J. Ferment. Bioeng.,* 73 : 31-6.

Tepfer, D. and Tempe, J. (1981). Production d'agropine pardes racines forme es sous l'action d' *Agrobacterium rhizogenes* : Souche A$_4$. *CR Acad. Sci. Paris, Ser.,* **III** : 292 : 153-156.

Tepfer, D. (1982). La transformation genetique de plants superieurens par *Agrobacterium rhizogenes* – In Ze Colleque sur les Researchers Fruitieres. *Centre Techniue Interprofessional des Fruits et legumes.* Pp. 47-59, Bordeaux.

Tepfer, D. (1983). The potential use of *Agrobacterium rhizogenes* in the genetic engineering of high plants : Nature got there first. In : *Genetic Engineering of Eukaryotes* (NATO ASI Series, A, Life Sciences), vol. 61 (Eds. P.E. Lurquin and A. Kleinhofs), pp. 153-164, Plenum Press, New York.

Tepfer, D. (1984). Transformation of several species of higher plants by *Agrobacterium rhizogenes* : sexual transmission of the transformed genotype and phenotype. *Cell,* 37 : 959-967.

Tepfer, D. (1989) R, TDNA from Agrobacterium rhizogenes a source of genes having applications in rhizosphere bilogy and plant development, ecology and evolurion in plant microbe interactions molecular ad genetic perspectives Vol. 3,. PP 294-342 (eds) T. Kosuge and E.W. Nester (USA, Mc Graw-Hill)

Thomas, M.R., Rose, R.J. and Nolan, K.E. (1992). Genetic transformation of *medicago truncatula* using *Agrobacterium* with genetically modified Ri and disarmed Ti plasmid. *Plant Cell reports,* 11 : 113-7.

Toivenon, L. (1993). Utilization of hairy root cultures for production of secondary metabolites. *Biotechnol. Prog.,* 9 : 12-20.

Toivonen, L., Balsevich, J. and Kurz, W.G.W. (1989). Indole alkaloid production by hairy root cultures of *Catharanthus roseus*. *Plant Cell Tiss. Org. Cult.,* 18 : 79-93.

Toivonen, L., Ojala, M., Kauppinen, V. (1990). Indole alkaloid production of hairy root cultures of *Cathranthus roseus*. Growth kinetics and fermentation. *Biotechnol. Lett.*, 12 : 519.

Trotin, F., Moumou, Y. and Vasseur, J. (1993). Flavanol production by *Fagopyrum esculentum* hairy and normal root cultures. *Phytochemistry,* 32: 929-931.

Trulson, A.J., Simpson, R.B. and Shahin, E.A. (1986). Transformation of cucumbe (*Cucumis sativus* L.) plants with *Agrobacterium rhizogenes*. *Theor. Appl. Gene.,* 73 : 11-5.

Trypsteen, M., Van-Lijsekettens, M., Van Severen, R. and Van Montagu, M. (1991). *Agrobacterium rhizogenes* mediated transformation of *Echinacea purpurea*. *Plant Cell Reports,* 10 : 85-9.

Tzfira, T., Vainstein, A. and Altman, A. (1999). *Rol*-gene expression in transgenic aspen (Popular tremula) plants result in accelerated growth and improved stem production. *Trees-Berlin Index,* 14 : 49-54.

Uozumi, N., Ohtake, Y., Nakashimada, Y., Morikawa, Y., Tanaka, N. and Kobayashi, T. (1996). Efficient regeneration from a GUS transformed ajuga hairy roots. *J. Ferment Bioeng.,* 81 : 374-8.

Van Altvorst, A.C., Bino, R.J., Van Dijk, A.J., Lamers, A.M.J., Lindhout, W.H., Mark, F.V. and Dons, J.J.M. (1992). Effects of the introduction of *Agrobacterium rhizogenes* rol genes on tomato plant and flower development. *Plant Science,* 83 : 77-85.

Van der Salm, T.P.M., Hanisch ten Gate, C.H. and Dons, H.J.M. (1996). Prospects for applications of *rol* genes for crop improvement. *Plant Mol. Biol. Rep.,* 14 : 207-228.

Vanhala, L., Hiltunen, R. and Oksman-Caldentey, K.M. (1995). Virulence of different *Agrobacterium* strains on hairy root formation of *Hyoscyamus muticus*. *Plant Cell Reports,* 14 : 236-40.

Vargas, V.M.L. and Ham, M.L.M. (1995). Root cultures as source of secondary metabolites of economic importance. In : Phytochemistry of Medicinal Plants (eds.), Arnason, J.T., Mata, R. and Romeo, J.T., 29, Plenum Press, New York and London, p. 225.

Vazques-Flota F., V.O. Moreno, et al. (1994). Catharanthin and ajmalicine synthesis in Catharanthus roséus hairy root cultures: Medium optimization and elicitatiion. Plant cell Tissue Organ Culture 38(2-3): 273-279.

Verdejo, S. and Jaffee, B.A. (1988). Reproduction of *Pasteuria penetrans* in a tissue culture system containing *Meloidogyne javanica* and *Agrobacterium rhizogenes* transformed roots. *Phytopathology,* 78 : 1284-1286.

Visser, R.F.G., Hesseling-Meinders, A., Jacobsen, E., Nijdam, H., Witholt, B. and Feenstra, W.J. (1989). Expression and inheritance of inserted markers in binary vectors carrying *Agrobacterium rhizogenes* transformed potato (*Solanum tuberosum* L.). *Theor. Appl. Genet.,* 78 : 705-14.

Visser, R.G.F., Jocobsen, E., Witholt, B. and Feenstra, W.J. (1989). Efficient transformation of potato (*Solanum tuberosum* L.) using a binary vector in *Agrobacterium rhizogenes*. *Theor. Appl. Genet.,* 78 : 594-600.

Walton, N.J., Peerless, A.C.J., Robins, R.J., Rhodes, M.J.C., Boswell, H.D. and Robins, D.J. (1994). Purification and properties of Putrescine N-methyltransferase from transformed roots of *Datura stramonium* L. *Planta,* 193 : 9-15.

Walton, N.J. and Belshaw, N.J. (1998). The effect of cadaverine on the formation of cnabasine from lysine in hairy root cultures of *Nicotiana hesperis*. *Plant Cell Rep.,* 7 : 115-118.

Washida, D., Shimomura, K., Nakajima, Y., Takido, M. and Kitanaka, S. (1998). Ginsenosides in hairy roots of a *Panax* hybrid. *Phytochemistry,* 49 : 2331-5.

Waugh,R and Brown, J.W.S (1991) Plant gene structure and espression. In: Plant genetic engineering (ed) Grierson, D.PP1-32 Blackie Glasgow and London

Wei, Z.M., Kamada, H. and Harada, H. (1986). Transformation of *Solanum nigrum* L. Protoplasts by *Agrobacterium rhizogenes*. *Plant Cell Reports,* 5: 93-96.

Weidle, U.H., Borgya, A., Mattes, R., Lenz, H. and Buckel, P. (1987). Reconstitution of functionally active antibody directed against creatine kinase from separately expressed heavy and light chains in non-lymphoid cells. *Gene.,* 51 : 21-29.

Weiler, E.W. and Westekemper, P. (1979). *Phytochemistry,* 20 : 2009-2016.

Westcott, R.J. (1988). Thiophene production from hairy roots of *Tagetes*. *In : Manipulating secondary metabolism in Plant Cell Culture*. (Eds. R.J. Robins and M.J.C. Rhodes), pp. 233-237, Cambridge University Press, Cambridge.

White, F.F., Ghidossi, G., Gordon, M.P. and Nester, E.W. (1982). Tumour induction by *Agrobacterium rhizogenes* involves the transfer of plasmid DNA to the plant genome. *Proc. Nat. Acad. Sci. USA.,* 79 : 3193-3197.

White, F.F., Taylor, G.H., Huffmann, G.A., Gordon, M.P. and Nester, E.W. (1985). Molecular and genetic analysis of the transferred DNA of the root inducing plasmid of *Agrobacterium rhizogenes. J. Bacter.,* 164 : 33-44.

Whitney, P.J. (1992). Novel bioreactors for the growth of roots transformed by *Agrobacterium rhizogenes. Enzyme Microbial. Technology,* 14 : 13-17.

Wielanek, M and Urbanek, H. (1999). Glucotropaeolin and myrosinase production in hairy root cultures of Tropaeolim majus plant cell, tissue and organ culture 57 : 39.45.

Williams, R.D. and Ellis, B.E. (1993). Alkaloids from *Agrobacterium rhizogenes* transformed *Papaver somniferum* cultures. *Phytochemistry,* 32 : 719-23.

Wilson, P.D.G., Hilton, M.G., Robins, R.J. and Rhodes, M.J.C. (1987). Fermentation studies of transformed root cultures. In : *Bioreactors and Biotransformations* (Eds. G.W. Moody and P.B. Baker), pp. 38-51. Elsevier Publishers, London.

Wilson, P.D.G., Hilton, M.G., Meehan, P.T.H., Waspe, C.R. and Rhodes, M.J.C. (1990). The cultivation of transformed roots from laboratory plant to pilot scale. In : *Abstracts of the VII International IUPAc Congress on Plant Tissue and Cell Culture,* pp. 338. Amsterdam, June 24-29, Kluwer Academic Publishers : Dordrecht.

Wongsamuth, R. and Doran, P.M. (1997). Production of monoclonal antibodies by tobacco hairy roots. *Biotechnol. Bioeng.,* 54 : 401-405.

Wu, T., Wittkamper, J. and Flores, H.E. (1999). Root herbivory *in vitro* : interactions between roots and aphids frown in aseptic coculture. *In vitro Cell Dev. Biol. Plant,* 35 : 259-264.

Yamanaka, M., Ishibashi, K., Shimomura, K. and Ishimaru, K. (1996). Polyacetylene glucosides in hairy root cultures of *Lobelia cardinalis. Phytochemistry,* 41(1) : 183-5.

Yonemitsu, H., Shimomura, K., Satake, M., Mochida, S., Tanaka, M., Endo, T. and Kaji, A. (1990). Lobeline production by hairy root cultures of *Lobelia inflata* L. *Plant Cell reports,* 9 : 307-10.

Yoshikawa, T. and Furuya, T. (1987). Saponin production by cultures of *Panax ginseng* transformed with *Agrobacterium rhizogenes. Plant Cell Reports,* 6: 449-453.

Yoshikawa, T. Asada, Y. and Furuya, T. (1993). Continuous production of glycosides by a bioreactor using ginseng hairy root culture. *Appl. Microbial. Biotechnol.* 39 : 460-464.

Yoshimatsu, K. and Shimomura, K. (1992). Transformation of opium poppy (*Papaver somniferum* L.) with *Agrobacterium* rhizogenes MAFE 03-01724. *Plant Cell Reports,* 11 : 132-136.

Yoshimatsu, K., Yamaguchi, H. and Shimomura, K. (1996). Traits of *Panax ginseng* hairy roots after Cold storage and cryopreservation. *Plant Cell Reports,* 15 : 555-560.

Yu, S., Kwok, K.H. and Doran, P.M. (1996). Effect of sucrose, exogenous product concentration and other culture conditions on growth and steroidal alkaloid production by *Solanum aviculare* hairy roots. *Enzyme Microbiol. Technol.,* 18 : 238-43.

Zambryski, P., Tempe, J. and Schell, J. (1989). Transfer and function of T-DNA genes from *Agrobacterium* Ti and Ri plasmids in plants. *Cell.,* 56 : 193-201.

Zhan, X.C., Jones, D.D. and Kerr, A. (1988). Regeneration of flax plants transformed by *Agrobacterium rhozogenes. Plant Mol. Biol.,* 11 : 551-60.

Zheng, Z.R., Song, C.Q., Liu, D., Hu, Z.B. (1998). Determination of 6-isoflavonoids in the hairy root cultures of *Astragalus membranaceus* by HPLC. *Acta Pharmaceutica Sinica,* 33(2) : 148-151.

Zhou, Y., Hirotani, M., Yoshikava, T. and Furuya, T.(1997) Flavonoids and phenylethanoids from hairy root cultures of *Scutellaria baicalensis. Phytochemistry,* 44(1) : 83-7.

Chapter 20

Large Scale Cultivation of Plant Cells Using Bioreactors

INTRODUCTION

The fermentation of microbial cells to produce primary metabolites (e g amino acids and organic acids) and secondary metabolites (e.g antibiotics) has the long history, where as the application of same techniques for the production of plant metabolites using plant tissue culture for use as pharmaceuticals, flavors, colors is of recent origin. Several new *in vitro* plant products have just reached semi-commercial production levels (Table 20.1). Large-scale commercial production of secondary compounds from plants *in- vitro* is recognized for Shikonin, Berberine, Ginseng saponins and Taxol.

Table 20.1. Commercial and emerging/ semi-commercial large-scale secondary metabolite production schemes for plant cells in vitro.

Product	Cell culture source	Bioreactor/maximum volumes	Producer (s)
Berberine	*Thalictrum minus, Coptis japonica*	Batch & contionus flow; impeller driven 4000 L	Mitsui Petrochemical Industries, Japan
Shikonin	*Lithospermum erythrorhizon*	Batch 750 ml	Mitsui Petrochemical Industries , Japan
Ginseng saponins	*Panax ginseng*	Cell & root cultures 20,000 L	Nitto Denko, Japan
Vanillin	*Vanilla planifolia*	Impeller driven reactor 72 L	ESCAgenetics. USA
Taxol	*Taxus brevifolia: Taxus sp.*	2- stage system impeller driven and airlift reactor systems 2500 L/ 75000L	ESCAgenetics. USA Phyton Inc. USA, Nippon Oil Company, Japan, Samyang Genex Co. Ltd., Korea
Anthocyanins	*Euphorbia millii*	Rotary culture system	Nippon paint Company, Japan
Sanguinarine	*Papaver sommiferum*	Airlift system 300 l	Vipont Research Labs, USA
Digoxin	*Digitalis lanata*		Boehringer, Germany
Rosmarinic acid	*Coleus blumei*		Nattermann, Germany

One factor that is particularly important to note is that plant cell cultures are "totipotent". It is possible to produce unorganized, dedifferentiated callus or suspension cultures from differentiated whole plant material; it is also possible to regenerate whole plants from this dedifferentiated plant material. With animals terminally differentiated cells (e.g hepatocytes) can not be used to regenerate other cell types. The ability of plant cells to be continually able to respond to development clues is an important factor to consider in bioreactor strategy.

Related factors are the possibilities of cell-to-cell communication. Plant cells often grow in large aggregates and diffusional limitations may result in concentration gradients of key nutrients or metabolic by-products. Such gradients may result in microenvironments that foster cellular response to developmental clues. Further, the cells in a plant tissue are interconnected with plasmodesmata which are channels that interconnect the cytoplasm of plant cells in a tissue; the plasmodesmata allow the more rapid exchange of low molecular weight metabolites (<800 kDa) than would be possible by simple diffusion (Table 20.2). Plant cells have a large central vacuole and typically secrete products of interest within such vacuoles, while animal cells typically secrete products of interest into the extracellular compartment.

Table 20.2. Comparsion of key culture characteristics among plant, mammalian, and insect cultures (Shuler 1993)

Chracteristic	Plant	Insect	Mammalian
Is transformation necessary to establish continous cell lines?	No	No	Yes
Contact inhibition?	3-D Growth	Monolayer typically	Monolayer for normal cells
Cryopreservation	Difficult	Fairly routine	Fairly routine
Is differentiation of cells Irreversible?	No	Yes	Yes
Aggregate formation in Suspension?	Yes	Unusual	Unusual
Large central vacuole present?	Yes	No	No
Cell wall?	Yes	No	No
Intercell connections	Plasmodesmata	Gap-junctions	Gap-junctions
Size	15-150 µm	10-20 µm	10 µm
Doubling time	24-100 h	14-26 h	10-30 h
Typical attainable cell mass in suspension	10-50 g dw/L	<1.0 dw/L	<0.5 g dw/L

Another key point of differentiation between plant and animal cell culture is the level of cell density that can be attained. The high cell number densities that can be achieved by plant cells increases product concentration and possibly volumetric productivities. It also complicates the delivery of oxygen. Plant cells contain large amounts of water so in many cases more than 50% of the bioreactor volume is cellular volume even at dry weight cell densities of 20 to 30 g/L. This large amount of cells alters the fluid dynamics in a reactor. Although the specific O_2 requirement for plant cells is low, the volumetric requirements are relatively high due to the high cell density and constraints on mixing due to shear sensitivity. Because plant cells have a cell wall, they are less sensitive to shear than animal cells. Nonetheless, their large size makes them more vulnerable to turbulent shear since their cell size is on the order of the size of the smallest eddies (Table 20.3).

Table 20.4 summarizes key aspects of nutrition and medium formulation. The media used for plant cell culture are typically chemically defined, autoclavable and much less expensive than media for animal cells. The requirements with respect to carbon/energy and nitrogen sources are much simpler for plant cells, and their metabolism yields fewer problems with build-up of metabolic by-

Table 20.3. Comparison of key environmental parameters for plant, insect, and mammalian cells (Shuler 1993)

Parameter	Plant	Insect	Mammalian
Temperature optimum	25-30ºC	27-30ºC	37ºC
pH optimum	5.5-5.8	6.0-6.5	7.0-7.5
CO_2 incubator needed?	No	No	Typical because of bicarbonate buffer
Shear toleranace	Moderate	Low	Low
O_2 requirements (m mol O_2/g dw h)	0.5	0.4	0.2

Table 20.4. Comparsion of media for plant, insect, and mammalian cells (Shuler 1993)

Nutrient	Plant	Insect	Mammalian
Chemically-defined?	Yes	Possible	Not usually
Autoclavable?	Yes	No (filter sterilize)	No (filter sterilize)
Typical	Low to moderate	High	High
Carbon/energy source	Usually not photosynthetic; typically sucrose @ 30 g/L	Complex mixture of sugars and amino acids	Glucose, glutamine and other amino acids; less complex that for insect cells
Nitrogen sources	NO_3^- and NH_4^+	Amino acids	Amino acids

products (alanine, lactate, and ammonia) that limits animal cell culture. For plant cells lactate and ammonia are not waste products. Lactate build-up can be a problem for insect cells, but ammonia can be converted to uric acid. With mammalian cells both lactate and ammonia accumulation are important.

These differences between plant and animal cells presents unique considerations in developing appropriate bioreactor strategies.

The heart of the fermentation process is the fermenter. It is a container in which, maintained an environment favorable to the operation of a desired biological process. The difference between a bioreactor and fermenter is the type of cells being grown. The fermenter grows prokaryotic cells such as bacteria, fungi, etc. and a bioreactor is used for the mass cultivation of eukaryotic cells, such as insects, mammalian and plant cells. The key difference between the fermentor and bioreactor is in the agitation and aeration system. The difference in the characteristics of plant cell and microbial cultures from large-scale cultivation of point view is depicted in Table 20.5.

20.1. PROPERTIES OF PLANT CELL CULTURES IN RELATION TO MASS CULTURE

a) Cell Cultures

Plant cells are large (50-100 µM) by microbial standards and this, in itself, makes the application of the technology of fermentation developed for microorganisms difficult to apply to plant systems. The formation of aggregates affects the bioengineering properties interms of cellular physiology and gene expression. Plant cell cultures are heterogeneous in number of characteristics. To cultivate plant cell in a bioreactor requires high level of plant growth substances, which may be mutagenic and induces disorganized growth. This may lead to genetic variation in cells present in cultivation (D'Amato,

Table 20.5. The characteristics of plant cell and microbial cultures (Scragg 1992)

Characteristics	Microbial cell	Plant cell suspension	Consequences for plant cells
Size	2-10 μm	50-100 um	Possibly shear sensitive; rapid sedimentation
Individual cells	Often	Not often generally aggregated up to 1 mm in diameter	Rapid sedimentation possible micro-environment; low viscosity; large sample ports
Doubling time (growth rate)	1 h (rapid)	Days (slow)	Long culture runs, sterility maintenance
Inoculation	Small	5-20%	Large inoculation vessels required
Shear stress	Insensitive	Sensitive	Slow stirred speeds required, may eliminate STR bioreactor
Aeration	High 1-2 vol/vol min	Low 0.2-0.3 VVM	Low aeration required low K_la adequate
Variability	Can be stable	Can vary greatly	Culture characteristics not stable; may require continual selection
Product formation	Often into medium	Into vacuole	Cells need harvesting, no continual extraction from the medium

1978). With colored products such as anthocyanins and shikonin it is easy to observe the biochemical variation. Fujita *et al.* (1982) found significant differences in the levels of shikonins in individual protoplasts derived from callus culture of *L. erythrorhizon*. Holden *et al.* (1988) showed that the factors which stimulated anthocyanins production in cell suspension cultures of *C. roseus* tended to increase the propagation of pigmented cells rather than to increase the concentration of anthocyanins in individual cells. Variation in aggregate size is another source of heterogenecity in plant cell cultures. Plant cell cultures generally consist of aggregates of a few several thousand cells in particles of a few 100 μ to mm in diameter. The formation of aggregates affects the bioengineering and biochemical properties. The mixing and mass transfer in a bioreactor becomes difficult. These may also influence the pattern of gene expression.

Plant cells in culture, in addition to being large, frequently have thinner cell walls than cells in intact plants. It is often observed that plant cells secrete pectinaceous material and glycoproteins into the culture during growth. The combination of large size and thin yet rigid cell wall makes plant cell relatively fragile.

The rate of growth of plant cells in culture is low by microbial standard. *Nicotiana tabacum* cell suspension culture has 15 hr doubling time compared with 1-2 hr for typical microbial system. This requires maintenance of long term sterility. Plant cells have a low demand for oxygen relative to microorganisms. Plant cell respiration rate 10^{-4} MnO_2 (g dry weight). The underlying heterogenecity of plant cell culture of the genetic biochemical and cell aggregates levels provides the major challenge in the design of fermenters to maximize the particular metabolites (Table 20.5).

b) Immobilized Cell cultures

Immobilization of plant cells will be carried out using polymers such as alginate, agar, k-carrageenan, gelatins, polyacrylamide. The factors involved in the immobilization of plant cells onto polymer surfaces are (i) adhesion (ii) electrostatic factor. *Adhesion* is related to interfacial tension between cell and support, cells and suspending medium, support and medium. Electrostatic factors include pH, ionic strength and composition of medium (see Chapter 11).

c) Plant Organ Cultures

Both in cell aggregates in free suspensions and in cells in an immobilization matrix, a degree of cell organization may occur which may be beneficial to product formation. In root and shoot cultures in which growth and differentiation are ordered, cell separation does not occur except in rare situations and the full intercellular and vascular systems within the organ are intact. Unlike cell suspension cultures, organ cultures do not show the genetic variation in the dividing cells (Deus-Neuman and Zenk, 1984) and causes increase in ploidy. The expression of secondary metabolism in organ culture is under endogenous control within the normal programme of differentiation. The cultures show a high degree of stability, both genetic and biochemical, over extended periods.

The transformed root ('hairy root') culture has very rapid growth and highly dependent on the maintenance of root integrity. The high concentrations of auxins (endogenous) and mechanical damage induces root disorganization leading to loss in secondary metabolite formation. The hairy root cultures have high O_2 demand compared to cell suspension cultures. The difference in the properties of cell culture, organ cultures Vs hairy root culture is shown in Table 20.6.

Table 20.6. Characteristics of plant cell suspensions and organized cultures (Scragg 1994)

Characteristics	Suspension	Root, shoot, embryo	'Hairy' roots
Size	10-200 μm in length, often aggregated, up to 2 mm in diameter	Large organized structures, up to 2 cm in length	Long, highly branched roots; negatively geotropic
Growth rate	Doubling times 2-5 days	Doubling times long (5-15 days)	Rapid growth
Aeration requirement	Low	Low, may require certain levels of CO_2 or ethylene	Low
Shear stress sensitivity	In a number of cases, tolerant	Probably sensitive	Sensitive

20.2. HISTORICAL DEVELOPMENTS OF BIOREACTORS FOR MASS CULTURE OF PLANT CELLS

The mass culture of plant cell will be done by three methods:

a) Batch culture system
b) Semi-continuous culture system
c) Continuous culture system.

Tulecke and Nickell (1959) reported the first truly large-scale culture system for plant cell. A 20-1 carboxy with rubber stoppered fitted with four tubes – air in, air out "medium in" and medium out. Filtered compressed air supplied both aeration and agitation. They successfully grown cell lines derived from ginkoo and rose. Cultures were carried in liquid medium (100 ml/300 ml shake flasks). For scale up of the inoculum, 1 week old cultures was transferred to 900 ml of medium in a 2 l Fernbach Shake flask which after cultivation for 1 week was used to inoculate 9 l of medium in the carboxy (Fig 20.1).

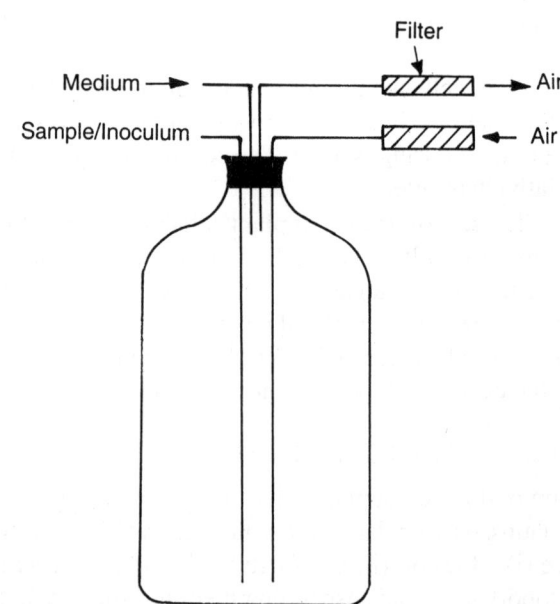

Fig. 20.1. Carboxy system of Tulecke & Nickell (1959).

Wang and Staba (1963) used dual carboxy system (Fig 20.2) for the culturing of spearmint cell. Half unit is carboxy unit like of Tulecke and Nickell. However, an aeration disc, stirrer, and a condenser on the exhaust line were added. The inoculum was 15% v/v of 5-day-old culture. The poor growth in 2nd vessel is one of the drawbacks.

Lamport (1964) described a roller-bottle system of 10-l flask fitted with cotton plug for aseptic sampling (Fig 20.3). Graebe and Novelli (1966) reported the described apparatus for large-scale cultivation of plant cell (Fig 20.4). Culture vessel was either 6 or 12 L florence flask, the neck of which was fitted with a large standard taper joining below which was located an outlet for exhaust gas on to the standard taper was fitted with a glass cap with transfer and aeration tubes. A magnetic stirr bar, placed in the culture vessel, for agitation. The culture was agitated and aerated by passing sterile air through the fitted sparger.

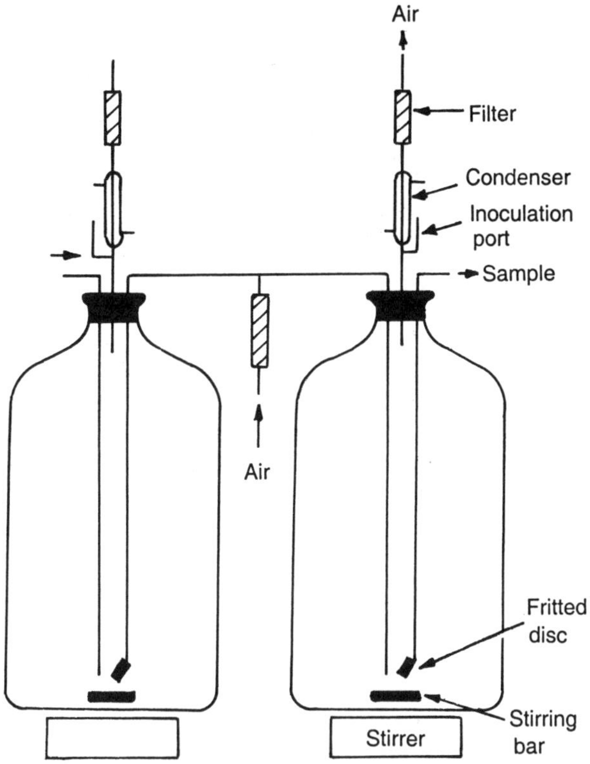

Verma and van Huystee (1971) described a culture system (Fig 20.5) which reduces the contamination during scanning of the tissue and culture of the medium. 5 L three necked culture vessel was fitted with exhaustion gas, finger action stirrer, and F shaped glass fitting for air and medium inlets. Lack of flexibility in design and complexity or operation was the major disadvantages for this bioreactor.

Veliky and Martin (1970) separated a culture vessel of inverted Erlenmeyer flask (Fig 20.6). In 1962 Byrne and Koch reported on the successful use of New Brunswick fermentor for growing plant cell. Rose and Martin (1975) used 7.5-l New Brunswick microform fermenters for the Ipomoea sp. Kato *et al.* (1972) examined the influence of agitation speed on the growth of tobacco cells in stirred-Jar fermenter (Fig 20.7).

Fig. 20.2. Dual carboxy system of Wang and Staba (1963).

b) Semi Continuous Systems

A semi continuous system is a simply a batch system provided with the facility to remove at intervals a large volume of culture and replace it with fresh sterile

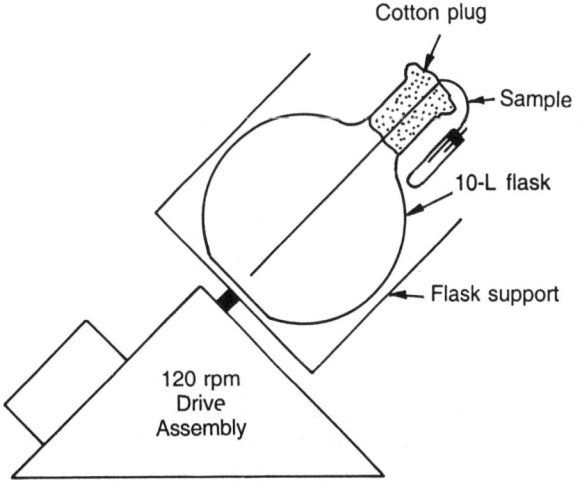

Fig. 20.3. Roller flask system of Lamport (1964).

medium. The usual operating procedure is to allow the culture to develop until near maximum yield is attained. The every day or two a portion (up to 50%) of the culture is harvested and replaced with fresh medium. The only difference between semi continuous system with that or Batch culture is harvest size.

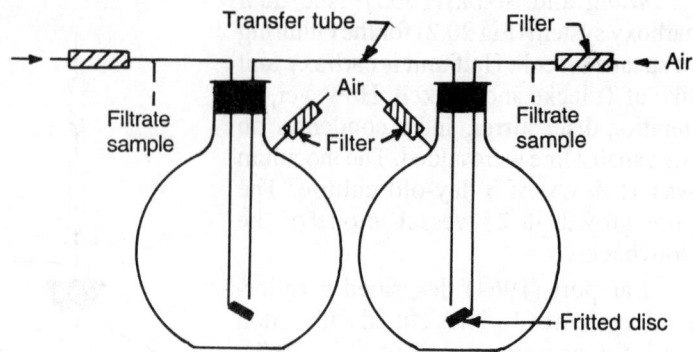

Fig. 20.4. System of Graebe and Novelli (1966).

c) Continuous Systems

Plant cells go through a growth cycle analogous to microorganism in Batch culture, where as in continuous culture cell division taken place at a defined rate and in uniform environment. The principle of continuous culture system is based on Monod formula

$$\mu = \frac{S}{max \ K + S}$$

μ = specific growth rate
K = saturation constant
S = substrate concentration

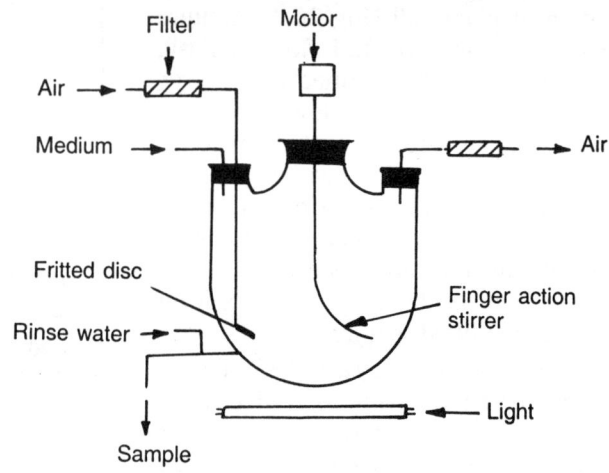

Fig. 20.5. System of Verma and van Huystee (1971).

Miller *et al.,* (1968) reported a continuous system containing a glass sparger and agitation by magnetic stirrer placed at the bottom of the vessel, automatic sampling valve for continuous withdrawals of culture at the interval of 20 min. Medium was fed continuously through a tube (Fig 20.8).

Kurz (1975) culture vessel contains a thin-walled glass tube, bottom of tube as part for aeration, and other parts for sampling and harvesting. A peristaltlic pump for continuos flow medium and harvest (Fig 20.9) used Wilson *et al.* (1971) developed chemostat or turbidostat modes for continuous culture. The 4 l vessel was stirred by a magnetic stirrer held of the bottom, and temperature was controlled by tempered water flowing through a glass coil. Fresh medium was introduced into the vessel via variable speed pump. Downstream of the pump was an optical density monitor, while upstream was a side arm leading via solenoid valve to the sample collector.

In operation, the vessel was inoculated, the circulating pump started, and the fresh medium flow begun. When the medium touched the upper electrode, the solenoid valve opened, and culture flowed by gravity into the sample collector. When the medium level dropped, the solenoid valve closed. Then by using a limiting nutrient, growth was controlled by a flow rate, i.e. the system operated as a chemostat. In turbidostat mode, the fresh medium outlet was connected to the circulating pump via a double action solenoid valve. When optical density of the circulating culture reached the upper set point, the circulating line closed, the medium inlet opened, and fresh medium was pumped into the system. As fresh medium forced into the density monitor, the lower set point was passed, the medium supply was cut off, and the culture circulated. The constant level device functioned as before to maintain the volume of the culture (Fig 20.10).

20.3. BIOREACTOR CHARACTERISTICS REQUIRED FOR LARGE-SCALE PRODUCTION

A bioreactor or fermenter is a vessel made up of glass or steel in which plant cells are cultivated. It should have the following characteristics :

i) A sterile environment

ii) Supply of air

iii) Mixing

iv) Temperature control, addition or removal of heat

v) Provision for sampling

vi) Access for environmental monitoring e.g pH, Dissolved oxygen, temperature etc.

The basic function of a bioreactor or fermenter is, it should maintain controlled environment for the growth of microbial cells or plant cells to obtain a desired product. The basic parts of a bioreactor are

1. A culture vessel

2. Associate supply and environmental systems

3. Measurement and control system

20.3.1. Design of a bioreactor (Fig 20.11)

Points to be considered :

1. The vessel should be capable of being operated for long term (days) aseptically

2. Adequate aeration and agitation

3. Power consumption should be low

4. A system for temperature control

5. A system for pH

6. Sampling port

7. Minimization of evaporation losses

8. Culture vessel should be useful for wide range of processes

9. Vessel should be constructed with smooth internal surfaces

10. The material for construction should be available at cheaper rate

20.3.2. Body construction

In the simplest cases, plant cells may be cultured in Erlenmeyer flasks, baffle flasks or Car-

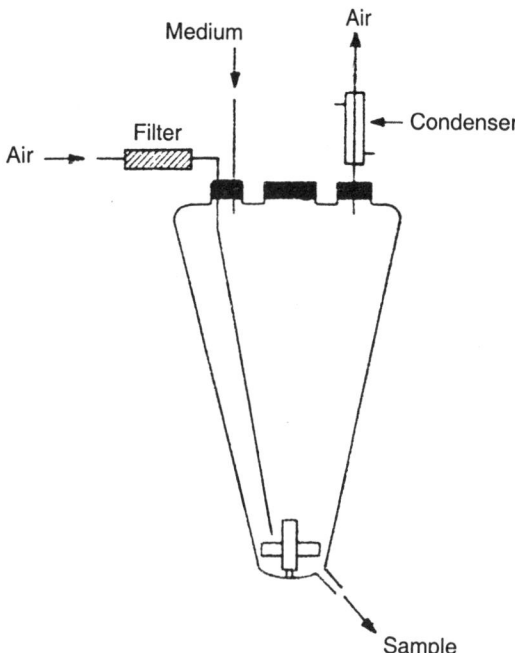

Fig. 20.6. System of Veliky and Martin (1970).

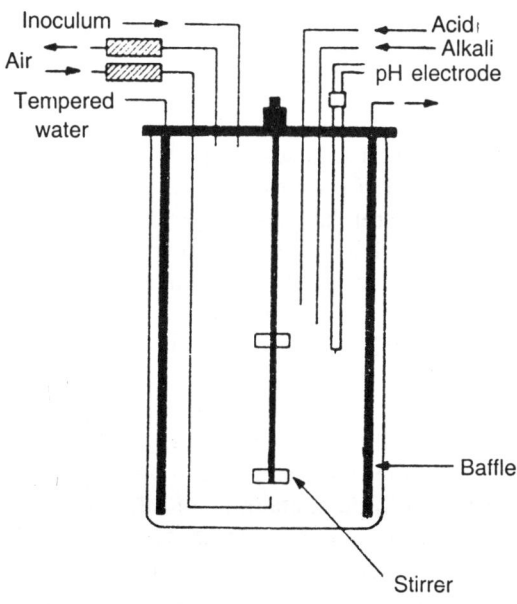

Fig. 20.7. Stirrer-Jar Fermenter (Kato *et al.*, 1972).

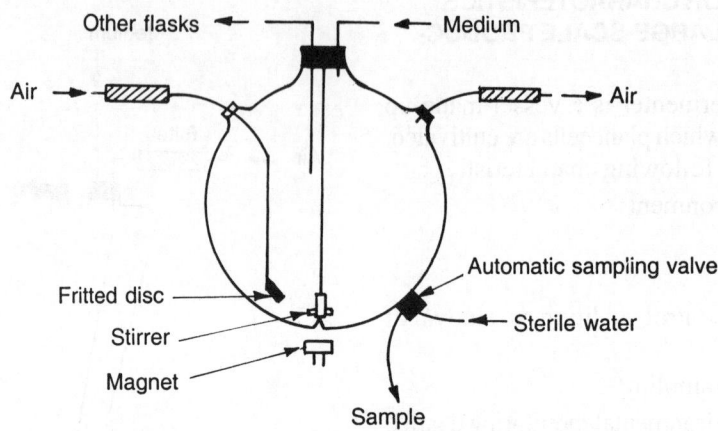

Fig. 20.8. Continuous Fermenter (Miller *et al.*, 1967).

boxy bottles (20l) (Tulecke & Nickell 1959), or in special, closed tanks (Willemot & Durand 1977). If the main purpose is biomass production, they are commonly called fermenters, but if they serve mainly to produce secondary compounds, they are often called bioreactors (Hashimoto & Azechi 1988).

Suitable materials should be inert and sterilizable. At a laboratory scale of up to 2 l, plastic foils (polyamide, polypropylene) are as well as glass are used. For capacities up to 20 l, steel and stainless steel, usually used for larger volumes, may be used in addition to glass (Markl 1989).

On pilot and large scale, any materials used will have to be assessed on their ability to withstand pressure sterilization and corrosion and their potential toxicity and cost. Walker and Holdsworth (1958), Solomons (1969) have discussed the suitability of various materials used in the construction of fermenters. Pilot-scale and industrial-scale vessels are normally constructed of stainless steel or at least have a stainless-steel cladding to limit corrosion. Mild steel coated with glass or phenolic epoxy materials has occasionally been used (Gordon *et al.,* 1947; Fortune et al 1950, Buelow and Jonson, 1952, Irving, 1968). The thickness of the construction material will increase with scale.

Normally in the design and construction of a fermenter there must be adequate provision for temperature control which will affect the design of the vessel body. Heat will be produced in bioreactor through the mechanical agitation, should be removed from the system.

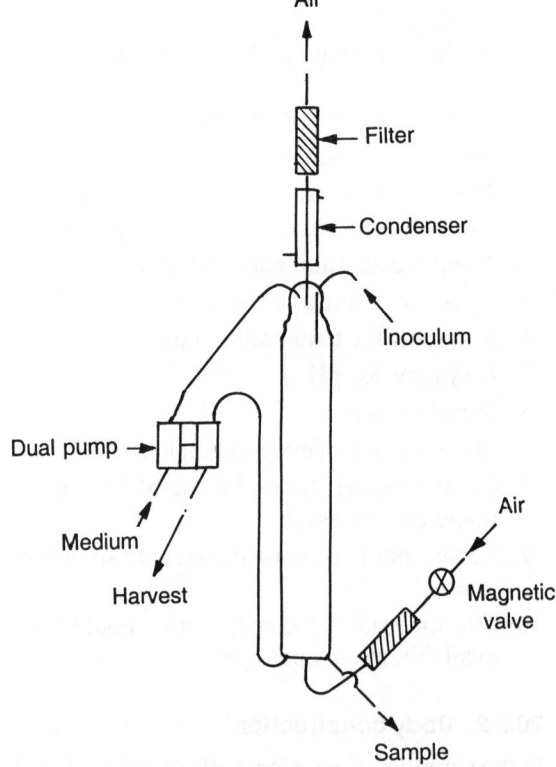

Fig. 20.9. Continuous Fermenter (Kurz, 1975).

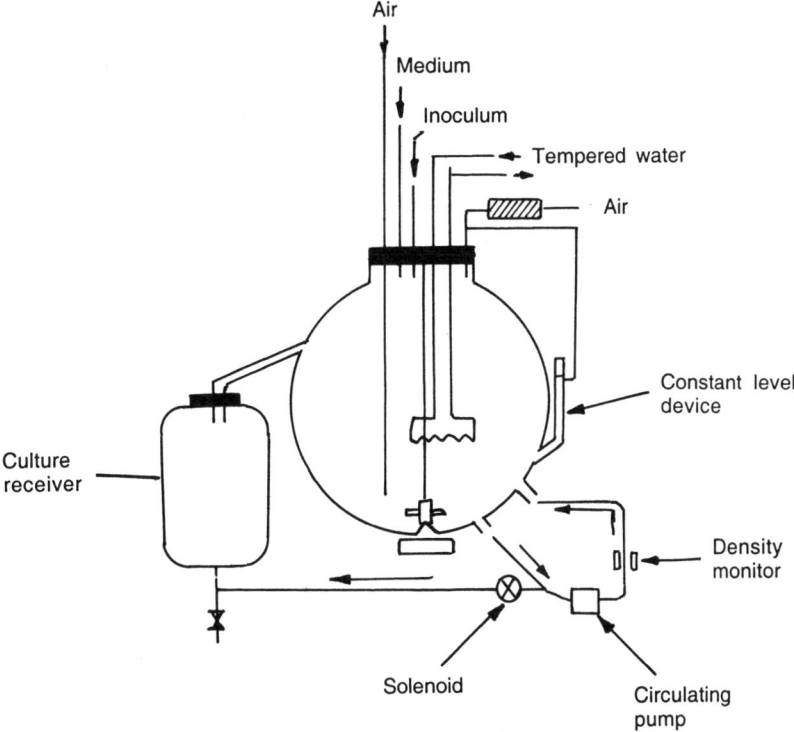

Fig. 20.10. Continuous fermenter (Wilson *et al.*, 1971).

20.3.3. Aeration and Agitation

The primary purpose of aeration should be to provide plant cells/microbial cells in fermenter with sufficient oxygen for metabolic requirements, while agitation should ensure that a uniform suspension of microbial cells/plant cells is achieved in a homogenous nutrient medium. The type of aeration – agitation system depends on characteristics of the fermentation process.

Aeration and Agitation system contains

1. The impeller (agitator)
2. Stirrer glands and bearings.
3. Baffles
4. The aeration system (sparger).

20.3.3.1. Impeller

The functions of the impeller are :

1. To diminish the size of air bubbles to give a bigger interfacial area of oxygen transfer and to decrease the diffusion path.
2. To maintain uniform environment throughout the vessel contents.

Impellers may be classified as disc turbines, vaned discs, and open turbines of variable pitch and properties (Fig 20.12). The most commonly used necessary to consider the size of the imepller and where to position it in the vessel. In tall vessels more than one impeller is needed if adequate aeration–agitation is to be obtained. Ideally the impeller should be one third to one-half of the vessel diameter above the base of the vessel.

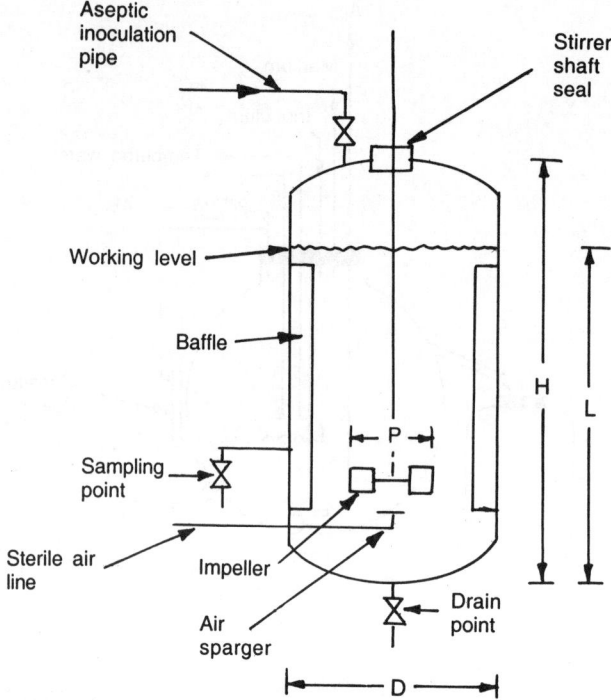

Fig. 20.11. Typical design of bioreactor.

20.3.3.2. Stirrer Glands and Bearings

The satisfactory sealing of the stirrer shaft assembly has been one of the most difficult problems to overcome in the contruction of fermentation equipment, which can be operated aseptically for long periods. A number of different designs have been developed to obtain aseptic seals. The stirrer shaft can enter the vessel from the top, side (Richards, 1968) or bottom of the vessel. Top entry is most commonly used but bottom entry may be advantageous if more space is needed on the top plate for entry ports, and the shorter shaft permits highly stirrer speeds to be used by eliminating the problem of the shaft whipping at high speeds. In general, bottom entry stirrers would be undesirable, as the bearing would be submerged, although Chain *et al.* (1952) has successfully operated vessels of this type. The various aseptic seals used for this purpose are packed gland seal, bush seal, mechanical seal and magnetic drives.

20.3.3.3. Baffles

Four baffles are normally incorporated into agitated vessels of all sizes to prevent a vortex and to improve aeration efficiency. They are metal stripes roughly one-tenth of the vessel diameter and attached radially to the wall. Walker and Holdsworth (1958) recommended that baffles should be installed so that a gap existed between them and the vessel wall, so that there was a scouring action around and behind the baffles thus minimizing microbial growth on the baffles and the fermenter walls.

20.3.3.4. The aeration system (sparger)

A sparger may be defined as a device for introducing air into the liquid in a fermenter. It is important

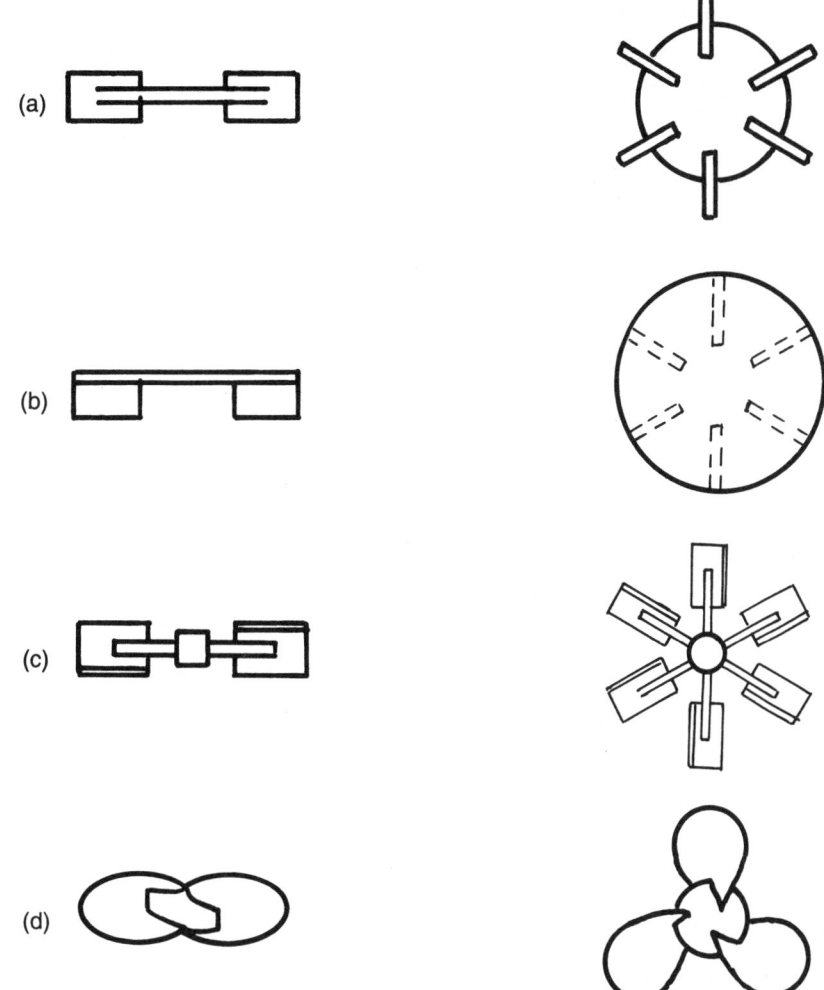

Fig. 20.12. Types of Impeller. a) disc turbine, b) Vaned disc, c) Open turbine d) Marine propeller (Solomons, 1969).

to know whether the sparger is to be used in its own or with mechanical agitation as this can influence equipment design to determine initial bubble size. Three basic types of sparger have been used and may be described as the porous sparger, the orifice sprager (a perforated pipe) and the nozzle sparger (an open or partially close pipe). A combined sparger agitator may be used in laboratory fermenters.

20.4. THE ACHIEVEMENT AND MAINTENANCE OF ASEPTIC CONDITIONS

The following operations may have to be performed to achieve and maintain aseptic conditions during fermentation.

20.4.1. Sterilization of the fermenter

The fermenter should be so designed that it may be steam sterilized under pressure. The medium may be sterilized in the vessel or separately, and subsequently added aseptically, If the medium is sterilized *in situ* its temperature should be raised prior to the injection of live steam to prevent the formation of large amount of condense.

All pipes should be constructed as simply as possible and slope towards drainage points to make sure steam reaches all parts of the equipment and is not excluded by siphons or pocket of condense or mash.

20.4.2. Sterilization of the air supply

Sterile air will be required in very large volumes in many aerobic fermentation processes. Although there are a number of ways of sterilizing air, only three have found permanent application. These are heat, filtration through fibrous material and filtration through granular material.

20.4.3. Aeration and Agitation (see details on 20.3.3)

20.4.4. The addition of inoculum, nutrients and other supplements

To prevent contamination during an addition it is essential that both the addition vessel and the fermenter are maintained under a positive pressure and the addition port is equipped with a steam supply.

20.4.5. Sampling

Sampling port is kept in 40 % formalin or a suitable substitute for aseptic conditions.

20.4.6. Foam control

In any fermentation it is very important to minimize foaming. When foaming becomes excessive, there is a danger that filters become wet resulting in contamination. There is also the possibility of siphoning developing leading to the loss of all or part of the contents to the fermenter. It is common practice to add antifoam to a fermenter when the culture starts foaming above a certain predetermined level. Valves attached to fermenters and ancillary equipment are used for controlling the flow of liquids and gases e.g. Ball valves, Needle valves, piston valves etc.

20.5. IMPORTANT FACTORS FOR BIOREACTORS

20.5.1. Gas-Liquid Mass Transfer

Most bioreactors are rated in terms of their oxygen mass transfer coefficient. K_{La} scale – up of many bioprocesses is based on maintaining a constant K_{La}. The goal for culturing plant cell is often to maximize the oxygen transfer coefficient. This is achieved by using mechanical agitators at high rotational speeds and by spraying large volume of gas (e.g. 1 volume of gas per minute per volume of liquid). Since plant cells have lower respiration rates. The oxygen transfer requirements are considerably less. If cells which respire at a rate of 0.2 m mol /g-h are to grown to 10g/l without allowing the dissolved oxygen concentration to fall below 20% of saturation. The equation suggest that the K_{La} should be

$$K_{La} \frac{Qo\ max\ X}{(CO_2 - CO_2\ crit)} = \frac{0.2\ mmol}{g\text{-}h} = (10\text{-}g/l)\,/\,[0.25\ mmol/l - 0.2\,(0.25\ mmol/l)] - 10h^{-1}$$

Plant cell bioreactors are typically operated at K_{La} values of 10-30 hr^{-1}. Operation at higher K_{La} values has often been observed to result in poorer plant cell growth. (Tanaka 1981, 1987, Smart and Fowler 1984) or product synthesis (YamaKawa *et al.,* 1983). This poor performance at high K_{La} values may be due to either increased shear associated with the high-K_{La} conditions or to enhanced carbondioxide stripping from the liquid (Hegarty *et al.,* 1986)

20.5.2. Shear Sensitivity and Rheology

Because of their large size, extensive vacuole, and rigid cell wall, plant cells have been regarded as sensitive to shear stress. It is difficult to define what is meant by shear. Shear refers to forces exerted on the surface of a body in a direction parallel to the surface. This is in contrast to the normal forces, which are exerted on a surface but perpendicular to the surface.

Shear is described by the equation

$T = \eta r$

Where T = shear stress (a force per unit surface area)

R = shear rate (a change in velocity across a distance)

η = Viscosity (a coefficient which describes the resistance to flow)

To describe the shear damage to eukaryotic cells in mechanically agitated systems, shear damage resulted from "highly localized velocity gradients" (Midler and Finn 1966) and these localized gradients or the maximum shear rates could be related to the speed with which the outer edges of the impeller blade traveled (Oldshue 1966). Impeller tip speed is given by the equation

$$\text{Tip speed} = \pi \, N d_i$$

Where N is the Impeller speed (rpm)

d_i is the impeller diameter

In addition to hydrodynamic shear, recent studies reveal that cell damage may result from cell-cell and cell-impeller collisions (Cherry and Papoutsakis 1988, Prokop and Rosenberg 1989). Shear can also results from gas sparging, even in the absence of mechanical agitation.

The rheological behaviour of certain plant cell cultures have been determined, for example *Morinda citrifolia* showed shear thinning and thixotrophic behaviour whereas *Cudrania tricuspidata*, *Vinca rosea* and *Agrostemma githago* were non-Newtonian and pseudoplastic. In these studies the rheological properties of the cultures could be estimated using normal rotational viscometers such as the cup-and-bob system. However, plant cell suspensions are in general highly aggreated. As a consequence of this particulate nature, the normal methods of measuring viscosity, i.e., the cone-and-plant or cup-and-bob viscometers are usually inadequate, because the aggregates settle rapidly or become trapped in the narrow gaps in the viscometer. Alternative designs of viscometer using either anchor and turbine impellers that eliminate the problems of settling have been used but they do not give a direct measurement of viscosity because of the turbulent flow produced. Using both alternative types of viscometer, plant cell suspensions have been shown to be non-Newtonian and pseudoplastic having a low viscosity in the order of 1-10 mpa/s.

20.5.3. Mixing

Mixing refers to the convective transport of matter (e.g. transfer of solutes associated with bulk fluid motion). The mixing of dissolved nutrients (e.g. salts and sugars) has generally not been a problem in suspension culture. There are three important mixing problems in suspension culture systems.

1. Because of the large size of plant cells and especially cell aggregates which lead to settling at the bottom of the bioreactor. These cells can then settled into dead zones or unmixed regions of the bioreactor. These dead zones can become depleted of key nutrients (e.g., dissolved oxygen).
2. Attachment of cells onto surface above the level of the liquid. Since these cells are not bathed in the liquid medium, they can become deprived of nutrients.
3. Because of the desire to culture cells at high concentrations. With moderate cell concentration (packed cell volumes less then 50%), bringing cells in contact with the nutrient-containing liquid

is often not a problem. However, incomplete mixing can be a serious problem when plant cell is cultured of high concentrations.

20.6. TYPE OF BIOREACTORS

The objective in culturing plant cells, whether in shake flask or in bioreactors is to provide conditions to the culture which allow it perform in a reproducible and optimal manner. Various measures of culture performance exist, and appropriateness of a measure depends on the desired results. For e.g. if efficient cell growth is desired, then bioreactor performance is evaluated on the basis of doubling times and maximum cell concentrations. If secondary metabolites are desired, then an appropriate performance measure would be the overall productivity, or the amount of product produced per total cultivation time. The bioreactors are classified into 3 categories based on the mode of aeration and agitation 1) Mechanically agitated bioreactor (2) Pneumatically agitated 3) Special class for e.g., Mist bioreactor. The second classification is based on the shape and size of the culture vessel, they are divided into 1) Stirred tank 2) Air-lift 3) Trickling film reactor 4) Mist 5) Rotating drum 6) Bubble column 7) Immobilized plant cell fermenter . The third classification is based on mode of operation they are divided into 1) Batch suspension reactor 2) Continuous flow stirred tank reactor (CFSTR) 3) Semi-continuous (Multistage (CFSTR) 4) Immobilized tank reactor.

20.6.1. Mechanically agitated bioreactor (Stirred tank-bioreactor)

Stirred tank reactor is a classical representative of bioreactor types where air is dispersed by mechanical agitation and it is most common type of reactor for culturing plant cells (including microbial and animal cells). This type of reactor employs impellers and mechanical energy for gas-liquid mass transfer and mixing. Various type of impellers have been used,(See Fig 20.12) like flat-blade turbine impeller and marine propellor etc. (Fig 20.13). In stirred tank reactor rotational motion of impeller can lead to the formation of a vortex. Vortex formation is eliminated by the use of baffles. Since baffles also provide a surface onto which the cells can attach they can aggravate problems associated with the adhesion of cells onto surfaces above the liquid level.

Stirrers have been invented which can be used to mix plant cell cultures grown in reactors. Various types of stirrers i.e., spin, helix, bladed, paddle have been designed for this purpose. Tanaka (1987) has cultured *Maclura tricupidata* cells in bioreactor equipped with several types of stirrers. In this best result were obtained when a paddle impeller was used to mix the plant cells.

Ulbrich *et al*. (1985) compared various agitation modes in standard vessels and found the rosmarinic acid production with *Coleus blumei* cells was best when a helical stirrer, termed "module spiral stirrer" was used. Cardenolide biotransformation was improved when performed in stirred-tank reactors.

The ratio of stirrer/shear flow is dependent on both the shape and size of the impeller and is sometimes chosen as the scale-up criterion. Small diameter impeller rotating rapidly generate high shear force but are not effective system, where large diameter impeller moving at low speed have a low shear-flow ratio. Still higher stirrer velocities resulted in a reduced cell growth and hence reduced productivity. Several stirrer types (spin stirrer, bladed stirrer, helical stirrer were tested and it was found that maximum alkaloid production is achieved if a bladed stirrer is used together with baffles. It is thus seen that plant cells may be cultivated in conventional stirred-tank bioreactors. High growth rates as well as good production rate were achieved with either spin stirrers. Helical stirrers or bladed stirrers. Stirred reactors are more versatile systems. Although there are many indications that the effect of shear on cell viability may vary with the cell line.

The first industrial plant cell culture based process of shikonin production from *L. erythrorhizon* used stirred tank reactors of 200 and 500 litre capacity. Inspite of the fact that plant cells have great

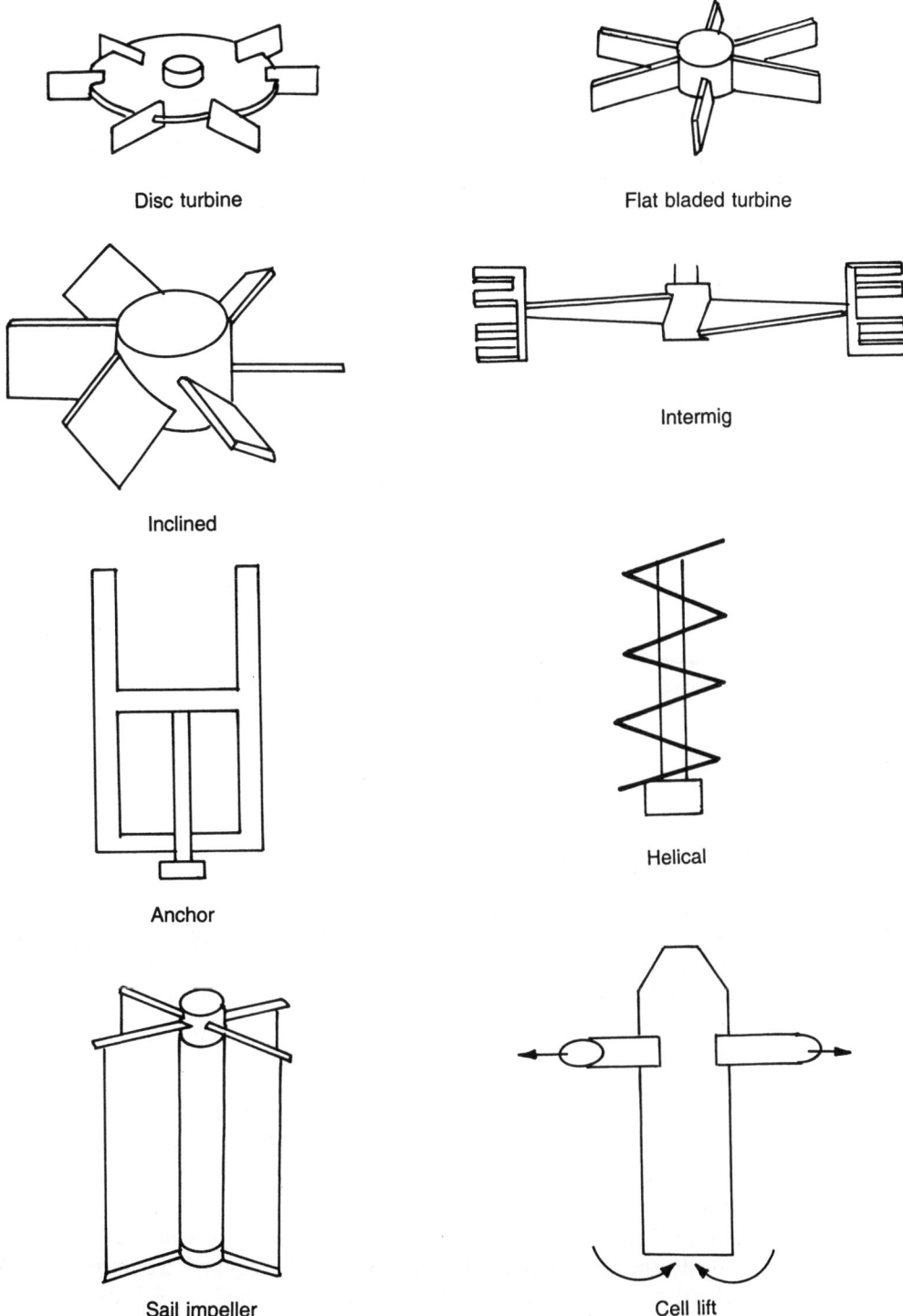

Disc turbine

Flat bladed turbine

Inclined

Intermig

Anchor

Helical

Sail impeller

Cell lift

Fig. 20.13. Impeller designs for stirred-tank reactor (Hooker *et al.*, 1989).

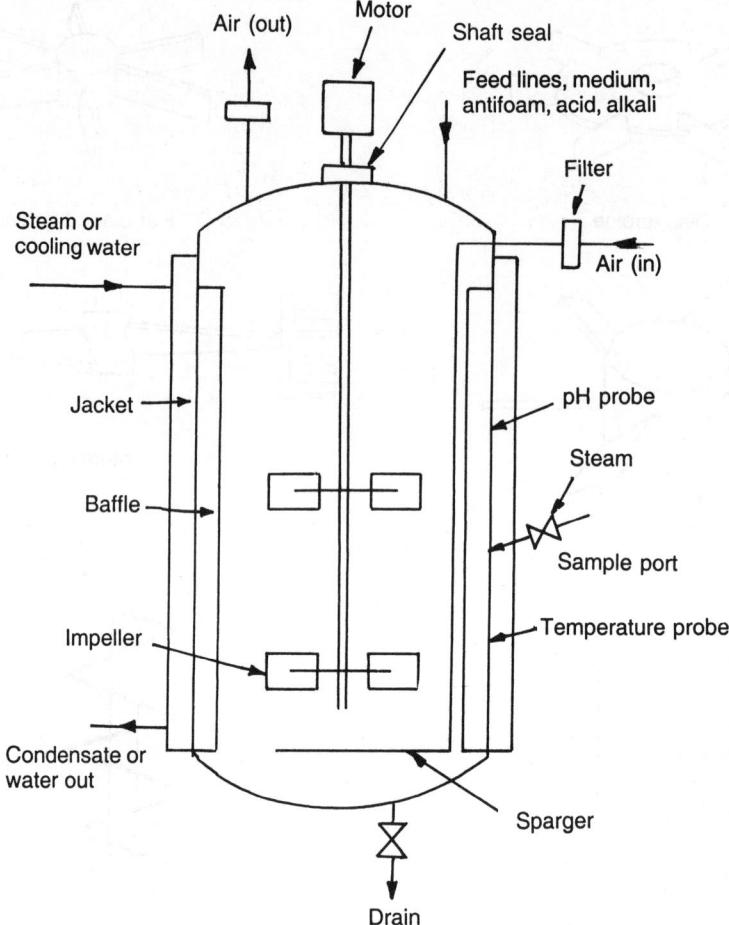

Fig. 20.14. A stirred-tank bioreactor.

potential when used at low agitation speed, high energy requirement, complexity of construction and difficulty in scaling up the volume, are a few disadvantages associated with stirred tank reactors. Cells of *Nicotiana tabacum* (Kato *et al.*, 1977; Knoop *et al.*, 1988; Noguchi *et al.*, 1977) *Catharanthus roseus* (Kozai, 1991), *Dioscorea* sp. (Fowler, 1981; Ducos and Pareilleux, 1986) *D. detoidea* (Spetler *et al.*, 1985) have been cultivated successfully in this type of reactor(Fig 20.14)

Wagner and Vogelmann (1977) reported influence of different types of reactors on natural product synthesis in *Morinda citrifolia* and recognized that low productivities achieved in reactors with turbine impellers resulted mainly from agitation problems. Oxygen limitation in stagnant zones together with mechanical damage to the cells in the turbulent region was the key reason. These results add to the development of airlift systems.

A CFSTR (continuous flow stirred tank reactor) is a device where a nutrient feed stream or streams is fed continuously and at least one effluent stream is also remove a continuously. Microbiologists describe such a device as a "chemostat" and the term is applied to CFSTRs with plant cells (Dougall 1983). Chemostat theory demonstrates that at steady-state operation the population-averaged growth rate of the culture (i) must equal the dilution rate, D (where D equals the volumetric flow rate into the reactor divided by the reactor volume). Thus the chemostat allows the investigator to independently set the growth rate. At steady state the culture response has, in principle, no dependence on the inoculum history. The chemostat is an ideal device to probe a culture response to a variety of stimuli under well-controlled conditions where the environment remains constant.

The application of chemostat theory to plant cell culture requires the investigator to be aware of the basic assumptions in developing the theory. The key assumption is that the contents of the vessel are perfectly mixed. Thus the concentration of cells and nutrients etc. must be the same at any position in the reactor and the same as the concentration in the exit line.

Reactor with low agitation is unlikely to keep the plant cells evenly suspended and often wall growth is significant in plant cell reactors. In a chemostat the cells must be dividing at a sufficiently rapid to replace the cells washed out in the effluent stream. Thus any product formation that takes place must be at least partially growth associated.

A multistage CFSTR is better suited to systems where product formation is non-growth associated. In the first stage conditions must be maintained to support cellular replication at rate equal to washout. In the second and subsequent stages replication is no longer vital since the previous reactor acts as a continuous source of new cellular material. Thus conditions in the second and subsequent stages can be used to direct the cells toward forming non-growth-associated products. The system can be used to direct the culture through a sequence of physiological states.

20.6.2. Pneumatically agitated Bioreactors

In addition to mechanical agitation, it has been suggested that the motion of the rising gas stream should be capable of providing the energy for fluid mixing. There are two type pneumatically agitated bioreactors, the bubble column and the airlift.

The pneumatically agitated reactors differ from the mechanically agitated systems not only by the absence of an impeller but also the pneumatically agitated systems are tall and thin. Typically the height to diameter ratio of pneumatically agitated bioreactor is 10 while this ratio is closer to 1 for mechanically agitated bioreactor. The advantages of pneumatically agitated bioreactors are 1) simple 2) no moving parts and 3) low hydrodynamic shear (Table 20.7).

Table 20.7. Respective advantages and drawbacks of airlift or stirred reactors for plant cell cultivation (Pareilleux 1987)

Characteristics	Airlift reactor	Stirred reactor
Device Mixing	Simple Correct problem at high cell density	More complex Efficient
Hydrodynamic shear forces	Low	High
Oxygen transfer	Correct (Except at high cell density; often insufficient in large volumes)	Efficient even at high cell density
Overventilation	Possible and frequent	Not frequent
Scaling	Up to medium volumes	Up to large volumes

20.6.2.1. Air-lift bioreactor (Fig 20.15)

It is a gas-liquid bioreactor where compressed air used for aeration and agitation and is based on the draught tube principle . Wagner and vogelmann (1977) analyzed the relation between bioreactor configuration and anthraquinone production from the cells of *Morinda citrifolia*. Eight fold higher yield of tropdiolide (anti HIV, antitumour agent) from the cells of *Triptergium wilfordii* was obtained in an airlift bioreactor when compared to a stirred tank reactor.

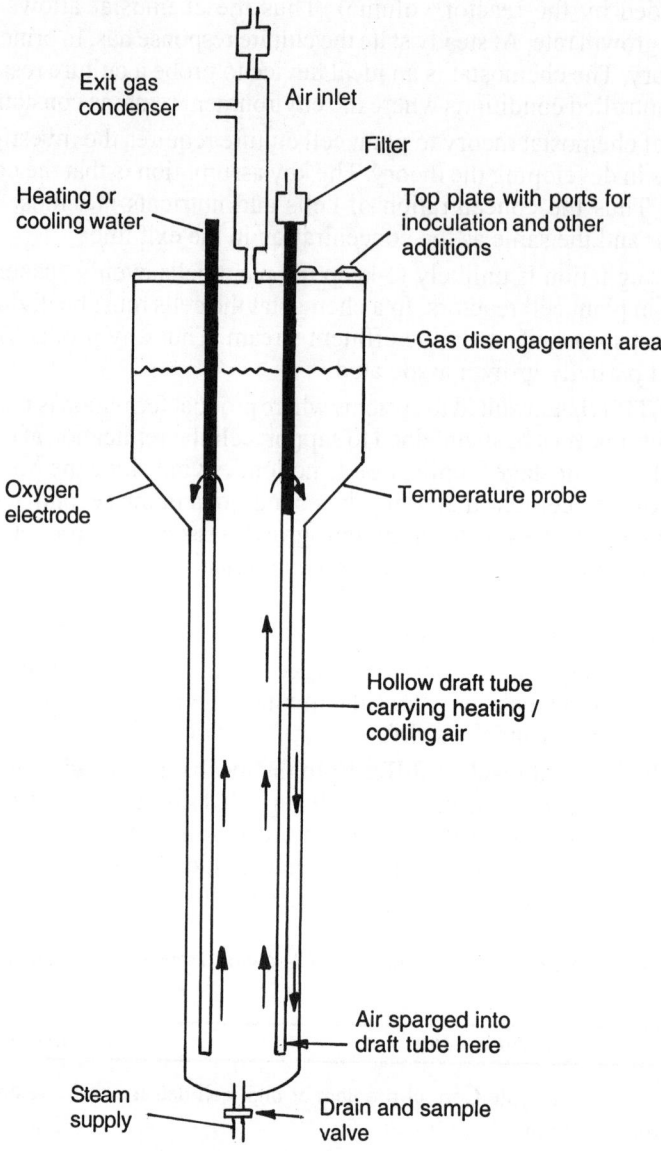

Fig. 20.15. An air-lift bioreactor.

Airlift reactors upto 100 L capacity have used extensively for mass cultivation of cells of *C.roseus* (Smart and Fowler 1984, Fowler 1981). The cells of *D. lanata*, (Spetler *et al.,* 1985, Alfermann *et al.,* 1977), *Berberis wilsone* (Breuling *et al.,*1985), *Cinchona ledgeriana* (Scragg and Fowler (1986), and *Morinda citrifolia* (Spetler *et al.,*1985) have also been cultured in air-lift bioreactors. In all these cases pneumatically agitated bioreactors are more suitable for the production of secondary metabolites than stirred tank reactors.

Oxygen transfer at low shear, less contamination due to non-movable parts, low operation cost because of simple design are a few merits of air-lift type reactors. The demerits are the development of dead zones inside the vessel due to insufficient mixing and nutrient supply caused by high cell densities.

20.6.2.2. *Bubble column bioreactor*

Several bioreactor configurations have been investigated for the growth of plant cells and the bubble column reactor is one of the simplest types of gas-liquid bioreactors for the growth of the cells of *N. tabacum* using a 155 litre culture vessel. However, insufficient mixing reduced the specific growth rate of cells. *Cudrania tricuspidata* cells were also grown in this type of reactor successfully. Undefined fluid flow pattern inside the reactor and non-uniform mixing are the main disadvantages associated with this configuration (Fig 20.16).

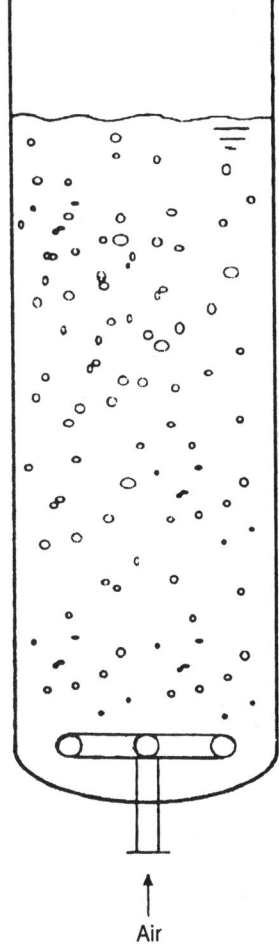

The major difference between the bubble column and airlift designs is that the airlift provides for liquid circulation. Where the gas is spared the fluid flows upward. After the gas disengages at the top of the column. The fluid then flows downward in the down corner section. These sections may be separated using a baffle a concentric cylinder or an external loop. The liquid circulation in the airlift promotes better top-to-bottom mixing and therefore has advantages in suspending cells and clumps. The airlift bioreactors are usually inferior to the bubble column with respect to an overall K_La. This mixing problem is only expected to occur in larger scale airlifts. It should be emphasized that the performance of an airlift is strongly dependent on the geometry of the system. Because of simplicity, efficiency for oxygen transfer and limited shear characteristics, pneumatically agitated bioreactors appear better suited for plant cell culture than traditional mechanically agitated bioreactors.

20.6.3. Rotating drum fermenter

Rotating drum reactors used in biotechnological processes consist of a horizontal rotating drum on rollers. The rotating motion of the drum facilitates proper mixing of the gas-liquid in reactors, thereby, promoting efficient oxygen transfer to cells at high densities. The rotating drum reactor imparts less hydrodynamic stress overcoming problems of shear stress, caused in the other types of reactors. The rotating drum

Fig. 20.16. A bubble column bioreactor.

reactor has been used for the cultivation of *Vinca rosea* cells preferably over the stirred tank reactor. It has been shown to be superior to airlift and modified stirred tank reactors for the cultivation of *L. erythrorhizon* cells. The major disadvantage of this type of cell reactor is its dependence on high-energy requirement for large-scale operation (Fig 20.17).

20.7. BIOREACTOR FOR DIFFERENT PLANT TISSUE CULTURES

20.7.1. Bioreactor for suspension cultures (Fig 20.18)

The main characteristics of plant cell suspensions, which impact bioreactor design, are shown in Table 20.6. Since suspension cultures require good mixing, moderate shear conditions, but has a low oxygen demand and thus requires a lower K_La. This is in contrast to microbial cultures, where oxygen supply is often critical and mixing is needed for bubble break-up to increase oxygen supply. The cultures grown using different bioreactors are depicted in Table 20.8 and the reactors size and different impeller for the large-scale cultivation of various cultures are included in Table 20.8. The airlift has been used to grow a wide range of plant cell suspensions. The slow growth rate of plant cells causes the extended bioreactors runs. The checks are required to ensure sterility. These checks involve the testing of the bioreactor for leaks, the inoculum and medium for contamination, and pre-inoculation incuba-tion to determine whether the whole system is sterile. The highly aggregated nature of the cultures and meringue formation can cause problems of sampling, non-homogeneous cul-tures, loss of biomass. In such cases, modifi-cations to the bioreactor should be made, for e.g. enlargement of the sample port. Meringue formation occurs when the culture begins to foam due to accumulation of proteins in the medium. Plant cell aggregates get trapped in this foam to produce a mixture of cells, poly-saccharides, and proteins which can extend across the surface of the bioreactor, reach 10 cm or more in depth, and block exit and sample ports. This is a particular problem in airlift bioreactors with their high air input and high

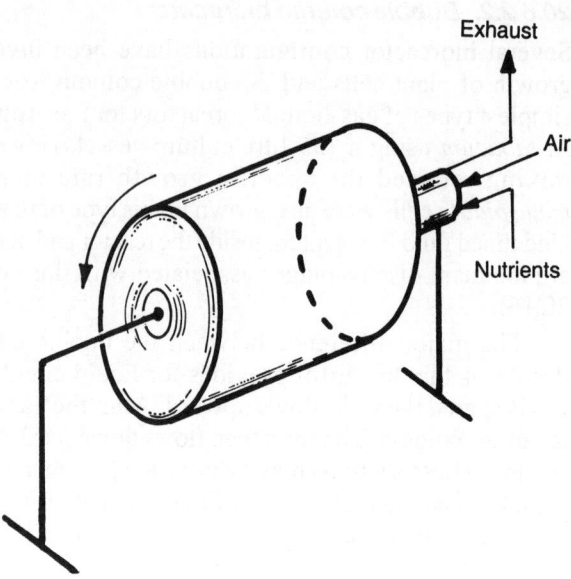

Fig. 20.17. A rotary drum bioreactor.

aspect ratio. Meringue formation can be inhibited by addition of antifoam agents, but these should be checked for inhibitory effects on the culture before use. Use of mechanical scrapers may reduce meringue, and for some small bioreactors a solid shake each day is often sufficient to dislodge meringue. For larger vessels (e.g. 20 litres) meringue does not appear to be a problem. In extreme cases of meringue formation the loss of biomass is very evident with a virtual clearing of the medium.

Table 20.8. Bioreactors used for plant cell suspension

Name	Reactor type	Size(L)	Reference
Mentha Sp.	STR	5 L	Tal et al.,1983
Dioscorea Sp	STR	5 L	
C. roseus	STR	3.5 L	Pareilleux and Vinas 1984
N. tabacum	STR	20,000 L	Azechi *et al.,*1983
Solanum demissum N.tabacum,	Total	800 L	Berlin *et al.,* 1985
C. roseus	Spindle	750 L	
	Tuber		
	Viroid		
Thalictrum rugosum	STR	20 L	Piehl *et al.,*1988
C. roseus	bubble column	0.5 L	Smart and Flower 1984
C. roseus	Air -lift	30 L	Payne *et al.,*1988
C. roseus	Air-lift bioreactor	4.5 L	Smart and Fowler 1984
Berberis wilsonae	Air-lift bioreactor	20 L	Breuling *et al.,*1987
Eschscholtzia californica	Bubble column reactor	10 L	Moo-young and Chish (1988)
Tripterygium wilfordii	air lift	10 L	Townsley and Webster, 1983
Lithospermum erythrorhizon	Rotary Drum	1000 L	Tanaka 1987
Echinacea purpurea	STR	75000	Rittershaus *et al.,* 1990
Cudriana Tricluspidata	STR	5 L	Tanaka 1987
Hyoscyamus muticus	STR	10	Curtis and Emery (1993)
Morinda citrifolia	Air lift	10	Wagner and Vogelmann, 1977

Even though airlift bioreactors have advantages over stirred tank reactors, (STR) plant cell suspensions continue to be grown in STR only. This may be due to general availability of the stirred tank rather than a specific preference (Phyton Inc. USA uses 75,000 liter bioreactors for Taxol production using a STR can be modified to satisfy low oxygen and shear requirement (see Table 20.1).

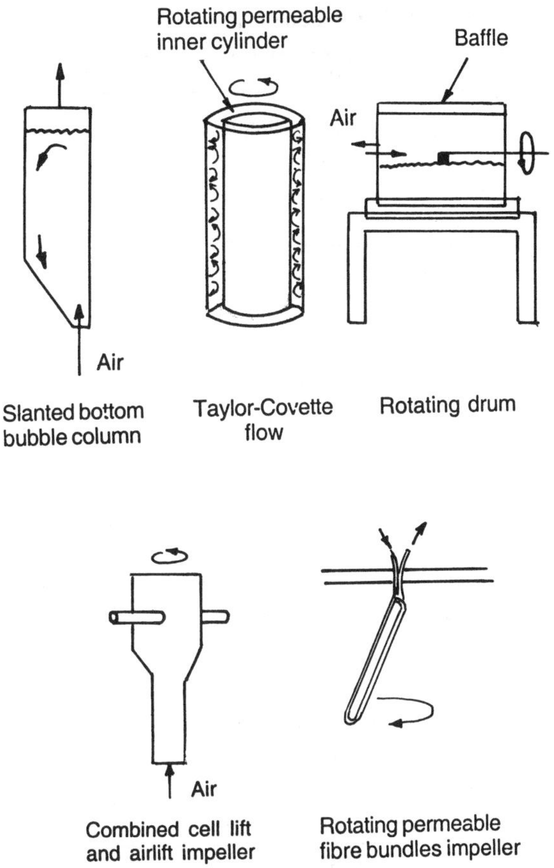

20.7.2. Bioreactors for organized cultures

Organized cultures consist of roots, shoots, or embryos, which are large structures and are prone to damage due to their size. To date, no one has determined their true shear sensitivity, but it is self-evident that they must be shear sensitive. Thus, if these cultures are to be grown in bioreactors, some form of low shear system will have to be used, and in the case of shoot cultures light may have to be provided. A number of bioreactors can be used for the growth of such cultures (Table 20.9 and Fig 20.19).

20.7.3. Bioreactors for hairy root cultures

Fig. 20.18. Some of the bioreactors used for large-scale cultivation of plant cells (Adopted from Scragg, 1994).

The vast potentiality of the hairy root cultures as a stable source of biologically active chemicals had focussed the attention of researchers towards the exploitation of this system through up-scaling in novel bioreactors which would provide the best conditions for optimum growth and secondary metabolite production comparable to or higher than the native roots (Payne *et al.*, 1992). The number of reports of hairy root culture in bioreactors has increased significantly within the last few years (Table 20.10). The reactor designing depends on the product location, which allow continuous harvest of the product without destruction of the biomass. In case of intracellular products, the root tissue is the product and the reactor design should facilitate free harvest. Roots have been cultivated in a wide range of reactor configurations varying from novel design to simple modifications of existing ones. The various bioreactors used for the large-scale cultivation are.

20.7.3.1. *Stirred tank reactor*

This type of bioreactor includes impeller or turbine blades which facilitate mass transfer, and is not usually suitable for hairy root cultures because of the wound response and callus formation that results from the shear stress caused by the impeller rotation (Taya *et al.*, 1989, Hilton and Rhodes 1990). However, recently some modified stirred tank bioreactors have been developed. These modified STRs have large impellers and baffles that are agitated at a very low speed (Fig 20.19).

Table 20.9. Bioreactor designs used for organized cultures (Adopted from Scragg, 1994)

Structure	Plant	Bioreactor design	Reference
Embryo	*Daucus carota*	4 litre, STR	Kessel and Carr 1972
Embryo	*Medicago sativa*	1 litre spin filter	Stuart *et al.*, 1987
Roots	*Atropa belladonna*	10 litre modified STR (1)	Akita and Takayema 1988
Embryo	*Mediacago sativa*	airlift	Stuart *et al.*, 1987
Embryo	*Medicago sativa*	2 litre airlift (4) 2 litre STR (5) 2 litrevibro-mixer (6)	Chen *et al.*,1987
Plantlets	*Artemisia annua*	2 litre squate airlift	Park *et al.*, 1989
Embryo	*Digitalis lanata*	5 litre airlift	
Mini-corms	*Gladiolus* sp.	Bubble column	Takayama 1992
Mini-tubers	*Solanum tuberosum*		
Plantlets	*Fragaris* sp.		
Shoots	*Nephrolepis exaltata* (fern)	1 litre bubble column	Ziu and Hodar 1991
Embryo	*Daucus carota*	1 litre spin filter (2)	Styer 1985
Embryo	*Panax ginseng*	Spin filter (2)	Takayama 1992
Embryo/shoots		Rotating drum	Takayama 1992
Embryo	*Euphorbia pulcherrima* (poinsettia)	Silicone tubing	Preil *et al.*, 1988
Shoots	*Musa* (banana), *Cordyline, Nephrolepis* (fern)	Mist bioreactor (3)	Weathers *et al.*, 1988

Bioreactors (1), (2), and (3) are shown in Fig 20.19. Sources of the other bioreactors are: (4) L.H. Engineering Ltd., Stoke Poges, UK; (5) New Brunswick, Edison, USA; (6) Pegasus Industrial Specialities Ltd, Agincourt, Canada.

Table 20.10. Bioreactors for cultivation of hairy roots

S.No	Plants species	Bioreactor configuration	Reference
1.	*Armoracia rusticana*	Air lift, batch	Taya *et al.*, 1989
2.	*Atropa belladonna*	Air lift, batch	Jung and Tepfer 1987
3.	*A. belladonna*	Turbine blade	
4.	*Beta vulgaris*	Turbine blade	Dilorio *et al.*, 1992
5.	*B. vulgaris*	Trickle bed	Jung and Tepfer 1987
6.	*Calystegia sepium*	Turbine blade	Dilorio *et al.*, 1992
7.	*Carthamus tinctorius*	Trickle bed	Dilorio *et al.*, 1992
8.	*Catharanthus roseus*	Air lift	Toivonen *et al.*, 1993
9.	*D. carota*	Mist	Kondo *et al.*, 1989
10.	*Datura stramonium*	Airlift continous flow	Hilton *et al.*, 1988
11.	*D. stramonium*	Stirred tank	Hilton *et al.*, 1988

(Contd.)

S.No Plants species	Bioreactor configuration	Reference
12. *D. stramonium*	Stirred tank	Hilton *et al.*, 1988
13. *Duboisia leichhardtii*	Air lift	Muranaka *et al.*, 1992
14. *D. leichhardtii*	Turbine blade	Hilton *et al.*, 1988
15. *Hyoscyamus mutics*	Trickle bed	Mckelvey 1992
16. *H. muticus*	Trickle bed	Flores & Curtis 1992
17. *H. muticus*	Ebb and Flow	Cuello *et al.*, 1991
18. *H. muticus*	Bubble column	Mckelvey 1992
19. *Lithospermum erythorhizon*	Air lift packed column with Amberlite XAD-2	Shimomura *et al.*, 1991
20. *L. erythorhizon*	Bubble column	Sim and Chang 1993
21. *Nicotiana rustica*	Air lift batch followed by continous flow	Rhodes *et al.*, 1986
22. *N. rustica*	Air lift	Rhodes *et al.*, 1986
23 *Panax ginseng*	Turbine blade	Inomata *et al.*, 1993
24. *Tagetes patula*	Stirred tank	Buitelaar 1991
25. *Trigonella foenum graecum*	Stirred tank	Rodrigues *et al.*, 1991
26. *Solanum tuberosum*	Bubble Column	Hilton and Rhodes 1990

20.7.3.2. Air lift bioreactors

Plant cells have large vacuole and slow growth so hairy roots require comparatively low oxygen supply of about 0.05-0.4 vol of air/vol of liquid/min.

20.7.3.3. Bubble column bioreactor

In a Bubble column bioreactor the bubbles create less shear stress, so that it is useful for hairy root cultures in mass scale. It has beeen successfully used to culture hairy roots *Tagetes p atula, H. muticus* and *Lithospermum erythrorhizon* etc (Buitelaar *et al.*, 1991, Flores and Curtis 1992, Shimomura *et al.*, 1991). However, their performance was rather poor in comparison to other reactors. This may be due to low oxygen supply because of channel-ing tissue damage caused by sparging.

20.7.3.4. Turbine blade reactor

This is a combination of air lift and STR. It has been successfully used for hairy roots of *Duboisia leichhardtii* (Muranaka *et al.*, 1992), *Panax ginseng* (Inomata *et al.*,1993).

20.7.3.5. Mist bioreactor (Trickle bed reactor)

Hairy roots may be oxygen limited unless cultures are aerated with oxygen enriched air (Yu and Doran 1994). Oxygen depletion triggers anaerobiosis and production of ethy-lene. Since it is in soluble in water, the formation of thin films at the surface of roots can cause a localized build-up to concentrations for in excess of normal. Overall, adequate moisture cycling and use of droplet sizes small enough to avoid excessive film build up are critical for proper control of gases in root culture. The mist bioreactors are the best-suited bioreactor for hairy root cultures. (Fig. 20.19).

In mist bioreactor the medium trickles over a whatman filter paper containing the biomass, then spent medium is drained from the bottom of the bioreactor to a reservoir and is recirculated at a specific rate. The degree of distribution of liquid varies according to the mechanism of liquid delivery

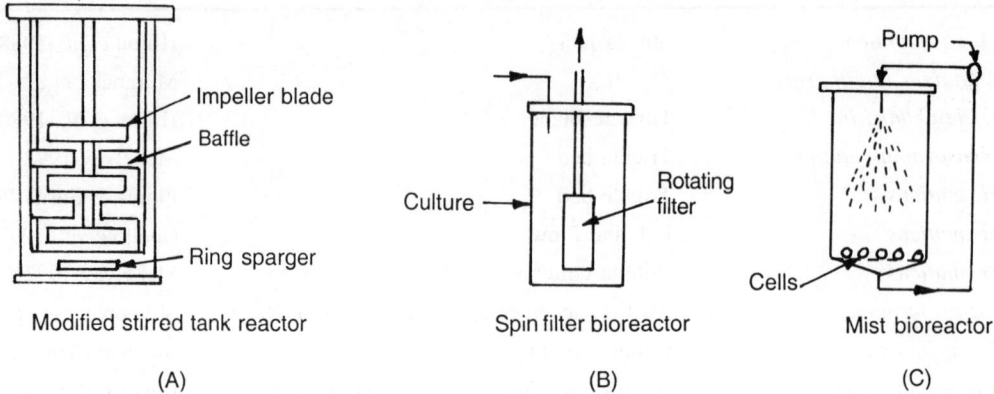

Fig. 20.19. Bioreactors used for large scale cultivation of hairy root cultures. (A–Akita and Takayama 1988, B–Styer 1985, C–Dilorio 1991).

at the top of the reactor chamber. For better dispersion spraying is done by mixing humidified air with medium that creates the mist (Dilorio *et al.,* 1992 a,b Whitney 1992).

20.7.3.6. Rotating Drum reactor

Kondo *et al.,* (1989) used this system for hairy roots from carrot. Hairy roots adhere to the walls of the reactor and as the drum rotates the roots tend to break up. To overcome this problem, a polyurethane foam sheet was fixed onto the surface of the drum, to which the hairy roots get attached. This resulted in higher growth without any detachment.

20.7.3.7. Other bioreactor

In spin filter bioreactor the rotating filter mixes the cultures and simultaneously allows for spent medium removal and fresh medium addition (see Fig 20.19). The Drip tube bioreactor has been successfully used for the cultivation of hairy roots of *Datura stramonium* and *N. glauca* (Holmes *et al*1997).

20.7.4. Bioreactors for immobilized cultures:

The major problems associated with the development of large scale immobilized plant cell system are
- Large scale aseptic immobilization procedures must be developed.
- Mass transfer limitations imposed by separating the cells from the nutrients may significantly affect cell metabolism.
- Products must be produced by non-growing cells.
- Products must be released from the cell into the medium.
- Experience in the scale-up of immobilized cell systems is limited.

Immobilized systems can promote more cell to cell interactions than suspension cultures. Cell to cell interactions are apparently required for morphological differentiation (Walker *et al.,*1979) and may be responsible for the enhanced capsaicin production by immobilized *Capsicum frutescens* cells (Lindsey and Yeoman 1984). In a specially designed reactor (Hallsby and Shuler 1986) significant differences have been observed in the response of tobacco cells to the absence or presence of strong diffusional limitations in an immobilized cell reactor. When slow, diffusional mechanisms were responsible for nutrient supply and product removal, dense, hard cohesive cells clumps were observed. These clumps, which are presumably some primitive form of organization due to radial alignment of cell, were observed to grow rapidly when transferred to hormone-free solid media. These clumps were not observed in a similar reactor when the media was forced to flow through the cell layer. These differences in morphogenic response were attributed to differences in mass transfer characteristics of the two operating conditions. Thus the design and operation of immobilized cell reactors may control the access of nutrients to the cells and greatly alter cellular responses.

Immobilized cell systems may present technical problems in scale-up. The susceptibility of soft polysaccharide gels to compressive and shearing forces may require changes in either immobilization techniques or reactor design.

20.7.4.1. Packed bed reactors

In this reactor, the cells can be immobilized either in the surface of, or throughout the support, and fluid-containing S flows past the support particles. When the support, the packed bed can accommodate a large number of cells per reactor volume (Fig 20.20).

However, there are several disadvantages for packed bed immobilized cell reactors. The characteristic low degree of mixing leads to difficulties when it is necessary to transfer oxygen, control temperature and pH, and when gaseous products (e.g., CO_2) must be removed.

20.7.4.2. Well Mixed Reactors

The advantages of a well-mixed reactor for immobilized cells are that gaseous substrates can be directly sparged, the temperature and pH can be carefully controlled, and mass transfer rates can be

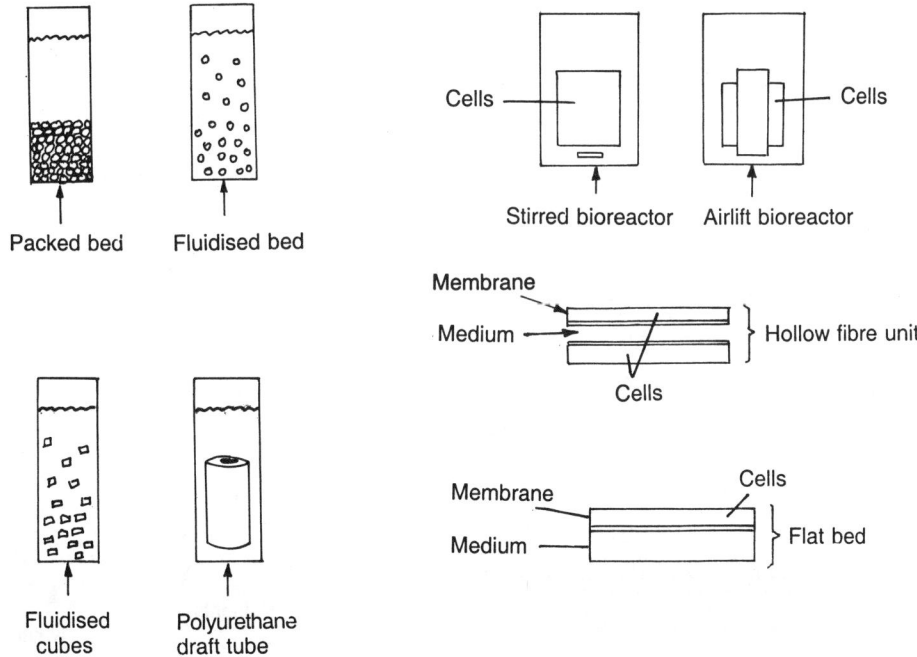

Fig. 20.20. Bioreactors used for large scale cultivation of immobilized plant cells.

improved over packed bed reactors. The suspension culture with complete cell recycle is essentially a well-mixed "immobilized" cell reactor.

The disadvantages of a well-mixed reactor for immobilized cell systems are the possibility of fragmenting the support due to particle collisions and shear. Using gentle pneumatic agitation can reduce this problem.

20.7.4.3. Fluidized Bed Reactors

Typical fluidized bed reactors utilize the energy of the flowing fluid (liquid and / or gas) to suspend the particles. Because the energy required for fluidization increases with increased particle size, small-immobilized particles are often employed (Fig 20.20). The mass transfer advantages of these small particles is one of the major benefits of the fluidized bed reactor. To obtain adequate conversions of substrate to product, the fluid must often be retained in the reactor for an extended period. This may be incompatible with the requirement that the fluid flow rate be sufficient to suspend the particles. To satisfy these two requirements, large gas volumes can be used to suspend the immobilized cells while maintaining low liquid conditions lead to a large degree of fluid mixing. Large recirculation rates were used in the fluidized bed reactors of Prenosil and Pedersen (1983) and Morris *et al.,*(1983). Major disadvantages of the immobilized supports and the complex fluid dynamics of such reactors make scale-up difficult.

20.7.4.4. Membrane bioreactor

The most commonly considered membrane reactors are the hollow fiber and spiral wound reactors shown in (Fig 20.21). In the hollow fiber reactor, the cells are retained either within the tubes or in the outer region. Preliminary studies to evaluate the use of these reactors have been reported by Shuler *et al.,* (1983), Jose *et al.,* (1983) and Prenosil and Pedersen, (1983). The spiral wound membrane reactor is essentially a flat plate reactor rolled into a cylindrical shape and although no reports have appeared on the use of a spiral wound reactor, Hallsby and Shuler (1986) has studied that plate reactor with tobacco cells. These studies have demonstrated that in addition to the membrane, the thickness of cell layer is also important. The inner portions of thick cell layers are generally characterized by substrate deficiencies and product accumulation and these conditions may prove beneficial for the morphological and chemical differentiation of plant cells.

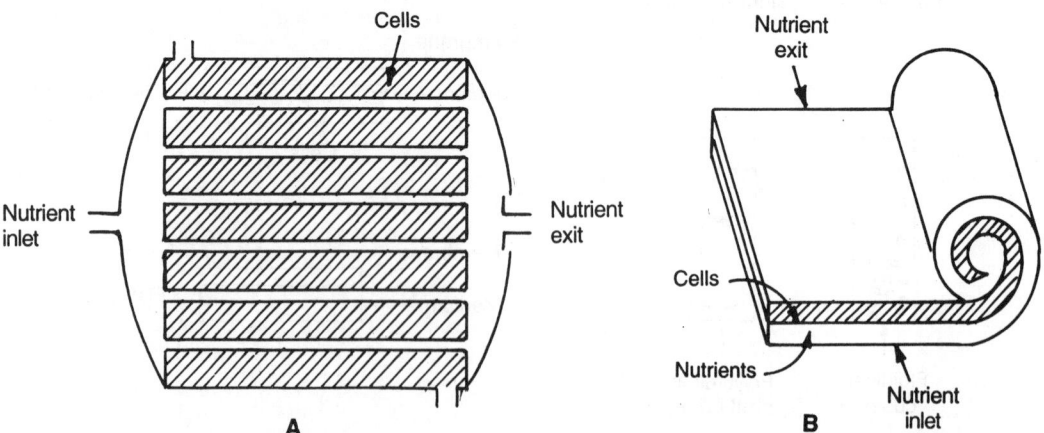

Fig. 20.21. Membrane bioreactors used for immobilized plant cells hollow fibre and B. Spiral.

An advantage of membrane reactors is the possibilities that these membranes can be reused. Thus despite the high capital costs, the reusability may make this form of immobilization economical. Also because plant cells are metabolically less active than microbes, thicker cell layers can be employed. Since reactor inoculation, or cell loading is easier with larger cell layer thickness it may be possible to develop plant cell reactors, which can be easily inoculated.

20.7.5. Bioreactors for Micropropagation

In order to achieve efficient and automated production in plant tissue culture by controlling the various chemical and / or physical properties in culture vessels, plant production systems have evolved from a small research scale to a large volume and high-yield culture system and liquid media are preferably used to facilitate handling (Leathers *et al.,* 1995). For the tissue culture of African violet and lettuce, the regenerated multiple shoot clumps need to be separated by hands (Pack and Hahn, 1999). Therefore, it is important to develop new propagation strategies to overcome the limitations of conventional micropropagation techniques.

The basic function of a bioreactor is to provide optimum growth conditions by regulating various chemical and/or physical factors. The fact that organogenic and embryogenic cultures of many economically important species have the potential for continuous proliferation and virtually unlimited production of propagules has stimulated the use of bioreactor culture for mass propagation (Table 20.11).

Table 20.11. Plants propagated by shake and/or bioreactor culture technique(Adopted from Kukreja and Ahuja 1994)

Plant species	Bioreactor configuration	Propaguic cultured
Digitalis purpurea	Air lift	shoot bud
Solanum tuberosum	Two stage air	microtuber
	Linear	mucrotuber
Fragaria ananasa	Air lift	shoot
Colocasia esculenta	Air-driven	shoot/microcom
Pinetha ternaia	Air-driven	shoot/microcom
Hyppeastrum hybridum	Shake culture	shoot/bulb
Gladiolus	Shake culture	shoot/microcom
Hyacinthus	Shake culture	shoot/bulbscale
Lilium species	Air- lift	shoot/bulbscale
Begonia	Air- lift	shoot
Saintpaulia	Air- lift	shoot
Gloxtnias	Air- lift	shoot
Orchids	Jar fermentor	protocorm
Colocasia esculenta	Shake culture	microcorm
Pinellia ternata	Shake culture	microcorm

In bioreactor culture for micropropagation, major progress can be found in the area of large-scale liquid culture and in the development of bioreactor process control systems (Fowler *et al.,* 1987, Heyerdahl *et al.,* 1995). Fujita *et al.*(1982) reported an efficient production of *Lithospermum erythrorhizon* cells using cell suspension culture. It is one of the examples in which cell suspension culture has been industrialized with a 750-L fermenter. Plant cell cultivation has been industrially accomplished up to a 75,000 L bioreactor. Subsequently, bioreactors were scaled-up to industrial standard in the Korea Forest Research Institute for large scale production of Ginseng adventitious roots, and more than 400 kg of the roots were harvested in a 6-week production using a 5000-liter bioreactor (Azechi *et al.,* 1983, Ritershaus *et al.,*1989).

Bioreactor design has been primarily developed for specific requirements of embryogenic or organogenic suspensions. The optimum bioreactor design for maximum growth and proliferation of different species must be evaluated systematically (Son *et al.,* 1999a,b). The various types of bioreactors employed for micropropagation are Airlift and bubble column-type bioreactors, Balloon-type bubble bioreactor (BTBB), Stirred tank bioreactor (STR), Ebb and flood bioreactor, etc.

7.5.1. Various advantages and disadvantages of bioreactor culture systems for micropropagation

Adapting bioreactors with liquid media for micropropagation is favorable due to the ease of scaling-up (Preil, 1991), the ability to prevent the physiological disorders of shoot and leaf hyperhydricity high control of culture factors such as physical and chemical environments, scheduled shipping and reduction of the production costs as a result. To adopt bioreactors may be beneficial for the development of new regeneration systems that are free form physiological disorders and the year-round production of the useful plants can also be achieved by adapting the bioreactors for micropropagation. Computers can be used as control systems for automation, saving labor and production costs (Preil *et al.,* 1988). However, only a few bioreactors are presently used in practice for plant propagation. This is mainly due to the lack of systematic and factorial experiments on propagation using bioreactors needed to reveal the complex interactions between plant physiology and physical parameters of bioreactors. Widely spread skepticism that large somaclonal variation is expected in the suspension culture-derived progenies (Armstrong and Phillips, 1988) is another reason why few bioreactors are applied for plant propagation. However, there is a high probability of finding genetically stable genotypes, which are suitable for bioreactor propagation, among vegetatively propagated agricultural crops.

However, large initial investment for equipment includes substantial costs in repair, replacement parts and upkeep of bioreactor facilities. If contamination is introduced to a large batch of propagules, the cost and lost time can be devastating (Leathere *et al.,* 1995). Organogenic and embryogenic propagules are more difficult to handle in bioreactors because the units are comprised of a variety of cell types, which complicates optimization of the process.

20.7.6. Shoot cultures

20.7.6.1. Introduction

The proliferation of shoots *in vitro* without intervening roots or undifferentiated tissue is referred to as a shoot culture.

Shoot cultures are potential sources of plant secondary metabolites that are not produced by undifferentiated cultures or that are not easily available from whole plants in nature (Table 20.12). Shoot cultures are also used for the clonal propagation of many horticultural plants. However, because of the labor-intensive nature of the current solidified-media-based shoot multiplication, there is interest in automated, large-scale, liquid-based culture for micropropagation purposes (Hussey, 1986). Finally, shoot cultures are also used for the production and long-term preservation of virus-free germplasm (Kartha, 1981; Withers, 1985).

Shoot cultures obviously differ from root cultures, but they do possess some similarities based on their common differences with undifferentiated cultures, for example, enhanced genetic stability. The other similarity, namely, their gross morphological characteristics, calls for a similar approach when it comes to reactor design.

Table 20.12. Examples of cases in which shoot differentiation is necessary for secondary metabolism (Payne et al., 1992)

Plant	Compound	Reference
Catharanthus roseus	Vindoline, Vinblastine	Constabel *et al.* (1982)
Chrysanthemum sp.	Pyrethrins	Staba *et al.* (1984)
Citrus sp.	Limonin, naringin	Barthe *et al.* (1987)
Digitalis sp.	Cardenolides	Hagimori *et al.* (1982a)
Heimia salicifolia	Quinolizidine alkaloids	Pelosi *et al.* (1985)
Lupinus sp.	Quinolizidine alkaloids	Wink and Hartmann (1980)
Mentha sp.	Essential oils	Hirata *et al.* (1990)
Papaver somniferum	Morphinan alkaloids	Constabel (1985)
Pelargonium spp.	Essential oils	Charlwood *et al.* (1989)
Withania somnifera	Withanolides	Heble (1985)

20.7.6.2. Initiation and Maintenance of Shoot Cultures

A shoot culture is typically initiated from a sterile seedling or a whole plant by dissecting the shoot apical meristem, or a leaf, or a stem onto solidified medium.

Shoot cultures proliferate by continuously forming new meristems may be axillary that is those arising at the leaf or stem. The greater the rate of formation of new meristems per unit mass of tissue, the greater will be the overall growth rate.

Though certain shoot cultures are capable of rapid growth in darkness (Hagimori et al 1982 a), light us usually required for normal greening and leaf development. The shoot cultures are genetically stable. Roja et al (1987) reported the consistent high production of alkaloids in shoot cultures of *Rauwolfia serpentina* without major changes in the alkaloid profile and levels for 5 years. Miura *et al.* (1988) reported the stable maintenance of a relatively fast growing high producing cultures for 30 months with little change in cultures in the growth rates, morphological characteristics or alkaloid productivity.

20.7.6.3. Biosynthetic characteristics of shoot cultures

20.7.6.3.1. Biosynthetic differences between shoot cultures and undifferentiated cultures.

Shoot differentiation enable the expression of several biosynthetic pathways that are expressed at very low levels, or at all, in undifferentiated cultures (see Table 20.12). The technology for large-scale culture of undifferentiated cultures is well established where for shoot cultures it is not. Therefore, much initial effort is usually spent in confirming that shoot differentiation is really necessary for secondary metabolism and that undifferentiated cultures cannot be easily coaxed to produce the secondary metabolite.

Many secondary metabolites are synthesized and stored in specialized cell in the plant. Since leaf formation allows cell specialization, it is perhaps not surprising that shoot cultures allow the expression of certain pathways not normally expressed in undifferentiated cultures. For e.g. key enzymes in the monoterpene biosynthetic pathway have been localized in the glandular trichomes of spearmint (*Mentha spicata*) leaves (Gershenzon *et al.*, 1989). Charlwood *et al.* (1989) was able to correlate essential oil accumulation with glandular hairy density in shoot cultures of *Pelargonium fragrans*.

Many studies have reported a positive correlation between light, chlorophyll content, and secondary metabolite production in shoot culture systems (Hagimori *et al.*, 1982a; Pelosi *et al.*, 1985; De Luca *et al.*, 1988; Saito *et al.*, 1989b) suggesting that certain key steps in the pathway may be chloroplast-associated. In some of these cases, chloroplast development is sufficient for expression of secondary metabolism, and shoot differentiation is not necessary, as evidenced by the ability of chlorophyllous suspensions (i.e., suspensions containing chloroplasts) to produce the secondary metabolite e.g. *Catharanthus roseus, Lupinus* Sp. *Heimla salicifolia, Digitalis purpurea* etc.

20.7.6.3.2. Differences between shoot cultures and whole plants

It may often be incorrect to assume that a shoot culture will produce all the compounds that are present in the leaves of a whole plant. The site of synthesis may be different from the site of storage (Wink, 1987). For example, the commercially important purine and tropane alkaloids are accumulated in the leaves of *Nicotiana* and *Duboisia* respectively, but are synthesized by the roots. Accordingly, root cultures of both species are capable of alkaloid biosynthesis (Rhodes *et al.*, 1986; Yoshioka *et al.*, 1989), whereas shoot cultures of both species do not accumulate alkaloids. Both shoot cultures do, however, take up exogenously added alkaloids and perform certain biotransformations (Saito *et al.*, 1989a; Hashimoto and Yamada, 1988).

Other factors affecting the whole plant such as leaf maturity, juvenility, flowering and adaptation to climatic stresses and seasonal variations no doubt affect secondary metabolite production. Since shoot cultures are not subjected to the same factors *in vitro*, their secondary metabolite profile may vary qualitatively and quantitatively from that of the leaves of the whole plant. Shoot cultures of *Catharanthus roseus*, for example accumulate significantly lower levels of the dimeric indole alkaloids than do leaves of a whole plant (Endo *et al.*, 1987; Miura *et al.*, 1988). Whereas the essential oil component profile of shoot cultures of *Pelargonium fragrans* was markedly different from that of the corresponding leaf oil with respect to the oxygenated monoterpenes, the essential oil components of the leaf oil and shoot cultures of *Pelargonium tomentosum* (Charlwood and Moustou, 1988) and of *Mentha tricata* (Spencer *et al.*, 1990) were very similar. The significant differences in the withanolide profiles between shoot cultures and whole plants of the *Withania somnifera* prompted Heble (1985) to suggest that shoot cultures may provide a source for obtaining novel secondary metabolites.

20.7.6.4. Bioreactors for shoot cultures (Fig 20.22)

The kinetics of product formation determines the mode of operation of a bioreactor to be used. The shoot culture bioreactor should serve two purposes. First, it should provide an environment that supports rapid growth. The growth rate achieved in a bioreactor should approach or, better still, exceed that achievable in a shake flask. Moreover, the maximum growth rate should be sustained over several doublings, and the final density of shoots should be maximized.

If the product is non-growth-associated and continuous operation with continuous product removal is desired, then the bioreactor should be designed to sustain the viability and productivity of tissue high densities for prolonged periods. The constraints for doing so may be even more stringent than for providing rapid growth. For example, higher oxygen concentrations or light intensities may be required during secondary metabolite production than during growth. Most studies, so far, have focused on bioreactor design for growth of shoot cultures.

The following factors will influence the design of bioreactor for shoot cultures.
1. Shoot morphology
2. Effect of submersion
3. Light requirement

Shoot morphology determines the susceptibility of tissue to mechanical stress associated with fluid flow, aeration and agitation. It also influences nutrient mass transfer. Shoot is aerial part of the plant and has evolved to grow and develop in air.

The light requirements of shoot cultures, as well as of other tissue, provide a unique challenge for bioreactor design and scale up. The mechanical constraints for adequate illumination are extreme. The transparent reactor walls are required. Takayama and Misawa (1981) used 3-L bubble column and a 10-L jar fermenter in order to scale up growth of *Begonia* shoot cultures.

The maximum growth rate of shoot cultures of *Digitalis purpurea* in a 3-L bioreactor (Doubling time - 3.2 days) actually exceeded that in a shake-flask. A final tissue density of 15.4 g dry wt/L was achieved. At these high densities, however, nonuniform light exposure was a problem, with chlorophyll and digitoxin contents varying with the position of shoots in the reactor. Lowering the final tissue density resulted in more uniform greening of the shoots.

Park *et al.* (1989) reported the growth of *Artemisia* shoot cultures in a sparged rectangular tank with no mechanical agitation. The maximum growth rate in shake flasks (Doubling time = 4.3-5.4 days) was superior to that in the reactor (Doubling time = 5.9 days). Shoots were successfully rooted in the bioreactor.

Bringi (1991) used a 1-L sparged vessel with mechanical agitation for cultivating tobacco shoot cultures. A stainless steel screen protected the shoots from the magnetic stirrer as shown in figure 9.5. Average-doubling times of 3.9 ± 0.3 days were achieved, which corresponds to slower growth than the same cultures are capable of in the recycle reactors (Fig 20.22). They also used a simple roller

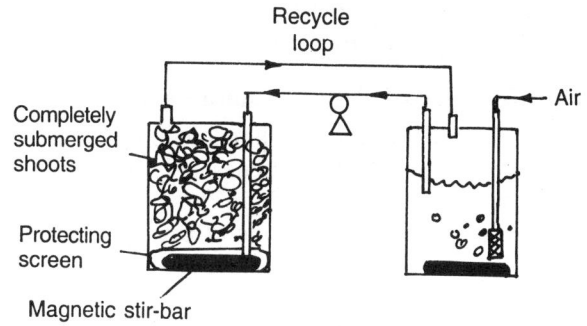

A recycle bioreactor

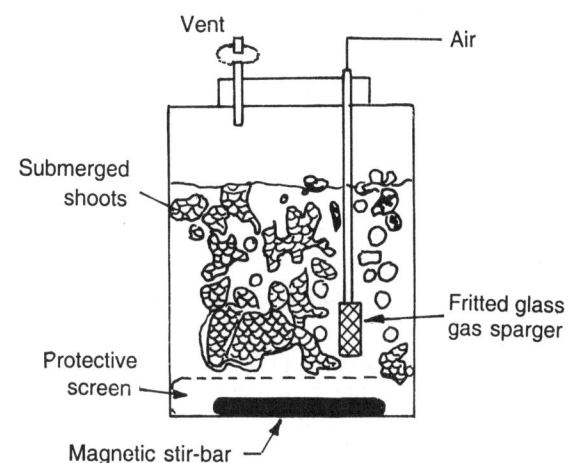

Air-sparged bioreactor

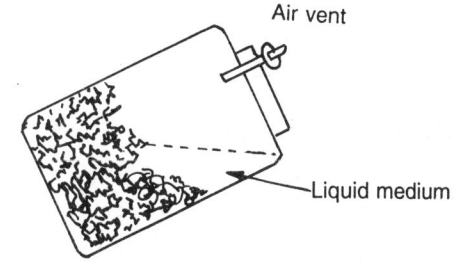

Tilted rotating bottle bioreactor

Fig. 20.22. Bioreactors used for shoot cultures (Adopted from Payne et al., 1992).

bottle to demonstrate the feasibility of alternative bioreactor design for shoot cultures of tobacco. Weathers and Giles (1988) have used the nutrient mist bioreactors for the culture of shoots, roots and other plant tissues.

20.8. SCALE-UP AND PROCESS DEVELOPMENT FOR PLANT CELL CULTURES FOR LARGE-SCALE CULTURE

20.8.1. Scale up of plant cell cultures

Many of the considerations applicable to scale-up of microbial systems are also applicable to plant cell cultures. These include nutrient supply and sterilization, uniformity of mixing, the nature of nutrient regime, handling of bulk medium flow, adequate gas transfer and culture homogeneity. Added to these considerations must be the effect of shear intensity and air bubble dispersion. The general methodology used of the large-scale production is shown in Fig. 20.23.

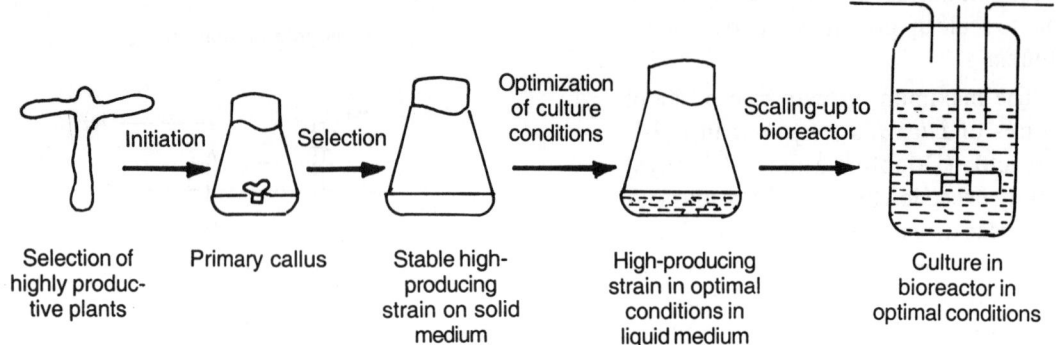

Fig. 20.23. General methodology used for scale-up of the plant cells.

When the optimum process conditions are found at the laboratory scale, there is a need to translate these findings for use in large bioreactors. The scale-up methods that have been most often proposed are given below :

1. Scale-up based on fixed power input
2. Scale-up based on fixed mixing time
3. Scale-up based on fixed oxygen transfer coefficient
4. Scale-up based on fixed environment (e.g., dissolved oxygen)
5. Scale-up based on fixed impeller tip speed

However, the details of commonly used methods is described in detail.

20.8.1.1. Scale-Up at Fixed $K_L a$

If physical mass transfer rates drop below certain values, growth is hampered or destroyed. Thus, $K_L a$ is frequently used as a basis for scaling-up, especially anaerobic systems. Table 20.13. gives an example of scale-up based on constant $K_L a$ in both reactors. From this table, it is seen that it may not be possible to maintain equal volumetric gas flow rates (vvm) since the linear gas velocity (V_s) through the vessel will increase differently with the scale and in fact would be impractical if the upper limit of liquid blow-out action is reached. However, it may be possible to reduce the volume of gas per volume of liquid per minute (vvm) on scale-up while increasing power input by changing the reactor geometry and/or power input per unit volume (P/V) as shown in the Table 20.13. $K_L a$ will remain constant in both cases.

It appears that $K_L a$ is often, but not always, a reasonable design approach. An increase in $K_L a$ can sometimes have an adverse effect because of damage to organisms in highly turbulent fermentation broth and/or oxygen poisoning. Other problems, such as gross bubble coalescence, are also important in nonmechanically stirred fermenters.

Table 20.13. Scaling-up Based on Constant $k_L a$ for Gas-Liquid Contacting in a Sparged Stirred-Tank Reactor. Effect of Scale-up Parameters on Common Operating Parameters (Blanch 1990)

Operating parameter	Laboratory reactor 75l		Plant reactor 10 m³	
H_l/T	1	1	1	2.8
P/V	1	1	Z	1
vvm	1	1	0.2	0.1
V_s	0.1	0.5ᵃ	0.1	1
k_L a	1	--	1	1

* Impractical liquid 'blow-out' conditions.

It is important to realize that $K_L a$ is not necessarily constant during fermentation. In highly viscous fermentation's, or in fermentations producing surfactants, large variations in $K_L a$ can occur. Thus, using the oxygen uptake rate throughout the fermentation as a scale-up parameter is often more successful.

20.8.1.2. Scale-up on flow Basis (Constant Power Input)

Another common design approach is based on constant input of agitator power per unit reactor volume,i.e.,

$$P_1/V_1 = P_2/V_2$$

where subscripts 1 and 2 refer to small- and large-scale vessels, respectively. Under turbulent conditions, the agitator power input is given by :

$$P\alpha\ P^{N3}D^5$$

For constant power input in geometrically similar vessels, we can therefore write :

$$\frac{PN^3_1 D^5_1}{V_1} = \frac{PN^3_2 D^5_2}{V_2}$$

and this becomes

$$N_2 = N_1 = \left(\frac{V_2}{V_1}\right)^{Y3} \left(\frac{d_1}{D_2}\right)^{5/3}$$

Given the impeller speed in the large fementer for successful scale-up.

It must be noted that this scale-up method relies on correlation for power input in ungassed systems and the demand for geometric similarity also limits its applicability.

20.8.2. Process development

As with microbial technology, plant cell process system can be separated into three areas
1. Up stream processing
2. The production or reactor phase
3. Down stream processing

The up stream phase relates to culture development, culture storage and the provision of seed cultures to initiate the main production reactor. The production or reactor phase is obviously central to the whole operation. This can be done in Batch, fed-Batch and continuous phase depending upon the nature of secondary metabolite from plant cell culture. In Batch culture, it is usually a compromise between optimizing growth to achieve short run times and, maximizing the yield of product. In continuous culture, secondary metabolite yield is typically very low. Tal and Goldberg (1982) obtained good cell growth with little steroid synthesis in continuous fermentations of *Dioscorea deltoidea*. Dougall *et al.*, (1983) in culturing *D. carota* on a continuous basis demonstrated that factors, which improve biomass productivity, either do not affect product formation or inhibit it.

Another approach as a process strategy is through the use of semi continuos culture, in which the culture is first set up in a batch-wise mode (Fowler 1984). By closely following the production of biomass and the desired product it should them feasible to select an optimum time point when the cells are still dividing, albeit slowly, in late exponential phase and when there is a reasonable level of product present.

Down stream processing include the recovery of products from plant cell cultures. The classical approaches have been used., where the products released into the nutrient broth. Plant cell processes face the same high cost problems of all fermentation processes, a small amount of product in a large volume of liquid.

20.9. INDUSTRIAL APPLICATION OF PLANT CELL BIOREACTOR SOME EXAMPLES

20.9.1. Smoking material-Tobacco cells

Matsumoto *et al.*, (1971) grow tobacco cells (*N. tabacum* By-2-cells) at a high growth rate in auxin supplemented medium to produce smoking material. These cells were grown as cell suspensions as a semi-continuous culture in 30 litre jar fermentor and growth conditions were standardized (Kato *et al.*,1976, 1977, Noguchi *et al.*,1977). A 5-days continuous culture achieved using a 500 fermenter (Kato *et al.*, 1976). Proved best to produce smoking material industrially in terms of high productivity, stability of process. These successful results were applied to the continuous culture of By-2 cells in 20,000 litre jar fermenter to confirm their suitability and conditions for stable continuous cultures cells had a characteristic and unfavorable taste on smoking and industrial production of tobacco by using tobacco cultured cells could not be undertaken. However, the plant cell fermenters have proved their utility for up scaling of cell culture for possible industrialisation in future.

20.9.2. Shikonin production from *Lithospermum erythrorhizon*

Shikonin and its derivatives such as acetyl shikonin and isobutyl shikonin accumulated in roots of *Lithospermum erythrorhizon* Sieb. and Zucc. are reddish purple pigments and have been used in traditional dyeing. The plant has also been used as an herbal medicine. The compound shikonin, used in Japan traditionally for its antibacterial and anti-inflammatory effects. The first commercial process for producing a plant natural product by plant cell culture was developed in Japan for the production of shikonin. Development of cell culture production methods was stimulated by the value of shikonin and difficulties encountered in meeting the demand using field-grown material (Tabata and Fuijita 1985, Table 20.13). Shikonin is biosynthesized from the shikimic acid and Acetate-MVA pathway (Fig 20.24). The high producing cell line of *L. erythrorhizon* was selected by visual inspection methods (Mizukami *et al.*, 1978).

In medium development studies, it was determined that cytokinins, copper (Fujita *et al.*, 1981a,b), high levels of sucrose, and low levels of nitrogen (Mizukami *et al.*, 1978) were beneficial for shikonin production. The synthetic auxin 2,4-D (Tabata *et al.*, 1974), gibberellin (Yoshikawa *et al.*, 1986), ammonium (Fujita *et al.*, 1981a), and glutamine inhibit the production (Yazaki *et al.*, 1987).

Table 20.14. A comparison of Shikonin production in intact plants and plant tissue culture (Brodelius, 1985).

	Time before harvest	Shikonin (% dry wt.)
Intact plant	2-3 yrs	1-2
Plant tissue culture	3 weeks	14

Since the medium (designated M-9) which supports high levels of shikonin production did not result in good growth (Fujita *et al.,*1981b), a two-stage process was developed. As shown in Fig 20.26, the cells are first grown in a growth medium (designated MG-5) for 9 days, then filtered and pumped to a second vessel in which the M-9 production medium is added. The cells are cultured for 14 additional days, after which they are recovered by filtration, and the shikonin is extracted from the cells. Shikonin levels at that time were reported to be 4 g/L (Tabata and Fujita 1985). They also studied the effect of oxygen on shikonin production.

Tanaka (1987) reported that in a fermenter with a paddle type impeller increased K_La for values up to $10\ h^{-1}$ enhanced shikonin production. Also, Tanaka (1987) reported that scale-up of the shikonin

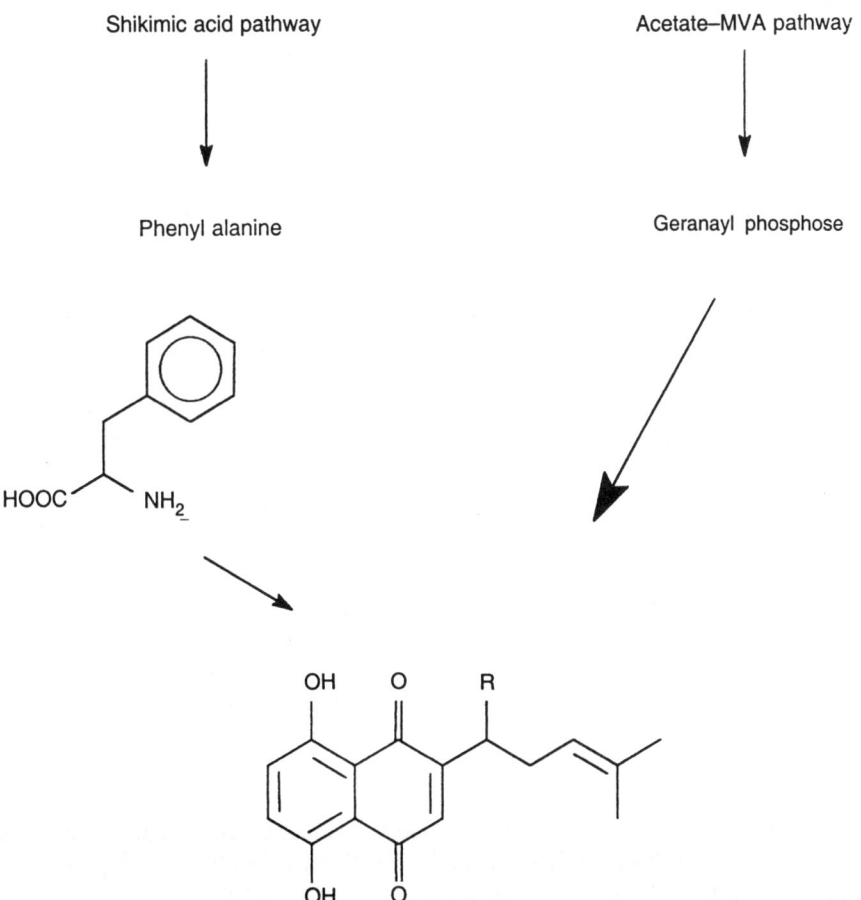

Fig. 20.24. Biosynthetic pathway for shikonin derivatives (Adopted from Tabata and Fujita 1985).

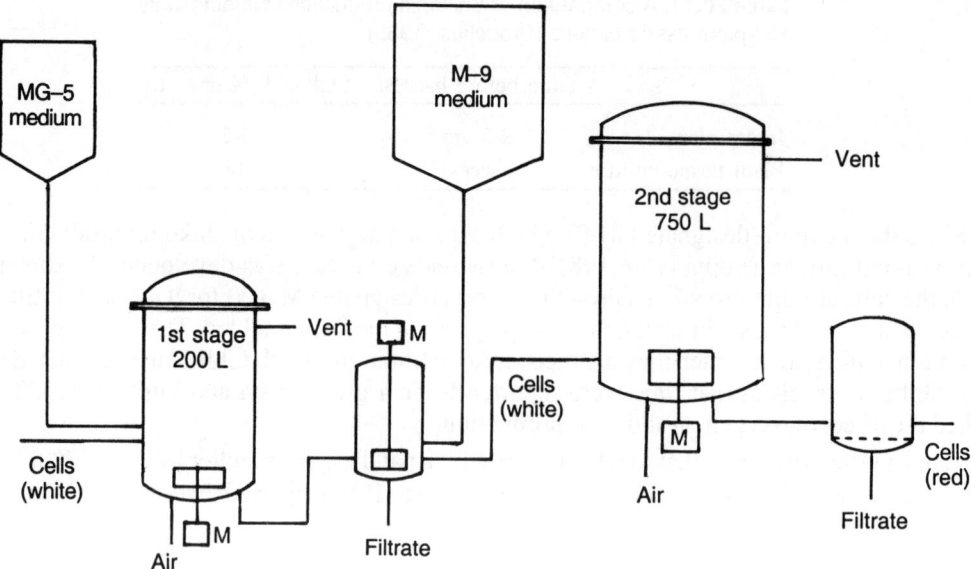

Fig. 20.25. The large-scale production of shikonin-Mitsui Process (Adopted from Tabata and Fujita 1985).

process was successful when scale-up was based on maintaining constant $K_L a$. In the two-stage batch process shikonin productivities were reported to be highest when the dissolved oxygen concentrations were 6.4 and 6 ppm in the individual stages (Tabata and Fujita 1985).

Fujita and Tabata (1987) reported some details on culturing *L.eryhrorhizon* cells in bioreactors. When a paddle impeller was used, good cell growth was observed, however, shikonin production was reduced compared to that by shake-flask-cultured cells. Also, increases in the impeller speed further reduced shikonin concentrations. Payne *et al* (1992) attributed this reduction in production at the higher agitation to cell injury. Efforts to use an airlift type reactor were frustrated by the "bubbling-up" of the cells and the subsequent adherence of the cells to the tank wall. Tanaka *et al.,* (1981) then adopted a rotary drum type tank which revolved slowly (Fig 6.6), thus washing the cells from the wall. This system has been scaled-up from 5 to 1,000 L in either mechanically agitated or airlift systems was reported to result in reductions in shikonin porduction (Tanaka 1987).

Shimomura *et al.,* (1991) established a hairy root culture of *L. erythrorhizon* with *Agrobacterium rhizogenes*. The hairy root culture did not produce shikonin on solid MS medium but produced the pigment in the root culture medium and also secreted it into the medium. Addition of adsorbents, XAD-2, XAD-4, charcoal and so on increased the concentration of shikonin produced. The roots were cultivated in a 2L air-lift type fermenter connected to a XAD-2 column (25 g) through a peristaltic pump and 5 mg/day of shikonin was continuously produced during a period of more than 220 days.

20.9.3. Berberine production

Berberine is an isoquinoline alkaloid, which has also been considered for commercial production and has been scaled up to the 4000-L scale by Mitsui Petrochemical Industries, Japan. The biosynthetic pathway for this alkaloid is shown in Fig. 20.26 (Zenk *et al.,* 1985).

There are various producing cultures, and interestingly, berberine is stored within *Coptis japonica* cells, whereas this alkaloid is released into the medium with *Thalictrum minus* cells (Yamamoto *et*

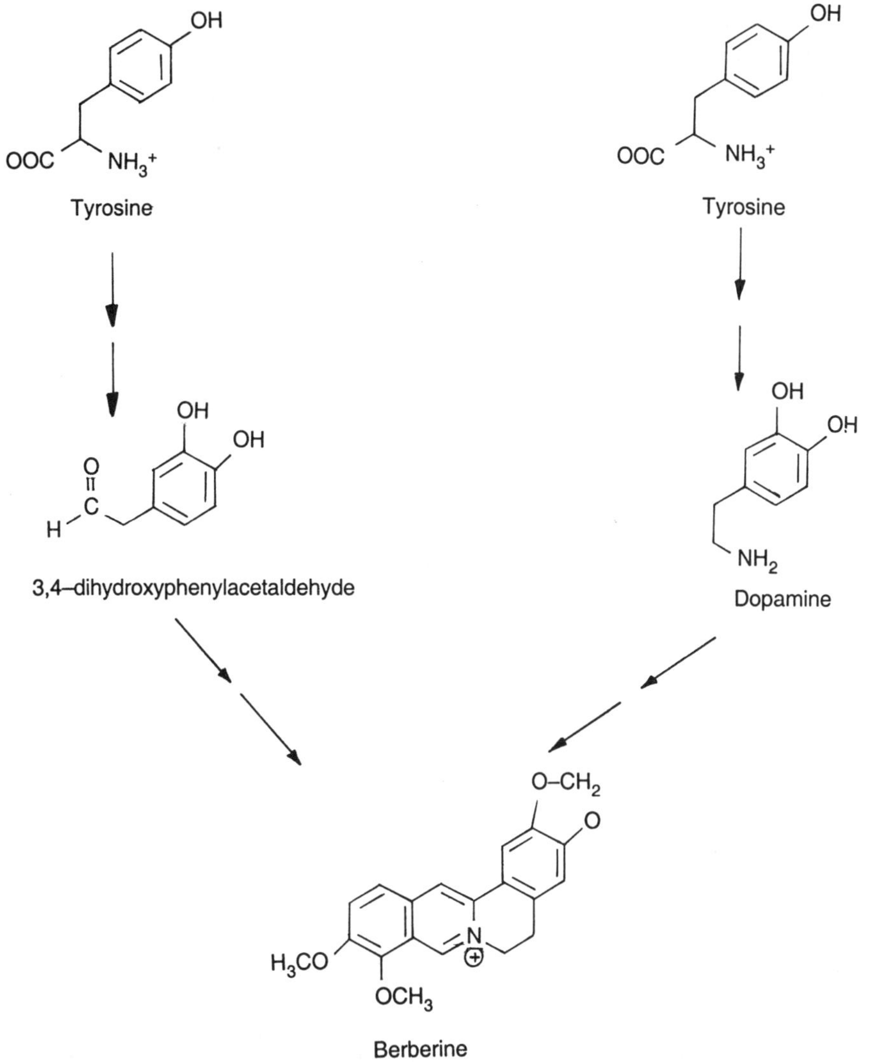

Fig. 20.26. Biosynthetic pathway of berberine (Adopted from Amann et al 1986).

al., 1986). The efforts were made to obtain high-producing cell lines (Yamada and Sato 1981, Sato and Yamada 1984, Yamada and Morikawa 1985). High berberine producers were identified using fluorescence microscopy, and these cells were selected from small aggregate protoplasts (Yamada and Morikawa 1985).

Nakagawa *et al.,*(1986) observed that although the synthetic auxin 2,4-D supported good growth of *T. minus* cells, berberine production was low. In contrast, when an auxin and a cytokinin were added, growth was inferior although berberine production was enhanced. Since optimal conditions for growth and product synthesis were different, these results suggested that a two-stage process would be appropriate. However, Fujita and Tabata (1987) reported that for *Coptis japonica* cultures, optimal conditions for growth and production were so similar that a single-stage process was appropriate.

Hara *et al.,* (1988) reported that gibberellic acid stimulated berberine production in *C. japonica.* Brodelius and co-workers (Funk *et al.,* 1987, Marques and Brodelius 1988, Gugler *et al.,* 1988) reported that a polysaccharide preparation from yeast was able to elicit both berberine production as well as activities of a key biosynthetic enzyme, tyorsine decarboxylase, from *Thalictrum rugosum.*

Breuling *et al.,* (1985) reported the use of a 19-L airlift bioreactor for berberine production by *Berberis wilsonae.* Although aeration between 0.1 and 0.5 volumes air per volume liquid per minute (dissolved oxygen levels between 20 and 50% saturation) had a small effect on growth, berberine production was increased at the higher dissolved oxygen levels. Kreis and Reinhard (1989) also discussed the dissolved oxygen requirements in mechanically agitated bioreactors.

20.9.4. Rosmarinic acid

The production of rosmarinic acid is a model process yielding highest reproduction rates and served as a model system for the process optimization in the field of plant cell culture. The rosmarinic acid has antiphlogistic activity (Berlin 1988). The biosynthesic pathway for rosmarinic acid is shown in Fig 20.27. The production of Rosmarinic acid (RA) from the cell cultures of *Coleus blumei* was reported. (Razzaque and Ellis 1977). De-Eknamkul and Ellis (1985a,b,1984) reported the production of RA from the cell cultures of *Anchusa officinalis.* The cell culture group of the A. Nattermann Company, (Koln, Germany) developed a process for the production of rosmarinic acid on a large scale using *Coleus blumei* suspension cultures. A two stage process was developed. *C. blumei* cells were grown in 42 litter airlift bioreactor containing growth medium. When the appropriate cell density was recorded about 50% of suspension was inoculated in a stirred tank bioreactor. Using this two-stage production process needs only 14 days to produce in total 200 gms rosmarinic acid of 97% purity in two parallel production batches. Ulbrich *et al.,* (1985) reported serious mixing problems in a 32-L (liquid volume) airlift system. They thus opted to use a mechanically agitated vessel with a helical impeller. (5.5 g of RA/L in two-stage batch process see detail chapter-21 on two-stage process).

20.9.5. Indole alkaloids from *Catharanthus roseus*

Fowler (1984), observed that a shear tolerant *C. roseus* cell line grew in a bioreactor at agitation speeds in excess of 300 rpm. In contrast, Wagner and Vogelmann (1977) reported that because of the increased shear sensitivity of older *C. roseus* cultures agitation rates as low as 28 rpm were inappropriate for alkaloid production. Apparently as a result of this shear sensitivity, many groups have used pneumatically agitated bioreactors for culturing *C. roseus* cells. Both external loop (Smart and Fowler *et al* 1984) and draft tube (Smart and Fowler 1984b) airlifts as well as bubble column (Payne *et al* 1988) systems have been used. In contrast, Schiel and Berlin (1987) used a mechanically agitated bioreactor was less than that observed in shake flasks-even when the shake-flask culture was derived from the same inoculum as was used in the bioreactor.

Indole alkaloid synthesis is often induced by transferring culture from a growth-promoting medium to a medium, which supports less growth but enhanced product synthesis. To exchange media, an external filtration step was performed in the shikonin process. By coupling filtration to aeration, it was possible to filter the cell from the medium without the need for operators outside the bioreactor (Payne *et al.,* 1988). This internal filtration operation should reduce contamination risks and permits the use of a single bioreactor for both the growth and production phases. A similar system- the spin filter-couples agitation with filtration to achieve cell separation within a mechanically agitated bioreactor (Ammirato and syter 1985, Styer 1985). Sedimentation has also been used for internal cell separation with *C. roseus* cultures (Pareilleux and Vinas 1984).

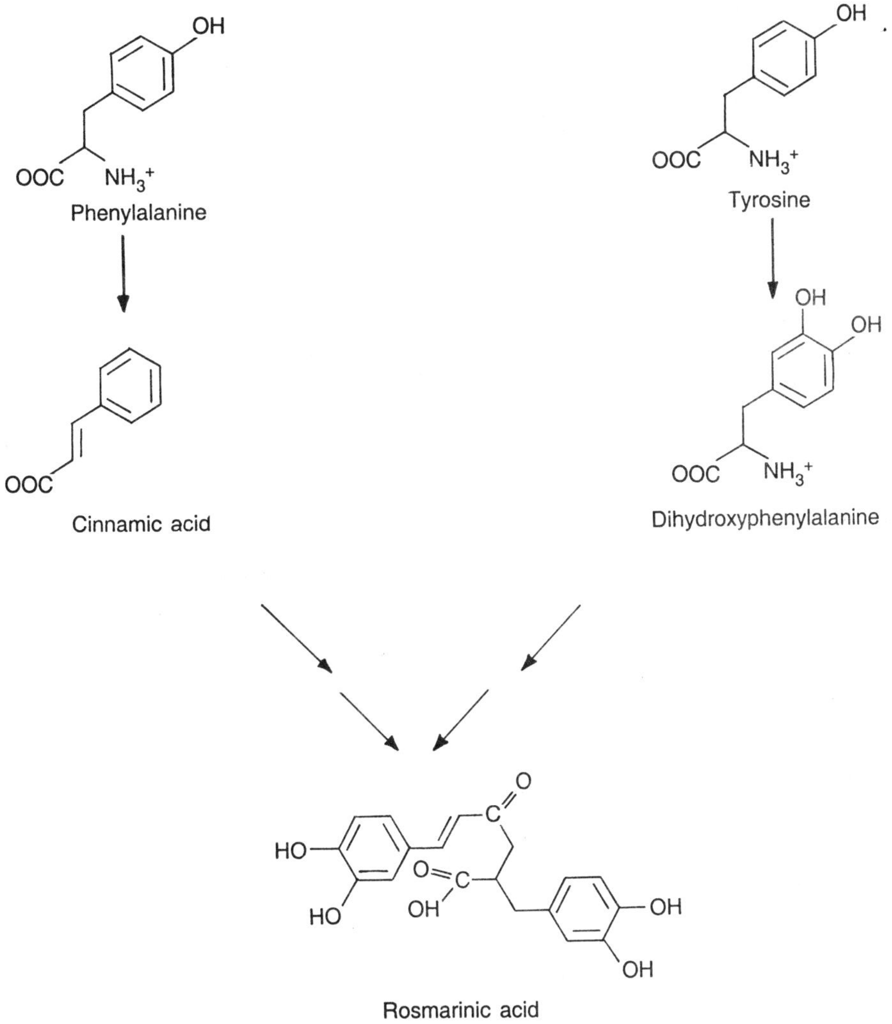

Fig 20.27. Biosynthetic pathway for Rosmarinic acid (Adapted from Ellis and Towers 1970).

20.9.6. Paclitaxel (Taxol) production from cell cultures

Taxol, a diterpenoid secondary product of the yew species, has been recognized as the best anticancer drug to emerge in the last 20 years. The supply of taxol for clinical use is still limited and dependent on extraction from the yew plant (*Taxus* sp.), as its bark is the only commercial source. It is difficult to synthesize chemically on economically viable method. The *Taxus* species is a slow growing tree, which compel to look into alternative sources of supply. Plant cell cultures method production for Taxol is the best alternative available for the problem of supply crisis (See chapter 6). There are several reports on the production of Taxol and its analogs from cell cultures of *Taxus* sp. (Srinivasan *et al.*, 1995, 1996, Veeresham *et al.*, 1995; Fett-Neto *et al.*, 1994; Prasad Babu *et al.*, 2001). However yields need to be improved. There are also several patents on the improvement of Taxol yield from cell cultures of *Taxus* sp. (Bringi *et al.*, 1995; Choi *et al.*, 1999, 2001, Chattopadhyay *et al.*, 1999, 2000; Kulkarni and Krishna Murthy, 2002; Yukimune *et al.*, 2002). The organizations and companies are developing/developed methods for commercial scale production using plant cell cultures process for Taxane/Taxol are U.S. Department of Agriculture, USA (Christen *et al.*, 1991), Nippon Steel Corp, Japan (Saito *et al.*, 1992), Escagenetics Inc (Smith *et al.*, 1993), USA, Phyton Incorporated (Bringi *et al.*, 1995) USA, Penn State Research Foundation (Arteca and Wickremesinhe, 1993) USA, Mitsui Petro Chemical Industries Ltd. (Yukimune *et al.*, 1995) Japan, E.R. Squibb and Sons, Inc. Princeton, USA (Cino *et al.*, 1997) and Samyang Genex Co., Ltd., Seoul, Korea (Choi *et al.*, 1999, 2001).

Srinivasan *et al.* (1995) reported the large scale cultivation of *Taxus baccata* cell culture in three bioreactor configurations, i.e. 250 ml Erlenmeyer flasks, a 1-L working volume pneumatically mixed glass bioreactors (PMB) and a 1-L working volume glass stirred tank bioreactor (STB). The temperature was maintained at 25°C for all three bioreactors and the reactors were kept in darkness. Air was filtered by first passing it through a Lab clear gas filter (Fischer Scientific, Pittsburgh, PA) to remove any oil and particulates and then sterilized by passing through an on-line 0.22 μm Gelman filter unit. The STB had a height to diameter ratio of 2.5. Agitation was achieved by means of a magnetically driven flat bladed turbine rotating at 120 rpm. The culture was aerated by filtered air at a flow rate of 280 ml/min. A K_La of ~ $9h^{-1}$ was obtained under these conditions. The PMB was constructed with a thick walled pyrex glass. The bottom was tapered to permit better suspension of larger aggregates. A water jacket around the bioreactor was used to maintain temperature. The air flow rate was maintained at 370 ml/min. Dimethylpolysiloxane (Sigma) in water (20 v/v) was added as an antifoaming agent not exceeding a total addition of 0.2% v/v.

The biomass accumulation and nutrient uptake profiles in all three bioreactors were very similar. The maximum biomass accumulation was on day 25. The maximum amount of taxol from *T. baccata* cell cultures from large-scale cultivation using the bioreactors are STB (0.1 mg/l), PMB (1.5 mg/l) and Erlenmeyer flasks (0.3 mg/l). Since pneumatically mixed operation generally provides adequate gas transfer and low mechanical shear, the better yield of taxol observed in the PMB. The lower yield in the STB could be related to differences in hydrodynamic shear.

Seki et al (1997) reported a continuos production of taxol (paclitaxel) using free and immobilized cells of *Taxus cuspidata*. Callus tissues were induced from the needles and young stems of the yew tree, *T. cuspidata* Sieb & Zucc. and its variations *T. cuspidata* varnana. The medium used for callus initation was B5 medium with a NAA (5 mg/l). Sucrose (20g/l) and casamino acids (3g/l). A considerable amount of paclitaxel was released into the suspension culture medium. The perfusion culture of *T. cuspidata* cell cultures using with a 300 ml Erlenmeyer flask with a mesh net cell separator shaken at 100 strokes/min at 300k in the dark. Fresh medium was continuously fed into the flask using a peri-staltic pump. Air exchange was done through a 50-mm membrane filter (Millipore) with 0.2 μm pore diameter. The medium in the flask was pumped out through a cell-separation unit that held nylon mesh netting with ca. 0.15.mm openings. The culture volume was maintained at 100 ml. About 1.5 g of wet cells from liquid subcultures was used for 10-14 days to inoculate the liquid medium in each flask.(Fig 20.28). The specific productivity of paclitaxel released into the medium was 3.2 mg/g of dry cells by the 11-day cultivation.

Seki and Furusaki (1996) demonstrated a system of continuous paclitaxel production using immobilized cells of *T.cuspidata*. Cells were entrapped in calcium alginate gel particles by dripping a concentrated cell suspension mixed with sodium alginate into a $CaCl_2$ solution. The perfusion cultures of immobilized flow out medium was ~ 0.2 mg/l (Fig 20.29).

Son *et al.* (2000) reported the large-scale growth and taxane production in cell cultures of *Taxus cuspidata* (Japansese yew) using a novel bioreactor Rapidly growing cell lines were selected from callus cultures derived from immature embryos of yew. The cells were inoculated in 20-l capacity bioreactors of different types to test the growth performance. The models of small scale bioreactors incorporated in this study included a ballon type bubble bioreactor (BTBB), a bubble column bioreactor (BCB),a BCB with a split plate internal loop a BCB with a concentric draught tube internal loop, a BCB with a fluidized bed bioreactor, and two different models of stirred tank reactors. Among the reactors, BTBB appeared to be the most efficient in promoting cell growth. The doubling time of cell growth in BTBB was 12 days with 30 % inoculation cell density. The optimum time for medium replacement or feeding was 12-15 days after inoculation as determined by monitoring both the levels of sugars and medium conductivity. When yew tree cells were grown in different sizes (100-500 l) of BTBBS, more than 70 % cell viability was recorded at the time of harvest. The growth pattern of the cells in the pilot-scale BTBB appeared to be the same as that of cells in the 20-l bioreactors. Approximately 3 mg/l of taxol and 74 mg/l total taxanes were obtained after 27 days of culture.

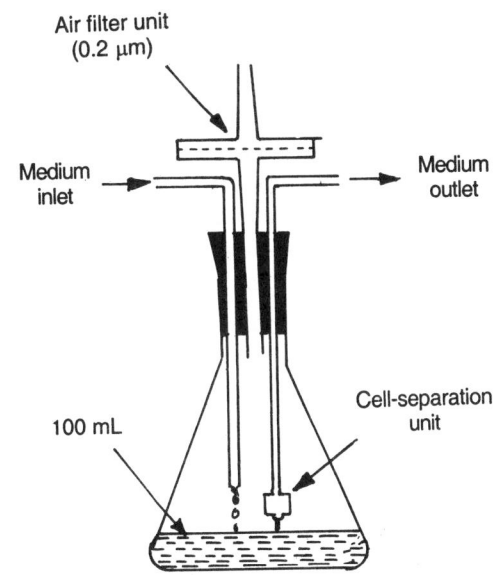

Fig. 20.28. Continuous production of taxol using *T. cuspidata* immobilized cells (adopted from Seki et al., 1997).

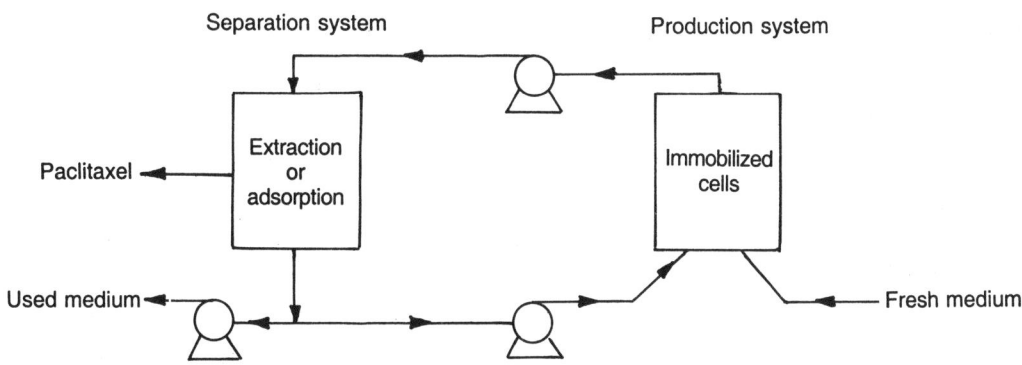

Fig. 20.29. Economical production of Paclitaxel (Adopted from Seki and Furusaki, 1996).

Osorio et al (2002) reported the taxol and baccatin III production in suspension cultures of *T. baccata* and *T. wallichiana* in an air lift bioreactor. Suspension cultures of *Taxus baccata* var fastigiata and *Taxus wallichiana* were grown in a 20-l air lift bioreactor in B5 medium with 2% sucrose (w/v) and NAA (1.86 mg/l). The bioreactor vessel, 40 cm wide and 100 cm long (with a height to diameter ratio 2.5), was constructed from borosilicate glass. A ring sparger with 15 holes was placed exactly in the centre nearly touching the bottom of the bioreactor vessel, which creating a bubble column which provided both agitation and mixing. The working volume was kept at 18-L culture medium, with an aeration rate of 2.5 l min^{-1} and a temperature of 25°C. The inoculum consisted of 100 g cell fresh weight L^{-1}. The *T.baccata* culture grew moderately, reaching a maximum of 25.98 g dry weight L^{-1} after 20 days of culture, whereas *T. wallichiana* grew more actively reaching the 25.98 g dry weight L^{-1} after only 12 days. After 24 days culture, the total taxol content in *T. wallichiana* culture (21.04 mg/l) was 73 % higer than the respective content of *T.baccata* culture (12.04 mg/l). The total bacctin III production from cell cultures of *T. wallichiana* was found to be 25.67 mg/l after 28 days of cultivation.

To date, there is only one industry in Germany using the plant cell culture process for commerical production of taxol using 75,000 L bioreactor (Phyton Incorporated, Ithaca, NY. USA).

Table 20.15. Products of interest of Industry from cell cultures

Name	*Compounds*
Papaver somnifera, P. bracteatum	Morphinan Alkaloids Codeine, thebaine, Sanguinarine (anti-plague agent)
Coptis japonica, Phellondendron amurense, Thalictrum minus, T. rugosum	Berberine
D. myoporoides, D. leichhardtii, Scopolia Sp. *Atropa belladonna, Hyoscyamus niger Datura stramonium*	Tropane alkaloid (Atropine, Hyoscyamine, Scopolamine)
Digitalis lanata, D. purpurea	Cardiac glycorides (Digoxin etc)
Mucuna pruriens, M. hassjoo M. deeringiana	L- DOPA
Nardostachys jatamansi Valeriana wallichii, V. officinalis, N. chinensis	Valeopotriates
Taxus brevifolia, T. baccata, Taxus Sp.	Taxol
C. acucminata, Nothapodytes foetida	Camptothecine
Cephalotaxus harringtonia	Harringtonine
Podophyllum peltatum, P. hexandrum, Callitris drummondii,Linum album	Podophyllotoxin
Catharanthus roseus	Indole alkaloids, Ajmalicine, serpentine, Vincristine, vinblastine, vincamine
Panax ginseg	ginseng saponins
Coleus blumei, Anchusa officinalis Lithospermum erythrorhizon	Rosmarinic acid
Vaccinium vitisidaea, Arctostaphylos uva-ursi, Pyrus communis	Arbutin
Food Additives	
Shikonin	*L. erythrorhizon*
Anthocyanins	*Euphorbia millii*
Safflower yellow	*Carthamus tinctorius*
Saffron	*Crocus sativus*
Madder colorants	*R. tinctorum*
Chicle (raw material for chewing gum)	*Achras sapota*
Mucliage	*Astragalus gummifera*
Hernandulcin	*Lippia dulcia*
Azadirachtian(Biopesticide)	*Azadirachta indica*

REFERENCES

Akita, M.and Takayama, S. (1988). Mass propagation of potato tubers using jar fermentor techniques. *Acta Hortic.,* 230 : 5-61.

Alfermann, A.W., Boy, H.M., Doller, P.C., Hagedorn, W., Heins, M., Wahl, J. and Reinhard, E. (1977). Biotransformation of cardiac glycosides by plant cell cultues. In : Plant Tissue Culture and its Biotechnological Applications (Barz, W., Reinhard, E. and Zenk, M.H. (eds.)), pp. 125-141, Springer-Verlag, New York.

Allan, E.J. and Scragg, A.H. (1986). Comparison of the growth of *Cinchona ledgeriana* suspension cultures in shake flasks and seven litre air lift bioreactors. *Biotechnol. Lett.,* 8 : 635-638.

Amann, M., Wanner, G. and Zenk, M.H. (1986). Intracellular compartmentalization of two enzymes of berberine biosynthesis in plant cell cultures. *Planta,* 167 : 310-320.

Ammirato, P.V. and Styer, D.J. (1985). Strategies for large-scale manipulation of somatic embryos in suspension culture. In "Biotechnology in Plant Science : Relevance to agriculture in the Eighties", M. Zaitlin, P. Day and A. Hollaender (eds.), Academic Press, New York, pp. 161-178.

Armstrong, C.L and Philips R .L. (1988). Genetic and cytogenetic variation in plants regenerated from organogenic and friable, embryogenic tissue cultures of maize. *Crop Sci* 28: 363-369

Arteca, R.N. and Wickremesinhe,E. PCT Int. Appl. WO93/23555, 25 Nov. 1993.

Azechi, S., Hashimoto, T., Yuyama, T., Nagatsuka, S., Nakashizuka, M., Nishiyama, T. and Murata, A. (1983). Continuous cultivation of tobacco plant cells in an industrial scale plant. Hakkokogaku Kaishi, 61 : 117-128.

Bahakov, A.V., Bartova, L.M., Dridze, I.L., Maisuryan, A.N., Margulis, G.U., Organin, R.R., Voblikova, V.D. and Muromisev, G.S. (1995). Culture of transformed horseadish roots as source of Fusicoccin-like ligands. *J. Plant Growth Reg.,* 14 : 163-167.

Barthe, G.A., Jourdan, P.S., Melntosh, C.A. and Mansell, R.L. (1987). Naringin and limonin production in callus cultures and regenerated shoots from *Citrus* spp. *J. Plant Physiol.,* 127 : 55-65.

Berlin, J. (1985). The use of immobilised plant cells, An Evaluation. *Int. Ass. Plant Tissue Culture Newslett.,* 46: 8-14.

Berlin, J. (1988). Formation of secondary metabolites in Cultured Plant Cells and Its Impact on Pharmacy. In "Biotechnology in Agriculture and Forestry." Vol. 4, Y.P.S. Bajaj (ed.) Springer-Verlag, Berlin, pp. 37-59.

Bhadra, R., Vani, S. and Shanks, J.V. (1993). Production of indole alkaloids by selected hairy root lines of *Catharanthus roseus. Biotechnol. Bioeng.,* 41 : 581-592.

Blanch, H.W.(1990) Introduction to bioreactor engineering In; Biotechnology and Food Process engineering, Schwartzberg,H.G. and Rao, M.A. (eds) Marcel decker,Inc. Newyork,USA. PP-41.

Breuling, M., Alfermann, A.W. and Reinhard, E. (1985). Cultivation of cell cultures of *Berberis wilsone* in 20-1 airlift bioreactor. *Pl. Cell Rep.,* 4 : 220-223.

Breuling, M., Spieler, H., Schwantag, D., Alferman, A.W. and Reinhard, E. (1987). Large scale cultivation of plant cells for production of natural products. In Chmiel, H. Hammes, W.P.and Bailey, J.E.(eds.), *Biochemical Engineering - A Challenge for Interdisciplinary Cooperation,* Gustav FischerVerlag, Stuttgart, pp. 443-449.

Bringi, V. (1991). Differentiation in plant cultures (1) Tracheary element induction in immobilized tobacco cells: (2) Bioreactor considerations for shoot culture cultivation. Ph.D. thesis, Cornell University, Ithaca, New York.

Bringi, V., Kadkade, P.g., Prince, C.L., Schubmehl, B.F., Kane, E.J., Roach, B. (1995). Enhanced production of taxol and taxanes by cell cultures of *Taxus* species. United States Patent No. 5,407, 816.

Brodelius, P. (1985). The potential role of immobilization in plant cell biotechnology. *Trends Biotechnology,* 3 : 280-285.

Buelow, G. H. and Johnson, M. J. (1952) Effect of separation on citric acid production in 50 gallon tanks. *Ind. Eng. Chem.* 44, 2945-2946.

Buitelaar, R.N., Langehoff, A.A.N., Heidstra, R. and Tramper, J. (1991). Growth and thiophene production by hairy root cultures of *Tagetus patula* in various two liquid phase bioreactors. *Enzyme Microbiol. Technol.,* 13 : 487-494.

Byrne, A.F. (1962). Food production by submerged culture of Plant Tissue Cells. Activities Report 14, Armed Forces Food and Container Institute, Chicago, 177.

Byrne, A.F. and Koch, M.B. (1962). Food production by submerged culture of plant tissue cells. *Science,* 135 : 215.

Chain, E. B., Gualandi, G. and Morisi, G. (1966) Aeration studies. IV. Aeration conditions in 3,000 litre submerged fermentations with various micro-organisms. *Biotech. Bioeng.* 8:595-619.

Chain, E.B., Palabino, S., Callow, D.S., Vgolini, F. and Van Der Sluis, J. (1952). Studies on aeration I. *Bull. World. Health Org.,* 66 : 83-98.

Charlwood, B.V. and Moustou, C. (1988). Essential oil accumulation shoot-proliferation cultures of *Pelarganium* spp. In "Manipulating Secondary Metabolism in Culture", R.J. Robins and M.J.C. Rhodes (eds.), Cambridge University Press, Cambridge, U.K., pp. 187-194.

Charlwood, B.V., Moustou, C., Brown, J.T., Ilegarry, P.K. and Charlwood, K.A. (1989). The regulation of accumulation of lower isoprenoids in plant cell cultures. In "Primary and Secondary Metabolism of Plant Cell Cultures", W.G.W. Kurz, (eds.) Springer-Verlag, Berlin, pp. 73-84.

Chattopadhyay, Sunil Kumar, S., Ram Prakash, R., Sushil Kumar (2000). Process for the production of important taxol analogues 10-deacetyl taxol A, B and C. United States Patent No. 6028206.

Chattopadhyay, Sunil Kumar, S., Ram Prakash, R., Sushil Kumar, Padmanabhan, K.M. (1999). Process for the production of taxol. United States Patent No. 5, 856,532.

Chen, T.H.H., Thompson, B.E. and Gerson, D.F.(1987). *In vitro* production of alfalfa somatic embryos in fermentation systems. *J. Ferment Technol.,* 65 : 353-357.

Cherry, R.S. and Papoutsakis, E.T. (1988). Physical Mechanisms of Cell Damage in Microcarrier Cell Culture Bioreactors. *Biotechnol. Bioeng.,* 32 : 1001-1014.

Choi, H.K., Adams, T.L., Stahlhut, R.W., Kim, S.I., Yun, J.H., Song, B.K., Kim, J.H., Song, J.S. Hong, S.S. and Lee, H.S. (1999). Method for mass production of taxol by semi-continuous culture with *Taxus chinensis* cell cultures, United States Patent No. 5,871,979.

Choi, H.K., Adams, T.L., Stahlhut, R.W., Kim, S.I., Yun, J.H., Song, B.K., Kim, J.H., Song, J.S. Hong, S.S. and Lee, H.S. (2001). Production of taxol from *Taxus* plant cell culture adding silver nitrate, United States Patent No.: 6,248,572.

Christen, A.A., Gibson, D.M. and Bland, J. (1991). Production of *Taxol* or Taxol like compounds in cell cultures. U.S. Patent No. : 5019504.

Christen, P. *Centranthus* species : *In vitro* culture and the production of valepotriates and other secondary metabolites. In : Bajaj, Y.P.S., editor. Biotechnology in Agriculture and Forestry, Medicinal and Aromatic Plants, XI, vol. 43, Berlin : Springer-Verlag, pp. 42-56.

Christen, P., Roberts, M.F. Phillipson, J.D. Evans, W.C. (1991). Alkaloids of hairy root cultures of a *Datura candida* hybrid. *Pl. Cell Rep.,* 9 : 101-104.

Cino, P.M., Schwarz, S.R. and Cazzulino, D.L. (1997). Callus cell induction and the preparation of taxanes, United States Patent No. 5,665,576.

Constabel, F., Gauder-La Prairie, P., Kurz, W.G.W. and Kutney, J.P. (1982). Alkaloid production in *Catharanthus roseus*. XII. Biosynthetic capacity of callus from original explants and regenerated shoots. *Plant Cell Rep.,* 1 : 139-142.

Constabel, F., (1985) Morphinan alkaloids from plant tissue cultures. In: The chemistry and bilogy of isoquinoline alkaloids. Phillipson, J.D. Roberts, M.F. and Zenk, M.H., (eds) Springer-Verlag, Berlin PP257-264.

Cuello, J.L., Walker, P.N. and Curtis, W.R. (1991). Ebb and flow : Bioreactor for hairy root cultures. International winter Meeting. Chicago : IL Dec. 17-20, Paper No. 917528 Chicago. *American Society for Agricultural Engineering.*

Curtis, W.R. and Emery, A.H. (1993). Plant cell suspension culture theology. *Biotechnol. Bioeng.,* 42 : 520.

D'Amato, F. (1978). Chromosome number variation in cultured cells and regenerated plants. In : Thorpe, T.A. (ed.), *Frontiers of Plant Tissue Culture,* IAPTC, University of Calgary, Canada, pp. 287-295.

Davioud, E., Kan, C., Hamon, J.T. and Hussan, H.P. (1989). Production of indole alkaloids by *in vitro* cultures from *Cantharanthus trichophyllus. Phytochem.,* 32 : 2675.

De Luca, A., Balsevich, J., Tyler, R.T. and Kurz, W.G.W. (1988). Development regulation of enzymes of indole alkaloid biosynthesis in *Catharanthus roseus. Plant Physiol.,* 86 : 447-450.

De-Eknamkul, W. and Ellis, B.E. (1984). Rosmarinic acid production and growth characteristics of *Anchusa officinalis* cell suspension cultures. *Planta Med.,* 51 : 346-350.

De-Eknamkul, W. and Ellis, B.E. (1985a). Effects of Macronutrients on Growth and Rosmarinic acid formation in cell suspension cultures of *Anchusa officinalis. Plant Cell Rep.,* 4 : 50-53.

De-Eknamkul, W. and Ellis, B.E. (1985b). Effects of macronutrients on growth and rosmanic acid formation in cell suspension cultures of *Anchusa officinalis. Plant Cell Rep.,* 4 : 46-49.

Deus-Neuman, B. and Zenk, M.H. (1984). Instability in indole alkaloid production in *Catharanthus roseus* cell suspension cultures. *Planta Medica,* 50 : 427-431.

Dilorio,A.A., (1991). Betacyanin Production and efflux from Transformed Roots of *Beta vulgaris* in a Nutrient Mist Bioreactor. Ph.D thesis, Worcester Polytechnic Institute, USA.

Dilorio, A.A., Cheetham, R.D. and Weathers, P.J. (1992a). Carbon dioxide improves the growth of hairy roots cultured on solid medium and in nutrient mists. *Appl. Microbiol. Biotechnol.,* 37 : 463-467.

DiIorio, A.A.,Cheetham,R.D., and Weathers,P.J., (1992b) Growth of transofrmed roots in a nutrient mist bioreactor; reactor performance and evaluation. *Appl. Microbial. Biotechnol* 37 : 457-462.

Dougall, D.K., LaBrake, S. and Whitten, G.H. (1983). The effects of limiting nutrients, dilution rate, culture pH, and temperature on the yield constant and Anthocyanin accumulation of carrot cells in semi-continuous chemostat cultures. *Biotechnol. Bioeng.,* 25 : 569-579.

Ducos, J.P. and Pareilleux, A. (1986). *Appl. Microbiol. Biotechnol.,* 25 : 101-105.

Ellis, B.E. and Towers, G.H.N. (1970). Biogenesis of Rosmarinic acid in *Mentha. Biochem. J.,* 118 : 291-297.

Endo, T., Goodbody, A. and Misawa, M. (1987). Alkaloid production in root and shoot cultures of *Catharanthus roseus. Planta Med.,* 53 : 479-182.

Fett-Neto, A.G., Zhang, W.Y., DiCosmo, F. (1994). Kinetics of taxol production, growth and nutrient uptake in cell suspension cultures of *Taxus cuspidata. Biotechnol. Bioeng.,* 44 : 205-210.

Flores, H.E and Curtis, W.R. (1992). Approaches to understanding and manipulating the biosynthetic potential of plant roots. *Ann. NY Acad. Sci.,* 665 : 188.

Flores, H.e. Pickard, J.J. and Hoy, M.W. (1988). Production of polyacetylenes and thiophenes in heterotrophic and photosynthetic root cultures of *Asteraceae.* In : Lam J., Hreheler, H., Amason, T., Hansen, L. editors. Chemistry and Biology of Natural Occurring Acetylenes and Related Compounds (NOARC). *Bioactive Molecules,* 7 : 233-254.

Fortune W, B., McCormick, S. L., Rhodehamel, H. W. and Stefaniak, J. J. (1950) Antibiotics development. *Ind. Eng. Chem.* 42, 191-198.

Fowler MW, Bond P and Scragg AH (1987). Developments in plant cell culture technology. In: Biochemical engineering. Eds. Chemiel H, Hammes WP and Bailey JE. 33-341 Gustave Fischer, New York

Fowler, M.W. (1981). The large-scale cultivation of plant cells. *Prog. Ind Microbiol* 16. 207-229.

Fowler, M.W. (1984). Plant cell culture : Natural products and industrial application. *Biotechnol* and *Genet Engneerign Rev* 2: 41-67.

Fowler, M.W. (1984). Large scale cultures of cells in suspension In Cell Cultures and somatic Cell Genetics of Plants: Vol. 1 Laboratory Procedures and their Applications Vasil IK pp. 167-74. Academic press. New York.

Fowler, M.W. (1984). Problems in commercial exploration of plant cell cultures. applications of Plant Cell and Tissue culture. Ciba Foundation Syposimm. pp 239-53.

Fujita, Y. and Tabata, M. (1987). Secondary metabolites from Plant Cells – Pharmaceutical Applications and Progress in Commercial Prouction. In "Plant Tissue and Cell Culture," Green, C.E., Somers, D.A., Hackett, W.P. and Biesboer, D.D. (eds.), Alan R. Liss Inc., New York, pp. 169-185.

Fujita, Y., Hara, Y., Ogino, T. and Suga, C. (1981a). Production of Shikonin Derivatives by Cell Suspension Cultures of *Lithospermum erythrorhizon* I. Effect of Nitrogen Source on the Production of Shikonin Derivatives. *Plant Cell Rep.,* 1 : 59-60.

Fujita, Y., Hara, Y., Suga, C. and Morimioto, T. (1981b). Production of Shikonin Derivatives by Cell Suspension Cultures of *Lithospermum erythrohizon* I. A New Medium for the Production of Shikonin Derivatives. *Plant Cell Rep.,* 1 : 61-63.

Fujita, Y., Tabata, M., Nishi, A. and Yamada, Y. (1982). In : Fukiwara, A. (ed.) New Medium and Production of Secondary Compounds with the Two State Culture Medium in Plant Tissue Culture, *Japanese Association of Plant Tissue Culture,* Tokyo, pp. 399-400.

Funk, C., Gugler, K. and Brodelius, P. (1987). Increased secondary product formation in plant cell suspension cultures after treatment with a yeast carbohydrate preparation (Elicitor). *Phytochemistry,* 26 : 401-405.

Gershenzon, J., Maffei, M. and Croteau, R. (1989). Biochemical and histochemical localization of monoterpene biosynthesis in the glandular trichomes of spearmint (*Mentha spicata*). *Plant Physiol.,* 89 : 1351-1357.

Giri, A., Banerjee, S., Ahuja, P.S., Giri,C.C. (1997). Production of hairy roots in *Aconitum heterophyllum* wall. Using *Atrobacterium rhizogenes. In vitro Cell Dev. Biol-Plant,* 33 : 280-284.

Giulietti, A.M., Parr, A.J. and Rhodes, M.J.C. (1993). Tropane alkaloids production in transformed root cultures of *Brugmansia candida. Plant Medica.,* 59 : 428-31.

Gordon, J. J., Grenfell, E., Knowles, E., Legge, B. J., McAllister, R. C. A. and White, T. (1947) Methods for penicillin production in submerged culture on a pilot scale. *J. Gen. Microbiol.* 1 : 171-186.

Graebe, J.E. and Novelli, G.D. (1966). A practical method for large-scale plant tissue culture. *Exp. Cell Res.,* 41: 509.

Gugler, K., Funk, C. and Brodelius, P. (1988). Elicitor-induced tyrosine decarboxylase in Berberine-synthesizing suspension cultures of *Thalictrum rugosum. Eur. J. Biochem.,* 170 : 661-66.

Hagimori, M., Matsumoto, T. and Obi, Y. (1982a). Studies on the production of *Digitalis purpurea* Cardenolides by Plant Tissue Culture. *Plant Physiol.,* 69 : 653-656.

Hallsby, G.A. and Shuler, M.L. (1986). Altering Fluid Flow Patterns Changes Pattern of Cellular Associations in Immobilized Tobacco Tissue Cultures. *Biotechnol. Bioeng. Symp. Ser.,* 17 : 741-746.

Hamill, J.D., Parr, A.J., Robins, R.J. and Rhodes, M.J.C. (1986). Secondary production formation by cultures of *Beta vulgaris* and *Nicotiana rustica* transformed with *Agrobacterium rhizogenes. Pl. Cell Rep.,* 5 : 111-114.

Hara, Y., Yoshioka, T., Morimoto, T., Fujita, Y. and Yamada, Y. (1988). Enhacement of berberine production in suspension cultures of *Coptis japonica* by Gibberellic acid treatment. *J. Plant Physiol.,* 133 : 12-15.

Hashimoto, T. and Azechi, S. (1988). Bioreactors for the large scale culture of plant cells. In : Biotechnology in Agriculture and forestry. Y.P.S. Bajaj (ed.), Vol. 4, Springer-Verlag, Berlin, pp. 104-122.

Hashimoto, T. and Yamada, Y. (1988). Biosynthesis of Scopolamine from (7-â-H-2)6-â-Hydroxyhyoscyamine in *Dubosia* shoot cultures. *Agric. Biol. Chem.,* 53 : 863-864.

Heble, M.R. (1985). Multiple shoot cultures. A viable alternatives *In vitro* system for the production of known and new biologically active plant constituents. In : Primary and Secondary Metabolism of Plant Cell Cultures", K.H. Neumann, W. Barz, and E. Reinhard (eds.), Springer-Verlag, Berlin, pp. 281-289.

Hegarty, P.K., Smart, N.j., Scragg, A.H. and Fowler, M.W. (1986). The aeration of *Catharanthus roseus* L. G. Don suspension cultures in airlift bioreactors : The inhibitory effect at high aeration rates on culture growth. *J. Exp. Bot.,* 37 : 1911-1920.

Heyerdahl, P.H., Olsen, O.A.S., and Hvoslef-Eide, A.K., (1995). Engineering aspects of plant propagation bioreactors. In: Automation and environmental control in plant tissue culture. Eds. Aitken-Christie J, Kozai T and Smith ML. 87-123 Kluwer Acad. Publ, Dordrecht, Boston, London

Hilton, M.G. and Rhodes, M.J.C. (1990). Growth and hyoscyamine production of hairy root cultures of *Datura stramonium* in a modified stirred tank reactor. *Appl. Microbiol Biotechnical.,* 33 : 132-138.

Hilton, M.G., Wilson, P.D.G., Robins, R.J., Rhodes, M.J.C. (1988). Transformed root culture-fermentation aspects. In : Manipulating secondary metabolism in culture. Cambridge University Press, Cambridge, PP 239-245.

Hirata, T., Murakami, S., Ogihara, K. and Suga, T. (1990). Volatile monoterpenoid constituents of the plantlets of *mentha spicata* produced by shoot tip culture. *Phytochemistry,* 29 : 493-495.

Holden, P.R., Holden, M.A. and Yeoman, M.M. (1988). Variation in the secondary metabolism of cultured plant cells. In Robins, R.J. and Rhodes, M.J.C. (eds.) *Manipulating Secondary Metabolism in Culture,* Cambridge University Press, Cambridge, pp.15-30.

Holmes, P., Li, S.L., Green, K.D., Lloyd, F.V.B. and Thomas, N.H. (1997). Drip tube technology for continuous culture of Hairy roots with Integrated Alkaloids extraction. In :Hairy root cultures and Applications Doran, P.M. (eds.) Harwood Academic Publishers, Australia, Canada, India, p. 201-209.

Hooker,B.S.,Lee,J.M., and An,G., (1989) Response of plant tissue culture to a high shear environment, *Enzyme.Microbial technology.* 11, 484-490.

Hussey, G. (1986). Problems and prospects in the *In vitro* propagation of Herbaceous Plants in "Plant Tissue Culture and its Agricultural Applications" Withers, L.A. and Alderson, P. (eds.), Butterworths, Boston, p. 69-84.

Inomata, S., Yokoyama, M., Gozu, Y., Shimizu, T. and Yanagi, M. (1993). Growth pattern and ginsenoside production of *Agrobacterium* transformed *Panax ginseng* roots. *Pl. Cell Rep.,* 12 : 681-686.

Ionkava, I., Kartnig, T. and Alfermann, W. (1997). Cycloartane saponin production in hairy root culture of *Astragalus mongholicus. Phytochemistry,* 45(8) : 1596-1600.

Irving, G.M. (1968). Construction materials for breweries. *Chem. Eng.* (New York), 75(14) : 100, 102-104.

Jaziri, M., Shimomura, K., Yoshimatsu, K., Fauconnier, M.L., Marlier, M. and Homes, J. (1995). Establishment of normal and transformed root cultures of *Artemisia annua* L. for artemisinin production. *J. Plant Physiol.,* 145 : 175-177.

Jose, U., Pedersen, H. and Chin, Ch.K. (1983). Immobilization of plant cells in a hollow-fiber reactor. *Ann. Ny. Acad. Sci.,* 413 : 409.

Jung, G. and Tepfer, D. (1987). Use of genetic transformation by the Ri T-DNA of *Agrobacterium rhizogenes* to stimulate biomass and tropane alkaloid production in *Atropa belladonna* and *Calystegia sepium* roots grown *in vitro. Plant Sci.,* 50 : 145-151.

Kartha, K.K. (1981). Meristem culture and cryopreservation methods and applications in "Plant tissue culture and applications in Agriculture" T.A. Thorpe (ed.) Academic Press, New York, pp. 181-211.

Kato A, Shiozawa Y. Yamada A, Nishida K and Noguchi M. (1977). A jar fermenter culture of *Nicottana tabacum* L. Cell suspensions. *Agric. Biol Chem* 36: 899-904

Kato, A., Kawazole, S., Lizima, M. and Shimzu, Y. (1976). Continuous culture of tobacco cells. *J. Ferment Technol* 54 : 82-87.

Kato, K., Shiozawa, Y., Yamada, A., Nishida, K. and Noguchi, M. (1972). A jar fermenter culture of *Nicotiana tabacum* L. cell suspensions. *Agric. Biol. Chem.,* 36 : 899.

Kessel, R.H.J. and Carr, A.H. (1972). The effect of Dissolved oxygen concentration on growth and differentiation of carrot (*Daucus carota*) tissue. *J. Exp. Bot.,* 3 : 996-1007.

Knoop E. Struass A and Wehrli W. (1988). Root induction on several solanacae species by *Agrobacterum rhizogenes* and the setermanation of root tropane alkaloid content. *Pl Cell Rep.,* 7 : 590-93

Kondo, O., Honda, T.M. and Kobayashi, T. (1989). Comparison of growth properties of carrot hairy root in various bioreactors. *Appl. Microbiol. Biotechnol.,* 32 : 291-294.

Kozai T. (1991). Controlled environments in conventional and automated microprogation In : Scale Upto and Automation in Plant Propagation. [Vasit IK (ed)], pp.213-30. Academic Press. New York.

Kreis, W. and Reinhard, E. (1989). The production of secondary metabolites by plant cells cultivated in bioreactors. *Planta Med.,* 55 : 409-416.

Kuboi, T. and Yamada, Y. (1976). Caffeic acid-O-methyltransferase in a suspension of cells aggregates of tobacco. *Phytochemistry,* 15 : 397-400.

Kukreja, A.K. and Ahuja, P.S. (1994). Plant cell fermenters potential for Industrial Applications, Current Research in Medicinal and Aromatic Plants, (JMAPS), 16(3) : 53-74.

Kulkarni, A.A. and Krishna Murthy, K.V. (2002). Culture medium composition useful for induction and proliferation of *Taxus calli*, United States Patent No. 6,365,407.

Kurz, W.G.W. (1971). A chemostat for growing higher plant cells in single cell suspension cultures. *Exp. Cell Res.,* 64 : 477.

Kurz, W.G.W. (1975). A fermenter system for the continuous culture of plant cells, in *Plant Tissue culture Methods,* Gamborg, O.L. and Wetter, L.R. Eds. National Research Council, Saskatoon.

Lamport, D.T.A. (1964). Cell suspension cultures of higher plants : isolation and growth energetics. *Exp. Cell Res.,* 33 : 195.

Leathers RR, Smith MAI and Aitken-Christie J (1995). Automation of the bioreactor process for mass propagation and secondary metabolism. In: Automation and environmental control in plant tissue culture. Eds. Aiken-Christie J, Kozai T, Smith MAI. 187-214 Kluwer Acad. Publ, Dordrecht, Boston, London

Lindsey, K. and Yeoman, M.M. (1984a). The synthetic potential of immobilised cells of *Capsicum frutescens* Mill. cv. *annum. Planta,* 162: 495-501.

Lindsey, K. and Yeoman, M.M. (1984b). The viability and biosynthetic activity of cells of *Capsicum frutescens* Mill. cv. *annuum* Immobilized in Reticulate Polyurethane. *J. Exp. Bot.,* 35 : 1684-1696.

Markl, H. (1989). Folien and membranen also neue Elemente in Fermentabau Forum Mikrobiol., 12 : 234.

Marques, I.A. and Brodelius, P.E. (1988). Elicitor-induced L-tyrosine decarboxylase from plant cell suspension cultures I. Induction and purification. *Plant Physiol.,* 88 : 46-51.

Matsumoto T. Okumnshi K. Nishida K. Npguchi M and Tamaki E. (1971). Studies in the culture conditions In : Plant Tissue Culture conditions of higher plant cells in suspension cultures. II Effect of nutritional factors on the growth. *Agric. Biol. Chem.* 35: 543-57

Matsumoto, T. and Tanaka, N. Production of phytoecdysteroides by hair root cultures of *Ajuga reptans* var *atropurpurea. Agric. Biol. Chem.,* 55 : 10-25.

McKelvey, S.A. (1992). Effect of bioreactor design on growth in *Agrobacterium* transformed hairy root cultures of *Hyoscyamus muticus*, B.S. Honors Thesis Pennsylvania : Pennsylvania State University.

Midler, M. and Finn, R.K. (1966). A model system for evaluating shear in the design of stirred fermentors. *Biotechnol. Bioeng.,* 8 : 71-84.

Miller, R.A., Shyluk, J.P., Gamborg, O.L. and Kirkpatricl J.W. (1968). Phytostat for continuous culture and automatic sampling of plant-cell suspension. *Science,* 159 : 540.

Miura, Y., Hirata, K., Kurano, N., Miyamoto, K. and Uchida, K. (1988). Formation of vinblastine in multiple shoot culture of *Catharanthus roseus. Planta Med.,* 54 : 18-20.

Mizukami, H., Konoshima, M. and Tabata, M. (1978). Variation in pigment production in *Lithospermum erythrorhizon* callus cultures. *Phytochemistry,* 17 : 95-97.

Moo-Young, M. (1988). Bioreactors immobilized enzymes and cells. Fundamentals and Applications. Elsevier applied science, London and New York, p. 175.

Moo-Young, M. and Chisti, Y. (1988). Considerations for designing bioreactors for shear-sensitive culture. *Biotechnology,* 1291-1296.

Morris, P., Smart, N.J. and Fowler, M.W. (1983). A fluidised bed vessel for the culture of immobilised plant cells and its application for the continuous production of the fine cell suspensions. *Plant Cell Tissue Organ Culture,* 2 : 207-216.

Mukundan, U., Rai, A., Dawda, H., Ratnaparkhi, S. and Bhinde, V. (1998). Secondary metbaolites in *Agrobacterium rhizogenes* mediated transformed root cultures. In : Srivastava PS, editor. *Plant Tissue Culture and Molecular Biology Applications and Prospects*, New Delhi : Narosa Publishing House, pp. 302-331.

Muranaka, T., Ohakawa, H. and Yamada, Y. (1992). Scopolamine release into media by *Duboisia leichhardtii* hairy root clones. *Appl. Microbiol. Biotechnol.,* 37 : 554-559.

Nakagawa, K., Fukui, H. and Tabata, M. (1986). Hormonal Regulation of berberine Production Cell suspension Cultures of *Thalictrum minus*. *Plant Cell Rep.* 5: 69-71.

Nin, S., Bennici, A., Roselli, G., Mariotti, D., Schiff, S. and Magherini, R. (1997). Agrobacterium medited transformation of *Artemisia absinthum* L. (worm wood) and production of secondary metabolites. *Plant Cell Reports,* 16 : 725-730.

Noguchi M, Matsumoto, T. Hiarata Y. Yamamoto K. Katsuyana A. Kato A. Azechi, S. and Kato, K. (1977). Improvement of growth rates of plant cells cultures In: Plant tissue cultures and Biotechnolgical Applications [Barz W. Reinbard E and Zenk M.H. (eds)] ,pp. 85-94. Springer-Verlag New York.

Oldshue, J.Y. (1966). Fermentation Mixing Scale-up Techniques. *Biotechnol. Bioeng.* 8, 3-24.

Osorio, A.N., Garden, H., Cusido, R.M., Palazon, J., Alfermann, A.W. and Pinol, T.M. (2002). *Taxol* and baccatin III production in suspension cultures of *Taxus baccata* and *Taxus wallichiana* in an air lift bioreactor. *J. Plant Physiol.,* 159 : 97-102.

Pack,K.Y., and Hahn, E.J., (19990. Variations in African violet "Crimson Frost" micropropagated by homogenized leaf tissue culture. Hort. Technol 9: 625-628

Panda, A.K., Bisaria, V.S. and Mishra, S. (1992). Alkaloid production by plant cell culture of *Holarrhena antidysentrica* II. Effect of precursor feeding and cultivation in stirred tank bioreactor. *Biotechnol. Bioeng.,* 39 : 1052.

Pareilleux, A. (1987). The large scale cultivation of plant cells. In : Proceedings of the NATO Advanced Study Institute on Plant Cell Biotechnology held in Albuferia, Algarve, Portugal, March 29- April 10, 1987, Pais, M.S. and Mavituna, F. (eds.), Springer-Verlag, Berlin, Heidelberg, New York, p. 313-328.

Pareilleux, A. and Vinas, R. (1984). A Study on the Alkaloid Production by Resting Cell Suspensions of *Catharanthus roseus* in a Continuous Flow Reactor. *Appl. Microbiol. Biotechnol.* 19:316-320.

Park, J.M., Hu, W.S. and Staba, F.J. (1989). Cultivation of *Artemisia annua* L. Plantlets in a bioreactor containing a single carbon and a single nitrogen source. *Biotechnol. Bioeng.,* 34 : 1209-1213.

Parr, A.J. and Hamill, J.D. (1987). Relationship between *Agrobacterium rhizogenes* transformed hairy roots and intact, uninfected *Nicotiana* plants. *Phytochem.,* 26 : 3241-3245.

Parr, A.J., Payne, J., Eagles, J., Chapmau, B.T., Robins, R.J. and Rhodes, M.J.C. (1990). Variation in tropane alkaloid accumulation with in the solonaceae and strategies for its exploitation. *Phytochem.,* 29 : 2545-2550.

Payne, G.F., Bringi, V., Prince, C. and Shuler, M.L. (1992). Plant cell tissue culture in liquid systems, Hanser Publishers, Munich, Vienna, New York, Barcelona, p. 147-171.

Payne, G.F., Payne, N.N. and Shuler, M.L.(1988). Bioreactor considerations for secondary metabolite production from plant cell tissue culture : indole alkaloids from *Catharanthus roseus*. *Biotechnol. Bioeng.,* 31 : 905-912.

Pelosi, L.A., Rother, A. and Edwards, M. (1985). The effect of light on the production of *Heimia* alkaloids. *Phytochemistry,* 24 : 2215-2218.

Piehl, G.W., Berlin, J., Mollenschott, C. and Lehmann, J. (1988). Grown and alkaloid production of a cell suspension culture of *Thalictrum rugrsum* in shake flasks and membrane- stirrer reactors with bubble free aeration. *Appl. Microbiol. Biotechnol.* 29:456-461.

Prasad Babu, Ch., Mamatha, R., Kokate, C.K. and Veeresham, C. (2001). Elicitation of *Taxus wallichiana* (Himalayan yew) cell cultures for the production of Taxanes. *Indian Drugs,* 38(10) : 502-505.

Preil W, Florek P, Wix U and Beck A (1988). Towards mass propagation by use of bioreactors. Acta. Hort 226: 99-105

Preil E, Showalter SD and Roberts M (1991). The topisomerase I inhibitor, camptothecin, inhibits Equine infectious anemia virus replication in chronically infected CF2th cells. *J. Virol* 65(8): 4137-4141

Prenosil, J.E. and Pedersen, H. (1983). Immobilized Plant Cell Reactors. *Enzyme Microb. Technol.,* 5 : 323-331.

Prokop, A. and M. Z. Rosenberg. (1989). Bioreactor for Mammalian Cell Culture. *Adv. Biochem. Eng. Biotechnol.* 39: 29-71.

Rao, K.V., Venkanna, N. and Lakshmi Narasu, M. (1998). *Agrobacterium rhizogenes* mediated transformation of *Artemisia annua*. *J. Sci. Ind. Res.,* 57 : 773-776.

Razzaque, A. and Ellis, B. E. (1977). Rosmarinic Acid Production in *Coleus* Cell Cultures. *Planta.* 137, 287-291

Rhodes, M. J.C.., Hilton, M., Parr A. J., Hamill, J.D., Robins, R.J. (1986). Nicotine production by "hairy root" cultures of *Nicotiana rustica :* fermentation and product recovery. *Biotechnol, Lett* 8: 415-420.

Richards, J. W. (1968) Design and operation of aspetic fermenters. In: Introduction to Industrial Sterilization, pp. 107-122. Academic Press, London.

Rittershaus, E., Ulrich, J., Weiss,A., and Westphal,K., (1989). Large scale industrial fermentation of plant cells: experiences in cultivation of plant cells in a fermentation cascade up to a volume of 75,000 L. Bioenig 5: 28-34

Rittershaus, E., Ulrich, J. and Westphal, K. (1990). Large-scale production of plant-cell cultures. *Int. Assoc. Plant Tissue Cult. News,* 61 : 2.

Rodrigues- Mendiola, M.A,. Stafford,A.,Cresswell,R., and Arioas-Castro,C., (1991). Bioreactors for growth of plant roots. *Enzyme. Microb. Technol.* 13 : 697-702.

Roja, P.C., Benjamin, B.D., Heble, M.R. and Chadha, M.S. (1985). Indole alkaloids from multiple shoot cultures of *Rauwolfia serpentina. Planta Med.,* 51 : 73-74.

Roja, P.C., Sipahimalani, A.T., Heble, M.R. and Chadha, M.S. (1987). Multiple shoot cultures of *Rauwolfia serpentina. J. Nat. Prod.,* 50 : 872-875.

Rose, D. and Martin, S.M. (1975). Growth of suspension cultures of plant cells (Ipomoeas) at various temperatures. *Can. J. Bot.,* 53 : 315.

Saito, K., Murakoshi, I., Inze, D. and Van Montagu, M. (1989a). Biotransformation of Nicotine alkaloids by tobacco shooty teratomas induced by a Ti plasmid mutant. *Plant Cell Rep.,* 7 : 607-610.

Saito, K., Ohashi, H.,Hibi, M. and Tahara, M. PCT Int. Appl. WO92/13961, 20 Aug.1992.

Saito, K., Yamazaki, M., Yamakawa, K., Fujisawa, S., Takamatsu, S., Kawaguchi, A. and Murakoshi, I. (1989b). Lupin alkaloids in tissue culture of *Sophora flavescens* var. *angustifolia.* Greening induced production of matrine. *Chem. Pharm. Bull.,* 37 : 3001-3004.

Sato, F. and Yamada, Y. (1984). High Berberine-Producing Cultures of *Coptis japonica* Cells. *Phytochemistry.* 23:281-285.

Schiel, O. and Berlin, J. (1987). Large Scale Fermentation and Alkaloid Production of Cell Suspension Cultures of *Catharanthus roseus. Plant Cell Tissue Organ Culture* 8: 153-161.

Scragg A. H. and Fowler, M.W. (1986). The mass culture of palnt cells. In : Cell Culture and Somatic Cell Genetics of Plants. Vol. 2. [Vasil I.K. (ed)] pp. 103-128 Academic Press, New York.

Scragg, A.H. (1992). Bioreactors for the mass cultivation of plant cells. In : Fowler, M.W. and Warren, G.S. (eds.). Plant Biotechnology Suppl. Comprehensive Biotechnology, Pergamon Press, Oxford, p. 45-62.

Scragg, A.H. (1994). Secondary products from cultured cells and organs II. Large scale culture. In : Plant Cell Culture – A Practical Approach : 2nd edition, Dixon, R.A., Gonzales, R.A. (eds.) IRL Press, Oxford University Press, Oxford, New York, Tokyo, p. 200-222.

Seki, M. and Furusaki, S. (1996).An immobilized cell system for *Taxol* production, *Chemtech,* 41-44.

Seki, M., Ohzora, C., Takeda, M. and Furusaki, S. (1997). *Taxol* (paclitaxel) production using free and immobilized cells of *Taxus cuspidata. Biotechnology and Bioengineering,* 53 : 214-219.

Sevon, N., Drager, B., Hiltunen, R. and Oksman-Caldentey, K.M. (1997). Characterization of transgenic plants derived from hairy roots of *Hyoscyamus muticus. Plant Cell Reports,* 16 : 605-611.

Shimomura, K., Sataka, M. and Kamada, H. (1986). Production of useful secondary metabolites by hairy roots transformed with Ri plasmid In : International Congress of Pl. Tiss. and Cell Cult., Univ. of Minnesota, Minneapolis, pp. 155.

Shimomura, K., Sudo, H., Saga, H. and Kamada, H. (1991). Shikonin production and secretion by hairy root cultures of *Lithospermum erythrorhizon. Plant Cell Reports,* 10 : 282-285.

Shuler, M.L. (1993). Strategies for improving productivity in plant cell, tissue, and organ culture in bioreactors. In : Bioproducts and Bioprocesses 2, Yoshida, Tanner (eds.) Springer-Verlag, Berlin, Heidelberg, p. 235-245.

Shuler, M.L., Shahai, O.P. and Hallsby, G.A. (1983). Entrapped Plant Cell Cultures. *Ann. N.Y. Acad. Sci.,* 413 : 373-382.

Sim, S.J. and Chang, H.N. (1993). Increased shikonin production by hairy roots of *Lithospermum erythrorhizon* in two phase column reaction. *Biotechnol. Lett.,* 15 : 145.

Smart, N.J. and Fowler, M.W. (1984a). An air lift column bioreactor suitable for large-scale cultivation plant cell suspenstons *J .Exp Bot* 35 : 531-535.

Smart, N. J. and Fowler, M. W. (1984b). Mass Cultivation of *Catharanthus roseus* Cells Using a Nonmechanically Agitated Bioreactor. *Appl. Biochem. Biotechnol.* 9, 209-216.

Smart, D.A., Strickland, S.G. and Water, K.A. (1987). Bioreactor production of alfalfa somatic embryos. *Hort. Sci.,* 22 : 800-803

Smith, R.,Adams, T.L., Barton, C.R., Stahlut, R.W.(1993) PCT Int. Appl. WO93/10253, 27 may 1993.

Solomons, G. L., (1969) *Materials and Methods in Fermentation.* Academic Press, London.

Son,S.H., Choi, S.M., Choi, K.B., Lee, Y.H., Lee, D.S.,, Choi, M.S., and Park Y.G., (1999a). Selection and proliferation of rapid growing cell lines from embryo derived cell cultures of yew tree (Taxus cuspidata Sieb et zucc). Biotechnol. Biopro. Engg. 4: 112-118

Son, S.H., Choi,S.M., Know, S.R., Lee,Y.H., Pack, K.Y (1999b). Large-scale culture of plant cell and tissue by bioreactor system. *J. Plant. Biotechnol* 1: 1-8

Son, S.H., Choi, S.M. Lee, Y.H., Chol, K.B., Yun, S.R., Kim, J.K., Park, H.J., Kwon, O.W., Noh, E.W., Seon, J.H., Park, Y.G. (2000). Large-scale growth and taxane production in cell cultures of *Taxus cuspidata* (Japanese yew) using a novel bioreactor. *Plant Cell Rep.,* 19(6) : 628-633.

Spencer, A., Hamill, J.D. and Rhodes, M.J.C., (1990). Production of terpenes by differentiated shoot cultures of *Mentha citrata* transformed with *Agrobacterium tumefaciens* T37. *Plant Cell Rep.,* 8 : 601-604.

Spetler, H., Alfermann, A.W. and Reinhard, E. (1985). Biotransformation of â-methyl-digitoxin by cell cultures of *Digitalis lanata* in air lift and stirred tank reactors. *Appl. Microbial Biotechnol.,* 23 : 1-4.

Srinivasan, V., Ciddi, V., Bringi, V., Shuler, M.L. (1996). Metabolic inhibitors, elicitors and precursors as tools for probing yield limitation in taxane production by *Taxus chinensis* cell cultures. *Biotechnol. Prog.,* 12 : 457-465.

Srinivasan, V., Pestchanker, L., Moser, S., Hirasuna, T., Tatlcek, R.A., Shuler, M.L. (1995). Taxol production in bioreactors : kinetics of biomass accumulation, nutrient uptake, and taxol production by cell suspensions of *Taxus baccata. Biotechnol. Bioeng.,* 47 : 666-676.

Staba, E.J., Nygaard, B.G. and Zito, S.W. (1984). Light effects on *Pyrethrum* shoot cultures. *Plant Cell Tissue Organ Cult.,* 3 : 211-214.

Stuart, D.A., Stickland,S.G., and Walker,K.A., (1987) *Hortic.Sci.* 22: 461-467.

Styer, D. J. (1985). Bioreactor Technology for Plant Propagation. In "Tissue Culture in Forestry and Agriculture," R. R. Henke, K. W. Hughes, M. P. Constaritin, and A. Hollaender (eds.), Plenum Press, New York, pp 117-130.

Styler, D.J. (1985). Bioreactor technology for plant propagation. *Basic Life Sci.,* 32 : 117-130.

Tabata, M. and Fujita, Y. (1985). Production of Shikonin by Plant Cell Cultures. In "Biotechnology in Plant Science: Relevance to Agriculture in the Eighties," M. Zaitlin, P. Daya, and A. Hollaender (eds), Academic Press, New York, pp 207-218.

Tabata, M., Mizukami, H., Hiraoka, N. and Konoshima, M. (1974). Pigment formation in callus cultures of *Lithospermum erythrorhizon. Phytochemistry,* 13 : 927-932.

Tada, H., Nakashima, T., Kuntake, H., Mori, K., Tanaka, M. and Ishimaru, K. (1996). Polyacetylenes in hairy root cultures of *Campanula medium* L. *J. Plant Physiol.,* 147 : 617-9.

Tahara, M., Sakamoto, T., Takami, M., Takigawa, K.(1995) PCT Int. Appl. WO95/04154, 9 Feb. 1995.

Takayama, S. and Misawa, M. (1981). Mass propagation of *Begonia X biemalis* plantlets by shake culture. *Plant Cell Physiol.,* 22 : 461-467.

Takayama, S (1992) In: Biotechnology in Agariculture and Forestry. Bajaj, Y.P.S. (eds). 17: 17-34. Springer-Verlag, Berlin.

Tal, B. and Goldberg, I. (1982). Growth and diosgenin production by *Dioscrorea deltoidea* cells in Batch and continuous culture. *Planta Medica,* 44 : 107-110.

Tal. B., Rokem, J.S, and Goldberg, I. (1983). Factors affecting growth and product formation in plant cells grown in cintinuous culture. Plant cell Reports, 2: 219- 222

Tanaka, H. (1981). Technological Problems in Cultivation of Plant Cells at High Density. *Biotechnol. Bioeng.* A Review. *Process Biochem.,* August 106-113. 23, 1203-1218.

Tanaka, H. (1987). Large-scale cultivation of plant cells at high density : A review, *Process Biochemistry*, 106-113.

Taya M, M, Kino-Oka, S Tone and T Kobayshi (1989a). A kinetic model of branching growth of palnt hairy root. *I Chem Eng japan* 22 (6), 698-700

Taya M,A Yoyama, Kondo,O., Kobayashi,T., and Masturi,C., (1989b). Growth chracteristics of plant hairy roots and their cultures in bioreactors. *J chem Eng Japan* 22 (1), 84-89

Taya, M., Mine, K., Kinoka, M., Tone, S. and Ichi, T. (1992). Production and release of pigments by cultures of transformed hairy roots of red beet. *J. Ferment. Bioeng.,* 73 : 31-36.

Taya, M., Yoyama, A., Nomura, R., Kondo, O., Matsui, C. and Kobayashi, T. (1989). Production of peroxidase with horse radish hairy root cells in a two step culture s ystem. *J. Ferment. Bioeng.,* 67 : 31-34.

Teoh, K.H., Weathers, P.J., Cheetham, R.D. and Walcerz, D.B. (1996). Cryopreservation of transformed (hairy) roots of *Artemisia annua. Cryobiology,* 33 : 106-117.

Toivonen, L., (1993) Utilization of hairy root cultures for the production of secondary metabolites. Biotechnology Progress. 9:12-20.

Toivonen, L and Rosenqvist, H., (1995). Establishment and growth chracteristics of *Glycyrrhiza glabra* haity root cultures. *Plant Cell Tiss Organ Cult* 41: 249-258.

Townsley, P. M. and Webster, F. (1983). The recycling air lift transfer fermenter for plant cells. *Biotechnol. Lett.* 5 : 13-18.

Tulecke, W. and Nickell, L.G. (1959). Production of large amounts of plant tissue by submerged culture. *Science,* 130 : 863.

Tulecke, W. and Nickell, L.G. (1960). methods, problems and results of growing plant cells under submerged conditions. *Trans. N.Y. Acad. Sci.,* 22 : 196.

Ulbrich, B., W. Wiesner, and H. Arens (1985). Large Scale Production of Rosmarinic Acid from Plant Cell Cultures of *Coleus blumei* Benth. In "Primary and Secondary Metabolism of Plant Cultures," K. H. Neumann, W. Barz, and E. Reinhard (eds.), Springer-Verlag, New York, pp. 293-303.

Veeresham, C., Srinivasan, V. and Shuler, M.L. (1995). Elicitation of *Taxus* sp. cell cultures for production of *Taxol. Biotechnology Letters,* 17(12) : 1343-1346.

Veliky, I. and Martin, S.M. (1970). A fermentor for plant cell suspension cultures. *Can. J. Microbiol.,* 16 : 223.

Verma, D.P.S. and van Huystee, R.B. (1971). Derivation, characteristics and large-scale cultivation of a cell line from *Arachis hypogaea* L. cotyledons. *Exp. Cell Res.,* 69 : 402.

Wagner, F. and Vogelmann, H. (1977). Cultivation of plant tissue cultures in Bioreactors and formation of secondary metabolites. In : Plant Tissue Culture and its Biotechnological applications. W. Barz, E. Reinhard and M.H. Zenk (eds.), Springer-Verlag, New York, pp. 245-252.

Walker, J. A. H. and Holdsworth, H. (1958) Equipment design.In: *Biochemical Engineering.* Street, R. (eds) Heywood, London. pp.223-273

Walker, K.A., Wendeln, M.L. and Jaworski, E.G. (1979). Organogenesis in Callus Tissue of *medicago sativa.* The Temporal separation of Induction Processes from Differentiation Processes. *Plant Sci. Lett.,* 16: 23-30.

Wallner, S.J. and Nevin, D.J. (1973). Formation and dissociation of cell aggregates in suspension cultures of Paul's Scarlet rose. *Am. J. Bot.,* 60(3) : 255.

Wang, C.J. and Staba, E.J. (1963). Peppermint and spearmint tissue culture. II. Dual-carboy culture of spearmint t issues. *J. Pharm. Sci.,* 52 : 1058.

Weathers, P.J. and Giles, K. (1988). Regeneration of plants using nutrient mist culture. *In vitro*, 24 : 727-732.

Weathers, P.J., Cheetham, R.D., Follanskie, E. and Teoh, K. (1994). Artemisinin production by transformed roots of *Artemisia annua*. *Biotechnol. Lett.,* 16 : 2181-1286.

Whitney, P.J. (1992). Novel bioreactors for the growth of roots transformed by *Agrobacterium rhizogenes. Enzymes Microbial. Technol.,* 14 **:** 13-17.

Willemot, R.M. and Durand, G. (1977). Les reacteurs biologiques. La Recherche, 18(188) **:** 164.

Wilson, S.B., King, P.J. and Street, H.E. (1971). Studies of the growth in culture of plant cells XII. A versatile system for the large scale batch or continuous culture of plant cell suspension. *J. Exp. Bot.,* 22 : 177.

Wink, M. (1987). Physiology of the accumulation of secondary metbaolites with special reference to alkaloids. In : Cell Culture and Somatic Cell Genetics of Plants, vol. 4, Constabel, F. and Vasil, I.K. (eds.), Academic Press, New York, pp. 17-42.

Wink, M. and Hartmann, T. (1980). Production of quinolizidine alkaloids by photomixotrophic cell suspension culturres. *Biochemical and Biogenetic aspects. Planta Med.,* 40 : 149-155.

Withers, L.A. (1985). Cryopreservation of cultured cells and meristems. In : Cell culture and somatic cell genetics of plants, Vol. 2, I.K. Vasil (ed.), Academic Press, New York, pp. 253-316.

Yamamoto, H., K. Kakagawa, H. Fukui, and M. Tabata. (1986). Cytological Changes Associated with Alkaloid Production in Cultured Cells of *Coptic japonica* and *Thalictrum minus. Plane Cell Rep.* 5: 65-68.

Yamada, Y. and F. Sato (1981). Production of Berberine in Cultured Cells of *Coptis japonica. Phytochemistry* 20 : 545-547.

Yamada, Y. and H. Morikawa (1985). Proptoplast Fusion of Secondary Metabolite-Producing Cells. In "Primary and Secondary Metbolism of Plant Cell Cultures," K.-H. Neumann, W. Barz, and E. Reinhard (eds.), Springer-Verlag, New York, pp. 255-217.

Yamakawa, T., F. Sato, K. Ishida , T. Kodama, and Y. Minoda (1983). Production of Anthocyanins by *Vitis* Cells in Suspension Culture. *Agri. Biol. Chem.* 47, 2185-2191.

Yazaki, K., H. Fukui,M. Kikuma, and M. Tabata (1987). Regulation of Shikonin Production by Glutamine in *Lithospermum erythrorbizon* Cell Cultures. *Plant Cell Rep.* 6, 131-134.

Yoshikawa, N., H. Fukui, and M. Tabata (1986). Effect of Gibberellin A3 on Shikonin Production in *Lithospermum* Callus Cultures. *Phytochemistry* 25: 612-622.

Yoshikawa, T. and Furuya, T. (1987). Saponin production by cultures of *Panax ginseng* transformed with *Agrobacterium rhizogenes. Plant Cell reports,* 6 : 449-453.

Yoshioka, T., Yamagata, H., Ithoh, A., Deno, H., Fujita, Y. and Yamada, Y. (1989). Effects of Exogenous polyamines on tropane alkaloid production by root culture of *Duboista myoporoides. Plant Med.,* 55 : 523-524.

Yu, S. and Doran, P.M. (1994). Oxygen requirements and mass transfer in hairy root culture. *Biotechnol. Bioeng.,* 44 **:** 880-887.

Yukimune, T., Hara, Y., Higashi, Y.; Ohnishi, N., Tabata, H., Suga, C. and Matsubara, K. PCT Int. Appl. WO95/14103, 26 May 1995.

Yukimune, Y., Hara, Y., Higashi, Y., Ohnishi, N., Tabata, H ., Suga, C. and Matsubara, K. (2002). Method of producing a taxane type diterpene and a method of obtaining cultured cells which produce the taxane type diterpene at a high rate. United States Patent No. 6,403,343.

Zenk, M.H., M. Ruffer, M. Amann and B. Deus-Neumann (1985). Benzylisoquinone Biosynthesis by Cultivated Plant Cells and Isolated Enzymes. *J. Nat Prod.* 48:725-738.

Ziu, M and Hoddar, A (1991) J.Bot (Isarel) 40: 7.

Chapter 21

In-situ Adsorption, Two-Phase and Two-Stage Cultures

INTRODUCTION

Secondary metabolites derived from plant cells cultured in liquid suspensions have been of commercial interest for the forty years. There are about 15 products produced commercially through the use of plant cell culture method (See Table 20.1). The major hurdle facing for the production of high value phytopharmaceuticals by plant tissue culture is the lack of economic feasibility due to low yields and productivities from suspension cultures. The mechanisms of control are far from being understood (Zenk 1982) and the reasons for these failures have been extensively discussed (Teuscher 1973, Berlin 1983).

If low but measurable amounts of secondary substances are produced, it is expected that genes coding for the enzymes of the secondary substance pathway are expressed, but the products is kept very low by one of the following events:

a) Feed-back inhibition of the enzymes and/or the membrane transport by secondary metabolites (Luckner 1980).

b) Degradation of secondary substances by enzymic or nonenzymic processes in the medium and/or cells.

c) Volatile secondary products evaporate through the gas phase.

The secondary metabolites from plant cells are secreted into surrounding medium by natural secretion mechanisms (passive secretion and active secretion). In order to excrete the desired compounds, which are usually stored in the vacuoles, the permeability of two membranes must be changed: that of the tonoplast and that of the plasmalemma. The processes used for this purpose can be divided into those using biotic components and those using exclusively abiotic tools (Table 21.1). An essential condition for their biosynthetic capacity remains "unlimited" in time. However, the elimination of compartmentalization also releases potentially harmful compounds such as phenols and catabolic enzymes.

21.1. OSMATIC SHOCK. pH, IONIC STRENGTH

The osmotic shock method successfully used in *Beta vulgaris* to release betalain glycoside (betanine) was proven to be inadequate because of the resulting microscopically visible damage. In *Catharanthus roseus* and *Acer pseudoplatamus,* cycle change in the pH of the medium was highly successful. Media with high ionic strength used in cell cultures of *Catharanthus roseus* did not damage or influence cell respiration or growth capacity.

Table 21.1. Components inducing variations in cell permeability

Abiotics	Biotics
Electroporation	Variation in the
Immobilization	- Medium composition
Application of	- ionic strength
- specific chemicals	- pH
Application of elicitors	
- inert secondary phases	Osmotic shocks

21.2. ELICITORS

Co- cultures of plant cells with suitable elicitors proved to be especially useful for enhanced production of secondary metabolites (see chapter 12).

21.3. ELECTROPORATION

In an electric field, cellular membranes behave like condensors. If the threshold specific to the cell type is exceeded during a strong (k V/cm) but short (ns to is) or a weak (V/cm) but long (ms) electric pulse, the membrane breaks down locally. Essentially, it is locally perforated. These pores allow compounds such as indole alkaloids which are usually stored as cations inside the vacuole due to thier basic properties, to pass through the membrane (Joersbo and Brunstedt 1991). Throughout the persistence of an electric filed, the secondary compounds, which are usually positively charged, travel towards the cathode by ionophoresis. Uncharged susbstances are released by diffusion.

21.4. ADDITION OF CHEMICALS

The addition of certain chemical compounds (Table 21.2) can induce a short-term change in membranes properties which allow secondary products to be released.

Membranes affected by such permeabilization differ in their sensitivity to the applied chemicals. Tonoplasts are usually relatively insensitive. Therefore, dimethyl-sulphoxide (DMSO) only becomes effective at the tonoplast level at concentrations that are toxic for plasmalemmas. By means of a 5-10% (v/v) DMSO solution the permeability of plasmalemmas can be selctively increased (Meravy *et al.*1988)

Table 21.2. Some chemicals for permeabilization of cells acting via cell membranes

Ca^{+2} Chelating agents
    EDTA and Nystain
Detergents
    Cetyl-trimethyl-ammonium bromide, Lyso-lecithin, Triton-X-100
Organic solvents
    Chloroform, Dimethylsulphoxide, diethylEther, Methanol
    n-Propanol, Polyethyleneglycol (PEG), Toluene
Polycations
    Polylsine, Chitosan

Table 21.3. The effect of various treatments on betanin release from cells of *Beta Vulgaris* (Parr et al. 1986)

Treatment		Percentage of betanin released (cells in phase of rapid growth)
None		
1 % (v/v)Toluene	1h	1
10 % DMSO	1h	86
10 % Ethanol	1h	26
0.5 % Phenethyl alcohol	1h	60
5 % Tween 20	20 h	40
0.1 % Triton X-100	1h	25
0.1 % Lysolecithin	6 h	50
100 µg/ml Nystatin	7 h	5
100 µg/ml Polylysine	24 h	3
0.8 M Mannitol	24 h	4
0.8M Mannitol	2h	10
followed by return to isotonic medium		

However, permeabilization using DMSO is usually deadly for most of the cells in a culture. The originally postulated reversibility of the process could not be verified for all cultures. The described restoration process is therefore probably due to cells that were only partially or not at all permeabillized. Thus *Catharanthus roseus* cell cultures always resume growth and ajmalicine production during the DMSO free interphases of permeabilization. On the other hand cells of *C. ledgeriana* suspension cultures which are difficult to culture even under normal conditions are completely destroyed by DMSO (Felix 1982, Meravy *et al* 1988).

21.5. IMMOBILIZATION (See for detail Chapter 11)

On considering only the secondary metabolites excreted into the medium, (the feed-back inhibition of the enzymes and/or the membrane transport by secondary substance by enzymic or non enzymic process in the medium and/or cells), the continuous removal of secondary substances from the medium should allow the permanent production and /or conservation of secondary substances. This may be done by the addition of a second phase to the culture medium (Two phase culture) (Beiderbeck 1982).

The idea of creating an accumulation and conservation site is derived from the situation in the intact plant, where secondary substances are often harmful or even toxic to the protoplast. Although they are synthesized by the protoplast itself, they are separated from it by different types of compartmentalization (Luckner 1980). They are secreted either through the tonoplast into the vacuole often as glycosides (mustard oils, cyanogenic glycosides, tannins), accumulated in liposomes of specialized oil cells and idioblasts, or secreted through the plasmalemma into the cell walls (lignin), extracellular spaces (schizogenic oil ducts, resin ducts or subcuticular spaces, Denffer *et al.* 1978). A similar compartmentalization could be observed in few known cases of tissue cultures capable of producing essential oils : oil production or its accumulation is restricted to idioblasts with in the cultured tissues (Reinhard *et al.* 1968; Nagel and Reinhard 1975; Bisson *et al.* 1983) .

An artificial accumulation site could therefore mimic similar relationships in cultures not able to differentiate such specialized cells, An ideal secondary substance accumulating material should have the following characteristics.

a) be stable under sterilizing conditions, preferably autoclaving

b) be nontoxic to the cultivated cell and unable to extract nutrients and hormones from the medium,

c) have the ability to dissolve or bind the secondary subtances of interest and stabilize them

d) be well disperse with the culture

e) allow elution and recovery of the secondary substances bound and accumulated during the preceding culture.

21.6.1. Two Phase Culture

21.6.1.1. Liquid Lipophilic Materials

In fully differentiated plants, lipophilic compounds in particular are secreted into structures with corresponding chemical properties in suspension cultures can be mimicked by adding solid or liquid lipophilic substances. Miglyol 812, a water insoluble triglyceride of low viscosity composed of fatty acids of C_8-C_{10} chain length, was addded (10-12%) to flasks containing a modified MS medium and different cell lines of *Matricaria* (Beiderbeck 1982, Bisson *et al* 1983) or cultures of *Nicotiana tabacum* and *Thuja occidentalis* (Berlin *et al* 1984). Miglyol 812 meets almost all the conditions characterstics of an ideal second phase substance, except the recovery of secondary metabolites from the miglyol is some what diffcult. The cell cultures of *Matricaria* together with a miglyol phase, the oil alcumulated uv-absorbing products. Upon GC/MS. analysis of miglyol phase was identified as the sesquiterpene alcohol α bisabolole, one of the flower oil of *Matricaria* which could not be found in these cultures before (Bisson *et al* 1983). A cell suspension culture of *Thuja orientalis* produced upto 3 mg monoterpenoids (α-Pinene, β-Pinene, myrcene, limonene and terpinolene) per g. dry weight/day in a miglyol two phase system (Berlin *et al*. 1984).

Instead of the triglyceride, miglyol, liquid parraffin have also been used to isolate uv-absorbing secondary metabolites from cell suspension cultures of *Matricaria chamomilla* (Beiderbeck 1982). Later, silicone fluid was found to successfully remove and enhance production of benzophenanthridine alkaloids in cell suspension cultures of *Eschscholtzia californica* (Byun *et al* 1990, Byun and Pedersen 1994). The use of silicone oil as second phase in the hairy root cultures of *Catharanthus roseus* ,ingnificantly enhanced the production of tabersonine and lochnericine (Tikhomiroff *et al* 2002). Tricaprylin, a glycerol derivative is second phase for extracting sanguinarine from *E. californica* (Dutta *et al* 1994). Tricaprylin is autoclavable and non - toxic. It removes the compound of interest from the medium with a minimal amount of other medium components required for normal growth and, since its specific gravity is 0.95, it exists as a distinct, immiscible second phase, facilitating its removal and sampling from the aqueous liquid. This is especially important in a continuous bioreactor system..

Tricaprylin at 20% v/v as second phase has no effect on growth of cells and enchanced the production of sanguinarine from cell cultures of *E. californica*. The addition of tricaprylin as second phase to the cell cultures of *Taxus brevifolia* enhanced the extracellular accumulation of taxol, while no effect on growth (Collins-pavao *et al* 1996).

21.6.1.2. Solid lipophilic susbstances

A cell culture with this type of second phase consists of a liquid nutrient phase with cells in which the lipophilic phases are suspended as a powder or beads.

Lichroprep . RP-8. RP-8 is a silica gel (particle size 40 -60 μm) where the outer SiOH groups are covalently bound to C8 hydrocarbons. These hydrocarbons coat the silica gel particles as a monomolecular lipophilic surface layer. The RP-8 material fulfills almost all conditions of an ideal second phase. The only difficult is to separate the fine powder from the cultured cells.

Using a cell culture of *Pimpinella anisum,* small amounts of the phenylpropanoid anethole could be isolated from RP-8 material, whereas it could not be found in the conventional single phase culture (Bisson *et al* 1983).

Similarly, in RP-8 phases of cell cultures of *Valeriana wallichii,* valepotriates could be detected. One cell line which did not release any valepotriate during conventional culture yielded a low amount of these substances in the presence of RP-8. Another cell line normally containing 0.5% valepotriates produced increased amount in the presence of RP-8. These valepotriates were partly adsorbed to the RP-8 material (Becker and Herold 1983).

21.6.1.3. Solid polar Substances

A cell suspension culture with this type of second phase consists of liquid nutrient phase in which the cells and a solid phase of an adsorbent are suspended.

21.6.1.3.1. Activated Charcoal (AC)

Activated charcoal is an adsorbent which is very widely used in science and industry, although the binding principles are not fully understood (Mattson and Mark 1971). AC of 2-3 mm diameter beads are most suitable for cell culture purposes, since they can be easily removed from cells.

Knoop and Beiderbeck (1983) reported that the crown gall cell cultures of *Matricaria chamomilla* which releases coniferyl aldehyde into the medium due to the addition of a 0.8% -4.1% w/v of AC. 20-60 fold amount of coniferyl aldehyde can be recovered from cell cultures of *M. chamomilla* as compared to the free medium. The increased secondary metabolite production is partly a consequence to the AC and partly a consequence of growth reduction . The growth reduction by AC is may be due to the withdrawl of essential nutrients from the culture medium (Constantin *et al* 1977).

21.6.1.3.2. Other Adsorbents

As a disadvantage, AC has a pronounced catalytic surface activity which may lead to chemical reactions of secondary substances released by the cells. Therefore a series of other adsorbents was tested with regard to a possible use in cell cultures. Adsorbents available only as fine powders have been enclosed in droplets of alginate to improve removal from the cells after culture; adsorbents available as beads or granules were used directly (Mbanaso and Roscoe 1982).

XAD-4 is an adsorber resin binding a great variety of diverse susbtances. It is available as beads of 0.3-1 mm diameter, can be separated from cell cultures by repeated decanting, and is extractable with acetone. The addition of different XAD-4 (0.8%-12.5% w/v) to a cell suspension of *Nicotiana tabacum* (Beiderbeck and Knopp 1988), drastically enhanced the production of secondary metabolite compared to the adsorbent free control. The addition of XAD-7 resin to the cultures of *Catharanthus roseus* stimulated alkaloid production without affecting glucose consumption. This may be due to feed back inhibition of alkaloid production (Payne *et al* 1988). The addition of wofatit ES and Amberlite XAD-2 to the cell cultures of *Galium vernum,* stimulated the production of anthraquinones (Strobel *et al.* 1991). Similarly the addition of XAD-7 (5% w/v) to the cell cultures of *Papaver somniferum* to actively growing cultures led to increase in sanguinarine production and release of 30 % to 40 % and 60 % respectively (Williams *et al* 1992). The addition of XAD-7 greatly enhanced the release of catharanthine and ajmalicine from hairy root cultures of *C. roseus,* with an increase in total production (Sim 1994).

The cultivation of plant cells in aqueous two-phase polymer systems will provide a method of cell immobilization which avoid many of the problems associated with other immobilization techniques.

Using this method, culture growth is unhindered since cells remain in a free suspension than being bound within a matrix or between barrier membranes. In addition, when mixed these systems achieve a high degree of phase contacting which diminishes mass transfer limitations.

Aqueous two-phase systems are formulated by mixing two different water soluble polymers, such as polyethylene glycol and dextran, in water (Albertsson *et al*.1987) and have been used in a variety of applications including protein extraction (Shanbhag and Johansson1974), alcohol fermentation (Kuhn 1980), bioconversion (Andersson *et al.* 1984) and Microbial cultivation (Persson *et al.* 1984). These systems provide an outstanding medium for biological processing because of the high water content of each phase (Albertsson 1971). More importantly these systems have the ability to partition components such as nutrient, metabolites, proteins, cell particles and whole cells unevenly between the two polymer phases.

Suspension cultures of *Nicotiana tabacum* have been successfully grown in aqueous two-phase systems comprised of polyethylene glycol (PEG) and dextran in a modified LS Medium. Aqueous two-phase systems may be advantageous for plant tissue cultivation since cells can be immobilized in one phase while secondary products are collected and withdrawn in the other phase, thus enchancing productivity. Culture growth rate was compared in a variety of two-phase systems, covering a range of both polymer molecular weight and concentration. Systems exhibiting relatively higher phase miscibility yielded increased growth rates as compared to less miscible phase formulations. The highest observed growth rate occurred in 3 % PEG 20000/ 5 % crude dextran and approached growth rates and cell densities of cultures grown in standard LS medium (Hooker and Lee 1990). Robins and Rhodes (1986) observed that in situ removal by resin (adsorbent) stimulated anthraquinone production by *Cinchona ledgeriana* cell cultures.

In-situ extraction technique can be coupled with other strategies of yield improvements of secondary metabolite from cell cultures.

21.6.7. In-situ extraction with bioreactor for large -scale cultivation

21.6.7.1. Shikonin production

Large-scale cultivation of plant tissue and organ is hampered by the fragile nature of the cells, low growth rate and complex cellular metabolic pathways. The successful exploitation of plant cell culture technology requires an understanding of the cellular information (biochemical, physical and genetic factors affecting metabolite production) and reactor design variables (Concerning bioreactor performances including oxygen transfer, mixing and rheological chracteristics of cultured cells). Most common bioreactor types like stirred tank and the bubble column can be used as two phase bioreactors.

A packed bed reactor was designed to continuously produce secondary metabolites from immobilized plant cell cultures (Kim and Chang 1990) to provide the oxygen necessary for the maintenance through the packed cell beads.Continuous extraction of produced shikonin was performed at the medium reservoir using 50 mL *n*-hexadecane. Shikonin production continuously increased during 13 days of the experiments. Specific production was 2 mg/g cell at day 13 (Fig. 21.1).

Fig. 21.2 Show a bubble column bioreactor for hairy root culture with in situ separation. A stainless steel mesh was isolated inside the reactor to immobilize hairy roots. During the bioreactor operation cell growth was indirectly estimated by conductivity measurement (Taya *et al* 1989). The hairy roots grew much better and the reactor was almost packed with hairy roots. Shikonin was produced at a constant level of 10.6 mg/l day during the culture period of 54 days.

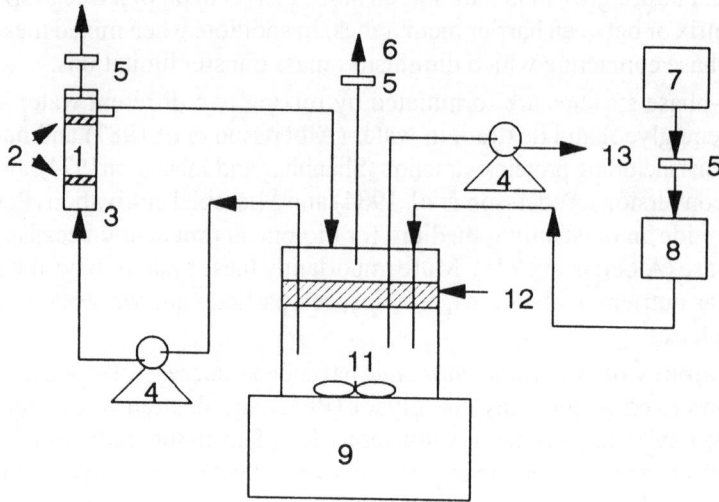

Fig. 21.1. Schematic diagram of experiment setup for packed bed bioreactor operation (1. Alginate beads, 2. Glass bead, 3. Packed-bed reactor, 4. Sampling, adopted from Kim and Chang 1990).

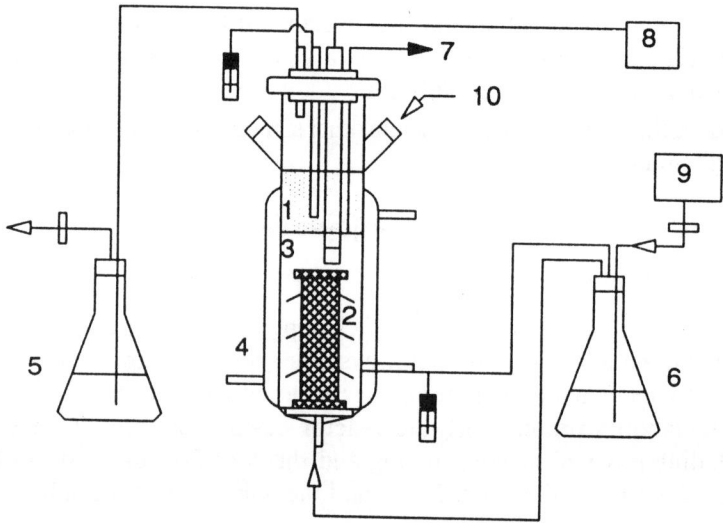

Fig. 21.2. Schematic diagram of experimental set-up for two phase bubble column bioreactor (1. solvent, 2. stainless steel structure, 3. medium, 4. water jacket, 5. air oulet adopted from Chang and Sim 1994).

21.6.7.2. Indole alkaloids

A new reactor was used to culture *Catharanthus roseus* cells, in which the draft tube was made up of polyurethane foam and acted as the immobilizing matrix. The reactor was connected in series to an adsorbent column with a neutral polymeric resin (NKA-9 resin, Nankai University, Japan, 20-50 mesh size) which absorbs these alkaloids. The synthesis of alkaloid was stimulated by adding the resin column and the total content of alkaloid secreted by cells reached 380 mg/L, which was 4.5 times of that in the control experiment. Meanwhile, most of the intracellular alkaloid produced by *Catharanthus roseus* was secreted into the medium. (Fig 21.3) (Yuan *et al* 1999).

21.6.8. Two Stage Cultures

For the production of secondary metabolites by plant cell cultures a two stage batch operation has been widely advocated, because cell growth and product formation are nega-tively correlated in most cases. In the first stage, optimal conditions for fast growth of biomass are created and in the second stage product formation is induced often by the application of a high sugar concentration. There are numerous reports on the optimization of both the growth and production stage (Berlin *et al* 1987, Knobloch and Berlin 1980, Morris 1986, Payne *et al* 1991, Toivonen *et al.*, 1991, Zenk *et al* 1977). Surprisingly, there is hardly any infor-

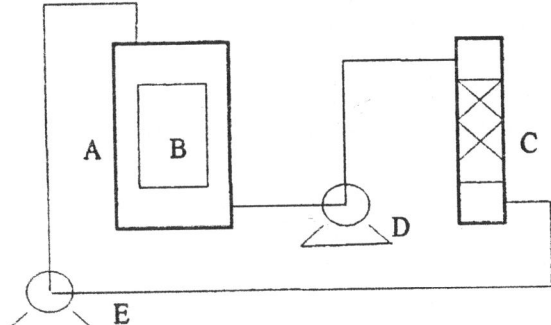

Fig. 21.3. Experimental scheme for two phase culture of C. roseus A. Air left bioreactor (2L), B. Draft tube, C. Resin column (Adopted from Yuan et al., 1999).

mation about the relationship between both stages, and what is the best moment to start the production stage.

From an economical point of view, the growth stage should be as short as possible, producing the highest biomass possible. Therefore, the most suitable moment to start the production stage seems to be end of the growth. However, the biomass compositon intracellular concentrations of biologically important substances such as carbon, phosphate (VanGulik *et al.*,1993), and nitrogen as well as enzyme activities of the ajmalicine biosynthetic pathway (Moreno *et al.*, 1993) changes continuously during the stationary phase. In the production stage, the initial biomass composition may have an effect on secondary metabolism and the capacity of the cell to withstand the high sugar concentration in the induction medium. Hence there is possibly an optimum time to induce secondary metabolism.

Schlatmann *et al* (1995) described a two stage batch process for production of ajmalicine by *Catharanthus roseus*. In the production stage, ajmalicine production by *C.roseus* in a 3 L stirred tank reactor was induced with a high glucose concentration (80 g/L). Ajmalicine production in cultures started with cells from the late stationary phase was five times higher than in cultures started with cells from the early stationary phase. After transfer to the production stage cells from the early stationary phase showed a transient increase in respiration and enzyme induction, followed by culture browning. In contrast, cells in the late stationary phase showed a typical induction pattern constant respiration, and permanent enzyme induction. A striking similarity between the geraniol-10 hydroxylase (G10 H) activity and the ajmalicine accumulation profile could be observed in all cultures suggesting that G 10H regulated ajmalicine production in this investigation. The intracellular nitrate concentration was significantly higher in the inoculum showing a high ajmalicine production than in the inoculum with a low production. Consequently, nitrate may act as a marker for the start of the production stage: as soon as the nitrate is depleted in the growth medium secondary metabolism can be induced. The biosynthetic pathway for ajmalicine is shown in Fig. 21.4.

Jung *et al.*, (1994) reported the effect of the concentrations of inorganic salts in Schenk and Hildebrand (SH) medium on catharanthine production in hairy root culture of *Catharanthus roseus*. The inorganic salt components could be categorized into four groups. The first group (nitrate) supported both the growth and catharanthine production by hairy roots with incremental increase in the concentration. The second (ammonium and phosphate) yielded contradictory effects with respect to growth and production. The third (borate and molibdate) inhibited both growth and production while the fourth (potassium iodide, sulfate and iron) did not exhibit any significant effects. Through optimization of the concentrations of inorganic salts in the medium, a two stage process of hairy root cultures with different media for growth and production was developed which enabled, to enhance the volumetric yield of cathranthine up to 60.5 mg/L. This productivity was 5.4 times higher than that of a one stage culture in the original SH medium.

Cultured cells of *Carthamus tinctorius* released a red pigment into culture medium (Hanagata *et al.*, 1992; Saito *et al.*, 1988) which is different from the red pigment called carthamin produced in the petals of the intact plant (Hanagata *et al.*, 1992; Saito *et al*., 1988). Kobayashi *et al.* (1992) determined the structure of this red pigment and named it kinobenon A. Hanagata *et al.* (1992) previously determined the optimum production medium for red pigment formation in *C. tinctorius* suspension culture. No cell growth occured in the production of red pigment by cultured cells of *C. tinctorius*. In general, during the first culture stage cell growth is achieved whereas the second culture stage is for production of the secondary metabolite.

Hanagata and Karube (1994) reported a two stage culture system of *Carthamus tinctorius* for the production of a red pigment (Kincbenon). The growth levels of cells in an Erlenmeyer flask on a rotary shaker and

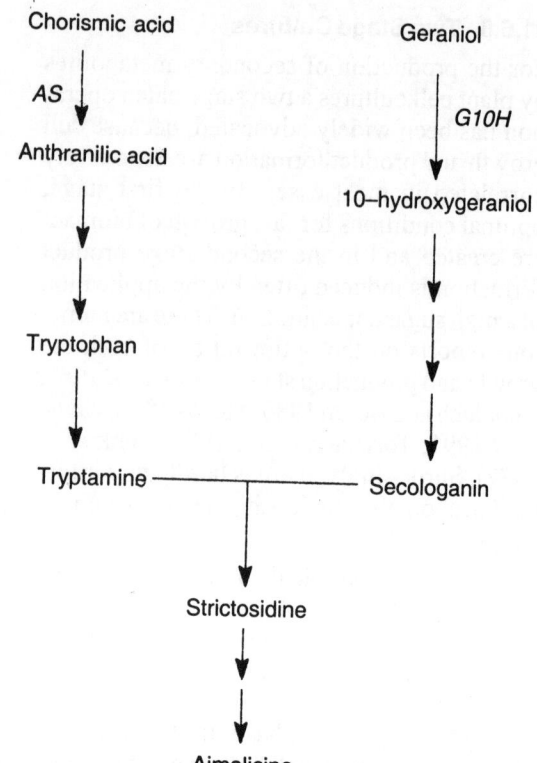

Fig. 21.4. Some intermediates and key enzymes in the biosynthesis of ajmalicine (Adopted from Schlatmann et al., 1995. As = Anthranilic acid synthase; G10H = Geraniol-10-hydroxylase.

a see saw type reactor were higher than those of STR and bubble column. The medium used for growth was MS with NAA (10^{-6}M) and Kinetin (10^{-6} M) for 10 days. Product inhibition was observed for pigment formation in the second culture stage. High pigment production levels were achieved by periodical removal of pigment from the reactor. Production medium in the second culture stage could be reused by removal of the pigment formed. The production medium (Hanagata *et al* 1992) contains autoclaved blue greenalga, *Nostoclinckla* (150 mg/L) as elicitor (Hanagata *et al* 1994). The production of secondary metabolite was enhanced because of decoupling the growth and production stages for e.g.Shikonin (See chapter 20) and also Taxol from Taxus sp. cell cultures.

REFERENCES

Albertsson, P.A. (1971). Partition of cell particles and macromolecules. Almqvist and Wiksell, Stockholm.

Albertsson, P.A., Cajarville, A., Brooks, D.E. and Tjerneld, F. (1987). *Biochim. Biophys. Acta.*, 926 : 87-93.

Andersson, E.,Mattiasson, B. and Hahn-Hagerdal, B. (1984). *Enzyme Microb. Technol.*, 6 : 301-306.

Becker, H. and Herold, S (1983) RP-8 als Hilfsphase zur Akkumulation von Valeporiaten aus Zellsuspensionskulturen von *Valeriana wallichii*, *Planta Med* 49: 191-192

Beiderbeck, R. and Knoop, B. (1988). Enhanced production of secondary substances : Addition of artificial accumulation sites to culture. In : Biotechnology in Agriculture and Forestry 4. Medicinal and Aromatic Plants I, Y.P.S. Bajaj (eds.), Springer-Verlag, Berlin,p. 123-135.

Beiderbeck, R. (1982). Zweiphasenkultur - ein Weg zur Isolierung lipophiler Substanzen aus pflanzilchen Suspensionskulturen. *Z. Pflanzenphysiol.,* 108: 27-30.

Berlin, J. (1983). Naturstoffe aus pflanzlichen Zelkulturen. *Chiu Z.,* 17 : 77-84.

Berlin, J., Mollenschott, C. and DiCosmo, F. (1987). Comparison of various strategies designed to optimize indole alkaloid accumulation of a cell suspension culture of *Catharanthus roseus. Z. Naturforsch.,* 42C : 1101-1108.

Berlin, J., Witte L., Schubert, W. and Wray, V. (1984). Determination and quantification of monoterpenoids secreted into the medium of cell cultures of *Thuja occidentalis. Phytochemistry,* 23 : 1277-1279.

Bisson, W., Beiderbeck, R. and Reichling, J. (1983). Die Produktion atherischer Ole durch Zellsuspensionen der Kamille in einem Zweiphasensystem. *Planta Med.,* 47 : 164-168.

Byun, S.Y. and Pedersen, H. (1994). Two-phase airlift fermentor operation with elicitation for enhanced production of benzophenanthridine alkaloids in cell suspensions of *E. californica. Biotechnol. Bioeng.,* 44 : 14-20.

Byun, S.Y., Pedersen, H. and Chin, C.K. (1990). Two-phase culture for the enhanced production of benzophenanthridine alkaloids in cell suspensions of *Eschscholtzia californica. Phytochemistry,* 29 : 3135-3139.

Chang, H.N. and Sim, S.J. (1994). Production of plant secondary metabolites by extractive cultivation. In : Advances in Plant Biotechnology, Studies in Plant Science, 4, Ryu, D.D.Y. and Furusaki, S. (eds), pp. 355-369, Elsevier, Amestradam, New York.

Collins-Pavao, M., Chin, C.K. and Pedersen,H. (1996). Taxol partitioning in two-phase plant cell cultures of *Taxus brevifolia. J. Biotechnol.,* 49 : 95-100.

Constantin, M.J., Henke, R.R. and Mansur, M.A. (1977). Effect of activated charcoal on callus growth and shoot organogenesis in tobacco. *In vitro,* 13 : 293-296.

Corry, P.J., Reed, W.L. and Curtis, W.R. (1993). Enhanced recovery of solavetivone from Agrobacterium transformed root cultures of *Hyoscyasmus muticus* using integrated product extraction. *Biotechnol. Bioeng.,* 42 : 503-508.

Denffer, D. Von, Ehrendorfer, F., Maegdefrau, K. and Ziegler, H. (1978). Lehrbuch der Botanik, 31st edn. Gustav Fischer, Stuttgart.

Dutta, A., Pedersen, H. and Chin, C.K. (1994). Two-phase culture system for plant cells. *Ann. NY Acad. Sci.,* 745: 251-260.

Felix, H.(1982). Permeabilized cells. *Anal. Biochem.* 120 : 211.

Hanagata, N and Karube, I. (1994). Red pigment production by *Carthamus tinctorius* cells in a two stage culture system. *J. Biotechnol.* 37: 59-65

Hanagata, N., Ito, A., Fukuju, Y. and Murate, K. (1992) Red pigment formation in cultured cells of *Carthamus tinctorius L.Biosci. Biochem.* 56,44-47.

Hanagata,N ., Uchera , H.,Ito, A, Takeuchhi, T. and Karube, I. (1994) Elicitor for red pigment formation in *Carthamus tinctorius* cultured cells. *J. Biotechnol.* 34: 71-77.

Hooker, S.B. and Lee, J.M. (1990). Cultivation of plant cells in aqueous two-phase polymer systems. *Plant Cell Rep.,* 8 : 546-549.

Joersbo, M. and Brunstedt, J. (1991) Electroporation mechanism and transient expressions, table transformation and biological effects in plant protoplats. *Physiol. Plant.,* 81 : 256.

Jung, H.K., Kwak, S.S., Choi, C.Y. and Liu, J.R. (1994). Development of two stage culture process by optimization of inorganic salts for improving catharanthine production in hairy root cultures of *C. roseus. J. Ferm. and Bioprocess.* 77(1) : 57-61.

Kim, D.J. and Chang, H.N. (1990). Enhanced shikonin production from *L. erythrorhizon* by in situ extraction and calcium alginate immobilization. *Biotechnol. Bioeng.,* 36 : 460-466

Knobloch, K.H. and Berlin, J. (1980). Influence of medium composition on the formation of secondary compounds in cell suspension cultures of *Catharanthus roseus* (L). G. Don *Z. Naturforsch.* 35C : 551-556.

Knoop, B. and Beiderbeck, R. (1983). Adsorbenskultur-ein weg Zur Steigerung der Sekundarstolff production in pflanzlichen suspension cultures. *Z. Naturforsch.,* 38C : 484-486.

Kobayashi, A., Kawazu, K., Hisasaka, K.,Wakayama, Y., Nakagawa, N. and Terui, S. (1992). Abstract, Annual meeting of Japan Society for Bioscience, Biotechnology and Agrochemistry, 1-3 April 1992, Toko, Japan, p. 223.

Kuhn, I. (1980). Biotechnology and Bioengineering, 22 : 2393-2398.

Luckner, M. (1980). Expression and control of secondary metabolism. In : Bell, E.A., Charlwood B.V. (eds.) Encyclopedia of plant physiology, New Ser, vol. 8 Springer, Berlin Heidelberg New York, pp 22-63.

Mattson, J.S. and Mark, H.B. (1971). Activated carbon, Dekker, New York.

Mbanaso, E.N.A. and Roscoe, D.H. (1982). Alginate : an alternative to agar in plant protoplast culture. *Plant Sci. Lett.,* 25 : 61-66.

Meravy, L., Cvikrova, M. and Hrubeova, M. (1988). The effect of DMSO on extraction of proteins and phenolic substrates by *Nicotiana tabacum* cells into culture medium. *Biol. Plant.,* 30(4) : 241.

Moreno, P.R.H., Schlatmann, J.E., Van der Heijden, R., van Gulik, W.M., ten Hoopen, H.J.G., Verpoorte, R. and Heijnen, J.J. (1993). Induction of ajmalicine formation and related enzyme activities in *Catharanthus roseus* cells : Effect of inoculum density. *Appl. Microbiol. Biotechnol.* 39: 42-47.

Morris, P. (1986). Regulation of product synthesis in cell cultures of *catharanthus roseus.* II. Comparison of production media. *Planta Med.,* 50 : 121-126.

Nagel, M. and Reinhard, E. (1975). Das atherische Ol der Kalluskulturen von *Ruta graveolens.* II. Physiologie zur Bidung des atherischen Oles. *Planta Med.,* 27 : 264-274.

Parr, A.J., Robins, R.J. and Rhodes, M.J.C. (1986). Product release from plant cells grown in culture. In : Morris, P. Scragg, A.H., Stafford, A., Fowler, M.W. (eds.). Secondary metabolites in plant cell cultures, University Press, Cambridge, p. 173.

Payne, G.F., Payne, N.N., Shuler, M.L. and Asada, M. (1988). *In situ* adsorption for enhanced alkaloid production by *Catharanthus roseus. Biotechnol. Lett.,* 88 : 187-191.

Payne, G.F., Prince, C. and Shuler., M.L. (1991). Plant cell and tissue culture in liquid systems Hanser, Munich

Persson, I., Tjerneld, F. and Hahn-Hagerdal, B. (1984). *Enzyme Microb. Technol.,* 6 : 415-418.

Reinhard, E., Corduan, G. and Volk, O.H. (1968). Uber Gewebekulturen von *Ruta graveolens. Plant Med.,* 1 : 8-16.

Robins, R. and Rhodes, M.J.C. (1986). The stimulation of anthraquinone production by *Cinchona ledgeriana* cultures with polymeric absorbents. *Appl. Microbiol. Biotechnol.,* 24 : 35-41.

Saito, K., Daimon, E., Kusaka, K., Wakaama, S. and Sekino, (1988). Accumulation of a novel red pigment in cell suspension cultures of floral meristem tissue from *Catharanthus tinctorius* L. *Z. Naturforsch.,* 43C : 862-870.

Schlatmann, J.E., Moreno, P.R.H., Selles, M., Vinke, J .L., Hoopenten, H.J.G., Verpoorte, R. and Heijnen, J.J. (1995). Two-stage batch process for the production of Ajmalicine by *Catharanthus roseus* : the link between growth and production stage. *Biotechnol. Bioeng.,* 47 : 54-59.

Shanbhag, V.P. and Johansson, G. (1974). *Biochem. Biophys. Res. Comm.,* 61 : 1141-1146.

Sim, S.J. (1994). Production and secretion of indole alkaloids in hairy root cultures of *Catharanthus roseus* : effects of *in situ* adsorption, cell permeabilization. *J. Ferment. Bioeng.,* 78 : 229-234.

Strobel, J., Hieke, M. and Groger, D. (1991). Increased anthraquinone production in *Galium vernum* cultures induced by polymeric adsorbents. *Plant Cell Tiss. Org. Cult.,* 24 : 207-210.

Taya, M., Yoyoma, A., Kondo, O., Kobayashi, T. and Matsui, C. (1989). Growth characteristics of plant hairy roots and their cultures in bioreactors. *J. Ferment. Bioeng.,* 72 : 457-460

Teuscher, E. (1973). Probleme der Produktion sekundarer Pflanzenstoffe mit Hilfe von Zellkulturen. *Pharmazie,* 28 : 6-18.

Tikhomiroff, C., Allais, S. and Jolicoeur,M. (2002). Continuouss elective extraction of secondary metabolites from *C. roseus* hairy root with silicone oil in a two-liquid phase bioreactor. *Biotechnol. Prog.* (in press).

Toivonen, L., Ojala, M. and Kauppinen, V. (1991). Studies on the optimazation of growth and indole alkaloid production by hairy root cultures of *Catharanthus roseus. Biotechnol. Bioeng.* 37 : 673-680

Van Gulik, W.M., ten Hoopen, H.J.G., and Heijnen, J.J. (1993). A structured model describing carbon and phosphate limited growth of *Catharanthus roseus* plant cell suspensions in batch and hemostat culture. *Biotechnol. Bioeng.*, 41 : 771-780.

Williams, R.D., Chauret, N., Bedart, C. and Archambault, J. (1992). Effects of polymeric adsorbents on the production of sanguinarine by *Papaver somniferum* cell cultures. *Biotechnol. Bioeng.*, 40 : 971-977.

Yuan, Q., Xu, H. and Hu, Z.(1999). Two-phase culture for enhanced alkaloid synthesis and release in a new airlift reactor by *Catharanthus roseus*. *Biotechnol. Techniques*, 13 : 107-109.

Zenk, M.H. (1982). Pflanzliche Zellkulturen in der Arzneimittelforschung. *Naturwissenschaft.*, 69 : 534-536.

Zenk, M.H., El-Shagi, H. Arens, H., Stockigt, J., Weiler, E.W. and Deus, B. (1977). Formation of the indole alkaloids serpentine and ajmalicine in cell suspension cultures of *Catharanthus roseus*. pp 27-44. In : W. Barz, E. Reinhard, and M.H. Zenk (eds), Plant tissue culture and its biotechnological application, Springer, Berlin.

In situ Regeneration: Two-Phase and Two-Stage Cultures. 468

Vancura, W.D., Jin Haapala, T.L., and Heinonen, T., (1988). A structured model demonstrating the phenomenon filled growth of *Catharanthus roseus* plant cell suspensions with batch and continuous ... *Biotechnol. Bioeng.* 31:77-180.

Williams, P.D., Granier, M., De Luca V., and Anhan Conti, J. (1992). Effects of polymeric adsorbents on the ...

Yeon, T.-Y., Jin, T.L. (1990). Two-phase culture for enhanced alkaloids synthesis with different ... reactor by a new culture ... *Biotechnol. Bioeng.* 36:369-378.

Zenk, M.H. (1987). Biosynthesis of alkaloids ...

Zenk, M.H. (1989). ...

Chapter 22

Cryopreservation of Germplasm

INTRODUCTION

In the recent years, with the tremendous increase in the population, pressure on the forest and land resources have increased, which in turn caused decrease in the population of medicinal and aromatic plant species. Even some of the plant species are at the verge of vanishing from the forest. The list of endangered species is growing day by day (Table 22.1). The conventional methods of germplasm preservation are prone to possible catastrophic losses because of :

1. Attack by pests and pathogens
2. Climatic disorders
3. Natural disasters and
4. Political and economic causes.

In addition, the seeds of many important medicinal plants lose their viability in a short time under conventional storage system. The conservation of germplasm can be done by two methods.

1. *In-situ* preservation : Preservation of the germplasm in their natural environment by establishing biospheres, national parks etc.
2. *Ex-situ* preservation : In the form of seed or *in vitro* cultures.

Seeds form the most common material to conserve plant germplasm, their method has the following disadvantages :

- Some plants do not produce fertile seeds.
- Loss of seed viability
- Seed destruction by pests, etc.
- Poor germination rate.
- This is only useful for seed propagating plants.

In vitro preservation by tissue culture has several advantages over seed preservation.

- Small areas can store large amount of material.
- Protection from environmental methods.

Disadvantages:

- It is a costly process.

Table 22.1. List of Threatened (Endangered) Medicinal Plants

S.No.	Name	Present Status
1	Aconitum deinorrhizum	Almost extinct
2	A. heterophyllum	Greatly threatened
3	Angelica glauca	Threatened
4	Arnebia benthemii	Threatened
5	Artemisia brevifolia	Likely to be threatened
6	A. maritima	Likely to be threatened
7	Atropa acuminata	Threatened
8	Berberis aristata	Threatened
9	Bunium persicum	Greatly threatened
10	Colchium luteum	Threatened
11	Corydalis govaniana	Likely to be threatened
12	Dactylorhiza hatagirea(orchis)	Threatened
13	Dioscorea deltoidea	Threatened
14	Ephedra gerardiana	Likely to be threatened
15	Ferula jaeschkeana	Threatened
16	Gentiana kurroa	Threatened
17	Hedychium spicatum	Likely to be threatened
18	Jurinea dolomiaea (Dhoop)	Likely to be threatened
19	Nardostachys jatamansi	Threatened
20	Orchis latifolia	Threatened
21	Picrorhiza kurroa	Likely to be threatened
22	Podophyllum emodi	Threatened
23	Rheum emodi	Threatened
24	Swertia chirata	Threatened
25	Valeriana wallichii	Likely to be threatened
26	Zanthoxylum alatum	Likely to be threatened

The use of large-scale culture of plant cells for the storage and the production of phytopharmaceuticals and other products have been emphasized. The purpose of cryopreservation is to retain the valuable cell line for the required purpose. The routine techniques of culture maintenance are expensive and time consuming. Typically, a cell suspension culture would need to be transferred every 7-10 days and callus cultures every 14-30 days. Furthermore, the maintenance for cultures in the fast-growing state bears the risk of possible loss through contamination or equipment failure. Besides these technical aspects, it should be kept in mind that plant cells growing in the dedifferentiated stage are genetically unstable. The factors influencing conservation by subculturing is shown in Table 22.2.

The preservation of cell, tissue and organs in liquid nitrogen is called cryopreservation and the science pertaining to this activity is known as cryobiology.

Cryopreservation is the non-lethal storage of biological material at ultra-low temperature. At the temperature of liquid nitrogen (-196°C) almost all the metabolic activities of cells are ceased and the sample can then be preserved in such state for extended periods. However, only few biological materials can be frozen to (-196°C) without affecting the cell viability.

Cryopreservation comprises many steps, of which freezing is only one, successful recovery is dependent on the combined effects of cryogenics and pre and post freeze treatments.

22.1. METHOD

The technique of cryopreservation involves the following steps:

1. Selection of plant material
2. Pre culture or pregrowth
3. Cryoprotective treatment
4. Freezing and storage
5. Thawing

Table 22.2. Factors influencing conservation by subculturing

Somaclonal variation
Mass invasion by microbes
Destroying the whole parent culture.
Subculturing leads to selection for good growth, neglecting productivity.

22.1.1. Selection of Plant Material

Tissues must be selected from healthy plants and in case of *in vitro* material culture parameters should be optimized before cryopreservation. The morphological and physiological conditions of the plant material influence the ability of an explant to survive freezing at – 196ºC. Different types of tissues can be used for cryopreservation such as plant organs (ovules, anther/pollen, embryos and endosperm), plant cells somatic embryos, and protoplasts, etc. In general, small, young, richly cytoplasmic, meristematic cells survive better than the larger, highly vacuolated cells. A number of cell suspensions of medicinal plants have been successfully frozen (Table 22.4). The uniform suspensions composed of small groups of cytoplasmic, meristematic callus are more suitable than highly vacuolated cells which have water content and they should be late lag phase or exponential phase actively growing suspension are able to withstand freezing much better than relatively old culture. A high density to cells per ampoule yield better results.

Callus derived from tropical plants is more resistant to freezing damage. A rapidly growing stage of callus shortly after 1 or 2 weeks of subculture is best for cryopreservation. The old cells at the top of callus and blackened area should be avoided. Cultured cells are not ideal for freezing. Instead, organized structures such as shoot apices, embryos or young *plantlets* are preferred.

22.1.2. Pregrowth

Pre-growth involves cold hardening treatments or the application of additives known to enhance plant stress tolerance e.g., abscisic acid, proline, trehalose. Partial tissue dehydration can be achieved by the application of osmotically active compounds. The addition of low concentration of DMSO (1-5%) during pre-growth often improves shoot tip recovery, e.g. :

(i) *C. roseus* cells were precultured in a medium containing 1M sorbitol before freezing (Chen *et al.,* 1984).

(ii) *Digitalis* cells were precultured on 6% mannitol medium for 3 days before freezing (Seitz *et al.,* 1983).

(iii) *Nicotiana sylvestris* with 6% sorbitol for 2-5 days before freezing (Maddox *et al.,* 1983).

22.1.3. Cryoprotective Treatments

There are two potential sources of cell damage during cryopreservation.

(i) Formation of large ice crystals, inside the cells, leading to rupture of organelle and the cell itself.

(ii) Intracellular concentration of solutes increases to toxic levels before or during freezing as a result of dehydration.

Cryoprotectants are categorized as :

Penetrating : exert their protective colligative action

Non-penetrating : effect through osmotic dehydration.

(a) *Vitrification* : Avoidance of ice formation in biological tissues exposed to low and ultra low temperature reduces damage. This can be achieved through vitrification, a process in which ice formation cannot take place because the aqueous solution is too concentrated to permit ice crystal nucleation. Instead, water solidifies into an amorphous 'glassy' state.

(b) *Cryoprotective dehydration* : If cells are sufficiently dehydrated they may be able to withstand immersion in liquid nitrogen without further application of traditional cryoprotectant mixtures. Dehydration can be achieved by growth in the presence of high concentration of osmotically active compounds (sugars, polyols) and / or air desiccation in a sterile flow cabinet or over silica gel. Dehydration reduces the amount of water available for the ice formation.

(c) *Encapsulation and dehydration* : This involves the encapsulation of tissues in calcium alginate beads which are pre-grown in liquid culture media containing high concentrations of sucrose. The beads are transferred to a sterile airflow in a laminar cabinet and desiccated further. After these treatments the tissues are able to withstand exposure to liquid nitrogen without application of chemical cryoprotectants.

Addition of cryoprotectants controls the appearance of ice crystals in cells and protects these cells from the toxic solution effect. A large number of heterogeneous groups of compounds have been shown to possess cryoprotective properties with different efficiencies (Table 22.3). e.g. Glycerol, DMSO, etc. Cryoprotectants depress both the freezing point and the super cooling point of water, i.e., the temperature at which the homogeneous nucleation of ice occurs thus retarding the growth of ice crystal formation. DMSO has excellent cryoprotective properties and it is very widely employed for cryopreservation of cultured cells.

Table 22.3. Cryoprotectants used in cryopreservation

Alcohols	Sulphur containing compounds	Polymers
Ethylene glycol	Amino acids	Hydroxyethyl amidon
Glycerol	Dimethyl sulphoxide	Polyethylene glycol
Propylene glycol	*Sugar* (glucose, trehalose, saccharose)	Polyvinyl pyrrolidine
Sorbitol		
Mannitol		

22.1.4. Freezing and Storage

The type of crystal water within stored cells is very important for survival of the tissue. Three different types of freezing procedures have been developed.

22.1.4.1. Rapid Freezing

The plant material is placed in vials and plunged into liquid nitrogen and decrease of –300 to –1000°C/min or more occurs. The quicker the freezing is done, the smaller the intracellular ice crystals are. A somewhat slower temperature decrease is achieved when the vial containing plant material is put in the atmosphere over liquid nitrogen (-10 to –70°C/min). Dry ice (CO_2) instead of nitrogen can also be used in a similar manner. It is also possible that ultra rapid cooling may prevent the growth of intracellular ice crystals by rapidly passing the cells through the temperature zone in which lethal ice crystal growth occurs. This method is technically simple and easy to handle. It has been successfully used for the cryopreservation of shoot tips of potato, strawberry, Brassica sps, and somatic embryos.

22.1.4.2. Slow Freezing

The tissue is slowly frozen with a temperature decrease of 0.1 – 10°C/min from 0°C to –100°C and then transferring to liquid nitrogen. Survival of cells frozen at slow freezing rates of –0.1 to –10°C/ min may involve some beneficial effects of dehydration, which minimizes the amount of water that freezes intracellularly. Slow cooling permits the flow of water from the cells to the outside, thereby promoting extracellular ice formation instead of lethal intracellular freezing. It is generally agreed that upon extracellular freezing the cytoplasm will be effectively concentrated and plant cells will survive better when adequately dehydrated. This method has been successfully employed for cryopreservation of meristems of peas, potato, cassava and strawberry, etc.

22.1.4.3. Stepwise Freezing

A slow freezing procedure down to –20 to –40°C, a stop for a period of time (approximately 30 min) and then additional rapid freezing to –196°C is done by plunging in liquid nitrogen. A slow freezing procedure initially to –20 to –40°C permits protective dehydration of the cells. An additional rapid freezing in liquid nitrogen prevents the growing of big ice crystals in the biochemically important structures.

22.1.5. Storage

The frozen cells/tissues are immediately kept for storage at temperature ranging from –70°C to – 196°C. A liquid nitrogen refrigerator running at 150°C in the vapor phase or –196°C in the liquid phase is ideal for this purpose. The temperature should be sufficiently low for long term storage of cells to stop all metabolic activities and prevent biochemical injury. Long term storage is done at – 196°C. Thus as long as regular supply of liquid nitrogen is ensured in liquid nitrogen refrigerator, it is possible to maintain the frozen material with little further care.

22.1.6. Thawing

The temperature and rate at which tissues are thawed is dependent on the freezing method. Thawing is particularly critical in vitrified tissues as ice crystallization can occur during re-warming. In conventional freezing methods, thawing is usually achieved by plunging the cryovials into sterile water maintained at 40 – 45°C. Once all the ice has melted the samples are removed. Tissues which have been frozen by encapsulation and/or dehydration are frequently thawed at ambient temperatures. Vitrified and excessively dehydrated tissues may require a sequential lowering of media osmoticum during thawing.

22.2. DETERMINATION OF VIABILITY

As mentioned in the chapter of immobilization (11.8) and growth measurement in chapter (5.2).

22.3. CRYOPRESERVATION EQUIPMENT

Simple freezing chambers utilizing a cooled solvent system may be used. However, a typical, modern programmable freezer comprises a liquid nitrogen cooled sample chamber, to which nitrogen is pumped from an accessory Dewar or storage tank. Cooling rates, holding times, and terminal transfer temperatures are programmed, and chamber and sample temperatures are monitored by thermocouples. Models can be purchased from several companies (Planar UK, Cryomed USA, L'Air Liquide, France).

The following equipment is required for cryopreservation.

1. a reliable source of liquid nitrogen
2. safety equipment (gloves, apron, face shield, pumps for dispensing liquid nitrogen from a large storage dewars, trolleys for the transport of dewars)
3. small (1-2 litre) liquid nitrogen resistant dewar(s)
4. dewar(s) for the routine storage of liquid nitrogen
5. dewar(s) for the long-term storage of specimens-these may be equipped with an alarm system which is activated if the liquid nitrogen level falls below a critical level
6. cryovials, straws, boxes, canes, racks
7. a refrigerator (–20°C)
8. a programmable freezer with dewar and pump
9. a water-bath for thawing at 40–50°C

22.4. CRYOPRESERVATION OF CALLUS AND CELL SUSPENSIONS

The list of medicinal and aromatic plant tissue culture successfully cryopreserved is shown in Table 22.4. The cell suspensions of *Datura stramonium* frozen at the rate of 1°C/min in the presence of 7% DMSO gave the best results. Periwinkle (*Catharanthus roseus*) cell cultures subjected to cooling at the rate of 0.5°C/min to –40°C and then immersed in liquid nitrogen showed viability of about 60% of the controls (Chen *et al.*, 1984). The cell cultures of *Digitalis lanata* survived upto 50%, and showed stability of biotransformation potential after cryostorage. The capacity to transform α-methyldigitoxin to β-methyldigoxin remained unchanged (Diettrich *et al.*, 1982; Seitz *et al.*, 1983). Likewise callus and cell suspensions of *Anisodus acutangulus* retained their biosynthetic ability of hyoscyamine and scopolamine after freezing. The storage at –196°C was better than at –20°C, and callus cultures gave better results than cell suspensions. Lactalbumin hydrolysate used as cryoprotectant was similar to DMSO, but glycerine was ineffective (Zheng Guangzhi *et al.*, 1983).

Table 22.4. Cryopreservation of medicinal and aromatic plants (Adopted from Bajaj, 1988a)

Plant	Culture	Cryoprotectant	Freezing	Thawing (°C)	Reference
Anisodus acutangulus	Cell suspension	DMSO or lactalbumin hydrolysate	37	–196°C	Zheng Guang-zhi *et al.* (1983)
Atropa belladonna	Cell suspension	DMSO (5%)	2°C/min, –100, –196°C	37	Nag and Street (1975)
A. belladonna	Protoplasts	DMSO + mannitol	Exposed to vapors, immersed in LN	35	Bajaj (1988)
Catharanthus roseus	Cell suspension	DMSO (5%) + IM sorbitol	0.5-3°C/min to –35°C, 196°C	40	Chen et al. (1984)
Citrus sp.	Ovule/nucellar embryo	DMSO 7% + Sucrcse 7%	–196°C	35-40	Bajaj (1984)
Cinchona ledgeriana	Callus	Mannitol + DMSO	Slow freezing	Rapid thawing	Hunter (1986)

(Contd.)

Plant	Culture	Cryoprotectant	Freezing	Thawing (oC)	Reference
Datura stramonium	Cell suspension	DMSO 7%	1°C/min, −196°C	36 - 38	Bajaj (1976)
D. innoxia	Protoplasts	DMSO + mannitol	Exposed to vapors, immersed in LN	35	Bajaj (1988)
D. innoxia	Cell suspension	1 M Sorbitol	1°C/min, −196°C	33	Weber et al. (1983)
Dioscorea deltoidea	Cell suspension	7% DMSO	Hardening at 2-10, −30, −70, then −196°C	20 or 40	Butenko et al. (1984)
Digitalis lanata	Cell suspension	DMSO + glycerol + sucrose	1°C/min, −35°C, −196oC	40	Diettrich et al. Seitz et al. (1983)
Lavandula vera	Cell suspension	DMSO (10%) + glucose (20%)	1°C/min, −40°C, -196°C	Rapid	Watanabe et al. (1983)
Nicotiana tabacum	Cell suspension	DMSO, sucrose or glycerol	1-2°C/min, or −20, −70, −196°C	36 - 38	Bajaj (1976) Hauptmann and Widholm (1982), Maddox et al. (1983)
N. plumba- ginifolia N. sylvestris	Cell suspension	DMSO + glycerol	-40 to −70 for 4h, then −196°C	45	Maddox et al. (1983)
N. tabacum	Protoplasts	DMSO + mannitol	Exposed to vapors, immersed in LN	35	Bajaj (1988)
Panax ginseng	Cell suspension	Sucrose/glycerol	Hardening, −30, −70, then −196°C	20 or 40	Butenko et al. (1984)

Dioscorea deltoidea and *Panax ginseng* are two other species of medicinal value for the production of diosgenin and saponin respectively, and which have yielded excellent results (Butenko *et al.,* 1984). Preculture and hardening of *Panax* cells in the presence of 7% - 25% sucrose at 2 – 10°C resulted in considerable increase in their survival. As a cryoprotectant, sucrose was best for *Panax*, whereas 7% DMSO gave optimal results in *Dioscorea*, and they retained biosynthetic potential for diosgenin, sitosterol, stigmasterol, and other metabolites.

Cinchona callus precultured on 3 days in a medium containing 5% PVP, and cryoprotected with CB5 + mannitol + DMSO, frozen slowly in liquid nitrogen vapors, transferred to liquid nitrogen and then rapidly thawed, gave rise to slowly growing calli on agar medium. However, the meristems did not survive after freezing (Hunter, 1986).

REFERENCES

Bajaj, Y.P.S. (1976) Regeneration of plants from cell suspensions frozen at −20, -70 and −196°C. *Physiol. Plant,* 37 : 263-268.

Bajaj, Y.P.S. (1984). Induction of growth in frozen embryos of coconut and ovules of citrus. *Curr. Sci.,* 53 : 1215-1216.

Bajaj, Y.P.S. (1988). Regeneration of plants from frozen (-196°C) protoplasts of *Atropa belladonna, Datura innoxia* and *Nicotiana tabacum. Indian J. Exp. Biol.,* 26 : 289-292.

Bajaj, Y.P.S. (1988a). Cryopreservation and the retention of biosynthetic potential in cell cultures of medicinal and alkaloid-producing plants. In : Biotechnology in Agriculture and Forestry, Vol. 4, Medicinal and Aromatic Plants, Y.P.S. Bajaj (eds.), Springer-Verlag-Berlin, Hieldelberg, pp. 172-173.

Butenko, R.G., Popov, A.S., Vokova, L.A., Chernyak, N.D. and Nosov, A.M. (1984). Recovery of cell cultures and their biosynthetic capacity after storage of *Dioscorea deltoidea* and *Panax ginseng* cells in liquid nitrogen. *Plant Sci. Lett.,* 33 : 285-292.

Chen, T.H.H., Kartha, K.K., Leung, N.L., Kurz, W.G.W., Chatson, K.B. and Constabel, F. (1984). Cryopreservation of alkaloid producing cell cultures of periwinkle (*Catharanthus roseus*). *Plant Physiol.,* 75 : 726-731.

Diettrich, B., Popov, A.S., Pfeiffer, B., Neumann, D., Butenko, R. and Luckner, M. (1982). Cryopreservation of *Digitalis lanata* cell cultures. *Planta Med.* 46 : 82-87.

Hauptmann, R.M. and Widholm, J.M. (1982). Cryostorage of cloned amino acid analog resistant carrot and tobacco suspension culture. *Plant Physiol.* 70 : 30-37.

Hunter, C.S. (1986). *In vitro* propagation and germplasm storage of *Cinchona*. In : Alderson P, Withers LA (eds) Plant tissue culture and its agricultural applications. Butterworth, London, pp 291-301.

Maddox, A.D., Gonsalves, F. and Shields, R. (1983). Successful preservation of suspension cultures of three *Nicotiana* species at the temperature of liquid nitrogen. *Plant Sci. Lett.,* 28 : 157-162.

Nag, K.K. and Street, H.E. (1975). Freeze-preservation of cultured plant cells II. The freezing and thawing phases. *Physiol. Plant,* 34 : 261-265.

Seitz, U., Alfermann, W. and Reinhard, E. (1983). Stability of biotransformation capacity in *Digitalis lanata* cell cultures after cryogenic storage. *Plant Cell Rep.,* 2 : 273-276.

Watanabe, K., Mitsuda, H. and Yamada, Y. (1983). Retention of metabolic and differentiation potential of green *Lavandula vera* callus after freeze-preservation. *Plant Cell Physiol.,* 24 : 119-122.

Weber, G., Roth, E.J., and Schweiger, H.G. (1983). Storage of cell suspensions and protoplasts of *Glycine max* (L.) Merr., *Brassica napus* (L.), *Datura innoxia* (Mill.), and *Daucus carota* (L.) by freezing. *Z. Pflanzenphysiol.,* 109 : 29-39.

Zheng Guang-zhi, He Jing-bo and Wang Shi-ling (1983). Cryopreservation of calli and their suspension culture cells of *Anisodus acutangulus. Acta Bot. Sin.* 25(6) : 512-517 (in Chinese).

Bajaj, Y.P.S. (1985), Cryopreservation and the retention of biosynthetic potential in cell cultures of medicinal and alkaloid-producing plants. In: Biotechnology in Agriculture and Forestry, Vol. 1, Medicinal and Aromatic Plants I, Bajaj, Y.P.S. (ed.), Springer-Verlag, Berlin Heidelberg, pp. 393-413.

Butcher, D.N., Cocking, E.C., Hennerty, M.J. and Power, J.B. (1988), Recovery of cell cultures and their morphogenetic ability after freezing and thawing of cryopreserved cells in European Phaseolus vars. Plant Sci. 54, 285-295.

Chen, T.H.H., Kartha, K.K., Finn, Kurz, L.C., Stushnoff, C.K. and Constabel, F. (1984), Cryopreservation of alkaloid-producing cell cultures of periwinkle (Catharanthus roseus). Plant Physiol. 75, 726-731.

Dereuddre, J., Scottez, C., Arnaud, Y., Fauré, M., Bottineau, F. and Leddet, C. (1982), Cryopreservation of meristems, cell cultures. Plant Cell Rep. 4, 29-32.

Haquimann, R.N. and Withers, L.A. (1981), Cryopreservation of cloned in-line technology. Improvement in mature suspension cultures. Plant Physiol. 70, 1032.

Küster, E. (1989), Low-temperature photographs and germplasm storage in the cryopreserved Withers, L.A. (ed.), EUCARPIA, University of Birmingham, Birmingham, Edgbaston, pp. 29-36.

Kartha, K.K., Leung, N.L. and Chen, P.K. (1988), Cryopreservation and plant regeneration cultures of strawberry, species-related temperature responses. Plant Cell Rep. 1, 135-138.

Nag, K.K. and Street, H.E. (1975), Freeze-preservation of cultured plant cells, I. The freezing and thawing phases. Physiol. Plant. 34, 254-260.

Index